Imaging: Sensors and Technologies

Special Issue Editor
Gonzalo Pajares Martinsanz

MDPI

Guest Editor
Gonzalo Pajares Martinsanz
University Complutense of Madrid
Spain

Editorial Office
MDPI AG
St. Alban-Anlage 66
Basel, Switzerland

This edition is a reprint of the Special Issue published online in the open access journal *Sensors* (ISSN 1424-8220) from 2015–2017 (available at: http://www.mdpi.com/journal/sensors/special_issues/imaging-sensors-technologies).

For citation purposes, cite each article independently as indicated on the article page online and as indicated below:

Author 1; Author 2; Author 3 etc. Article title. *Journalname*. **Year**. Article number/page range.

ISBN 978-3-03842-360-7 (Pbk)
ISBN 978-3-03842-361-4 (PDF)

Table of Contents

About the Guest Editor

Gonzalo Pajares received his Ph.D. degree in Physics from the Distance University, Spain, in 1995, for a thesis on stereovision. Since 1988 he has worked at Indra in critical real-time software development. He has also worked at Indra Space and INTA in advanced image processing for remote sensing. He joined the University Complutense of Madrid in 1995 on the Faculty of Informatics (Computer Science) at the Department of Software Engineering and Artificial intelligence. His current research interests include computer and machine visual perception, artificial intelligence, decision-making, robotics and simulation and has written many publications, including several books, on these topics. He is the co-director of the ISCAR Research Group. He is an Associated Editor for the indexed online journal *Remote Sensing* and serves as a member of the Editorial Board in the following journals: *Sensors, EURASIP Journal of Image and Video Processing, Pattern Analysis and Applications.* He is also the Editor-in-Chief of the *Journal of Imaging.*

Preface to "Imaging: Sensors and Technologies"

This book contains high-quality works demonstrating significant achievements and advances in imaging sensors, covering spectral electromagnetic and acoustic ranges. They are self-contained works addressing different imaging-based procedures and applications in several areas, including 3D data recovery; multispectral analysis; biometrics applications; computed tomography; surface defects; indoor/outdoor systems; surveillance. Advanced imaging technologies and specific sensors are also described on the electromagnetic spectrum (ultraviolet, visible, infrared), including airborne calibration systems; selective change driven, multi-spectral systems; specific electronic devices (CMOS, CCDs, CZT, X-Ray, and fluorescence); multi-camera systems; line sensors arrays; video systems. Some technologies based on acoustic imaging are also provided, including acoustic planar arrays of MEMS or linear arrays.

The reader will also find an excellent source of resources, when necessary, in the development of his/her research, teaching or industrial activity, involving imaging and processing procedures.

This book describes worldwide developments and references on the covered topics—useful in the contexts addressed.

Our society is demanding new technologies and methods related to images in order to take immediate actions or to extract the underlying knowledge on the spot, with important contributions to welfare or specific actions when required.

The international scientific and industrial communities worldwide also benefit indirectly. Indeed, this book provides insights into and solutions for the different problems addressed. It also lays the foundation for future advances toward new challenges. In this regard, new imaging sensors, technologies and procedures contribute to the solution of existing problems; conversely, they contribute where the need to resolve certain problems demands the development of new imaging technologies and associated procedures.

We are grateful to all those involved in the edition of this book. Without the invaluable contribution of the authors together with the excellent help of the reviewers, this book would not have seen the light of day. More than 150 authors have contributed to this book.

Thanks to *Sensors* journal and the whole team involved in the edition and production of this book for their support and encouragement.

Gonzalo Pajares Martinsanz
Guest Editor

Article

Depth Errors Analysis and Correction for Time-of-Flight (ToF) Cameras

Ying He [1,*], Bin Liang [1,2], Yu Zou [2], Jin He [2] and Jun Yang [3]

[1]   Shenzhen Graduate School, Harbin Institute of Technology, Shenzhen 518055, China;
      bliang@tsinghua.edu.cn
[2]   Department of Automation, Tsinghua University, Beijing 100084, China;
      y-zou10@mails.tsinghua.edu.cn (Y.Z.); he-j15@mails.tsinghua.edu.cn (J.H.)
[3]   Shenzhen Graduate School, Tsinghua University, Shenzhen 518055, China;
      yangjun603@mails.tsinghua.edu.cn
*   Correspondence: heying@hitsz.edu.cn; Tel.: +86-755-6279-7036

Academic Editor: Gonzalo Pajares Martinsanz
Received: 2 September 2016; Accepted: 9 December 2016; Published: 5 January 2017

Abstract: Time-of-Flight (ToF) cameras, a technology which has developed rapidly in recent years, are 3D imaging sensors providing a depth image as well as an amplitude image with a high frame rate. As a ToF camera is limited by the imaging conditions and external environment, its captured data are always subject to certain errors. This paper analyzes the influence of typical external distractions including material, color, distance, lighting, etc. on the depth error of ToF cameras. Our experiments indicated that factors such as lighting, color, material, and distance could cause different influences on the depth error of ToF cameras. However, since the forms of errors are uncertain, it's difficult to summarize them in a unified law. To further improve the measurement accuracy, this paper proposes an error correction method based on Particle Filter-Support Vector Machine (PF-SVM). Moreover, the experiment results showed that this method can effectively reduce the depth error of ToF cameras to 4.6 mm within its full measurement range (0.5–5 m).

Keywords: ToF camera; depth error; error modeling; error correction; particle filter; SVM

---

1. Introduction

ToF cameras, which have been developed rapidly in recent years, are a kind of 3D imaging sensor providing a depth image as well as an amplitude image with a high frame rate. With its advantages of small size, light weight, compact structure and low power consumption, this equipment has shown great application potential in fields such as navigation of ground robots [1], pose estimation [2], 3D object reconstruction [3], identification and tracking of human organs [4] and so on. However, limited by its imaging conditions and influenced by the interference of the external environment, the data acquired by a ToF camera has certain errors, among which is the fact it has no unified correction method for any non-systematic errors caused by the external environment. Therefore, different depth errors must be analyzed, modeled and corrected case by case according to the different causes.

ToF camera errors can be divided into two categories: systematic errors, and non-systematic errors. A systematic error is triggered not only by its intrinsic properties, but also by the imaging conditions of the camera system. The main characteristic of this kind of error is that their form is relatively fixed. These errors can be evaluated in advance, and the correction process is relatively convenient. Systematic errors which can be reduced by calibration under normal circumstances [5] and can be divided into five categories.

A non-systematic error is an error caused by the external environment and noise. The characteristic of this kind of error is that the form is not fixed and random, and it is difficult to establish a unified model to describe and correct such errors. Non-systematic errors are mainly divided into four categories: signal-to-noise ratio, multiple light reception, light scattering and motion blurring [5].

Signal-to-noise ratio errors can be removed by the low amplitude filtering method [6], or an optimized integration time can be decided by using a complex algorithm as per the area to be optimized [7]. Other ways generally reduce the impact of noise by calculating the average of data to determine whether it exceeds a fixed threshold [8–10].

Multiple light reception errors mainly exist at surface edges and depressions of the target object. Usually, the errors in surface edges of the target object can be removed by comparing the incidence angle of the adjacent pixels [7,11,12], but there is no efficient solution to remove the errors of depressions in the target object.

Light scattering errors are only related to the position of a target object in the scene; the closer it is to the target object, the stronger the interference will be [13]. In [14], a filter approach based on amplitude and intensity on the basis of choosing an optimum integration time was proposed. Measurements based on multiple frequencies [15,16] and the ToF encoding method [17] both belong to the modeling category, which can solve the impact of sparse scattering. A direct light and global separation method [18] can solve mutual scattering and sub-surface scattering among the target objects.

In [19], the authors proposed detecting transverse moving objects by the combination of a color camera and a ToF camera. In [20], transverse and axial motion blurring were solved by an optical flow method and axial motion estimation. In [21], the authors proposed a fuzzy detection method by using a charge quantity relation so as to eliminate motion blurring.

In addition, some error correction methods cannot distinguish among error types, and uniformly correct the depth errors of ToF cameras. In order to correct the depth error of ToF cameras, a fusion method with a ToF camera and a color camera was also proposed in [22,23]. In [24], a 3D depth frame interpolation and interpolative temporal filtering method was proposed to increase the accuracy of ToF cameras.

Focusing on the non-systematic errors of ToF cameras, this paper starts with the analysis of the impacts of varying external distractions on the depth errors of ToF cameras, such as materials, colors, distances, and lighting. Moreover, based on the particle filter to select the parameters of a SVM error model, an error modeling method based on PF-SVM is proposed, and the depth error correction of ToF cameras is realized as well.

The reminder of the paper is organized as follows: Section 2 introduces the principle and development of ToF cameras. Section 3 analyzes the influence of lighting, material properties, color and distance on the depth errors of ToF cameras through four groups of experiments. In Section 4, a PF-SVM method is adopted to model and correct the depth errors. In Section 5, we present our conclusions and discuss possible future work.

2. Development and Principle of ToF Cameras

In a broad sense, ToF technology is a general term for determining distance by measuring the flight time of light between sensors and the target object surface. According to the different measurement methods of flight time, ToF technology can be classified into pulse/flash, continuous wave, pseudo-random number and compressed sensing [25]. The continuous wave flight time system is also called ToF camera.

ToF cameras were firstly invented at the Stanford Research Institute (SRI) in 1977 [26]. Limited by the detector technology at that time, the technique wasn't used widely. Fast sampling of receiving light didn't come true until the lock-in CCD technique was invented in the 1990s [27]. Then, in 1997 Schwarte, who was at the University of Siegen (Germany), put forward a method of measuring the phases and/or magnitudes of electromagnetic waves based on the lock-in CCD technique [28]. With this technique, his team invented the first CCD-based ToF camera prototype [29]. Afterwards, ToF cameras began to develop rapidly. A brief development history is shown in Figure 1.

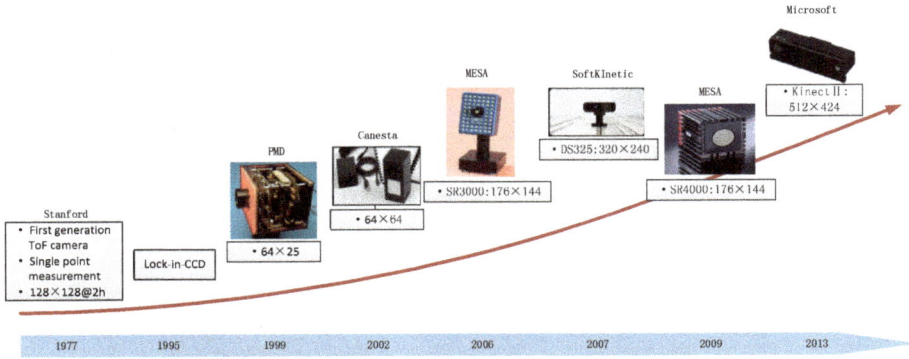

Figure 1. Development history of ToF cameras.

In Figure 2, the working principle of ToF cameras is illustrated. The signal is modulated on the light source (usually LED) and emitted to the surface of the target object. Then, the phase shift between the emitted and received signals is calculated by measuring the accumulated charge numbers of each pixel on the sensor. Thereby, we can obtain the distance from the ToF camera to the target object.

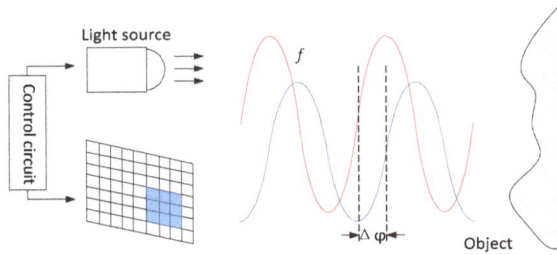

Figure 2. Principle of ToF cameras.

The received signal is sampled four times at equal intervals for every period (at 1/4 period). From the four samples (φ_0, φ_1, φ_2, φ_3) of phase φ, offset B and amplitude A can be calculated as follows:

$$\varphi = \arctan\left(\frac{\varphi_0 - \varphi_2}{\varphi_1 - \varphi_3}\right), \tag{1}$$

$$B = \frac{\varphi_0 + \varphi_1 + \varphi_2 + \varphi_3}{4} \tag{2}$$

$$A = \frac{\sqrt{(\varphi_0 - \varphi_2)^2 + (\varphi_1 - \varphi_3)^2}}{2} \tag{3}$$

Distance D can be derived:

$$D = \frac{1}{2}\left(\frac{c\Delta\varphi}{2\pi f}\right), \tag{4}$$

where D is the distance from ToF camera to the target object, c is light speed and f is the modulation frequency of the signal, $\Delta\varphi$ is phase difference. More details on the principle of ToF cameras can be found in [5].

We list the exterior and parameters of several typical commercial ToF cameras on the market in Table 1.

Table 1. Parameters of typical commercial ToF cameras.

ToF Camera		Maximum Resolution of Depth Images	Maximum Frame Rate/fps	Measurement Rage/m	Field of View/°	Accuracy	Weight/g	Power/W (Typical/Maximum)
MESA-SR4000		176×144	50	0.1–5	69×55	±1 cm	470	9.6/24
Microsoft-Kinect II		512×424	30	0.5–4.5	70×60	±3 cm@2 m	550	16/32
PMD-Camcube 3.0		200×200	15	0.3–7.5	40×40	±3 mm@4 m	1438	-

3. Analysis on Depth Errors of ToF Cameras

The external environment usually has a random and uncertain influence on ToF cameras, therefore, it's difficult to establish a unified model to describe and correct such errors. In this section, we take the MESA SR4000 camera (Zurich, Switzerland, a camera with good performance [30], which has been used in error analysis [31–33] and position estimation [34–36]) as an example to analyze the influence of the external environment transformation on the depth error of ToF cameras. The data we get from the experiments provide references for the correction of depth errors in the next step.

3.1. Influence of Lighting, Color and Distance on Depth Errors

During the measurement process of ToF cameras, it seems that the measured objects tend to have different colors, different distances and may be under different lighting conditions. Then, the following question arises: will the difference in lighting, distances and colors affect the measurement results? To answer this question, we conduct the following experiments.

As we know, there are several natural indoor lighting conditions, such as light-sunlight, indoor light-lamp light and no light. This experiment mainly considers the influence of these three lighting conditions on the depth errors of the SR4000. Red, green and blue are three primary colors that can be superimposed into any color. White is the color for measuring error [32,37,38], while reflective papers (tin foil) can reflect all light. Therefore, this experiment mainly considers the influence of these five conditions on the depth errors of the SR4000.

As the measurment target, the white wall is then covered by red, blue, green, white and reflective papers, respectively, as examples of backgrounds with different colors. Since the wall is not completely flat, laser scanners are used to build a wall model. Then we used a 25HSX laser scanner from Surphaser (Redmond, WA, USA) to provide a reference value, because its accuracy is relatively high (0.3 mm). The SR4000 camera is set on the right side of the bracket, while the 3D laser scanner is on the left. The bracket is mounted in the middle of two tripods and the tripods are placed parallel to the white wall. The distances between the tripods and the wall are measured with two parallel tapes. The experimental scene is arranged as shown in Figure 3 below.

The distances from the tripods to the wall are set to 5, 4, 3, 2.5, 2, 1.5, 1, 0.5 m respectively. At each position, we change the lighting conditions and obtain one frame with the laser scanner and 30 frames with the SR4000 camera. To exclude the influence of the integral time, the SR_3D_View software of the SR4000 camera is set to "Auto".

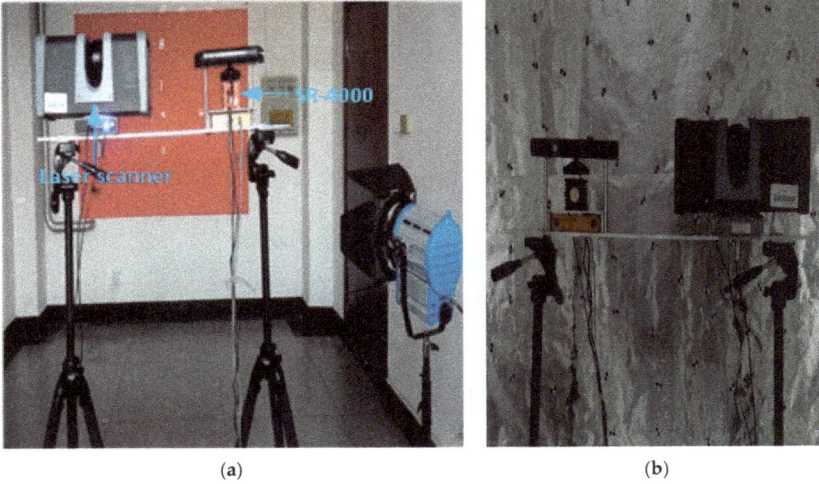

(a)   (b)

Figure 3. Experimental scene. (**a**) Experimental scene; (**b**) Camera bracket.

In order to analyze the depth error, the acquired data are processed in MATLAB. Since the target object can't fill the image, we select the central region of 90 × 90 pixels of the SR4000 to be analyzed for depth errors. The distance error is defined as:

$$h_{i,j} = \frac{\sum\limits_{f=1}^{n} m_{i,j,f}}{n} - r_{i,j}, \tag{5}$$

$$g = \frac{\sum\limits_{i=1}^{a} \sum\limits_{j=1}^{b} h_{i,j}}{s} \tag{6}$$

where $h_{i,j}$ is the mean error of pixel i,j, f is the frame number of the camera, $m_{i,j,f}$ is the distance measured at pixel i,j in Frame f, $n = 30$, $r_{i,j}$ is the real distance, a and b are the row and column number of the selected region respectively and s is the total number of pixels. The real distance $r_{i,j}$ is provided by the laser scanner.

Figure 4 shows the effects of different lighting conditions on the depth error of the SR4000. As shown in Figure 4, the depth error of the SR4000 is on slightly affected by the lighting conditions (the maximum effect is 2 mm). The depth error increases approximately linearly with distance, and the measurement error value complies with the error test of other Swiss Ranger cameras in [37–40]. Besides, as seen in the figure, SR4000 is very robust against light changes, and can adapt to various indoor lighting conditions for the lower accuracy requirements.

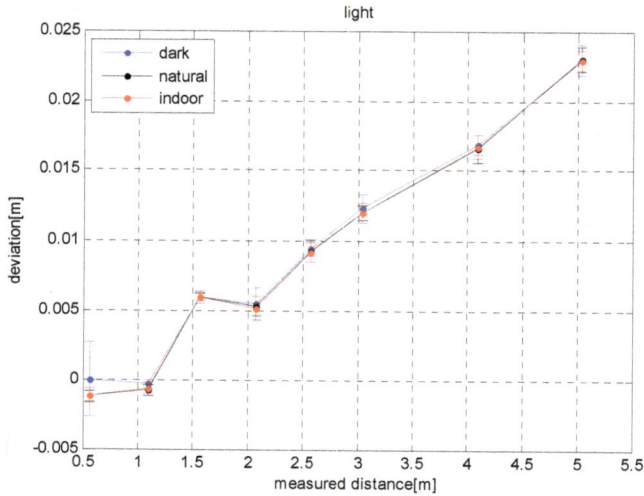

Figure 4. Influence of lighting on depth errors.

Figure 5 shows the effects of various colors on the depth errors of the SR4000 camera. As shown in Figure 5, the depth error of the SR4000 is affected by the color of the target object, and it increases linearly with distance. The depth error curve under reflective conditions is quite different from the others. When the distance is 1.5–2 m, the depth error is too large, while at 3–5 m, it is small. When the distance is 5 m, the depth error is 15 mm less than when the color is blue. When the distance is 1.5 m, the depth error when the color is white is 5 mm higher than when the color is green.

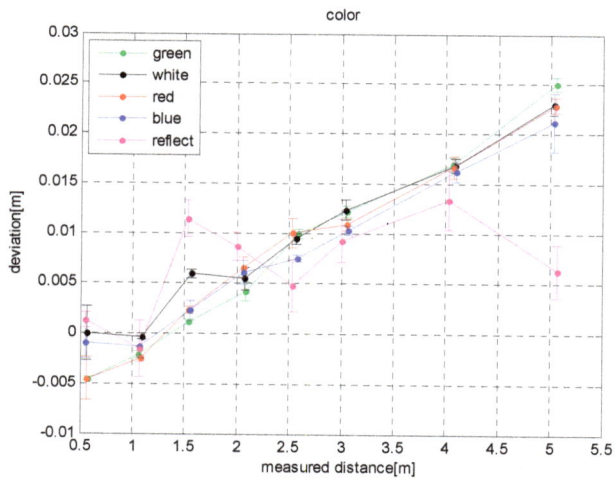

Figure 5. Influence of color on depth errors.

3.2. Influence of Material on Depth Errors

During the measurement process of ToF cameras, it seems that the measured objects tend to be of different materials. Then, will this affect the measurement results? For this question, we conducted the following experiments: to analyze the effects of different materials on the depth errors of the SR4000, we chose four common materials in the experiment: ABS plastic, stainless steel, wood and glass. The tripods are arranged as shown in Figure 3 of Section 3.1, and the targets are four 5-cm-thick boards of the different materials, as shown in Figure 6. The tripods are placed parallel to the target and the distance is set to about 1 m, and the experiment is operated under natural light conditions. To differentiate the boards on the depth image, we leave a certain distance between them. Then we acquire one frame with the laser scanner and 30 consecutive frames with the SR4000 camera. The integral time in the SR_3D_View software of the SR4000 camera is set to "Auto".

Figure 6. Four boards made of different materials.

For the SR4000 and the laser scanner, we select the central regions of 120 × 100 pixels and 750 × 750 pixels, respectively. To calculate the mean thickness of the four boards, we need to measure the distance between the wall and the tripods as well. Section 3.1 described the data processing method and Figure 7 shows the mean errors of the four boards.

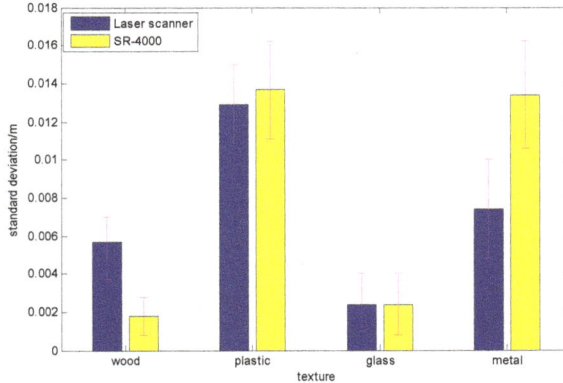

Figure 7. Depth data of two sensors.

As shown in Figure 7, the material affects both the depth errors of the SR4000 and the laser scanner. When the material is wood, the absolute error of the ToF camera is minimal and only 1.5 mm. When the target is the stainless steel board, the absolute error reaches its maximum value and the depth error is 13.4 mm, because, as the reflectivity of the target surface increases, the number of photons received by the light receiver decreases, which leads to a higher measurement error.

3.3. Influence of a Single Scene on Depth Errors

The following experiments were conducted to determine the influence of a single scene on depth errors. The tripods are placed as shown in Figure 3 of Section 3.1, and as shown in Figure 8, the measuring target is a cone, 10 cm in diameter and 15 cm in height. The tripods are placed parallel to the axis of the cone and the distance is set to 1 m. The experiment is operated under natural light conditions. We acquire one frame with the laser scanner and 30 consecutive frames with the SR4000 camera. The integral time in the SR_3D_View software of the SR4000 camera is set to "Auto".

Figure 8. The measured cone.

As shown in Figure 9, we choose one of the 30 consecutive frames to analyze the errors, extract point cloud data from the selected frame and compare it with the standard cone to calculate the error. The right side in Figure 9 is a color belt of the error distribution, of which the unit is m. As shown in Figure 9, the measurement accuracy of SR4000 is also higher, where the maximal depth error is 0.06 m. The depth errors of the SR4000 mainly locate in the rear profile of the cone. The measured object deformation is small, but, compared with the laser scanner, its point cloud data are sparser.

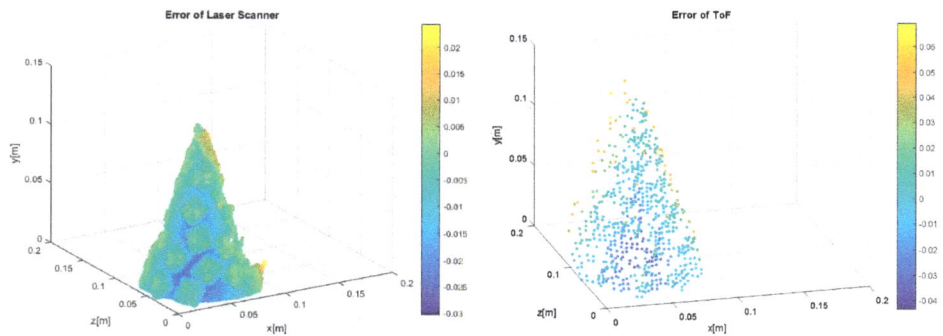

Figure 9. Measurement errors of the cone.

3.4. Influence of a Complex Scene on Depth Errors

The following experiments were conducted in order to determine the influence of a complex scene on depth errors. The tripods are placed as shown in Figure 3 of Section 3.1 and the measurement target is a complex scene, as shown in Figure 10. The tripods are placed parallel to the wall, and the distance is set to about 1 m. The experiment is operated under natural light conditions. We acquire one frame with the laser scanner and 30 consecutive frames with the SR4000 camera. The integral time in the SR_3D_View software of the SR4000 camera is set to "Auto".

Figure 10. Complex scene.

We then choose one of the 30 consecutive frames for analysis and, as shown in Figure 11, obtain the point cloud data of the SR4000 and the laser scanner. As shown in Figure 11, there is a small amount of deformation in the shape of the target object measured by the SR4000 compared to the laser scanner, especially on the edge of the sensor where the measured object is clearly curved. However, distortion exists on the border of the point cloud data and artifacts appear on the plant.

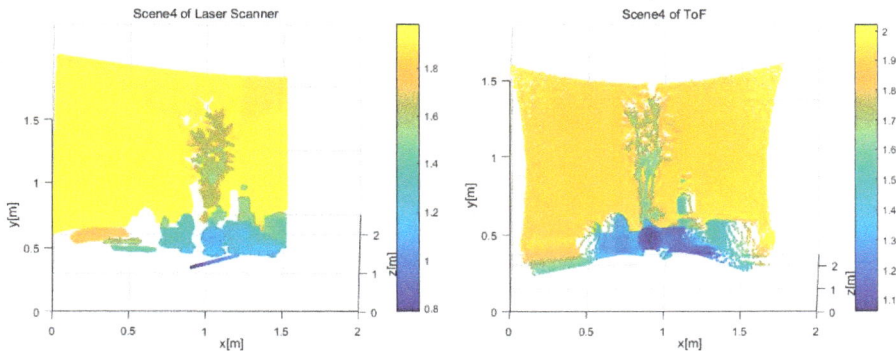

Figure 11. Depth images based on the point cloud of depth sensors.

3.5. Analysis of Depth Errors

From the above four groups of experiments, the depth errors of the SR4000 are weakly affected by lighting conditions (2 mm maximum under the same conditions). The second factor is the target object color. Under the same conditions, this affects the depth error by a maximum of 5 mm. On the other hand, the material has a great influence on the depth errors of ToF cameras. The greater the reflectivity of the measured object material, the greater the depth error, which increases approximately linearly with the distance between the measured object and ToF camera. In a more complex scene, the depth error of a ToF camera is greater. Above all, lighting, object color, material, distance and complex backgrounds could cause different influences on the depth errors of ToF cameras, but it's difficult to summarize this in an error law, because the forms of these errors are uncertain.

4. Depth Error Correction for ToF Cameras

In the last section, four groups of experiments were conducted to analyze the influence of several external factors on the depth errors of ToF cameras. The results of our experiments indicate that different factors have different effects on the measurement results, and it is difficult to establish a unified model to describe and correct such errors. For a complex process that is difficult to model mechanically, an inevitable choice is to use actual measurable input and output data to model. Machine learning is proved to be an effective method to establish non-linear process models. It maps the input space to the output space through a connection model, and the model can approximate a non-linear function with any precision. SVM is a new generic learning method developed on the basis of a statistical learning theory framework. It can seek the best compromise between the complexity of the model and learning ability according to limited sample information so as to obtain the best generalization performance [41,42]. Also in the last section of this paper, we learn and model the depth errors of ToF cameras by using a LS-SVM [43] algorithm.

Better parameters generate better SVM recognition performance to build the LS-SVM model. We need to determine the penalty parameter C and Gaussian kernel parameter γ. Cross-validation [44] is a common method which suffers from large computation demands and long running times. A particle filter [45] can be used to approximate the probability distribution of parameters in the parameter state space by spreading a large number of weighted discrete random variables, based on which, this paper puts forward a parameter selection algorithm, which can fit the depth errors of ToF cameras quickly and meet the requirements of correcting the errors. The process of the PF-SVM algorithm is shown in Figure 12 below.

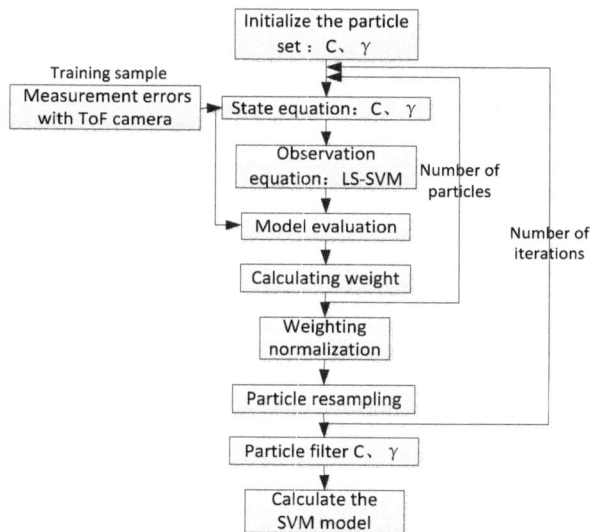

Figure 12. Process of PF-SVM algorithm.

4.1. PF-SVM Algorithm

4.1.1. LS-SVM Algorithm

According to statistical theory, during the process of black-box modeling for non-linear systems, training set $\{x_i, y_i\}$, $i = 1, 2, \ldots, n$ is generally given and non-linear function f is established to minimize Equation (8):

$$f(x) = w^T \varphi(x) + b, \tag{7}$$

$$\min_{w,b,\delta} J(w, \delta) = \frac{1}{2} w^T w + \frac{1}{2} C \sum_{i=1}^{n} \delta^2, \tag{8}$$

where $\varphi(x)$ is a nonlinear function, and w is the weight. Moreover, Equation (8) satisfies the constraint:

$$y_i = w^T \varphi(x_i) + b + \delta_i, i = 1, 2 \cdots n, \tag{9}$$

where $\delta_i \geq 0$ is the relaxation factor, and $C > 0$ is the penalty parameter.

The following equation introduces the Lagrange function L to solve the optimization problem in Equation (8):

$$L = \frac{1}{2}\|w\|^2 + \frac{1}{2}C\sum_{i=1}^{n}\delta_i^2 - \sum_{i=1}^{n}\alpha_i(\varphi(x_i) \cdot w + b + \delta_i - y_i), \tag{10}$$

where α_i is a Lagrange multiplier.

For $i = 1, 2, \ldots n$ by elimination of w and δ, a linear equation can be obtained:

$$\begin{bmatrix} 0 & e^T \\ e & GG^T + C^{-1}I \end{bmatrix}_{(n+1)\times(n+1)} \begin{bmatrix} b \\ \alpha \end{bmatrix} = \begin{bmatrix} 0 \\ y \end{bmatrix}, \tag{11}$$

where e is an element of one n-dimensional column vector, and I is the $n \times n$ unit matrix:

$$G = \begin{bmatrix} \varphi(x_1)^T & \varphi(x_2)^T & \cdots & \varphi(x_n)^T \end{bmatrix}^T, \tag{12}$$

According to the Mercer conditions, the kernel function is defined as follows:

$$K(x_i, x_j) = \varphi(x_i) \cdot \varphi(x_j), \tag{13}$$

We substitute Equations (12) and (13) into Equation (11) to get a linear equation from which α and b can be determined by the least squares method. Then we can obtain the non-linear function approximation of the training data set:

$$y(x) = \sum_{i=1}^{n}\alpha_i K(x, x_i) + b, \tag{14}$$

4.1.2. PF-SVM Algorithm

The depth errors of ToF cameras mentioned above are used as training sample sets $\{x_i, y_i\}$, $i = 1, 2, \ldots n$, where x_i is the camera measurement distance, and y_i is the camera measurement error. Then the error correction becomes a black-box modeling problem of a nonlinear system. Our goal is to determine the nonlinear model f and correct the measurement error with it.

The error model of ToF cameras obtained via the LS-SVM method is expressed in Equation (14). In order to seek a group of optimal parameters for the SVM model to approximate the depth errors in the training sample space, we put this model into a Particle Filter algorithm.

In this paper, the kernel function is:

$$k(x, y) = \exp\left(\frac{-\|x - y\|^2}{2\gamma^2}\right), \tag{15}$$

(1)   Estimation state.

The estimated parameter state x at time k is represented as:

$$x_0^j = \begin{bmatrix} C_0^j & \gamma_0^j \end{bmatrix}^T, \tag{16}$$

where x_0^j is j-th particle when $k = 0$, C is the penalty parameter and γ is the Gauss kernel parameter.

(2) Estimation Model.

The relationship between parameter state x and parameter α,b in non-linear model $y(x)$ can be expressed by state equation $z(\alpha,b)$:

$$z(\alpha, b) = F(\gamma, C),$$ (17)

$$
\begin{bmatrix} b \\ \alpha_1 \\ \vdots \\ \alpha_n \end{bmatrix} = \begin{bmatrix} 0 & 1 & \cdots & 1 \\ 1 & K(x_1, x_1) & \cdots & K(x_1, x_n) \\ \vdots & \vdots & \ddots & \vdots \\ 1 & K(x_n, x_1) & \cdots & K(x_n, x_n) + \frac{1}{C} \end{bmatrix}^{-1} \begin{bmatrix} 0 \\ y_1 \\ \vdots \\ y_n \end{bmatrix},
$$ (18)

where Equation (17) is the deformation of Equation (11).

The relationship between parameter α,b and ToF camera error $y(x)$ can be expressed by observation equation f:

$$y(x) = f(\alpha, b),$$ (19)

$$y(x) = \sum_{i=1}^{n} \alpha_i K(x, x_i) + b,$$ (20)

where Equation (20) is the non-linear model derived from LS-SVM algorithm.

(3) Description of observation target.

In this paper, we use y_i of training set $\{x_i, y_i\}$ as the real description of the observation target, namely the real value of the observation:

$$z = \{y_i\},$$ (21)

(4) The calculation of the characteristic and the weight of the particle observation.

This process is executed when each particle is under characteristic observation. Hence the error values of the ToF camera are calculated according to the sampling of each particle in the parameter state x:

$$\bar{z}^j\left(\bar{\alpha}^j, \bar{b}^j\right) = F\left(\gamma^j, C^j\right),$$ (22)

$$\bar{y}^j(x) = f\left(\bar{\alpha}^j, \bar{b}^j\right),$$ (23)

Here we compute the similarity between the ToF camera error values and the observed target camera values of each particle. The similarity evaluation *RMS* is defined as follows:

$$RMS = \sqrt{\frac{1}{n}\sum_{i=1}^{n}\left(\bar{y}^j - y_i\right)^2},$$ (24)

where $\bar{y}^j$ is the observation value of particle j and y_i is the real error value. The weight value of each particle is calculated according to the Equation (24):

$$w(j) = \frac{1}{\sqrt{2\pi}\sigma}e^{-\frac{RMS^2}{2\sigma}},$$ (25)

Then the weight values are normalized:

$$w^j = \frac{w^j}{\sum_{j=1}^{m} w^j},$$ (26)

(5)   Resampling

Resampling of the particles is conducted according to the normalized weights. In this process, not only the particles with great weights but also a small part of particles with small weights should be kept down.

(6)   Outputting particle set $x_0^j = \begin{bmatrix} C_0^j & \gamma_0^j \end{bmatrix}^T$. This particle set is the optimal LS-SVM parameter.

(7)   The measurement error model of ToF cameras can be obtained by introducing the parameter into the LS-SVM model.

4.2. Experimental Results

We've performed three groups of experiments to verify the effectiveness of the algorithm. In Experiment 1, the depth error model of ToF cameras was modeled with the experimental data in [32], and the results were compared with the error correction results in the original text. In Experiment 2, the depth error model of ToF cameras was modeled with the data in Section 3.1, and the error correction results under different test conditions were compared. In Experiment 3, the error correction results under different reflectivity and different texture conditions were compared.

4.2.1. Experiment 1

In this experiment, we used the depth error data of the ToF cameras which was obtained from Section 3.2 of [32] as the training sample set. The training set consists of 81 sets of data, where x is the distance measurement of the ToF camera and y is the depth error of the ToF camera, as shown in Figure 13 by blue dots. In the figure, the solid green line represents the error modeling results by using the polynomial given in [32]. It shows that the fitting effect is better when the distance is 1.5–4 m, and the maximum absolute error is 8 mm. However, when the distance is less than 1.5 m or more than 4 m, the error model deviated from the true error values. By using our algorithm, we can obtain the results $C = 736$ and $\gamma = 0.003$. By substituting these two parameters into the abovementioned algorithm, we can also obtain the depth error model of the ToF camera as shown in the figure by the red solid line. For this, it can be seen that the error model can match the real errors well.

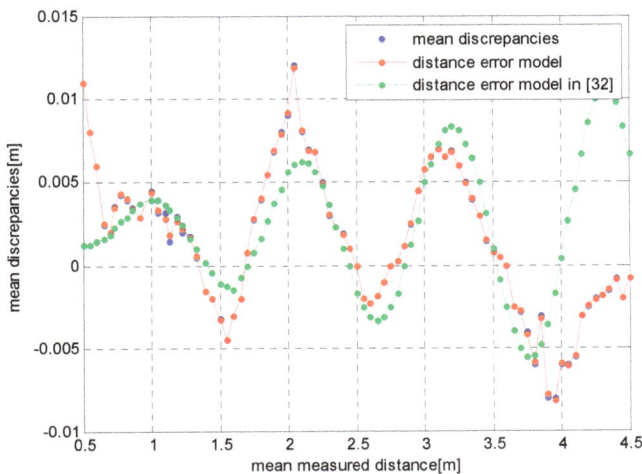

Figure 13. Depth error and error model.

In order to verify the validity of the error model, we use the ToF camera depth error data obtained from Section 3.3 of [32] as a test sample set (the measurement conditions are the same as Section 3.2

of [32]). The test sample set consists of 16 sets of data, as shown in Figure 14 by the blue line. In the figure, the solid green line represents the error modeling results by using the polynomial in [32]. It shows that the fitting effect is better when the distance is 1.5–4 m, and the maximum absolute error is 8.6 mm. However, when the distance is less than 1.5 m or more than 4 m, the error model has deviated from the true error value. The results agree with the fitting effect of the aforementioned error model. The model correction results obtained by using our algorithm are shown by the red solid line in the figure. It shows that the results of the error correction are better when the distance is in 0.5–4.5 m, and the absolute maximum error is 4.6 mm. Table 2 gives the detailed performance comparison results of these two error corrections. From Table 2, we can see that, while expanding the range of error correction, this method can also improve the accuracy of the error correction.

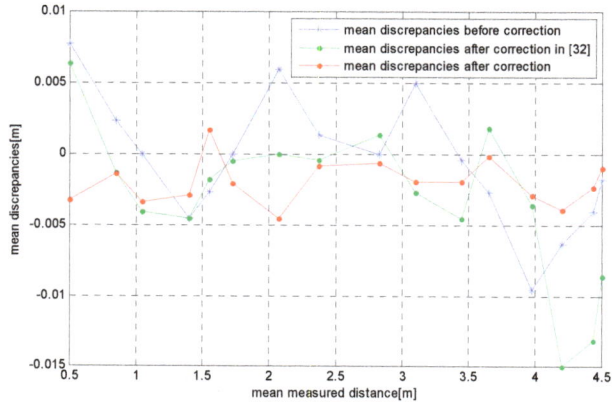

Figure 14. Depth error correction results.

Table 2. Analysis of depth error correction results.

Comparison Items	Maximal Error/mm		Average Error/mm		Variance/mm		Optimal Range/m	Running Time/s
	1.5–4	0.5–4.5	1.5–4	0.5–4.5	1.5–4	0.5–4.5		
This paper's algorithm	4.6	4.6	1.99	2.19	2.92	2.4518	0.5–4.5	2
Reference [32] algorithm	4.6	8.6	2.14	4.375	5.34	29.414	1.5–4	-

4.2.2. Experiment 2

The ToF depth error data of Section 3.1 on the condition of blue background is selected as the training sample set. As shown in Figure 15 by blue asterisks, the training set consists of eight sets of data. The error model established by our algorithm is shown by the blue line in Figure 15. The model can fit the error data well, but the training sample set should be as rich as possible in order to build the accuracy of the model. To verify the applicability of the error model, we use white, green and red background ToF depth error data as test samples, and the data after correction is shown in the figure by the black, green and red lines. It can be seen from the figure that the absolute values of the three groups of residual errors is less than the uncorrected error data after the application of the blue distance error model. The figure also illustrates that this error model is very applicable to the error correction of ToF cameras for different color backgrounds.

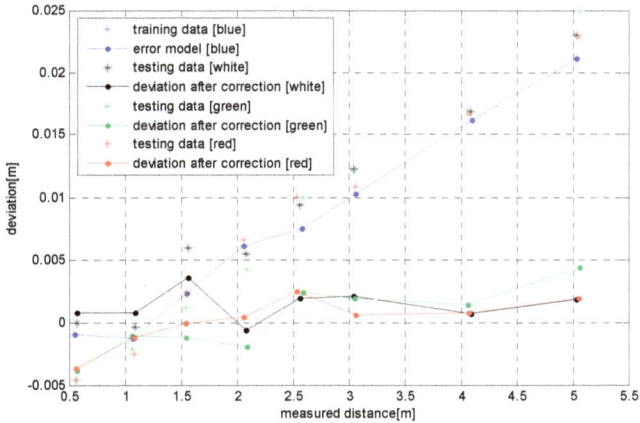

Figure 15. Depth error correction results of various colors and error model.

4.2.3. Experiment 3

The experimental process similar to that of Section 3.1 hereof was adopted in order to verify the validity of the error modeling method under different reflectivity and different texture conditions. The sample set, including 91 groups of data, involved the depth errors obtained from the white wall surfaces photographed with a ToF camera at different distances, as shown with the blue solid lines in Figure 16. The error model established by use of the algorithm herein is shown with the red solid lines in Figure 16. The figure indicates that this model fits the error data better. With a newspaper fixed on the wall as the test target, the depth errors obtained with a ToF camera at different distances are taken as the test data, as shown with the black solid lines in Figure 16, while the data corrected through the error model created here are shown with the green solid lines in the same figure. It can be seen from the figure that the absolute values of residual errors is less than the uncorrected error data after the application of the distance error model. The figure also illustrates that this error model is very applicable to the full measurement range of ToF cameras.

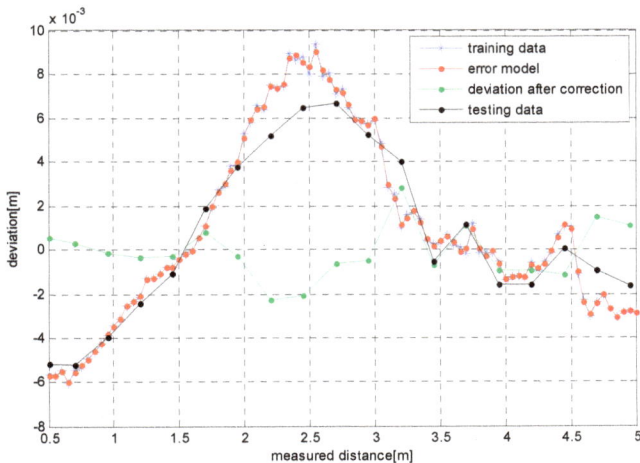

Figure 16. Depth error correction results and error model.

5. Conclusions

In this paper, we analyzed the influence of some typical external distractions, such as material properties and color of the target object, distance, lighting and so on on the depth errors of ToF cameras. Our experiments indicate that lighting, color, material and distance could cause different influences on the depth errors of ToF cameras. As the distance becomes longer, the depth errors of ToF cameras increase roughly linearly. To further improve the measurement accuracy of ToF cameras, this paper puts forward an error correction method based on Particle Filter-Support Vector Machine (PF-SVM). Then, the best parameters with particle filter algorithm on the basis of learning the depth errors of ToF cameras are selected. The experimental results indicate that this method can reduce the depth error from 8.6 mm to 4.6 mm within its full measurement range (0.5–5 m).

Acknowledgments: This research was supported by National Natural Science Foundation of China (No. 61305112).

Author Contributions: Ying He proposed the idea; Ying He, Bin Liang and Yu Zou conceived and designed the experiments; Ying He, Yu Zou, Jin He and Jun Yang performed the experiments; Ying He, Yu Zou, Jin He and Jun Yang analyzed the data; Ying He wrote the manuscript; and Bin Liang provided the guidance for data analysis and paper writing.

Conflicts of Interest: The authors declare no conflict of interest.

References

1. Henry, P.; Krainin, M.; Herbst, E.; Ren, X.; Fox, D. RGB-D mapping: Using kinect-style depth cameras for dense 3D modeling of indoor environments. *Int. J. Robot. Res.* **2012**, *31*, 647–663. [CrossRef]
2. Brachmann, E.; Krull, A.; Michel, F.; Gumhold, S.; Shotton, J.; Rother, C. *Learning 6D Object Pose Estimation Using 3D Object Coordinates*; Springer: Heidelberg, Germany, 2014; Volume 53, pp. 151–173.
3. Tong, J.; Zhou, J.; Liu, L.; Pan, Z.; Yan, H. Scanning 3D full human bodies using kinects. *IEEE Trans. Vis. Comput. Graph.* **2012**, *18*, 643–650. [CrossRef] [PubMed]
4. Liu, X.; Fujimura, K. Hand gesture recognition using depth data. In Proceedings of the IEEE International Conference on Automatic Face and Gesture Recognition, Seoul, Korea, 17–19 May 2004; pp. 529–534.
5. Foix, S.; Alenya, G.; Torras, C. Lock-in time-of-flight (ToF) cameras: A survey. *IEEE Sens. J.* **2011**, *11*, 1917–1926. [CrossRef]
6. Wiedemann, M.; Sauer, M.; Driewer, F.; Schilling, K. Analysis and characterization of the PMD camera for application in mobile robotics. *IFAC Proc. Vol.* **2008**, *41*, 13689–13694. [CrossRef]
7. Fuchs, S.; May, S. Calibration and registration for precise surface reconstruction with time of flight cameras. *Int. J. Int. Syst. Technol. App.* **2008**, *5*, 274–284. [CrossRef]
8. Guomundsson, S.A.; Aanæs, H.; Larsen, R. Environmental effects on measurement uncertainties of time-of-flight cameras. In Proceedings of the 2007 International Symposium on Signals, Circuits and Systems, Iasi, Romania, 12–13 July 2007; Volumes 1–2, pp. 113–116.
9. Rapp, H. Experimental and Theoretical Investigation of Correlating ToF-Camera Systems. Master's Thesis, University of Heidelberg, Heidelberg, Germany, September 2007.
10. Falie, D.; Buzuloiu, V. Noise characteristics of 3D time-of-flight cameras. In Proceedings of the 2007 International Symposium on Signals, Circuits and Systems, Iasi, Romania, 12–13 July 2007; Volumes 1–2, pp. 229–232.
11. Karel, W.; Dorninger, P.; Pfeifer, N. In situ determination of range camera quality parameters by segmentation. In Proceedings of the VIII International Conference on Optical 3-D Measurement Techniques, Zurich, Switzerland, 9–12 July 2007; pp. 109–116.
12. Kahlmann, T.; Ingensand, H. Calibration and development for increased accuracy of 3D range imaging cameras. *J. Appl. Geodesy* **2008**, *2*, 1–11. [CrossRef]
13. Karel, W. Integrated range camera calibration using image sequences from hand-held operation. *Int. Arch. Photogramm. Remote Sens. Spat. Inf. Sci.* **2008**, *37*, 945–952.
14. May, S.; Werner, B.; Surmann, H.; Pervolz, K. 3D time-of-flight cameras for mobile robotics. In Proceedings of the 2006 IEEE/RSJ International Conference on Intelligent Robots and Systems, Beijing, China, 9–15 October 2006; Volumes 1–12, pp. 790–795.

15. Kirmani, A.; Benedetti, A.; Chou, P.A. Spumic: Simultaneous phase unwrapping and multipath interference cancellation in time-of-flight cameras using spectral methods. In Proceedings of the IEEE International Conference on Multimedia and Expo (ICME), San Jose, CA, USA, 15–19 July 2013; pp. 1–6.

16. Freedman, D.; Krupka, E.; Smolin, Y.; Leichter, I.; Schmidt, M. Sra: Fast removal of general multipath for ToF sensors. In Proceedings of the European Conference on Computer Vision, Zurich, Switzerland, 6–12 September 2014.

17. Kadambi, A.; Whyte, R.; Bhandari, A.; Streeter, L.; Barsi, C.; Dorrington, A.; Raskar, R. Coded time of flight cameras: Sparse deconvolution to address multipath interference and recover time profiles. *ACM Trans. Graph.* **2013**, *32*, 167. [CrossRef]

18. Whyte, R.; Streeter, L.; Gree, M.J.; Dorrington, A.A. Resolving multiple propagation paths in time of flight range cameras using direct and global separation methods. *Opt. Eng.* **2015**, *54*, 113109. [CrossRef]

19. Lottner, O.; Sluiter, A.; Hartmann, K.; Weihs, W. Movement artefacts in range images of time-of-flight cameras. In Proceedings of the 2007 International Symposium on Signals, Circuits and Systems, Iasi, Romania, 13–14 July 2007; Volumes 1–2, pp. 117–120.

20. Lindner, M.; Kolb, A. Compensation of motion artifacts for time-of flight cameras. In *Dynamic 3D Imaging*; Springer: Heidelberg, Germany, 2009; Volume 5742, pp. 16–27.

21. Lee, S.; Kang, B.; Kim, J.D.K.; Kim, C.Y. Motion Blur-free time-of-flight range sensor. *Proc. SPIE* **2012**, *8298*, 105–118.

22. Lee, C.; Kim, S.Y.; Kwon, Y.M. Depth error compensation for camera fusion system. *Opt. Eng.* **2013**, *52*, 55–68. [CrossRef]

23. Kuznetsova, A.; Rosenhahn, B. On calibration of a low-cost time-of-flight camera. In Proceedings of the Workshop at the European Conference on Computer Vision, Zurich, Switzerland, 6–12 September 2014; Lecture Notes in Computer Science. Volume 8925, pp. 415–427.

24. Lee, S. Time-of-flight depth camera accuracy enhancement. *Opt. Eng.* **2012**, *51*, 527–529. [CrossRef]

25. Christian, J.A.; Cryan, S. A survey of LIDAR technology and its use in spacecraft relative navigation. In Proceedings of the AIAA Guidance, Navigation, and Control Conference, Boston, MA, USA, 19–22 August 2013; pp. 1–7.

26. Nitzan, D.; Brain, A.E.; Duda, R.O. Measurement and use of registered reflectance and range data in scene analysis. *Proc. IEEE* **1977**, *65*, 206–220. [CrossRef]

27. Spirig, T.; Seitz, P.; Vietze, O. The lock-in CCD 2-dimensional synchronous detection of light. *IEEE J. Quantum Electron.* **1995**, *31*, 1705–1708. [CrossRef]

28. Schwarte, R. Verfahren und vorrichtung zur bestimmung der phasen-und/oder amplitude information einer elektromagnetischen Welle. DE Patent 19,704,496, 12 March 1998.

29. Lange, R.; Seitz, P.; Biber, A.; Schwarte, R. Time-of-flight range imaging with a custom solid-state image sensor. *Laser Metrol. Inspect.* **1999**, *3823*, 180–191.

30. Piatti, D.; Rinaudo, F. SR-4000 and CamCube3.0 time of flight (ToF) cameras: Tests and comparison. *Remote Sens.* **2012**, *4*, 1069–1089. [CrossRef]

31. Chiabrando, F.; Piatti, D.; Rinaudo, F. SR-4000 ToF camera: Further experimental tests and first applications to metric surveys. *Int. Arch. Photogramm. Remote Sens. Spat. Inf. Sci.* **2010**, *38*, 149–154.

32. Chiabrando, F.; Chiabrando, R.; Piatti, D. Sensors for 3D imaging: Metric evaluation and calibration of a CCD/CMOS time-of-flight camera. *Sensors* **2009**, *9*, 10080–10096. [CrossRef] [PubMed]

33. Charleston, S.A.; Dorrington, A.A.; Streeter, L.; Cree, M.J. Extracting the MESA SR4000 calibrations. In Proceedings of the Videometrics, Range Imaging, and Applications XIII, Munich, Germany, 22–25 June 2015; Volume 9528.

34. Ye, C.; Bruch, M. A visual odometry method based on the SwissRanger SR4000. *Proc. SPIE* **2010**, *7692*, 76921I.

35. Hong, S.; Ye, C.; Bruch, M.; Halterman, R. Performance evaluation of a pose estimation method based on the SwissRanger SR4000. In Proceedings of the IEEE International Conference on Mechatronics and Automation, Chengdu, China, 5–8 August 2012; pp. 499–504.

36. Lahamy, H.; Lichti, D.; Ahmed, T.; Ferber, R.; Hettinga, B.; Chan, T. Marker-less human motion analysis using multiple Sr4000 range cameras. In Proceedings of the 13th International Symposium on 3D Analysis of Human Movement, Lausanne, Switzerland, 14–17 July 2014.

37. Kahlmann, T.; Remondino, F.; Ingensand, H. Calibration for increased accuracy of the range imaging camera Swissranger^{TM}. In Proceedings of the ISPRS Commission V Symposium Image Engineering and Vision Metrology, Dresden, Germany, 25–27 September 2006; pp. 136–141.

38. Weyer, C.A.; Bae, K.; Lim, K.; Lichti, D. Extensive metric performance evaluation of a 3D range camera. *Int. Soc. Photogramm. Remote Sens.* **2008**, *37*, 939–944.

39. Mure-Dubois, J.; Hugli, H. Real-Time scattering compensation for time-of-flight camera. In Proceedings of the ICVS Workshop on Camera Calibration Methods for Computer Vision Systems, Bielefeld, Germany, 21–24 March 2007.

40. Kavli, T.; Kirkhus, T.; Thielmann, J.; Jagielski, B. Modeling and compensating measurement errors caused by scattering time-of-flight cameras. In Proceedings of the SPIE, Two-and Three-Dimensional Methods for Inspection and Metrology VI, San Diego, CA, USA, 10 August 2008.

41. Vapnik, V.N. *The Nature of Statistical Learning Theory*; Springer: New York, NY, USA, 1995.

42. Ales, J.S.; Bernhand, S. A tutorialon support vector regression. *Stat. Comput.* **2004**, *14*, 199–222.

43. Suykens, J.A.K.; Vandewalle, J. Least squares support vector machine classifiers. *Neural Process. Lett.* **1999**, *9*, 293–300. [CrossRef]

44. Zhang, J.; Wang, S. A fast leave-one-out cross-validation for SVM-like family. *Neural Comput. Appl.* **2016**, *27*, 1717–1730. [CrossRef]

45. Gustafsson, F. Particle filter theory and practice with positioning applications. *IEEE Aerosp. Electron. Syst. Mag.* **2010**, *25*, 53–82. [CrossRef]

sensors

MDPI

Article

Expanding the Detection of Traversable Area with RealSense for the Visually Impaired

Kailun Yang, Kaiwei Wang *, Weijian Hu and Jian Bai

College of Optical Science and Engineering, Zhejiang University, Hangzhou 310027, China;
elnino@zju.edu.cn (K.Y.); huweijian@zju.edu.cn (W.H.); bai@zju.edu.cn (J.B.)
* Correspondence: wangkaiwei@zju.edu.cn; Tel.: +86-571-8795-3154

Academic Editor: Gonzalo Pajares Martinsanz
Received: 13 September 2016; Accepted: 8 November 2016; Published: 21 November 2016

Abstract: The introduction of RGB-Depth (RGB-D) sensors into the visually impaired people (VIP)-assisting area has stirred great interest of many researchers. However, the detection range of RGB-D sensors is limited by narrow depth field angle and sparse depth map in the distance, which hampers broader and longer traversability awareness. This paper proposes an effective approach to expand the detection of traversable area based on a RGB-D sensor, the Intel RealSense R200, which is compatible with both indoor and outdoor environments. The depth image of RealSense is enhanced with IR image large-scale matching and RGB image-guided filtering. Traversable area is obtained with RANdom SAmple Consensus (RANSAC) segmentation and surface normal vector estimation, preliminarily. A seeded growing region algorithm, combining the depth image and RGB image, enlarges the preliminary traversable area greatly. This is critical not only for avoiding close obstacles, but also for allowing superior path planning on navigation. The proposed approach has been tested on a score of indoor and outdoor scenarios. Moreover, the approach has been integrated into an assistance system, which consists of a wearable prototype and an audio interface. Furthermore, the presented approach has been proved to be useful and reliable by a field test with eight visually impaired volunteers.

Keywords: RGB-D sensor; RealSense; visually impaired people; traversable area detection

1. Introduction

According to the World Health Organization, 285 million people were estimated to be visually impaired and 39 million of them are blind around the world in 2014 [1]. It is very difficult for visually impaired people (VIP) to find their way through obstacles and wander in real-world scenarios. Recently, RGB-Depth (RGB-D) sensors revolutionized the research field of VIP aiding because of their versatility, portability, and cost-effectiveness. Compared with traditional assistive tools, such as a white cane, RGB-D sensors provide a great deal of information to the VIP. Typical RGB-D sensors, including light-coding sensors, time-of-flight sensors (ToF camera), and stereo cameras are able to acquire color information and perceive the environment in three dimensions at video frame rates. These depth-sensing technologies already have their mature commercial products, but each type of them has its own set of limits and requires certain working environments to perform well, which brings not only new opportunities but also challenges to overcome.

Light-coding sensors, such as PrimeSense [2] (developed by PrimeSense based in Tel Aviv, Israel), Kinect [3] (developed by Microsoft based in Redmond, WA, USA), Xtion Pro [4] (developed by Asus based in Taipei, Taiwan), MV4D [5] (developed by Mantis Vision based in Petach Tikva, Israel), and the Structure Sensor [6] (developed by Occipital based in San Francisco, CA, USA) project near-IR laser speckles to code the scene. Since the distortion of the speckles depends on the depth of objects, an IR CMOS image sensor captures the distorted speckles and a depth map is generated through triangulating algorithms. However, they fail to return an efficient depth map in sunny environments because projected speckles are submerged by sunlight. As a result, approaches for VIP with light-coding sensors are just proof-of-concepts or only feasible in indoor environments [7–15].

ToF cameras, such as CamCube [16] (developed by PMD Technologies based in Siegen, Germany), DepthSense [17] (developed by SoftKinetic based in Brussels, Belgium), and SwissRanger (developed by Heptagon based in Singapore) [18] resolve distance based on the known speed of light, measuring the precise time of a light signal flight between the camera and the subject independently for each pixel of the image sensor. However, they are susceptible to ambient light. As a result, ToF camera-based approaches for VIP show poor performance in outdoor environments [19–21].

Stereo cameras, such as the Bumblebee [22] (developed by PointGrey based in Richmond, BC, Canada), ZED [23] (developed by Stereolabs based in San Francisco, USA), and DUO [24] (developed by DUO3D based in Henderson, NV, USA) estimates the depth map through stereo matching of images from two or more lenses. Points on one image are correlated to another image and depth is calculated via shift between a point on one image and another image. Stereo matching is a passive and texture-dependent process. As a result, stereo cameras return sparse depth images in textureless indoor scenes, such as a blank wall. This explains why solutions for VIP with stereo camera focus mainly on highly-textured outdoor environments [25–28].

The RealSense R200 (developed by Intel based in Santa Clara, CA, USA) uses a combination of active projecting and passive stereo matching [29]. IR laser projector projects static non-visible near-IR patterns on the scene, which is then acquired by the left and right IR cameras. The image processor generates a depth map through an embedded stereo-matching algorithm. In textureless indoor environments, the projected patterns enrich textures. As shown in Figure 1b,c, the texture-less white wall has been projected with many near-IR patterns which are beneficial for stereo matching to generate depth information. In sunny outdoor environments, although projected patterns are submerged by sunlight, the near-IR component of sunlight shines on the scene to form well-textured IR images as shown in Figure 1g. With the contribution of abundant textures to robust stereo matching, the combination allows the RealSense R200 to work under indoor and outdoor circumstances, delivering depth images though it has many noise sources, mismatched pixels, and black holes. In addition, it is possible to attain denser depth maps pending new algorithms. Illustrated in Figure 1, the RealSense R200 is quite suitable for navigational assistance thanks not only to its environment adaptability, but also its small size.

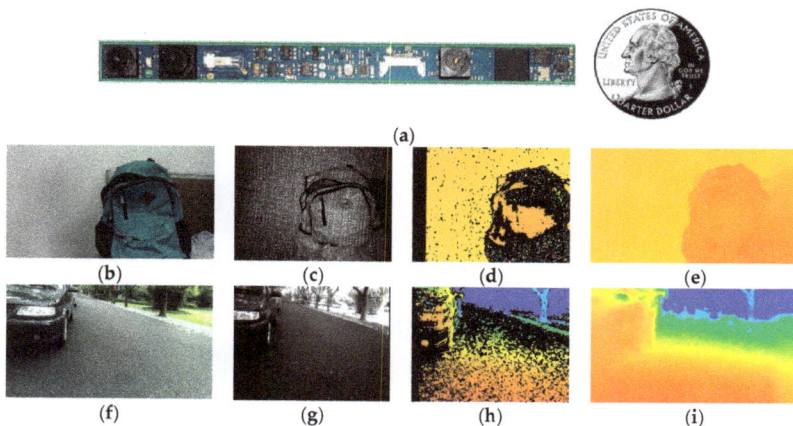

Figure 1. (a) The RealSense R200; (b,f) color image captured by the RealSense R200; (c,g) IR image captured by the right IR camera of the RealSense R200; (d,h) the original depth image from the RealSense R200; and (e,i) the guided filtered depth image acquired in our work.

However, the depth range of the RGB-D sensor is generally short. For the light-coding sensor, the speckles in the distance are too dark to be sensed. For the ToF camera, light signals are overwhelmed by ambient light in the distance. For stereo-cameras, since depth error increases with the increase of the depth value, stereo-cameras are prone to be unreliable in the distance [30]. For the RealSense R200, on the one hand, since the power of IR laser projector is limited, if the coded object is in the distance, the speckles are too dark and sparse to enhance stereo matching. On the other hand, depth information in the distance is much less accurate than that in the normal working distance ranging from 650–2100 mm [31]. As shown in Figure 2, the original depth image is sparse a few meters away. In addition, the depth field angle of RGB-D sensor is generally small. For the RealSense R200, the horizontal field angle of IR camera is 59°. As we know, the depth image is generated through stereo matching from overlapping field angles of two IR cameras. Illustrated in Figure 3, though red and green light are within the horizontal field angle of the left IR camera, only green light is within the overlapping field angle of two IR cameras. Thus, the efficient depth horizontal field angle is smaller than 59°, which is the horizontal field angle of a single IR camera. Consequently, as depicted in Figure 2, both the distance and the angle range of the ground plane detection with the original depth image are small, which hampers longer and broader traversable area awareness for VIP.

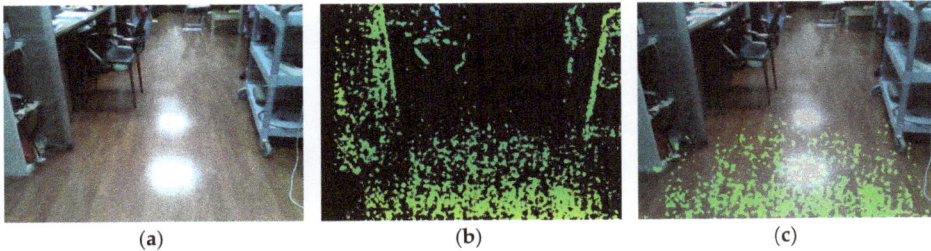

Figure 2. (**a**) Color image captured by the RealSense R200; (**b**) the original depth image captured by the RealSense R200; (**c**) traversable area detection with original depth image of the RealSense R200, which is limited to short range.

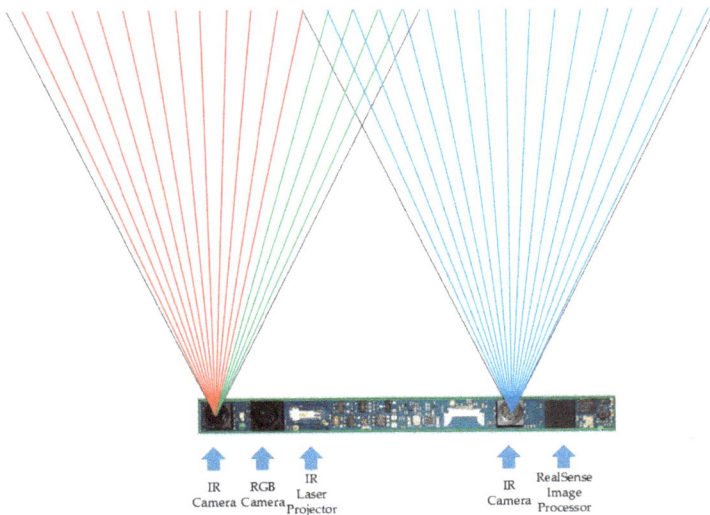

Figure 3. Horizontal field angle of IR cameras.

In this paper, an effective approach to expand the traversable area detection is proposed. Since the original depth image is poor and sparse, two IR images are large-scale matched to generate a dense depth image. Additionally, the quality of the depth image is enhanced with the RGB image-guided filtering, which is comprised of functions, such as de-noising, hole-filling, and can estimate the depth map from the perspective of the RGB camera, whose horizontal field angle is wider than the depth camera. The preliminary traversable area is obtained with RANdom SAmple Consensus (RANSAC) segmentation [32]. In addition to the RealSense R200, an attitude sensor, InvenSense MPU6050 [33], is employed to adjust the point cloud from the camera coordinate system to the world coordinate system. This helps to eliminate sample errors in preliminary traversable area detection. Through estimating surface normal vectors of depth image patches, salient parts are removed from preliminary detection results. The highlighted process of the traversable area detection is to extend preliminary results to broader and longer ranges, which fully combines depth and color images. On the one hand, short-range depth information is enhanced with long-range RGB information. On the other hand, depth information adds a dimension of restrictions to the expansion stage based on seeded region growing algorithm [34]. The approach proposed in this paper is integrated with a wearable prototype, containing a bone-conduction headphone, which provides a non-semantic stereophonic interface. Different from most navigational assistance approaches, which are not tested by VIP, eight visually impaired volunteers, three in whom are suffering from total blindness, have tried out our approach.

This paper is organized as follows: in Section 2, related work that has addressed both traversable area detection and expansion are reviewed; in Section 3, the presented approach is elaborated in detail; in Section 4, extensive tests on indoor and outdoor scenarios demonstrate its effectiveness and robustness; in Section 5, the approach is validated by the user study, effected by real VIP; and in Section 6, relevant conclusions are drawn and outlooks to future work are depicted.

2. Related Work

In the literature, a lot of approaches have been proposed with respect to ground plane segmentation, access section detection, and traversable area awareness with RGB-D sensors.

In some approaches, ground plane segmentation is the first step of obstacle detection, which aims to separate feasible ground area from hazardous obstacles. Wang adopted meanshift segmentation to separate obstacles based on the depth image from a Kinect, in which planes are regarded as feasible areas if two conditions are met: the angle between the normal vector of the fitting plane and vertical direction of the camera coordinate system is less than a threshold; and the average distance and the standard deviation of all 3D points to the fitting plane are less than thresholds [35]. Although the approach achieved good robustness under certain environment, the approach relies a lot on thresholds and assumptions. Cheng put forward an algorithm to detect ground with a Kinect based on seeded region growing [15]. Instead of focusing on growing thresholds, edges of the depth image and boundaries of the region are adequately considered. However, the algorithm is unduly dependent on the depth image, and the seed pixels are elected according to a random number, causing fluctuations between frames, which is intolerable for assisting because unstable results would confuse VIP. Rodríguez simply estimated outdoor ground plane based on RANSAC plus filtering techniques, and used a polar grid representation to account for the potential obstacles [25]. The approach is one of the few which have involved real VIP participation. However, the approach yields a ground plane detection error in more than ten percent of the frames, which is resolvable in our work.

In some approaches, the problem of navigable ground detection is addressed in conjunction with localization tasks. Perez-Yus used the RANSAC algorithm to segment planes in human-made indoor scenarios pending dense 3D point clouds. The approach is able to extract not only the ground but also ascending or descending stairs, and to determine the position and orientation of the user with visual odometry [36]. Lee also incorporated visual odometry and feature-based metri-topological simultaneous localization and mapping (SLAM) [37] to perform traversability analysis [26,38]. The navigation system extracts ground plane to reduce drift imposed by the head-mounted RGB-D sensor and the paper demonstrated that the traversability map works more robustly with a light-coding sensor than with a stereo pair in low-textured environments. As for another indoor localization application, Sánchez detected floor and navigable areas to efficiently reduce the search space and thereby yielded real-time performance of both place recognition and tracking [39].

In some approaches, surface normal vectors on the depth map have been used to determine the accessible section. Koester detected the accessible section by calculating the gradients and estimating surface normal vector directions of real-world scene patches [40]. The approach allows for a fast and effective accessible section detection, even in crowded scenes. However, it prevents practical application for user studies with the overreliance on the quality of 3D reconstruction process and adherence to constraints such as the area directly in front of the user is accessible. Bellone defined a novel descriptor to measure the unevenness of a local surface based on the estimation of normal vectors [41]. The index gives an enhanced description of the traversable area which takes into account both the inclination and roughness of the local surface. It is possible to perform obstacle avoidance and terrain traversability assessments simultaneously. However, the descriptor computation is complex and also relies on the sensor to generate dense 3D point clouds. Chessa derived the normal vectors to estimate surface orientation for collision avoidance and scene interpretation [42]. The framework uses a disparity map as a powerful cue to validate the computation from optic flow, which suffers from the drawback of being sensitive to errors in the estimates of optical flow.

In some approaches, range extension are concerned to tackle the limitations imposed by RGB-D sensors. Muller presented a self-supervised learning process to accurately classify long-range terrain as traversable or not [43]. It continuously receives images, generates supervisory labels, trains a classifier, and classifies the long-range portion of the images, which complete one full cycle every half second. Although the system classifies the traversable area of the image up to the horizon, the feature extraction requires large, distant image patches within fifteen meters, limiting the utility in general applications with commercial RGB-D sensors, which ranges mush closer. Reina proposed a self-learning framework to automatically train a ground classifier with multi-baseline stereovision [44]. Two distinct classifiers include one based on geometric data, which detects the broad class of ground, and one based on color data, which further segments ground into subclasses. The approach makes predictions based on past observations, and the only underlying assumption is that the sensor is initialized from an area free of obstacles, which is typically violated in applications of VIP assisting. Milella features a radar-stereo system to address terrain traversability assessment in the context of outdoor navigation [45,46]. The combination produces reliable results in the short range and trains a classifier operating on distant scenes. Damen also presented an unsupervised approach towards automatic video-based guidance in miniature and in fully-wearable form [47]. These self-learning strategies make feasible navigation in long-range and long-duration applications, but they ignore the fact that most traversable pixels or image patches are connected parts rather than detached, which is fully considered in our approach, and also supports an expanded range of detection. Aladrén combines depth information with image intensities, robustly expands the range-based indoor floor segmentation [9]. The overall diagram of the method composes complex processes, running at approximately 0.3 frames per second, which fails to assist VIP at normal walking speed.

Although plenty of related works have been done to analyze traversable area with RGB-D sensors, most of them are overly dependent on the depth image or cause intolerable side effects in navigational assistance for VIP. Compared with these works, the main advantages of our approach can be summarized as follows:

- The 3D point cloud generated from the RealSense R200 is adjusted from the camera coordinate system to the world coordinate system with a measured sensor attitude angle, such that the sample errors are decreased to a great extent and the preliminary plane is segmented correctly.
- The seeded region, growing adequately, considers the traversable area as connected parts, and expands the preliminary segmentation result to broader and longer ranges with RGB information.

- The seeded region growing starts with preliminarily-segmented pixels other than according to the random number, thus the expansion is inherently stable between frames, which means the output will not fluctuate and confuse VIP. The seeded region growing is not reliant on a single threshold, and edges of the RGB image and depth differences are also considered to restrict growing into non-traversable area.
- The approach does not require the depth image from sensor to be accurate or dense in long-range area, thus most consumer RGB-D sensors meet the requirements of the algorithm.
- The sensor outputs efficient IR image pairs under both indoor and outdoor circumstances, ensuring practical usability of the approach.

3. Approach

In this section, the approach to expand traversable area detection with the RealSense sensor is elaborated in detail. The flow chart of the approach is show in Figure 4. The approach is described in terms of depth image enhancement, preliminary ground segmentation, and seeded region expansion, accordingly.

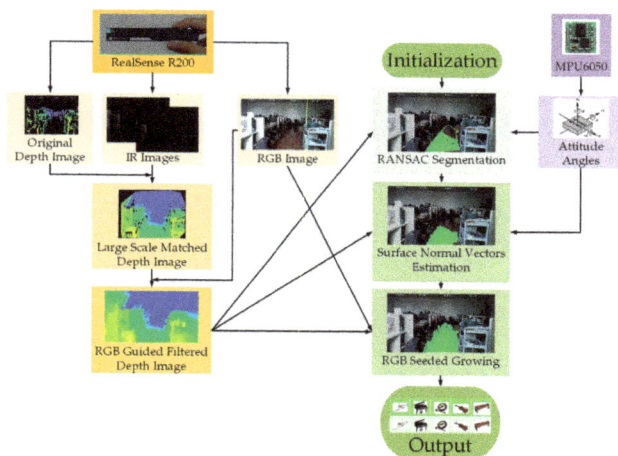

Figure 4. The flowchart of the approach.

3.1. Depth Image Enhancement

The original depth image from the RealSense R200 is sparse and there are many holes, noises, and mismatched pixels. Besides, the embedded stereo-matching algorithm in the processor is fixed, which is unable to be altered. The embedded algorithm is based on local correspondences, and parameters are fixed with the algorithm, such as the texture threshold and uniqueness ratio, limiting the original depth map to be sparse. Typical original depth images are shown in Figure 1d,h. Comparatively, IR images from the RealSense are large-scale matched in our work.

To yield a dense depth map with calibrated IR images, original efficient depth pixels are included in the implementation of efficient large-scale stereo matching algorithm [48]. Support pixels are denoted as pixels which can be robustly matched due to their textures and uniqueness. Sobel masks with fixed size of 3×3 pixels and a large disparity search range are used to perform stereo matching and obtain support pixels. As Sobel filter responses are good, but still insufficient, for stereo matching, original depth image pixels are added to the support pixels. In addition, a multi-block-matching principle [49] is employed to obtain more robust and sufficient support matches from real-world textures. Given the resolution of IR images is 628×468, the best block sizes found with IR pairs are

$41 \times 1, 1 \times 41, 9 \times 9$, and 3×3. Then, the approach estimates the depth map by forming triangulation on a set of support pixels and interpolating disparities. As shown in Figure 5, the large-scale matched depth image is much denser than the original depth map, especially in less-textured scenarios, even though these original depth images are the denser ones acquired with the sensor.

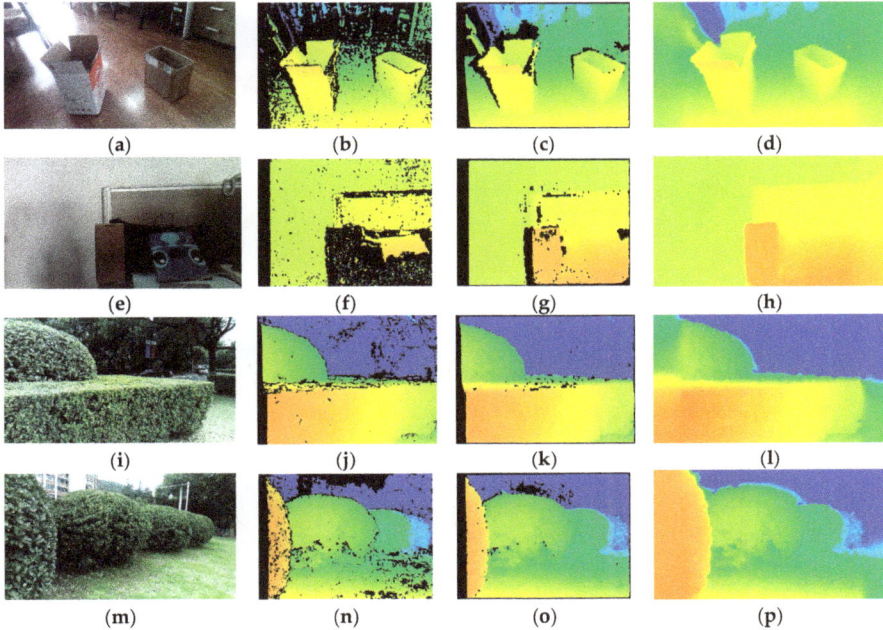

Figure 5. Comparison of depth maps under both indoor and outdoor environments. (**a,e,i,m**) Color images captured by the RealSense sensor; (**b,f,j,n**) original depth image from the RealSense sensor; (**c,g,k,o**) large-scale matched depth image; and (**d,h,l,p**) guided-filter depth image.

However, there are still many holes and noises in the large-scale depth image. Moreover, the horizontal field view of the depth image is narrow, which hampers broad navigation. In order to take advantage of available color images acquired with the RealSense R200 instead of filling invalid regions in a visually plausible way using only depth information, we incorporate a color image and apply the guided filter [50] to refine and estimate the depth of unknown areas. In this work, we implement a RGB guided filter within the interface of enhanced photography algorithms [51] to improve the depth image, which is to fill holes, de-noise and, foremost, estimate the depth map from the field view of the RGB camera. The color image, depth image, and calibration data are input to the post-process, within which the original depth image is replaced by a large-scale matched depth image. Firstly, depth information from the perspective of one IR camera is projected onto the RGB image with both IR cameras and the RGB camera calibration parameters. In this process, depth values are extracted from the large-scale matched depth image instead of original depth image. Secondly, a color term is introduced so that the weighting function in the guided filter is able to combine color information for depth inpainting. This color-similarity term is based on an assumption that neighboring pixels with similar color are likely to have similar depth values. In addition, there are filter terms which decide that the contribution of depth values to an unknown pixel varies according to geometric distance and direction. Additionally, the pixels near the edges of the color image are estimated later than the pixels which are far away from them to preserve fine edges. Overall, the interface of enhanced photography algorithms is hardware accelerated with OpenCL, so it is computationally efficient to be used in the approach to obtain smoother and denser depth images, which are beneficial for both the detection and the expansion of the traversable area. Shown in Figure 5, the presented approach remarkably smooths and improves the density of the original depth image from the RealSense sensor: firstly, the horizontal field angle of depth image has increased from 59° to 70°, which is the field angle of the color camera, allowing for broader detection; secondly, the filtered depth image has far less noise and fewer mismatches than the original depth image; lastly, the guided filtered depth image achieves 100% density.

3.2. Preliminary Ground Segmentation

In order to detect the ground, a simple and effective technique is presented. Firstly, 3D coordinates of the point cloud are calculated. Given the depth Z of pixel (u, v) in the depth image, the calibrated focal length f, and (u_0, v_0) the principal point, the point cloud in the camera coordinate system can be determined using Equations (1) and (2):

$$X = Z \times \frac{u - u_0}{f} \tag{1}$$

$$Y = Z \times \frac{v - v_0}{f} \tag{2}$$

On the strength of the attitude sensor, X, Y, and Z coordinates in the camera coordinate system can be adjusted to world coordinates. Assume a point in the camera coordinate system is (X, Y, Z) and the attitude angles acquired from the attitude sensor are (a, b, c). This means the point (X, Y, Z) rotates about the x-axis by $\alpha = a$, then rotates about the y-axis by $\beta = b$ and rotates about z-axis by $\gamma = c$ in the end. Shown in Equation (3), multiplying the point (X, Y, Z) by the rotation matrix, and the point (X_w, Y_w, Z_w) in world coordinates is obtained:

$$\begin{bmatrix} X_w \\ Y_w \\ Z_w \end{bmatrix} = \begin{bmatrix} \cos\gamma & -\sin\gamma & 0 \\ \sin\gamma & \cos\gamma & 0 \\ 0 & 0 & 1 \end{bmatrix} \begin{bmatrix} \cos\beta & 0 & \sin\beta \\ 0 & 1 & 0 \\ -\sin\beta & 0 & \cos\beta \end{bmatrix} \begin{bmatrix} 1 & 0 & 0 \\ 0 & \cos\alpha & -\sin\alpha \\ 0 & \sin\alpha & \cos\alpha \end{bmatrix} \begin{bmatrix} X \\ Y \\ Z \end{bmatrix} \tag{3}$$

The ground plane detection is based on the RANdom SAmple Consensus (RANSAC) algorithm [32]. By using the plane model, the RANSAC algorithm provides a robust estimation of the dominant plane parameters, performing a random search to detect short-range ground preliminarily, which is assumed to be the largest plane in the scenario. Although the assumption is violated in some real-world scenes, attitude angles of the camera and real vertical heights are employed to restrict the sampling process. The plane model is shown in Equation (4), and the inlier points of ground are determined with Equation (5). Firstly, a set of 3D points are randomly chosen from the point cloud to solve for the initial parameters A, B, C, and D. Secondly, the remaining 3D points are validated to count the number of inliers. After m computations, the ground plane is determined, which is the plane with the most inlier points. For the RANSAC algorithm, shown in Equation (6), if P is the probability of not failing the computation of outliers, p is the dimension of the model (three in our case), and η is the overall percentage of outliers, the number of computed solutions m can be selected to avoid overall sampling error:

$$AX_w + BY_w + CZ_w + D = 0 \tag{4}$$

$$d(X_w, Y_w, Z_w) = \frac{|AX_w + BY_w + CZ_w + D|}{\sqrt{A^2 + B^2 + C^2}} < T \tag{5}$$

$$m = \frac{\log(1 - P)}{\log(1 - (1 - \eta)^p)} \tag{6}$$

Rather than generate ground plane segmentation with the original point cloud, points are adjusted from the camera coordinate system to the world coordinate system in consideration of three respects:

- The inclination angle θ of the sampled plane can be calculated using Equation (7). This allows for dismissing some sample errors described in [25]. For example, if inclination angle of a sampled plane is abnormally high, the plane could not be the ground plane.
- Since the incorrect sampled planes are dismissed directly, the validation of inlier 3D points can be skipped to save much computing time.
- Given points in the world coordinate system, we obtain a subset of 3D points which only contains points whose real height is reasonable to be ground according to the position of the camera while the prototype is worn. Points which could not be ground points, such as points in the upper air are not included. As a result, η the percentage of outliers is decreased, so m, the number of computations, is decreased and, thereby, a great deal of processing time is saved.

$$\theta = arccos\frac{|B|}{\sqrt{A^2 + B^2 + C^2}} \tag{7}$$

After initial ground segmentation, some salient parts, such as corners and little obstacles on the ground may be included in ground plane. Salient parts should be wiped out of the ground for two reasons: little obstacles may influence VIP; these parts may extend out of the ground area in the stage of seeded region growing. In this work, salient parts are removed from the ground based on surface normal vector estimation. Firstly, the depth image is separated into image patches; secondly, the surface normal vector of each patch is estimated through principal component analysis, the details of which are described in [14]; lastly, patches whose normal vector has a low component in the vertical direction are discarded. In the sampling stage, the number of iterations m equals 25, and inclination angle threshold of the ground plane is empirically set to $10°$. Figure 6 depicts examples of short-range ground plane segmentation in indoor and outdoor environments, both of them achieving good performance, detecting the ground plane and dismissing salient parts correctly.

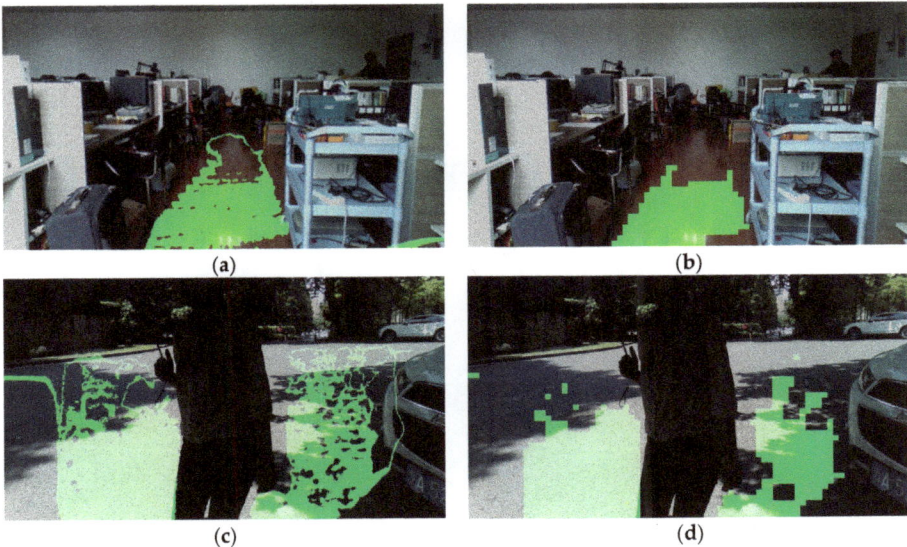

Figure 6. Ground plane segmentation in indoor and outdoor environments. (**a,c**) Ground plane detection based on the RANSAC algorithm; (**b,d**) salient parts in the ground plane are dismissed with surface normal vector estimation.

3.3. Seeded Region Growing

In order to expand traversable area to longer and broader range, a seeded region growing algorithm is proposed, combining both color images and filtered depth images. Instead of attaching importance to thresholds, edges of the color image are also adequately considered to restrict growth to other obstacle regions.

Firstly, seeds are chosen according to preliminary ground detection. A pixel is set as a seed to grow if two conditions are satisfied: the pixel is within the ground plane; four-connected neighbor pixels are not all within the ground plane. The seeds are pushed into the stack.

Secondly, a seed is valid to grow when it meets two conditions: the seed has not been traversed before, which means each seed will be processed only once; the seed does not belong to the edges of the color image.

Thirdly, we assume the growing starts from pixel G, whose depth value is d and hue value is v. One of the four-connected neighbors is Gi, whose depth value is d_i and hue value is v_i. Whether Gi belongs to G's region and be classified as traversable area depends on the following four growing conditions:

- Gi is not located at Canny edges of color image;
- Gi has not been traversed during the expansion stage;
- Real height of Gi is reasonable to be included in traversable area; and
- $|v - v_i| < \delta_1$ or $\begin{aligned} |v - v_i| &< \delta_2 \\ |d - d_i| &< \delta_h \end{aligned}$, where δ_1 is the lower hue growing threshold, and δ_2 is the higher growing threshold, while δ_h the height growing threshold, limits the expansion with only the color image.

If all four conditions are true, Gi is qualified for the region grown from G, so Gi is classified as a traversable area. Each qualified neighbor pixel is put into the stack. When all of G's four-connected pixels have been traversed, pop G out of the stack and let Gi be the new seed and repeat the above process. When the stack is empty, the seeded growing course finishes. After the seeded growing stage, the short-range ground plane has been enlarged to a longer and broader traversable area. Figure 7 depicts examples of expansion based on seeded region growing under indoor and outdoor situations, both expanding the traversable area to a great extent and preventing growth into other non-ground areas.

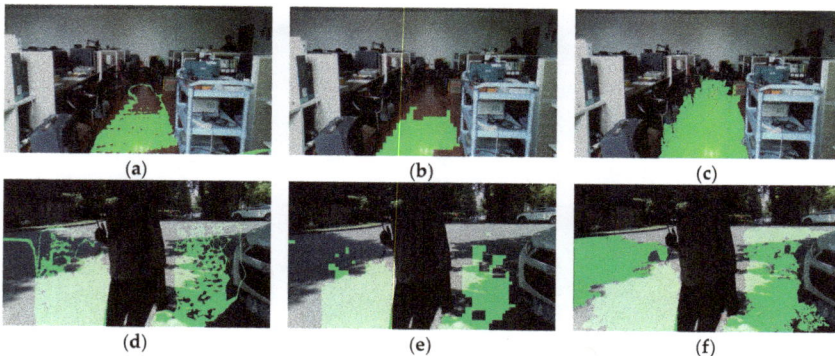

Figure 7. Traversable area expansion in indoor and outdoor environments. (**a**,**d**) Ground plane detection based on the RANSAC algorithm; (**b**,**e**) salient parts in the ground plane are dismissed with surface normal vector estimation; and (**c**,**f**) preliminary traversable area are expanded greatly with seeded region growing.

4. Experiment

In this section, experimental results are presented to validate our approach for traversable area detection. The approach is tested on a score of indoor and outdoor scenarios including offices, corridors, roads, playgrounds, and so on.

Figure 8 shows a number of traversable area detection results in the indoor environment. Largely-expanded traversable area provides two superiorities: firstly, longer range allows high-level path planning in advance; and, secondly, broader range allows precognition of various bends and corners. For special situations, such as color image blurring and image under-exposing, the approach still detects and expands the traversable area correctly, as shown in Figure 8g,h. Additionally, the approach is robust regardless of continuous movement of the cameras as the user wanders in real-world scenes.

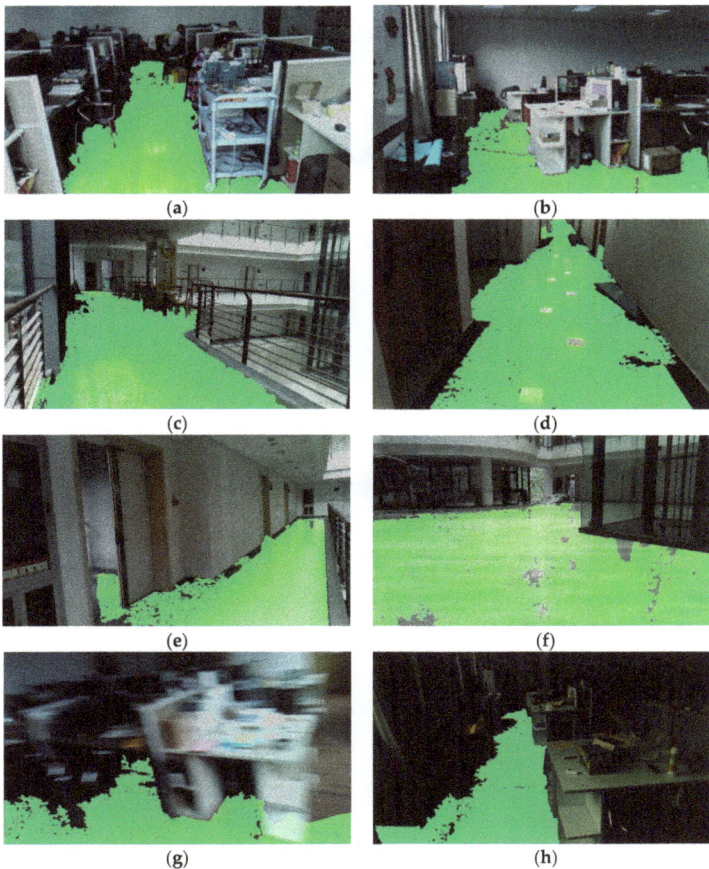

Figure 8. Results of traversable area expansion in indoor environment. (**a**,**b**) Traversable area detection in offices; (**c–e**) traversable detection in corridors; (**f**) traversable area detection in an open area; (**g**) traversable area detection with color image blurring; abd (**h**) traversable area detection with color image under-exposing.

Figure 9 shows several traversable area detection results under outdoor circumstances. It can be seen that traversable area has been enlarged greatly out to the horizon. Rather than the short-range ground plane, the expanded traversable area frees the VIP to wander in the environment.

Figure 9. Results of traversable area expansion in outdoor environment. (**a–g**) Traversable area detection on roads; (**h**) traversable area detection on a platform; and (**i**) traversable area detection on a playground.

To compare the performance of traversable area detection with respect to other works in the literature, the results of several traversable detection approaches on a typical indoor scenario and outdoor scenario are shown in Figure 10. Given the depth image, the approach proposed by Rodríguez estimated the ground plane based on RANSAC plus filtering techniques [25]. Figure 10n is a correct result of detecting the local ground, but the wall is wrongly detected as the ground plane in Figure 10e, which is one type of sample error mentioned in the paper. This kind of error is dissolvable in our work with consideration of the inclination angle of the plane. The approach proposed by Cheng detected the ground with seeded region growing of depth information [15]. The approach in [15] projects RGB information onto the valid pixels of depth map, so the detecting result shown in Figure 10f,o has many noises and black holes, and the detecting range is restricted since the depth information is discrete and prone to inaccuracy in long range. However, the main problem of the algorithm lies in that the seed pixels are elected randomly, thereby causes intolerable fluctuations to confuse VIP. In our previous works, we only employed depth information delivered by the light-coding sensor of the Microsoft Kinect [14,15]. However, the sensor outputs a dense 3D point cloud (ranges from 0.8 m to 5 m) indoors and fails in sunny outdoor environments. As a result, the algorithms are unable to perform well when the sensor could not generate a dense map. In Figure 10g,p, the idea of using surface normal vectors to segment ground presented in [14,40] is able to segment the local ground plane but fails to segment the long-range traversable area robustly as the estimation of normal vectors asks the sensor to produce dense and accurate point clouds. In this paper, we fully combine RGB information and depth information to expand the local ground plane segmentation to long range. In the process, IR image large-scale matching and RGB image guided filtering are incorporated to enhance the depth images. Although the computing time improves from 280 ms to 610 ms per frame on a 1.90 GHz Intel Core Processor, within which the RGB image-guided filtering is hardware accelerated with the HD4400 integrated graphics, the range of traversable detection has been expanded to a great extent and the computing time contributed in this process endows VIPs to perceive traversability at long range and plan routes in advance so the traversing time eventually declines. Figure 10h,q shows the results of traversable area detection without IR image large-scale matching and RGB image-guided filtering. The seeded region growing process is unable to enlarge the local ground segmentation based on RANSAC to long-range as the depth map is still discrete and sparse in the distance. Comparatively, in Figure 10i,r, after IR image large-scale matching and RGB image-guided filtering, the segmented local ground plane largely grows to a longer and broader traversable area. The set of our images is available online at Kaiwei Wang Team [52].

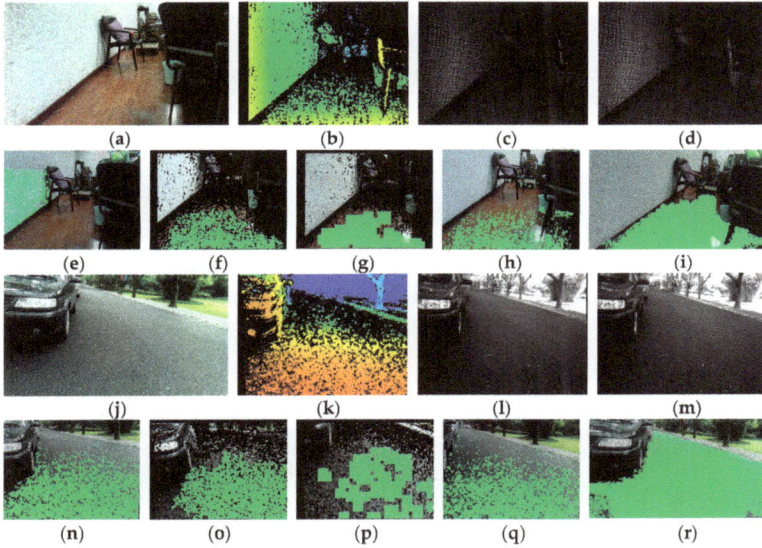

Figure 10. Comparisons of results of different traversable area detection approaches. (**a–d**) The set of images of a typical indoor scenario including color image, depth map, and calibrated IR pairs; (**e–i**) the results of different approaches on the indoor scenario; (**j–m**) the set of images of a typical outdoor scenario; and (**n–r**) the results of different approaches on the outdoor scenario.

Figure 11. An example of expansion error. The ground has been unexpectedly expanded to a part of the car.

The approach creates a multithreaded program including a thread for image acquisition and depth enhancement, a thread for traversable area detection and expansion, as well as a thread for audio interface generation for the VIP. Together, the average processing time of a single frame is

610 ms on a 1.90 GHz Intel Core 5 processor, making the refresh rate of the VIP audio feedback 1.6 times per second. In addition, detection rate and expansion error for indoor and outdoor scenarios are presented to demonstrate the robustness and reliability of the approach. Indoor scenarios, including a complicated office room and a corridor are analyzed, while outdoor scenarios, including school roads and a playground, are evaluated. Typical results of the four scenarios are depicted in Figures 8a,c and 9c,i. As depicted in Figure 11, part of the car has been classified as traversable area, which is a typical example of expansion error.

In order to provide a quantitative evaluation of the approach, given Equations (8) and (9), detection rate (*DR*) is defined as the number of frames which ground has been detected correctly (*GD*) divided by the number of frames with ground (*G*). Meanwhile, expansion error (*EE*) is defined as the number of frames which traversable area has been expanded to non-ground areas (*ENG*) divided by the number of frames with ground (*G*):

$$DR = GD/G \qquad (8)$$

$$EE = ENG/G \qquad (9)$$

Shown in Table 1, detection rates of the four scenarios are all above 90%, demonstrating the robustness of the approach. For the scene of the corridor, it yields an expansion error of 15.9%. This is mainly due to inadequate lighting on the corners in the corridor, so the edges of the color image are fuzzy and the traversable area may be grown to the wall. Overall, the average expansion error is 7.8%, illustrating the reliability of the approach, which seldom recognizes hazardous obstacles as safe traversable area.

Table 1. Detection rate and expansion error of the approach.

Scenario	Frames with Ground (*G*)	FRAMES Detected Ground Correctly (*GD*)	Detection Rate (*DR*)	Frames Expanded to Non-Ground Areas (*ENG*)	Expansion Error (*EE*)
An office	1361	1259	92.5%	44	3.2%
A corridor	633	614	97.0%	101	15.9%
School roads	837	797	95.2%	81	9.7%
A playground	231	228	98.7%	13	5.6%
All	3062	2898	94.4%	239	7.8%

Additionally, the average density of depth images of four different scenarios is calculated to prove that IR image large-scale matching and RGB image guided filtering remarkably improve the density of the original depth image from the RealSense sensor. The density of the depth image is defined as the number of valid pixels divided by the resolution. As shown in Table 2, the average density of the large-scale matched depth image is much higher than the original depth image and the guided-filtered depth image achieves 100% density.

Table 2. Average density of depth images including the original depth image, large-scale matched depth image and guided-filtered depth image.

Scenario	Original Depth Image (Resolution: 293,904)	Large Scale Mathced Depth Image (Resolution: 293,904)	Guided Filtered Depth Image (Resolution: 360,000)
An office	68.6%	89.4%	100%
A corridor	61.4%	84.5%	100%
School roads	76.2%	91.2%	100%
A playground	79.5%	92.0%	100%

5. User Study

In this section, a user study is elaborated in terms of assisting system overview, non-semantic stereophonic interface, and assisting performance study.

5.1. Assisting System Overview

The approach presented has been integrated in an assisting system. As shown in Figure 12, the system is composed of a RGB-D RealSense R200sensor, an attitude sensor MPU6050, a 3D-printed frame which holds the sensors, a processor Microsoft Surface Pro 3, as well as a bone-conducting headphone AfterShokz BLUEZ 2S [53], which transfers non-semantic stereophonic feedback to the VIP. Since the RealSense R200 only uses part of the USB 3.0 interface to transmit data, spare interfaces which are compatible with USB 2.0 are employed to transmit attitude angles from the MPU6050. Additionally, the processor communicates with the headphone through Bluetooth 4.0. Thereby, the system only needs a USB 3.0 cord to transfer images and data from sensors to the processor. As we know, VIP rely on voices from the environment great deal. For example, they use the sounds from cars to understand the orientation of streets. The assisting prototype is not only wearable but also ears-free, because the bone-conducting interface will not block VIP's ears from hearing environmental sounds.

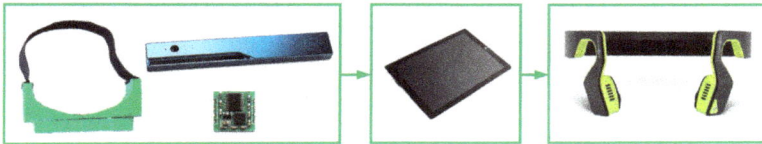

Figure 12. The assisting system consists a frame which holds the RealSense R200 and the attitude sensor, a processor, and a bone-conducting headphone.

5.2. Non-Semantic Stereophonic Interface

The assisting system uses a non-semantic stereophonic interface to transfer traversable area detection results to the VIP. The generation of the non-semantic stereophonic interface follows rules below:

- Divide the detection result into five directions, since the horizontal field view has been enlarged from 59° to 70°, so each direction corresponds to traversable area with a range of 14°.
- Each direction of traversable area is represented by a musical instrument in 3D space.
- In each direction, the longer the traversable area, the greater the sound from the instrument.
- In each direction, the wider the traversable area, the higher the pitch of the instrument.

To sum up, the directions of traversable area are differentiated not only by sound source locations in 3D space, but also by musical instruments, whose tone differs from each other. As shown in Figure 13, five instruments, including trumpet, piano, gong, violin, and xylophone, produce sounds simultaneously which last for 0.6 s, notifying the user the traversable area. Additionally, we also implemented a simple obstacle detection method to warn against walking on the ground under obstacles in the air (e.g., Figure 8g). The 3D points which are not within traversable area and are within close range (1 m in our case) are counted in respectively five directions. If the number of points in one direction exceeds a threshold, it means there is one obstacle in the close range. In this case, the audio interface generates a friendly prompt to help VIP to be aware of close obstacles. Since that is not the major topic of this paper, specific parameters of the audio feedback are not discussed here.

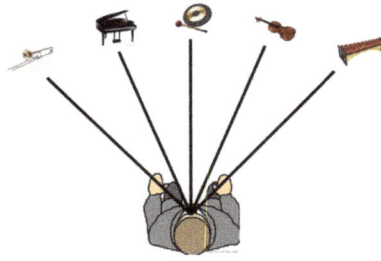

Figure 13. Non-semantic stereophonic interface of the assisting system. Sounds of five directions of traversable area are presented by five musical instruments in 3D space, including trumpet, piano, gong, violin, and xylophone.

5.3. Assisting Performance Study

Eight visually impaired volunteers including three suffering from total blindness participated in the user study. Figure 14 are the moments of the assisting study. During the assisting study, participants first learned the audio feedback. The working pattern of the system and signals from the headphone was introduced. Each one of them has ten minutes to learn, adapt to the audio interface, and wander around casually. After that, participants were asked to traverse through obstacles without collisions, and finally find the person standing at the end point. A contrary test is designed to compare its performance under two conditions: the signal from the audio interface is generated according to the original ground detection, and the audio interface is generated according to the traversable area expansion.

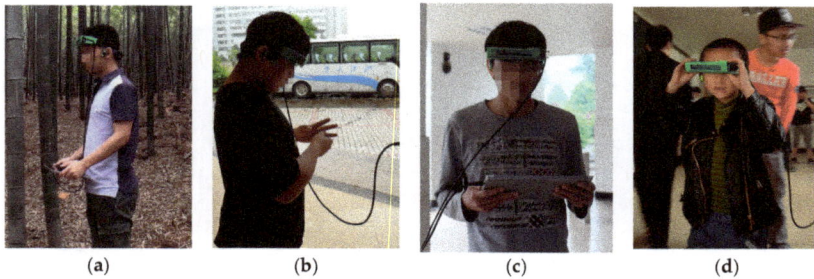

(a)　　　　　　(b)　　　　　　(c)　　　　　　(d)

Figure 14. Eight visually impaired volunteers took part (**a–d**). The moments of the assisting study. Participants' faces are blurred for the protection of the privacy (we have gotten the approval to use the assisting performance study for research work).

After the learning stage, eight visually impaired participants were required to travel through obstacles. Shown in Figure 15, six different white boards including large columns were employed as obstacles. Five different obstacle arrangements were generated by arranging the position of obstacles differently. Firstly, they were asked to complete the course with traversable area expansion, and a typical detection example is shown in Figure 7. Secondly, they were asked to complete the course with original ground detection, which is shown in Figure 2. All visually impaired participants completed the test and found the person standing at the end point. The average number of collisions, average time and average number of steps to complete a single test were recorded. Collisions include collisions with obstacles and walls. The timer starts when a participant is sent to the start region and stops when the participant completes a single test. The distance between the start region and the end point is the same for all tests. However, the number of steps to complete a single test varies. As shown in Table 3, average number of collisions to complete a single test with traversable area expansion is 78.6% less than that with original ground detection. Most of the collisions occurred when the user did not know which direction to walk as the original ground detection is at short-range. Additionally, the average time to complete a single test with traversable area expansion is 29.5% less than that with original ground detection. Moreover, the number of average steps to complete a single test with traversable area expansion is 43.4% less than that with original ground detection. It is the expansion of traversable area which endows VIP the ability to plot routes farther ahead and, therefore, reduce traversing time and the number of steps. Each participant completes the tests with different obstacle arrangements in a random order. As a result, the participants have no idea about the arrangement of obstacles each time. It is ruled out that the decrease of collisions and traversing time after traversable area expansion is due to variation of familiarity with the prototype. Since the test was taken with traversable area expansion first and the taken with original ground detection afterwards, if it was due to the variation of familiarity, it would enhance rather than weaken the performance of navigational assistance, such as the number of collisions and the traversing time taken with traversable expansion would be more than with original ground detection. It can be proved convincingly that traversable area expansion improves the performance dramatically. In other word, the safety and robustness enhances navigation.

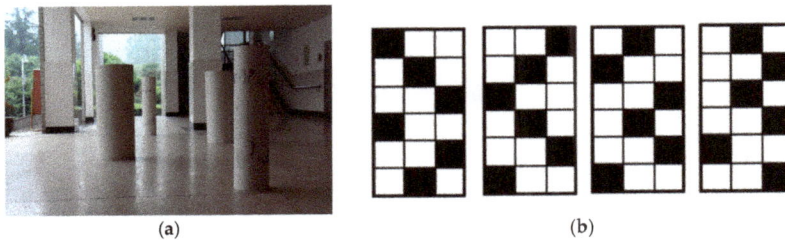

(a)   (b)

Figure 15. Obstacle arrangements. (**a**) An image of obstacle arrangement; (**b**) Four other obstacle arrangements.

Table 3. Number of collisions and time to complete tests in two conditions: the audio interface transferred to the VIP is generated according to original ground detection or traversable area expansion.

Detection Result Transfered to VIP	Total Number of Collisions	Average Number of Collisions of Each Time	Total Time to Complete Tests	Average Time to Complete a Single Test	Total Number of Steps	Average Number of Steps to Complete a Single Test
Original ground deteciton	103	2.58	733 s	18.33 s	1850	46.25
Traversable area expansion	22	0.55	517 s	12.93 s	1047	26.18

After the test, eight participants were asked two simple questions including whether the prototype is easy to wear and whether the system provides convenient assistance to travel in an unfamiliar environment. Shown in the questionnaire (Table 4), all users answered that the system is useful and can offer help in unknown or intricate environments. It not only gave us significant confidence, but also demonstrated usefulness and reliability of the approach. In addition, some users gave some advice on adding functions, such as face recognition or GPS navigation and a user hopes that the prototype be designed in a hat.

Table 4. A questionnaire. After the test, eight participants were asked two simple questions.

User	Total Blind or Partially Sighted	Easy to Wear?	Useful?	Advice
User 1	Partially sighted	Yes	Yes	
User 2	Partially sighted	Yes	Yes	Add face recognition
User 3	Total blind	Yes	Yes	Design the prototype in a hat
User 4	Partially sighted	Yes	Yes	
User 5	Partially sighted	No	Yes	Add GPS navigation
User 6	Total blind	Yes	Yes	
User 7	Total blind	No	Yes	
User 8	Partially sighted	Yes	Yes	

6. Conclusions

RGB-D sensors are a ubiquitous choice to provide navigational assistance for visually impaired people, with good portability, functional diversity, and cost-effectiveness. However, most assisting solutions, such as traversable area awareness, suffer from the limitations imposed by RGB-D sensor ranging, which is short, narrow, and prone to failure. In this paper, an effective approach is proposed to expand ground detection results to a longer and broader range with a commercial RGB-D sensor, the Intel RealSense R200, which is compatible with both indoor and outdoor environments. Firstly, the depth image of the RealSense is enhanced with large scale matching and color guided filtering. Secondly, preliminary ground segmentation is obtained by the RANSAC algorithm. The segmentation is combined with an attitude sensor, which eliminates many sample errors and improves the robustness of the preliminary result. Lastly, the preliminary ground detection is expanded with seeded region growing, which fully combines depth, attitude, and color information. The horizontal field angle of the traversable area has been increased from 59° to 70°. Additionally, the expansion endows VIP the ability to predict traversability and plan paths in advance since the range has been enlarged greatly to a large extent. The approach is able to see smoothly to the horizon, being acutely aware of the traversable area at distances far beyond 10 m. Both indoor and outdoor empirical evidences are provided to demonstrate the robustness of the approach, in terms of image processing results, detection rate, and expansion error. In addition, a user study is described in detail, which proves the approach to be usable and reliable.

In the future, we aim to incessantly enhance our navigational assistance approach for the visually impaired. Especially, the implementation of the algorithm is not yet optimized, so we are looking forward to speeding it up. Additionally, a cross-modal stereo-matching scheme between IR images and RGB images would also be interesting and useful to inherently improve the detecting range and ranging accuracy of the camera.

Author Contributions: Kailun Yang conceived the method, performed the traversable area detection and wrote the paper. Kaiwei Wang designed the experiments. Weijian Hu developed the audio interface. Jian Bai analyzed the experimental results. All authors participated in the assisting performance study and approved the paper.

Conflicts of Interest: The authors declare no conflict of interest.

References

1. World Health Organization. Available online: www.who.int/mediacentre/factsheets/fs282/en (accessed on 7 November 2016).
2. PrimeSense. Available online: www.en.wikipedia.org/wiki/PrimeSense (accessed on 7 November 2016).
3. Kinect. Available online: www.en.wikipedia.org/wiki/Kinect (accessed on 7 November 2016).
4. Xtion Pro. Available online: www.asus.com.cn/3D-Sensor/Xtion_PRO (accessed on 7 November 2016).
5. Mantis Vision. Available online: www.mv4d.com (accessed on 7 November 2016).
6. Structure Sensor. Available online: structure.io (accessed on 7 November 2016).

7. Zöllner, M.; Huber, S.; Jetter, H.; Reiterer, H. *NAVI—A Proof-of-Concept of a Mobile Navigational Aid for Visually Impaired Based on the Microsoft Kinect. Human-Computer Interaction—INTERACT 2011*; Springer: Berlin/Heidelberg, Germany, 2007; Volume 6949, pp. 584–587.

8. Hicks, S.L.; Wilson, I.; Muhammed, L.; Worsfold, J.; Downes, S.M.; Kennard, C. A depth-based head-mounted visual display to aid navigation in partially sighted individuals. *PLoS ONE* **2013**, *8*, e67695. [CrossRef] [PubMed]

9. Aladren, A.; Lopez-Nicolas, G.; Puig, L.; Guerrero, J.J. Navigational assistance for the visually impaired using rgb-d sensor with range expansion. *IEEE Syst. J.* **2014**, *99*, 1–11.

10. Takizawa, H.; Yamaguchi, S.; Aoyagi, M.; Ezaki, N.; Mizuno, S. Kinect cane: An assistive system for the visually impaired based on three-dimensional object recognition. *Pers. Ubiquitous Comput.* **2012**, *19*, 740–745.

11. Park, C.H.; Howard, A.M. Real-time haptic rendering and haptic telepresence robotic system for the visually impaired. In Proceedings of the World Haptics Conference, Daejeon, Korea, 14–18 April 2013; pp. 229–234.

12. Khan, A.; Moideen, F.; Lopez, J.; Khoo, W.L.; Zhu, Z. KinDetect: Kinect Detecting Objects. In *Computers Helping People with Special Needs*; Springer: Berlin/Heidelberg, Germany, 2012; pp. 588–595.

13. Ribeiro, F.; Florencio, D.; Chou, P.A.; Zhang, Z. Auditory augmented reality: Object sonification for the visually impaired. *IEEE Int. Workshop Multimed. Signal Proc.* **2012**, *11*, 319–324.

14. Yang, K.; Wang, K.; Cheng, R.; Zhu, X. A new approach of point cloud processing and scene segmentation for guiding the visually impaired. In Proceedings of the IET International Conference on Biomedical Image and Signal Processing, Beijing, China, 19 November 2015.

15. Cheng, R.; Wang, K.; Yang, K.; Zhao, X. A ground and obstacle detection algorithm for the visually impaired. In Proceedings of the IET International Conference on Biomedical Image and Signal Processing, Beijing, China, 19 November 2015.

16. PMD. Available online: www.pmdtec.com (accessed on 7 November 2016).

17. SoftKinetic. Available online: www.softkinetic.com (accessed on 7 November 2016).

18. HEPTAGON. Available online: hptg.com/industrial (accessed on 7 November 2016).

19. Zeng, L.; Prescher, D.; Webber, G. Exploration and avoidance of surrounding obstacles for the visually impaired. In Proceedings of the 14th International ACM SIGACCESS Conference on Computers and Accessibility, Boulder, CO, USA, 22–24 October 2012; pp. 111–118.

20. Tamjidi, A.; Ye, C.; Hong, S. 6-DOF pose estimation of a Portable Navigation Aid for the visually impaired. In Proceedings of the IEEE International Symposium on Robotics and Sensors Environments, Washington, DC, USA, 21–23 October 2013; pp. 178–183.

21. Lee, C.H.; Su, Y.C.; Chen, L.G. An intelligent depth-based obstacle detection for visually-impaired aid applications. In Proceedings of the IEEE 2012 13th International Workshop on Image Analysis for Multimedia Interactive Services (WIAMIS), Dublin, Ireland, 23–25 May 2012; pp. 1–4.

22. PointGrey Bumblebee. Available online: www.ptgrey.com/bumblebee2-firewire-stereo-vision-camera-systems (accessed on 7 November 2016).

23. Stereolabs. Available online: www.stereolabs.com (accessed on 7 November 2016).

24. DUO. Available online: duo3d.com (accessed on 7 November 2016).

25. Rodríguez, A.; Yebes, J.J.; Alcantarilla, P.F.; Bergasa, L.M.; Almazán, J.; Cele, A. Assisting the visually impaired: Obstacle detection and warning system by acoustic feedback. *Sensors* **2011**, *12*, 17476–17496. [CrossRef] [PubMed]

26. Lee, Y.H.; Medioni, G. RGB-D camera based navigation for the visually impaired. In Proceedings of the RSS RGBD Advanced Reasoning with Depth Camera Workshop, Los Angeles, CA, USA, 27 June 2011.

27. Martinez, J.M.S.; Ruiz, F.E. Stereo-based aerial obstacle detection for the visually impaired. In Proceedings of the Workshop on Computer Vision Applications for the Visually Impaired, Marseille, France, 18 October 2008.

28. Lin, K.W.; Lau, T.K.; Cheuk, C.M.; Liu, Y. A wearable stereo vision system for visually impaired. In Proceedings of the 2012 International Conference on Mechatronics and Automation (ICMA), Chengdu, China, 5–8 August 2012; pp. 1423–1428.

29. Intel RealSense R200. Available online: software.intel.com/en-us/realsense/r200camera (accessed on 7 November 2016).

30. Kytö, M.; Nuutinen, M.; Oittinen, P. Method for measuring stereo camera depth accuracy based on stereoscopic vision. *Proc. SPIE Int. Soc. Opt. Eng.* **2011**. [CrossRef]

31. Getting Started with the Depth Data Provided by Intel RealSense Technology. Available online: software. intel.com/en-us/articles/realsense-depth-data (accessed on 7 November 2016).

32. Fischler, M.A.; Bolles, R.C. Random sample consensus: A paradigm for model fitting with applications to image analysis and automated cartography. *Comm. ACM* **1981**, *24*, 381–395. [CrossRef]

33. InvenSense MPU-6050. Available online: www.invensense.com/products/motion-tracking/6-axis/mpu-6050 (accessed on 7 November 2016).

34. Adams, R.; Bischof, L. Seeded Region Growing. *IEEE Trans. Pattern Anal. Mach. Intell.* **1994**, *16*, 641–647. [CrossRef]

35. Wang, T.; Bu, L.; Huang, Z. A new method for obstacle detection based on Kinect depth image. *IEEE Chin. Autom. Congr.* **2015**. [CrossRef]

36. Perez-Yus, A.; Gutierrez-Gomez, D.; Lopez-Nicolas, G.; Guerrero, J.J. Stairs detection with odometry-aided traversal from a wearable RGB-D camera. *Comput. Vis. Image Underst.* **2016**, in press. [CrossRef]

37. Wang, Z.; Huang, S.; Dissanayake, G. *Simultaneous Localization and Mapping*; Springer: Berlin/Heidelberg, Germany, 2008.

38. Lee, Y.H.; Medioni, G. RGB-D camera based wearable navigation system for the visually impaired. *Comput. Vis. Image Underst.* **2016**, *149*, 3–20. [CrossRef]

39. Sánchez, C.; Taddei, P.; Ceriani, S.; Wolfart, E.; Sequeira, V. Localization and tracking in known large environments using portable real-time 3D sensors. *Comput. Vis. Image Underst.* **2016**, *139*, 197–208. [CrossRef]

40. Koester, D.; Schauerte, B.; Stiefelhagen, R. Accessible section detection for visual guidance. In Proceedings of the IEEE International Conference on Multimedia and Expo Workshops, San Jose, CA, USA, 15–19 July 2013; Volume 370, pp. 1–6.

41. Bellone, M.; Messina, A.; Reina, G. A new approach for terrain analysis in mobile robot applications. In Proceedings of the IEEE International Conference on Mechatronics, Wollongong, Australia, 9–12 July 2013; Volume 307, pp. 225–230.

42. Chessa, M.; Noceti, N.; Odone, F.; Solari, F.; Sosa-García, J.; Zini, L. An integrated artificial vision framework for assisting visually impaired users. *Comput. Vis. Image Underst.* **2015**, *149*, 209–228. [CrossRef]

43. Hadsell, R.; Sermanet, P.; Ben, J.; Erkan, A.; Scoffier, M.; Kavukcuoglu, K. Learning long-range vision for autonomous off-road driving. *J. Field Robot.* **2009**, *26*, 120–144. [CrossRef]

44. Reina, G.; Milella, A. Towards autonomous agriculture: Automatic ground detection using trinocular stereovision. *Sensors* **2012**, *12*, 12405–12423. [CrossRef]

45. Milella, A.; Reina, G.; Underwood, J.; Douillard, B. Visual ground segmentation by radar supervision. *Robot. Auton. Syst.* **2014**, *62*, 696–706. [CrossRef]

46. Reina, G.; Milella, A.; Rouveure, R. Traversability analysis for off-road vehicles using stereo and radar data. In Proceedings of the IEEE International Conference on Industrial Technology, Seville, Spain, 17–19 March 2015.

47. Damen, D.; Leelasawassuk, T.; Mayol-Cuevas, W. You-Do, I-Learn: Egocentric unsupervised discovery of objects and their modes of interaction towards video-based guidance. *Comput. Vis. Image Underst.* **2016**, *149*, 98–112. [CrossRef]

48. Geiger, A.; Roser, M.; Urtasun, R. *Efficient Large-Scale Stereo Matching. Asian Conference on Computer Vision*; Springer: Berlin/Heidelberg, Germany, 2010; Volume 6492, pp. 25–38.

49. Einecke, N.; Eggert, J. A multi-block-matching approach for stereo. In Proceedings of the 2015 IEEE Intelligent Vehicles Symposium (IV), Seoul, Korea, 28 June–1 July 2015; pp. 585–592.

50. He, K.; Sun, J.; Tang, X. Guided Image Filtering. *IEEE Trans. Softw. Eng.* **2013**, *35*, 1397–1409. [CrossRef] [PubMed]

51. Intel RealSense Depth Enabled Photography. Available online: software.intel.com/en-us/articles/intel-realsense-depth-enabled-photography (accessed on 7 November 2016).

52. Kaiwei Wang Team. Available online: wangkaiwei.org (accessed on 7 November 2016).

53. AfterShokz BLUEZ 2S. Available online: www.aftershokz.com.cn/bluez-2s (accessed on 7 November 2016).

sensors

MDPI

Article

A 3D Optical Surface Profilometer Using a Dual-Frequency Liquid Crystal-Based Dynamic Fringe Pattern Generator

Kyung-Il Joo [†], Mugeon Kim [†], Min-Kyu Park, Heewon Park, Byeonggon Kim, JoonKu Hahn and Hak-Rin Kim *

School of Electronics Engineering, Kyungpook National University, Daegu 41566, Korea; kijoo@knu.ac.kr (K.-I.J.); im2781@gmail.com (M.K.); mkpark@ee.knu.ac.kr (M.-K.P.); heewonpark@ee.knu.ac.kr (H.P.); bgkim@knu.ac.kr (B.K.); jhahn@knu.ac.kr (J.H.)
* Correspondence: rineey@knu.ac.kr; Tel.: +82-53-950-7211; Fax: +82-53-940-8622
† These authors contributed equally to this work.

Academic Editor: Gonzalo Pajares Martinsanz
Received: 23 September 2016; Accepted: 24 October 2016; Published: 27 October 2016

Abstract: We propose a liquid crystal (LC)-based 3D optical surface profilometer that can utilize multiple fringe patterns to extract an enhanced 3D surface depth profile. To avoid the optical phase ambiguity and enhance the 3D depth extraction, 16 interference patterns were generated by the LC-based dynamic fringe pattern generator (DFPG) using four-step phase shifting and four-step spatial frequency varying schemes. The DFPG had one common slit with an electrically controllable birefringence (ECB) LC mode and four switching slits with a twisted nematic LC mode. The spatial frequency of the projected fringe pattern could be controlled by selecting one of the switching slits. In addition, moving fringe patterns were obtainable by applying voltages to the ECB LC layer, which varied the phase difference between the common and the selected switching slits. Notably, the DFPG switching time required to project 16 fringe patterns was minimized by utilizing the dual-frequency modulation of the driving waveform to switch the LC layers. We calculated the phase modulation of the DFPG and reconstructed the depth profile of 3D objects using a discrete Fourier transform method and geometric optical parameters.

Keywords: optical surface profilometry; interference; phase modulation; liquid crystal; dynamic fringe pattern generator

1. Introduction

Recently, the demand for acquiring 3D information has increased, accompanied by improvements in several 3D applications like 3D displays, 3D printing technologies, 3D medical or dental imaging systems, and 3D vision modules for robot or vehicle applications [1–6]. For these applications, 3D depth information together with conventional 2D images is critically needed, and cannot be obtained with conventional 2D vision systems. Optical 3D vision systems can measure 3D information for an object with a wide scope in a relatively short time because they obtain the 2D coordinate information together with the depth information optically in parallel using a Charge-Coupled Device (CCD) or Complementary Metal Oxide Semiconductor (CMOS) image sensor.

The optical 3D surface profilometer can optically measure the surface morphology of an object and compute the depth information without direct contact. In general, optical surface profilometers, which extract 3D depth information from distortions of the projected optical beam patterns, need an optical beam pattern generator module to utilize several structured beam. To measure a surface profile of 3D objects, N-bits of binary-coded stripe patterns could be projected [7], where the number of the binary-coded projection beam patterns needed to be increased to improve the measurement resolution. By utilizing gray-level beam patterns, the number of projection patterns could be effectively reduced [8], but this approach needs more complex spatial light modulators (SLMs) with a higher pixel resolution and a larger pixel density to generate the gray-level patterns. To effectively reduce the measurement time, a one-shot scanning method using periodic grid patterns was proposed [9]. Using a coherent light source, interference fringe patterns could be projected [10–14]. To avoid the optical phase ambiguity from the fringe patterns distorted by the surface morphology, multiple sets of interference patterns were needed, and phase unwrapping or geometrical parameter methods were applied to reconstruct surface depth profiles. The depth extraction method based on the geometrical parameters could provide the depth and position value informations in the absolute coordinate system without the phase unwrapping process [10]. These optical techniques do not damage the surface morphology of an object even for a soft surface.

In most cases of optical surface profilometries, to acquire more precise 3D depth information, several sets of specific beam patterns are needed. These are generated by beam pattern projecting modules such as mechanically moving wedge plates [15], tunable gratings [16], switchable gradient index lenses [17], polymer dispersed liquid crystal (LC) techniques [18,19], and several types of phase modulators using SLMs [20,21]. Recently, to obtain color information of an object in addition to the 3D depth information, time-sequential projection of structured and RGB-colored beam patterns was also proposed using fast switching digital light processing (DLP) projectors [22]. However, conventional projection units based on commercial SLMs and DLP are too bulky and not cost-effective for industrial field applications, especially dental imaging, which needs an elaborate and complex semiconductor manufacturing process for the preparation of backplanes to control the 2D phase or intensity patterns with matrix driving schemes [23–25].

In this study, we propose an LC-based dynamic fringe pattern generator (DFPG) for a more compact 3D optical surface profilometer system, which can generate multiple fringe patterns to enhance the 3D depth extraction and avoid the optical phase ambiguity in analyzing the 3D depth profile from distorted fringe patterns induced by the surface morphologies. Sixteen interference patterns are generated with four-step phase shifting and four-step spatial frequency varying schemes by the proposed LC-based DFPG without a mechanically moving part. In our DFPG, the 16 sets of fringe patterns can be generated by a single, compact LC cell, which has one common slit operated by an electrically controllable birefringent (ECB) LC mode and four switching slits operated by a twisted nematic (TN) LC mode. Four different moving fringe patterns are controlled by the phase value of the ECB LC layer behind one common slit. The interferometric fringe patterns with four different spatial frequencies are controlled by electrically selecting one of the switching slits operated by the TN mode. The switching time of the DFPG module, required for projecting the 16 sets of fringe patterns, is minimized by utilizing the dual-frequency-based LC switching scheme. We present the optical and material design of the LC-based DFPG and its manufacturing process obtaining multiple sets of fringe patterns using a compact module with a fast switching time. The electrical switching properties of the phase modulation and the slit spacing controls for the interference patterns of the DFPG are characterized and the 3D depth profile reconstruction from the distorted fringe patterns using the discrete Fourier transform (DFT) method and geometric optical parameters is presented.

2. Operation Principle of the DFPG and the Theory of Depth Extraction

2.1. Schematic and Operation Principle of the DFPG

2.1.1. Schematic of the DFPG

In this study, the DFPG was designed to develop a 3D optical surface profilometer that can exhibit multiple sets of interference fringe patterns with four steps of the spatial frequency and four steps of the phase shifting properties without any mechanical translation stages. As shown in Figure 1, the DFPG consists of a multi-directional LC alignment layer on the top substrate to provide the ECB LC and TN LC modes for the phase-shifting common slit and four switching slits, respectively. On the other side of the multi-alignment layer of the top substrate, five Al slits were fabricated by the lift-off process. The indium tin oxide (ITO) layer under the multi-directional alignment layer was also patterned using photolithography to control one ECB LC layer and four TN LC layers individually by applying appropriate voltages to them. The positions of the patterned ITO layer, patterned LC alignment layers, and Al slits were precisely aligned with each other with alignment marks. The spacings between one common slit with the ECB LC mode and the other switching slits with the TN LC modes were $\Delta g = 200, 400, 600$, and $800\ \mu m$. The widths of the Al slits and the patterned ITO electrodes were 52 and $50\ \mu m$, respectively. A uni-directionally rubbed LC alignment layer was prepared on the non-patterned ITO glass substrate as the bottom substrate. Parallel polarizers were attached on both sides of the two substrates.

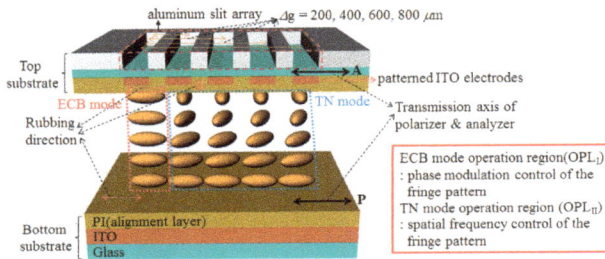

Figure 1. Schematic of the LC-based dynamic fringe pattern generator for projecting multiple sets of interference patterns with phase-shifting and spatial-frequency-varying properties.

The multi-spatial frequency properties of the interference fringe patterns were achieved by applying a turn-on voltage on one of the switching slits prepared with the TN LC mode. The four-step phase shifting properties were developed by controlling the applied voltage through the ECB LC mode layer used for the common slit. As a result, the DFPG can generate 16 sets of the multiple fringe patterns that exhibit four different spatial frequencies together with four-step phase-shifted fringe patterns for each selected spatial frequency.

The DFPG was fabricated with a dual-frequency LC (MLC-2048, Merck Ltd., Seoul, Korea) to enhance its switching response time required for generating the 16 sets of fringe patterns. The inversion frequency of the dielectric anisotropy of MLC-2048 is 50 kHz. Its dielectric anisotropy values are $\Delta\varepsilon = 3.2$ at the low frequency AC driving of 1 kHz and $\Delta\varepsilon = -3.4$ at the high frequency AC driving of 100 kHz. The extraordinary refractive index of the dual-frequency LC is 1.7192, and the ordinary refractive index is 1.4978. Therefore, the cell gap of the DFPG was calculated as $2.48\ \mu m$ to realize over $3\pi/2$ phase modulation of the LC layer. We dropped a mixture of an optical adhesive polymer (NOA 65, Norland Products, Inc., Cranbury, NJ, USA) with ball spacers on the four edges of the patterned ITO/Al substrate, and the top substrate was covered with the bottom electrode substrate. The ball spacers uniformly supported the cell gap of the DFPG required for reliable phase modulation. In our DFPG, the cell gap was about $4\ \mu m$. The empty DFPG cell was filled with MLC-2048 by the capillary force over the nematic-isotropic phase transition temperature ($T_{NI} = 106.2\ ^\circ C$) of the LC. After slowly cooling the LC cell to room temperature, the LC layer was well aligned multi-directionally along the patterned rubbing directions of the alignment layer of the top substrate, showing the two LC domains of the ECB and TN LC modes.

2.1.2. Operation Principle of the DFPG

The operation principles of our DFPG and field-dependent LC orientations on each patterned slit are shown in Figure 2, where the projected fringe patterns, which are measured under the far-field interference conditions using a CCD camera, are co-plotted according to the applied voltage conditions. As shown in Figure 2, the common slit is aligned with the ECB LC mode that can modulate the phase shifting and the four switching slits are aligned with the TN LC mode that can modulate the spatial frequencies of the fringe pattern. Under the parallel polarizer condition, the LC alignment direction of the bottom substrate is parallel with the transmission axes of the top and bottom polarizers. Therefore, the incident polarization state after the bottom polarizer does not change irrespective of the applied voltage within the ECB LC mode. The incident beam passing through the ECB LC layer always transmits the top polarizer after passing though the common slit without intensity loss irrespective of the applied voltage.

However, the polarization states of the incident beams passing through the TN LC layers, initially polarized by the bottom polarizer, are rotated by 90° owing to the polarization rotating effect of the twisted LC structure in the field-off state. Thus, the beams after four switching slits are blocked by the second polarizer without an applied voltage in our parallel polarizer scheme. The dielectric anisotropy of the LC used in our experiment is positive under low frequency AC driving conditions, and the LCs are reoriented along the applied field direction. When a voltage sufficient to fully reorient the LCs along the vertical field direction is applied to the TN LC layer under one of the switching slits by the patterned ITO electrode, the LC molecules in the selected local area can be fully reoriented to the vertically aligned geometry, as shown in Figure 2a,b. Thus, the beam passing through the selected switching slit can be transmitted through the second polarizer. Therefore, depending on the applied voltages of the patterned ITO electrodes under the four switching slits, the distance between two interference slits can be controlled, which enables four different spatial frequencies of the projected fringe patterns, as shown in Figure 2a,b. In Figure 2a,b, the fringe patterns measured under the shortest and the longest slit distances in our DFPG are shown for the example cases of the lowest and the highest spatial frequencies of the projected fringe patterns, respectively. By using these methods, the multi-spatial frequency schemes were developed in the DFPG without any mechanical translation stage part.

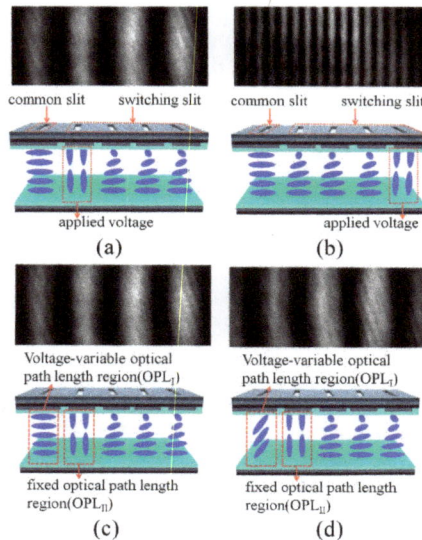

Figure 2. Operation principles of the DFPG switched by field-induced patterned LC orientations and the resulting projected fringe patterns. (**a**,**b**) show the LC orientations and spatial frequency variations of the projected fringe patterns by control of the selected switching slit with applied voltages: (**a**) Δg = 200 μm slit spacing and (**b**) Δg = 800 μm slit spacing; (**c**,**d**) show the field-induced LC orientations and the moving fringe patterns by control of the phase shifting at the common slit with applied voltages under the fixed slit spacing condition (Δg = 200 μm slit spacing).

Under the fixed spatial frequency attained by selecting the switching slit of one TN LC layer, the four-step phase-shifting schemes can be achieved by applying voltages to the ECB LC layer of the common slit for an appropriate phase difference with the phase of the selected switching slit. The phase shift is decided by the optical path length difference between the ECB LC layer and the field-applied TN LC layer under two slits of one common slit and the selected switching slit, respectively. Figure 2c,d show the initial and the relatively moved fringe patterns according to the applied voltages in the ECB LC layer for the same selected switching slit condition, where the spatial frequencies of two fringe patterns maintain each other. The effective refractive index of the ECB LC layer can be varied by applying voltages because the incident polarization is parallel to the rubbing direction of the LC alignment layer, and the LC layer operated with the ECB LC mode, which has a field-dependent tilting redistribution of the LC layer, does not exhibit any LC twisting distortion irrespective of the applied voltages. Thus, the optical path length passing though the common slit can be modulated according to the voltage applied to the ECB LC layer. For 3D depth extraction from the projected fringe patterns, which are distorted by an object, four-step phase shifting, especially over $3\pi/2$ phase shifting, is needed, and the thickness of the DFPG LC cell is designed considering the birefringence of the LC used in our experiment.

2.1.3. Fabrication Process of the DFPG

To electrically control the spatial-frequencies and phase modulation of the fringe patterns, the ITO electrodes were patterned by a photolithography and chemical etching process, as shown in Figure 3a. The first step for developing the patterned ITO electrodes was the spin-coating of a positive photoresistor (GXR-601, AZ Electronic Materials Co., Wiesbaden, Germany) on the ITO glass substrate. The GXR-601 photoresistor was coated under the conditions of 2500 rpm for 5 s, 3500 rpm for 30 s, and 2500 rpm for 5 s. After the spin-coating process, the coated ITO glass was heated to 90 °C on a hotplate for 90 s. Then, the coated photoresistor was projected through the photo-mask pattern by ultraviolet light and was etched with the photoresistor developer (AZ300, AZ Electronic Materials Co.) for 18 s.

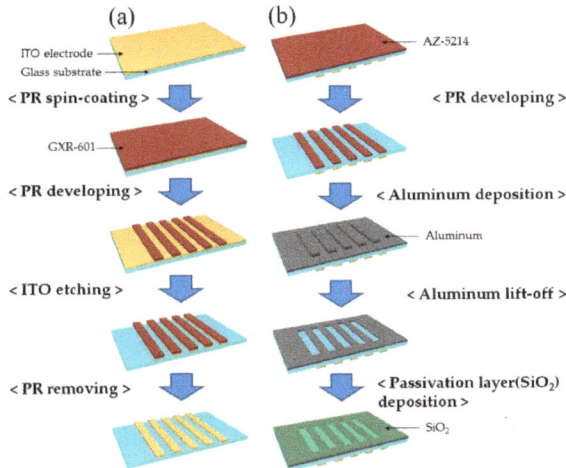

Figure 3. Fabrication process of the top substrate for the multiple slits and the aligned ITO patterns of the LC-based DFPG device: (**a**) ITO electrode patterning and (**b**) Al slit array patterning using the lift-off process.

The patterned photoresistor coated on the ITO glass was heated again on the hotplate at 120 °C for 3 min. Finally, to define the ITO electrode pattern, the ITO electrode was etched by the ITO etchant

(LCE-12K, Cyantek Co., Fort Worth, TX, USA) for 20 min and the photoresistor, which might have remained on the substrate, was removed by an acetone cleaning process.

To generate the interference patterns, the Al slits were prepared on the backside of the patterned ITO surface of the glass substrate. In this study, the Al slits were fabricated using the metal lift-off method to avoid chemical damage on the ITO patterns during our Al slit manufacturing prepared at the same substrate. For the Al slit patterning process, the patterned ITO glass substrate was turned over and the photoresist was patterned with a photo-mask, as shown in Figure 3b. The Al slits must be aligned precisely with the patterned ITO electrodes to control each slit independently. The positions of the patterned ITO electrodes and the Al slits were aligned with the alignment marks. A negative photoresist (AZ-5214, AZ Electronic Materials Co.) was used for the Al slit patterning, which was suitable for the metal lift-off process. After the process of the negative photoresist pattering, the Al layer was deposited using vacuum thermal evaporation equipment with a 500 nm thickness. The Al slit patterns were defined clearly by removing the photoresist with acetone. To protect the Al slits from a physical scratch damage, the SiO_2 layer was deposited on the patterned Al layer as a passivation layer with a 500 nm thickness.

To realize the optical 3D surface profilometry system without any mechanical stage, we implemented an orthogonally aligned LC sample, where the two LC alignment directions were precisely aligned with two regions of the common slit and four switching slits. Those were also aligned with the patterned ITO electrodes, as shown in Figures 1 and 4. As shown in Figure 4, the orthogonally aligned LC sample was implemented using the multi-rubbing method on the patterned ITO/Al substrate. After rubbing the LC alignment layer with a soft rubbing cloth attached on the roll-based rubbing machine, the LC molecules can be unidirectionally aligned on the LC alignment surface along the rubbing direction owing to the rubbing-induced surface morphology change and the alignment effect of the LC-interactive side chains of the LC alignment surface material [26]. However, with this conventional rubbing process, the multi-directional patterned LC alignment condition, required in our DFPG for coexisting two LC modes of the ECB LC mode and the TN LC mode in single LC cell, cannot be obtained. To develop the DFPG with the multi-directional LC alignment, we suggested two steps of the rubbing process supported by the photolithography process between each rubbing step. First, the LC alignment layer (polyimide, PI, SE-5811, Nissan Chemical Industries Co., Tokyo, Japan) was spin-coated on the patterned ITO electrode under the conditions of 1000 rpm for 5 s, 3000 rpm for 30 s, and 1000 rpm for 5 s.

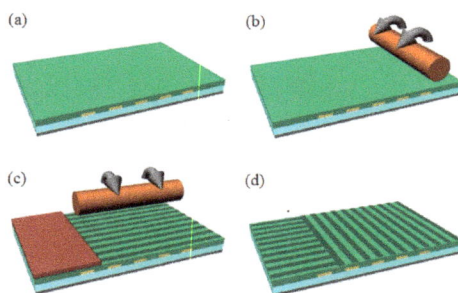

Figure 4. Schematic of the multi-rubbing process for two domains of the initial LC alignments on patterned ITO substrates to enable the ECB LC mode on the common slit for the phase changing of the interference patterns and the TN LC mode on the switching slits for changing the spatial frequency of the projected fringes: (**a**) spin-coating of the LC alignment layer; (**b**) the first rubbing process over the whole area; (**c**) the second rubbing process after the formation of the patterned passivation layer with photoresistor where the rubbing direction is orthogonal to the first rubbing direction; and (**d**) the final multi-rubbed LC alignment layer for the top substrate of the DFPG device, prepared after removing the passivation layer.

The coated PI layer was baked at 230 °C for 30 min and then it was uni-directionally rubbed with the roll-based rubbing machine at a roller speed of 300 rpm and a substrate speed of 10 mm/s in the horizontal direction, as shown in Figure 4b. Before the second rubbing process, the rubbed PI layer was protected by the photoresist (GXR-601) layer, and it was partially patterned by the photolithography process using the photo-mask, as shown in Figure 4c, where the LC alignment layer prepared for the ECB LC mode operation under the common slit was locally protected for the second rubbing process. With the partially passivated PI layer, the uncovered PI surface was rubbed orthogonal to the first rubbing direction, and then the residual GXR-601 layer was completely removed using acetone, as shown in Figure 4d, which schematically shows the final LC alignment directions of the top substrate of our DFPG, which has two orthogonal LC alignments for two LC domains of the ECB and TN LC modes. For our two domains of LC alignment, the photoresist process and its removing process should not physio-chemically degrade the LC alignment capability produced by the first rubbing process. Thus, the types of photoresist and their etchants should be carefully chosen, as mentioned in our experimental procedure.

Figure 5a,b show the photo-mask used for the ITO etching and the Al slit patterning process, respectively. In Figure 5a, five wide ITO patterns, directly connected to each ITO line pattern, can be seen, which were prepared for the wire-bonding process to control the electro-optic properties of the common slit and five switching slits individually with the driving signals. Figure 5c shows the finally implemented top substrate of the DFPG with the patterned ITO electrodes and the Al slit arrays. The substrate size was 15×20 mm^2 in our sample implementation. The size of the actual area, which optically acts as the DFPG, was almost 8×1 mm^2.

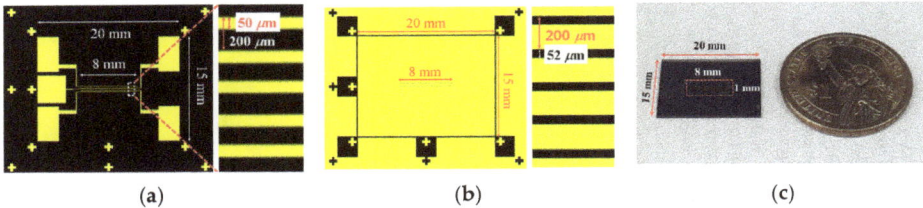

Figure 5. CAD images of photo-lithographic masks with the alignment marks for (**a**) ITO patterning and (**b**) Al slit patterning; (**c**) Image of the fabricated DFPG device.

2.2. Theory of Depth Extraction

2.2.1. Calculation of the Phase Modulation

When using our four-step phase-shifting scheme for the 3D depth extraction, the phase-shifting in our DFPG according to an applied voltage to the common slit needs to be precisely measured. The optical set-up of the phase modulation measurement is shown in Figure 6a. The coherent light source of a He-Ne laser ($\lambda = 632.8$ nm) was used to generate interference fringe patterns and to measure the field-dependent phase modulation of the LC layer operated by the ECB LC mode. The speckle noise and high order fringe visibility could be minimized to be a negligible level as shown in Figures 2 and 6 because the coherence length of the He-Ne laser used in our experiment was 20 cm that was much smaller than those of coherent light sources used for conventional holographic interference experiments. The fringe visibility of the dynamic fringe patterns was about 0.9 in our system. This fringe visibility was enough to obtain the 3D depth extraction with our four-step phase-shifting and four steps of multi-spatial frequency scheme [21]. The beam from the He-Ne laser was expanded by a beam expander and was passed through an iris to reduce the optical noise so that a suitable spot size with a uniform beam intensity can cover all of the five Al slits. The light polarized with the x-axis polarizer was passed through the DFPG. The DFPG generated fringe patterns from two slits and the fringe patterns expanded by a projection lens were projected on a flat-panel screen. The projected fringe patterns on the flat-panel screen were captured by a CCD (FL2-14S3H, Pointgrey, Richmind, BC, Canada) with 4.4 μm pixel pitch and the field of view (FOV) of the CCD lens module was 12°.

We used the DFT method to measure the phase modulation. The fringe pattern by the DFPG was calculated using the equation $M_{phase} = angle[X(pN/\Lambda + 1)]$, as explained in our previous study, obtained using a mechanical moving slit system [21]. The values of the phase modulation were repeatedly calculated depending on the change of the applied voltage with an increase of 0.01 V per step. As shown in Figure 6b, the phase modulation over $3\pi/2$ was achieved at 5.6 V. The voltage values for the four-step phase modulation were measured as 2.1, 3.1, 3.8, and 5.6 V, each for the phase shift of $\Delta\phi = 0$, $\pi/2$, π, and $3\pi/2$, respectively, as shown in Figure 6b. For each step of increasing the applied voltage to the common slit, the projected fringe patterns with the same spatial frequency were moved spatially by a quarter of the periodicity of the fringe patterns, as shown in the inset picture of Figure 6b.

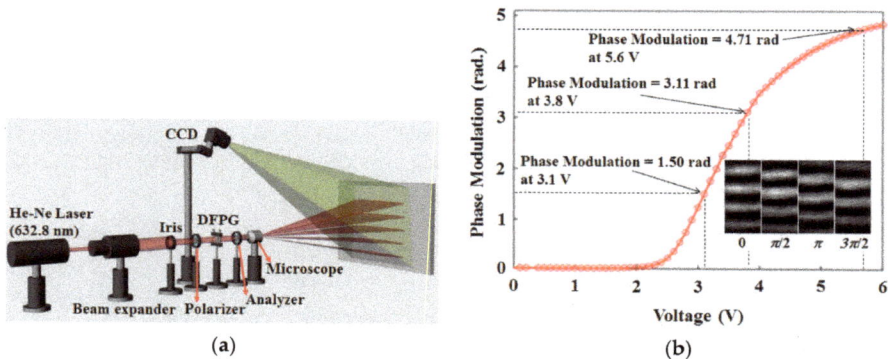

Figure 6. (a) Optical system for measuring phase modulation values of the common slit according to an applied voltage and their moving fringe patterns; (b) Measured phase modulation curve according to an applied voltage and the fringe patterns captured under $\Delta\phi = 0$, $\pi/2$, π, and $3\pi/2$ conditions.

Figure 7 shows the time sequence of the driving waveforms applied to the common slit part and four switching slit parts used for our DFPG. The driving waveforms for the common slit were changed every 200 ms and the amount of the applied voltage was increased to 2.1 V, 3.1 V, 3.8 V, and 5.6 V at 1 kHz to achieve phase shifting of $\Delta\phi = 0$, $\pi/2$, π, and $3\pi/2$, respectively, as shown in Figure 6. At a given phase-shifting amount condition, four-step spatial-frequency-varying fringe patterns were projected by applying dual-frequency-driving waveforms to the TN LC parts individually operated by the four patterned ITO electrodes. The turn-on waveform for each TN LC part of the switching slits was applied with 5 V of the applied voltage at 1 kHz for 50 ms, and then the turn-off waveform was applied with 5 V of the applied voltage at 100 kHz for 150 ms. These frequency-modulating AC waveforms were applied in sequence to the ITO electrode pattern of each switching slit every 50 ms to obtain four sets of fringe patterns with four different spatial frequencies. When applying the turn-off waveform with high frequency AC driving to the switching slit IV, the voltage required for the next step of the phase-shifting was applied to the ECB LC layer of the common slit and then, frequency-modulating waveforms were sequentially applied to the four TN LC layers to select four different spatial frequencies of the projected fringe patterns. In this manner, the whole scanning time required for the four-step multi-spatial frequencies and four-step phase shifting schemes could be completed in less than 800 ms. This means that the optical surface profilometry can capture 16 sets of fringe patterns distorted by an object in less than 800 ms. Our profilometry using the DFPG can exhibit quite fast scanning and capturing speeds suitable for hand-held dental applications, compared with our previous study of a 3D optical surface profilometry system constructed with a mechanically moving optical element [21].

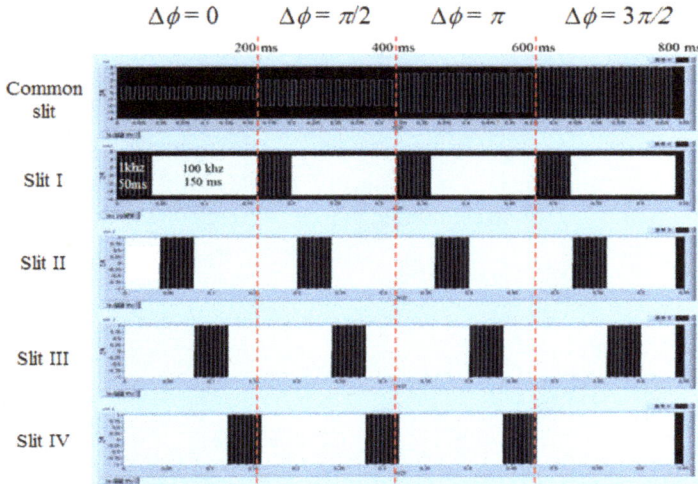

Figure 7. Driving waveforms of voltages applied to the LC layer on each slit of the DFPG utilizing the dual-frequency modulation for projecting 16 fringe patterns with the four different multi-spatial frequencies and the four-step phase shifting within the fast switching time.

2.2.2. Theory of 3D Depth Extraction

Figure 8a shows the optical setup used to reconstruct a 3D depth profile with the DFPG device. The multiple sets of fringe patterns generated by the DFPG were projected though the projection lens onto the surface of an object. The fringe patterns distorted by the object surface were captured by a CCD. Figure 8b shows the geometrical parameters used to calculate the depth and position information from the CCD images. The optical system is composed of the DFPG, projection lens, CCD camera, and testing object. The DFPG and CCD are on the *y-z* plane. The center points of the projection lens, CCD, and object are defined by (y_M, z_M), (y_C, z_C) and (y_{POI}, z_{POI}), respectively. The fringe patterns are expanded through the projection lens, and the projected fringe patterns are distorted depending on the surface profile of an object. The optical axis of the DFPG is parallel to the *z*-axis, and the optical axis of the CCD is tilted at an angle of ϕ_C to the *z*-axis. A point of interest (POI) on the object is expressed using geometrical parameters. As shown in Figure 8, a POI on the object is defined as (y_{POI}, z_{POI}). The line between the center of the projection lens and the POI is tilted at an angle of α with respect to the optical axis of the DFPG. The line between the center of the CCD and the POI is tilted at an angle of θ_{POI} with respect to the optical axis of the CCD. In this case, the triangular method, defined by the geometrical parameters, leads to the following relationships:

$$z_{POI} - z_C = -(y_{POI} - y_C)\cot(\phi_C - \theta_{POI}) \text{ and} \tag{1}$$

$$z_{POI} - z_M = (y_{POI} - y_M)\cot\alpha \tag{2}$$

From Equations (1) and (2):

$$z_{POI} = \frac{(y_C - y_M) - (z_C - z_M)\tan\alpha}{\tan\alpha - \tan(\phi_C - \theta_{POI})} + z_C \text{ and} \tag{3}$$

$$y_{POI} = \frac{(z_C - z_M) - (y_C - y_M)\cot\alpha}{\cot\alpha + \cot(\phi_C - \theta_{POI})} + y_C \tag{4}$$

are obtained. Finally, Equations (3) and (4) show the depth and position information of the object calculated from the geometrical parameters [21].

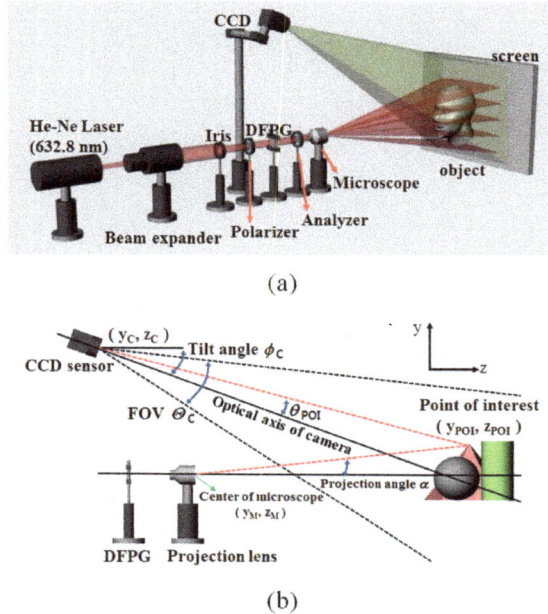

(a)

(b)

Figure 8. (**a**) Optical set-up to reconstruct the depth profile of a 3D object with the DFPG device; (**b**) Geometrical parameters of the 3D vision system used for 3D depth extraction.

3. Experimental Results and Discussion

3.1. Dynamic Phase Changing and Multiple Spatial Frequency Modulation Properties of the DFPG

In our DFPG for 3D optical surface profilometry, the transmission axes of the two polarizers and the rubbing direction of the LC alignment layer of the bottom substrate are aligned to be parallel to each other. In this device configuration, Figure 9a shows the polarization optical microscope (POM) images of the DFPG cell measured with applying the turn-on voltage to one of the TN LC layers used for switching the switching slits with varying the spacing for two beam interference, $\Delta g = 200, 400, 600$ and 800 μm. As shown in Figure 9a, the beam transmitted through the common slit, which is used as one beam spot for two-beam interference, is always under the turn-on state without intensity variation. However, the spacing between the two slits for interference can be controlled by the applied voltage conditions of the TN LC layers under the four switching slits. In all cases, three switching slit areas, which were not selected by the turn-on voltage, showed a dark texture without light transmittance.

To observe the field-dependent birefringence change in the ECB LC layer under the common slit, POM images of the DFPG cell were obtained between the crossed polarizers after detaching the parallel polarizers from the DFPG cell, as shown in Figure 9b. In this measurement, the POM images were captured by rotating the rubbing direction of the bottom substrate of the DFPG cell by 45° with respect to the crossed polarizers. On the common slit area, light transmittance or a color change could be observed owing to a variation of the phase modulation with increasing applied voltage to the ECB LC layer.

Sensors 2016, 16, 1794

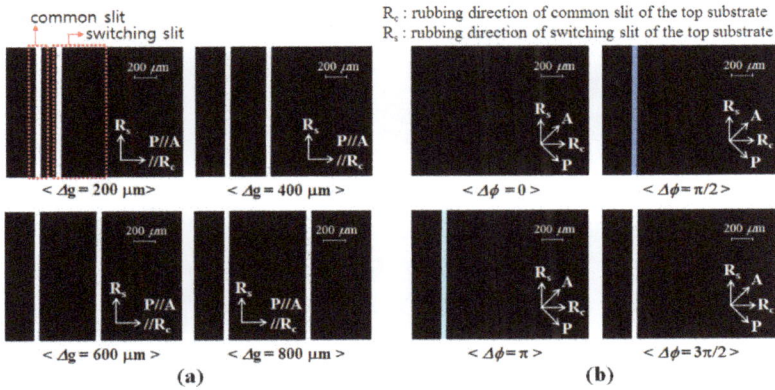

Figure 9. POM images depending on applied voltage conditions at each slit. (**a**) POM images measured between the parallel polarizers with varying selected switching slits for interference, which creates fringe patterns with four different spatial frequencies; (**b**) POM images showing the four-step phase shifting of the LC layer on the common slit captured between the crossed polarizers to show field-dependent birefringence variation.

It is important to obtain the fast response property of the DFPG required in projecting 16 sets of multiple fringe patterns. The 3D depth profile is calculated from the 16 fringe pattern images distorted by surface morphology. The fast response time of each LC slit is positively necessary to improve the operating time of the 3D vision system. In this study, the dual-frequency LC was used to improve the response time of the LC layer switching. A dual-frequency LC exhibits the frequency-dependent dielectric anisotropy where $\Delta\varepsilon$ is positive under the low frequency operation and $\Delta\varepsilon$ is negative under the high frequency operation. That means that the LC molecules are reoriented along the applied field direction under the low frequency operation, whereas the LC molecules are reoriented perpendicular to the applied field direction under the high frequency operation.

The response times of the DFPG were measured by electro-optic characterization equipment (LCMS200, Sesim Photonics Technology Co. Ltd., Uiwang, Korea). The common slit was blocked to measure the response times of the switching slits operated by the TN LC mode. Figure 10a shows that the turn-on response time switched by the low Hz operation of 5 V at 1 kHz was about 1.6 ms, sufficient for projecting multiple fringe patterns. However, the turn-off response time, measured by the field-off condition, showed a relatively slow response with a value of about 36.1 ms, as shown in Figure 10b. The natural field-off LC relaxation to the initial TN state is supported only by the LC elastic property and the LC surface anchoring, which was too slow to be used in our optical surface profilometry system. In our experiment, the turn-off response of the switching slit was improved by applying the high frequency operating field to obtain the field-driven turn-off property instead of the natural field-off LC relaxation. Figure 10c shows that the turn-off response was improved to about 8.1 ms by applying a high frequency AC field of 5 V at 100 kHz. Consequentially, the total response time of the TN slit was about 9.7 ms. As shown in Figure 7, the dual-frequency modulation scheme was used to switch four switching slits in sequence at a given phase-shifting amount of the common slit.

49

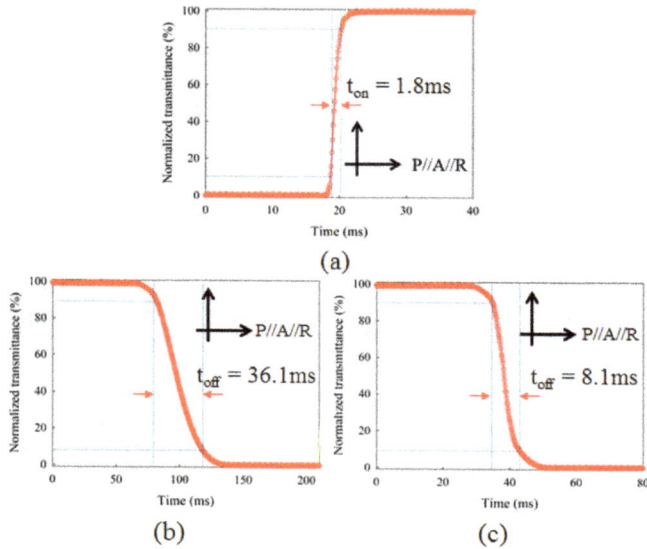

Figure 10. On/Off response times of the LC layer of the DFPG. (**a**) Field-on response time; (**b**) Field-off response time obtained by the natural relaxation of the LC layer due to the surface alignment effect without applying any electric field; (**c**) Off response time obtained by applying a high frequency field to the LC layer utilizing the frequency inversion effect of the LC dielectric anisotropy.

3.2. Depth Extraction from 3D Optical Profilometry

Figure 11a shows a the photographic image of the object used for the 3D depth extraction with the presented 3D optical surface profilometry system, which was deliberately chosen to have a continuously slanted surface together with an abrupt depth change discontinuity from the background reference surface. The smallest height at the abrupt side edges of the slanted object was over 40 mm, which was much higher than the optical wavelength used in our experiment. The regions marked with the dotted red box were reconstructed after projecting multiple sets of fringe patterns with our DFPG. Figure 11b shows the 16 sets of fringe patterns distorted on the slanted object. Four sets of laterally moved fringe patterns were obtained by changing $\Delta\phi$ via the common slit control. At a given $\Delta\phi$ condition, multiple fringe patterns having four different spatial frequencies were sequentially projected by selecting the Δg condition of the switching slits. Because there was a high depth discontinuity between the slanted surface and the background surface, some bright (or dark) fringe lines on the slanted surface were continuous with those on the background surface. This optical phase ambiguity could be solved using the four-step phase shifting scheme during the depth reconstruction [21,27].

The upper image of Figure 12a shows the enlarged CCD image of one of the projected fringe patterns, where we can observe some optical noises within the fringe patterns, which might be produced by some particles from our optical components. However, the synthesized phase map obtained after applying the four-step phase shifting algorithm, presented in the lower image of Figure 12a, shows that this type of the background optical noise within the fringe patterns can be successfully eliminated after the 3D depth reconstruction.

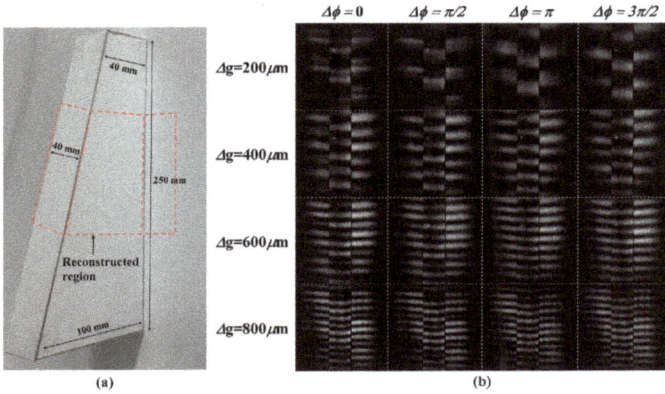

Figure 11. (**a**) Photographic image of the slanted object used for the 3D depth extraction; (**b**) CCD images of the distorted fringe patterns projected on the slanted object.

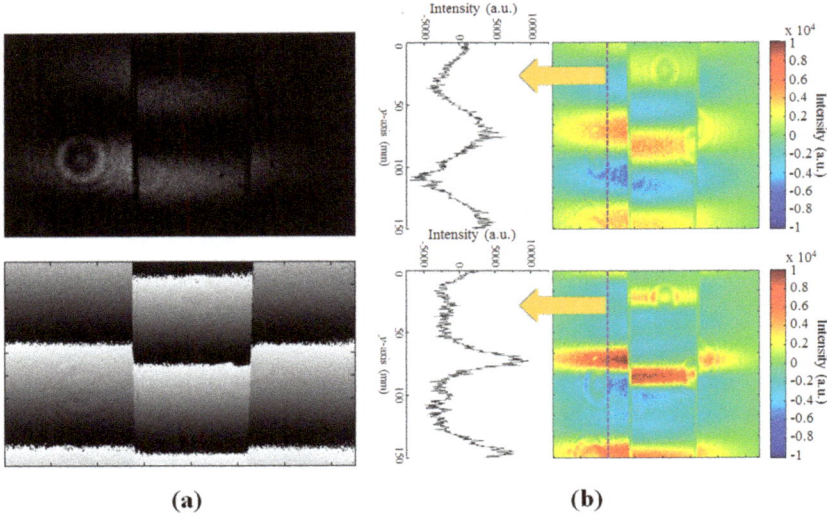

Figure 12. (**a**) Original fringe optical pattern image (upper image) projected on the slanted surface of Figure 11a using the interference slit of $\Delta g = 200$ μm and the synthesized phase map (lower image) obtained using the four-step phase shifting algorithm; (**b**) Fictitious scanning pattern images generated from two spatial frequencies ($\Delta g = 200$ and 800 μm) of the projected fringe patterns (the upper images) and four spatial frequencies ($\Delta g = 200, 400, 600,$ and 800 μm) of the projected multiple fringe patterns (lower images).

In our depth reconstruction, the fictitious scanning pattern is generated through multiple spatial frequencies of the interference fringes for the 3D depth reconstruction using the geometrical optical parameters [21,27]. In Figure 12b, the upper and the lower images show the fictitious scanning patterns generated from the two spatial frequencies of $\Delta g = 200$ and 800 μm and generated from full sets of four spatial frequencies ($\Delta g = 200, 400, 600,$ and 800 μm), respectively. Compared with the upper image, the fictitious scanning pattern shown in the lower image can make much sharper peaks owing to more sinusoidal sets of the spatial frequencies used in the synthesis of the fictitious scanning pattern [21,27]. This could result in the enhanced depth resolution and precision after the 3D depth reconstruction.

Figure 13 shows the 3D depth profile reconstructed from the slanted object using the 16 sets of the dynamic fringe patterns. Compared with conventional phase unwrapping methods [28], one of the merits of the depth reconstruction based on the geometrical optical parameters used in our experiment is that the absolute depth values together with the 2D positional values, not just the relative phase values, can be obtained, as shown in Figure 13.

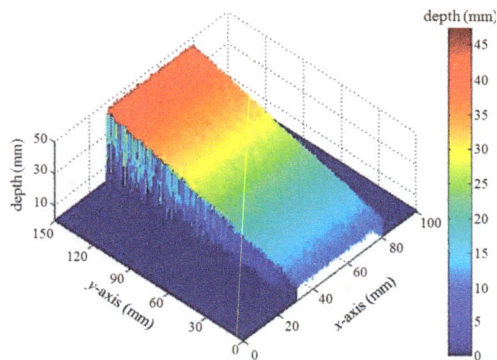

Figure 13. 3D depth profile reconstructed from the slanted object.

When the size of an object to be measured with projecting the dynamic fringe patterns and to be reconstructed with the 3D depth extraction increases with approaching the values of the numerical aperture (NA) of the projection lens and the FOV of the CCD lens module, the calibrations of the fringe images captured by the CCD camera would be needed to reduce and to compensate the image distortions which might become more severe especially in the captured image boundaries owing to like the vignetting effect of the lens-based projection and imaging system. In our experiment, the NA of the projection lens was 0.4 and the FOV of the CCD lens module was 12°. Considering the projection distance of 1 m, the object size which can be measured in our experiment would be about 25×25 cm^2. However, our object size was limited under 20×20 cm^2 because we had a problem in enlarging the detectable object size because of the highly weakened fringe intensity with increasing the fringe pattern size as shown in Figures 2 and 11. In our projection system, the much parts of the incident optical beams were blocked by the five slits and the intensity profiles projected by the projection lens decrease from the center to the image edges. We expect that this problems would be solved by improving the collimating projection lens part and by introducing the optical coupling module between the light source and the individual slits in the future work. However, this nonuniform intensity profile did not affect the depth extraction process in our four-step phase shifting and four steps of multi-spatial frequency scheme as shown in Figure 12. In our experiment set-up, we used the surface-distorted dynamic fringe patterns captured by the CCD lens module without the positional image calibration for the 3D depth extraction process. We checked the effect of the image distortion by our lens system by placing a grid pattern at the object plane, the image distortion was negligible under our experimental conditions of the FOV and the measurement distance. In Figure 13, the measurement error in the absolute value was less than 2 mm. This might originate from additional image distortions produced during the depth reconstruction process based on the geometric parameters, which are dependent on the perspective imaging condition expressed by the angle values of α and θ_{POI} of Equations (3) and (4). We expect that the measurement accuracy can be further improved after calibration of these geometric parameters by using the periodic test patterns and the size of the measurable object can be enlarged after improving the projection parts and improving the CCD camera module.

Figure 14a shows the CCD images of the distorted fringe patterns projected on a square box object ($80 \times 80 \times 80$ mm^3). The perspective view of the 3D depth profile obtained after the 3D depth reconstruction is presented in Figure 14b with the absolute coordinate axis values. For each *x*, *y*, and *z* coordinate axis, the measurement errors of the extracted 3D depth profile were 1.6, 0.9, 1.2 mm, respectively, which were slightly different depending on the coordinate axis. This coordinate-dependent measurement error might originated in the different perspective viewing conditions between the projection lens part and the image-capturing CCD camera part in our optical system as shown in Figure 8. In the previous approaches, most of the optical surface profilometry systems are based on the phase unwrapping algorithm for the depth reconstruction [28–30] and need more complex optical elements like the time-sequentially switching multi-wavelength light sources [31,32], additional optical components [33,34], and a mechanical translation state [21,35] for the projection to generate multi-spatial-frequency or/and phase-shifting fringe patterns. Our optical surface profilometry system based on the electrically switching single DFPG cell can generate dynamic fringe patterns in a fast switching time and can extract the depth and 2D positional information of a 3D object even for cases having large surface depth discontinuities.

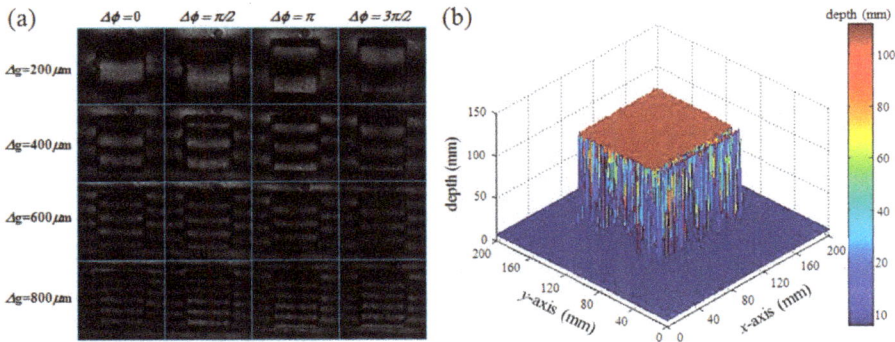

Figure 14. (a) CCD images of the distorted fringe patterns projected on the square box object; (b) 3D depth profile reconstructed from the square box object.

4. Conclusions

3D optical surface profilometry systems have been applied to measure the depth profile of a 3D object and their application fields are much growing recently. To enhance the accuracy of the 3D depth extraction and/or to avoid the phase ambiguity problem, multiple sets of fringe pattern projections are essential in the optical surface profilometry system. We suggested a single LC cell DFPG operated by a simple passive driving scheme, which can be fabricated into a compact optical module. To project 16 sets of dynamic fringe patterns having the optical features of four-step phase shifting and four different spatial frequencies without any mechanical parts and any complex and expensive SLM parts, two types of LC modes—the ECB LC mode and the TN LC mode—were included in the single LC cell by preparing two orthogonal LC alignments on one of the LC alignment layers. Five optical slits were prepared using photolithography and aligned with the patterned LC alignment layer, where one common slit with the ECB LC layer was used for the four-step phase-shifting control and four switching slits with the TN LC layers were used for the generation of four different spatial frequencies. Moreover, to improve the scanning time required for generating the 16 sets of dynamic fringe patterns, dual-frequency switching LC was employed, where the switching time—including the turn-on and the turn-off times—controlled by the frequency modulation increased four-fold about four times compared with the switching time controlled by the conventional single frequency. As a result, the 16 sets of fringe patterns could be projected and captured within 800 ms. For the depth reconstruction procedures, the phase shifting in the common slit of the DFPG according to the applied voltages was accurately measured. The resolution of the depth reconstruction obtained by applying the DFT method and the optical geometric parameters was improved by using the four different sets of spatial frequencies of the dynamic fringe patterns. Our 3D optical surface profilometry system yields 3D depth profiles with respect to the absolute value 3D coordinates, not just the relative phase depth maps, even for objects with abruptly changing surface profiles. Owing to the compactness and fast switching properties of the DFPG, we expect that the presented 3D optical surface profilometry system can be applied to a hand-held 3D vision module, which can be widely used for several mobile applications requiring 3D depth information.

Acknowledgments: This work was supported by 'The Cross-Ministry Giga KOREA Project' grant from the Ministry of Science, ICT and Future Planning, Korea.

Author Contributions: Kyung-Il Joo and Mugeon Kim equally contributed to this paper. Hak-Rin Kim conceived and designed the experiments. Kyung-Il Joo and Mugeon Kim performed the optical experiments. Min-Kyu Park and Heewon Park prepared the LC device. Mugeon Kim and Byeonggon Kim designed the optical module and packaged it. Mugeon Kim, Kyung-Il Joo and Hak-Rin Kim analyzed the data and wrote the paper. JoonKu Hahn contributed the reconstruction algorithm. Hak-Rin Kim proofread and revised the manuscript.

Conflicts of Interest: The authors declare no conflict of interest.

Abbreviations

The following abbreviations are used in this manuscript:

DFPG	Dynamic Fringe Pattern Generator
LC	Liquid Crystal
TN	Twisted Nematic
ECB	electrically controlled birefringence
ITO	indium thin oxide
He-Ne Laser	Helium-Neon Laser
DFT	Discrete Fourier Transform
POM	Polarizing Optical Microscope
FOV	Field of View

References

1. Salvi, J.; Fernandez, S.; Pribanic, T.; Llado, X. A state of the art in structured light patterns for surface profilometry. *Pattern Recognit.* **2010**, *43*, 2666–2680. [CrossRef]
2. Su, X.; Chen, W. Fourier transform profilometry: A review. *Opt. Lasers Eng.* **2001**, *35*, 263–284. [CrossRef]
3. Park, C.-S.; Park, K.-W.; Jung, U.; Kim, J.; Kang, S.-W.; Kim, H.-R. Dynamic fringe pattern generation using an electrically tunable liquid crystal Fabry-Perot cell for a miniaturized optical 3-D surface scanning profilometer. *Mol. Cryst. Liq. Cryst.* **2010**, *526*, 28–37. [CrossRef]
4. Han, X. 3D Shape Measurement Based on the Phase Shifting and Stereovision Methods. Ph.D. Thesis, Stony Brook University, New York, NY, USA, 2010.
5. Yoneyama, S.; Morimoto, Y.; Fujigaki, M.; Yabe, M. Phase-measuring profilometry of moving object without phase-shifting device. *Opt. Lasers Eng.* **2003**, *40*, 153–161. [CrossRef]
6. Mermelstein, M.S.; Feldkhun, D.L.; Shirley, L.G. Video-rate surface profiling with acousto-optic accordion fringe interferometry. *Opt. Eng.* **2000**, *39*, 106–113.
7. Ishii, I.; Yamamoto, K.; Doi, K.; Tsuji, T. High-speed 3D Image Acquisition Using Coded Structured Light Projection. In Proceedings of the IEEE/RSJ International Conference on Intelligent Robots and Systems, San Diego, CA, USA, 29 October–2 November 2007; pp. 925–930.
8. Waddington, C.; Kofman, J. Analysis of measurement sensitivity to illuminance and fringe-pattern gray levels for fringe-pattern projection adaptive to ambient lighting. *Opt. Lasers Eng.* **2010**, *48*, 251–256. [CrossRef]
9. Ulusoy, A.O.; Calakli, F.; Taubin, G. One-Shot Scanning Using De Bruijn Spaced Grids. In Proceedings of the IEEE International Conference on Computer Vision Workshops, Kyoto, Japan, 27 September–4 October 2009; pp. 1786–1792.

10. Sansoni, G.; Carocci, M.; Rodella, R. Calibration and Performance Evaluation of a 3-D Imaging Sensor Based on the Projection of Structured Light. *IEEE Trans. Instrum. Meas.* **2000**, *49*, 628–636. [CrossRef]
11. Su, X.; Chen, W. Reliability-guided phase unwrapping algorithm: A review. *Opt. Laser Eng.* **2004**, *42*, 245–261. [CrossRef]
12. Chen, L.-C.; Yeh, S.-L.; Tapilouw, A.M.; Chang, J.-C. 3-D surface profilometry using simultaneous phase-shifting interferometry. *Opt. Commun.* **2010**, *283*, 3376–3382. [CrossRef]
13. Su, X.; Xue, L. Phase unwrapping algorithm based on fringe frequency analysis in Fourier-transform profilometry. *Opt. Eng.* **2001**, *40*, 637–643. [CrossRef]
14. Tian, J.; Peng, X.; Zhao, X. A generalized temporal phase unwrapping algorithm for three-dimensional profilometry. *Opt. Laser Eng.* **2008**, *46*, 336–342. [CrossRef]
15. Saldner, H.O.; Huntley, J.M. Profilometry using temporal phase unwrapping and a spatial light modulator-based fringe projector. *Opt. Eng.* **1997**, *36*, 610–615.
16. Yu, C.J.; Jang, E.; Kim, H.-R.; Lee, S.D. Design and fabrication of high-performance liquid crystal gratings. *Mol. Cryst. Liq. Cryst.* **2006**, *454*, 765–778. [CrossRef]
17. Kraan, T.C.; Bommel, T.V.; Hikmet, R.A.M. Modeling liquid-crystal gradient-index lenses. *J. Opt. Soc. Am. A* **2007**, *24*, 3467–3477. [CrossRef]
18. Lucchetta, D.E.; Karapinar, R.; Manni, A.; Simoni, F. Phase-only modulation by nanosized polymer-dispersed liquid crystal. *J. Appl. Phys.* **2002**, *91*, 6060–6065. [CrossRef]
19. Ren, H.; Lin, Y.-H.; Fan, Y.-H.; Wu, S.-T. Polarization independent phase modulation using a polymer-dispersed liquid crystal. *Appl. Phys. Lett.* **2005**, *86*, 141110. [CrossRef]
20. Hu, L.; Xuan, L.; Liu, Y.; Cao, Z.; Li, D.; Mu, Q.Q. Phase-only liquid-crystal spatial light modulator for wave-front correction with high precision. *Opt. Express* **2004**, *12*, 6403–6409. [CrossRef] [PubMed]
21. Joo, K.-I.; Park, C.-S.; Park, M.-K.; Park, K.-W.; Park, J.-S.; Seo, Y.; Hahn, J.; Kim, H.-R. Multi-spatial-frequency and phase-shifting profilometry using a liquid crystal phase modulator. *Appl. Opt.* **2012**, *51*, 2624–2632. [CrossRef] [PubMed]
22. Zhang, Z.H.; Towers, C.E.; Towers, D.P. Phase and Colour calcualation in Colour fringe projection. *J. Opt. A Appl. Opt.* **2007**, *9*, S81–S86. [CrossRef]
23. Geng, J. Structured-light 3D surface imaging: A tutorial. *Adv. Opt. Photonics* **2011**, *3*, 128–160. [CrossRef]
24. Lanman, D.; Taubin, G. Build Your Own 3D Scanner: 3D Photography for Beginners. In Proceedings of the SIGGRAPH 2009 Course Notes, New Orleans, LA, USA, 3 August–7 August 2009.
25. Hahn, J.; Kim, H.; Lee, B. Optimization of the spatial light modulation with twisted nematic liquid crystals by a genetic algorithm. *Appl. Opt.* **2008**, *47*, 87–95. [CrossRef]
26. Van Aerle, N.A.J.M.; Barmentlo, M.; Hollering, R.W.J. Effect of rubbing on the molecular orientation within polyimide orienting layers of liquid-crystal displays. *J. Appl. Phys.* **1993**, *74*, 3111–3120. [CrossRef]
27. Kim, E.-H.; Hahn, J.; Kim, H.; Lee, B. Profilometry without phase unwrapping using multi-frequency and four-step phase-shift sinusoidal fringe projection. *Opt. Express* **2009**, *17*, 7818–7830. [CrossRef] [PubMed]
28. Wu, F.; Zhang, H.; Lalor, M.J.; Burton, D.R. A novel design for fiber optic interferometric fringe projection phase-shifting 3-D profilometry. *Opt. Commun.* **2001**, *187*, 347–357. [CrossRef]
29. Jeught, S.V.D.; Sijbers, J.; Dirckx, J.J. Fast Fourier-Based Phase Unwrapping on the Graphics Processing Unit in Real-Time Imaging Applications. *J. Imaging* **2015**, *1*, 31–44. [CrossRef]
30. Pedraza-Ortega, J.C.; Gorrostieta-Hurtado, E.; Delgado-Rosas, M.; Canchola-Magdaleno, S.L.; Ramos-Arreguin, J.M.; Fernandez, M.A.A.; Sotomayor-Olmedo, A. A 3D Sensor Based on a Profilometrical Approach. *Sensors* **2009**, *9*, 10326–10340. [CrossRef] [PubMed]
31. Mehta, D.S.; Dubey, S.K.; Hossain, M.M.; Shakher, C. Simple multifrequency and phase-shifting fringe-projection system based on two-wavelength lateral shearing interferometry for three-dimensional profilometry. *Appl. Opt.* **2005**, *44*, 7515–7521. [CrossRef] [PubMed]
32. Hagen, K.M.; Burgarth, V.; Zeid, A.A. Profilometry with a multi-wavelength diode laser interferometer. *Meas. Sci. Technol.* **2004**, *15*, 741–746. [CrossRef]
33. Li, E.B.; Peng, X.; Xi, J.; Chicharo, J.F.; Yao, J.Q.; Zhang, D.W. Multi-frequency and multiple phase-shift sinusoidal fringe projection for 3D profilometry. *Opt. Express* **2005**, *13*, 1561–1569. [CrossRef] [PubMed]

Sensors **2016**, *16*, 1794

34. Enzberg, S.V.; Al-Hamadi, A.; Ghoneim, A. Registration of Feature-Poor 3D Measurements from Fringe Projection. *Sensors* **2016**, *16*, 283. [CrossRef] [PubMed]
35. Xiao, S.; Tao, W.; Zhao, H. A Flexible Fringe Projection Vision System with Extended Mathematical Model for Accurate Three-Dimensional Measurement. *Sensors* **2016**, *16*, 612. [CrossRef] [PubMed]

sensors

MDPI

Article

Robust Depth Image Acquisition Using Modulated Pattern Projection and Probabilistic Graphical Models

Jaka Kravanja [1], Mario Žganec [1], Jerneja Žganec-Gros [1], Simon Dobrišek [2] and Vitomir Štruc [2,*]

[1]   Alpineon d.o.o., Ulica Iga Grudna 15, Ljubljana SI-1000, Slovenia; jaka.kravanja@alpineon.si (J.K.); mario.zganec@alpineon.si (M.Ž.); jerneja.gros@alpineon.si (J.Ž.-G.)

[2]   Faculty of Electrical Engineering, University of Ljubljana, Tržaška cesta 25, Ljubljana SI-1000, Slovenia; simon.dobrisek@fe.uni-lj.si

*   Correspondence: vitomir.struc@fe.uni-lj.si; Tel.: +386-1-4768-839; Fax: +386-1-4768-316

Academic Editor: Gonzalo Pajares Martinsanz
Received: 9 August 2016; Accepted: 10 October 2016; Published: 19 October 2016

Abstract: Depth image acquisition with structured light approaches in outdoor environments is a challenging problem due to external factors, such as ambient sunlight, which commonly affect the acquisition procedure. This paper presents a novel structured light sensor designed specifically for operation in outdoor environments. The sensor exploits a modulated sequence of structured light projected onto the target scene to counteract environmental factors and estimate a spatial distortion map in a robust manner. The correspondence between the projected pattern and the estimated distortion map is then established using a probabilistic framework based on graphical models. Finally, the depth image of the target scene is reconstructed using a number of reference frames recorded during the calibration process. We evaluate the proposed sensor on experimental data in indoor and outdoor environments and present comparative experiments with other existing methods, as well as commercial sensors.

Keywords: depth imaging; modulated acquisition; structured light; triangulation; probabilistic graphical models; 3D reconstruction

---

1. Introduction

Over the last few years, we have witnessed the rapid growth of 3D imaging technologies in various application areas, ranging from autonomous navigation of robots, drones or cars [1–3], medical applications [4,5], consumer electronics [6] and surveillance systems [7,8] to object reconstruction [9,10], biometrics [11–13], and others [14–16]. Especially, with the introduction of low-cost commercial 3D imaging sensors, such as Microsoft's Kinect (Microsoft, Redmond, WA, USA) [6], depth sensing has become a popular research direction with new applications and use cases being presented on a regular basis. Several major corporations have since introduced their own depth imaging technology (e.g., Intel (Santa Clara, CA, USA) recently announced the Euclid sensor; Sony (Tokio, Japan) introduced the PlayStation Camera with the PS4 console; and Infineon (Neubiberg, Germany) developed the Real3 sensor) with the goal of participating in this rapidly growing depth-sensor market.

Existing 3D imaging techniques can be divided into two main categories: (i) active and (ii) passive. Active techniques utilize an active source of illumination to project a suitably-devised pattern of structured light onto the target scene and then perform 3D reconstruction based on temporal or spatial distortions of the projected pattern caused by interactions with the target scene. Examples of active techniques include standard structured light approaches [17–19], time-of-flight methods [20,21] or interferometry [22]. A comprehensive review of existing techniques from this group can be found in [23,24]. Passive techniques, on the other hand, do not rely on active illumination, but commonly require only a calibrated pair of cameras. Typical examples of passive techniques represent stereo-vision [25], shape-from-focus [26], shape-from-shading [27] and other related shape-from-X approaches [28]. The reader is referred to [29] for more detailed coverage of this topic.

When applied in outdoor environments, 3D imaging techniques are expected to provide accurate depth information regardless of the external lighting and atmospheric conditions, which are known to negatively affect the existing range measurement techniques. This property is crucial for various outdoor applications that require reliable depth information to function properly. This paper addresses the problem of outdoor depth imaging with structured light approaches and presents a novel sensor designed specifically for outdoor deployment [30]. The sensor exploits the recently-proposed concept of modulated pattern projection, introduced by our group in [31], which facilitates the acquisition of spatial distortion maps in real-world environments (even in the presence of strong incident sunlight), where other existing approaches often fail or at least struggle with their performance. Correspondences between the projected pattern and the acquired distortion map are established based on a novel probabilistic approach relying on graphical models (inspired by [32,33]) and are used in conjunction with prerecorded reference frames to compute the depth information for each point of the projected structured light pattern. All components of the sensor presented are designed for outdoor deployment and contribute to the overall performance, as demonstrated in the experimental section.

We make the following three contributions in this paper: (i) we present a novel 3D imaging sensor that supports robust acquisition of spatial distortion maps and is able to generate accurate depth maps in challenging outdoor environments; (ii) we describe the complete hardware and software (algorithmic) design of the sensor; and (iii) we present a comprehensive experimental evaluation, as well as comparative results with competing techniques and existing commercial sensors.

The paper is organized as follows: Section 2 presents the main components of the depth sensor and outlines their characteristics. The individual components are discussed in Sections 3, 4 and 5. Experimental results and comparative evaluations are described in detail in Section 6. The paper concludes with some final remarks and directions for future work in Section 7.

2. Sensor Overview

This section presents a short overview of a novel sensor designed for depth image acquisition. The sensor presented was developed as part of our research efforts with respect to an active triangulation system (ATRIS) capable of capturing depth images in difficult settings; for example, under exposure to strong incident sunlight. The sensor relies on the established concept of depth image acquisition based on structured light, in which a light pattern is first projected onto a target scene, and the shape (i.e., depth information) of the scene is then inferred based on the spatial distortions of the projected pattern and the (known) geometrical properties of the prototype.

A schematic representation of the three key components of the ATRIS sensor is shown in Figure 1.

- The image acquisition procedure uses specialized hardware (comprised of a laser projector and a high-speed camera) to project a structured light pattern onto the target scene with the goal of capturing an image of the distorted pattern (i.e., a spatial distortion map). The procedure is based on the recently-introduced concept of modulated pattern projection [31,34], which ensures that spatial distortion maps of good quality can be captured in challenging conditions; for example, in the presence of strong incident sunlight or under mutual interference caused by other similar sensors directed at the same scene.

- The light plane-labeling procedure establishes the correspondence between all parts of the projected light pattern and the detected pattern that has been distorted due to the interaction with the target scene. The procedure uses loopy-belief-propagation inference over probabilistic graphical models (PGMs) as proposed in [33] to solve the correspondence problem and, differently from other existing techniques in the literature, exploits spatial relationships between parts of the projected pattern, as well as temporal information from several consecutive frames to establish correspondence.

- The 3D reconstruction procedure reconstructs the depth image of the target scene based on (i) the reference frames of the light pattern projected onto a planar surface at different distances from the camera and (ii) the established correspondence between parts of the projected pattern and the detected distortion map.

A detailed description of all three key ATRIS components is presented below.

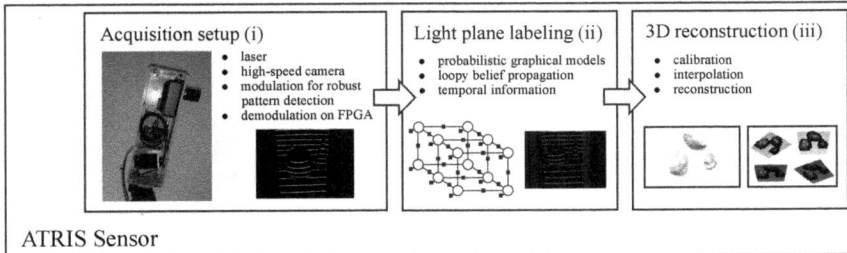

Figure 1. Schematic representation of our active triangulation system (ATRIS) sensor. The sensor comprises three key components: (i) an image acquisition procedure, which captures an image of the projected light pattern; (ii) a light plane labeling technique, which establishes the correspondence between all parts of the projected and detected patterns; and (iii) a 3D reconstruction procedure, which constructs a depth image from the detected light pattern.

3. The Acquisition Procedure

This section describes the modulated pattern acquisition procedure used in the ATRIS sensor. The section starts by presenting the hardware setup used in the sensor and then proceeds by describing the pattern acquisition procedure and its characteristics. Note that the underlying concept of the acquisition procedure was originally introduced in [31].

The hardware setup (shown in Figure 2a) used in the ATRIS sensor comprises a high-speed industrial camera with an integrated FPGA processing core and a modulable 650-nm LED laser. The high-speed (Velociraptor [35]) camera (Optomotive, Ljubljana, Slovenia) is capable of operating at a frame rate of 480 fps, which allows the ATRIS sensor to capture images of the structured light pattern at a speed of several frames per second. The sensor can therefore be used with static or dynamic scenes.

To acquire one image (or better said, a single frame) of the projected light pattern, the camera first captures a series of images of the target scene (we refer to these images as sub-frames). Every time an image (sub-frame) is taken, the laser projector is turned either on or off depending on the current value of the pseudo-random binary control/modulation sequence $c \in \{0,1\}$ that is cyclically shifted in the FPGA modulation register [31]. If the control/modulation sequence takes a value of $c = 1$, the laser projector is turned on, and the captured sub-frame contains a snapshot of the illuminated target scene. Similarly, when the control/modulation sequence takes a value of $c = 0$, the laser projector is turned off, and the captured sub-frame contains a snapshot of the scene without the structured light pattern (see Figure 2b). Based on these sub-frames, the final image of the projected pattern is generated as a normalized superposition of all sub-frames captured during one cycle of the control sequence.

As illustrated in Figure 2b, all sub-frames captured during the on state of the laser are added to the superposition, and all sub-frames captured during the off state are subtracted. This demodulation procedure is implemented in FPGA and removes most information about the appearance of the target scene from the generated image/frame, thus significantly emphasizing the projected pattern. A thresholding step is ultimately applied to the demodulated image to remove all remaining (scene-related) artifacts and to produce the final binary image of the projected pattern needed for the light plane labeling.

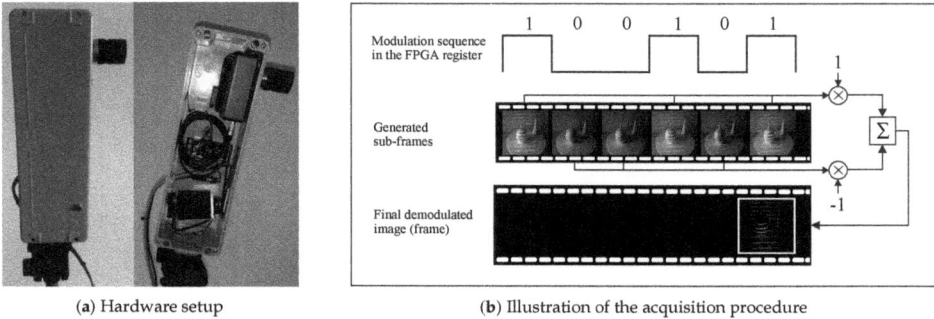

(a) Hardware setup   (b) Illustration of the acquisition procedure

Figure 2. The acquisition procedure. (**a**) Visual appearance of the hardware setup of the ATRIS sensor. The left side of the image shows the casing of the sensor and the right side the arrangement of the camera (above) and the laser projector (below) in the casing. (**b**) Illustration of the modulated acquisition procedure.

The acquisition procedure presented exhibits several desirable characteristics that are also experimentally validated in Section 6.1 (for a formal theoretical argumentation of the characteristics, the reader is referred to [31]):

- Noise suppression: The modulated acquisition procedure is robust for various types of noise. If information related to the visual appearance of the target scene is treated as "background noise," then the procedure presented obviously removes the background noise as long as the control/modulation sequence employed in the FPGA register is balanced (a balanced modulation sequence is defined as a sequence with an equal number of zero- and one-valued bits). Because demodulation is a pixel-wise operation, the acquisition procedure presented also suppresses sensor noise (typically assumed to be Gaussian) caused, for instance, by poor illumination or high temperatures, where a simple pair-wise sub-frame subtraction would not suffice.

- Operation under exposure to incident sunlight: Even if the illumination of the target scene by incident sunlight is relatively strong, the modulation sequence is capable of raising the level of "signal" pixels sufficiently to recover a good-quality image of the projected pattern. This characteristic is related to the noise suppression property discussed above, because incident sunlight behaves very much like background noise under the assumption that the intensity level of the sunlight is reasonably stable.

- Mutual interference compensation: With the modulated acquisition procedure, it is possible to compensate for the mutual interference typically encountered when two or more similar sensors operate on the same target scene. This can be done by constructing the control/modulation sequences based on cyclic orthogonal (Walsh–Hadamard) codes, in which the cross-correlation properties of the modulation codes are exploited to compensate the mutual interference (see [31] for more information). Similar concepts are used in other areas, as well; for example, for synchronized CDMA (code division multiple access) systems [36] or sensor networks [37], for which mutual interference also represents a major problem.

It should be noted that in the current implementation of the ATRIS sensor, a diffractive optical element (DOE) mounted in front of the laser is used to split the laser beam and produce a structured light pattern comprising 11 parallel light planes (see Figure 3a). For the modulation sequence, a 16-bit long modulation sequence is used, which results in a stable pattern acquisition rate of 30 fps given a camera frame rate of 480 fps (i.e., 480 fps/16 = 30 fps).

(a) Original structured light pattern　　(b) Sample target scene　　(c) Image of deformed pattern

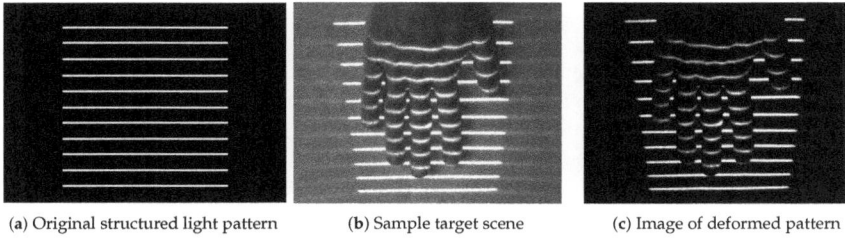

Figure 3. The correspondence problem: (**a**) an illustrative image of the structured light pattern produced by the ATRIS sensor, without distortions; (**b**) an example of a target scene illuminated by the light pattern; (**c**) an example of the spatial distortion map captured with the sensor. To be able to reconstruct a depth image of the target scene, each pixel comprising the light pattern in (**c**) needs to be assigned a label corresponding to one of the light planes in the pattern shown in (**a**).

4. Light-Plane Labeling

The modulated acquisition procedure presented in the previous section results in a binary image (or frame) of the projected pattern that forms the basis for depth image reconstruction. As pointed out in Section 2, the depth image is computed based on the geometrical properties of the acquisition setup and the distortions of the structured light pattern caused by projecting the light pattern onto the target scene [33]. Although the geometrical properties of the acquisition setup are commonly known in advance, the pattern distortions need to be quantified before a depth image can be constructed. Typically, this is achieved by establishing the correspondence between all parts of the original structured light pattern (Figure 3a) and all parts of the captured distortion map (Figure 3c). Because solving this correspondence problem (illustrated in Figure 3) is crucial for the success of the depth-image-construction step, an efficient procedure based on probabilistic graphical models (PGMs) was developed for the ATRIS sensor. A detailed description of the procedure is given below in this section.

4.1. Problem Statement

The binary distortion map captured with the ATRIS acquisition setup contains a large number of binary regions. The 11 light planes that constitute the structured light pattern are usually not detected as large connected binary regions in the captured image, but in the form of shorter, potentially discontinuous line fragments (we refer to any connected binary region (using eight-adjacency) in the image as a line fragment), as shown in Figure 4. In addition, small binary regions not corresponding to any of the projected light planes can also appear in the captured image due to the presence of noise.

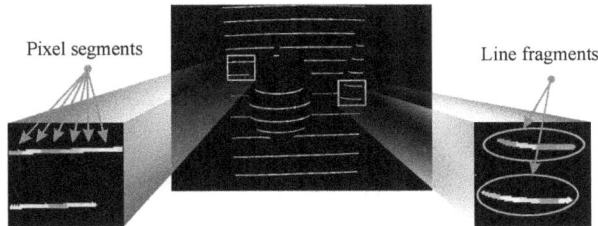

Figure 4. Visual illustration of some terminology used in this paper. Connected binary regions (using eight-adjacency) are referred to as line fragments. Smaller parts of the line fragments of fixed width are referred to as pixel segments.

The structured light pattern used in the ATRIS sensor consists of 11 parallel light planes. Solving the correspondence problem, therefore, amounts to finding the correct light plane label for each of the connected binary regions in the captured binary distortion map [33]. Although this labeling problem could be approached for each non-zero pixel individually, we group the non-zero pixels into pixel segments (i.e., parts of the line fragments with a fixed width; see Figure 4) and try to assign each pixel segment one of the light plane labels to reduce the computational burden of the labeling procedure.

The illustrated labeling problem can be formally defined as follows: assume that the detected light pattern is represented in the form of the binary distortion map I, that the scene points illuminated by the projected pattern are encoded with a pixel value of one and that all other pixels are encoded with a value of zero. Furthermore, assume that the non-zero pixels that form the line fragments are grouped into pixel segments of fixed width (i.e., spanning a predefined number of image columns). Let us now denote the set of all pixel segments in the distortion map I as $\mathcal{P} = \{p_1, p_2, \ldots, p_N\}$, where p_i stands for the i-th pixel segment (for $i = 1, 2, \ldots, N$) and N represents the number of all pixel segments in I. Moreover, let us denote the set of indices of the light planes constituting our pattern as $\mathcal{L} = \{1, 2, \ldots, M\}$, where M stands for the number of light planes in the structured light pattern ($M = 11$ in our case), and Index 1 represents the light plane that is closest to the bottom of I. The correspondence (or labeling) problem can then be defined as the mapping ψ that assigns each pixel segment from $\mathcal{P}$ an index (or label) from $\mathcal{L}$:

$$\psi : p_i \to \mathcal{L}, \text{ for } i = 1, 2, \ldots, N. \tag{1}$$

4.2. Labeling with Graphical Models

We follow the ideas presented in [32,33] and formulate the correspondence problem as an inference problem over probabilistic graphical models (PGMs). The formalism associated with PGMs allows us to break down complex problems into (smaller) simpler parts that can easily be modeled. For the labeling problem in the ATRIS sensor, these simpler parts correspond to geometrical relationships between pixel segments and their relative positions in a series of consecutive frames (captured by our ATRIS sensor).

Graphical models $\mathcal{G}$ are defined by a set of vertices $\mathcal{V}$ and a set of edges $\mathcal{E}$ connecting the vertices; that is, $\mathcal{G} = (\mathcal{V}, \mathcal{E})$. To represent the labeling problem in the form of a graphical model, the pixel segments in the detected pattern are represented as vertices $v \in \mathcal{V}$, and the dependencies between the pixel segments are represented as edges $e \in \mathcal{E}$ of the graph. Each pixel segment (and in turn each vertex) is associated with a discrete random variable X from $\mathcal{X}^t = \{X_1^t, X_2^t, \ldots, X_N^t\}$, where the set of all N random variables $\mathcal{X}^t$ is defined by the binary distortion map I taken at time instance t. Similarly, each edge is associated with a factor that models the functional relationship between the vertices (random variables) connected by the edge. Solving the labeling problem defined in Equation (1) amounts to finding the most likely value (from $\mathcal{L}$) for each random variable in $\mathcal{X}^t$ given the dependencies (and relationships) between the pixel segments.

The PGM-modeling procedure used for the ATRIS sensor is illustrated in Figure 5. Here, the left side of Figure 5 depicts two sample frames, each containing two line fragments and a total of four pixel segments. The two frames are assumed to have been captured at two consecutive time instances, $t - 1$ and t, and the color-coded pixel segments are assumed to be reasonably well aligned in the vertical (y axis), horizontal (x axis) and "temporal" (t axis) directions. The right side of Figure 5 shows the corresponding PGM constructed based on the two frames. As can be seen, the state (or value) of each random variable (i.e., each pixel segment) depends on the state of its horizontal, vertical and temporal neighbors. The dependencies between the neighboring pixel segments are defined by so-called factors (illustrated by squares), which model the relationships/dependencies between random variables and are for the case of horizontal, vertical and temporal neighbors denoted as ϕ_h, ϕ_v and ϕ_t, respectively. So-called unary factors are also used in our modeling procedure to construct the graph. These factors act only on a single variable (vertex) at a time and, in our case, encode the prior knowledge about the structure of the projected pattern [33]. They are denoted as ϕ_p in Figure 5.

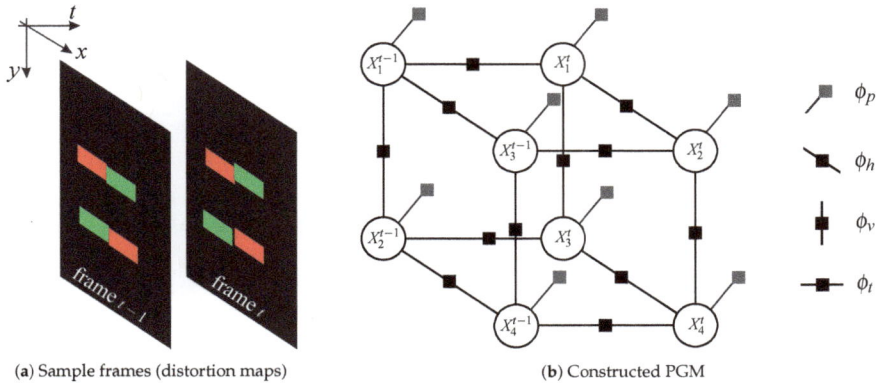

(a) Sample frames (distortion maps)   (b) Constructed PGM

Figure 5. Illustration of the probabilistic graphical model (PGM)-based modeling procedure: (a) simplified distortion map; (b) corresponding PGM.

With the illustrated modeling approach, the joint probability distribution of the PGM used in the ATRIS sensor can be written as a factor product:

$$p(\mathcal{X}^{t-1}, \mathcal{X}^t) \;=\; \frac{1}{Z} \prod_{\substack{t'=t-1 \\ (i,j)\in\mathcal{E}_h}}^{t} \phi_h(X_i^{t'}, X_j^{t'}) \prod_{\substack{t'=t-1 \\ (i,j)\in\mathcal{E}_v}}^{t} \phi_v(X_i^{t'}, X_j^{t'}) \prod_{(i,j)\in\mathcal{E}_t} \phi_t(X_i^{t-1}, X_j^t) \prod_{\substack{t'=t-1 \\ i=1}}^{t,N} \phi_p(X_i^{t'}), \quad (2)$$

where Z denotes the partitioning function and the sets $\mathcal{E}_h$, $\mathcal{E}_v$ and $\mathcal{E}_t$ correspond to subsets of all edges $\mathcal{E}$, over which the horizontal, vertical and temporal factors are defined, respectively. In the above equation, N can in general also take different values at different time instances. The joint distribution is defined only for the case of two consecutive frames (from time instances t and $t-1$), but the extension to a longer sequence is trivial [33].

Inference over the constructed model can be conducted using various inference algorithms (e.g., [38] or [39]), where the goal is to find the most likely value (label) for each random variable in the constructed graph. A detailed description of the inference algorithm used for the ATRIS sensor is given in Section 4.2.3.

4.2.1. Graph Construction

Unlike the toy example in Figure 5, where all line fragments are more or less parallel and the pixel segments are near perfectly aligned in all directions, building a PGM from real sensory data is a more complex task. Because no specific topology (e.g., nodes arranged in a grid) is present in the light pattern that is projected onto the target scene with the ATRIS sensor, it is necessary to formulate criteria for identifying vertical, horizontal and temporal neighbors. Based on these criteria, dependencies (i.e., factors) between neighboring pixel segments can be defined, and inference over the constructed graph can be conducted.

For the ATRIS graph construction procedure, horizontal neighbors are defined as connected pixel segments (here, eight-adjacency is used [40] to probe for the connectivity). On the left side of Figure 6, where four pixel segments (labeled a, b, c and d) are presented, only segment pairs a-b and d-b represent horizontal neighbors, whereas the segment pair b-c does not, because b and c are not connected. The main motivation for introducing horizontal neighbors in the PGM construction procedure is to "encourage" horizontally-connected pixel segments to take the same label.

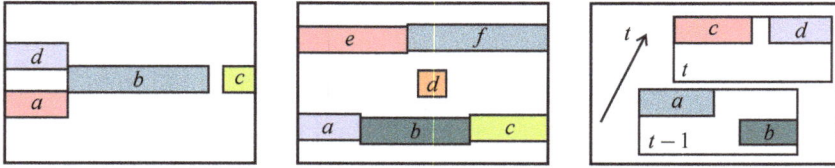

Figure 6. Defining the neighbors: horizontal neighbors (left image), valid neighbors are a-b, d-b; vertical neighbors (middle image), valid neighbors are a-e, b-e, b-d, b-f, c-f, d-f; (right image) temporal neighbors, valid neighbors are a-c.

Vertical neighbors in the ATRIS sensor are defined as pairs of pixel segments that are not connected, but share at least one pixel at the same x-coordinate (and different y-coordinates). In the middle image of Figure 6, pixel-segment pairs a-e, b-e, b-d, b-f, c-f and d-f represent vertical neighbors according to this definition. Note that a pixel segment can easily have several vertical neighbors. Vertical neighbors are needed in the PGM to ensure that the detected light planes tend to be labeled consecutively. Due to this fact, vertical dependencies between pixel segments are extremely important for the modeling procedure.

Finally, temporal neighbors are defined as pixel segments belonging to detected patterns recorded at two consecutive time instances, $t - 1$ and t, that share at least one non-zero pixel at the same spatial coordinates. This definition requires no tracking of the pixel segments over time and is extremely simple to implement. On the right side of Figure 6, only pixel segments a and c represent temporal neighbors, whereas all other segment pairs do not. Temporal neighbors are included in the graphical model to exploit additional temporal information when labeling the light planes of the projected pattern. As shown in the experimental section, the addition of temporal neighbors contributes to the accuracy of the labeling procedure.

The definitions presented define the topology of the PGM (i.e., vertices and edges) constructed from the given input image I. However, to be able to conduct inference on the graph, factors between pairs of neighboring vertices (or on a single vertex) that model the dependencies between the random variables associated with the vertices need to be defined, as well. The procedure for defining the factors used in this paper is described in the next section.

4.2.2. Factor Definition

Factors represent functions of random variables. Typically, factors model the dependencies (relationships, constraints) between neighboring vertices and, hence, represent functions of two random variables. Alternatively, they relate only to a single vertex and act as functions of a single random variable. In the modeling procedure used for the ATRIS sensor, factors are used to model the relationships between horizontally-, vertically- and temporally-neighboring pixel segments, and unary factors are added to include knowledge about the structure of the projected light pattern.

Horizontal factors ϕ_h describe the relationship between horizontally-neighboring pixel segments and return a fitness score with respect to the labels assigned to the neighboring segments. The factor returns a score of one when both pixel segments are assigned the same label and some small score f_c if they are assigned different labels. This definition reflects the structure of the projected pattern and encourages horizontal neighbors to take the same light plane label [33]. The fitness score returned by the horizontal factor is defined as:

$$\phi_h(X_i^t = k, X_j^t = k') = \begin{cases} 1, & k = k', \\ f_c, & \text{else} \end{cases}, \tag{3}$$

where $k, k' \in \mathcal{L}$ and f_c $(0 < f_c < 1)$ denotes the fraction-cost parameter that penalizes horizontal neighbors that are labeled differently.

Vertical factors ϕ_v are assigned between random variables identified as vertical neighbors. The factor returns a high fitness score when the two vertically-neighboring pixel segments are labeled in an ascending manner and a fitness score of zero otherwise. This definition encourages the vertical neighbors to take consecutive light plane labels and prevents the assignment of labels in a non-ascending order. The relationship between vertically-neighboring segments in the ATRIS sensor is modeled as follows:

$$\phi_v(X_i^t = k, X_j^t = k') = \begin{cases} f(k - k'), & k > k' \\ 0, & \text{else} \end{cases}, \tag{4}$$

where $k, k' \in \mathcal{L}$ and f denotes a linear function of the difference of two labels. The function f decreases monotonically with the label difference:

$$f(\delta) = \begin{cases} g(1 - (\delta - 1)h), & \delta \neq 0 \\ o_c, & \text{else} \end{cases}. \tag{5}$$

The parameter h defines the slope of the linear part of the function f; o_c (overlap cost) stands for a parameter that penalizes vertical neighbors with the same variable value; and the function $g(.)$ represents a function that truncates all negative values to zero.

Temporal factors ϕ_t are assigned between pixel segments identified as temporal neighbors in two consecutive frames (of distortion maps). Under the assumption of a sufficiently high frame rate, the spatial location of most pixel segments can be considered constant. Pixel segments originating from two consecutive frames having approximately the same spatial location should therefore be assigned the same light plane label. The temporal factors defined for our modeling approach are functions that assign a fitness score of one if the pixel segments are assigned the same label and a fitness score of zero if the labels differ; that is [33]:

$$\phi_t(X_i^{t-1} = k, X_j^t = k') = \begin{cases} 1, & k = k' \\ 0, & \text{else} \end{cases}, \tag{6}$$

where $k, k' \in \mathcal{L}$.

Finally, the prior factors ϕ_p are assigned to all vertices and operate on a single random variable at a time. They are used to incorporate prior knowledge about the spatial structure of the projected pattern into the modeling procedure and in a sense carry information about the most likely range of values a random variable can take with respect to the vertical position of the pixel-segments and the number of its vertical neighbors above and below. The prior factors are computed based on the pseudo-procedure presented in Algorithm 1. Below, we outline the algorithm for a single pixel segment based on the toy example shown in Figure 7. However, the procedure is identical for all pixel segments.

Assume that our goal is to compute the prior factor for the red pixel segment and that all other pixel segments in image I are shown in white and gray (Figure 7b). To compute the prior factor, we first scan over all x-coordinates of the red segment and for each x-coordinate search for (at most M) line fragments above and below the current x-position of the red segment. We then label the pixel segments found at the current x-coordinate consecutively from the bottom of the image up and increase the likelihood of the label assigned to the red-segment by some arbitrary constant q. If we are able to find M pixel segments at the given x-coordinate, only the likelihood of a single label is increased (shown by the graph in Figure 7a), whereas the likelihood of several labels is increased if fewer than M pixel segments were found (shown by the graph in Figure 7c). The procedure aggregates the likelihoods over all x-coordinates of the red pixel segment and in the final step normalizes the likelihoods to the unit L_1 norm over all light plane labels to produce the final prior factor for the corresponding random variable.

Algorithm 1 Calculating prior factors

1: **for** all pixel-segments (i.e., random variables X_i) in the image I **do**

2:     **Init:** Initialize **p** as an M-dimensional vector of all zeros

3:     **Result:** Normalized distribution (prior factor) $\phi_p(X_i)$

4:     **for** all x-coordinates of the pixel-segment corresponding to X_i **do**

5:         ▷ find (at most) M biggest line fragments in I having a pixel segment at the current x-coordinate

6:         ▷ record the position, k, of the pixel-segment (corresponding to X_i) among the found m line fragments counting from the bottom of image I up

7:         **if** the number of found line fragments m equals M **then**

8:             ▷ increase the k-th element of **p** by some positive constant q

9:         **else**

10:             ▷ increase all elements of **p** from position k to $k + (M - m)$ by some positive constant q

11:         **end if**

12:     **end for**

13:     ▷ normalize the vector **p** to unit L_1 norm; $\phi_p(X_i) = \mathbf{p}$

14: **end for**

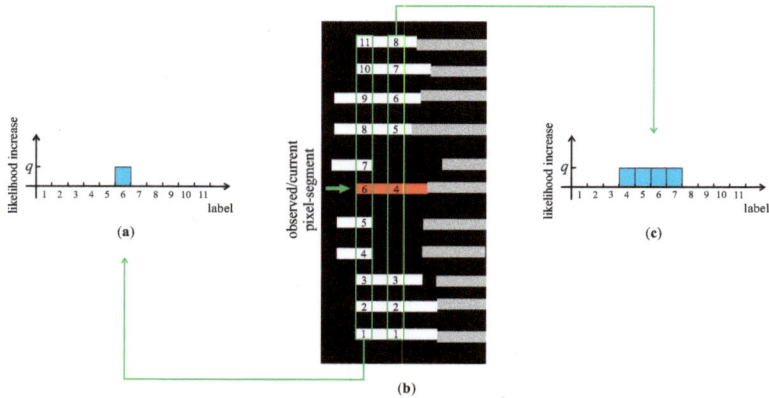

Figure 7. Computing prior factors: illustration of the procedure with a simple example (shown in (**b**)). For each random variable, the prior factor represents a probability distribution over all light-plane labels. Estimates of the probabilities are obtained by labeling the pixel segments and increasing the likelihood (shown in (**a**) and (**c**)) of the label assigned to the observed pixel segment at each x-coordinate.

4.2.3. Inference

To solve the labeling problem using the constructed PGM, a value needs to be assigned to each random variable (or vertex) constituting the graph, for which the range of possible values is given by the set of light plane labels $\mathcal{L}$ (see Equation (1)). The assignment is computed based on maximum a posteriori probability (MAP) estimation:

$$\hat{\mathcal{X}}^t = \arg\max_{\mathcal{X}^t} p(\mathcal{X}^{t-1}, \mathcal{X}^t), \qquad (7)$$

where $p(\mathcal{X}^{t-1}, \mathcal{X}^t)$ is the joint probability distribution of the PGM defined by Equation (2) and $\hat{\mathcal{X}}^t$ is the most likely configuration of random-variable assignments for the PGM at time instance t.

MAP estimation can be conducted using different inference techniques; for example, [38,39,41]. In the case of acyclic graphs, an exact solution can be found; on the other hand, with cyclic graphs, as in our case, the problem is NP hard and only an approximate solution can be computed. Thus, for the ATRIS sensor, we use loopy-belief propagation for the inference on the PGM as described in [38].

5. Depth Image Reconstruction

In order to reconstruct a depth image of the target scene based on the labeled distortion map, we use a simple processing approach involving reference frames of the projected pattern taken at various distances from the camera.

To capture the reference frames, we start by placing a planar surface parallel to the XY plane of the camera's coordinate system (see Figure 8a) at some initial distance z_1 from the camera. We project our light pattern onto the surface and capture the first reference frame, R_1, corresponding to the distance z_1. We then move the surface by some depth increment Δz away from the camera and take another reference frame, R_2, at the distance z_2 from the camera. We repeat the procedure for the entire measurement range of our ATRIS sensor and thus generate a set of reference frames that are later used for depth calculation. The procedure for capturing the reference frames is illustrated in Figure 8a, and some sample frames are shown in Figure 8b. Here, the fourth and fifth light plane are labeled in each frame to demonstrate how the position of the detected light planes changes in accordance with the distance at which the reference frames are recorded.

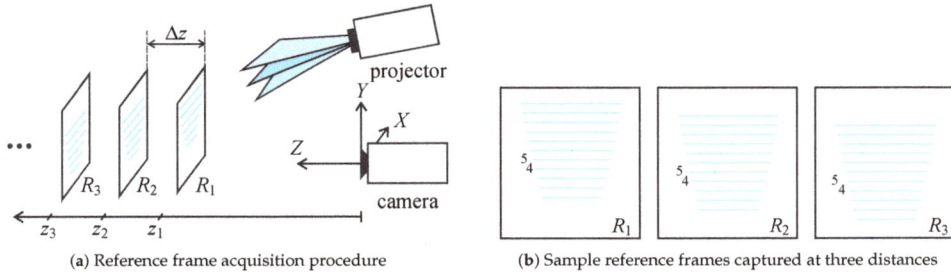

(a) Reference frame acquisition procedure (b) Sample reference frames captured at three distances

Figure 8. Illustration of the reference frame acquisition procedure: (a) the setup; (b) sample reference frames captured at distances z_1, z_2 and z_3. The reference frames are used to compute the depth value of each pixel segment in the labeled distortion map.

Let us denote the distances at which the reference frames, R_s, were captured with:

$$z_s = z_0 + s \cdot \Delta z, \text{ for } s = 1, \ldots, S, \tag{8}$$

where z_0 denotes some minimum distance from the camera and the measurement range of the sensor lies between z_1 and z_S. The depth increment Δz defines the depth resolution of the ATRIS sensor and may be selected arbitrarily. Each reference frame, R_s, contains at most M line fragments $f_i^{(s)}$ (for $i = 1, \ldots, m \leq M$) with associated light plane labels $k_i^{(s)} \in \mathcal{L}$.

Consider a non-zero pixel (note that depth calculation is conducted for each pixel separately and not for the entire pixel segment at once) from the distortion map I located at image coordinates $\mathbf{p} = [x_{pix}, y_{pix}]^T$ and associated with some light plane label $k \in \mathcal{L}$ assigned during the labeling procedure. To compute the $[x, y, z]^T$ position of the pixel (in camera coordinates), we first find the reference frame, $R_{\hat{s}}$, that contains the line fragment (at the same x-coordinate, i.e., x_{pix}) with the same label as the given non-zero pixel and is closest in terms of its y coordinate; that is:

$$\hat{s} = \arg\min_s |y_{f_i^{(s)}} - y_{pix}|, \text{ subject to } k = k_i^{(s)}, \tag{9}$$

where $y_{f_i^{(s)}}$ stands for the y-coordinate of the i-th line fragment in the reference frame R_s. We then assign the distance, at which the frame $R_{\hat{s}}$ was taken, as the z-coordinate of the given pixel with respect to the camera's coordinate system:

$$z = z_0 + \hat{s} \cdot \Delta z. \tag{10}$$

The x and y coordinates of the pixel (in camera coordinates) are computed using the hardware's intrinsic parameters, which can be estimated with standard techniques [42]. The procedure presented is applied to all non-zero pixels of the spatial distortion map I and results in a sparse depth map, which is interpolated during the last processing step to fill in the missing values.

6. Experiments

This section describes the experiments conducted to demonstrate the merits of the developed sensor and to evaluate its performance on experimental data using quantitative performance metrics. We start the section by presenting experiments related to the characteristics of the sensor and the proposed acquisition procedure, proceed by providing results on the performance of the proposed light plane labeling technique and conclude the section with some examples of 3D reconstructions of scenes generated with our ATRIS sensor.

6.1. Characteristics of the Acquisition Procedure

One of the main merits of our pattern acquisition procedure is the fact that it is possible to deploy several depth sensors exploiting our procedure in the same environment. In fact, we demonstrated in [31] that it is possible to completely compensate for the mutual interference usually encountered when deploying several identical depth sensors in the same environment by constructing the modulation sequence of our acquisition procedure based on cyclic orthogonal (Walsh–Hadamard) codes. (A detailed discussion on the construction of the modulation sequence is beyond the scope of this paper. The reader is referred to [31] for detailed coverage of this topic.) An illustrative example of this characteristic is presented in Figure 9 on a simple indoor toy scene. Here, the images in the upper row correspond to our ATRIS sensor, and the images in the lower row correspond to images captured with the first generation Kinect sensor, which also exploits structured light [6]. The image in the upper left corner presents a sample scene with two of our sensors directed at it; the second image shows a demodulated image with non-cyclic orthogonal codes; and the last image in the upper right shows the demodulated image based on cyclic orthogonal codes. Note how the projected pattern can be recovered despite the presence of more than one active sensor operating on the same scene. In the lower row, the left most image depicts the acquisition setup using a pair of Kinect sensors. The middle image shows the depth map acquired when only one sensor is active, and the third image in the lower row demonstrates the effect of two Kinects capturing depth images of the same target scene. In the latter case, white areas appear in the image where depth information cannot be computed. This effect demonstrates the effect of the mutual interference of the two Kinects and is not present with the ATRIS sensor. As a consequence of the interaction of the Kinects' light patterns, the shape of the objects comprising the scene is distorted, and part of the depth information is missing.

Another important aspect of the developed pattern acquisition procedure is its robustness to ambient illumination and the presence of incident sunlight. To demonstrate this characteristic, we again provide a few (qualitative) illustrative examples. We first present sample results for a simple indoor scene imaged in three distinct illumination conditions: (i) under ambient lighting with no additional illumination directed at it (first row of Figure 10), (ii) under ambient lighting and with the room lights turned on (second row of Figure 10) and (iii) under ambient lighting, with room lights turned on and with a flashlight directed at the scene (third row of Figure 10). The first and third columns of Figure 10 show gray-scale images of the scene with the ATRIS prototype and Kinect sensor (taken at the same time instance), respectively, and the second and fourth columns show the corresponding distortion maps (for ATRIS) and depth images (for Kinect). Our acquisition procedure produces stable results with minor differences in the intensities of the distortion maps, but information is missing from the depth images generated by Kinect when the imaging conditions become more challenging. Note that missing information corresponds to white areas in the depth images.

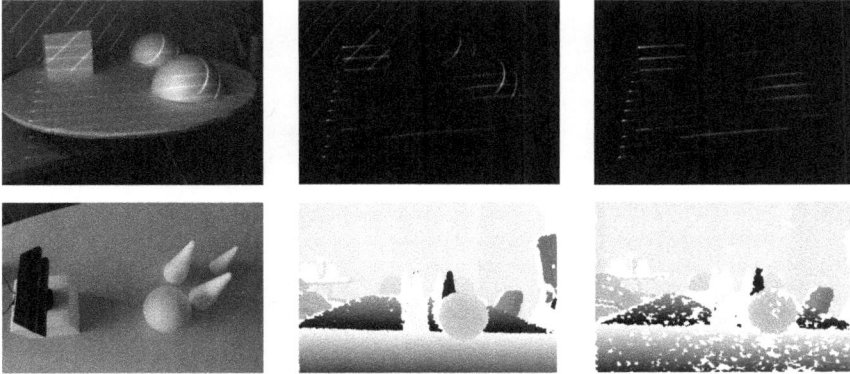

Figure 9. An illustrative example of the behavior of the developed acquisition procedure when two identical sensors are directed at the same scene and a comparison with a commercial sensor. Upper row: a sample scene illuminated with two ATRIS sensors (**left**); the demodulated images with non-cyclic orthogonal codes (**middle**); the demodulated images with cyclic orthogonal codes (**right**). Lower row: acquisition setup with two Kinect (v1) sensors (**left**); captured depth image when one sensor is active (**middle**); captured depth image when both sensors are active (**right**). Observe how the ATRIS sensor is able to compensate for the mutual interference, to recover a spatial distortion map and is unaffected by the pattern projected by the second ATRIS sensor.

Figure 10. Qualitative examples of the performance of the developed acquisition procedure under various ambient lighting conditions. The first column shows a sample scene in different illumination conditions (from **top** to **bottom**): no additional illumination (**top**), with room lights on (**middle**) and with room lights on and a flashlight directed at the scene (**bottom**). The second column of images shows the distortion maps captured with the ATRIS sensor for the different illumination conditions. The third column depicts the same scene captured at the same time instance as the images in the first column, but with the Kinect sensor (v1). The last column presents the corresponding depth images generated by the Kinect sensor. Here, white areas indicate that no depth information could be computed.

In our next experiment, we deploy our sensor outdoors and again provide comparative results with the first-generation Kinect sensor that uses the same imaging technology (i.e., active structured light) as our ATRIS sensor. The results of this experiment are shown in Figure 11. Here, the first column shows images of our scene with the projection pattern barely visible due to the incident sunlight; the second column shows images of the acquired distortion maps; and the third column shows images taken with the Kinect sensor. The upper row of images was taken under moderate incident sunlight, and the lower row of images was taken under relatively stronger sunlight. Note that the Kinect sensor is unable to acquire a complete depth map of the observed scene due to deployment outdoors (white areas in the output image of the Kinect sensor indicate that no data are available for that area), but our acquisition procedure produces stable (though noisy) distortion maps that can be used with our light plane labeling procedure. Although there are obvious differences in the number of pixels in which the two sensors interact with the scene, it is clear from the images presented that the ATRIS sensor is capable of operating outdoors in a robust manner.

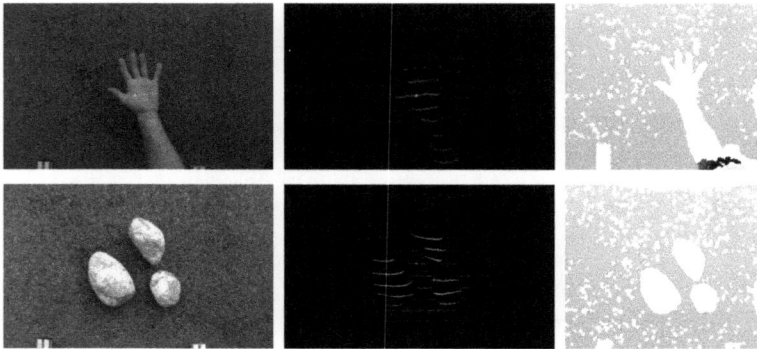

Figure 11. Illustrative example of the behavior of the developed acquisition procedure when deployed in outdoor environments. The first column shows gray-scale images of the target scene, the second column shows the distortion maps captured with the ATRIS sensor; and the last column of images presents the output of the Kinect sensor. The upper row presents images taken under exposure to moderate sunlight, and the lower row shows images taken under exposure to strong incident sunlight.

For our third and last experiment with the acquisition procedure, we set up another outdoor scene and compared the performance of our ATRIS sensor with the first and second generation Kinect sensor. The second generation Kinect (v2) uses time of flight (ToF) technology and requires a full-fledged GPU supporting DirectX 11 to produce depth maps. Due to the required computing resources and different technology, Kinect v2 is not a direct competitor to our ATRIS sensor (which runs on a simple FPGA), but is included in our comparison to demonstrate the performance of a state-of-the-art depth sensor. The qualitative comparison is presented in Figure 12. Here, the first column of images represents the outdoor scene and corresponding distortion map captured with the ATRIS sensor; the second column depicts the scene and the depth image acquired with the first generation Kinect; and the third column shows the scene and the depth map captured with the second generation Kinect. As can be seen, both Kinect v2 and our ATRIS sensor produce solid, though noisy, results for all measured pixels, whereas the first generation Kinect struggles with its performance outdoors.

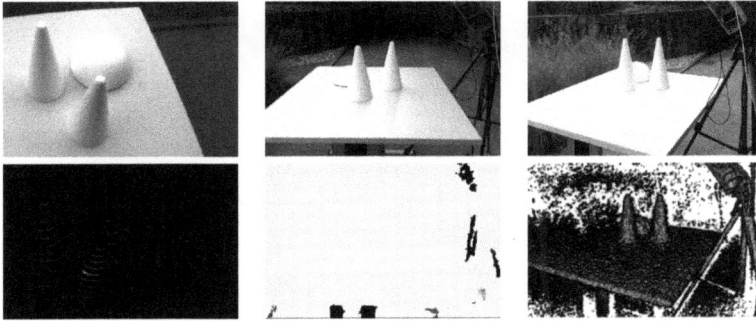

Figure 12. Qualitative comparison of the ATRIS sensor and both generations of the Kinect sensor on an outdoor scene. The first column shows the scene and distortion map from the perspective of the ATRIS sensor; the second column presents images from the first generation Kinect; and the last column of images shows the results produced by the second generation Kinect. Note that the second generation Kinect that uses time of flight technology, and our ATRIS sensor produces good results in all pixels measured, whereas the first generation Kinect performs less well outdoors.

6.2. Characteristics of the Light-Plane Labeling Technique

The distortion maps (see Figures 9–12) acquired with our ATRIS sensor form the basis for the light plane labeling procedure presented in Section 4. To evaluate the performance of the proposed procedure, we construct two datasets of spatial distortion maps.

The first dataset serves as our development set and is used in the experiments for tuning the open-hyper parameters of the labeling procedure (e.g., the values of o_c, f_c, h, etc.). In practice, it is necessary to fix the open-hyper parameters in such a way that the labeling technique exhibits the best possible performance. We therefore construct the first dataset from 152 images of a simple indoor scene, which is suitable for our purposes, because images taken indoors contain very little noise. The indoor scene comprises three objects positioned over a rotating table that change position backwards, forward, left and right, thus creating different depth discontinuities. The second dataset used for our experiments is a more realistic dataset of outdoor images. Here, we record 15 images of a scene containing a moving vehicle and a person passing between the vehicle and our ATRIS sensor. The images in this dataset contain objects with more complex geometry and are used to evaluate the performance of the labeling technique with fixed hyper-parameters.

All images from the two datasets are manually annotated to provide the ground truth for our experiments, in which we measure the accuracy of the labeling procedure using (what we refer to as) the correct labeling rate (CLR):

$$\text{CLR} = \frac{n_c}{N_a}, \tag{11}$$

where n_c denotes the number of correctly-labeled non-zero pixels and N_a stands for the number of all non-zero pixels (the term "correctly" in this context stands for "being the same as the ground truth"). The correct labeling rate (CLR), as defined above, measures the fraction of correctly-labeled pixels among all pixels that have to be labeled. Note that the CLR in all graphs and tables presented below is computed over all images of the given dataset.

A few sample images from both datasets and color-coded examples of the ground truth are shown in Figure 13. The upper group of images presents sample images from the (first) indoor dataset, and the lower group of images presents images from the (second) outdoor dataset. The imaging conditions outdoors are more challenging than the conditions indoors, which results in a higher level of noise in the demodulated images of the second dataset (observe the difference between the second and fourth row of images on the left). The images on the right side of Figure 13 represent annotated samples from both datasets.

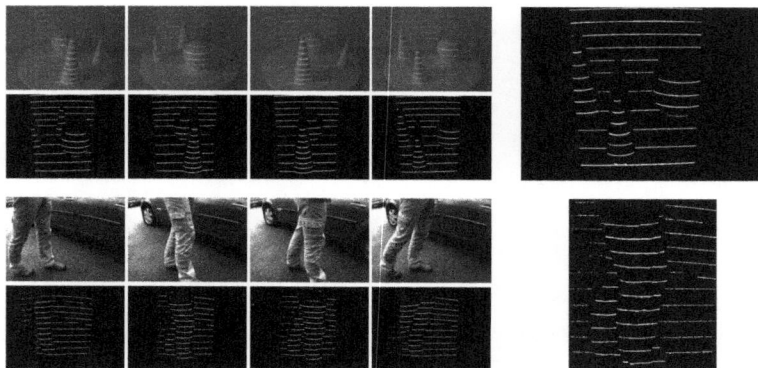

Figure 13. Sample images from the two datasets used in the experiments. The upper group of images shows sample images from the indoor dataset, and the lower group of images shows samples from the outdoor dataset. Both visible-spectrum and demodulated images are shown. The images on the right show the color-coded ground truth (best viewed in color).

As indicated above, the goal of the first series of experiments is to examine the impact of various hyper-parameters of the proposed techniques on the labeling accuracy. Towards this end, we first set the values of all hyper-parameters to a default value, then change a single parameter at a time and observe how the labeling accuracy changes with respect to the varying parameter. Even though the hyper-parameters of the proposed labeling technique are generally not mutually independent, we can, nevertheless, obtain a rough impression of the performance of the proposed method with respect to the varying parameter. We only make use of the first, the indoor dataset, in this series of experiments.

Figure 14 shows that the fraction cost f_c (see Equation (3)) and overlap cost o_c (see Equation (5)) have only a little effect on the labeling accuracy (graphs in the top row), whereas the function drop rate h (see Equation (5)), on the other hand, has a significantly larger impact on the labeling accuracy (graph in the lower left corner of Figure 14). These results suggest that the parameters f_c and o_c can be selected over a wide range of values with no significant performance loss, whereas h needs to be kept sufficiently small to ensure good performance. The most interesting observation of this series of experiments, which supports our working hypothesis that temporal information can improve labeling accuracy, can be made from the graph in the lower right corner of Figure 14. Note how the accuracy of the labeling procedure improves when two consecutive images from a sequence are used for constructing the PGM instead of only one. Adding additional images to the sequence further improves the labeling performance, albeit to a lesser extent. The best labeling accuracy we manage to achieve on the indoor dataset is a CLR of 0.9755 given a sequence length of five.

Based on the results of this series of experiments, the following parameter values are selected for the subsequent experiments on the outdoor dataset: $o_c = 1e^{-6}$, $f_c = 1e^{-5}$ and $h = 0.1$. Note that it is not our goal to find values of the hyper-parameters that result in the best possible performance on the indoor dataset because this could lead to over-fitting and poor generalization abilities of the final labeling approach. We therefore make no further effort to find a better set of parameters for our technique and run a second series of experiments on the outdoor dataset with the hyper-parameter values listed above.

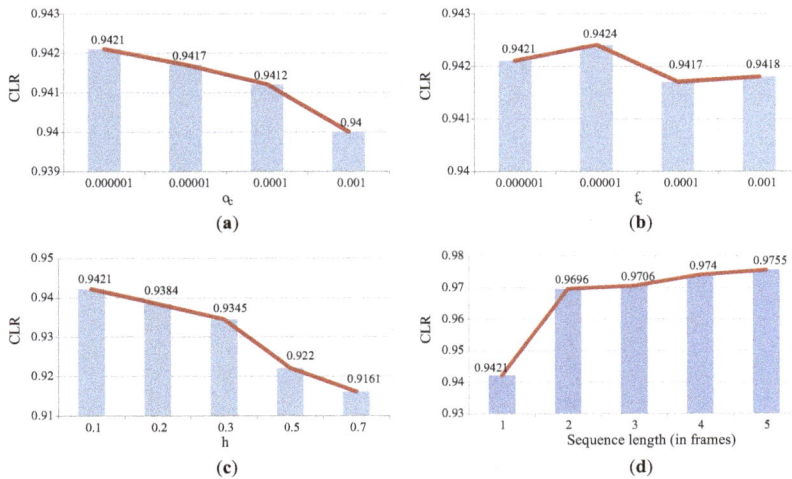

Figure 14. Impact of various hyper-parameters of the proposed technique on the labeling accuracy (measured in terms of the CLR). The results were generated on the indoor dataset and, except for the graph in the lower right corner, were computed using a sequence length of one. (**a**) impact of the overlap-cost parameter o_c, (**b**) impact of fraction-cost parameter f_c, (**c**) impact of the function drop rate h, (**d**) impact of the sequence length.

To gain insight into the characteristics of the proposed labeling approach and examine its behavior on more challenging data, we run several tests in the second series of experiments. These tests are conducted on the outdoor dataset and aim at: (i) examining the rationale behind defining the PGMs using horizontal, vertical and prior factors, (ii) evaluating the importance of temporal information on more challenging data and (iii) comparing the proposed technique to the existing labeling techniques.

To demonstrate the importance of each of the factors in the PGM, we conduct several tests. During each test, we remove a single factor and keep the rest. (For example, we remove the horizontal factors ϕ_h, which pull horizontally-neighboring pixel segments towards the same label, and keep only the vertical, prior and temporal factors. This case is denoted as "no ϕ_h". Other cases follow a similar notation.) We run the tests two times, first with a single image of the scene (i.e., $q = 1$) and then with two consecutive images (i.e., $q = 2$), to directly demonstrate the importance of the temporal factors, as well. The results of these tests are shown in Figure 15a. Several observations can be made from the results presented. First of all, the results indicate that prior factors are the most important component of the PGM. If the information about the structure of the projected patterns encoded in the prior factors is removed, the labeling accuracy drops significantly, to a value of CLR = 0.346 (in the case of a single image), as shown by the graph labeled "no ϕ_p". Without the prior information, the labeling accuracy becomes even worse when a second image is added to the sequence; that is, when temporal factors are introduced. However, when the prior factors are considered during the construction of the PGM, temporal information always improves the labeling performance. Similarly, both the horizontal and vertical factors also add to the overall labeling accuracy, as noticeable from the graphs labeled "no ϕ_h", "no ϕ_v" and "all". Here, "all" stands for the case when all four factor types are considered. All in all, the results of these tests suggest that all factors are important for the labeling procedure and contribute to the overall performance of the proposed labeling technique.

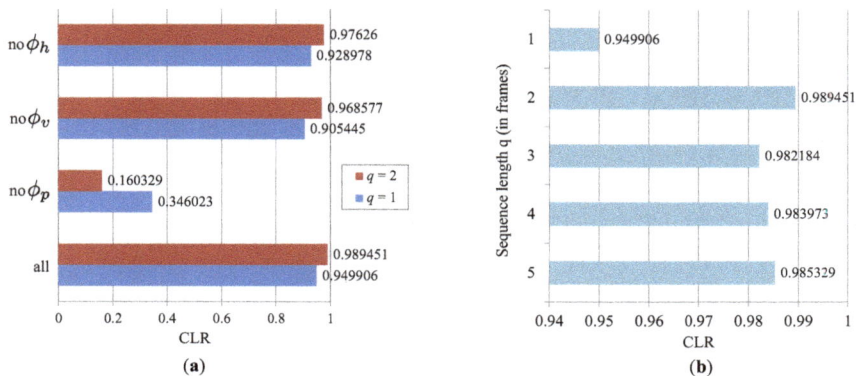

Figure 15. Results obtained on the outdoor dataset: (**a**) the results illustrate the importance of the selected structure of the PGM and (**b**) the impact of the sequence length on the correct labeling rate (CLR).

The graphs in Figure 15b show how the number of consecutive images used for constructing the PGM affects the labeling accuracy. As can be seen, the biggest increase in performance is noticeable when two consecutive images are used during construction of the PGM instead of one. The performance jump here is in fact a little larger than in the case of the indoor dataset, which can be attributed to the fact that the acquired distortion maps from the outdoor dataset are noisier, and hence, using more than one image from a sequence helps reduce the noise and determine the right labels. Interestingly, using more than two images from a sequence does not increase the performance further, but keeps it more or less stable. These results suggest that temporal information is useful for the labeling technique.

Next, we demonstrate the performance of the proposed labeling technique in comparison with other techniques that can be used for labeling the structured light pattern used in the ATRIS sensor. The implemented reference techniques are related to other structured light approaches from the literature that exploit light patterns comprised of parallel stripes (e.g., [43–46]), but, differently from these techniques, do not rely on coding strategies (in terms of color, intensity, geometry or time; see [18] for information on existing coding strategies) to solve the correspondence problem. Our comparison is therefore limited to techniques capable of handling uncoded structured light. Specifically, we implement the following reference techniques and include them in our comparison presented in Table 1:

- The naive labeling approach (NLA), which assigns light plane labels to the detected non-zero pixels in a consecutive manner. The first non-zero pixel at the given x-coordinate (looking from the bottom of the image up) is assigned the label 1; the second detected non-zero pixel at the given x-coordinate is assigned the label 2, and so on; until all 11 labels have been assigned.
- The labeling approach based on prior information (PR), which assigns light plane labels to the detected non-zero pixels by constructing a PGM based on prior factors only. This approach represents a refined version of the naive labeling technique introduced above.
- The reference approach from Ulusoy et al. (RUL) [32], which also exploits probabilistic graphical models, but relies only on spatial information to assign light plane labels to the detected non-zero pixels in the distortion map.

Table 1. Quantitative comparison with other labeling techniques (higher is better; 1 indicates a perfect score). NLA, naive labeling approach; PR, prior information approach; RUL, reference approach from Ulusoy et al.

Method	Outdoor Dataset (Noisy)			
	NLA	PR	RUL	Ours
CLR	0.888	0.912	0.950	0.989

Note that even the naive labeling approach results in a relatively high labeling accuracy with a CLR of 0.888. This approach is expected to work well in simple conditions, where there is no noise in the detected distortion maps and no large depth discontinuities are present in the scene. All other techniques improve the performance of the NLA technique with the proposed approach resulting in a CLR of 0.989.

All in all, the results of our experimental assessments suggest that exploiting spatio-temporal information for determining light plane labels in our ATRIS sensor is a feasible approach that results in state-of-the-art performance. The PGM approach is capable of assigning the correct label to most pixel segments of the detected light pattern even if large depth discontinuities are present in the scene observed. To visually demonstrate the efficacy of our approach, a few illustrative results of the labeling procedure are presented in Figure 16. Here, the first row depicts sample images from the outdoor dataset; the second row shows the color-coded ground truth; and the third row shows the color-coded results of the labeling procedure. Note how most of the assigned labels correspond to the ground truth, while there are, of course, a few errors, as well (right side of the image: the errors are marked with arrows). These errors typically introduce artifacts in the reconstructed depth images, but can easily be removed through simple post-processing of the depth images if they are not too frequent.

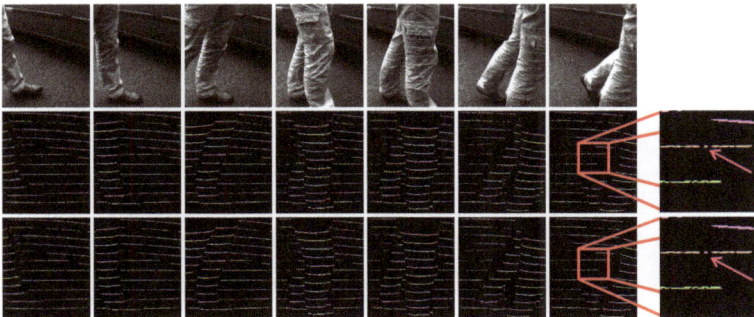

Figure 16. Visual examples of the results of the proposed light plane labeling procedure. The first row shows sample images from the outdoor dataset; the second row shows the (color-coded) manually-annotated ground truth; and the third row shows color-coded results of the labeling procedure. The images on the right side show an example of a labeling error, which is highlighted by arrows (best viewed in color).

6.3. Constructing Depth Maps: 3D Reconstruction

Once the light plane labels have been assigned to all parts of the detected light pattern, depth images of the observed scene can be reconstructed using the procedure presented in Section 5. To demonstrate the result of this process for our ATRIS sensor, we again present a couple of illustrative examples. The first example, which is shown in Figure 17, shows a number of stones with relatively simple geometry. The stones were placed on a street outdoors, and an image was captured using a commercial camera (upper left corner of Figure 17), as well as our sensor. Note that, despite the exposure to relatively strong incident sunlight, the sensor was able to capture an image with the projected pattern clearly visible (lower left corner of Figure 17) and reconstruct the depth images quite well (right side of Figure 17).

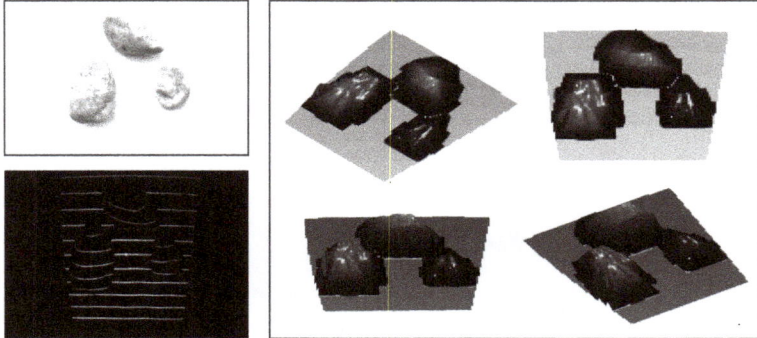

Figure 17. A visual example of the 3D reconstruction capabilities of the developed sensor. The image in the upper left corner shows a (visible light) image of the observed scene captured under strong incident sunlight; the image in the lower left corner shows the image of the detected light pattern; and the image on the right shows the reconstructed depth image from various angles.

The second example in Figure 18 shows a gray-scale image of a hand captured with a commercial camera (upper left corner of Figure 18), the labeled spatial distortion map generated by our ATRIS sensor (lower left corner of Figure 18) and the 3D reconstruction from various viewing angles (right side of Figure 18). Note that, despite the more challenging geometry of the hand (compared to the stones in the first example), our ATRIS sensor successfully captures all parts of the hand and recovers a good-quality depth image. The results show that usable depth images can be obtained with our sensor in difficult imaging conditions, as well as with relatively complex geometry of the target scene. This makes the sensor applicable in outdoor applications that require reliable depth information regardless of the external imaging conditions.

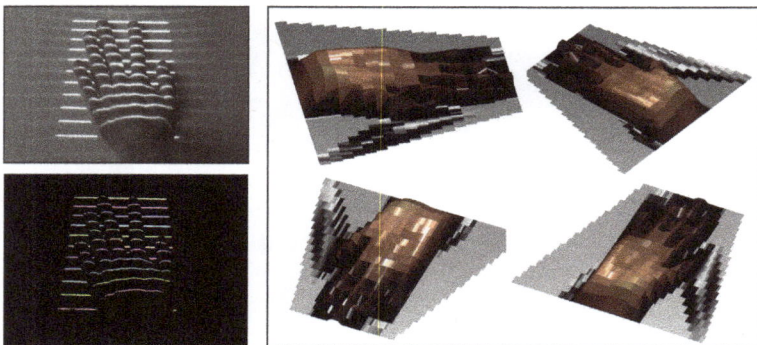

Figure 18. A visual example of the 3D reconstruction capabilities of the developed sensor. The image in the upper left corner shows a (visible light) image of the observed scene (i.e., a hand); the image in the lower left corner shows the labeled distortion map; and the image on the right shows the depth image from our ATRIS sensor from various viewing angles.

7. Conclusions and Future Work

We have presented and experimentally demonstrated the merits of a novel sensor for depth image acquisition. The sensor presented is based on the recently-introduced concept of modulated pattern projection [31], which ensures that the procedure of detecting the projected light pattern is robust with respect to various factors, such as background noise, background illumination or the mutual inference of similar systems operating on the same scene. The procedure for determining the correspondence between the projected and detected light patterns, which forms the basis for depth image reconstruction, is implemented with an approach based on probabilistic graphical models and, in addition to spatial information, also exploits temporal information when solving the correspondence problem. As demonstrated in the experimental section, the proposed procedure performs well even when large depth discontinuities are present in the scene. The experimental results also show that the sensor presented is capable of acquiring stable distortion maps when competing commercial systems struggle with their performance.

As part of our future work, we plan to further improve the sensor presented. One of the main drawbacks of the current implementation is the structure of the projected light pattern, which affects the resolution and quality of the acquired depth image. To address this issue, we intend to explore structured light patterns that can be used with the PGM-based labeling procedure presented. The goal here is to devise a pattern that ensures an even better quality of the captured depth images compared to what is possible with the current sensor. A possible way to achieve this is by using more lines in the light pattern or combining the existing pattern with line scanning techniques capable of generating dense depth maps.

The current implementation of the ATRIS sensor is suitable for outdoor applications, such as collision avoidance or autonomous navigation, where approximate depth maps need to be acquired as reliably as possible and the resolution of the depth maps is not of major concern. Another application domain for our ATRIS sensor is computer vision applications exploiting action recognition [47,48], pose estimation [49,50], facial expression recognition [51,52] or motion analysis technology [53]. These applications are commonly deployed outdoors and could benefit from robust depth imaging technology.

Acknowledgments: This research was supported by the European Union, European Social Fund, within the scope of the framework of the Operational Programme for Development of Human Resources in the period 2007–2013, Contract No. 321111000492 (ATRIS) and the ARRS Research Programme P2-0250(B) Metrology and Biometric Systems.

Author Contributions: M. Žganec and J. Žganec-Gros conceived of the acquisition procedure for the sensor presented. J. Kravanja, S. Dobrišek and V. Štruc devised the labeling procedure. J. Kravanja implemented the software part of the prototype and conducted the experiments. S. Dobrišek and V. Štruc drafted and polished the paper.

Conflicts of Interest: The authors declare no conflict of interest.

References

1. Weingarten, J.W.; Gruener, G.; Siegwart, R. A state-of-the-art 3D sensor for robot navigation. In Proceedings of the 2004 IEEE/RSJ International Conference on International Conference on Intelligent Robots and Systems (IROS), Sendai, Janpan, 28 September–2 October 2004; pp. 2155–2160.
2. Gutmann, J.S.; Fukuchi, M.; Fujita, M. 3D perception and environment map generation for humanoid robot navigation. *Int. J. Robot. Res.* **2008**, *27*, 1117–1134.
3. Ranft, B.; Dugelay, J.L.; Apvrille, L. 3D perception for autonomous navigation of a low-cost MAV using minimal landmarks. In Proceedings of the International Micro Air Vehicle Conference and Flight Competition (IMAV2013), Toulouse, France, 17–20 September 2013.
4. Hohne, K.H.; Fuchs, H.; Pizer, S. *3D Imaging in Medicine: Algorithms, Systems, Applications;* Springer Science & Business Media: Berlin/Heidelberg, Germany, 2012.
5. Udupa, J.K.; Herman, G.T. *3D Imaging in Medicine*; CRC Press: Boca Raton, FL, USA, 1999.

6. The Kinect Sensor. Available online: http://msdn.microsoft.com/en-us/library/hh438998.aspx (accessed on 8 August 2016).
7. Wallhoff, F.; Rub, M.; Rigoll, G.; Gobel, J.; Diehl, H. Surveillance and activity recognition with depth information. In Proceedings of the International Conference on Multimedia and Expo (ICME), Beijing, China, 2–5 July 2007; pp. 1103–1106.
8. Lim, S.N.; Mittal, A.; Davis, L.S.; Paragios, N. Uncalibrated stereo rectification for automatic 3D surveillance. In Proceedings of the 2004 International Conference on Image Processing (ICIP), Singapore, 24–27 October 2004; pp. 1357–1360.
9. Nguyen, T.T.; Slaughter, D.C.; Max, N.; Maloof, J.N.; Sinha, N. Structured light-based 3D reconstruction system for plants. *Sensors* **2015**, *15*, 18587–18612.
10. Zhang, Y.; Teng, P.; Shimizu, Y.; Hosoi, F.; Omasa, K. Estimating 3D leaf and stem shape of nursery paprika plants by a novel multi-camera photogrphy system. *Sensors* **2016**, *16*, 874.
11. Krizaj, J.; Struc, V.; Dobrisek, S. Towards robust 3D face verification using Gaussian mixture models. *Int. J. Adv. Robot. Syst.* **2012**, *9*, 1–11.
12. Savran, A.; Alyuz, N.; Dibeklioglu, H.; Celiktutan, O.; Gokberk, B.; Sankur, B.; Akarun, L. Bosphorus database for 3D face analysis. In *European Workshop on Biometrics and Identity Management;* Springer: Berlin, Germany, 2008.
13. Krizaj, J.; Struc, V.; Dobrisek, S. Combining 3D face representations using region covariance descriptors and statistical models. In Proceedings of the 2013 IEEE International Conference and Workshops on Automatic Face and Gesture Recognition Workshops, Shanghai, China, 22–26 April 2013.
14. Sansoni, G.; Trebeschi, M.; Docchio, F. State-of-the-art and applications of 3D imaging sensors in industry, cultural heritage, medicine, and criminal investigation. *Sensors* **2009**, *9*, 568–601.
15. Soutschek, S.; Penne, J.; Hornegger, J.; Kornhuber, J. 3D gesture-based scene navigation in medical imaging applications using time-of-flight cameras. In Proceedings of the 2008 IEEE Computer Society Conference on Computer Vision and Pattern Recognition Workshops (CVPRW), Anchorage, AK, USA, 23–28 June 2008; pp. 1–6.
16. Natour, G.E.; Ait-Aider, O.; Rouveure, R.; Berry, F.; Faure, P. Toward 3D reconstruction of Outdoor Scenes using an MMW radar and a monocular vision sensor. *Sensors* **2015**, *15*, 25937–25967.
17. Scharstein, D.; Szeliski, R. High-accuracy stereo depth maps using structured light. In Proceedings of the 2003 IEEE Computer Society Conference on Computer Vision and Pattern Recognition (CVPR), Madison, WI, USA, 18–20 June 2003.
18. Salvi, J.; Pages, J.; Batlle, J. Pattern codification strategies in structured light systems. *Pattern Recognit.* **2004**, *37*, 827–849.
19. Batlle, J.; Mouaddib, E.; Salvi, J. Recent progress in coded structured light as a technique to solve the correspondence problem: A survey. *Pattern Recognit.* **1998**, *31*, 963–982.
20. Kawahito, S.; Halin, I.A.; Ushinaga, T.; Sawada, T.; Homma, M.; Maeda, Y. A CMOS time-of-flight range image sensor with gates-on-field-oxide structure. *IEEE Sens. J.* **2007**, *12*, 1578–1586.
21. Lange, R.; Seitz, P. Solid-state time-of-flight range camera. *IEEE J. Quantum Electron.* **2001**, *37*, 390–397.
22. Thiebaut, E.; Giovannelli, J.F. Image reconstruction in optical interferometry. *IEEE Signal Process Mag.* **2010**, *1*, 97–109.
23. Besl, P.J. Active optical range imaging sensors. In *Advances in Machine Vision*; Springer: New York, NY, USA, 1989; pp. 1–63.
24. Blais, F. Review of 20 years of range sensor development. *J. Electron. Imaging* **2004**, *13*, 231–240.
25. Faugeras, O. *Three-Dimensional Computer Vision: A Geometric Viewpoint*; The MIT Press: Cambridge, MA, USA, 1993.
26. Nayar, S.K.; Nakagawa, Y. Shape from focus. *IEEE Trans. Pattern Anal. Mach. Intell.* **1994**, *16*, 824–831.
27. Prados, E.; Faugeras, O. Shape from shading. In *Handbook of Mathematical Models in Computer Vision*; Springer: New York, NY, USA, 2006; pp. 375–388.
28. Aggarwal, J.K.; Chien, C.H. 3D Structures from 2D Images. In *Advances in Machine Vision*; Sanz, J.L.C., Ed.; Springer: New York, NY, USA, 1989.
29. Forsyth, D.A.; Ponce, J. *Computer Vision: A modern Approach*, 2nd ed.; Pearson: Upper Saddle River, NJ, USA, 2012.

30. Kravanja, J. Analiza Projiciranih Slikovnih Vzorcev Za Pridobivanje Globinskih Slik. Ph.D. Thesis, University of Ljubljana, Ljubljana, Slovenia, 2016. (In Slovene)

31. Volkov, A.; Zganec-Gros, J.; Zganec, M.; Javornik, T.; Svigelj, A. Modulated acquisition of spatial distortion maps. *Sensors* **2013**, *13*, 11069–11084.

32. Ulusoy, A.O.; Calakli, F.; Taubin, G. Robust one-shot 3D scanning using loopy belief propagation. In Proceedings of the IEEE International Conference on Computer Vision and Pattern Recognition Workshops (CVPRW), San Francisco, CA, USA, 13–18 June 2010; pp. 15–22.

33. Kravanja, J.; Zganec, M.; Zganec-Gros, J.; Dobrisek, S.; Struc, V. Exploiting Spatio-Temporal Information for Light-Plane Labeling in Depth-Image Sensors Using Probabilistic Graphical Models. *Informatica* **2016**, *27*, 67–84.

34. Zganec, M.; Zganec-Gros, J. Active 3D Triangulation-Based Method and Device. US Patent 7,483,151 B2, 27 January 2009.

35. Optomotive Velociraptor Camera Datasheet. Available online: http://www.optomotive.com/products/velociraptor-hs (accessed on 8 August 2016).

36. Amadei, M.; Manzoli, U.; Merani, M.L. On the assignment of Walsh and quasi-orthogonal codes in a multicarrier DS-CDMA system with multiple classes of users. In Proceedings of the IEEE Global Telecommunications Conference, Taipei, Taiwan, 17–21 November 2002; pp. 841–845.

37. Tawfiq, A.; Abouei, J.; Plataniotis, K.N. Cyclic orthogonal codes in CDMA-based asynchronous wireless body area networks. In Proceedings of the 2012 IEEE International Conference on Acoustics, Speech and Signal Processing (ICASSP), Kyoto, Japan, 25–30 March 2012; pp. 1593–1596.

38. Kschischang, F.R.; Frey, B.J.; Loeliger, H.A. Factor Graphs and the Sum-Product Algorithm. *IEEE Trans. Inf. Theory* **2001**, *14*, 498–519.

39. Wiegerinck, W.; Heskes, T. Fractional belief propagation. In *Neural Information Processing Systems (NIPS)*; The MIT Press: Cambridge, MA, USA, 2003; pp. 438–445.

40. Gonzales, R.C.; Woods, R.E. *Digital Image Processing*, 3rd ed.; Prentice Hall: Upper Saddle River, NJ, USA, 2008.

41. Weiss, Y.; Freeman, W.T. On the optimality of solutions of the max-product belief-propagation algorithm in arbitrary graphs. *IEEE Trans. Inf. Theory* **2001**, *47*, 736–744.

42. Hartley, R.I.; Zisserman, A. *Multiple View Geometry in Computer Vision*, 2nd ed.; Cambridge University Press: Cambridge, UK, 2004.

43. Posdamer, J.; Altschuler, M. Surface measurement by space-encoded projected beam system. *Comput. Graph. Image Proc.* **1982**, *18*, 1–17.

44. Ishii, I.; Yamamoto, K.; Doi, K.; Tsuji, T. High-speed 3D Image Acquisition Using Coded Structured Light Projection. In Proceedings of the 2007 IEEE/RSJ International Conference on Itenlligent Robots and Systems, San Diego, CA, USA, 29 October–2 November 2007; pp. 925–930.

45. Li, Z.; Curless, B.; Seitz, S.M. Rapid shape acquisition using color structured light and multi-pass dynamic programming. In Proceedings of the 2002 First International Symposium on 3D Data Processing Visualization and Transmission, Padova, Italy, 19–21 June 2002.

46. Horn, E.; Kiryati, N. Toward optimal structured light patterns. *Image Vision Comput.* **1999**, *17*, 87–97.

47. Chen, C.; Liu, M.; Zhang, B.; Han, J.; Jiang, J.; Liu, H. 3D Action Recognition Using Multi-Temporal Depth Motion Maps and Fisher Vector. Available online: https://www.researchgate.net/profile/Chen_Chen82/publication/300700290_3D_Action_Recognition_Using_Multi-temporal_Depth_Motion_Maps_and_Fisher_Vector/links/570ac58308ae8883a1fc05da.pdf (accessed on 6 October 2016).

48. Chen, C.; Zhang, B.; Hou, Z.; Jiang, J.; Liu, M.; Yang, Y. Action recognition from depth sequences using weighted fusion of 2D and 3D auto-correlation of gradients features. In *Multimedia Tools and Applications*; Springer: New York, NY, USA, 2016; pp. 1–19.

49. Ly, D.L.; Saxena, A.; Lipson, H. Pose estimation from a single depth image for arbitrary kinematic skeletons. **2011**, arXiv:1106.5341.

50. Qiao, M.; Cheng, J.; Zhao, W. Model-based human pose estimation with hierarchical ICP from single depth images. *Adv. Autom. Robot.* **2011**, *2*, 27–35.

51. Gajšek, R.; Štruc, V.; Dobrišek, S.; Mihelič, F. Emotion recognition using linear transformations in combination with video. In Proceedings of the 10th Annual Conference of the International Speech Communication Association, Brighton, UK, 6–10 September 2009.

Sensors **2016**, *16*, 1740

52. Gajšek, R.; Žibert, J.; Justin, T.; Štruc, V.; Vesnicer, B.; Mihelič, F. Gender and affect recognition based on GMM and GMM-UBM modeling with relevance MAP estimation. In Proceedings of the 11th Annual Conference of the International Speech Communication Association, Chiba, Japan, 26–30 September 2010.

53. Ye, M.; Zhang, Q.; Wang, L.; Zhu, J.; Yang, R.; Gall, J. A survey on human motion analysis from depth data. In *Time-of-Flight and Depth Imaging. Sensors, Algorithms, and Applications*; Springer: New York, NY, USA, 2013; pp. 149–187.

sensors

MDPI

Article

Are We Ready to Build a System for Assisting Blind People in Tactile Exploration of Bas-Reliefs?

Francesco Buonamici, Monica Carfagni, Rocco Furferi *, Lapo Governi and Yary Volpe

Department of Industrial Engineering, University of Florence, Florence 50139, Italy; rocco.furferi@unifi.it (F.B.); monica.carfagni@unifi.it (M.C.); lapo.governi@unifi.it (L.G.); yary.volpe@unifi.it (Y.V.)
* Correspondence: rocco.furferi@unifi.it; Tel.: +39-055-275-8741

Academic Editor: Gonzalo Pajares Martinsanz
Received: 6 July 2016; Accepted: 18 August 2016; Published: 24 August 2016

Abstract: Nowadays, the creation of methodologies and tools for facilitating the 3D reproduction of artworks and, contextually, to make their exploration possible and more meaningful for blind users is becoming increasingly relevant in society. Accordingly, the creation of integrated systems including both tactile media (e.g., bas-reliefs) and interfaces capable of providing the users with an experience cognitively comparable to the one originally envisioned by the artist, may be considered the next step for enhancing artworks exploration. In light of this, the present work provides a description of a first-attempt system designed to aid blind people (BP) in the tactile exploration of bas-reliefs. In detail, consistent hardware layout, comprising a hand-tracking system based on Kinect® sensor and an audio device, together with a number of methodologies, algorithms and information related to physical design are proposed. Moreover, according to experimental test on the developed system related to the device position, some design alternatives are suggested so as to discuss pros and cons.

Keywords: hand-tracking system; Kinect sensor; 3D reconstruction; blind people

1. Introduction

Lack of sight affects blind people's possibilities in many aspects of everyday life. Movements, tasks and actions that are simple, or even trivial, to sighted people become really challenging for blind people (BP). To support BP in a great number of situations, in the last decades many devices have been designed all over the world. In most cases, research has been focused on developing systems for assisting BP in their everyday activities such as walking [1], reading books, using computers [2] and so on. However, there are other blind people needs that, although not essential for living, contribute to the overall well-being of an individual. The possibility of enjoying artworks is probably one of the most relevant one since it helps BP in taking part, on an equal basis with others, in cultural life. Not surprisingly, some museums (e.g., the Omero Tactile Museum of Bologna or the Art Institute of Chicago) have created tactile exhibitions dedicated to blind people. The majority of museums, however, present touchable reproductions of sculptures or other 3D objects; tactile 3D reproductions of pictures or other 2D artworks are very rare and only a few institutions have them at their disposal. Usually, these models are handmade by artists thus offering artistic 3D interpretations of the 2D original artwork. To increase and speed-up this "translation" process, in the last few years a few computer aided approaches have been developed [3–5]. Recent studies [6,7] suggest that the mere tactile exploration of 3D models (even in case these are optimally reproducing the original painting) is not sufficient to fully understand, and enjoy, the artwork. Blind people understanding of the original artwork is, in fact, subject to a lot of factors (e.g., sensitivity and personal ability of the person, size and quality of the tactile model); for this reason a good quality verbal guidance is essential in order to appreciate even the best possible artistic relief reproduction of a given painting. Such a verbal description is usually provided by a sighted person (museum employee, accompanying person, etc.);

this allows the blind person to build a complete mental image of the tactile model, and to not be stopped by the lack of comprehension of a single element of the bas-relief. The presence of another person, however, could be perceived as a limiting factor in enjoying the artwork since the blind person is forced to discover in the perspective of someone else; art is through a language that requires autonomy and freedom to be fully apprehended!

The introduction of an automatic verbal guide could increase the autonomy of the user during the exploration, allowing him to lead the experience (e.g., autonomously establishing the time needed for a full appreciation, moving the hands freely, taking time to think, etc.), achieving the same freedom of sighted people. To fulfil this goal, the guide should not be merely automatic (e.g., audio-guide such as the ones already available for sighted people), but rather "active" i.e., capable of following the user's movements so as to provide information in form of verbal descriptions [7].

This ambitious objective is still far to be accomplished in scientific literature, not only for technical restrains: guiding a BP in the exploration of an artworks cannot be limited to a description of an artwork scene and/or of touched areas, but is rather a gradual help to acquire information and to organize it into a "mental scheme" that become progressively more and more complete and detailed. However, the design of a first-attempt system able to automatically provide verbal information of touched areas is still an advancement of the state of the art in this topic.

With this aim in mind, the authors of the present work presented a brief feasibility study [8] of a system to improve blind people tactile exploration of bas-reliefs, where a possible methodology for conceiving an active guide for BP was sketched.

Starting from such a preliminary work, the present paper provides a comprehensive description of the design phases required to build a first-attempt cost-effective system able to properly guide BP in exploring tactile paintings. Such a designed system consists of (1) a 3D Kinect® sensor + software package to track the user hands; (2) a number of algorithms capable of detecting the position of the bas-relief in the same reference frame defined by the acquisition sensor; (3) a number of algorithms aiming at detecting the position and the distance of the user hand/finger with respect to the model; (4) the complete knowledge of the digital 3D bas-relief model and (5) an appropriate verbal description linked to relevant objects/subjects in the scene. The designed system, integrating latest methodologies and algorithms, represents a first consistent step in building an assistive system (not obviously aimed in completely replacing human assistance) to help BP in tactile exploration.

The remainder of the paper is as follows: in Section 2 a brief description of the state of the art for most relevant previous works (related to the designed system) is provided. In Section 3 the system hardware layout is described. In Section 4 methods and algorithms implemented and tested to build the Kinect® sensor-based system are provided. In Section 5, physical layout alternatives of the system are analyzed. Finally, conclusions and future works are discussed in Section 6.

2. Background

The bas-relief exploration system (BES) relies on the implementation of well-known pre-existing methods to perform hand tracking, point cloud registration and 3D evaluation of the distance between two point clouds. Therefore, it's hereby presented a brief review of these techniques, focused on the most promising approaches in literature for the considered application.

2.1. Hand Tracking

Hand tracking (HT) techniques aim at identifying, in real-time, the 3D position of a human hand. This goal is tackled with various approaches in the state of the art, using different data inputs and strategies. HT has been extensively applied in a number of fields: gestural interfaces, virtual environments and videogames are only few examples of the areas where it is gradually becoming a key-factor [9–12]. Application of HT techniques to help impaired people, including BP, in a number of everyday life problems makes no exception [13–15]. Since for the present application the HT system should not limit the user's haptic sensitivity and his/her gestural freedom (allowing for a fulfilling tactile exploration), among all the different HT techniques available in literature, this state of the art focuses on the vision-based ones. This class, in fact, uses only optical sensors (i.e., cameras, 3D optical scanners and other unobtrusive devices) to obtain data. Vision-based techniques can be roughly classified in two great groups: appearance-based [16–18] techniques and model-based ones [19].

Model-based approaches are the most interesting for the present application since provide a full DOF hand pose estimation together with real-time 3D position of the hand. In detail, a digital model of the hand, comprising all joints and articulations, is used. Usually, the solution is retrieved performing a minimization of an objective function that describes (using data from a set of visual cues) the discrepancy between observed data (real position of the hand) and the solution obtained using the digital model. Accordingly, although computationally costly, model-based approaches are the best candidates for this application where the hand position must be determined continuously and entirely.

A first approach to model-based tracking is presented in [20]; a 27 DOF hand modelled by quadrics is used to generate the contours of the hand, which are then confronted with processed images of the real hand. De La Gorce et al. [19] instead propose an approach that takes advantage of shading and texture information as visual cues to compare the digital model and data observed from a single RGB camera. One of the most promising works using model-based approach is the one proposed in [17] where a Microsoft Kinect® is used to obtain 3D data from the scene.

The user hand in 2D and 3D is isolated from the background by means of a skin colour detection followed by depth segmentation. The hand model (palm and five fingers) is described by geometric primitives and parametrized encoding 26-DOF (i.e., is represented by 27 parameters). The optimization procedure is carried out by means of a Particle Swarm Optimization technique [21]. The procedure contemplates temporal continuity of subsequent frames, searching for a solution in the neighborhood of the one found for the last frame analysed. The authors further developed their work in [22–24], covering simultaneous tracking of two hands and tracking of a hand interacting with real objects.

2.2. Point Cloud Registration

As widely recognized [25–27], point cloud registration is a class of algorithms that perform the alignment of two partially or entirely overlapping sets of points by means of a roto-traslation, minimizing relative distances. Among the wide range of methods for point cloud registration, the present work focuses on rigid techniques i.e., the ones that perform the alignment of the two sets of point by means of a rigid transformation (without changing the relative position of the points belonging to the transformed point cloud).

Rigid registration is usually performed by means of a two-step procedure: a first coarse registration and a subsequent fine one. Coarse registration performs a rough alignment of the two point sets, minimizing the distance between correspondences, such as points, curves or surfaces (or other geometric entities) extracted from the dataset with different criteria [27]. A number of algorithms can be used to perform coarse registration: Point Signature, Spin Image, RANdom SAmple Consensus (RANSAC)-based, Principal Component Analysis (PCA) and genetic algorithms. Fine registration, on the other hand, uses the result obtained by coarse registration as starting point and searches, in its neighborhood, for a more refined solution.

Among the wide range of algorithms available in the scientific literature, the most relevant for the present work are: the Iterative Closest Point (ICP) (which has been implemented in many different ways in recent years), the Chen's method (a variation of ICP), the signed distance fields and genetic algorithms [28]. ICP and Chen's methods are, by far, the most common and used: presented at the beginning of 90s, such methods are now implemented in many software libraries.

The ICP method aims to obtain an accurate solution by minimizing the distance between point-correspondences, known as closest point. When an initial estimation is known, all the points are transformed to a reference system applying the Euclidean motion. Then, every point in the first image is taken into consideration to search for its closest point in the second image, so that the distance between these correspondences is minimized, and the process is iterated until convergence.

Chen's method is quite similar to ICP; the only difference is the use of point-to-plane distance instead of point-to-point.

The minimization function is defined by the distances between points in the first image with respect to tangent planes in the second. In other words, considering a point in the first image, the intersection of the normal vector at this point with the second surface determines a second point in which the tangent plane is computed. The algorithm is, in this formulation, usually less conditioned by local minima and by the presence of non-overlapping regions [27].

2.3. Distance Evaluation

A number of methods coping with distance evaluation between sets of 3D points can be found in the literature [29–32]. Specifically, this work deals with the so called nearest neighbour search (NNS) problem, (also known as "proximity search"), which addresses the goal of finding the nearest point, within a data set, to a given query point (and the consequent computation of its distance). Although very simple in its definition, this issue becomes complicate either when the data set consists of a huge number of points or when a high number of query points are provided. Due to its importance in a number of computer vision problems, over the years the NNS problem has been tackled with several different strategies, partially discussed in this section.

Basically, NNS methods exploit the construction of search trees among the inspected dataset (i.e., data structures that organize the information about points distribution in a convenient way), in order to increase the efficiency of the nearest point search. In one of the most used methods, i.e., the "KD-Tree" one [29] a k-dimensional tree-like structure is created by means of recursive binary partitions of the dataset resulting from regions circumscribed by k-dimensional hyper-planes.

Another known method is the so called "Ball Tree" (also known as "Metric Tree"): in this case, the dataset is described by a tree modelled using hyper-spheres; this kind of structure, although computationally costly to build, guarantees a faster search, especially with high-dimension problems.

3. System Layout

As depicted in Figure 1, the layout of the designed BES consists of:

(1) A physical bas-relief to be explored by BP and its digital counterpart (e.g., a high-definition point cloud/polygonal model describing it).
Even if, in principle, any kind of bas-relief could be used for developing the BES, in this work the used tactile models are the ones created by using the procedure described in [33], where shape from shading-based methods are devised to obtain both 3D polygonal models (e.g., STL) and a physical prototype of such a digital model starting from a shaded picture (for example a renaissance painting). In fact, by using such a procedure both the physical and digital 3D information are directly available. In any case, the proposed procedure can be applied to any kind of bas-relief (or in case the bas-relief is not allowed to be touched, to a replica) since the required initial information (polygonal model) can be easily achieved using a commercial 3D scanner.

(2) A 3D acquisition device capable of (i) tracking the user hands and (ii) detecting the position of the physical bas-relief in its reference frame.
The device used to build the system is the Microsoft Kinect®. As widely known, it consists of a projector-camera triangulation device furnished with a 43° vertical by 57° horizontal field of view that covers, at 1 m distance a visible rectangle of 0.8 m × 1.1 m. Such a field of view, to be considered as a plausible value for tracking according to [17], is required to cover the typical dimension of tactile bas-reliefs.

(3) A PC workstation, in control of the whole BES.
This element is responsible for the hand tracking, the required calculations (point clouds registration and distances computation, as previously described) and for the touch identification.

The hardware needs to be equipped for GPU computing, and with hand tracking performances comparable with [17], to assure satisfying results.

(4)   An Audio system.

Since the final outcome of the BES is, as already mentioned above, a verbal description of the scene and/or of touched objects or features, the system is equipped with headsets/headphones. Of course, to locate headsets could be difficult for unaccompanied BP; unfortunately, since the installation is specifically addressed to museum installations, the use of audio speakers could not represent a valid option.

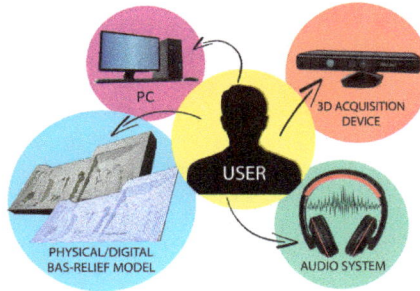

Figure 1. BES layout.

4. Materials and Methods

This section provides a step-by-step description of methods and algorithms implemented and tested to build such a system. All necessary procedures were developed using Matlab® that offers a number of embedded tools and algorithms useful for this application.

To help in understanding the devised system the overall method is described with reference to the tactile reproduction of "Guarigione dello storpio e resurrezione di Tabita" by Masolino da Panicale (see Figure 2). The physical model has size $900 \times 420 \times 80$ mm while its digital counterpart is described by 3.6 million points.

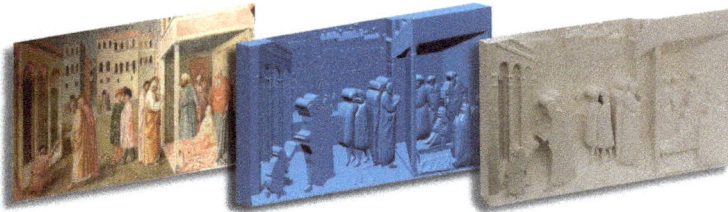

Figure 2. "Guarigione dello storpio e resurrezione di Tabita" by Masolino da Panicale: original artwork, digital 3D model and 3D-printed bas-relief.

4.1. Hand Tracking

The first step of the entire procedure consists of detecting the position of the bas-relief to be explored in the Kinect® reference frame and, contextually, to detect the areas touched by the user (whose hand position has to be expressed in the same reference system). In fact the knowledge of the position of the hand, together with the position of the tactile model, will allow to determine if the user is touching the bas-relief and in which area. This information, however, must be known in the same reference system.

Accordingly, the very first step of the proposed procedure consists of tracking the user hand. Among the several interesting works in literature exploring the use of the Kinect® sensor as a device for hand tracking [10] the HT system used for building the proposed system is the one developed in [17]. Such a real time HT system, working using Microsoft Kinect® as optical sensor, is characterized by a 20 fps framerate when running on modern architecture PCs and moreover it does not require any visual marker on the user's hand. Moreover, the system is delivered with a convenient ready-to-use library (developed by the Forth Institute in 2015). Accordingly, the use of the above mentioned HT system is a straightforward method to know the fingertip position directly in the acquisition device reference system. For the proposed application a high frame rate value is crucial. In fact, it directly affects the quality of the solution provided by the system: with higher frame rate values rapid movements of the hand are more easily registered. Moreover, a small time step between evaluated solutions increases the soundness of the last-known position, used as reference value for the location of the hand.

For this reason, in the proposed system, the procedures accomplishing the HT (i.e., the HT library), have been left free to run separately and independently to the rest of the algorithms (e.g., touch identification), which could slow down the HT. In fact, two main cycles run simultaneously: a hand tracking cycle (HTC), which evaluates continuously hand pose solutions and saves them, and a touch identification cycle (TIC). TIC consists of a number of procedures (extensively described in next steps) that rely on the latest hand pose solution stored in the system by the HTC to assess if and where the user is touching the bas-relief.

Moreover, to reduce the complexity of tracking problem, hand tracking has been performed with reference to a single fingertip (index). This choice is recommended for this application since the proposed BES is only a prototypal version of a future automatic verbal guide system to be installed into museum environments. Consequently, the final result obtained by using the HT system is to detect the coordinates of the extreme point of the index fingertip in the Kinect® reference frame.

4.2. Bas-Relief Positioning

To retrieve the position of the tactile bas-relief in the Kinect® reference frame, the simplest way is to use such a device as a sort of "traditional" 3D scanner; with a single placement a 3D scan of the scene in the Kinect® field of view is accomplishable simply using Kinect® Fusion library.

However, the quality of the Kinect® 3D scan is not good enough to obtain detailed information about the bas-relief or to identify the contact with the finger; especially with a single placement, obtained 3D polygonal model have a low resolution and is affected by high noise. Moreover, the Kinect® acquires the entire scene (not only the bas-relief), resulting in lot of undesired scan points or polygons. Nonetheless, despite the device provides low-definition (LD) scans almost useless for accurate reconstruction of the scene, the scanned points can be used as a provisional reference for registering the (available) high-definition (HD) model as explained in the next procedural step.

4.3. Registration

Once the LD model (correctly referred to the device reference frame) is available, the original high-definition model (HD) of the bas-relief is registered upon the LD one. With this strategy, a very refined 3D model correctly referenced in the Kinect® frame can be obtained. The registration is accomplished by using a two-steps procedure: first a coarse registration is performed to roughly align and over-impose the HD onto the LD points (belonging to the point cloud or polygon vertices in case polygonal models are used). Then, a fine registration, using the rough results obtained in the coarse registration as initial guess, is made to increase the quality of points' alignment.

4.3.1. Coarse Registration

As described in the introductory session, several methods for coarse registration are available in literature. However, in the present work, it is performed with an appositely devised interactive procedure, taking advantage of the hand tracking system implemented to obtain the required initial rough alignment. In effect, traditionally coarse registration algorithms search for geometric correspondences in the two point sets. This procedure, instead, imitates the common "point and click" procedure for coarse registration that is usually implemented in reverse engineering software (where the selection of correspondences is done by the user itself).

In detail an appositely developed point-and-click interface (Figure 3) is used to pick a number $N \geq 3$ of non-aligned pairs of equivalent points in both the LD and HD clouds (or polygonal models).

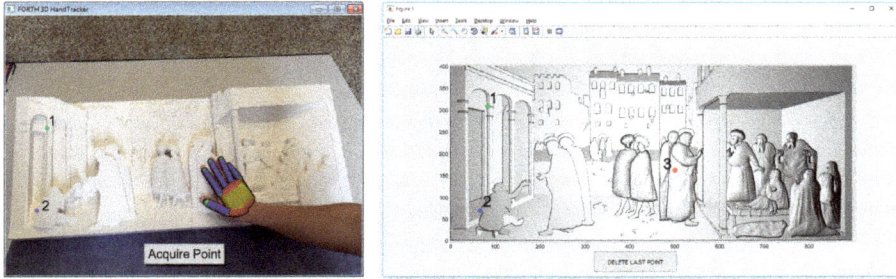

Figure 3. Custom-made coarse registration GUI.

To select the points in the LD model, the HT system is used as follows: first the user touches the desired point using the tracked fingertip. Once the contact between finger and physical bas-relief is established, the user click on the "acquire" pushbutton so that the coordinates of the fingertip are stored in a matrix P_{LD} (size $N \times 3$) and a numbered tag is attached in the touched point. Subsequently, the user is required to touch the correspondent point on the HD model, using the sequence defined by the numbered tags. The coordinates of these points are finally stored in a matrix P_{HD} (size $N \times 3$).

The P_{LD} and P_{HD} matrices, whose elements are the coordinates of, roughly the same points in, respectively, the reference frame of the LD and HD models, can be used for effectively registering the HD model onto the LD ones.

In fact, the reciprocal alignment between the two mentioned reference frames consists of the roto-translation described by the following equation:

$$P_{HD} = R \times P_{LD} + t \tag{1}$$

where R is the rotation matrix and t is the translational vector. Since in Equation (1) both R and t are unknown, a proper procedure for determining them is required. In particular, a singular value decomposition (SVD)-based method (Besl and McKay, [34]), covering three steps has been used.

Firstly, the centroids of both sets of points P_{LD} and P_{HD} are computed as follows:

$$\text{centroid}_{LD} = \frac{1}{N} \sum_{i=1}^{N} P_{LD}^i \tag{2}$$

$$\text{centroid}_{HD} = \frac{1}{N} \sum_{i=1}^{N} P_{HD}^i \tag{3}$$

where P_{LD}^i and P_{HD}^i are the 1×3 vectors describing the coordinates of the ith point belonging, respectively, to the set P_{LD} and P_{HD}.

Once the centroids are evaluated, it is possible to build the matrix H as follows:

$$H = \sum_{i=1}^{N} (P_{LD}^i - \text{centroid}_{LD})(P_{HD}^i - \text{centroid}_{HD})^T \tag{4}$$

The widely known SVD procedure can now be applied to matrix H allowing to determine the matrices U, W and V:

$$[U, W, V] = SVD(H) \tag{5}$$

As a consequence, the rotation matrix R is easily evaluable as follows:

$$R = VU^T \tag{6}$$

Once R is known, the translational vector can be evaluated using the following equation:

$$t = -R \times centroid_A + centroid_B \tag{7}$$

Finally, the knowledge of R and t allows to determine the rough alignment between HD model and Kinect® acquired LD one, according to Equation (2); in other words, to find the whole set of coordinates P'_{HD} of the HD model in the reference frame of the LD one (i.e., the Kinect® reference frame) it is sufficient to apply Equation (1) as follows:

$$P'_{HD} = R \times P_{HD} + t \tag{8}$$

This procedure showed good results on the point registration, especially when the points chosen by the user are well-separated and non-aligned.

4.3.2. Fine Registration

Fine registration is performed starting from the roughly aligned point sets resulted from Section 4.3.1 (Coarse Registration). Among the already mentioned algorithms proposed in scientific literature, ICP and Chen's method were tested so as to find the best one suited for this application.

Both methods perform an iterative minimization of properly defined distance functions. Given the two point sets P'_{HD} and P_{LD} to be aligned, ICP iteratively searches for each P'^{i}_{HD} point of set P'_{HD} the nearest point P^{i}_{LD} of set P_{LD} and apply to the original set P'_{HD} a proper roto-translation to minimize the distance between the two points. At the end of iterations (reached when a proper cost function is minimized) the set P''_{HD} represents the best HD aligned model.

As mentioned in Section 2, Chen's method is quite similar to the ICP one, with the difference of using point-to-plane instead of point-to-point distances. Also using this algorithm, the final result is the set P''_{HD} describing the aligned HD model.

Both these methods, easily implementable in the Matlab® environment, are reliable and show overall good results. Accordingly, ICP and Chen's methods were tested on the registration of the LD and HD scans and showed comparable results. Despite Chen's method being considered in the literature as the most reliable among the two analysed, for the proposed application tests demonstrated that it exhibit more sensitivity to local minima during iterations. Given that fine registration needs to be executed just once during the calibration of the models (i.e., before the bas-relief exploration starts), solution stability was considered as the most important factor. ICP was therefore chosen as preferred method to perform fine registration. Tests performed by authors demonstrated that the average time for convergence of ICP is in the range of 5–8 min, with model dimensions in the order of 10^5 points for the HD scan and 10^6 points for the LD scan (it has to be noticed that LD scans contains also points that are not belonging to the bas-relief). Iterations usually stop with a RMS error between 2–3 mm, value comparable with the Kinect® accuracy. A visual example of the final result obtained with this method is depicted in Figure 4.

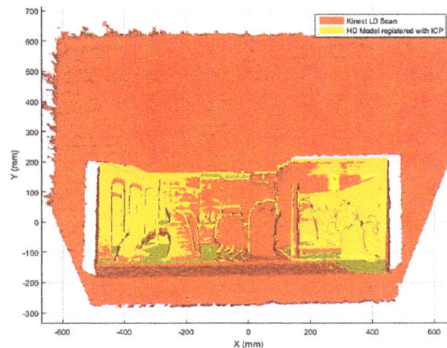

Figure 4. HD and LD bas-relief models after coarse and fine registration (ICP algorithm).

It is important to remark that the whole registration procedure (coarse + fine) should be performed only one time, before the exploration task starts, or at worst it has to be repeated in case the relative position between the bas-relief and the sensor changes for any reason.

4.4. Touch Identification

As already said, thanks to the strategies presented in Section 4.1, Section 4.2 and Section 4.3, the position of the index fingertip and of all the points composing the HD bas-relief model are known in the same reference system. The next step consists of identifying if and where the contact between the finger and the bas-relief occurs.

To perform touch identification, the most convenient method is to find, among all the points of the 3D model, the nearest to the fingertip. Moreover, the distance between such two points is compared to a given threshold to decide whether the finger is in contact or not with the bas-relief. To find the nearest point, the k-nearest neighbour algorithm method was tested against both the "N-D nearest point search" method and the "brute force" method.

Though the proposed prototypal application is based on a single query point (i.e., the index fingertip) algorithms were tested with up to 16 query points with the aim of simulating more complex versions of the system (i.e., with more hand points processed by the system at the same time and/or with more points taken in a single finger). In particular, tests were carried out increasing the number of query points (1–2–4–8–16 points) and the dimension of the dataset (50–100–200–400–800–1600 k points). K-nearest neighbour resulted as the best performing method in all the situations since its computing time is lower than 0.1 s even in the most challenging condition. Such a value guarantees a frame rate of approximately 10 fps and may be therefore considered acceptable for the touch identification task.

In Figure 5 the results of the test performed with one query point are presented. Computing time value, equal to about 0.05 s in the worst conditions, confirms that the implemented k-nearest neighbour algorithm performs perfectly for the considered application.

Accordingly, once the nearest point $P \in P''_{HD}$ to the query point Q is evaluated using the k-nearest neighbours (together with the distance value d) it is possible to identify the touching condition. In fact, if d is smaller than the threshold value d_{touch}, the finger is identified as in touch with the bas-relief; conversely, the devised algorithm considers that no contact occurred between the finger and the HD model. In this last case, the current touch identification cycle (TIC) is considered completed and the touch identification task starts again.

On the basis of a number of tests performed using the whole system of Figure 6, the threshold value d_{touch} was set to 5 mm. This value showed the best compromise between false positive and negative occurrences.

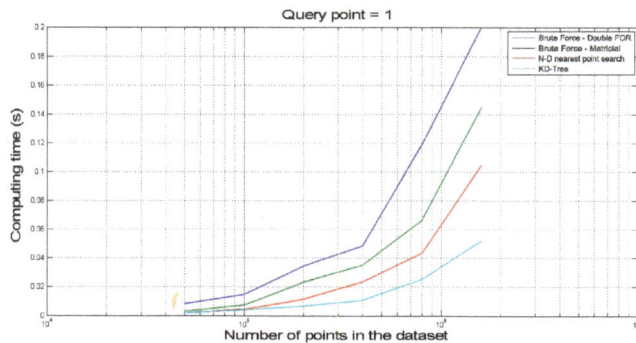

Figure 5. Nearest point methods comparison, 1 query point.

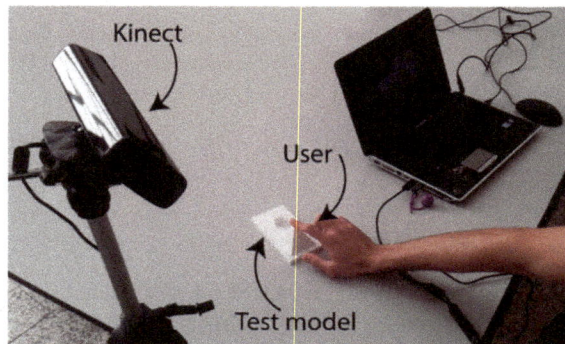

Figure 6. Hardware for the test of touch identification.

A touch identification test was performed using the setup pictured in Figure 6 to evaluate the performances of this phase. The Kinect® sensor was placed at a distance of 1 m from the test model (a plausible value to obtain good hand tracking results, according to [17]). The test began with the hand tracking calibration (required by the implemented method), which registered the hand model upon the user hand. Once that the tracking was stable, the user approached the test model with his hand, describing a roughly vertical movement, until his right index fingertip touched the tip of a pyramidal shape (i.e., the target point of the test), as in Figure 6. The test results (Table 1) showed 74 positive touch identifications on a total of 100 runs. Eighteen false negatives (situations where the touch condition was not recognized) occurred, partly caused by the complete loss of the tracking; eight false positives were registered, characterized by the identification of touch between the finger and the model that significantly anticipated the actual contact between the two.

Table 1. Touch identification test results.

Number of Tests	Positive Touch Identifications	False Negatives	False Positives
100	74	18	8

These values, although promising for a first test, represent a significant limit to an actual implementation of the whole system; this aspect, therefore, needs to be considered and addressed during the study of a first BES functional prototype.

4.5. 3D Segmentation and Region Identification

The last step to be performed consists on the identification of the touched bas-relief region (in such a region, in the future, it will be possible to convey associated verbal description). Region identification starts from a 3D segmentation of the HD digital bas-relief model. Different regions of the bas-relief are identified considering their significance in the original artwork and according to the desired level of detail. These m regions are easily segmentable using a reverse engineering software (e.g., Polyworks®); for the proposed procedure each segment is individually stored as an STL file.

Of course each segmented region consists of a number of points of the set P''_{HD}. As a consequence it is possible to associate to each point of the 3D model a label (from 1 to m) identifying the region containing the point itself. In other words, it is possible to build a matrix A (size m × 4) where the first three columns are the coordinates xyz and the last column is the associated label.

Once the touch identification cycle identifies the touching condition, the corresponding touched region is detected simply by searching in matrix A the label associated to the coordinates P.

4.6. Verbal Description

Each segmented region of the HD model can also be enriched by an audio file containing a verbal description; by a way of example it is sufficient to associate to each region a single wav file. Once the touched area is identified, such description can be transmitted to the user by means of a pair of headphones, to guide him/her in the exploration so to allow a full-immersion experience. Until the description is provided, the TIC is maintained in stand-by in order to avoid undesired interruptions due to a different position of the finger.

5. Physical Layout Alternatives

To select the best physical layout of the devised prototypal system, the relative positions among the user, the bas-relief and the Kinect® have to be investigated. In fact, the system should be capable of providing the best possible accessibility to the tactile bas-relief thus maximizing the comfort during the tactile exploration and, at the same time should guarantee the best system performances. To this purpose, CAD models of the 50th percentile male and of the bas-relief to be explored have been realized in order to get a first idea of the overall dimensions of the two elements (see Figure 7a).

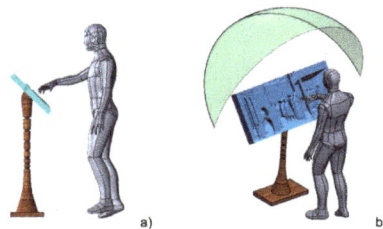

Figure 7. (**a**) CAD model of the bas-relief and the user; (**b**) CAD model of the plausible region (green) to host the Kinect®.

The bas-relief has been positioned at an approximate height of 1.2 m from the ground and with an inclination toward the user of 45°. This position was determined to be the most comfortable for the user, thanks to information gathered by the authors in tests performed together with a panel of blind persons, within the T-VedO project [33]. Starting from this configuration, it has been possible to determine the occlusions introduced in the scene by the users' hand (a key information to place the visual sensor) and, therefore, to identify the areas suitable for housing other elements of the system.

Of course, the most important element to be positioned is the Kinect®; according to [17], and supported by further tests performed by authors of the present work using the device library, an average distance of 1 m between the sensor and the user's hand may be considered among the best options to obtain good performances in terms of resolution and visibility. Moreover, besides the possible obstruction of the scene provided by users' hand, it has to be taken into account that some areas have to be left free for accessing the bas-relief. As a consequence a portion of spherical shell with radius equal 1 m, centred in the barycentre of the bas-relief and with an angle of 60° (Figure 7b) has been found as a plausible area to host the Kinect®.

The sensor position influences also the percentage of bas-relief directly visible by the acquisition device due to the optical occlusions created by parts of the bas-relief itself. Since the system needs to be functional independently from the specific bas-relief selected for exploration, it is not possible to determine a sensor position for avoiding all possible self-occlusions. However, by studying a set of representative case studies it is possible to select a position which is generally suitable though not optimal for each single case. In detail, the percentage of bas-relief visible from a discrete set of points taken on the previously defined spherical shell is evaluated. The results, depicted in Figure 8, show a common central region that maximises the visibility percentage (with an average value near to 90%).

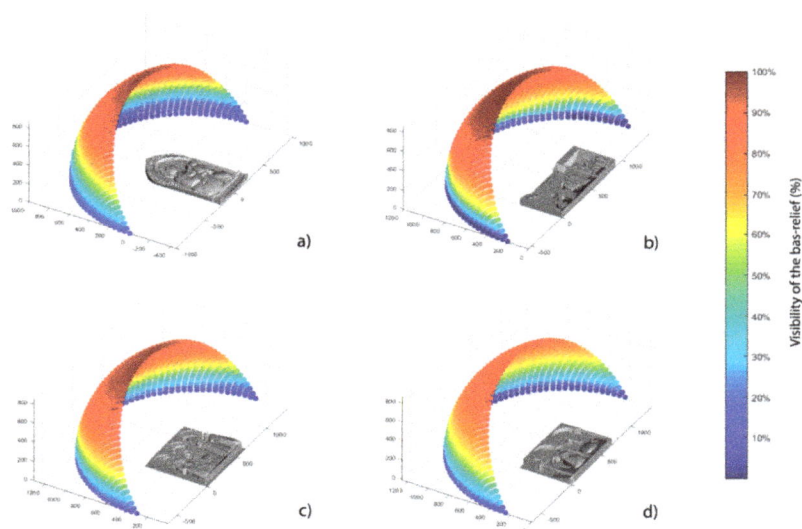

Figure 8. Visibility analysis for bas-reliefs; (**a**) "Madonna con Bambino e Angeli" by Niccolò Gerini di Pietro; (**b**) "Guarigione dello storpio e resurrezione di Tabita" by Masolino da Panicale; (**c**) "Pala di Santa Lucia de' Magnoli" by Domenico Veneziano; (**d**) "Annunciazione" by Beato Angelico.

To choose, among the set of positions that maximise the visibility of the bas-relief, the better solution, it has to be considered that the HT precision strongly depends also by the portion of users' hand acquired during the sensor acquisition. Test performed with different positions of the Kinect® (in the points with maximum percentage of visibility) showed that the best performance is obtained when the hand is placed perpendicular with respect to the line of sight of the Kinect® sensor (see the red hand in Figure 9a). This is probably due to the fact that, in this position, the hand shows to the sensor its most distinctive features (e.g., the silhouette shows all five fingers and the back of the hand is clearly visible). Conversely, when the acquisition device line of sight is inclined with respect to the hand (see the green hand in Figure 9a), it shows less features and accordingly the silhouette could be lost or ignored during skin segmentation.

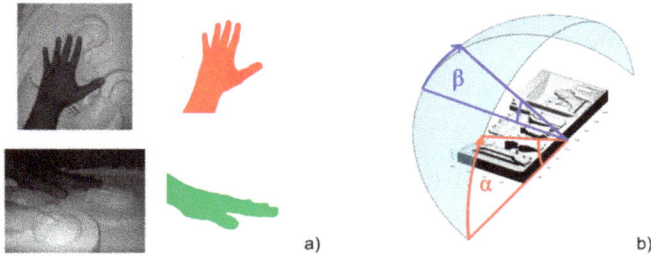

Figure 9. (**a**) Views and silhouettes of the user hand as seen by the Kinect® with different elevation angles; (**b**) Bas-relief reference frame for Kinect® positioning, where α is the azimuth angle and β the elevation angle.

Since on the basis of the mentioned tests performed by authors in the T-VedO project, BP usually explore bas-reliefs with palms placed approximately parallel to the explored surface, among the possible configurations showing better visibility results, the preferable options are the ones where the sensor is placed approximately perpendicular to the bas-relief. Finally, it is possible to state that the best positioning option, referring to Figure 9b, corresponds to the angles $\alpha = 90°$ (azimuth angle) and $\beta = 60°$ (elevation angle). In fact this configuration is the optimal compromise between bas-relief visibility and hand features recognition.

On the basis of the above considerations, the final layout of the proposed prototype is the one depicted in Figure 10 where a vertical structure, located directly under the bas-relief, is designed with the aim of (1) containing the computer hardware and control devices; (2) sustaining the tactile model in the desired position; (3) placing the acquisition sensor in the above mentioned optimal position. The structure, moreover, has a lectern-like shape, suitable for museum exhibits.

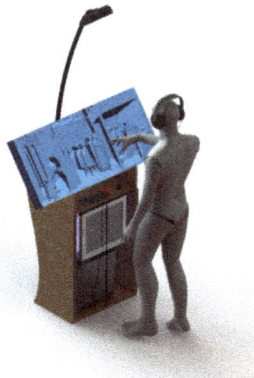

Figure 10. Final layout of the BES prototype.

6. Discussion and Conclusions

In this work, a full description of a first-attempt tactile bas-relief exploration system to improve blind people tactile exploration of artwork has been presented. The system, today in the form of prototype, consists of a bas-relief, a 3D scanner (Microsoft Kinect®) tracking the user's hands connected to a PC, and an audio device providing the user verbal descriptions in response to the hand movements relatively to the bas-relief.

The system functionalities are of course limited, mainly due to the fact it tracks only a single finger and the verbal guidance is only an embryonal idea. Moreover, the accuracy of the touch identification needs to be carefully improved and assessed. The main drawback of the proposed system, up-to-date, is related to the HT system; in fact while slow movements and limited hand rotations are excellently tracked, the hand position is lost from time to time when movement speed increases. Accordingly, at this time it is still premature to think of an implementation of the entire system at least until the issues related to HT and touch identification are coped with. In other words, the answer to the question posed at the beginning is: "probably not yet!".

Fortunately, additional refinements of the work in [22–24] are now under further development at the Forth Institute and the release of better HT algorithms is expected soon. Consequently, such improvements will be tested with the proposed system. Furthermore, the use of the new version of the Kinect® (Kinect® 2.0) with enhanced depth fidelity could improve the HT performance as well, together with an expectable higher resolution provided by the sensor.

Future work will be addressed to the implementation of the HT system to detect more fingers and even the two full hands. The introduction of multiple points, although challenging, could be useful to refine the identification logic of the regions (e.g., interpretation of the full hand position could resolve conflicts in the identification of two contiguous regions). Other issues that will be investigated to improve the system performances are: (a) lighting condition on the scene, which could affect skin segmentation performed by the HT system; (b) implementation of multiple visual sensors, to increase the number of viewpoints and strengthen the HT; (c) study of multi-cue strategies to increase the robustness of the hand identification by HT, appositely devised for this application and its features (e.g., computer vision techniques like background subtraction). Moreover, despite the fact the devised system is largely based on prior studies with a panel of BP, further studies will be assessed by involving more BP to highlight strengths and weaknesses of the proposed system as well to find possible improvements. Moreover, a detailed analysis of system performance on a panel of BP is still required to assess the effectiveness of the designed solution.

The implementation of a gestural interface (i.e., conferring different meanings to specifics gestures made by the user) could dramatically increase the autonomy of blind people, the interactivity of the system and, therefore, its potentiality. Different positions of the hand during exploration could, in effect, be interpreted to transmit different kind of information to the user (e.g., art-related, semantics of the touched regions). This particular issue is not trivial and accordingly more work is required prior to reach exploitable results. Issues such as cost of the entire system, industrial production feasibility, optimization of procedures at industrial scale constitute further future studies to be confronted with prior to effectively introduce the proposed system in museum environments.

Author Contributions: Rocco Furferi, Lapo Governi and Yary Volpe conceived the main steps for creating the system; Francesco Buonamici implemented the system and provided some procedures; he, moreover, carried out experimental tests. Monica Carfagni helped in drafting the procedures; Francesco Buonamici, Rocco Furferi, Yary Volpe and Lapo Governi wrote the paper.

Conflicts of Interest: The authors declare no conflict of interest.

References

1. Lahav, O.; Mioduser, D. Haptic-feedback support for cognitive mapping of unknown spaces by people who are blind. *Int. J. Hum.-Comput. Stud.* **2008**, *66*, 23–35. [CrossRef]
2. Loiacono, E.T.; Djamasbi, S.; Kiryazov, T. Factors that affect visually impaired users' acceptance of audio and music websites. *Int. J. Hum.-Comput. Stud.* **2013**, *71*, 321–334. [CrossRef]
3. Song, W.; Belyaev, A.; Seidel, H.P. Automatic generation of bas-reliefs from 3d shapes. In Proceedings of the IEEE International Conference on shape modeling and applications, Lyon, France, 13–15 June 2007; pp. 211–214.

4. Wang, M.; Chang, J.; Zhang, J.J. A review of digital relief generation techniques. In Proceedings of the 2nd International Conference on Computer Engineering and Technology (ICCET), Chengdu, China, 16–18 April 2010; pp. 198–202.
5. Wang, R.; Paris, S.; Popović, J. 6D hands: Markerless hand-tracking for computer aided design. In Proceedings of the 24th Annual ACM Symposium on User Interface Software and Technology, Santa Barbara, CA, USA, 16–19 October 2011; pp. 549–558.
6. Hayhoe, S. *Arts, Culture and Blindness: Studies of Blind Dtudents in the Visual Arts*; Teneo Press: Youngstown, NY, USA, 2008.
7. Carfagni, M.; Furferi, R.; Governi, L.; Volpe, Y.; Tennirelli, G. Tactile representation of paintings: An early assessment of possible computer based strategies. In *Progress in Cultural Heritage Preservation*; Springer: Heidelberg, Germany, 2012.
8. Buonamici, F.; Furferi, R.; Governi, L.; Volpe, Y. Making blind people autonomous in the exploration of tactile models: A feasibility study. In *Universal Access in Human-Computer Interaction*; Springer: Basel, Switzerland, 2015.
9. Qin, H.; Song, A.; Liu, Y.; Jiang, G.; Zhou, B. Design and calibration of a new 6 DOF haptic device. *Sensors* **2015**, *15*, 31293–31313. [CrossRef] [PubMed]
10. Arkenbout, E.; de Winter, J.; Breedveld, P. Robust hand motion tracking through data fusion of 5DT data glove and nimble VR Kinect camera measurements. *Sensors* **2015**, *15*, 31644–31671. [CrossRef] [PubMed]
11. Airò Farulla, G.; Pianu, D.; Cempini, M.; Cortese, M.; Russo, L.; Indaco, M.; Nerino, R.; Chimienti, A.; Oddo, C.; Vitiello, N. Vision-based pose estimation for robot-mediated hand telerehabilitation. *Sensors* **2016**, *16*, 208. [CrossRef] [PubMed]
12. Spruyt, V.; Ledda, A.; Philips, W. Robust arm and hand tracking by unsupervised context learning. *Sensors* **2014**, *14*, 12023–12058. [CrossRef] [PubMed]
13. Boccanfuso, L.; O'Kane, J.M. CHARLIE: An adaptive robot design with hand and face tracking for use in autism therapy. *Int. J. Soc. Robot.* **2011**, *3*, 337–347. [CrossRef]
14. Imagawa, K.; Lu, S.; Igi, S. Color-based hands tracking system for sign language recognition. In Proceedings of the 3rd IEEE International Conference on Automatic Face and Gesture Recognition, Nara, Japan, 14–16 April 1998; pp. 462–467.
15. Farulla, G.A.; Russo, L.O.; Pintor, C.; Pianu, D.; Micotti, G.; Salgarella, A.R.; Camboni, D.; Controzzi, M.; Cipriani, C.; Oddo, C.M.; et al. Real-time single camera hand gesture recognition system for remote deaf-blind communication. In *Augmented and Virtual Reality*; Springer: Basel, Switzerland, 2014.
16. Payá, L.; Amorós, F.; Fernández, L.; Reinoso, O. Performance of global-appearance descriptors in map building and localization using omnidirectional vision. *Sensors* **2014**, *14*, 3033–3064. [CrossRef] [PubMed]
17. Oikonomidis, I.; Kyriazis, N.; Argyros, A.A. Efficient model-based 3D tracking of hand articulations using Kinect. *BmVC* **2011**, *1*, 1–11.
18. Erol, A.; Bebis, G.; Nicolescu, M.; Boyle, R.D.; Twombly, X. Vision-based hand pose estimation: A review. *Comput. Vis. Image Und.* **2007**, *108*, 52–73. [CrossRef]
19. De La Gorce, M.; Paragios, N.; Fleet, D.J. Model-based hand tracking with texture, shading and self-occlusions. In Proceedings of the IEEE conference on Computer Vision and Pattern Recognition, Anchorage, AK, USA, 23–28 June 2008; pp. 1–8.
20. Stenger, B.; Thayananthan, A.; Torr, P.; Cipolla, R. Hand pose estimation using hierarchical detection. In *Computer Vision in Human-Computer Interaction*; Springer: Heidelberg, Germany, 2004.
21. Esmin, A.A.; Coelho, R.A.; Matwin, S. A review on particle swarm optimization algorithm and its variants to clustering high-dimensional data. *Artif. Intell. Rev.* **2015**, *44*, 23–45. [CrossRef]
22. Oikonomidis, I.; Kyriazis, N.; Argyros, A.A. Markerless and efficient 26-DOF hand pose recovery. In Proceedings of the 10th Asian Conference on Computer Vision, Queenstown, New Zealand, 8–12 November 2010; pp. 744–757.
23. Oikonomidis, I.; Kyriazis, N.; Argyros, A.A. Full DOF tracking of a hand interacting with an object by modeling occlusions and physical constraints. In Proceedings of the IEEE International Conference on Computer Vision (ICCV 2011), Barcelona, Spain, 6–13 November 2011; pp. 2088–2095.
24. Oikonomidis, I.; Kyriazis, N.; Argyros, A.A. Tracking the articulated motion of two strongly interacting hands. In Proceedings of the IEEE Conference Computer Vision Pattern Recognition, Providence, RI, USA, 16–21 June 2012; pp. 1862–1869.

25. Poreba, M.; Goulette, F. A robust linear feature-based procedure for automated registration of point clouds. *Sensors* **2015**, *15*, 1435–1457. [CrossRef] [PubMed]
26. Chen, L.; Hoang, D.; Lin, H.; Nguyen, T. Innovative methodology for multi-view point cloud registration in robotic 3D object scanning and reconstruction. *Appl. Sci.* **2016**, *6*, 132. [CrossRef]
27. Salvi, J.; Matabosch, C.; Fofi, D.; Forest, J. A review of recent range image registration methods with accuracy evaluation. *Image Vis. Comput.* **2007**, *25*, 578–596. [CrossRef]
28. Marichal, G.; Del Castillo, M.; López, J.; Padrón, I.; Artés, M. An artificial intelligence approach for gears diagnostics in AUVs. *Sensors* **2016**, *16*, 529. [CrossRef] [PubMed]
29. Kibriya, A.M.; Frank, E. An empirical comparison of exact nearest neighbour algorithms. In *Knowledge Discovery in Databases: PKDD*; Springer: Heidelberg, Germany, 2007.
30. Chiabrando, F.; Chiabrando, R.; Piatti, D.; Rinaudo, F. Sensors for 3D imaging: Metric evaluation and calibration of a CCD/CMOS time-of-flight camera. *Sensors* **2009**, *9*, 10080–10096. [CrossRef] [PubMed]
31. Keller, J.M.; Gray, M.R.; Givens, J.A. A fuzzy K-nearest neighbor algorithm. *IEEE Trans. Syst. Man Cybern.* **1985**, *SMC-15*, 580–585. [CrossRef]
32. Piegl, L.A.; Rajab, K.; Smarodzinava, V.; Valavanis, K.P. Point-distance computations: A knowledge-guided approach. *Comput. Aided Des. Appl.* **2008**, *5*, 855–866. [CrossRef]
33. Furferi, R.; Governi, L.; Volpe, Y.; Puggelli, L.; Vanni, N.; Carfagni, M. From 2D to 2.5D ie from painting to tactile model. *Graph. Models* **2014**, *76*, 706–723. [CrossRef]
34. Besl, P.J.; McKay, N.D. Method for registration of 3-D shapes. In Proceedings of the SPIE 1611, Sensor Fusion IV: Control Paradigms and Data Structures, Boston, MA, USA, 12–15 November 1991.

![sensors logo] *sensors*

MDPI

Article

Extracting Objects for Aerial Manipulation on UAVs Using Low Cost Stereo Sensors

Pablo Ramon Soria [1,*], Robert Bevec [2], Begoña C. Arrue [1], Aleš Ude [2] and Aníbal Ollero [1]

[1]   Robotics, Vision and Control Group, University of Seville, Camino de los Descubrimientos, s/n,
     Seville 41092, Spain; barrue@us.es (B.C.A.); aollero@us.es (A.O.)
[2]   Humanoid and Cognitive Robotics Lab, Department of Automatics, Biocybernetics and Robotics,
     Jožef Stefan Institute, Jamova cesta 39, Ljubljana 1000, Slovenia; robert.bevec@ijs.si (R.B.);
     ales.ude@ijs.si (A.U.)
*   Correspondence: pabramsor@gmail.com; Tel.: +34-697-513-380

Academic Editor: Gonzalo Pajares Martinsanz
Received: 8 March 2016; Accepted: 10 May 2016; Published: 14 May 2016

Abstract: Giving unmanned aerial vehicles (UAVs) the possibility to manipulate objects vastly extends the range of possible applications. This applies to rotary wing UAVs in particular, where their capability of hovering enables a suitable position for in-flight manipulation. Their manipulation skills must be suitable for primarily natural, partially known environments, where UAVs mostly operate. We have developed an on-board object extraction method that calculates information necessary for autonomous grasping of objects, without the need to provide the model of the object's shape. A local map of the work-zone is generated using depth information, where object candidates are extracted by detecting areas different to our floor model. Their image projections are then evaluated using support vector machine (SVM) classification to recognize specific objects or reject bad candidates. Our method builds a sparse cloud representation of each object and calculates the object's centroid and the dominant axis. This information is then passed to a grasping module. Our method works under the assumption that objects are static and not clustered, have visual features and the floor shape of the work-zone area is known. We used low cost cameras for creating depth information that cause noisy point clouds, but our method has proved robust enough to process this data and return accurate results.

Keywords: UAV; object detection; object recognition; SVM; manipulation

1. Introduction

Unmanned aerial vehicles (UAVs) have been the subject of much research [1] and attracted the interest of the public in recent years. Not only do they offer cost reductions in deployment and operation in a number of scenarios, but they also provide new capabilities in industrial and consumer applications. In order to reduce their dependence on an operator, UAVs have also gained several autonomous capabilities. They are able to autonomously plan paths, cooperate with each other, and even avoid obstacles while flying [2–4]. More recently, the interest in developing manipulation capabilities for UAVs has been spurred.

Such a robotic system, also called an aerial manipulator, merges the versatility of multirotor UAVs with the precision of robotic arms. However, the coupling effects between the aerial vehicle and the manipulator gives rise to several modeling and control problems. Several studies have focused on the control of the arm and how it influences the dynamics of UAVs [5–10], with the goal of developing a cooperative free-flying robot system for assembly structures. An extension of this idea is under way, which aims to develop the first aerial robots in the world with multiple arms.

Nevertheless, manipulation involves more than the control of an aerial manipulator. In order to grasp and manipulate objects, the robot must first be able to perceive them. The complementary required skill to aid manipulation is therefore object perception using arbitrary sensors. Methods using different types of markers for detecting objects have been developed, e.g., radio markers [11] or visual printed tags [11,12]. In this article, however, we focus on object detection methods that do not rely on additional visual cues but solely on the object's visual characteristics. Since drones are often used for surveillance tasks, many methods of detecting and tracking objects have been proposed. Some look for motion in images as a cue for object detection [13–16]. Others use color and intensity information [17,18]. These methods apply for objects that are relatively far away from a high flying drone. We want to solve the task of locating an object in close proximity to the drone, where it can reach it. In order to attempt to grasp an object, the UAV has to acquire some three-dimensional (3D) information about the scene; therefore, the methods mentioned above do not suffice.

Such 3D information is usually gathered in a map of an area. There are several approaches to creating general 3D maps, usually simultaneously localizing the viewer in the scene (SLAM). Some of these approaches use monocular systems [19–21], stereo cameras [19,22,23], depth cameras (or RGB-D sensors) [24], or even laser sensors that return very accurate distances to objects in the scene [25]. The task at hand, however, is not to accurately map a large area, but to return the objects pose relative to the drone, so that the drone can use that shape information in order to grasp the object.

There are several methods to describe objects in order to obtain grasping data. For the sake of the greater generality, we are interested in methods that do not require object CAD models (or computer-aided drafting models) to initiate the grasping. Using sparse depth information about an object a Gaussian processes can be used to describe its implicit surfaces [26,27]. The main advantage of this approach is that it provides a guess of the surface of the object and also offers a measurement of uncertainty of the shape, which can be used to decide where to further inspect the object. It has also been shown that when dealing with novel objects, a reactive grasping mechanism can be used to grasp objects using a humanoid robot by determining its dominant axis and centroid [28,29].

In this paper, we propose a method that extracts objects from a local map of the work-zone, generated using depth information. Candidate objects are extracted from this map by detecting areas different from our floor model. The candidates are then evaluated using their projections in the color images, where the object classification is executed. Our method builds a sparse cloud representation of each object and calculates the object's centroid and dominant axis. This information can be passed to a grasping module for a grasping attempt. Our code is suitable for on-board execution and ought to be initiated after the drone has approached the work-zone. The principle of how the drone comes to the actual pick-up location, within two meters of the objects, is beyond the scope of this paper. In our implementation, we use extremely low cost USB cameras, as seen in Figure 1, which capture images using a rolling shutter and do not have control over image triggering and focus. We show that our system is robust enough to successfully tackle the effects produced by these affordable cameras.

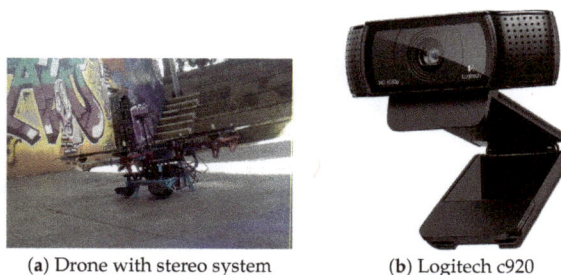

(**a**) Drone with stereo system   (**b**) Logitech c920

Figure 1. The UAV (**a**) used in the experiments with a stereo system comprised of low cost cameras; (**b**) More information about the system is shown is Section 3.

2. Methodology

2.1. System Description

The general principle of our method is best understood by following the flow chart in Figure 2. The system is initialized once the drone is within two meters above the work-zone floor for our cameras to return usable depth information. Our method requires input from the camera and the inertial measurement unit (IMU) with integrated compass data as explained in Section 2.2. The robot determines whether it is necessary to learn the floor appearance model as described in Section 2.4. The robot then predicts its current position in the map, depending on its previous movement (Section 2.6) and determines whether the images are blurry [30]. Due to motion blur or focus hunting, the images might be useless, and, therefore, the robot does not waste time processing them and goes straight to the Extended Kalman filter module to update the UAV's current pose (Section 2.6). If the images are in focus, the robot first excludes the floor from the images in case the floor appearance model has been learned (Section 2.4). Afterwards, a point cloud is generated as explained in Section 2.3 and aligned to the map (Section 2.5). The robot then extracts candidate objects from the map and attempts to classify them using a support vector machine (SVM) classifier, as described in Sections 2.7 and 2.8, respectively. Lastly, at the end of each loop, the object data is returned for a grasping attempt (Section 2.9).

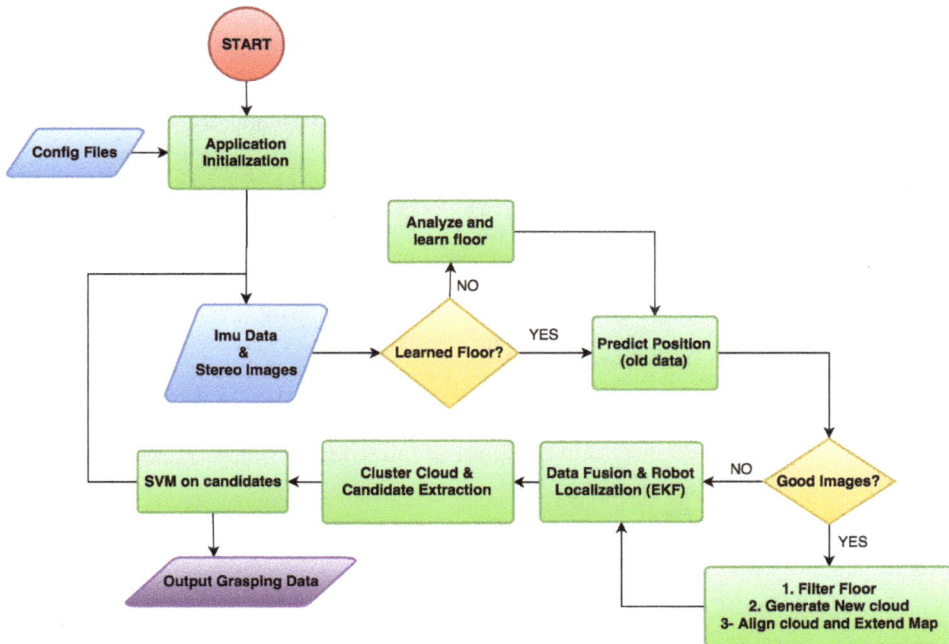

Figure 2. System flow chart.

2.2. Data Acquisition

Our system requires information from two different sensors:

1. Stereo cameras.
2. Inercial measure unit (IMU) module with compass data (Accelerometer, Gyroscope).

We focused on using cheap stereo cameras that do not have trigger control, which results in unsynchronized stereo images. In the execution loop, the cameras are prompted to return new images and the time difference between them can be anything up to $1/FPS$. The drone moves rather slowly above the objects while inspecting them, but larger time shifts are still noticeable and they result in poor point clouds. In Section 2.5, we describe how we deal with noisy clouds, and it is possible to see examples of good and bad clouds in Section 3. The IMU unit provides acceleration and orientation information. The latter can be used directly to estimate the robot's orientation, while the acceleration data is corrected for gravity and fed into an Extended Kalman filter for the motion model (Section 2.6).

2.3. Point Cloud Generation

In order to satisfy a broad spectrum of applications, an affordable depth sensor is required to gather 3D Rdata. Common off-the-shelf depth sensors, such as the Kinect, do not work well at short distances under 1 m [31,32], which makes them difficult to use for manipulation by drones with short arms. An alternative is to use stereoscopy to recover 3D information, while also acquiring 2D color information. Our proposed method works with any method of acquiring depth information and color images; however, we used low cost unsynchronized USB web cameras for the task.

In this implementation, the generation of the point clouds is divided into three steps:

1. Visual feature detection in the left image.
2. Template matching in the right image.
3. Triangulation.

Keypoints or visual features are distinctive points in an image that are invariant to small changes in view. Keypoints extracted from one stereo image should therefore also be distinctive in the other image. Our camera pair is calibrated, therefore we can use the constraints of epipolar geometry to look for keypoint matches. A template window is slid across the epipolar line and compared to the template of each corresponding keypoint. If the matching score is sufficient, a keypoint pair is then triangulated (Figure 3). All the triangulated keypoints make up the point cloud.

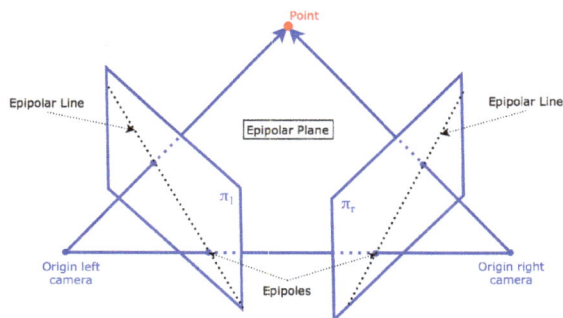

Figure 3. Epipolar geometry.

There are several feature detectors to choose from and several metrics for template matching. Often, features like scale invariant feature transform (SIFT) [33] or speeded up robust features (SURF) [34] are used. These features were designed to be robust in order to track them reliably over longer periods of time. However, we require features that are calculated quickly and we need to detect many of them in order to create a more dense point cloud. For this purpose, we chose the Shi–Tomasi corner detector [35] in combination with the squared sum of differences for template matching, but an arbitrary detector can be applied.

2.4. Floor Detection and Extraction

Floor extraction is analogous to the background subtraction problem, which is tackled frequently in surveillance tasks. The initialization of the background model is crucial to ensure foreground objects can be extracted effectively. Current state-of-the-art methods such as [36–39] propose different algorithms for the initialization and maintenance of the background model. However, these approaches assume static cameras. Some methods consider camera movement intrinsically as in [40,41]. Authors in [42] introduce a Bayesian filter framework to estimate the motion and the appearance model of background and foreground. Others like [43] tackle the problem using an optical flow algorithm to segment the foreground objects. All mentioned methods require moving objects in order to extract them from the background, which is not the case in our scenario, where a UAV tries to pick up a static object.

Our method works under the assumption that the floor shape model is known in advance. After the initial 3D map is generated from the point clouds, we use the random sample consensus (RANSAC) algorithm to find the best match of the floor model in the map [44]. Extracting the floor is very important, as it helps to segment the cloud into candidate objects. Only the floor shape model is assumed, but the robot also learns a color and texture model in order to extract it from the images. Floor extraction from the images is done due to the fact that visual features can appear overwhelmingly only on the floor. Our method of point cloud generation relies on visual features appearing on the objects in order to detect them. Distinctive keypoints appear predominantly on the floor, when it has small repetitive patterns, e.g., a gravel floor or a pebble floor. Our method looks for the maximum N best features for point cloud generation in order to satisfy the time constraints and the quality of the floor features can completely overwhelm the object features. We consider the following scenarios regarding the floor:

1. The floor in the scene is uniform so it has few features on it.
2. The floor has a texture that can be modeled/learned.
3. The floor has a texture that cannot be learned.

The first scenario is the simplest. The feature detector will mostly find keypoints that correspond to objects and will produce good and accurate 3D points. In this case, RANSAC will not detect a good floor match, but the pipeline will continue to work flawlessly and extract candidate objects.

In the second scenario, the floor has some textures that produce keypoints. RANSAC is able to detect the floor in this case and the robot tries to learn the floor appearance model. Repetitive small patterns in particular cause a lot of problems. However, the good thing is that these patterns can be learned and extracted from the images [45] before creating the 3D cloud. It is important to notice that cropping the floor at this stage will speedup the system as fewer features are detected in the remaining image, so the matching, triangulation and then aligning to the map takes less computational time.

Finally, the third scenario has the same problem with dominant features on the floor as the second. In this case, it is not possible to learn the floor pattern for some reason. This is a less likely event, but if a floor has great variance, it can occur and has to be considered. RANSAC extracts the floor model from the map, but at least some keypoints on the objects have to be found for the system to be able to extract objects. Since geometric parameters of floor are detected, we can exclude the points of the map that belong to the floor and pass the result on to the procedure for candidate extraction. In this scenario, the number of object keypoints will obviously be smaller due to most of them being part of the excluded floor; however, as shown later in Section 3, our system can handle these types of scenarios as well.

The system automatically detects whether the floor models can be learned in order to extract the floor either from the images (second scenario) or in 3D (third scenario). In all other cases, the system processes the entire map as in the first scenario. Figure 4 gives a the detailed flow chart of our floor learning method with the starred block representing all the other processes in the loop.

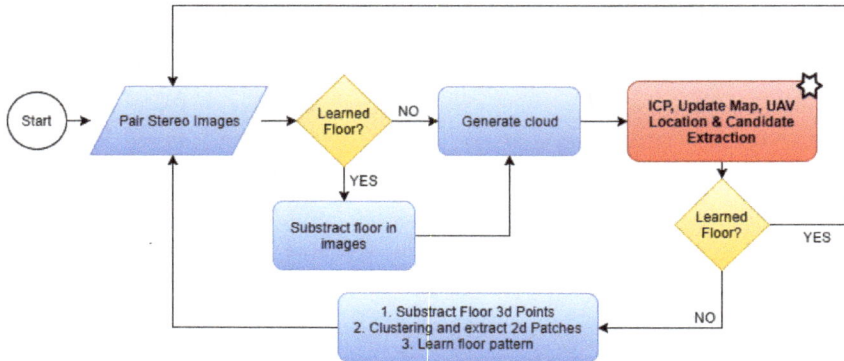

Figure 4. Floor detection and extraction flow chart.

2.5. Temporal Convolution Voxel Filtering for Map Generation

Point clouds generated by matching features from stereo cameras include noisy points due to bad matches, triangulation and calibration errors, mistimed stereo images (Unsynchronized stereo produces a delay between the captured frames), occasional rolling shutter effect because of vibrations, and even some partially blurry images that get through to this point. For this reason, it is necessary to process these clouds before adding them to the map. We developed a method that processes sequential point clouds in both spatial and time dimensions before adding the result to the map. Due to the movement of the UAV, sequential point clouds first need to be aligned properly using affine transformations. Section 2.6 will explain how this transformation is obtained. After the alignment of the point clouds, we filter out the bad points using our probabilistic map generation procedure based on a sequence of the N previous point clouds.

As mentioned, the point clouds are filtered in two steps: (1) Spatial filtering: Isolated particles or small clusters of particles are considered noise (using [46]) and the remaining points are transformed into a grid of cubic volumes of equal size, also called voxels, where a voxel is occupied if at least one point from the point cloud belongs to it [47]; (2) Time filtering: We propose a filtering method over time using sequential voxel point clouds stored into memory, also called history. The occupancy of each voxel is checked in each cloud in history, so that only voxels that have a higher probability of being occupied by a real point will be kept. We call this method Temporal Convolution Voxel Filtering (or TCVF).

Given a set of N consecutive point clouds PC_i, the goal is to obtain a realistic representation of the environment by filtering out incorrect points. The Algorithm 1 describes the process.

TCVF adds a new cloud to the history in each iteration and evaluates the clouds kept in the history at that moment. The result of this operation is then added to the map. By discretizing the space, the number of points for computation is reduced, which reduces the computational time. We use an occupancy requirement of 100% throughout the entire history, making this calculation a simple binary operation of occupancy check, which is very fast and is only evaluated on occupied voxels, making this method computationally light. The number of operations is $O(nk)$, with n the number of occupied voxels in the smallest cloud in the history and k the history size. The voxel size is predetermined and represents the resolution of our map. Figure 5 shows a schematic of a 2D example using a history size of three.

Algorithm 1 Probabilistic Map Generation.

$MAP \leftarrow empty$
for $i \in [0, N]$ **do**

    $PC_I \leftarrow \textbf{filter}(PC_I)$
    $PC_I \leftarrow \textbf{align}(PC_I)$
    $PC_I \leftarrow \textbf{voxel}(PC_I)$
    addToHistory(PC_I)
end for
for *point in* PC_i **do**

    if *point* $\exists$ *in* PC_i *with* $i \in [0, N]$ **then**

        MAP **add** *point*
    end if
end for

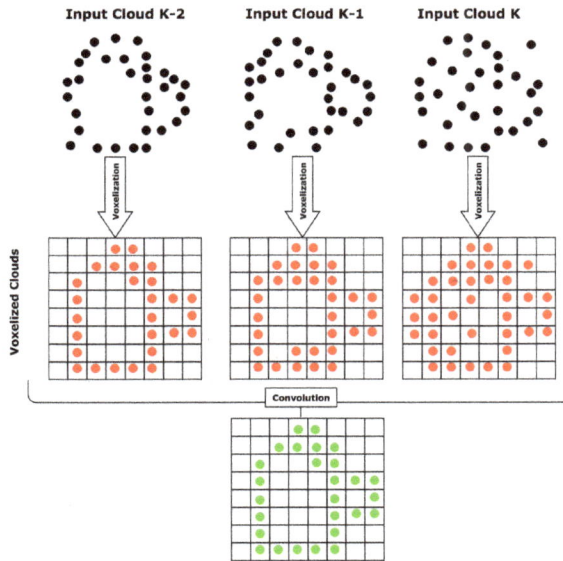

Figure 5. A 2D example of our Temporal Convolution Voxel Filtering for history size of 3. The method checks the occupancy of each voxel through the history of point clouds and only voxels occupied in the entire history are passed through the filter.

2.6. Drone Positioning and Cloud Alignment

At the start of the application, the drone acquires the first point cloud and initializes an empty local map. Since we do not want noise in our map, we use our TCVF algorithm to add points to the map. TCVF needs to first fill the entire history with sequential point clouds in order to determine whether specific points exist. However, the drone is not static, so from the camera's point of view, the points might move, even though they represent the same actual static point. The camera origin of a point cloud effectively represents the relative position of the drone to the detected scene. Obviously, the sequential clouds must be aligned in order for TCVF to work. The effect of aligning sequential clouds is also an assessment of the updated position of the drone. We use the iterative closest point (ICP) algorithm, which minimizes the distances of pairs of closest points in an iterative fashion, to align point clouds. However, ICP algorithms have difficulties detecting the correct transformation between two sequential point clouds if the change in pose between them is large.

This problem can be solved by using IMU data from the drone to provide an assessment of the pose change and feed this to the ICP algorithm. Unfortunately, it is not possible to rely solely on the IMU data for positioning in GPS-denied environments, because it tends to drift quickly. Luckily, the ICP result gives us an estimation of the drone position, so we implemented an algorithm to fuse the information from the IMU and the ICP result to estimate the position of the drone in the map.

Traditionally, an Extended Kalman Filter (EKF) is used to fuse the visual and inertial data. The result of using an EKF is a smoothed pose estimation. There are several implementations of this idea [48–51]. In particular, in [51], the effect of the biases in the IMU is studied and a solution provided. Suppose that the system's state is:

$$X_k = \{x_k^x, x_k^y, x_k^z, \dot{x}_k^x, \dot{x}_k^y, \dot{x}_k^z, \ddot{x}_k^x, \ddot{x}_k^y, \ddot{x}_k^z, b_k^{\ddot{x}}, b_k^{\ddot{y}}, b_k^{\ddot{z}}\} \tag{1}$$

and the observation's state:

$$Z_k = \{x_k^x, x_k^y, x_k^z, \ddot{x}_k^x, \ddot{x}_k^y, \ddot{x}_k^z\} \tag{2}$$

while the equations for the system and the observation are:

$$\begin{cases} x_k^i = x_{k-1}^i + \Delta t \dot{x}_{k-1}^i + \frac{\Delta t}{2} \ddot{x}_{k-1}^i, & i = x, y, z \\ \dot{x}_k^i = \Delta t \ddot{x}_{k-1}^i, & i = x, y, z \\ \ddot{x}_k^i = \ddot{x}_{k-1}^i, & i = x, y, z \\ bias_k^{\ddot{x}^i} = \frac{T}{T+\Delta t} bias_{k-1}^{\ddot{x}^i} + \frac{\Delta t + T}{T + \Delta t}(C_1 + C_2), & i = x, y, z \end{cases} \tag{3}$$

$$\begin{cases} X_k^i = Z_k^j, & i = j = 0.2 \\ X_k^i = Z_k^j, & i = 0.2, j = 3.5 \end{cases} \tag{4}$$

Introducing these equations into the EKF allows for predicting the current state of the system. This information is used to locate the cameras in the environment. It is also used to provide a guess in the next iteration of ICP, by taking the current state and assessing the drone's position after Δt. The orientation is taken directly from the IMU, since it is provided by the compass and does not drift.

Figure 2 shows the pipeline of the whole system and illustrates how the EKF information is used for drone positioning and cloud alignment:

1. The previous state X_{k-1} is used to obtain $\tilde{X}_k$, which is a rough estimation of the current position of the robot.
2. If the stereo system has captured good images, a point cloud is generated and aligned with the map using $\tilde{X}_k$ as the initial guess. The transformation result of the alignment is used as the true position of the drone $\hat{X}_k$. The obtained transformation is compared to the provided guess and discarded, if the difference exceeds a predefined threshold.
3. If the stereo system has not captured good images, it is assumed that $\tilde{X}_k$ is a good approximation of the state, so $\hat{X}_k = \tilde{X}_k$
4. The EKF merges the information from the ICP $\hat{X}_k$, with the information from the IMU, $\hat{\hat{X}}_k$, and the resulting X_k is the current filtered state.

2.7. Candidate Selection

As the robot builds the representation of the environment in the map, the search for candidate objects can be executed. The input cloud for this processing module has already had the floor points removed or has very few floor features as described in Section 2.4. Candidate objects are extracted from the cloud using a clustering algorithm based on Euclidean distances [52]. This clustering method extends each cluster if a point appears closer than a predetermined threshold to any current point in the cluster. An example of this procedure can be seen in Figure 6. Clusters that meet the minimum number of points requirement are selected as candidate objects. The idea behind this is that a tight cluster of features that is not part of the floor could represent an object. Assuming the objects are not cluttered, several candidate objects are extracted and passed to the object recognition module for validation.

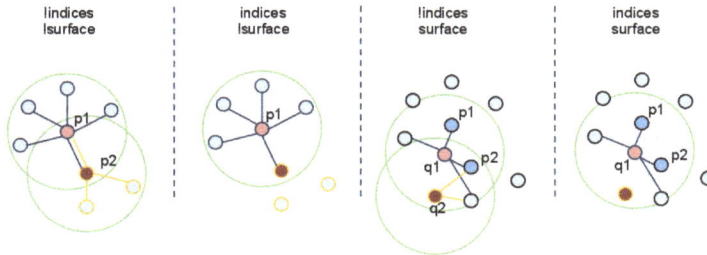

Figure 6. A visualization of the Euclidean cluster extraction [53].

2.8. Object Recognition

This is the final step of the object detection and recognition system. After the candidate extraction, their point clouds are projected onto the images from the cameras. Due to the accurate cloud alignment and drone positioning system described earlier, the projected points correspond well to their respective objects as seen in Figure 7. For each cluster of 2D points, a convex hull that envelops the object is generated, and a patch is extracted for object classification.

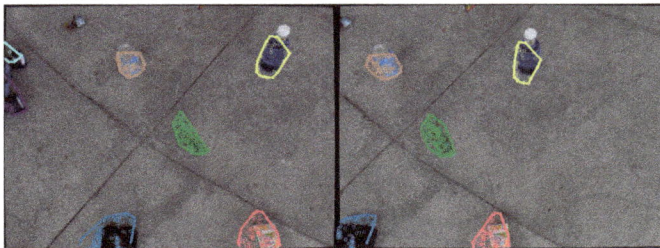

Figure 7. Reprojection of the points belonging to candidate objects, surrounded by a convex hull.

The recognition system is based on the *Bag of Visual Words* (BoW) model [54]. Each object is represented by a histogram of visual descriptors, computed by detecting features in the extracted image patch. We used the Shi–Tomasi corner detector in combination with the SIFT descriptor [33] to describe each object.

The BoW model requires a vocabulary, which is a set of representative descriptors that are used as a reference to quantify features in the images. The vocabulary is generated during the offline classifier training process, where the algorithm detects all features in the training input set and extracts a representative set of *words*. During the training process, an SVM model is trained using positive and negative object samples. The negative samples are used in order to reject false candidate objects, for example if a big enough cluster appears in the map due to noise or a bad floor model. The resulting object detection and recognition system returns the labels of the detected objects, which the drone users can use if a specific object must be picked up or located.

The BoW model has been chosen due to another useful characteristic. It has been shown that it is also good for learning more general representations, like object categories. By training the object to recognize object categories, a novel object can also be classified.

2.9. Grasping Data

Each object is represented by a cluster of points. The grasping module can process this information arbitrarily. However, for purposes of illustration and evaluation, we provide the grasping data in a similar form to that defined in [29]. A grasp is planned by calculating the mean position $\mathbf{p}$ and the dominant principal axis $\mathbf{a}$ of the object. The dominant axis is used for the robot to grasp the object on the narrow side and the centroid is the grasping approach target. Let $\mathbf{x}_i = [x_i, y_i, z_i]$ be the position of N cluster points:

$$\mathbf{p} = \frac{1}{N} \sum_{i=1}^{N} \mathbf{x}_i \tag{5}$$

By calculating the principal axes of the object's points, we estimate the object's greatest extent in each direction. First, the covariance matrix $\mathbf{\Sigma}$ is calculated as:

$$\mathbf{\Sigma} = \mathrm{cov}\left(\{\mathbf{x}_1, \mathbf{x}_2, \ldots \mathbf{x}_N\}\right) = \frac{1}{N-1} \sum_{i=1}^{N} (\mathbf{x}_i - \mathbf{p})(\mathbf{x}_i - \mathbf{p})^T \tag{6}$$

Next, we calculate the three eigenvalues $\lambda_1, \lambda_2, \lambda_3$ and eigenvectors $\mathbf{v}_1, \mathbf{v}_2, \mathbf{v}_3$ of the covariance matrix $\mathbf{\Sigma}$, which is done by solving the equation

$$(\mathbf{\Sigma} - \lambda \mathbf{I}) \mathbf{v} = 0 \tag{7}$$

The eigenvector associated with the biggest eigenvalue represents the dominant axis $\mathbf{a}$ of the object.

3. Experimental Validation

We performed several experiments to evaluate our proposed method. A pair of Logitech c920 cameras [55] were mounted on the bottom of a hexacopter as seen in Figure 1. The drone is equipped with a Pixhawk IMU [56] for the inertial measurements. The baseline of the cameras was approximately 20 cm and they were facing towards the floor at an angle of 70° to the horizon. The detailed specifications and parameters of the system can be found Appendix A. A human operator controlled the drone during flights and the initialization of our method was triggered manually. An Intel NUC [57] computer with a 5th generation i7 processor was on board for vision processing, and our method was able to run at about five frames per second.

The experiments were executed in four different scenarios, summarized in Table 1. In all examples, our floor model was a plane. The *Laboratory* scenario was executed on a white uniform floor with very few visual features, where the drone was hand-held and moved manually. Due to the white floor, this scenario does not trigger our floor extraction module, and there is no vibration from the motors to cause image acquisition problems.

The *Street 1* scenario was again hand-held, but executed outside on a gravel floor. This floor was full of visual features, an example where features are detected predominantly on the floor as described in Section 2.4. The floor extraction module detected the floor within the first three iterations and learned its appearance, removing it from the images before creating point clouds.

In the *Street 2* scenario (Figure 8), the drone was airborne, flown by an operator above a floor with similar texture as in *Street 1*. This scenario represents an example, where vibrations are generated by the motors and images may suffer from the rolling shutter effect. The effect results in bad point clouds.

Table 1. Description of the testing scenarios.

Scenario	Location	Movement	Floor Type
Laboratory	indoor	hand-held	white uniform
Street 1	outdoor	hand-held	gray textured
Street 2	outdoor	flight	gray textured
Testbed	indoor	flight	textured complex

Figure 8. The drone flying above the objects during an outdoor experiment.

Lastly, the scenario *Testbed* was executed in an indoor test facility for drones with a VICON [58] measuring system. The drone flew above a complex textured floor, where the ground truth of all of the objects' positions and the drone motion was measured by VICON. This represents the most difficult scenario for our method, where the floor has many visual features, but it is not possible to learn its appearance and the motors are running and causing vibrations. Our method has to deal with bad point clouds, where few visual features belong to actual objects.

Table 2 summarizes the results of our object extraction method after 15 s of flight. In our scenarios, the number of total points in the local map stopped, increasing significantly after this point, meaning the area had been mostly inspected and the results have converged. However, this is an empirically derived value that depends on the diversity of the observed scene and the flight path the UAV takes to inspect it.

Table 2. Results of the object extraction method.

Dataset	Categories			Objects		
	Precision	Recall	F-Score	Precision	Recall	F-Score
Laboratory	1	1	1	1	0.5	0.667
Street 1	0.429	0.6	0.6	0.333	0.4	0.36
Street 2	0.783	0.4	0.53	0.75	0.33	0.462
Testbed	0.429	0.5	0.462	0.2	0.167	0.182

The SVM classifier was trained using 16 individual objects, and, in the other case, the objects were grouped by categories (Figure 9): cans; juice boxes; circuits; cars; boxes; Some of the objects were very similar and hard to distinguish from certain angles, e.g., the original coke and generic copy, juice boxes of the same brand but different flavors, since they possess intentionally similar appearance. In the *Laboratory* scenario with a plain background, the results for individual object recognition were good and perfect when the objects were grouped in categories. It should be noted that we used no color information for training and recognition, although it could improve recognition results of individual objects. In the other scenarios, the recognition rates were lower, and we attribute this to our feature extraction implementation. We used a square bounding box around the objects to extract features instead of only the convex hull. In training, the background was plain, but for those examples with a textured floor, a lot of the background is included in the bounding box, and it is very rich with visual features. Still, we can see that the categories were detected better than the individual objects.

| (a) Objects of category *cans* | (b) Objects of category *juice* | (c) Objects of category *circuits* |

Figure 9. Our test objects were also grouped into categories. We see category examples belonging to *cans* in (**a**); *juice boxes* in (**b**) and *circuits* in (**c**).

Because our system works with low-cost unsynchronized USB cameras with auto-focus, some pairs of images are not useful. Blurring can occur due to motion or focus hunting, the timing difference between the pair can be too big and occasionally even distortions appear due to the vibrations producing the rolling shutter effect. There are two possible outcomes. If our blur detection module registers a blurry image, we omit it and get a new pair. However, in low light, when the shutter speed of the camera is decreased, motion blur can be present for a longer time. This poses a problem for the ICP algorithm, which requires a good guess for successful alignment. If the position of the drone is lost, the guess cannot be provided and the ICP fails. Our positioning system ensures that the position of the drone is estimated in the EKF using IMU data. The ICP is then able to align a new cloud after a longer period of blurry images and recover the drone's true position. Due to drift, the IMU based guess is not perfect, but, without using it, recovery is unlikely.

The other possible outcome is that the drone generates a point cloud from the bad images. In that case, the quality of the cloud decreases drastically as shown in Figure 10. However, the TCVF is able to handle such clouds and does not compromise the map with noisy data. A noisy point would have to appear in the same voxel throughout the entire history of k for it to appear in the map. In our experiments, we used empirically obtained parameters of history $k = 3$ and voxel size 5 mm.

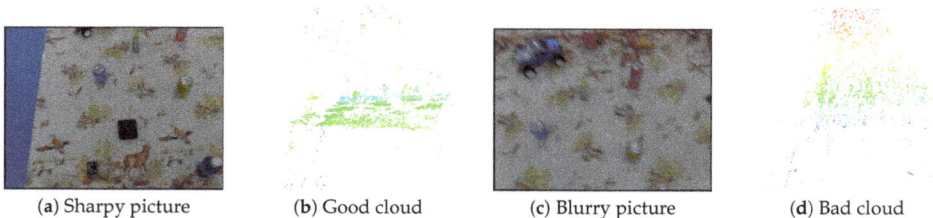

| (a) Sharpy picture | (b) Good cloud | (c) Blurry picture | (d) Bad cloud |

Figure 10. (**a**) shows a good input image. A point cloud generated from good images is shown in (**b**); where the floor plane is clear with objects protruding out; (**c**) shows a blurry image that generates a poor quality cloud (**d**); where the floor and objects are indistinguishable. A similarly bad cloud is also generated when images are mistimed too much.

The positioning system of the drone relative to the objects is one of the key issues. Figure 11 shows how the fusion of ICP information and IMU information gives more robust and accurate results than using only IMU data or ICP results separately. The RGB dotted lines refer to the XYZ estimation of position using only IMU information. As mentioned in Section 2.6, this tends to drift due to the accumulation of errors. The RGB dashed lines are the XYZ positions using only ICP. These results are initially good in so far as all the input clouds are confident. In iteration 120, the algorithm converges to a wrong solution, and then it does not recover. The RGB solid lines are the XYZ positions of the drone for the fusion algorithm. It returns stable and robust estimates of the position of the drone.

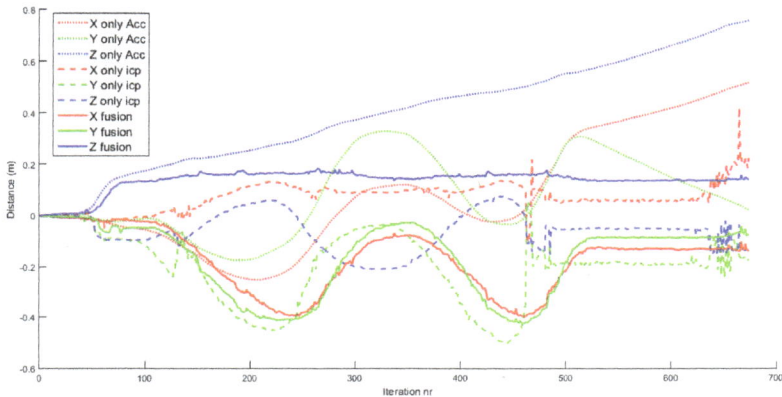

Figure 11. Comparison of drone positioning using the EKF with: only IMU data; only ICP results; fused IMU data and ICP results. Using only IMU data, the position drifts away quickly. Using only ICP results, the position has several bad discrete jumps and does not correspond to actual motion. Using the fused data, the position corresponds to actual motion.

The progress of building the local map of the scene can be seen in Figure 12. The number of features in the map grows and candidate objects appear defined by points of the same color. In order to analyze the quality of the object localization, we compared the results to the ground truth acquired using the dataset *Testbed*. Figure 13 shows the resulting position of the candidates in the scene after approximately 15 s of inspection. Each colored cluster represents a candidate with a PCA defined coordinate system of its pose in the center. The red circles represent the ground truth position of the objects. Three objects were not detected in this dataset.

In order to evaluate the grasping information returned by our system, we calculated the error between each candidate and the ground truth in position and orientation. Figure 14 shows how these errors evolve over time for each object. The position error decreases or remains very stable. When a candidate is first discovered, it has fewer features and might not contain features seen from different angles, rendering its centroid less accurate. As the object is inspected from different angles, this centroid improves accordingly. Similar behavior is noticeable with angles. When a complete representation of the object is acquired, the accuracy of the centroid and orientation is very good. Table 3 shows the average error and variance of the object centroids and orientation after observing the scene for about 50 s. An example of an extracted object can be seen in Figure 15.

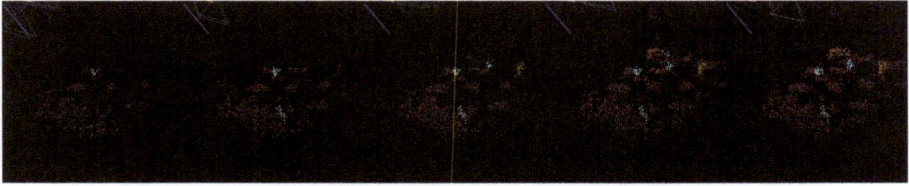

Figure 12. The local map in different iterations. The size of the map is increasing (**red** points) and new object candidates are discovered (seen in color).

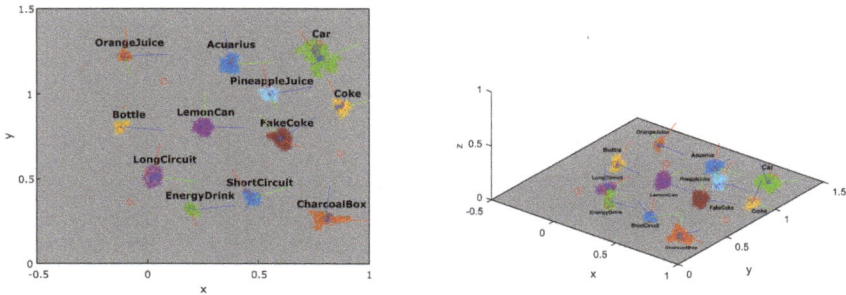

Figure 13. Reconstruction of the objects in the scene (**colored** clusters) compared to ground truth (**red** circles). The frame in each object represents the PCA (principal component analysis) results with the red axis representing the dominant axis.

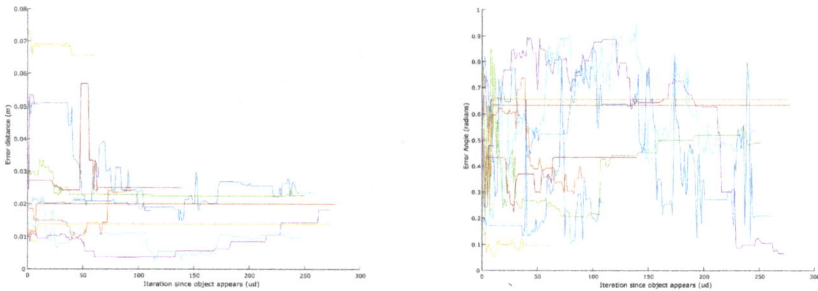

Figure 14. Error of extracted objects' centroid (**left**) and orientation (**right**). The moment of object discovery is aligned to zero on the abscissa. If an object disappears from the view, its plot ends but its last pose is kept.

Table 3. Centroid and orientation error of the extracted objects.

Error	Mean	σ
centroid (m)	0.0256	0.1356
angle (rad)	0.3831	0.4320

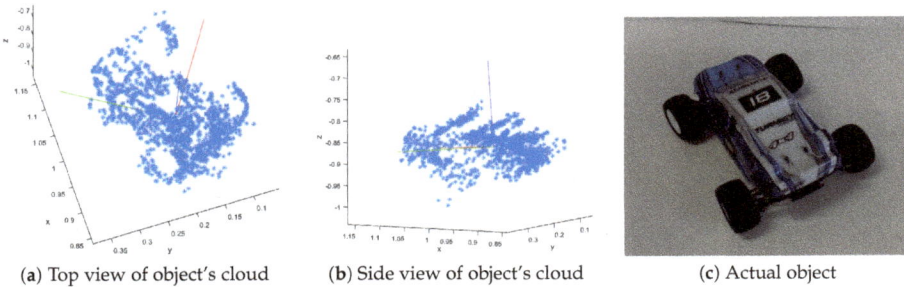

(a) Top view of object's cloud   (b) Side view of object's cloud   (c) Actual object

Figure 15. Example of an extracted object. (**a**,**b**) show the point cloud from different angles; (**c**) is a picture of the actual object.

4. Conclusions

We have developed an on-board object extraction method for rotary wing UAVs that calculates the information necessary for autonomous grasping of objects, without the need of providing a model of the object's shape. An SVM classification procedure is used to recognize specific objects or reject bad candidate objects in a work-zone populated with several objects. Our method works under the assumption that the objects are static and not clustered, have visual features, and that floor shape model of the work-zone area is known. The low cost cameras we have used for creating the depth information occasionally cause very noisy point clouds, but our method for creating a local map has proved robust enough to process this data and return accurate results.

There are several applications for our method to work under the previously mentioned assumptions, particularly industrial applications in partially unknown environments. A drop-off/pick-up zone for arbitrary objects can be selected and the drone ought to pick up objects autonomously without requiring any information about the exact location and shape of the object in advance. A limitation of our method is the time the UAV can handle without accurate point clouds due to either blurry images, the rolling shutter effect or unsynchronized stereo. Blurry images are not used for point cloud generation, while point clouds produced from mistimed or distorted images result in incorrect ICP alignment results. The UAV discards bad ICP results as described in Section 2.6 and must rely solely on the IMU data, causing a slow drift in position estimation. Eventually this drift is so large that the position cannot be recovered anymore due to ICP failure. We can tackle this limitation by improving the camera sensors, reducing the number of bad images, which improves the stability of the system for observing the scene.

A particular advantage of our system is that we extract objects using a bottom-up approach, where candidate objects are extracted from a stereo reconstructed local map of the scene and then recognized using 2D information from the images. Our method is therefore also applicable to the localization and grasping of completely novel objects, since no knowledge about the object is needed for the generation of the candidate objects. The manipulation module can then work in unison with our system to validate candidate objects with information acquired during grasping.

A lightweight arm must be designed and mounted on our UAV in order to test the grasping module that uses the information returned using our proposed method. We plan to design a control method for flying the UAV over objects in a way that will optimize building the local object map and the recognition of the objects. Currently, an operator either flies the UAV manually or predetermines a way-point flight plan. Ultimately, we want to enhance our method to handle mobile objects too, which represents a significant challenge.

Acknowledgments: We thank Robotics, Vision and Control Group and Humanoid and Cognitive Robotics Lab for supporting us during this work. This work has been developed in the framework of the project AEROARMS (SI-1439/2015) EU-funded project and the project AEROMAIN (DPI2014-59383-C2-1-R).

Author Contributions: Pablo Ramon Soria, Robert Bevec, Begoña C. Arrue, Aníbal Ollero and Aleš Ude conceived the methodology. Pablo Ramon Soria, Robert Bevec and Begoña C. Arrue designed the experiments; Pablo Ramon Soria and Robert Bevec performed the experiments; Pablo Ramon Soria, Robert Bevec and Begoña C. Arrue analyzed the data; Aníbal Ollero contributed materials and tools; Pablo Ramon Soria, Robert Bevec and Begoña C. Arrue wrote the paper.

Conflicts of Interest: The authors declare no conflict of interest.

Appendix   Parameters of the System

Parameter	Description	Value
Camera Parameters		
ROI	Usable part of images due to distortion	$[\sim 30, \sim 30, \sim 610, \sim 450]$
Blur Threshold	Max. value of blurriness (Section 2.2)	1.1–1.5
Disparity Range	Min-max distances between pair of pixels. This value determine the min-max distances	$[\sim 50, \sim 300]$ px $\rightarrow [\sim 0.3, \sim 2]$ m
Template Square Size	Size of the template for template matching	7–15
Max Template Score	Max. allowed score of template matching	0.004–0.05
Map Generation Parameters		
Voxel Size	Size of voxel in 3D space grid	0.005–0.03
Outlier Mean K	Outlier removal parameter	10–20
Outier Std Dev	Outlier removal parameter	0.05–0.1
ICP max epsilon	Max. allowed error in transformation between iterations in ICP	0.001
ICP max iterations	Number of iterations of ICP	10–50
ICP max corresp. dist	Max. initial distance for correspondences	0.1–0.2
ICP max fitting Score	Max. score to reject ICP result	0.05–0.1
Max allowed Translation	Max. translation to reject ICP result	0.05–0.15
Max allowed rotation	Max rotation to reject ICP result	10–20
History Size	History size of TCVF	2–4
Cluster Affil. Max. Dist.	Minimal distance between objects	0.035
EKF Parameters		
Acc Bias Calibration	Bias data from IMU X,Y,Z	$-0.14, 0.051, 6.935$
Acc Frequency	Mean frequency of data	1 KHz
Gyro Noise	Average magnitude of noise	$0.05\,rad/s$
Gyro Frequency	Mean frequency of data	30 KHz
Imu to cam Calibration	Transformation between camera and IMU	*Data from Calib*
Q System cov. Mat.	Covariance of System state variables	*Data from Calib*
R Observation cov. Mat.	Covariance of Data	*Data from Calib*
Recognition System		
Training params	Train parameters of SVM	---
Detector Descriptor	Feature detector and descriptor used	*SIFT*

References

1. Pajares, G. Overview and Current Status of Remote Sensing Applications Based on Unmanned Aerial Vehicles (UAVs). *Photogramm. Eng. Remote Sens.* **2015**, *81*, 281–329.
2. Budiyanto, A.; Cahyadi, A.; Adji, T.; Wahyunggoro, O. UAV obstacle avoidance using potential field under dynamic environment. In Proceedings of the International Conference on Control, Electronics, Renewable Energy and Communications, Bandung, Indonesia, 27–29 August 2015; pp. 187–192.
3. Santos, M.; Santana, L.; Brandao, A.; Sarcinelli-Filho, M. UAV obstacle avoidance using RGB-D system. In Proceedings of the International Conference on Unmanned Aircraft Systems, Denver, CO, USA, 9–12 June 2015; pp. 312–319.

4. Hrabar, S. 3D path planning and stereo-based obstacle avoidance for rotorcraft UAVs. In Proceedings of the International Conference on Intelligent Robots and Systems, Nice, France, 22–26 September 2008; pp. 807–814.

5. Kondak, K.; Huber, F.; Schwarzbach, M.; Laiacker, M.; Sommer, D.; Bejar, M.; Ollero, A. Aerial manipulation robot composed of an autonomous helicopter and a 7 degrees of freedom industrial manipulator. In Proceedings of the International Conference on Robotics and Automation, Hong Kong, China, 31 May–7 June 2014; pp. 2107–2112.

6. Braga, J.; Heredia, G.; Ollero, A. Aerial manipulator for structure inspection by contact from the underside. In Proceedings of the International Conference on Intelligent Robots and Systems, Hamburg, Germany, 28 September–2 October 2015; pp. 1879–1884.

7. Ruggiero, F.; Trujillo, M.A.; Cano, R.; Ascorbe, H.; Viguria, A.; Per, C. A multilayer control for multirotor UAVs equipped with a servo robot arm. In Proceedings of the International Conference on Robotics and Automation, Seattle, WA, USA, 26–30 May 2015; pp. 4014–4020.

8. Suarez, A.; Heredia, G.; Ollero, A. Lightweight compliant arm for aerial manipulation. In Proceedings of the International Conference on Intelligent Robots and Systems, Hamburg, Germany, 28 September–2 October 2015; pp. 1627–1632.

9. Kondak, K.; Krieger, K.; Albu-Schaeffer, A.; Schwarzbach, M.; Laiacker, M.; Maza, I.; Rodriguez-Castano, A.; Ollero, A. Closed-loop behavior of an autonomous helicopter equipped with a robotic arm for aerial manipulation tasks. *Int. J. Adv. Robot. Syst.* **2013**, *10*, 1–9.

10. Jimenez-Cano, A.E.; Martin, J.; Heredia, G.; Ollero, A.; Cano, R. Control of an aerial robot with multi-link arm for assembly tasks. In Proceedings of the International Conference on Robotics and Automation, Karlsruhe, Germany, 6–10 May 2013; pp. 4916–4921.

11. Fabresse, F.R.; Caballero, F.; Maza, I.; Ollero, A. Localization and mapping for aerial manipulation based on range-only measurements and visual markers. In Proceedings of the International Conference on Robotics and Automation, Hong Kong , China, 31 May–7 June 2014; pp. 2100–2106.

12. Heredia, G.; Sanchez, I.; Llorente, D.; Vega, V.; Braga, J.; Acosta, J.A.; Ollero, A. Control of a multirotor outdoor aerial manipulator. In Proceedings of the International Conference on Intelligent Robots and Systems, Chicago, IL, USA, 14–18 Septmber 2014; pp. 3417–3422.

13. Saif, A.; Prabuwono, A.; Mahayuddin, Z. Motion analysis for moving object detection from UAV aerial images: A review. In Proceedings of the International Conference on Informatics, Electronics & Vision, Dhaka, Bangladesh, 23–24 May 2014; pp. 1–6.

14. Sadeghi-Tehran, P.; Clarke, C.; Angelov, P. A real-time approach for autonomous detection and tracking of moving objects from UAV. In Proceedings of the Symposium on Evolving and Autonomous Learning Systems, Orlando, FL, USA, 9–12 December 2014; pp. 43–49.

15. Ibrahim, A.; Ching, P.W.; Seet, G.; Lau, W.; Czajewski, W. Moving objects detection and tracking framework for UAV-based surveillance. In Proceedings of the Pacific Rim Symposium on Image and Video Technology, Singapore, 14–17 November 2010; pp. 456–461.

16. Rodríguez-Canosa, G.R.; Thomas, S.; del Cerro, J.; Barrientos, A.; MacDonald, B. A real-time method to detect and track moving objects (DATMO) from unmanned aerial vehicles (UAVs) using a single camera. *Remote Sens.* **2012**, *4*, 1090–1111.

17. Price, A.; Pyke, J.; Ashiri, D.; Cornall, T. Real time object detection for an unmanned aerial vehicle using an FPGA based vision system. In Proceedings of the International Conference on Robotics and Automation, Orlando, FL, USA, 15–19 May 2006; pp. 2854–2859.

18. Kadouf, H.H.A.; Mustafah, Y.M. Colour-based object detection and tracking for autonomous quadrotor UAV. *IOP Conf. Ser. Mater. Sci. Eng.* **2013**, *53*, 012086.

19. Lemaire, T.; Berger, C.; Jung, I.K.; Lacroix, S. Vision-based SLAM: Stereo and monocular approaches. *Int. J. Comput. Vis.* **2007**, *74*, 343–364.

20. Davison, A.J.; Reid, I.D.; Molton, N.D.; Stasse, O. MonoSLAM: Real-time single camera SLAM. *IEEE Trans. Pattern Anal. Mach. Intell.* **2007**, *29*, 1–16.

21. Engel, J.; Schöps, T.; Cremers, D. LSD-SLAM: Large-scale direct monocular SLAM. *Eur. Conf. Comput. Vis.* **2014**, *8690*, 834–849.

22. Engel, J.; Stuckler, J.; Cremers, D. Large-scale direct SLAM with stereo cameras. In Proceedings of the International Conference on Intelligent Robots and Systems, Hamburg, Germany, 28 September–2 October 2015; pp. 1935–1942.

23. Fu, C.; Carrio, A.; Campoy, P. Efficient visual odometry and mapping for unmanned aerial vehicle using ARM-based stereo vision pre-processing system. In Proceedings of the International Conference on Unmanned Aircraft Systems, Denver, CO, USA, 9–12 June 2015; pp. 957–962.

24. Newcombe, R.A.; Molyneaux, D.; Kim, D.; Davison, A.J.; Shotton, J.; Hodges, S.; Fitzgibbon, A. KinectFusion: Real-time dense surface mapping and tracking. In Proceedings of the International Symposium on Mixed and Augmented Reality, Basel, Switzerland, 26–29 October 2011; pp. 127–136.

25. Fossel, J.; Hennes, D.; Claes, D.; Alers, S.; Tuyls, K. OctoSLAM: A 3D mapping approach to situational awareness of unmanned aerial vehicles. In Proceedings of the International Conference on Unmanned Aircraft Systems, Atlanta, GA, USA, 28–31 May 2013; pp. 179–188.

26. Williams, O.; Fitzgibbon, A. Gaussian Process Implicit Surfaces. Available online: http://gpss.cc/gpip/slides/owilliams.pdf (accessed on 13 May 2016).

27. Dragiev, S.; Toussaint, M.; Gienger, M. Gaussian process implicit surfaces for shape estimation and grasping. In Proceedings of the International Conference on Robotics and Automation, Shanghai, China, 9–13 May 2011; pp. 2845–2850.

28. Schiebener, D.; Schill, J.; Asfour, T. Discovery, segmentation and reactive grasping of unknown objects. In Proceedings of the International Conference on Humanoid Robots, Osaka, Japan, 29 November–1 December 2012; pp. 71–77.

29. Bevec, R.; Ude, A. Pushing and grasping for autonomous learning of object models with foveated vision. In Proceedings of the International Conference on Advanced Robotics, Istanbul, Turkey, 27–31 July 2015; pp. 237–243.

30. Marziliano, P.; Dufaux, F.; Winkler, S.; Ebrahimi, T. A no-reference perceptual blur metric. In Proceedings of the International Conference on Image Processing, Rochester, NY, USA, 22–25 September 2002.

31. Alhwarin, F.; Ferrein, A.; Scholl, I. IR Stereo Kinect: Improving Depth Images by Combining Structured Light with IR Stereo. In Proceedings of the PRICAI 2014: Trends in Artificial Intelligence: 13th Pacific Rim International Conference on Artificial Intelligence, Gold Coast, Australia, 1–5 December 2014; pp. 409–421.

32. Akay, A.; Akgul, Y.S. 3D reconstruction with mirrors and RGB-D cameras. In Proceedings of the 2014 International Conference on Computer Vision Theory and Applications (VISAPP), Lisbon, Portugal, 5–8 January 2014; Volume 3, pp. 325–334.

33. Lowe, D.G. Distinctive image features from scale-invariant keypoints. *Int. J. Comput. Vis.* **2004**, *60*, 91–110.

34. Bay, H.; Ess, A.; Tuytelaars, T.; van Gool, L. Speeded-up robust features (SURF). *Comput. Vis. Image Underst.* **2008**, *110*, 346–359.

35. Shi, J.; Tomasi, C. Good features to track. In Proceedings of the Conference on Computer Vision and Pattern Recognition, Seattle, WA, USA, 21–23 June 1994; pp. 593–600.

36. Huang, S.C. An advanced motion detection algorithm with video quality analysis for video surveillance systems. *IEEE Trans. Circuits Syst. Video Technol.* **2011**, *21*, 1–14.

37. Cheng, F.C.; Chen, B.H.; Huang, S.C. A hybrid background subtraction method with background and foreground candidates detection. *ACM Trans. Intell. Syst. Technol.* **2015**, *7*, 1–14.

38. Chen, B.H.; Huang, S.C. Probabilistic neural networks based moving vehicles extraction algorithm for intelligent traffic surveillance systems. *Inf. Sci.* **2015**, *299*, 283–295.

39. Colombari, A.; Fusiello, A. Patch-based background initialization in heavily cluttered video. *IEEE Trans. Image Process.* **2010**, *19*, 926–933.

40. Barnich, O.; van Droogenbroeck, M. ViBe: A universal background subtraction algorithm for video sequences. *IEEE Trans. Image Process.* **2011**, *20*, 1709–1724.

41. Yi, K.M.; Yun, K.; Kim, S.W.; Chang, H.J.; Choi, J.Y. Detection of moving objects with non-stationary cameras in 5.8 ms: Bringing motion detection to your mobile device. In Proceedings of the 2013 IEEE Conference on Computer Vision and Pattern Recognition Workshops, Portland, OR, USA, 23–28 June 2013; pp. 27–34.

42. Elqursh, A.; Elgammal, A. Computer Vision—ECCV 2012. In Proceedings of the 12th European Conference on Computer Vision, Florence, Italy, 7–13 October 2012.

43. Narayana, M.; Hanson, A.; Learned-Miller, E. Coherent motion segmentation in moving camera videos using optical flow orientations. In Proceedings of the 2013 IEEE International Conference on Computer Vision, Sydney, Australia, 1–8 December 2013; pp. 1577–1584.

44. Fischler, M.A.; Bolles, R.C. Random Sample Consensus: A paradigm for model fitting with applications to image analysis and automated cartography. *Graph. Image Process.* **1981**, *24*, 381–395.

45. Stauffer, C.; Grimson, W.E.L. Adaptive background mixture models for real-time tracking. *Comput. Visi. Pattern Recognit.* **1999**, *2*, 252.

46. Rusu, R.B.; Marton, Z.C.; Blodow, N.; Dolha, M.; Beetz, M. Towards 3D point cloud based object maps for household environments. *Robot. Auton. Syst.* **2008**, *56*, 927–941.

47. Roth-Tabak, Y.; Jain, R. Building an environment model using depth information. *Computer* **1989**, *22*, 85–90.

48. De Marina, H.G.; Espinosa, F.; Santos, C. Adaptive UAV attitude estimation employing unscented kalman filter, FOAM and low-cost MEMS sensors. *Sensors* **2012**, *12*, 9566–9585.

49. Lobo, J.; Dias, J.; Corke, P.; Gemeiner, P.; Einramhof, P.; Vincze, M. Relative pose calibration between visual and inertial sensors. *Int. J. Robot. Res.* **2007**, *26*, 561–575.

50. Nießner, M.; Dai, A.; Fisher, M. Combining inertial navigation and ICP for real-time 3D surface Reconstruction. In *Eurographics*; Citeseer: Princeton, NJ, USA, 2014; pp. 1–4.

51. Benini, A.; Mancini, A.; Marinelli, A.; Longhi, S. A biased extended kalman filter for indoor localization of a mobile agent using low-cost IMU and USB wireless sensor network. In Proceedings of the IFAC Symposium on Robot Control, Valamar Lacroma Dubrovnik, Croatia, 5–7 September 2012.

52. Rusu, R.B. Semantic 3D Object Maps for Everyday Manipulation in Human Living Environments. Ph.D. Thesis, Institut für Informatik der Technischen Universität München, Munich, Germany, 2010.

53. PCL - Point Cloud Library. An Open-Source Library of Algorithms for Point Cloud Processing Tasks and 3D Geometry Processing. Available online: http://www.pointclouds.org (accessed on 13 May 2016).

54. Csurka, G.; Dance, C.; Fan, L.; Willamowski, J.; Bray, C. Visual categorization with bags of keypoints. In *ECCV Workshop on Statistical Learning in Computer Vision*; Springer-Verlag: Prague, Czech Republic, 2004; Volume 1, pp. 1–22.

55. Logitech c920 Personal Eebcamera. *Logitech*. Available online: http://www.logitech.com/en-au/product/hd-pro-webcam-c920 (accessed on 13 May 2016).

56. Pixhawk. An Open-Source Autopilot System Oriented Toward Inexpensive Autonomous Aircraft. *3D Robotics* Available online: https://pixhawk.org/modules/pixhawk (accessed on 13 May 2016).

57. Intel NUC5i7RYH. Next Unit of Computing (NUC) a Small-form-factor Personal Computer. *Intel Corporation*. Available online: http://www.intel.com/content/www/us/en/nuc/products-overview.html (accessed on 13 May 2016).

58. VICON. Motion Capture System. *Vicon Motion Systems Ltd*. Available online: http://www.vicon.com/ (accessed on 13 May 2016).

sensors

Article

Reliable Fusion of Stereo Matching and Depth Sensor for High Quality Dense Depth Maps

Jing Liu [1,2,*], Chunpeng Li [1], Xuefeng Fan [1,2] and Zhaoqi Wang [1]

[1] The Beijing Key Laboratory of Mobile Computing and Pervasive Device, Institute of Computing Technology, Chinese Academy of Sciences, No.6 Kexueyuan South Road Zhongguancun, Haidian District, Beijing 100190, China; cpli@ict.ac.cn (C.L.); fanxuefeng@ict.ac.cn (X.F.); meifeng@ict.ac.cn (Z.W.)

[2] University of Chinese Academy of Sciences, No.19A Yuquan Road, Beijing 100049, China

* Correspondence: liujing01@ict.ac.cn; Tel.: +86-10-6260-0874

Academic Editor: Gonzalo Pajares Martinsanz

Received: 23 June 2015; Accepted: 17 August 2015; Published: 21 August 2015

Abstract: Depth estimation is a classical problem in computer vision, which typically relies on either a depth sensor or stereo matching alone. The depth sensor provides real-time estimates in repetitive and textureless regions where stereo matching is not effective. However, stereo matching can obtain more accurate results in rich texture regions and object boundaries where the depth sensor often fails. We fuse stereo matching and the depth sensor using their complementary characteristics to improve the depth estimation. Here, texture information is incorporated as a constraint to restrict the pixel's scope of potential disparities and to reduce noise in repetitive and textureless regions. Furthermore, a novel pseudo-two-layer model is used to represent the relationship between disparities in different pixels and segments. It is more robust to luminance variation by treating information obtained from a depth sensor as prior knowledge. Segmentation is viewed as a soft constraint to reduce ambiguities caused by under- or over-segmentation. Compared to the average error rate 3.27% of the previous state-of-the-art methods, our method provides an average error rate of 2.61% on the Middlebury datasets, which shows that our method performs almost 20% better than other "fused" algorithms in the aspect of precision.

Keywords: stereo matching; depth sensor; multiscale pseudo-two-layer model; segmentation; texture constraint; fusion move

---

1. Introduction

Depth estimation is one of the most fundamental and challenging problems in computer vision. For decades, it has been important for many advanced applications, such as 3D reconstruction [1], robotic navigation [2], object recognition [3] and free viewpoint television [4]. Approaches for obtaining 3D depth estimation can be distinguished into two categories: passive and active. The goal of passive methods like stereo matching is to estimate a high-resolution dense disparity map by finding corresponding pixels in image sequences [5]. However, these methods heavily rely on how the scene is presented and contain error matchings caused by the luminance variation. Passive methods fail in textureless and repetitive regions where there is not enough visual information to obtain the correspondence. On the contrary, active methods, like depth sensors (ASUS Xtion [6] and Microsoft Kinect [7]), do not suffer from ambiguities in textureless and repetitive regions, because they emit an infrared signal. Unfortunately, sensor errors and the properties of the object surfaces mean that depth maps from a depth sensor are often noisy [8]. Additionally, their resolution is at least an order of magnitude lower than common digital single-lens reflex (DSLR) cameras, which limits many applications. Moreover, they cannot satisfactorily deal with object boundaries and a wide range of distances. Therefore, fusing different kinds of methods using their complementary characteristics undoubtedly makes the obtained depth map more robust and improves the quality. Commonly-used consumer DSLR cameras have higher resolution and can record better texture information than depth sensors. Therefore, it is reasonable to fuse the depth sensor with DSLR cameras to yield a high resolution depth map. Note that for notational clarity, all values mentioned here are disparities (considering that depth values are inversely proportional to disparities).

Figure 1. The system used in our method. It consists of two Cannon EOS 700D digital single-lens reflex (DSLR) cameras and one Xtion depth sensor. All DSLR cameras are controlled by the wireless remote controller. We used an adjustable bracket to change the angle and height of the Xtion depth sensor.

In this paper, we propose a novel disparity estimation method for the system shown in Figure 1. It fuses the complementary characteristics of high resolution DSLR cameras and the Xtion depth sensor to obtain an accurate disparity estimate. Compared to the average error rate 3.27% of the previous state-of-the-art methods, our method provides an average error rate of 2.61% on the Middlebury datasets. It is clear that our method performs almost 20% better than other "fused" algorithms in the aspect of precision. The proposed method views a scene with complex geometric characteristics as a set of segments in the disparity space. It assumes that the disparities of each segment have a compact distribution, which strengthens the smooth variance of the disparities in each segment. Additionally, we assume that each segment is biased towards being a 3D planar surface. The major contributions are as follows:

1. We incorporated texture information as a constraint. The texture variance and gradient is used to restrict the range of the potential disparities for each pixel. In textureless and repetitive regions (which often cause ambiguities when stereo matching), we restrict the possible disparities for a neighborhood centered on each pixel to a limited range around the values suggested by the Xtion. This reduces the errors and strengthens the compact distribution of the disparities in a segment.
2. We propose the multiscale pseudo-two-layer image model (MPTL; Figure 2) to represent the relationships between disparities at different pixels and segments. We consider the disparities from the Xtion as the prior knowledge and use it to increase the robustness to luminance variance and to strengthen the 3D planar surface bias. Furthermore, considering the spatial structures of segments obtained from the depth sensor, we treat the segmentation as a soft constraint to reduce matching ambiguities caused by under- and over-segmentation. Here, pixels with similar colors, but on different objects are grouped into one segment, and pixels with different colors, but on the same object are partitioned into different segments. Additionally, we only retain the disparity discontinuities that align with object boundaries from geometrically-smooth, but strong color gradient regions.

Sensors **2015**, *15*, 20894–20924

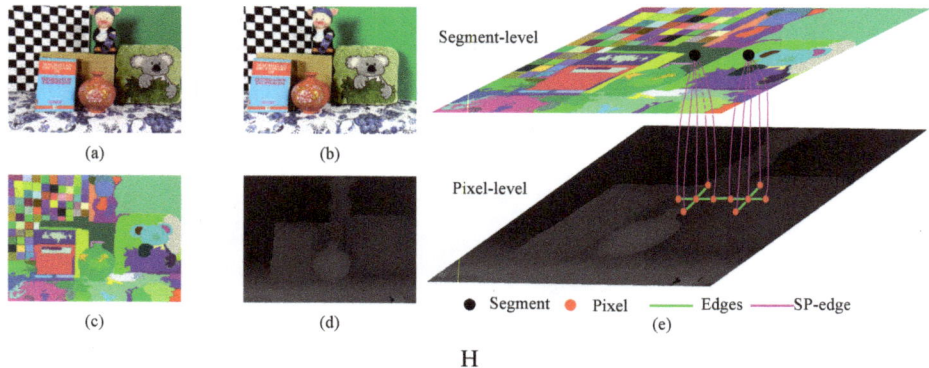

Figure 2. The illustration of the multiscale pseudo-two-layer (MPTL) model. The rectified left (**a**) and right (**b**) images from DLSRs, the segmentation; (**c**) as well as the depth map (**d**) from the depth sensor are taken as our inputs; (**e**) The conceptual structure of our MPTL. The MPTL captures the complementary characteristics of active and passive methods by allowing interactions between them. All interactions are defined to act in the segment-level component, the pixel-level component and the edges that connect them (Segment Pixel-edge, SP-edge). See Section 3.2 for the full details of the MPTL model and Section 4 for further results.

The remainder of this paper is organized as follows. Section 2 gives a summary of various methods used for disparity estimation. We present a pre-processing and some important notations of our model in Section 3.1. We discuss the details of the MPTL image model in Section 3.2, the optimization in Section 3.7 and the post-processing in Section 3.8. Section 4 contains our experiments, and Section 5 presents some conclusions with suggestions for future work.

2. Previous Work

There are many approaches to obtaining disparity estimation. They can generally be categorized into two major classes: passive and active. A passive method indirectly obtains the disparity map using image sequences captured by cameras from different viewpoints. Among the plethora of passive methods, stereo matching is probably the most well known and widely applied. Stereo matching algorithms can be divided into two categories [9]: local and global methods. Local methods [10] estimate disparity using color or intensity values in a support window centered on each pixel. However, they often fail around disparity discontinuities and low-texture regions. Global methods [11] use a Markov random field model to formulate the stereo matching as a maximum *a posteriori* probability energy function with explicit smoothness priors. They can significantly minimize matching ambiguities compared to local methods. However, the biggest disadvantage of them is the low computational efficiency. Segmentation-based global approaches [12,13] encode the scene as a set of non-overlapping homogeneous color segments. They are based on the hypothesis that the variance of the disparity in each segment is smooth. In other words, the segment boundaries are forced to coincide with object boundaries. Recently, the ground control point (GCP)-based methods [14] were used as prior knowledge to encode rich information on the spatial structure of the scene. Although a significant number of stereo matching methods have been proposed for obtaining dense disparity estimation, they heavily rely on radiometric variations and assumptions regarding the presentation of the scene. This means that stereo matching often fails in textureless and repetitive regions, where there is not enough visual information to obtain a correspondence. Furthermore, their accuracy is relatively low. Passive methods heavily rely on the luminance condition and how the scene is presented. They often fail in textureless and repetitive regions where there is not enough visual information to obtain the correspondence.

On the contrary, active methods like depth sensors, do not suffer from ambiguities in textureless and repetitive regions, because they emit an infrared signal. Three different kinds of equipments are used in active methods: a laser scanner device, a time-of-flight (ToF) sensor and an infrared single-based device (such as ASUS Xtion [6] and Microsoft Kinect [7]). The laser scanner device [15] can provide extremely accurate and dense depth estimation, but it is too slow to use in real time and too expensive for many applications. The ToF sensor and infrared single-based device can obtain real-time depth estimation and have recently become available from companies, such as 3DV [16] and PMD [17]. However, sensor errors and the properties of the object surfaces mean that depth maps from them are often noisy [8]. Additionally, their resolution is at least an order of magnitude lower than commonly-used DSLR cameras [18], which limits many applications. Moreover, they cannot satisfactorily deal with object boundaries.

It is clear that each disparity acquisition method is limited in some aspects where other approaches may be effective. Joint optimization methods that combine active and passive sensors have been used to make the obtained depth map more robust and to improve the quality. Zhu *et al.* please check throughout [19,20] fused a ToF sensor and stereo cameras to obtain better disparity maps. They improved the quality of the estimated maps for dynamic scenes by extending their fusion technology to the temporal domain. Yang *et al.* [21] presented a fast depth sensing system that combined the complementary properties of passive and active sensors in a synergistic framework. It relied on stereo matching in rich textured regions, while using data from depth sensors in textureless regions. Zhang *et al.* [22] proposed a system that addresses high resolution and high quality depth estimation by fusing stereo matching and a Kinect. A pixel-wise weighted function was used to reflect the reliabilities of the stereo camera and the Kinect. Wang *et al.* [23] presented a novel method that combined the initial stereo matching result and the depth data from a Kinect. Their method also considers the visibilities and pixel-wise noise of the depth data from a Kinect. Gowri *et al.* [24] proposed a global optimization scheme that defines the data and smoothness costs using sensor confidences and the low resolution geometry from a Kinect. They used a spatial search range to limit the scope of the potential disparities at each pixel. The smoothness prior was based on the available low resolution depth data from the Kinect, rather than the image color gradients.

Although existing disparity estimation methods have achieved remarkable results, they are typically performed using pixel-level cues, such as the smoothness of neighboring pixels, and do not consider the regional information (regarding, for example, 3D spatial structure, segmentation and texture) as a cue for the disparity estimation, which is the largest distinction between their method and ours. For example, occlusion cannot be precisely estimated using a single pixel, but a fitted plane-based filling occlusion in a segment can give good results. Additionally, if the spatial structure of neighboring segments is not known, matching ambiguities can arise at the boundaries of neighboring segments that physically belong to the same object, but have different appearances. Without texture information, we cannot be sure if the disparity from stereo matching is more confident than that from the depth sensor in textureless and repetitive regions (where stereo matching usually fails and the depth sensor performs well).

3. Method

The proposed method can be partitioned into four phases: pre-processing, problem definition, optimization and post-processing. Each phase will be discussed in detail later.

3.1. Pre-Processing

There are three camera coordinates involved in our system (Figure 1): the Xtion coordinate, the coordinates of the two DSLR cameras before the epipolar rectification and the DSLR camera coordinates after the epipolar rectification. During the pre-processing step, in order to combine the data from the Xtion and DSLR cameras, as shown in Figure 3, we firstly calibrated two DSLR cameras using the checkerboard-based method [25] and calibrated the DSLR camera pair with the Xtion sensor using the planar surfaces-based method [26], respectively. After the calibration, the depth image obtained from the Xtion is first transformed from the Xtion coordinate to the original DSLR cameras' coordinates, then rotated and up-sampled, so that it registers with the unrectified left image. Furthermore, according to the theory of epipolar geometric constraints, the registered depth image and original left image, as well as the original right image are rectified to be row-aligned, which means there are only horizontal disparities in the row direction. We denote the seed image (Π) as the map with disparities transferred from the rectified depth map. Each pixel $p \in \Pi$ is defined as a seed pixel when it is assigned a non-zero disparity. The initial disparity maps (D_L and D_R) of the rectified left and right images (I_L and I_R) are computed using a local stereo matching method [27]. I_L is partitioned into a set of segments using the edge-aware filter-based segmentation algorithm [28].

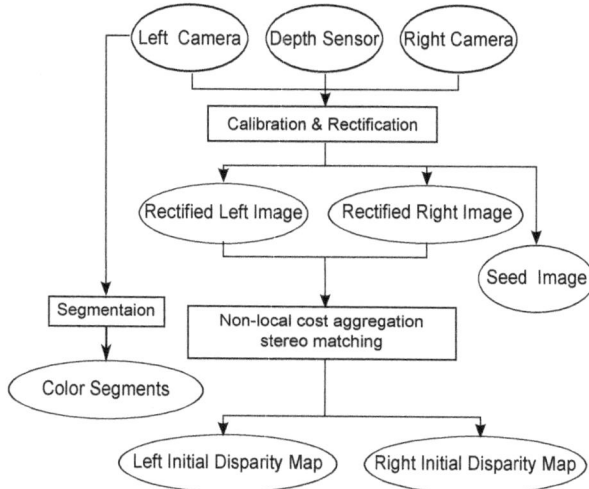

Figure 3. Conceptual flow diagram for the calibration and rectification phase.

In addition, as shown in Figure 4, all pixels and segments are divided into different categories. The occlusion judgment is used to find the occluded pixels with initial disparity maps (D_L and D_R of I_L and I_R, respectively) and to classify pixels into different categories: reliable and occluded. As we known, how to find occluded pixels accurately is always the challenging problem, because it often leads to error results that matching points might not even exist at all, especially in depth discontinuities. Pixels are defined as occluded when they are only visible from the left rectified view (I_L), but not from the right rectified view (I_R). Since image pairs have been rectified, we assume that occlusion only occurs in the horizontal direction. In early algorithms, cross-consistency checking is often applied to identify occluded pixels by enforcing a one-to-one correspondence between pixels. It is written as:

$$O(p) = \begin{cases} 0 & |D_L(p) - D_R(q)| < 1 \\ 1 & \text{otherwise} \end{cases} \quad p \in I_L, q \in I_R \tag{1}$$

$D_L(p)$ and $D_R(q)$ are the disparity of p and q, and q is the corresponding matching point of p. If p does not meet the cross-consistency checking, then it will be regard as an occluded pixel ($O(p) = 0$); otherwise, p is a reliable pixel ($O(p) = 1$). The cross-consistency checking states that a pixel of one image corresponds to at most one pixel of the other image. However, because of different sampling, the projection of a horizontal slant or a curved surface shows various lengths in the image pairs. Therefore, conventional cross-consistency checking that often identifies occluded pixels by enforcing a one-to-one correspondence is only suitable for a frontal parallel surface and cannot be true for a horizontal slant or curved surfaces. Considering the different sampling of image pairs, Bleyer *et al.* [29] proposed a new visibility constraint by extending the asymmetric occlusion model [30] that allows a one-to-many correspondence between pixels. Let p_0 and p_1 be neighboring pixels in the same horizontal line of I_L. Then, p_0 will be occluded by p_1 when they meet three conditions:

- p_0 and p_1 have the same matching point in I_R under their current disparity value;
- $D_L(p_0) \leq D_L(p_1)$;
- p_0 and p_1 belong to different segments.

In this paper, for each pixel p of I_L, if there is only one matching point in I_R, the conventional cross-checking is applied to obtain the occlusion Equation (1). Otherwise, if there are more than two matching points in I_R, pixels in I_L are marked as either reliable ($O(p_0) = 0$) or occluded ($O(p_0) = 1$), which satisfy or do not satisfy the Bleyer's asymmetric occlusion model. As shown in Figure 4, each segment belongs to the reliable segment (R) if it contains a sufficient amount of reliable pixels; otherwise, it belongs to the unreliable segment ($\overline{R}$). Furthermore, each segment $f_i \in R$ is denoted as a stable segment (S) when it contains a sufficient number of seed pixels. Otherwise, f_i belongs to the unstable segment ($\overline{S}$). We apply a RANSAC-based algorithm to approximate each stable segment $s_i \in S$ as a fitted plane Ψ_{s_i} using the image coordinates and known disparities of all seed pixels belonging to s_i. Table 1 lists important notifications used in this paper.

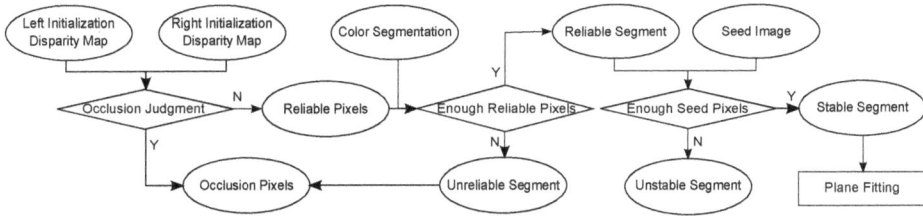

Figure 4. Conceptual flow diagram for the classification phase.

Table 1. Notations.

D:	Disparity map	$D(p)$:	Disparity value of pixel p	Π:	Seed image
D_L:	Initial disparity map of rectified left image	D_R:	Initial disparity map of rectified right image	I_L:	Rectified left DSLR image
I_R:	Rectified right DSLR image	R:	Reliable segment	$\overline{R}$:	Unreliable segment
S:	Stable segment	$\overline{S}$:	Unstable segment	f_i:	i-th segment
s_i:	i-th stable segment	Ψ_{s_i}:	Fitted plane of stable segment s_i	$f(p)$:	Segment that contains pixel p
f_i:	i-th segment	d_l:	Minimum disparity	d_u:	maximum disparity
$\wp$:	Segment boundary pixels	Λ_l^p:	Pixel's potential minimum disparity	Λ_u^p:	Pixel's potential maximum disparity
Ψ_i^l:	Minimum fitted disparity of the i-th stable segment	Ψ_i^u:	Minimum fitted disparity of the i-th stable segment		

3.2. Problem Formulation

In the problem formulation phase, we propose the MPTL model, which combines the complementary characteristics of stereo matching and the Xtion sensor. As shown in Figure 2, the MPTL model consists of three components:

- The pixel-level component, which improves the robustness against the luminance variance (Section 3.3) and strengthens the smoothness of disparities between neighboring pixels and segments (Section 3.4). Nodes at this level represent reliable pixels from stable and unstable segments. The edges between reliable pixels represent different types of smoothness terms.

- The edge that connects two level components (the SP-edge), which uses the texture variance and gradient as a guide to restrict the scope of potential disparities (Section 3.5).
- The segment-level component, which incorporates the information from the Xtion as prior knowledge to capture the spatial structure of each stable segment and to maintain the relationship between neighboring stable segments (Section 3.6). Each node at this level represents a stable segment.

Existing global methods have achieved remarkable results, but the capability of the traditional Markov random field stereo model remains limited. To lessen the matching ambiguities, additional information is required to formulate an accurate model. In this paper, the pixel-level improved luminance consistency term (E_l), the pixel-level hybrid smoothness term (E_s) and the SP-edge texture term (E_t), as well as the segment-level 3D plane bias term (E_p) are integrated as additional regularization constraints to obtain a precise disparity estimation (D) for a scene with complex geometric characteristics. According to Bayes' rule, the posterior probability over D given l, s, t and p is:

$$p(D|l,s,t,p) = \frac{p(l,s,t,p|D)p(D)}{p(l,s,t,p)} \tag{2}$$

During each optimization process, $P(l,s,t,p|D)$ is only dependent on l, s, t and p. Therefore, $P(D|l,s,t,p)$ can be rewritten as:

$$p(D|l,s,t,p) \propto p(l|D)p(s|D)p(t|D)p(p|D)p(D) \tag{3}$$

Because maximizing this posterior is equivalent to minimizing its negative log likelihood, our goal is to obtain the disparity map (D) that minimizes the following energy function:

$$E(D) = E_l(D) + E_s(D) + E_t(D) + E_p(D) \tag{4}$$

Each term will be discussed in detail in the following sections.

3.3. Improved Luminance Consistency Term

The conventional luminance consistency hypothesis is used to penalize the appearance dissimilarity between corresponding pixels in I_L and I_R, based on the hypothesis that the surface of a 3D object is Lambertian. Because it refers to a perfectly-diffuse appearance in which pixels originating from same 3D object have similar appearances in different views, its accuracy is heavily dependent on the lighting condition for which colors change substantially depending on the viewpoint. Furthermore, an object may appear to have different colors because different views have different sensor characteristics. In contrast, the Xtion sensor is more robust to the light condition and can be used as prior knowledge to reduce ambiguities caused by the non-Lambertian surface. Thus, the improved luminance consistency term is denoted as:

$$E_l = \sum_{p \in I_L} \lambda_l \cdot (1 - O(p)) \cdot \left[w_p^l \cdot C(p,q) + w_p^x \cdot X(p,q) \right] + O(p) \cdot \lambda_o \tag{5}$$

where q is the matching pixel of p in the other image. $O(p)$ is the asymmetric occlusion function described in Section 3.1, and λ_o is a positive penalty used to avoid maximizing the number of occluded pixels. $C(p,q)$ is defined as the pixel-wise cost function from stereo matching to measure the color dissimilarity.

$$C(p,q) = \alpha \cdot (1 - \exp\left(-\frac{C_{ssd}(p,q)}{r_{ssd}}\right)) + (1 - \alpha) \cdot (1 - \exp\left(-\frac{C_g(p,q)}{r_g}\right)) \tag{6}$$

where r_{ssd} and r_g are constant values defined by our experience. α is the scalar weight from zero to one. $C_{ssd}(p,q)$ and $C_g(p,q)$ are the color dissimilarity and gradient in three color channels as:

$$C_{ssd}(p,q) = \sqrt{\sum_{i=R,G,B}\left(I_L^i(p) - I_R^i(q)\right)^2} \tag{7}$$

$$C_g(p,q) = \sqrt{\sum_{i=R,G,B}\left(\nabla I_L^i(p) - \nabla I_R^i(q)\right)^2} \tag{8}$$

$X(p,q)$ is the components from the Xtion sensor, which are defined as:

$$X(p,q) = \min\{|D(p) - \Pi(p)|, T_\pi\} \tag{9}$$

T_π is the constant threshold, and $D(p)$ is the disparity value assigned to pixel p in each optimization. $\Pi(p)$ is the disparity of pixel $p \in \Pi$. w_p^x and w_p^l are pixel-wise confidence weights that are denoted as $w_p^x = 1 - w_p^l$. They are derived from the reliabilities of disparities obtained from stereo matching (m_p^l) and the Xtion (m_p^x) as:

$$w_p^l = \begin{cases} \frac{m_p^l}{m_p^l + m_p^x} & p \in \Pi \\ 1 & otherwise \end{cases} \tag{10}$$

where m_p^l is similar to the attainable maximum likelihood (AML) in [31], which models the cost for each pixel using a Gaussian distribution centered at the minimum actually achieved cost value for that pixel. The reliability of Xtion data m_p^x is the inverse of the normalized standard deviation of the random error [20]. The confidence of each depth value obtained from the depth sensor decreases with the increasing of the normalized standard deviation.

3.4. Hybrid Smoothness Term

The hybrid smoothness term strengthens the segmentation-based assumption that the disparity variance in each segment is smooth and reduces errors caused by under- and over-segmentation. It consists of four terms: the smoothness term for neighboring reliable pixels belonging to the unstable segment (E_{s_0}), the smoothness term for neighboring reliable pixels in the same stable segment (E_{s_1}), the smoothness term for neighboring reliable pixels in different stable segments (E_{s_2}) and the smoothness term for neighboring reliable pixels that belong to stable and unstable segments (E_{s_3}).

Because there is no prior knowledge about the spatial structure of unstable segments, we define the smoothness term E_{s_0} as the conventional second-order smoothness prior Equation (11), which can produce better estimates for a scene with complex geometric characteristics [11].

$$E_{s_0} = \sum_{\Phi_i^0 \in \Phi^0} \lambda_s^0 \cdot \left[1 - exp\left(-\frac{\nabla D(\Phi_i^0)}{\gamma_s}\right)\right] \{p_0, p_1, p_2\} \in \Phi_i^0 \tag{11}$$

where γ_s and λ_s^0 are the geometric proximity and the positive penalty. Φ^0 is the set of triple-cliques consisting of consecutive reliable pixels belonging to unstable segment. $\nabla D(\Phi_i^0)$ is the second derivative of the disparity map as:

$$\nabla D\left(\Phi_i^0\right) = D(p_0) - 2D(p_1) + D(p_2).\{p_0, p_1, p_2\} \in \Phi_i^0 \tag{12}$$

E_{s_0} captures richer features of the local structure and permits planar surfaces without penalty by setting $\nabla D(\Phi_i^0) = 0$. However, E_{s_0} only considers disparity information when representing the smoothness of neighboring pixels. This means that, in several cases, error matching can result in different disparity assignments, which correspond to the same second derivatives in the disparity map (see Figure 5b–d). Meanwhile, each stable segment can be represented as a fitted plane using the disparity data from the Xtion, which contains prior knowledge about the spatial structure of each stable segment. We can incorporate the spatial similarity weight with the prior knowledge from the Xtion into a conventional second-order smoothness prior. This term encourages constant disparity gradients for pixels in a stable segment and local spatial structures that are similar to the fitted plane of the stable segment. The smoothness term for neighboring reliable pixels in the same stable segment is as follows.

$$E_{s_1} = \sum_{\Phi_i^1 \in \Phi^1} \lambda_s^1 \cdot [2 - \delta\left(\Phi_i^1\right) - \exp\left(-\frac{\nabla D\left(\Phi_i^1\right)}{\gamma_s}\right)] \tag{13}$$

where Φ^1 is the set of triple-cliques defined by all 3×1 and 1×3 consecutive reliable pixels along the coordinate direction of the rectified image coordinate in each stable segment. λ_s^1 is a positive value penalty. As for the spatial 3D relationship shown in Figure 5, let s_i be the stable segment containing Φ_i^1 and Ψ_{s_i} be its corresponding fitted plane. Then, the spatial similarity weight $\delta\left(\Phi_i^1\right)$ is denoted as:

$$\delta\left(\Phi_i^1\right) = \begin{cases} 0 & \text{I}: \overline{p_0 p_1} \neq \overline{p_1 p_2} \\ 0.25 & \text{II}: \overline{p_0 p_1} = \overline{p_1 p_2} \text{ and } \overline{p_0 p_2} \cap \Psi_{s_i} \\ 0.5 & \text{III}: \overline{p_0 p_1} = \overline{p_1 p_2} \text{ and } \overline{p_0 p_2} // \Psi_{s_i} \\ 1 & \text{IV}: \overline{p_0 p_1} = \overline{p_1 p_2} \text{ and } \overline{p_0 p_2} \in \Psi_{s_i} \end{cases} \quad \{p_0, p_1, p_2\} \in \Phi_i^1 \tag{14}$$

- Case I: When $\overline{p_0 p_1} \neq \overline{p_1 p_2}$, the disparity gradients of pixels in Φ_i^1 are not constant ($\nabla D\left(\Phi_i^1\right) \neq 0$). This case violates the basic segmentation assumptions that the disparity variance of neighboring pixels is smooth, so a large penalty is added to prevent it from happening in our model (see Figure 5a).

- Cases II, III and IV: When $\overline{p_0 p_1} = \overline{p_1 p_2}$, the disparity gradients of pixels in Φ_i^1 are constant ($\nabla D\left(\Phi_i^1\right) = 0$). This means that the variance of the disparities is smooth. Furthermore, our model checks the relationship between all pixels in Φ_i^1 and Ψ_{s_i} (see Figure 5b–d). $\delta\left(\Phi_i^1\right)$ does not penalize the disparity assignment if all pixels in Φ_i^1 belong to Ψ_{s_i} (Case IV in Figure 5d), because it is reasonable to assume that the local structure of Φ_i^1 is the same as the spatial structure of Ψ_{s_i}. Note that we impose a larger penalty to Case II than to Case III to strengthen the similarity between the spatial structure of Φ_i^1 and Ψ_{s_i}.

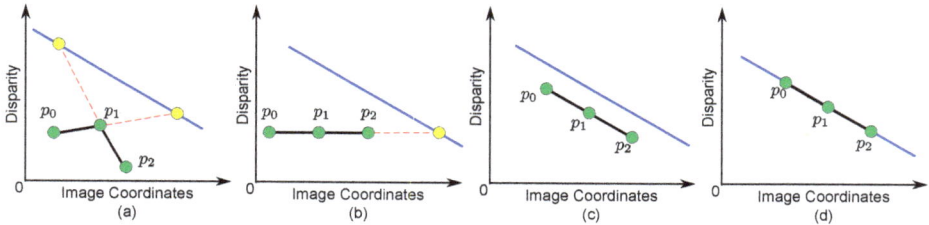

Figure 5. Smoothness term for pixels in the same stable segment (s_i) with the spatial similarity weight as the triple-clique $\Phi_i^1 = \{p_0, p_1, p_2\}$, under different disparity assignments. Given Ψ_{s_i} (blue line) as the fitted plane of s_i. Yellow nodes are the intersect points. (**a**) Case I, $D(p0) \neq D(p1) \neq D(p2)$ with $\nabla D\left(\Phi_i^1\right) \neq 0$; (**b**) Case II, $D(p0) = D(p1) = D(p2)$ with $\nabla D\left(\Phi_i^1\right) = 0$; (**c**) Case III, $D(p0) \neq D(p1) \neq D(p2)$ with $\nabla D\left(\Phi_i^1\right) = 0$; (**d**) Case IV, $D(p0) \neq D(p1) \neq D(p2)$ with $\nabla D\left(\Phi_i^1\right) = 0$. As shown in Cases II, III and IV, different disparity assignments correspond to the same second derivative ($\nabla D\left(\Phi_i^1\right) = 0$).

In some segmentation-based algorithms [32], the segmentation is implemented as a hard constraint by setting λ_s^0 and λ_s^1 to be positive infinity. This does not allow any large disparity variance within a segment. In other words, each segment can only be represented as a single plane model, and the boundaries of a 3D object must be exactly aligned with segment boundaries. Unfortunately, not all segments can be accurately represented as a fitted plane, and not all 3D object boundaries coincide with segment boundaries. The accuracy of the segmentation-based algorithms is easily affected by the initial segmentation. On the one hand, the initial segmentation typically contains some under-segmented regions (where pixels from different objects, but with similar colors are grouped into one segment). As a direct consequence of under-segmentation, foreground and background boundaries are blended if they have similar colors at disparity discontinuities. To avoid this, we use segmentation as a soft constraint by setting λ_s^0 and λ_s^1 to be positive finite, so that each segment can contain arbitrary fitted planes.

On the other hand, pixels with different colors, but on the same object are over-segmented into different segments in the initial segmentation, which causes computationally inefficiency and ambiguities on segment boundaries. In this paper, we considered the spatial structure of neighboring stable segments using disparities from the Xtion. Therefore, we apply the smoothness term for neighboring pixels belonging to different stable segments (E_{s_2}) to avoid errors caused by the over-segmentation. Let p and q be neighboring pixels belonging to stable segments s_i and s_j, respectively, Then, E_{s_2} can be expressed as:

$$E_{s_2} = \begin{cases} 0 & \text{I}: \Psi_{s_i} \neq \Psi_{s_j} \text{and} D(p) \neq D(q) \\ \lambda_s^2 & \text{II}:: \Psi_{s_i} \neq \Psi_{s_j} \text{and} D(p) = D(q) \\ \lambda_s^2 & \text{III}:: \Psi_{s_i} = \Psi_{s_j} \text{and} D(p) \neq D(q) \\ 0 & \text{IV}: \Psi_{s_i} = \Psi_{s_j} \text{and} D(p) = D(q) \end{cases} \tag{15}$$

As shown in Equation (15), for Cases I and II, if Ψ_{s_i} is not equal to Ψ_{s_j}, this means that s_i and s_j have different spatial structures, and the 3D object boundary coincides with the boundary between them. The disparity variance between p and q is allowed without any penalty (Case I); otherwise, a constant penalty λ_s^2 is added (Case II). In contrast, for Cases III and IV, if Ψ_{s_i} is equal to Ψ_{s_j}, this means that s_i and s_j have different appearances, but have similar spatial structures and belong to the same 3D object. In these two cases, the disparity variance between p and q is not allowed by adding a penalty. E_{s_2} reduces the ambiguities caused by over-segmentation and retains only the disparity discontinuities that are aligned with object boundaries from geometrically-smooth, but strong color gradient regions, where pixels with different colors, but from the same object are partitioned into different segments.

Because unstable segments do not have sufficient disparity information from the Xtion to regard their spatial plane models, the smoothness term for neighboring pixels that belong to the stable and unstable segments (E_{s_3}) encourages neighboring pixels to take the same disparity assignment. It takes the form of a standard Potts model,

$$E_{s_3} = \begin{cases} 0 & D(p) = D(q) \\ \lambda_s^3 & D(p) \neq D(q) \end{cases} \quad p \in S, q \in \overline{S} \tag{16}$$

Thus, let ω be the set of pixels belonging to segment boundaries, the hybrid smoothness term is:

$$E_s = \begin{cases} E_{s_0} & \{p_0, p_1, p_2\} \in \Phi_i \text{ and } \Phi_i \in \overline{S} \text{ and } \Phi_i \cap \omega = \varnothing \\ E_{s_1} & \{p_0, p_1, p_2\} \in \Phi_i \text{ and } \Phi_i \in S \text{ and } \Phi_i \cap \omega = \varnothing \\ E_{s_2} & \{p_0, p_1, p_2\} \in \Phi_i \text{ and } \Phi_i \cap \omega \neq \varnothing \text{ and } \{p_0, p_1, p_2\} \in S \\ E_{s_3} & \{p_0, p_1, p_2\} \in \Phi_i \text{ and } \Phi_i \cap \omega \neq \varnothing \text{ and } \{p_0, p_1\} \in S, \{p_2\} \in \overline{S} \end{cases} \tag{17}$$

3.5. Texture Term

Stereo matching often fails in textureless and repetitive regions, because there is not enough visual information to obtain a correspondence. However, the Xtion does not suffer from ambiguities in these regions. Therefore, the disparities from the Xtion are more reliable than those obtained from stereo matching on textureless and repetitive regions and should be closer to the range of potential disparities for pixels in these regions. In contrast, the disparities from the Xtion are susceptible to noise and problems caused by rich texture regions and have poor performance in preserving object boundaries. Therefore, the disparities obtained from stereo matching are more reliable than that of the Xtion and should be used to define the scope of potential disparities of pixels in those regions. Considering the complementary characteristics of stereo matching and the Xtion sensor, texture information can be used as a useful guide for disparities.

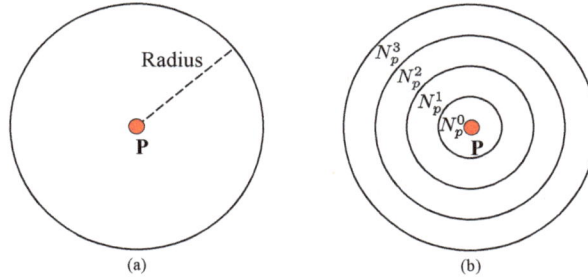

Figure 6. Surrounding neighborhood patch, N_p, for: (**a**) pixel p and (**b**) its corresponding sub-regions.

The texture variance and gradient are used as a cue to restrict the scope of potential disparities for pixels. This reduces errors caused by noise or outliers and makes the distribution of the disparity more compact. To do this, we first define a surrounding neighborhood patch N_p (with a radius of, for example, 20 pixels) centered at each pixel $p \in I_L$, as shown in Figure 6. Considering that the annular spatial histogram is translation and rotation invariant [33], N_p is evenly partitioned into four annular sub-regions. For each sub-region $N_p^i (i = 0 \cdots 3)$, we compute its normalized intensity 16-bin gray histogram $H_p^i = \left\{ h_p^{(i,j)}, j = 0 \cdots 15 \right\}$ to represent the annular distribution density of N_p as a 64-dimensional feature vector.

Finally, let L_p be a 1D line segment ranging from $(p - 10)$ to $(p + 10)$ in the same row of p in I_L. The texture variance and gradient of p is determined by the texture dissimilarity Γ_p Equation (18), using the Hamming distance Equation (19) between the annular distribution densities of p and its neighboring pixel q in L_p. That is,

$$\Gamma_p = min_{(q \in L_p, q \neq p)} \sum_{i=0}^{3} \sum_{j=0}^{15} H\left(h_p^{(i,j)}, h_q^{(i,j)} \right) \tag{18}$$

$$H\left(h_p^{(i,j)}, h_q^{(i,j)} \right) = \begin{cases} 1 & |h_p^{(i,j)} - h_q^{(i,j)}| \geq T_H \\ 0 & \text{otherwise} \end{cases} \tag{19}$$

Each pixel's disparity variance buffer (Ω_p) can be denoted as:

$$\Omega_p = 1 + \left[1 - exp\left(-\frac{\Gamma_p}{\gamma_H} \right) \right] \cdot \xi \xi = 0.2 \cdot (d_u - d_l) \tag{20}$$

d_l and d_u are the minimum and maximum disparities. Γ_p is small in the textureless and repetitive regions and is large in the rich texture regions or object boundaries. The scope of each pixel's potential disparities $[\Lambda_l^p, \Lambda_u^p]$ is denoted as:

$$Y_l = max \left\{ \left(\Psi_{f(p)}^l - \Omega_p \right), d_l \right\} \tag{21}$$

$$Y_u = \min\left\{\left(\Psi^l_{f(p)} + \Omega_p\right), d_u\right\} \tag{22}$$

$$\Lambda^p_l = \begin{cases} \max\left\{\left(\Theta^p_l - \Omega_p\right), Y_l\right\} & \chi_p \geq T_\Lambda \text{and} f(p) \in S \\ Y_l & \chi_p < T_\Lambda \text{and} f(p) \in S \\ d_l & f(p) \in \overline{S} \end{cases} \tag{23}$$

$$\Lambda^p_u = \begin{cases} \min\left\{\left(\Theta^p_u + \Omega_p\right), Y_u\right\} & \chi_p \geq T_\Lambda \text{and} f(p) \in S \\ Y_u & \chi_p < T_\Lambda \text{and} f(p) \in S \\ d_u & f(p) \in \overline{S} \end{cases} \tag{24}$$

where Θ^p_l and Θ^p_u are the minimum and maximum disparities from the Xtion in the region centered at p in I_L. $f(p)$ is the segment that contains p. $\Psi^l_{f(p)}$ and $\Psi^u_{f(p)}$ are the minimum and maximum fitted disparities of $f(p)$. χ_p is the number of seed pixels in the region centered at p. T_Λ is a positive value. As described in Equation (23), there are three cases for the definition of Λ^p_l:

-   When $f(p)$ is a stable segment ($f(p) \in S$) and contains sufficient seed pixels ($\chi_p > T_\Lambda$), Λ^p_l is equal to $\max\left\{\left(\Theta^p_l - \Omega_p\right), Y_l\right\}$. In this case, there are enough seed pixels from the Xtion to denote a guide for the variance of disparities of p. If p is in the textureless or repetitive region, Ω_p is small. This indicates that stereo matching may fail in these regions, and a small search range should be used around disparities from the Xtion. In contrast, if p is in the rich textured region or object boundaries, Ω_p is large. This indicates that disparities from the Xtion may be susceptible to noise and problems caused by rich texture regions where disparities obtained from stereo matching are more reliable. Then, a broader search range should be used, so that we can extract better results not observed by the Xtion.
-   When $f(p)$ is a stable segment ($f(p) \in S$), but there are not enough seed pixels around p ($\chi_p \leq T_\Lambda$), Λ^p_l is equal to Y_l. In this case, although there are some seed pixels from the Xtion, they are not enough to represent the disparity variance around p. On the other hand, because each stable segment is viewed as a 3D fitted plane, the search range for the potential disparities is limited by the fitted disparity of $f(p)$ and the disparity variance buffer (Ω_p).
-   When $f(p)$ is an unstable segment ($f(p) \in \overline{S}$), Λ^p_l is the minimum disparity (d_l).

Similarly, Λ^p_u can be obtained in the same way. Then, the SP-edge term (which defines the scope of pixel's potential disparities) is:

$$E_t = \begin{cases} 0 & \Lambda^p_l \leq D(p) \leq \Lambda^p_u \\ \lambda_t & \text{otherwise} \end{cases} \tag{25}$$

3.6. 3D Plane Bias Term

This 3D plane bias term focuses on strengthening the assumption that each stable segment has a 3D plane bias. It is denoted as:

$$E_p = \sum_{s_i \in S} \sum_{p \in s_i} \lambda_p \cdot \min\left\{|D(p) - \Psi_{s_i}(p)|, T_p\right\} \tag{26}$$

where $D(p)$ is the assigned value of pixel p in I_L. $\Psi_{s_i}(p)$ is the plane fitted value, and T_p is a threshold value. Note that for notation clarity, the traditional 3D bias assumption is a hard constraint that forbids any distinctive between $D(p)$ and $\Psi_{s_i}(p)$ by setting λ_p to be infinite. On the contrary, our 3D plane bias term is a soft constraint that a certain distinctive between $D(p)$ and $\Psi_{s_i}(p)$ is allowed by setting λ_p to be a finite positive value.

3.7. Optimization

The energy function defined in Equation (4) is a function of the real discrete disparity map. In this section, we describe how to optimize Equation (4) using the fusion move algorithm to obtain the disparity map D^*:

$$D^* = argmin_D E(D) \tag{27}$$

The fusion move approach [34] is an extended approach of the $\alpha - expansion$ algorithm [35], which allows arbitrary values for each pixel in the proposed disparity map. It generates a new result by fusing the current and proposed disparity maps with the energy either decreasing or remaining constant. Let D^c and D^p be the current and proposed disparity maps of I_L. Our goal is to optimally "fuse" D^c and D^p to generate a new depth map D^n, so that the energy $E(D^n)$ is lower than $E(D^c)$. This fusion move is achieved by taking each pixel in D^n from either D^c or D^p, according to a binary indicator map B. B is the result of the graph cut-based fusion move Markov random field optimization technique. During each optimization, each pixel either keeps its current disparity value ($B(p) = 0$) or changes it to proposed disparity value ($B(p) = 1$). That is,

$$D^n = (1 - B) \cdot D^c + B \cdot D^p \tag{28}$$

However, the fusion move is limited to optimizing the submodular binary fusion-energy functions that consist of unary and pairwise potentials. Because of the hybrid smoothness term, our binary fusion-energy functions are not submodular and cannot be directly solved using the fusion move [36]. Using the quadratic pseudo-Boolean optimization (QPBO) algorithm [37], we can obtain a partial solution for the non-submodular binary fusion-energy function by assigning either zero or one to partial pixels, and leaving the rest unassigned. The partial solution is a part of the global minimum solution, and its energy is not higher than that of the original solution. Because of the given lowest average number of unlabeled pixels, we used Quadratic Pseudo Boolean Optimization with Probing (QPBO-P) [38] and Quadratic Pseudo Boolean Optimization with Improving (QPBO-I) [39] as our fusion strategies. During the optimization, the pixel-level improved luminance consistency term (E_l), the SP-edge texture term (E_t) and the segment-level 3D plane bias term (E_p) are expressed as unary terms, respectively. We tackle the transformation problem of the pixel-level hybrid smoothness term (E_s) that contains triple-cliques using the decomposition method called Excludable Local Configuration (ELC) [40]. The essence of the ELC method is a QPBO-based transformation of a general higher-order Markov random field with binary labels into a first-order one that has the same minima as the original. It combines a new reduction with the fusion move and QPBO to approximately minimize higher-order multi-label energies. Furthermore, the new reduction technique is along the lines of the Kolmogorov-Zabih reduction that can reduce any higher-order minimization problem of Markov random fields with binary labels into an equivalent first-order problem. Each triple clique in E_s is decomposed into a set of unary or pairwise terms by ELC without introducing any new variables.

The choice of the proposed disparity maps in the fusion move approach is another crucial factor for the successful use and efficiency of the fusion move. Because there is not an algorithm that can be applied to all situations, our goal is to expect all proposed disparity maps to be correct in some parts and under some parameter setting. Here, we use the following schemes to obtain all proposed disparity maps:

- Proposal A: Uniform value-based proposal. All disparities in the proposal are assigned to a discrete disparity, in the range of d_l to d_u.
- Proposal B: The hierarchical belief propagation-based algorithm [41] is applied to generate proposals with different segmentation maps.
- Proposal C: The joint disparity map and color consistency estimation method [42], which combines mutual information, a SIFT descriptor and segment-based plane-fitting techniques.

During each optimization, the result of the current fusion move is used as the initial disparity map of the next iteration.

3.8. Post-Processing

The post-processing is composed of two steps: filling occlusions and refinement. Given that p is a occluded pixel in I_L, a two-step method is implemented to estimate disparities of occluded pixels. If $f(p) \in S$, the fitted plane value $\Psi_{f(p)}(p)$ is assigned as p's disparity. Otherwise, the disparity of p is the smaller disparity of its closet left and right seed pixels that belongs to the background.

After filling occlusions, in order to obtain an accurate disparity map and to remove ambiguities at object boundaries, the weighted joint bilateral filter with the slope depth compensation filter [43] is applied to refine the disparity map.

4. Results and Discussion

Here, a series of evaluations were performed to verify the effectiveness and accuracy of the proposed method. Results were composed of qualitative and quantitative analyses. The segmentation parameters for all experiment are the same: spatial bandwidth = 7, color bandwidth = 6.5, minimum region = 20. Other parameters are presented in Table 2. They were kept constant for all experiments and were typically empirically based.

Table 2. Parameter settings for all experiments.

T_H	T_Λ	T_π	T_p	γ_H	γ_s	λ_t	λ_l	λ_o	λ_s^0	λ_s^1	λ_s^2	λ_s^3	λ_p
0.1	50	3	5	35	1.5	200	12	200	40	40	15	15	10

4.1. Qualitative Evaluation Using the Real-World Datasets

We performed qualitative analyses of the proposed method using real-world datasets. In all evaluations, we captured the image pairs using the system in Figure 1 and regarded the left DSLR cameras as the target to be estimated by the disparity map. Notice that all scenes contain weakly-textured and repetitive regions, as well as a non-Lambertian surface.

In order to illustrate that our method combines the complementary characteristics of the various disparity estimation methods and outperforms using the conventional stereo matching or depth sensor alone, we evaluated the qualitative quality of the disparity estimates from three stereo matching methods, the Xtion depth sensor and the proposed method using several complex indoor scenes in Figure 7. As show in Figure 7c, although the local stereo matching with fast cost volume filtering (FCVF [44]) performed well by recovering the object boundaries using a color image as a guide, it was very fragile for noise and textureless regions (such as the uniformly-colored board in the yellow rectangle of Figure 7a). In contrast, the depth values obtained from the active sensor are more accurate (see the same regions in Figure 7f). Therefore, our method overcame these problems with the improved luminance consistency term and texture term by incorporating the prior depth information from the depth sensor. As shown in Figure 7d, segmentation-based global stereo matching with second order smoothness prior (SOSP [11]) overcame some of the problems caused by the noise and outliers, but did not solve the problems caused by the over-segmentation, which led to ambiguous matching when segment boundaries did not correspond to object boundaries (green rectangle in Figure 7a). Segment tree-based stereo matching (ST [13]) blended the foreground and background in the under-segmented region (as shown in Figure 7e), where different objects with similar appearances (such as the red rectangle in Figure 7b) were grouped into a segment. Comparing to our result and those of segmentation-based methods, it is clear that the proposed hybrid smoothness term helps reduce matching ambiguities causing by over-segmentation and under-segmentation with the indication of the depth sensor. The raw data from the depth sensor were noisy and had poor performance in preserving object boundaries (blue rectangle in Figure 7f); our result is more robust in this situation and can be used to improve the performance of the depth sensor by considering the color and segmentation information from stereo matching. Based on the above, we can safely draw the conclusion that the proposed method obtains accurate depth estimation by combining the complementary characteristics of stereo matching and the depth sensor. We also tested the proposed method on other real-world scenes to verify its robustness (see Figure 8).

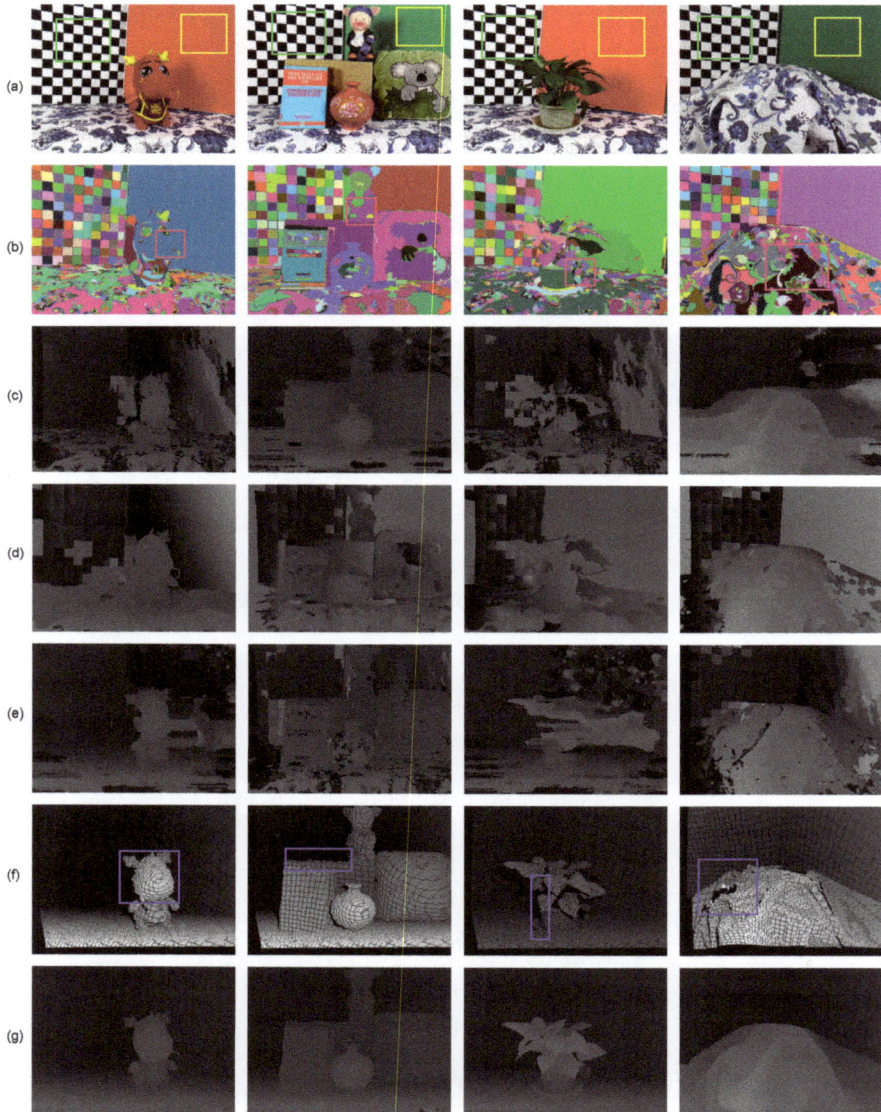

Figure 7. Results of the different methods applied to the real-world scenes: Dragon, Book, Plant and Tablecloth. Each column from up to down is: (**a**) the rectified left image; (**b**) the segmentation result; (**c**) the disparity map of FCVF [44]; (**d**) the disparity map of SOSP [11]; (**e**) the disparity map of segment tree (ST) [13]; (**f**) the seed image transformed from the Xtion data and (**g**) our result.

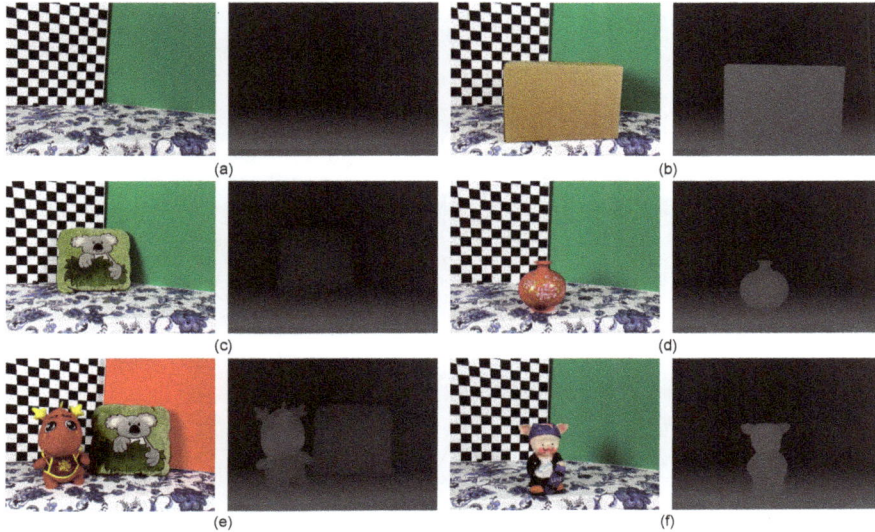

Figure 8. Comparative results for different real-world scenes: (**a**) Board; (**b**) Box; (**c**) Kola; (**d**) Vase; (**e**) Dragon and Kola; (**f**) Piggy. All scenes were approximately 0.5–1.5 m from the cameras, and the maximum disparity was 107 pixels. Each scene from left to right contains the rectified left image and its associated result of our method.

Furthermore, we implemented the post-processing processing introduced in Section 3.8 to assign valid disparities to pixels in the black regions of seed images. The seed image after assignment can be treated as the up-sampling disparity map of the target image captured by the Xtion alone. Then, we evaluated the quality of the 3D reconstruction from our method and that using only Xtion data (see Figure 9). The 3D point cloud reconstructions consist of the pixels' image coordinates and their associated disparities in disparity space. The blue rectangles highlight some regions where our method performed well. For example, the proposed method was more effective at retaining the boundaries of Piggy and Plant (Figure 9a,c) and correctly recovered the top of the head and beard of the dragon (Figure 9b). These comparisons illustrate that the stereo matching using the the depth sensor as the prior knowledge is more effective and accurate than using stereo matching or the depth sensor alone.

Figure 9. Comparative results for 3D reconstructions. (**a**) Book; (**b**) Dragon; (**c**) Plant; (**d**) Tablecloth; (**e**) Box; (**f**) Piggy; (**g**) Vase; (**h**) Dragon and Kola. Each scene from top to bottom contains the reconstruction using our method and the result using the up-sampled disparity map captured by the Xtion depth sensor.

4.2. Quantitative Evaluation Using the Middlebury Datasets

To quantitatively illustrate the validity of the proposed method, we also conducted evaluations on the Middlebury datasets [9,45] and focused on recovering the disparity map of the left image in each dataset. The evaluation is made by third-size resolution Views 1 and 5 of all image pairs. However, because there is nothing about the scanning depth information of this dataset, we used the method described in [14] to simulate the seed image transformed from the Xtion projected to View 1. This technique is based on a voting strategy and simply requires some disparity maps produced using several stereo methods [46–48]. Each pixel was labeled as a seed if its disparity in different maps was consistent (varied by less than a fixed threshold and was not near the intensity edge). Results on these datasets and their corresponding errors (compared with the ground truth) in non-occlusion regions are shown in Figure 10. As shown in Figure 11, our method ranks first among approximately 164 methods listed on the website [49]. It performs especially well on the Tsukuba image pairs, with minimum errors in non-occluded regions and near depth discontinuities.

On the other hand, as shown in Figure 12, we presented some evaluation results of the Middlebury extension datasets [50,51] to illustrate the robustness of the proposed method. Meanwhile, we also show the quality of 3D reconstruction of Middlebury datasets using the pixels' image coordinate and their corresponding disparities in disparity space (see Figure 13). The evaluation results in Figures 12 and 13 illustrate that our method is robust to different types of scenes and outperforms in slanted and highly curved surfaces.

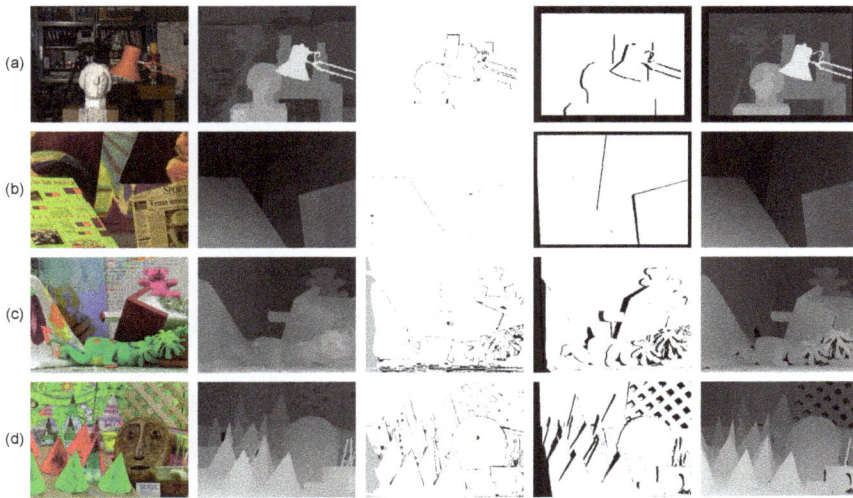

Figure 10. Evaluation results on the Middlebury standard data. (**a**) Tsukuba; (**b**) Venus; (**c**) Teddy; (**d**) Cones. Each row contains (from left to right): the left image, our results, the error map (error matching pixels whose absolute disparity errors are larger than one in non-occlusion and occlusion regions are marked in black and gray), the occlusion map (occluded pixels are marked black) and the ground truth map.

Error Threshold = 1		Sort by nonocc						Sort by all					Sort by disc	
Algorithm	Avg. Rank	Tsukuba ground truth			Venus ground truth			Teddy ground truth			Cones ground truth			Average Percent Bad Pixels
		nonocc	all	disc	nonocc	all	disc	nonocc	all	disc	nonocc	all	disc	
YOUR METHOD	10.5	0.79 2	1.21 4	4.30 2	0.10 7	0.21 12	1.27 9	3.49 15	9.04 31	10.9 17	2.06 5	7.05 18	5.80 4	3.85
IGSM [158]	10.8	0.93 11	1.37 13	5.05 13	0.07 3	0.17 4	1.04 3	4.08 20	5.98 8	11.4 21	2.14 10	6.97 15	6.27 9	3.79
TSGO [142]	13.0	0.87 5	1.13 1	4.66 7	0.11 10	0.24 13	1.47 13	5.61 43	8.09 19	13.8 38	1.67 3	6.16 3	4.95 3	4.06
LCU [155]	13.0	1.06 20	1.34 9	5.50 18	0.07 2	0.26 19	1.03 2	3.68 17	9.95 37	10.4 15	1.63 2	6.87 13	4.82 2	3.89
JSOSP+GCP [150]	14.8	0.74 1	1.34 10	3.98 1	0.08 4	0.16 1	1.15 4	3.96 18	10.1 38	11.8 22	2.28 20	7.91 35	6.74 23	4.18
ADCensus [82]	18.0	1.07 23	1.48 20	5.73 26	0.09 5	0.25 16	1.15 4	4.10 21	6.22 9	10.9 18	2.42 26	7.25 21	6.95 27	3.97
CoopRegion [39]	21.5	0.87 7	1.16 2	4.61 6	0.11 9	0.21 8	1.54 16	5.16 35	8.31 23	13.0 30	2.79 47	7.18 20	8.01 55	4.41
AdaptingBP [16]	21.9	1.11 26	1.37 12	5.79 28	0.10 8	0.21 11	1.44 12	4.22 23	7.06 17	11.8 23	2.48 30	7.92 37	7.32 36	4.23
CCRADAR [151]	26.2	1.15 28	1.42 18	6.23 41	0.15 21	0.27 20	1.89 26	5.39 38	10.6 43	14.7 47	2.01 4	7.37 23	5.88 5	4.75

Figure 11. Middlebury results of our method. All numbers are the percentage of error pixels whose absolute disparity error is larger than one. The blue number is the ranking in every column. Our method outperforms the conventional stereo matching algorithms and ranks first among approximately 164 methods according to the average of the sum of the rankings in every column (up to 20 April 2015).

Besides, we also compared our results with those produced by other "fused" schemes [20,23,52–56], and the compared results are listed in the Table 3. Our method provides an error rate of 2.61% on the Middlebury datasets, compared to the average error rate 3.27% of the previous state-of-the-art "fused" methods. It is clear that our method performs almost 20% better than other "fused" scheme-based algorithms in the aspect of precision. Furthermore, As shown in Figures 14 and 15, our method achieves comparable results in the following aspects:

- Noise and outliers are significantly reduced, mainly because of the improved luminance consistency term and the texture term.
- The method obtains precise disparities for slanted or highly-curved surfaces of objects with complex geometric characteristics, mainly because of the 3D plane bias term.
- Ambiguous matchings caused by over-segmentation or under-segmentation are overcome and disparity variances become smoother, mainly because of the hybrid smoothness term.

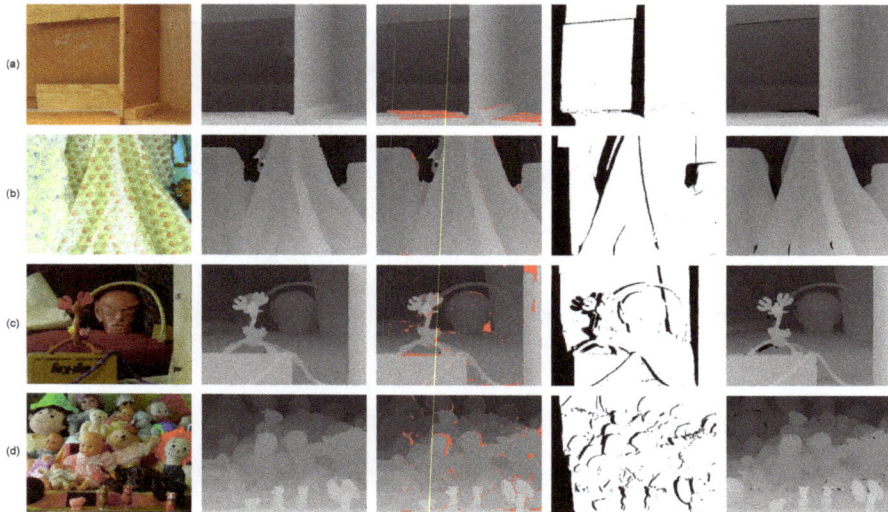

Figure 12. Evaluation results on the Middlebury extension datasets. (**a**) Wood1; (**b**) Cloth4; (**c**) Reindeer; (**d**) Dolls. Each row contains (left to right): the left image, our results, the error map (from top to bottom, the percentages of error pixels with absolute disparity error larger than one in non-occlusion regions are: 4.35%, 1.03%, 4.31%, 4.78%; error pixels are marked red), the occlusion map (occluded pixels are marked black) and the ground truth map.

Table 3. The percentages of error pixels (absolute disparity error larger than 1 in non-occlusion regions) of our method and other "fused" methods on the Middlebury datasets. "Averages" are the average percentages of error pixels over all images. Compared to the average error rate 3.27% of the previous state-of-the-art "fused" methods, our method provides a lower average error rate of 2.61% on the Middlebury datasets. It performs almost 20% better than other "fused" methods in the aspect of precision.

	The Percentages of Error Pixels (%)								
	Tsukuba	Venus	Teddy	Conse	Wood1	Colth4	Reimdeor	Dools	Averages
Zhu *et al.* [20]	1.16	0.14	2.83	3.47	5.38	3.74	5.83	5.46	3.50
Wang *et al.* [23]	0.89	0.12	6.39	2.14	4.05	3.81	3.55	2.71	2.96
Yang *et al.* [52]	0.94	0.26	5.65	7.18	1.76	2.60	4.43	4.13	3.37
Jaesik *et al.* [53,55]	2.38	0.56	5.59	6.28	3.72	2.88	4.04	4.69	3.77
James *et al.* [54]	2.90	0.29	2.12	2.83	2.74	2.32	5.02	4.02	2.78
Ours	0.79	0.10	3.49	2.06	4.35	1.03	4.31	4.78	2.61

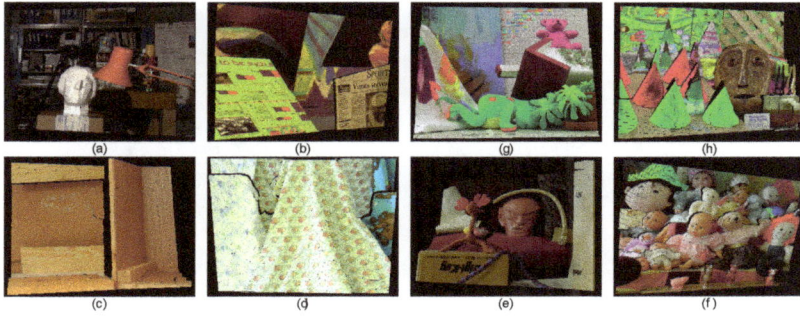

Figure 13. The results of 3D reconstructions. (**a**) Tsukuba; (**b**) Venus; (**c**) Teddy; (**d**) Cones; (**e**) Wood1; (**f**) Cloth4; (**g**) Reindeer; (**h**) Dolls.

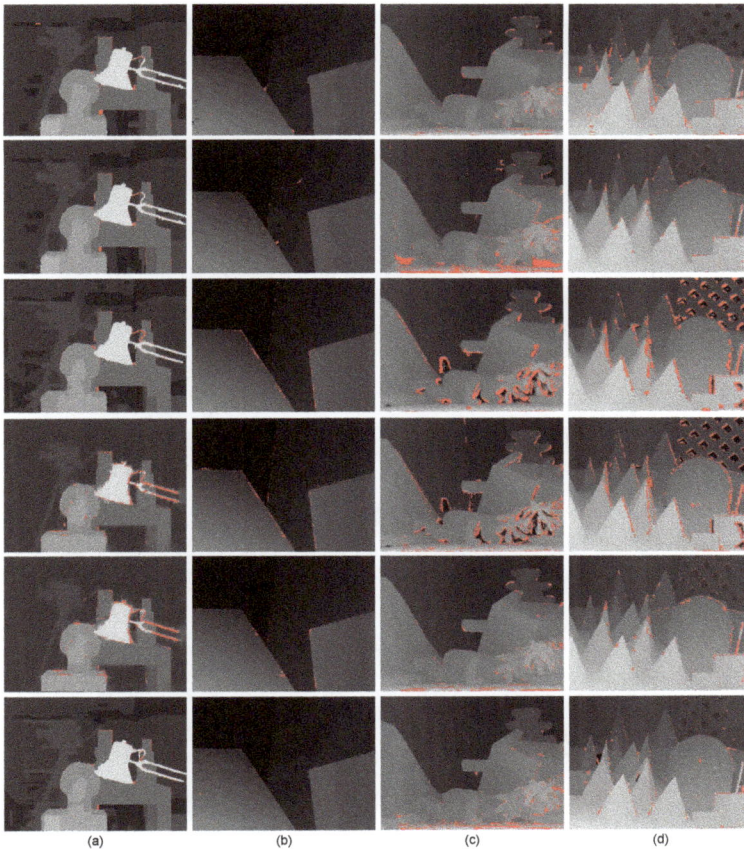

Figure 14. Evaluation results with the state-of-the-art "fused" scheme-based algorithms on the Middlebury datasets. (**a**) Tsukuba; (**b**) Venus; (**c**) Teddy; (**d**) Cones. Each column from top to bottom is the results obtained from: Zhu *et al.* [20], Wang *et al.* [23], Yang *et al.* [52], Jaesik *et al.* [53,55], James *et al.* [54] and our method. Error pixels with absolute disparity error larger than one in non-occlusion regions are marked red. The percentages of error pixels are listed in Table 3.

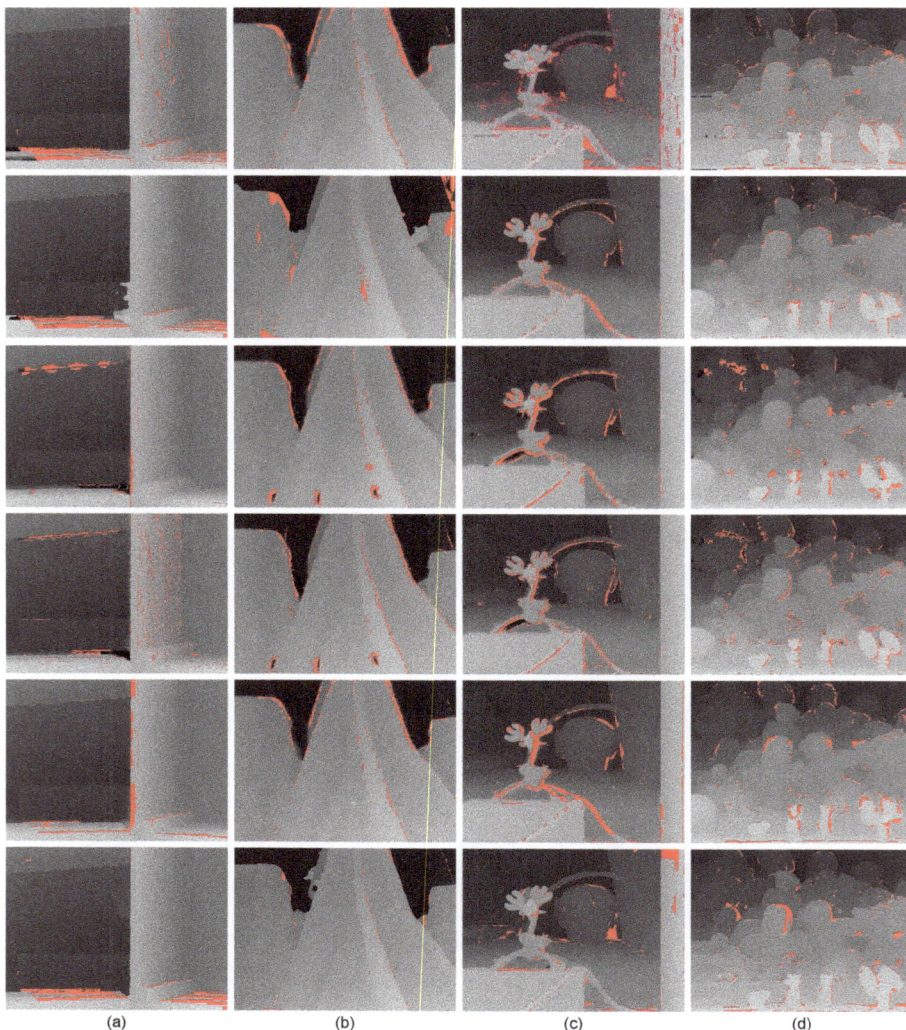

Figure 15. Evaluation results with the state-of-the-art "fused" scheme-based algorithms on the Middlebury extension datasets. (**a**) Wood1; (**b**) Cloth4; (**c**) Reindeer; (**d**) Dolls. Each column from top to bottom is the results obtained from: Zhu *et al.* [20], Wang *et al.* [23], Yang *et al.* [52], Jaesik *et al.* [53,55], James *et al.* [54] and our method. Error pixels with absolute disparity error larger than one in non-occlusion regions are marked red. The percentages of error pixels are listed in Table 3.

4.3. Evaluation Results for Each Term

We conducted evaluations to analyze the effect of the individual terms in Equation (4). In each experiment, one term was turned off and the others remained on. First, the texture term was turned off, which meant that the range of the potential disparities for each pixel was no longer restricted by the texture variance and gradient. Ambiguities occurred in textureless and repetitive texture regions without the prior restriction from the data of the depth sensor (see the yellow rectangle in Figure 16b). The average error rate of all images in non-occlusion regions sharply increased to 2.77%. Furthermore, the improved luminance consistency term was turned off by setting $w_p^x := 0$. Then, this term can be viewed as the conventional one that is easily affected by light variation and causes error matching on the non-Lambertian surface and rich texture regions (see the green rectangle regions in Figure 16c–e). The corresponding average error rate of all images in the non-occlusion regions is 2.37%. Thirdly, the hybrid smoothness term was turned out by replacing by the usual second-order smoothness term [11]. Some artifacts in the red rectangle in Figure 16f were caused by over-segmentation and under-segmentation. Its average error rate sharply increased to 3.02%. Finally, the 3D plane bias term was turned off by setting $\lambda_p := 0$. In that case, all 3D object surfaces are assumed as the frontal parallel ones, and the depth map is rather noisy, which makes it difficult to preserve the details at the boundary of objects (see the blue rectangle region in Figure 16g). Its average error rate is 2.14%. The corresponding error statistic analysis on the Middlebury datasets is listed in Table 4. It is clear that our method can obtain the lowest average error rate when all terms turn on (average error rate of 1.61% in non-occlusion regions).

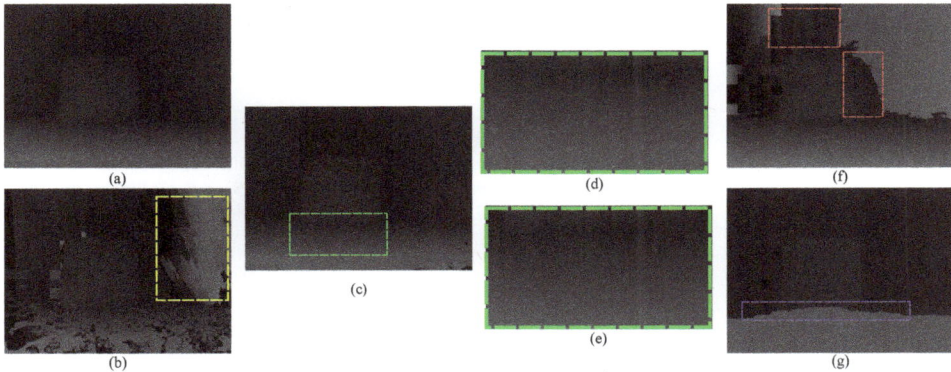

Figure 16. Evaluation results when turning off some terms. (**a**) Our result; (**b**) result without the texture term; (**c**) result without the improved luminance consistency term; (**d**) the detail with an enlarged scale in the green region of (c); (**e**) our result detail with enlarged scale in the same green region of (c); (**f**) result without the hybrid smoothness term; (**g**) result without the 3D plane bias term. Nonocc: non-occlusion regions.

Table 4. Error statistic for the Middlebury datasets with different constraint terms turned off. "Averages" are the average percentages of error pixels over all images in different regions.

	Tsukuba			Venus			Teddy			Cones			Averages		
	nonocc	all	disc	nonocc	all	disc	nonocc	all	disc	nonocc	all	disc	nonocc	all	disc
Texture term off	2.04	2.12	5.78	0.81	1.03	3.09	5.43	10.2	13.25	2.79	8.26	6.35	2.77	5.40	7.12
Luminance term off	1.01	1.65	4.78	0.11	0.25	1.54	5.49	11.20	14.92	2.88	8.47	7.74	2.37	5.39	7.25
Smoothness term off	1.39	2.14	4.94	0.85	0.93	2.02	6.57	13.01	14.80	3.28	7.50	7.13	3.02	5.90	7.22
Plane bias term off	0.88	1.49	4.86	0.23	0.65	2.27	4.53	9.30	10.63	2.90	7.96	8.96	2.14	4.76	6.68
All terms on	0.79	1.21	4.30	0.10	0.21	1.27	3.49	9.04	10.90	2.06	7.05	5.80	1.61	4.37	5.56

4.4. Computational Time Analyses

The proposed method was implemented on a PC with Core i5-2500 3.30 GHZ CPU and 4 GB RAM. Tables 5 and 6 list the running time of the proposed method for all experiments. It is obvious that the computational time is proportional to the image resolution and the scope of potential disparities. For example, it took approximately 1–9 mins to obtain results on Middlebury data and 19–25 mins on the real-world scene datasets. In the future, we aim to implement our method on a GPU to achieve a good balance between accuracy and efficiency.

Table 5. Running times for the real-world datasets. The disparity map resolution of all real-world datasets is 1024 × 960. The corresponding maximum disparity is 107.

	Dragon	Book	Plant	TableclotdBoard	Box	Kola	Vase	Piggy	Dragon and Piggy
Running Time (m):	22.15	20.54	22.16	24.31	21.51	24.26	23.43	22.30	19.44	23.04

Table 6. Running times for the Middlebury datasets.

	Tsukuba	Venus	Teddy	Cones	Wood1	Cloth4	Reindeer	Dolls
Running Time (m):	1.03	1.19	4.08	4.57	8.18	7.44	7.02	8.31
Disparity Map Resolution:	384 × 288	434 × 383	450 × 375	450 × 375	457 × 370	433 × 375	447 × 370	463 × 375
Maximum Disparity:	15	19	59	59	71	69	67	73

5. Conclusions

In this paper, we present an accurate disparity estimation fusion model that "fused" the advantages of the complementary nature of active and passive sensors. Our main contributions are the texture information constraint and the multiscale pseudo two-layer image model. The comparison results show that our method can reduce the error estimate caused by under- or over- segmentation and has good performance in keeping object boundaries compared to using the conventional stereo matching or the depth sensor alone. Furthermore, the proposed method provides an error rate of 2.61% on the Middlebury datasets, compared to the average error rate 3.27% of the previous state-of-the-art "fused" methods. It is clear that our method performs almost 20% better than other "fused" scheme-based algorithms in the aspect of precision. In the future, we will investigate a more accurate method for estimating the disparities of occluded pixels. We also intend to transform our method to a parallel GPU implementation.

Acknowledgments: This work is supported and funded by the National Natural Science Foundation of China (No. 61300131), the National Key Technology Research and Development Program of China (No. 2013BAK03B07) and the National High Technology Research and Development Program of China (863 Program) (No. 2013AA013902).

Author Contributions: All authors contributed to the understanding of the algorithm. Jing Liu and Zhaoqi Wang conceived of and designed the algorithm. Jing Liu and Xuefeng Fan performed the simulation analysis and the experiments. Jing Liu and Chunpeng Li wrote the paper. All authors read and approved the final manuscript.

Conflicts of Interest: The authors declare no conflict of interest.

References

1. Díaz-Vilariño, L.; Khoshelham, K.; Martínez-Sánchez, J.; Arias, P. 3D modeling of building indoor spaces and closed doors from imagery and point clouds. *Sensors* **2015**, *15*, 3491–3512. [CrossRef] [PubMed]
2. Pan, S.; Shi, L.; Guo, S. A kinect-based real-time compressive tracking prototype system for amphibious spherical robots. *Sensors* **2015**, *15*, 8232–8252. [CrossRef] [PubMed]
3. Yebes, J.J.; Bergasa, L.M.; García-Garrido, M. Visual object recognition with 3D-aware features in KITTI urban scenes. *Sensors* **2015**, *15*, 9228–9250. [CrossRef] [PubMed]
4. Tanimoto, M.; Tehrani, M.P.; Fujii, T.; Yendo, T. Free-viewpoint TV. *IEEE Signal Process. Mag.* **2011**, *28*, 67–76. [CrossRef]
5. Liu, J.; Li, C.; Mei, F.; Wang, Z. 3D entity-based stereo matching with ground control points and joint second-order smoothness prior. *Vis. Comput.* **2014**, *31*, 1–17. [CrossRef]
6. ASUS Xtion. Available online: www.asus.com/Multimedia/Xtion/ (accessed on 19 August 2015).

7. Microsoft Kinect. Available online: www.microsoft.com/zh-cn/kinectforwindows/ (accessed on 19 August 2015).
8. Song, X.; Zhong, F.; Wang, Y.; Qin, X. Estimation of kinect depth confidence through self-training. *Vis. Comput.* **2014**, *30*, 855–865. [CrossRef]
9. Scharstein, D.; Szeliski, R. A taxonomy and evaluation of dense two-frame stereo correspondence algorithms. *Int. J. Comput. Vis.* **2002**, *47*, 7–42. [CrossRef]
10. Yoon, K.J.; Kweon, I.S. Adaptive support-weight approach for correspondence search. *IEEE Trans. Pattern Anal. Mach. Intell.* **2006**, *28*, 650–656. [CrossRef] [PubMed]
11. Woodford, O.; Torr, P.; Reid, I.; Fitzgibbon, A. Global stereo reconstruction under second-order smoothness priors. *IEEE Trans. Pattern Anal. Mach. Intell.* **2009**, *31*, 2115–2128. [CrossRef] [PubMed]
12. Bleyer, M.; Rhemann, C.; Rother, C. Patch match stereo-stereo matching with slanted support windows. *BMVC* **2011**, *11*, 1–11.
13. Mei, X.; Sun, X.; Dong, W.; Wang, H.; Zhang, X. Segment-tree based Cost Aggregation for Stereo Matching. In Proceedings of the 2013 IEEE Conference on Computer Vision and Pattern Recognition, Portland, OR, USA, 23–28 June 2013; pp. 313–320.
14. Wang, L.; Yang, R. Global Stereo Matching Leveraged by Sparse Ground Control Points. In Proceedings of the IEEE Conference Computer Vision and Pattern Recognition, Providence, RI, USA, 20–25 June 2011; pp. 3033–3040.
15. LiDAR. Available online: www.lidarusa.com (accessed on 19 August 2015).
16. 3DV Systems. Available online: www.3dvsystems.com (accessed on 19 August 2015).
17. Photonix Mixer Device for Distance Measurement. Available online: www.pmdtec.com (accessed on 19 August 2015).
18. Khoshelham, K.; Elberink, S.O. Accuracy and resolution of kinect depth data for indoor mapping applications. *Sensors* **2012**, *12*, 1437–1454. [CrossRef] [PubMed]
19. Zhu, J.; Wang, L.; Gao, J.; Yang, R. Spatial-temporal fusion for high accuracy depth maps using dynamic MRFs. *IEEE Trans. Pattern Anal. Mach. Intell.* **2010**, *32*, 899–909. [PubMed]
20. Zhu, J.; Wang, L.; Yang, R.; Davis, J.E.; Pan, Z. Reliability fusion of time-of-flight depth and stereo geometry for high quality depth maps. *IEEE Trans. Pattern Anal. Mach. Intell.* **2011**, *33*, 1400–1414.
21. Yang, Q.; Tan, K.H.; Culbertson, B.; Apostolopoulos, J. Fusion of Active and Passive Sensors for Fast 3D Capture. In Proceedings of the IEEE International Workshop on Multimedia Signal Processing, Saint Malo, France, 4–6 October 2010; pp. 69–74.
22. Zhang, S.; Wang, C.; Chan, S. A New High Resolution Depth Map Estimation System Using Stereo Vision and Depth Sensing Device. In Proceedings of the IEEE 9th International Colloquium on Signal Processing and its Applications, Kuala Lumpur, Malaysia, 8–10 March 2013; pp. 49–53.
23. Wang, Y.; Jia, Y. A fusion framework of stereo vision and Kinect for high-quality dense depth maps. *Comput. Vis.* **2013**, *7729*, 109–120.
24. Somanath, G.; Cohen, S.; Price, B.; Kambhamettu, C. Stereo Kinect for High Resolution Stereo Correspondences. In Proceedings of the IEEE International Conference on 3DTV-Conference, Seattle, Washington, DC, USA, 29 June–1 July 2013; pp. 9–16.
25. Zhang, Z. A flexible new technique for camera calibration. *IEEE Trans. Pattern Anal. Mach. Intell.* **2000**, *22*, 1330–1334. [CrossRef]
26. Herrera, C.; Kannala, J.; Heikkilä, J. Joint depth and color camera calibration with distortion correction. *IEEE Trans. Pattern Anal. Mach. Intell.* **2012**, *34*, 2058–2064. [CrossRef] [PubMed]
27. Yang, Q. A Non-Local Cost Aggregation Method for Stereo Matching. In Proceedings of the IEEE Conference on Computer Vision and Pattern Recognition, Providence, RI, USA, 16–21 June 2012; pp. 1402–1409.
28. Christoudias, C.M.; Georgescu, B.; Meer, P. Synergism in low level vision. *IEEE Intern. Conf. Pattern Recognit.* **2002**, *4*, 150–155.
29. Bleyer, M.; Rother, C.; Kohli, P. Surface Stereo with Soft Segmentation. In Proceedings of the IEEE Conference on Computer Vision and Pattern Recognition (CVPR), San Francisco, CA, USA, 13–18 June 2010; pp. 1570–1577.
30. Wei, Y.; Quan, L. Asymmetrical occlusion handling using graph cut for multi-view stereo. *IEEE Intern. Conf. Pattern Recognit.* **2005**, *2*, 902–909.

31. Hu, X.; Mordohai, P. A quantitative evaluation of confidence measures for stereo vision. *IEEE Trans. Pattern Anal. Mach. Intell.* **2012**, *34*, 2121–2133. [PubMed]

32. Liu, Z.; Han, Z.; Ye, Q.; Jiao, J. A New Segment-Based Algorithm for Stereo Matching. In Proceedings of the IEEE International Conference on Mechatronics and Automation, Changchun, China, 9–12 August 2009; pp. 999–1003.

33. Rao, A.; Srihari, R.K.; Zhang, Z. Spatial color histograms for content-based image retrieval. In Proceedings of the 11th IEEE International Conference on Tools with Artificial Intelligence, Chicago, IL, USA, 9–11 November 1999; pp. 183–186.

34. Lempitsky, V.; Rother, C.; Blake, A. Logcut-Efficient Graph Cut Optimization for Markov Random Fields. In Proceedings of the IEEE 11th International Conference on Computer Vision, Rio de Janeiro, Brazil, 14–21 October 2007; pp. 1–8.

35. Boykov, Y.; Veksler, O.; Zabih, R. Fast approximate energy minimization via graph cuts. *IEEE Trans. Pattern Anal. Mach. Intell.* **2001**, *23*, 1222–1239. [CrossRef]

36. Kolmogorov, V.; Zabin, R. What energy functions can be minimized via graph cuts? *IEEE Trans. Pattern Anal. Mach. Intell.* **2004**, *26*, 147–159. [CrossRef] [PubMed]

37. Kolmogorov, V.; Rother, C. Minimizing nonsubmodular functions with graph cuts-a review. *IEEE Trans. Pattern Anal. Mach. Intell.* **2007**, *29*, 1274–1279. [CrossRef] [PubMed]

38. Boros, E.; Hammer, P.L.; Tavares, G. *Preprocessing of Unconstrained Quadratic Binary Optimization*; Technical Report RRR 10-2006, RUTCOR Research Report; Rutgers University: Piscataway, NJ, USA, 2006.

39. Lempitsky, V.; Roth, S.; Rother, C. FusionFlow: Discrete-Continuous Optimization for Optical Flow Estimation. In Proceedings of the Conference on Computer Vision and Pattern Recognition, Anchorage, AK, USA, 23–28 June 2008; pp. 1–8.

40. Ishikawa, H. Higher-Order Clique Reduction Without Auxiliary Variables. In Proceedings of the IEEE Conference on Computer Vision and Pattern Recognition, Columbus, OH, USA, 23–28 June 2014; pp. 1362–1369.

41. Yang, Q.; Wang, L.; Yang, R.; Stewénius, H.; Nistér, D. Stereo matching with color-weighted correlation, hierarchical belief propagation, and occlusion handling. *IEEE Trans. Pattern Anal. Mach. Intell.* **2009**, *31*, 492–504. [CrossRef] [PubMed]

42. Heo, Y.S.; Lee, K.M.; Lee, S.U. Joint depth map and color consistency estimation for stereo images with different illuminations and cameras. *IEEE Trans. Pattern Anal. Mach. Intell.* **2013**, *35*, 1094–1106. [PubMed]

43. Matsuo, T.; Fukushima, N.; Ishibashi, Y. Weighted Joint Bilateral Filter with Slope Depth Compensation Filter for Depth Map Refinement. *VISAPP* **2013**, *2*, 300–309.

44. Rhemann, C.; Hosni, A.; Bleyer, M.; Rother, C.; Gelautz, M. Fast Cost-volume Filtering for Visual Correspondence and Beyond. In Proceedings of the IEEE Conference on Computer Vision and Pattern Recognition, Providence, RI, USA, 20–25 June 2011; pp. 3017–3024.

45. Scharstein, D.; Szeliski, R. High-accuracy stereo depth maps using structured light. *IEEE Comput. Vis. Pattern Recognit.* **2013**, *1*, 195–202.

46. Chakrabarti, A.; Xiong, Y.; Gortler, S.J.; Zickler, T. Low-level vision by consensus in a spatial hierarchy of regions. 2014; arXiv:1411.4894.

47. Lee, S.; Lee, J.H.; Lim, J.; Suh, I.H. Robust stereo matching using adaptive random walk with restart algorithm. *Image Vis. Comput.* **2015**, *37*, 1–11. [CrossRef]

48. Spangenberg, R.; Langner, T.; Adfeldt, S.; Rojas, R. Large Scale Semi-Global Matching on the CPU. In Proceedings of the IEEE Intelligent Vehicles Symposium Proceedings, Dearborn, MI, USA, 8–11 June 2014; pp. 195–201.

49. Middlebury Benchmark. Available online: vision.middlebury.edu/stereo/ (accessed on 19 August 2015).

50. Hirschmuller, H.; Scharstein, D. Evaluation of Cost Functions for Stereo Matching. In Proceedings of the IEEE Conference on Computer Vision and Pattern Recognition, Minneapolis, MN, USA, 17–22 June 2007; pp. 1–8.

51. Scharstein, D.; Pal, C. Learning Conditional Random Fields for Stereo. In Proceedings of the IEEE Conference on Computer Vision and Pattern Recognition, Minneapolis, MN, USA, 17–22 June 2007; pp. 1–8.

52. Yang, Q.; Yang, R.; Davis, J.; Nistér, D. Spatial-depth super resolution for range images. In Proceedings of the IEEE Conference on Computer Vision and Pattern Recognition, Minneapolis, MN, USA, 17–22 June 2007; pp. 1–8.

53. Park, J.; Kim, H.; Tai, Y.W.; Brown, M.S.; Kweon, I.S. High-Quality Depth Map Upsampling and Completion for RGB-D Cameras. *IEEE Trans Image Process.* **2014**, *23*, 5559–5572. [CrossRef] [PubMed]
54. Diebel, J.; Thrun, S. An application of markov random fields to range sensing. *NIPS* **2005**, *5*, 291–298.
55. Park, J.; Kim, H.; Tai, Y.W.; Brown, M.S.; Kweon, I. High Quality Depth Map Upsampling for 3D-TOF Cameras. In Proceedings of the IEEE International Conference on Computer Vision, Barcelona, Spain, 6–13 November 2011; pp. 1623–1630.
56. Huhle, B.; Schairer, T.; Jenke, P.; Straßer, W. Fusion of range and color images for denoising and resolution enhancement with a non-local filter. *Comput. Vis. image Underst.* **2010**, *114*, 1336–1345. [CrossRef]

sensors

MDPI

Article

Simulated and Real Sheet-of-Light 3D Object Scanning Using a-Si:H Thin Film PSD Arrays

Javier Contreras [1,*], **Josep Tornero** [1], **Isabel Ferreira** [2], **Rodrigo Martins** [2,*], **Luis Gomes** [3] and **Elvira Fortunato** [2]

[1] Institute of Design and Manufacturing, Technical University of Valencia, CPI, Edif. 8G, 46022 Valencia, Spain; jtornero@idf.upv.es

[2] CENIMAT/I3N, Department of Material Science, Faculty of Science and Technology, FCT, New University of Lisbon and CEMOP/UNINOVA, 2829-516 Caparica, Portugal; imf@fct.unl.pt (I.F.); emf@fct.unl.pt (E.F.)

[3] Department of Electrical Engineering, Faculty of Science and Technology, FCT, New University of Lisbon, and CTS/UNINOVA, Campus da Caparica, 2928-516 Caparica, Portugal; lugo@uninova.pt

* Correspondence: jca8676@gmail.com (J.C.); rm@uninova.pt (R.M.); Tel.: +34-963877060 (J.C.); +351-212948525 (R.M.)

Academic Editor: Gonzalo Pajares Martinsanz

Received: 24 September 2015; Accepted: 9 November 2015; Published: 30 November 2015

Abstract: A MATLAB/SIMULINK software simulation model (structure and component blocks) has been constructed in order to view and analyze the potential of the PSD (Position Sensitive Detector) array concept technology before it is further expanded or developed. This simulation allows changing most of its parameters, such as the number of elements in the PSD array, the direction of vision, the viewing/scanning angle, the object rotation, translation, sample/scan/simulation time, *etc.* In addition, results show for the first time the possibility of scanning an object in 3D when using an a-Si:H thin film 128 PSD array sensor and hardware/software system. Moreover, this sensor technology is able to perform these scans and render 3D objects at high speeds and high resolutions when using a sheet-of-light laser within a triangulation platform. As shown by the simulation, a substantial enhancement in 3D object profile image quality and realism can be achieved by increasing the number of elements of the PSD array sensor as well as by achieving an optimal position response from the sensor since clearly the definition of the 3D object profile depends on the correct and accurate position response of each detector as well as on the size of the PSD array.

Keywords: three-dimensional sensing; arrays; three-dimensional image acquisition; optical sensing and sensors; thin film devices and applications; three-dimensional image processing

1. Introduction

Sheet-of-light range imaging is an interesting technique that is used in a number of 3D object rendering applications. Among the existing laser scanning techniques, a structured light triangulation method is considered to be fastest when acquiring 3D data information from an object in real time [1]. Generally, sheet-of-light systems use digital sensors such as CCDs (Charged Coupled Devices) or CMOS (Complementary Metal Oxide Semiconductors) [2], nevertheless, analog sensors such as arrays of PSDs (position sensitive detectors) with reported sizes of up to a maximum of 128 have also been employed [3,4]. The latter are fabricated using crystalline silicon, however, PSD arrays based on amorphous silicon nip or pin structures also exist and have already been described elsewhere [5,6]. In this work, results show for the first time the possibility of scanning an object in 3D when using an a-Si:H thin film 128 PSD array system and, in addition, a MATLAB/SIMULINK software simulation has been constructed in order to view and analyze the potential of PSD array concept technology before it is further expanded or developed. This study will enable the triangulation platform and system to be further developed and enhanced for future needs identified during simulations. Inspired by the simulation, examples of improvements and future adaptations could be the inclusion of suitable 360°rotation plates, faster translation tables, more precise optics and laser lines as well as larger sensor arrays depending of course on the limitations imposed by such a scenario. This simulation allows changing most of its parameters, such as the number of elements in the PSD array. The existing 128 PSD array sensor system, when mounted in a triangulation platform, is able to perform 3D object profile scans at high resolutions and speeds with a large number of frames, and this experiment was already presented here. Our previous works describe the implementation, characteristics and behavior of such 32/128 PSD array sensor systems, as well as the manner in which the scanned 3D object profile image is constructed thereby presenting a high speed sheet-of-light 3D object rendering platform [7] using as sensor an array of 32 amorphous silicon position sensitive detectors [6]. In the most recent work [8], we also exploited the 3D scanning optical characteristics of such an inspection system.

Analog versus Digital Technology

The foreseen PSD array technology roadmap proposes the use of amorphous silicon or other more suitable materials such as nanocrystalline silicon (which does not degrade as much), for the fabrication of analog 32/128/256/512/1024 PSD array sensors. The latter are used to scan and represent 3D profiles of objects in real-time [8], preferably at a greater speed than digital CCD or CMOS-based systems.

AT-Automation Technology GmbH [9], manufactures the most advanced CMOS sensor based sheet-of-light laser triangulation 3D cameras. They use digital sensors such as CMOS sensors (e.g., 1280 × 1024 pixels). The principle of application for these 3D sheet-of-light cameras is exactly the same as the one used for the 3D PSD sensor system and thereby they are the most advanced direct competitors in terms of speed, resolution and overall system performance, since they claim their cameras are the fastest in the world. One of their cameras (C4-1280) reaches 40,000 profiles/frames per second, when using 128 pixels and 128 rows of the sensors, representing only just a small part of the whole sensor (1280 × 1024 pixels).

Amorphous or nanocrystalline 32/128/256/512/1024 PSD array 3D sensors would be an alternative to these CMOS cameras in similar application scenarios. The overall performance between both technologies will be similar when using the 128 PSD sensor, however, amorphous or nanocrystalline 256/512/1024 PSD array 3D sensors are expected to outperform these 3D CMOS cameras when used in high speed 3D scanning applications. The reason for this is that the analog structure of the PSD sensor is different to the traditional discrete sensor and thereby it can process data much faster than a pixel-based structure. The resolution of these 3D CMOS cameras is quite good, depending on the application, and, in some applications, they claim to reach resolutions of about 35 μm or even 10 or 5 μm [9].

Therefore, it seems 3D cameras with digital sensors are already quite advanced and relatively cheap. A complete set generally includes a camera, laser and lens but not the software; nevertheless, third party software is readily available.

2. 128 PSD Array System 3D Object Profile Scanning

As already described in our previous work [7], an array of 32 amorphous silicon position sensitive detectors, was integrated inside a self-constructed machine vision system as the vision sensor component and the response was analyzed for the required application. Such work proposed the use of these sensors and relevant systems for 3D object profiling even at high speeds. A photograph of an amorphous silicon 128 element PSD linear array developed at CEMOP/UNINOVA is shown in Figure 1.

Figure 1. Amorphous silicon 128 PSD array glued and wire bonded to a suitable chip carrier.

This device can also be regarded as a three-dimensional PSD, being simply an array of one-dimensional PSDs mounted together in parallel to each other on a surface, where the separation between them and the detector width defines the minimum discrimination in one direction (discrete), while in the other (along the strip) it is continuous [6]. This type of structure has been specially designed for sheet-of-light 3D shape measurement.

The experimental setup and procedure for obtaining 3D object profiles with the 32 PSD sensor system were already described elsewhere [7,8] and these also apply for the 128 PSD sensor system. Here, we show results for the detection of the same object when using the 128 PSD array sensor system. The width of the rubber is small and it is only covered or detected by a few sensor channels. In addition, in these trials, the translation table (and object) moved from one side to the other so that the object (white color rubber), shown in Figure 2a, was scanned by the laser sheet-of-light system.

The object reflection was projected onto the active area of the 128 PSD sensor array. The dimensions of the rubber object, as measured by an electronic ruler, were the following: Length, 41.90 mm; Width, 16.70 mm; and Height, 11.74 mm.

The total distance travelled by the translation table (object) or scanned distance was about 69.37 mm at a speed of about 0.197 cm/s. As already reported elsewhere [8], here the incident scanning angle was fixed at 45° and the rate of acquisition was kept at 8 ms. The system integration time was maintained at 1ms for all scans and for low noise purposes, results were acquired using 128 sub-samplings (128 sample average). The light intensity recorded on the active area of the 128 PSD sensor array was 2.53 μW/cm².

The results obtained when channel 75 of a 128 PSD sensor array detects the light reflected by the white rubber as a function of the scanned distance are illustrated in Figure 2b. The response was acquired at 65 frames per second leading to 2289 frames in total.

The maximum resolution that can be obtained with an acquisition time of 8 ms derives from the number of maximum possible frames acquired in one second taking into account the lowest possible scanning speed. Therefore, $1/0.008$ s = 125 frames per second is the maximum possible acquisition frame rate at present (WINDOWS limitation) for this particular system configuration. Using the previously referred to acquisition time of 8ms and speed of 0.197 cm/s, a calculated scanning resolution of around 15.76 μm should be expected ($0.00197/125$ = 15.76 μm). However, in reality it takes more than 30 μm to acquire each frame, mainly due to software execution internal timings, *etc.*

The response shown in Figure 2b corresponds fairly well to the measured length of the rubber object being 41.90 mm as measured by an electronic ruler. Channel 75 of the sensor detects a signal for about 41mm. The difference in the position of the laser line reflected on the sensor active area [8] in Figure 2b (Y-axis) corresponds to the height of the rubber. The real measured height of the rubber being 11.74 mm is equivalent to a height of about 3 mm (from 0 mm to 3 mm on the Y axis) measured of course on the sensor active area after a lens reduction [8].

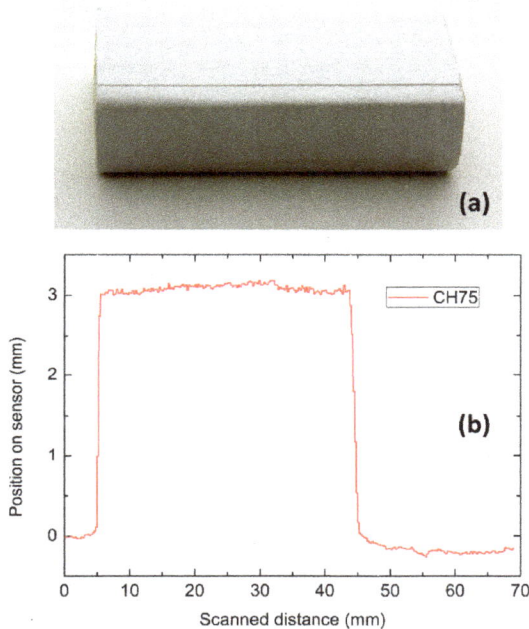

Figure 2. (**a**) Photograph of a white rubber object; (**b**) 3D object profile representation for the individual detection of channel 75 from the 128 PSD array sensor.

The stability of the profile detected is acceptable, even though a slight flickering is observed, which could be attributed to vibrations caused by the translation table during the scanning procedures as well as to the noise of the sensor.

3. 3D Sensor MATLAB/SIMULINK Simulation

A MATLAB/SIMULINK model has been constructed in order to analyze and view the possibilities of the PSD array sensor technology before it is further expanded or developed.

The simulation reads any "PLY" 3D data file located inside a predefined directory. In our case, the file "teapot.ply" is used. This file contains the raw 3D cloud data of the object that is to be scanned in 3D, and, in this case, it is a teapot; however any other predefined 3D object can be employed.

When the function "ply_read.m" is executed, the simulation reads the object. Two matrices, "[Tri, Pts]", are then recorded, one regarding the triangular connectivity information (triangle faces) and another regarding the VERTEX information (vertex or 3D points) of the object to be scanned in 3D within the simulation.

Figure 3 shows the structure of the simulation model with all its component blocks. Relevant blocks from this SIMULINK model are explained.

3.1. Initial Rotation

The object can be rotated before starting the scan. Any angle of initial rotation can be defined, e.g., (15°), however it should be entered/translated to radians (instead of degrees), so for 15° case, the term 15π/180 should be inserted. The default value is 0.

3.2. Rotation Velocity

The velocity of rotation is defined as the speed in which the object rotates per second. For example, a value of 36° per second could be inserted since the default value of the simulation is 10 s and this would result in the object rotating 360° in 10 s. The default value is 0. The value should be inserted in radians per second, instead of degrees, so therefore it would be 36π/180 or 2π/10. If we are only interested in rotating the object without translating, we can change the velocity of rotation to 2π/10 and change the translation velocity to 0 so that the object does not move. In addition, we can set the initial position to 0 so that the object is right below the lens and a better scan would be obtained in this particular case.

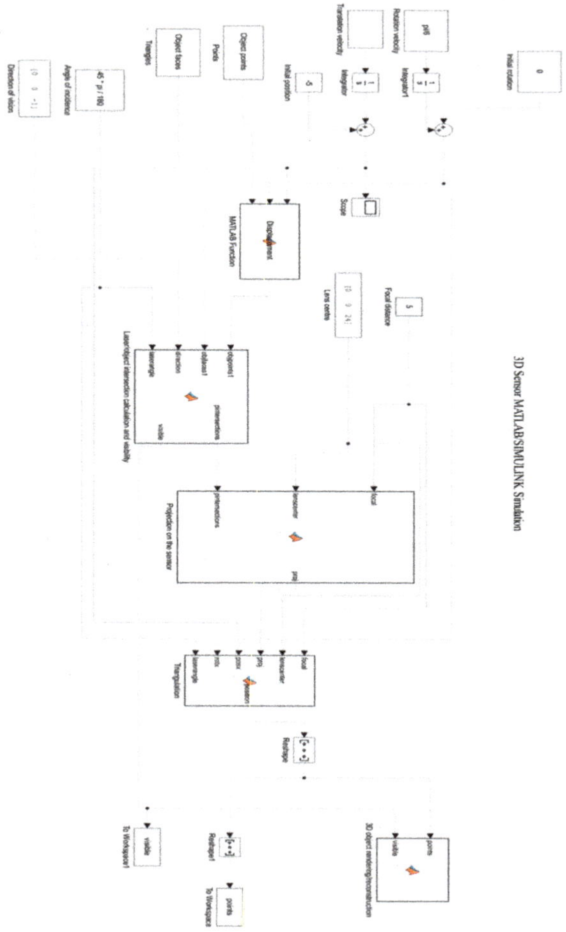

Figure 3. Structure of the SIMULINK simulation model with all its component blocks.

3.3. Translation Velocity

The translation velocity is defined as the speed in which the object translates, per second, in the X direction (X-axis). For example, a value of 1 cm/s could be inserted. The default value is 1. Thus, in order to rotate and translate the object at the same time, a value of rotation velocity of for example, $2\pi/10$ can be inserted in combination with the default value of 1cm/s for the translation velocity and the default value of initial position of -5 cm (see Section 3.4)).

3.4. Initial Position

This value determines the position (generally in the x-axis) where the object starts to be scanned in relation to the 0, 0, 0 (x, y, z) point or center of the system, perpendicular to the lens and 3D sensor. The default value is -5 cm in the X-axis, which is from where the object starts to be scanned along the X-axis. The center position is at the value x = 0, right below the lens.

3.5. Object Points

This value refers to the "Pts" matrix of points or (x, y, z) array of points returned by the ply_read.m function.

3.6. Object Faces

This value refers to the "Tri" matrix of faces or triangular order for the connectivity of the points returned by the ply_read.m function.

3.7. Angle of Incidence

This is the viewing or scanning angle, which exists between the laser and the normal to the lens/sensor system. A default value of $45\pi/180$ has been set ($45°$ in radians), however it can be changed in order to scan the object at various other angles. For example, a viewing/scanning angle of $15°$ will yield a different result to a $45°$ angle, since the laser would be projected differently and on different parts/sides of the object.

3.8. Direction of Vision

This parameter indicates the direction in which the lens/sensor system is viewing the object. This parameter is composed of three coordinates (x, y, z). The default value is [0, 0, -1], indicating that the direction of vision is downwards in the $-Z$ direction (negative Z-axis). In our system setup, this value is currently fixed since the object is passing and is being scanned below the lens/sensor system as it moves in the X-axis.

3.9. Displacement

This SIMULINK block calculates the displacement taking into account the parameters of translation and rotation velocity. The displacement is calculated as the simulation runs on every sample interval.

3.10. Focal Distance

This is the focal distance provided by the physical lens used in the real system setup, which, in this case, is set to a default value of 5 cm. Of course, in this simulation, it can also be changed to other values.

3.11. Lens Centre

This parameter defines where the center of the lens is located with reference to the 0, 0, 0 (x, y, z) point, which is located where the object meets the normal of the lens/sensor system. The default value is [0, 0, 24], so this means that the center of the lens is located 24 cm above the [0, 0, 0] reference point, so 24 cm in the Z-axis (upwards). This corresponds to the reality, since in the real system setup, the center of the lens is also located 24 cm above the [0, 0, 0] reference point.

3.12. Laser/Object Intersection Calculation and Visibility

This SIMULINK block considers the points of the object, the triangular object faces formed by those points from the object, the direction of vision, the viewing/scanning angle of the laser and the laser sheet of light itself in order to calculate the intersection of the object points/faces with the laser. The visibility of those intersecting points is also determined.

For a 32 PSD sensor, the physical number of detectors on the sensor is 32 and thereby the "numrays" variable defined in the code of the SIMULINK block is set to 32. This variable may be changed by accessing the code in the MATLAB workspace and by simply modifying the value of "numrays = 32" to any other value, taking into account that usually the physical number of detectors in these kind of PSD array sensors should vary from 32 through to 128, 256, 512 or 1024. The higher the number of detectors, the higher the resolution in the Y- axis.

Once the value is modified, the model should be "re-built" and thereafter updated using the icon "Build Model" on the MATLAB menu bar. The SIMULINK model should now be ready to "RUN" with the new value.

3.13. Projection on the Sensor

This SIMULINK block calculates and performs the 2D projection of the scanned object points or 2D frames on the sensor active area at each sample interval as the simulation runs.

3.14. Triangulation

This SIMULINK block uses the triangulation formulae and its relevant parameters in order to calculate the 3D coordinates (x, y, z) of each of the scanned points from the object at each sample interval.

3.15. 3D Object Rendering/Reconstruction

This SIMULINK block uses the previously calculated 3D coordinates and 2D frames to reconstruct or render the object shape in 3D. The 3D object mesh of points is plotted as a 3D map at each sample interval as the simulation runs.

3.16. Points

The matrix or array of points is returned to the MATLAB workspace at each sample interval.

3.17. Visible

The matrix or array of visible points is returned to the MATLAB workspace at each sample interval.

3.18. Running the Simulation

The simulation is compiled and run with the default parameters for each block; however, as already described, most of these can be modified to suit the needs of the required 3D scan.

Figure 4 shows the simulation running when the default parameters were used and Figure 5 presents the simulation results at the end of the simulation.

In Figures 4 and 5 the teapot was scanned and rendered using a 0.2 s sample time at a 45° scanning angle, a translation velocity of 1 cm/s, an initial position of –5 cm and no rotation velocity (0). At the end of the simulation, the relevant scanned 3D object generated data can be exported and, in that case, a file of "xyz" format called "puntos.xyz" is stored in a predefined directory. The file "puntos.xyz" can now be opened using a 3D mesh visualization program, such as MESHLAB. Figure 6a shows the generated file "puntos.xyz" opened in MESHLAB. Here, the teapot (object), which was scanned in the 3D sensor simulation, is illustrated and we can ZOOM IN, rotate and have a closer look at the resulting mesh of 3D points. Figure 6b shows the best 3D scanning results obtained when using a 32 PSD array sensor, starting at the default initial position of –5 cm, translating 1 cm/s, rotating 360° 10 times in 10 s, while using a sample time of 1/360 s.

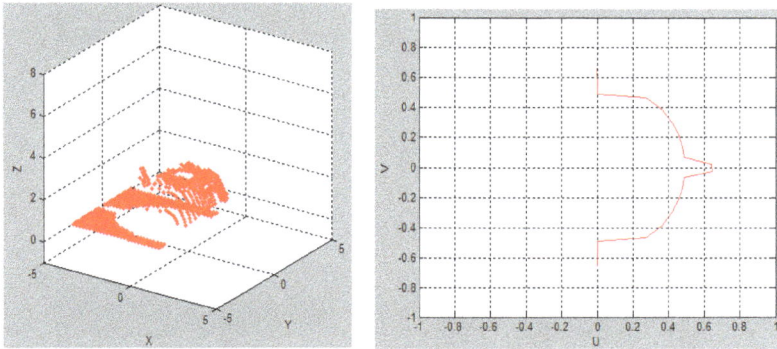

Figure 4. Simulation running when default parameters were used.

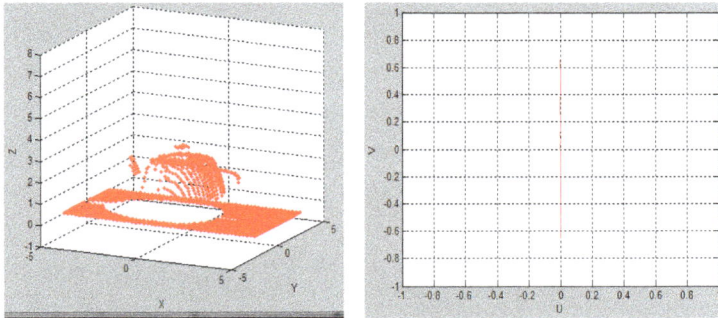

Figure 5. Simulation results at the end of the 3D object scan simulation.

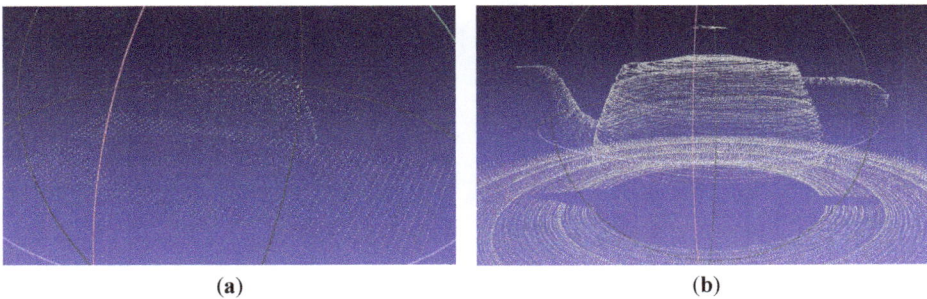

(a)   (b)

Figure 6. (**a**) Scanned teapot (3D point mesh) viewed in MESHLAB; (**b**) Best 3D scanning result (3D point mesh) viewed in MESHLAB, obtained with a 32 PSD sensor array.

3.19. Expansion of the PSD Array Size

As referred to in Section 3.12, the default number of simulated detectors on the sensor is 32 and can be expanded to any desired value, although the proposed sizes are 128, 256, 512 and 1024. The higher the number of detectors, the higher the resolution in the Y-axis and the higher the definition and quality of the 3D scanned object. Such a fact is clearly noticeable in Figure 7a–e, where we can see how the number of points or vertices of the scanned 3D object increases as we increase the number of detectors in the PSD sensor array.

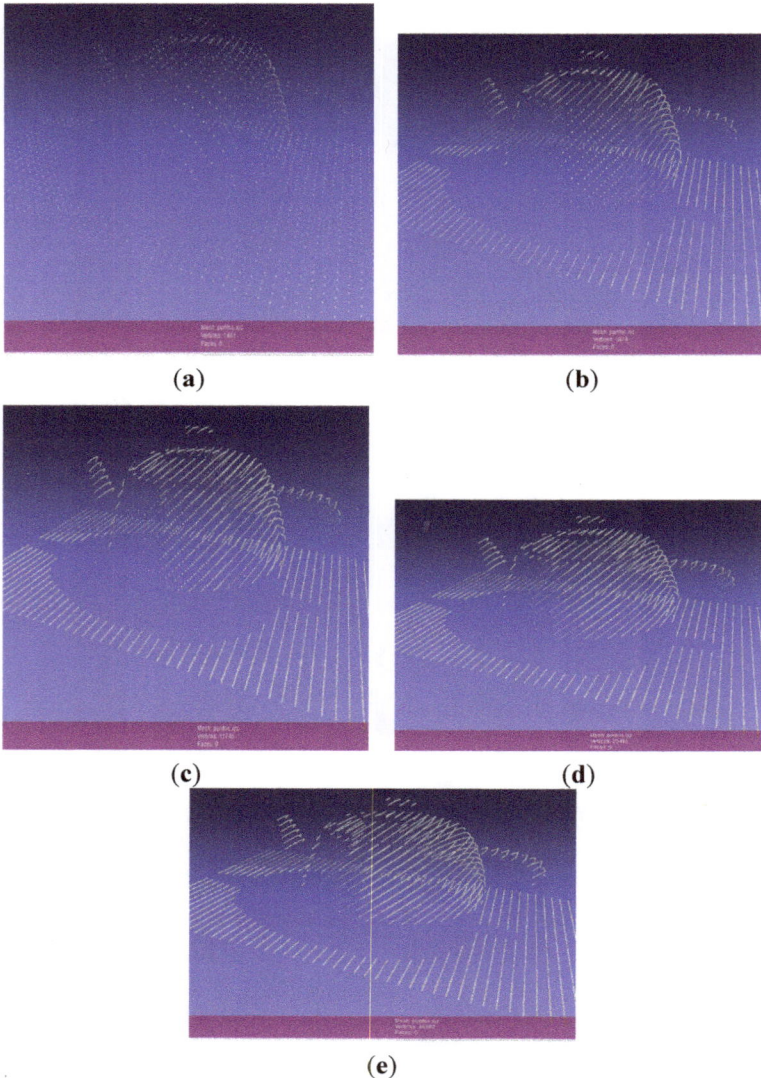

Figure 7. Expansion of the simulated PSD array size: (**a**) 32 PSD; (**b**) 128 PSD; (**c**) 256 PSD; (**d**) 512 PSD; and (**e**) 1024 PSD.

3.20. Object Resolution

The "teapot.ply" object file used in this simulation has 279 vertices and 500 faces. A lower resolution "teapot.ply" file does not affect the outcome of this research since the simulation calculates the point of intersection of the laser line with the surface of the object at each time frame and so it is scanning exactly what comes in the file. Even if the object's resolution is lower or higher, the simulation will still scan the object's surface at the intersection points between the laser line and the object. Therefore, the resolution does not depend on the number of vertices or faces and it will scan it as it is.

Since "teapot.ply" is composed of polygon faces that do not exactly represent the 3D profile of a real teapot composed of curved surfaces, a file with the highest possible resolution will of course best mimic the object being 3D scanned/simulated and provide the most precise estimate of a real teapot.

4. Triangulation Platform Configuration and Formulae

The triangulation configuration sketch and formulae derived by Park and DeSouza (see Figure 6.4 in reference [10]) was used to construct the simulation depicted in Figures 6 and 7 which corresponds to the real physical triangulation 3D sensor system scenario, which is being simulated in this SIMULINK model.

The coordinates of each of the 3D scanned points are calculated in accordance to the properties of similar triangles.

The Z-coordinate is calculated using the following expression [10]:

$$z = \frac{fb}{p + f \tan \theta} \tag{1}$$

where f is the focal length of the camera, p is the image coordinate of the illuminated point, θ is the incident/scanning angle, Xc and Zc (see Figure 6.4 in reference [10]) are two of the three principle axes of the camera coordinate system, respectively, and b (baseline) is the distance between the focal point and the laser along the Xc axis (see Figure 6.4 in reference [10]).

The X-coordinate is calculated using the following expression [10]:

$$x = b - z \tan \theta \tag{2}$$

The Y-coordinate is calculated using the following expression also from the principle of similar triangles:

$$y = \frac{Yq}{z} \tag{3}$$

where y is the distance along individual detectors on the sensor measured from the middle point of the sensor on the Y-axis on the sensor active area, Y is the distance measured from the middle point of the translation table (where object is placed) along the Y-axis within the scanning area, q is the distance between the sensor and the lens which is not the focal distance f, and z is the perpendicular distance between the lens and the object or the Z coordinate.

The error in the Z-coordinate measurement, Δz, is obtained by differentiating the equation of the Z-coordinate (Equation (1)), and the resulting expression is illustrated below [10]:

$$\Delta z = \frac{z^2}{fb} \Delta p + \frac{z^2 \sec^2 \theta}{b} \Delta \theta \tag{4}$$

where Δp and $\Delta \theta$ are the measurement errors of p and θ, respectively.

The results obtained by the physical sensor system (e.g., Figure 2b) when measuring object profiles in 3D conclude that the error of the sensor system when detecting objects in the existing triangulation platform setup is of ~5% in the X-axis and of ~5.6% in the Z-axis, which could be attributed to possible vibrations of the translation table (object) movement during the scanning process as well as to sensor noise. Signal fluctuations (noise in the signal) on the scanned profile define the object measurement error in the Z-axis and the maximum and minimum values of such fluctuations define such error interval.

5. Conclusions and Future Work

It has been demonstrated that a-Si:H 128 PSD sensor arrays and their corresponding systems work correctly and can be used as high speed and high resolution sheet-of-light 3D object scanning systems. The constructed simulation shows a huge potential for the proposed 3D sensor technology, which is clearly able to compete with the most advanced CMOS sensor-based sheet-of-light laser triangulation 3D cameras. Improvements are needed in order to achieve 100% correct sensor position response, especially for the 128 PSD array sensor and such a goal could be attained by fabricating a sensor using a different material, such as nanocrystalline silicon, which hardly degrades the overall position response of the structure over time. Other foreseen restrictions are expected when miniaturizing the sensor, for example to 256, 512 or 1024 elements, now that even if the technology for that purpose exists, problems may occur during material layer fabrication procedures, such as possible short circuiting in sensor channels, *etc*. Other limitations exist within the triangulation platform. Better optics and equipment can be used to improve the 3D detection setup overall. Anti-vibration and sensor noise reduction measures could be introduced, too. However, system hardware and software do not need to be enhanced.

The successful integration of amorphous silicon PSD array sensors into suitable sheet-of-light 3D object rendering systems is now possible and feasible. The quality and realism of the 3D object profiles depends on the correct and accurate position response of each detector from the sensor as well as on the size of the PSD array, meaning that the higher the number of PSDs integrated on the array, the higher the 3D object profile resolution on the discrete sensor axis, and, subsequently, the higher the number of total image 3D scan points.

Acknowledgments: This work was supported by IDF, Valencia, Spain as well as by *"Fundação para a Ciência e a Tecnologia"* (Foundation for Science and Technology, FCT-MEC) through a contract with CENIMAT/I3N and by the projects PEst-C/CTM/LA0025/2011 (Strategic Project—LA 25—2011–2012), PTDC/CTM/099719/2008, ADI—2009/003380, and ADI—2009/005610. The author is very grateful to Jordi Uriel for his relevant contribution to this work. The author would also like to thank his colleagues in CENIMAT for their valuable help on sensor fabrication. Javier Contreras would also like to thank FCT-MEC for providing him with the PhD fellowship SFRH/BD/62217/2009.

Author Contributions: Javier Contreras, Josep Tornero conceived and designed the experiments; Javier Contreras performed the experiments; Javier Contreras, Isabel Ferreira, Rodrigo Martins, Elvira Fortunato, Luis Gomes and Josep Tornero analyzed the data; Josep Tornero, Rodrigo Martins, Elvira Fortunato, Isabel Ferreira and Luis Gomes contributed with support/infrastructures/reagents/materials/analysis tools/know-how; Javier Contreras wrote the paper.

Conflicts of Interest: The authors declare no conflict of interest.

References

1. Raja, V.; Fernandes, K.J. *Reverse Engineering—An Industrial Perspective*; Springer: Berlin, Germany, 2008.
2. Johannesson, M. Can sorting using sheet-of-light range imaging and MAPP2200. In Proceedings of the International Conference on Systems, Man and Cybernetics, "Systems Engineering in the Service of Humans", Iowa, IA, USA, 17–20 October 1993; pp. 325–330.
3. Araki, K.; Shimizu, M.; Noda, T.; Chiba, Y.; Tsuda, Y.; Ikegaya, K.; Sannomiya, K.; Gomi, M. High speed and continuous 3-D measurement system. *Mach. Vis. Appl.* **1995**, *8*, 79–84. [CrossRef]
4. De Bakker, M.; Verbeek, P.W.; vanden Ouden, F.; Steenvoorden, G.K. High-speed acquisition of range images. In Proceedings of the 13th International Conference on Pattern Recognition, Vienna, Austria, 25–29 August 1996; pp. 293–297.

Sensors **2015**, *15*, 29938–29949

5. Martins, R.; Costa, D.; Aguas, H.; Soares, F.; Marques, A.; Ferreira, I.; Borges, P.; Pereira, S.; Raniero, L.; Fortunato, E. Insights on amorphous silicon nip and MIS 3D position sensitive detectors. *Mater. Sci. Forum.* **2006**, *514*, 13–17. [CrossRef]

6. Martins, R.; Fortunato, E.; Figueiredo, J.; Soares, F.; Brida, D.; Silva, V.; Cabrita, A. 32 Linear array position sensitive detector based on NIP and hetero a-Si:H microdevices. *J. Non Cryst. Solids* **2002**, *299*, 1283–1288. [CrossRef]

7. Contreras, J.; Ferreira, I.; Idzikowski, M.; Filonovich, S.; Pereira, S.; Fortunato, E.; Martins, R. Amorphous silicon position sensitive detector array for fast 3D object profiling. *IEEE Sens. J.* **2012**, *12*, 812–820. [CrossRef]

8. Contreras, J.; Gomes, L.; Filonovich, S.; Correia, N.; Fortunato, E.; Martins, R.; Ferreira, I. 3D scanning characteristics of an amorphous silicon position sensitive detector array system. *Opt. Exp.* **2012**, *20*, 4583–4602. [CrossRef] [PubMed]

9. AT-Automation Technology GmbH, Bad Oldesloe, Germany, 23843. Available online: www.automation technology.de (accessed on 6 October 2015).

10. Park, J.; DeSouza, G.N. 3D modelling of real-world objects using range and intensity images. In *Innovations in Machine Intelligence and Robot Perception*; Patnaik, S., Jain, L.C., Tzafestas, G., Bannore, V., Eds.; Springer-Verlag: Berlin, Germany, 2005.

sensors

MDPI

Article

An Indoor Obstacle Detection System Using Depth Information and Region Growth

Hsieh-Chang Huang [1,2], Ching-Tang Hsieh [2,*] and Cheng-Hsiang Yeh [2]

[1] Department of Information Technology, Lee-Ming Institute of Technology, New Taipei City 24346, Taiwan; sanmic@mail.lit.edu.tw
[2] Department of Electrical Engineering, Tamkang University, New Taipei City 25137, Taiwan; porkp617@hotmail.com
* Correspondence: hsieh@ee.tku.edu.tw; Tel.: +886-2-2621-5656; Fax: +886-2-2620-9814

Academic Editor: Gonzalo Pajares Martinsanz
Received: 16 June 2015; Accepted: 9 October 2015; Published: 23 October 2015

Abstract: This study proposes an obstacle detection method that uses depth information to allow the visually impaired to avoid obstacles when they move in an unfamiliar environment. The system is composed of three parts: scene detection, obstacle detection and a vocal announcement. This study proposes a new method to remove the ground plane that overcomes the over-segmentation problem. This system addresses the over-segmentation problem by removing the edge and the initial seed position problem for the region growth method using the Connected Component Method (CCM). This system can detect static and dynamic obstacles. The system is simple, robust and efficient. The experimental results show that the proposed system is both robust and convenient.

Keywords: obstacle detection; Kinect; depth map; travel aid

---

1. Introduction

According to new statistics [1], there are 285 million visually impaired people relying on the guide cane or guide dogs to move around freely in the world. However, not every visually impaired person can easily pair successfully with guide dogs and there is often a long wait for an animal.

Most visually impaired people use a cane to touch an obstacle, to assess the position of the obstacle and avoid it. Sometimes at the point when they touch the obstacle, the danger is unavoidable. These two methods for travel are neither convenient nor safe. Using computer vision technology reduces this problem. The efficient detection of obstacles is important. In recent years, there have been many developments in computer vision for this field. Many studies have proposed obstacle detection methods. In [2] Obstacle detection can be classified into three categories: Electronic travel aids (ETAs), electronic orientation aids (EOAs) and position locator devices (PLDs). However, this paper classifies obstacle detection into three categories. One uses non-depth information, a second uses depth information and the third uses neither.

There are many proposed methods for the first category, such as [3–8]. Ma *et al.* [3] proposed an object detection algorithm that uses edges and motion. The motion-information is used to determine the dynamic obstacles and the edge-information is used to determine obstacles. This information is combined with free space detection to determine the position of the obstacles. Zhang *et al.* [4] proposed an obstacle detection algorithm that uses a single camera. This uses edge detection to segment objects. However, these methods require a simple texture for the surface of the ground. Chen *et al.* [5] proposed an obstacle detection method that uses a saliency map. This uses a threshold value to determine the position of the obstacles. However, this method requires that there are few obstacles in the execution environment. Ying *et al.* [6] proposed an obstacle detection method that uses a gray-scale image. This method searches the region of interest (ROI) in the gray-scale image and then determines the location of obstacles. However, this method uses a gray-scale image, so it is easily affected by illumination. These methods are very robust if there is sufficient light, but not if there is insufficient light. The proposed system uses Kinect directly to capture the depth map, so it addresses these drawbacks.

The second category of methods for obstacle detection is been proposed in [9–24]. These methods detect obstacles using depth information. This is obtained from various capture devices, such as stereovision cameras, Leap Motion controllers [25], laser rangefinders [26], RealSense 3D Cameras [27] or Kinect sensors. Zollner *et al.* [8] just given a proof-of-concept idea of a mobile navigational aid, but the implementation of the proposed Kinect application was lacked. Filipe *et al.* [10] applied Neural Network to extract the features from the depth information captured by Kinect sensor and the extracted features are enabled to detect possible obstacles. In general, depth information of obstacles is really similar to the surrounding floor (ground plane) and the trained NN may be hard to separate the obstacles from the floor. Hotaka *et al.* [11] proposed Kinect cane system and tactile inform system, that is different from ours. Above three papers don't remove ground plane from depth map. However, our proposed system resolves the over-segmentation problem by removing the edge and the initialize seed position problem for the growth method (RGM) using the Connected Component Method (CCM). The RGM concept is simple. We only need a certain numbers of seed point to represent the property we want, then grow the region. The vocal inform system of our proposed system is more intuitive. And we do not change cane of visually impaired people. Zhang *et al.* [12] proposed an obstacle detection algorithm that uses a U-V disparity map analysis. This combines straight-line fitting and the standard Hough Transform [28] to determine the location of obstacles. However, the U-V disparity map is generated using two webcams, so the degree of illumination affects the performance of the system. In [13], Gao *et al.* use a 3D camera to obtain the depth map. This study combines straight-line fitting, the standard Hough Transform and a U-V disparity map to determine the location of obstacles. Choi *et al.* [14] used a Kinect sensor to obtain color images and depth maps (RGB-D images). This study uses edge detection for both color images and depth maps and then processes these edge images by morphology [29]. The results for the two images are then combined to determine the position of obstacles. However, the color image used in this study is still affected by illumination and the ground plane affects obstacle detection. The proposed system addresses these two problems.

For the third category of systems for obstacle detection, Brock *et al.* [30] used a vibrotactile belt to convey the position and distance to an obstacle using the position and strength of the vibrations. For more detail about a vibrotactile belt, please refer to [31]. The vOICe's Glasses for the Blind [32] are a wearable device that is equipped with a webcam and translates video data into a sound stream. Mann *et al.* [33] presented a novel head-mounted navigational aid that uses Kinect and vibrotactile devices built onto a helmet.

The method detailed in [34] does not process the ground, but segments object directly to calculate the standard deviation using an object's depth value and then determines whether it is an obstacle using the scale of the object's standard deviation. Although this detection method is simple, smaller objects on the ground are not detected. The proposed system filters the ground out before obstacle detection is begun, so this issue is eliminated. The system used in [35] is an autonomous navigation system that uses a finite state machine that is taught by an Artificial Neural Network (ANN) in an indoor environment. The system used in [36] uses machine learning for this field. The design goals for the proposed system are cost-efficiency, robustness and convenience. The system must address the ground plane problem, in order to detect rising stairs, descending stairs and static and dynamic obstacles.

The remainder of the paper is organized as follows. Section 2 gives a system overview and the details of the system. Section 3 gives the experimental results for different environments and the experimental results for two blind subjects and ten blindfolded subjects. Finally, a conclusion and details of future work are given in Section 4.

2. Proposed Methods

2.1. System Architecture

The proposed system flowchart is shown in Figure 1. Firstly, the morphology is dilated and eroded to remove the distracting noise of the depth map and the Least Squares Method (LSM) in a quadratic polynomial is used to approximate ground curves and to determine the ground height threshold in the V-disparity. The system then searches for dramatic changes in the depth value, depending on the ground height threshold, to determine stair-edge points. The Hough Transform is then used to determine the location of the drop line [37]. In order to strengthen the characteristics of the different objects and to overcome the drawbacks of the region growth method [38], edge detection is used to remove the edge. The ground height threshold and the features of the ground are then used to remove ground plane. The system then uses the region growth method to label the tags on different objects and analyzes each object to determine whether the object is a stair. Finally, the system allows users to navigate and gives them a vocal message about the distance to the obstacle and the obstacle category using Text To Speech (TTS).

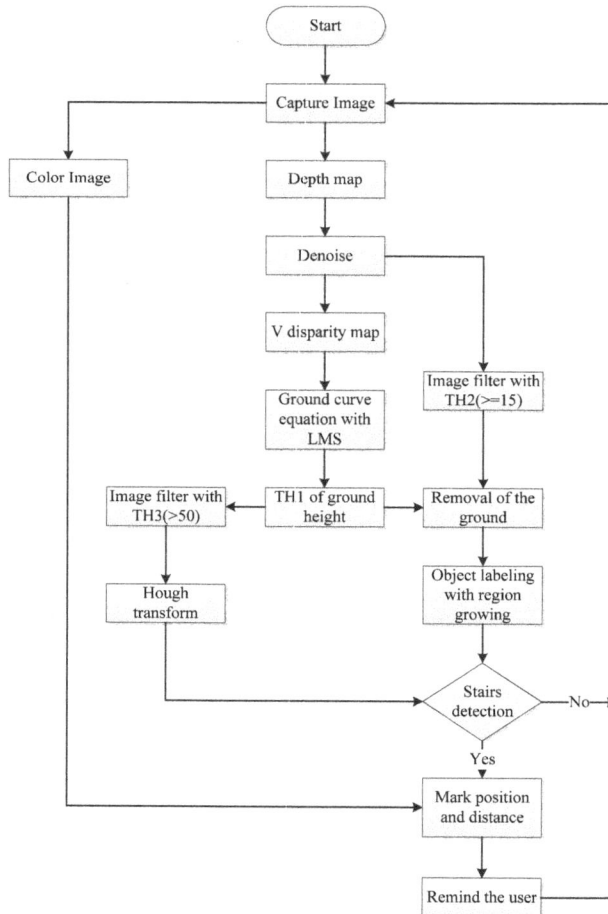

Figure 1. The system flowchart.

2.2. Noise Reduction

Because of the limitations of the Kinect hardware, a depth map can be broken. In order to make the depth map more complete, some simple morphology processing is used. This paper uses a closing operation for morphology to repair the black broken areas. Figure 2 shows that the processed depth maps are better than the original depth maps.

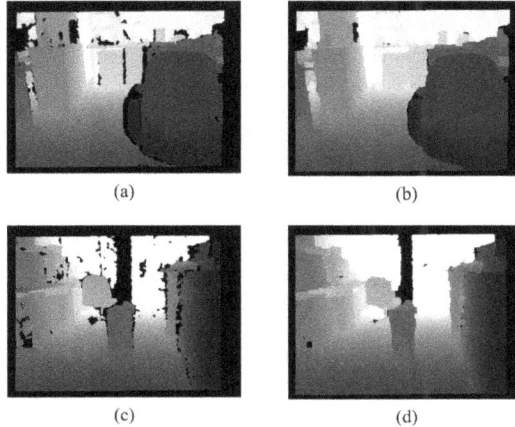

Figure 2. Noise Removal. (**a**) Original depth map; (**b**) Processing result; (**c**) Original depth map; (**d**) Processing result.

2.3. Ground Height Detection

A UV disparity map is composed of the U disparity map and the V disparity map from the depth map. Figure 3 shows that the V-disparity [39] concept simplifies the process of separating obstacles in an image, where "V" corresponds to the vertical coordinate in the (u, v) image coordinate system. Similarly, the U-disparity concept simplifies the process of separating obstacles in an image, where "U" corresponds to the vertical coordinate in the (u, v) image coordinate system.

A UV disparity map [40] is a statistical method that is similar to a histogram. However, the statistical target is different. The proposed system only uses V-Disparity because the effect is better. Figure 4a shows that this table is a depth map. The statistics for different depth values are gathered, row-by-row, and the results are shown in Figure 4b. For example, there are 15 zeros in row one in Figure 4a, so the position of Row 2 and Column 1 in Figure 4b records this value (15). This means that the depth value, 0, has an image height of 15.

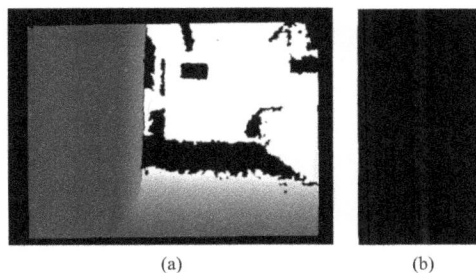

Figure 3. The relationship between the depth map and the V-disparity. (**a**) Depth map; (**b**) V-disparity.

Image Width

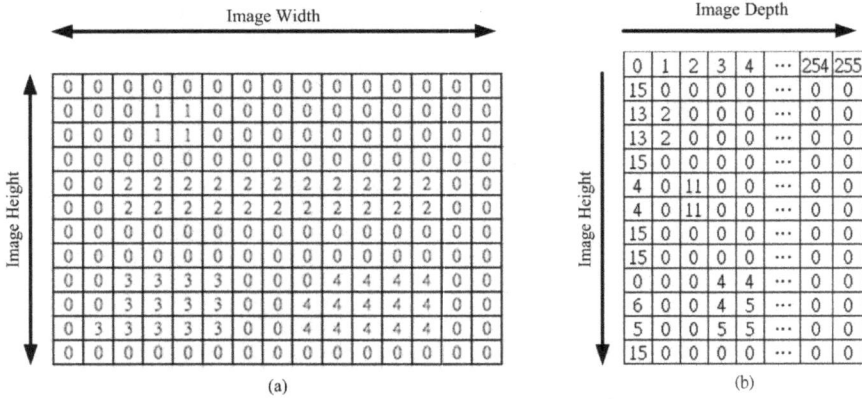

0	0	0	0	0	0	0	0	0	0	0	0	0	0	0
0	0	0	1	1	0	0	0	0	0	0	0	0	0	0
0	0	0	1	1	0	0	0	0	0	0	0	0	0	0
0	0	0	0	0	0	0	0	0	0	0	0	0	0	0
0	0	2	2	2	2	2	2	2	2	2	2	2	0	0
0	0	2	2	2	2	2	2	2	2	2	2	2	0	0
0	0	0	0	0	0	0	0	0	0	0	0	0	0	0
0	0	0	0	0	0	0	0	0	0	0	0	0	0	0
0	0	3	3	3	3	0	0	0	4	4	4	4	0	0
0	0	3	3	3	3	0	0	4	4	4	4	4	0	0
0	3	3	3	3	3	0	0	4	4	4	4	4	0	0
0	0	0	0	0	0	0	0	0	0	0	0	0	0	0

Image Height (vertical label, left side)

(a)

Image Depth

0	1	2	3	4	⋯	254	255
15	0	0	0	0	⋯	0	0
13	2	0	0	0	⋯	0	0
13	2	0	0	0	⋯	0	0
15	0	0	0	0	⋯	0	0
4	0	11	0	0	⋯	0	0
4	0	11	0	0	⋯	0	0
15	0	0	0	0	⋯	0	0
15	0	0	0	0	⋯	0	0
0	0	0	4	4	⋯	0	0
6	0	0	4	5	⋯	0	0
5	0	0	5	5	⋯	0	0
15	0	0	0	0	⋯	0	0

Image Height (vertical label, left side)

(b)

Figure 4. A schematic diagram of the V disparity map. (**a**) Depth map; (**b**) V disparity map.

The detection needs for subsequent steps require that noise must be removed from the captured depth map this must be projected into the V disparity map, as shown in Figure 5. The Y-axis height of the V disparity map corresponds to the Y-axis height of the depth maps, as shown in Figure 5, so the vertical length of an image represents the height of the actual object in the image. If the object is closer to the right side of the depth map, the distance between the object and the sensor is greater. The greater the pixel value in the V disparity map, the bigger is the object in the image. The normalization equation for the cumulative amount of depth is shown in the following equation. The cumulative value must be between 0 and 255. The cumulative value is statistical value of depth value of the row of the V disparity map image, and the Max cumulative value is image wide value of the depth map:

$$Depth\ cumulative\ value = \frac{cumulative\ value}{Max\ cumulative\ value} * 255 \tag{1}$$

According to [11], the ground is a rising curve in a V disparity map. The LSM is used to determine the equation of the curve, as shown in Figure 6 and Equation (2).

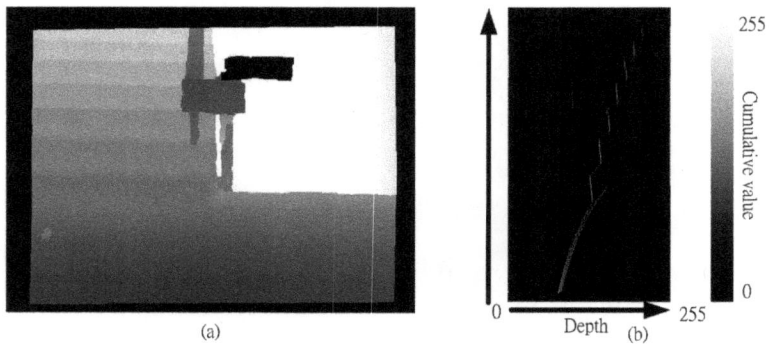

Figure 5. (**a**) The depth map with noise removed and (**b**) the V disparity map image.

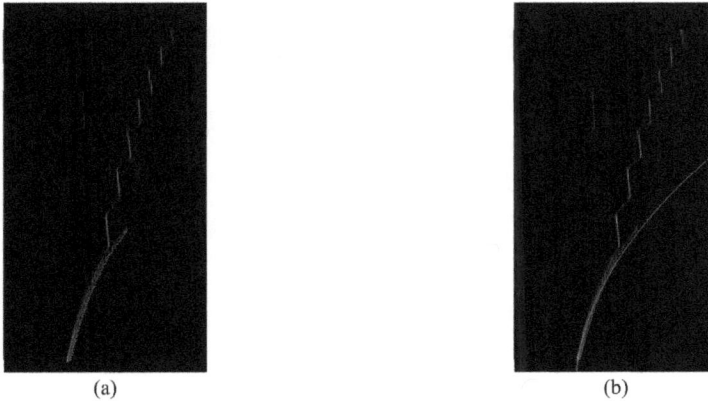

(a)   (b)

Figure 6. The ground curve in V disparity map. (a) Segment consisting of points (red); (b) The line of the equation (red).

$$ay^2 + by + c = d \tag{2}$$

where a, b and c respectively represent the parameters of the equation, y is the image height and d is the horizontal axis value (0 to 255) in the V-disparity map. However, we want to find out a quadratic equation to closer ground curve strip, then use it to remove ground plane. The ground plane is not only a simple line in the V-disparity map. Because pixels that are the same height in a depth map can have a different depth value, the curve becomes a strip, so several approximation targets, such as the minimum, the maximum, the mean and the specific value of every row of V-disparity map are used (the rightmost value of the strip, the leftmost value of the strip, the middle value of the strip on x-axis).When the obstacle is on the ground, these methods do not work. To address this problem, the proposed method uses the quadratic offset equation, which is shown as Equation (3):

$$TH1 = ay^2 + by + c - offset = d - offset \tag{3}$$

where *TH*1 is the shifted threshold depending on the ground height. The ground height threshold value indicates a height in the depth map and the minimum value cannot be less than *TH*1. The appropriate offset value is 35, which is obtained through experience. The offset value affects the removal of the ground, so several offset values, such as the minimum, the maximum, the mean and the specific value, are tried. The offset value controls the location of the approximation curve for the disparity map. The quadratic offset equation is the fastest and simplest method. Comparing the disparity map in Figure 7 with that in Figure 8, it is seen that the depth value of the ground plane (background) is greater than the depth value of the obstacle (foreground) for the same height. Figure 9 shows that the mean method (no offset) does not completely remove the ground plane. Therefore, the maximum method does not remove the ground plane either. In contrast, the minimum method is perhaps the best, but the depth of the obstacle interferes with this method. Because the depth value for the background is greater than the depth value for the foreground for the same height in the V-disparity, the minimum method cannot be used directly. Using the LSM to subtract the specific value is the best method, as shown in Figure 10. Figure 11 shows that Equation (3) improves the robustness of the system.

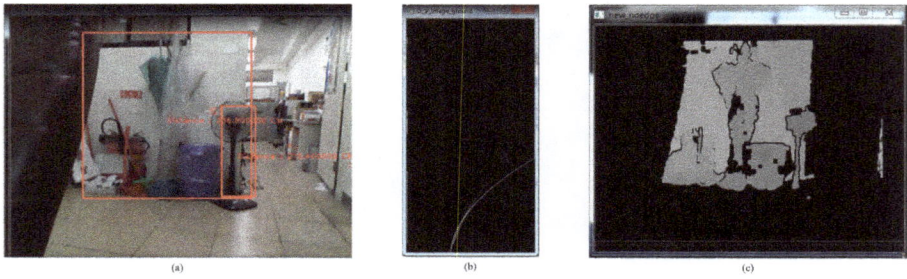

Figure 7. The scene without people. (**a**) Real scene; (**b**) V-disparity; (**c**) Depth map.

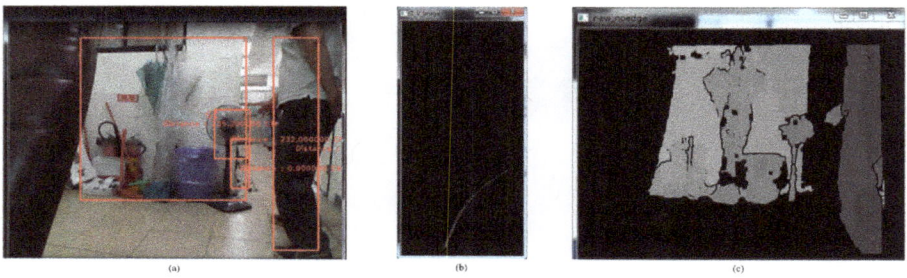

Figure 8. The scene with people. (**a**) Real scene; (**b**) V-disparity; (**c**) Depth map.

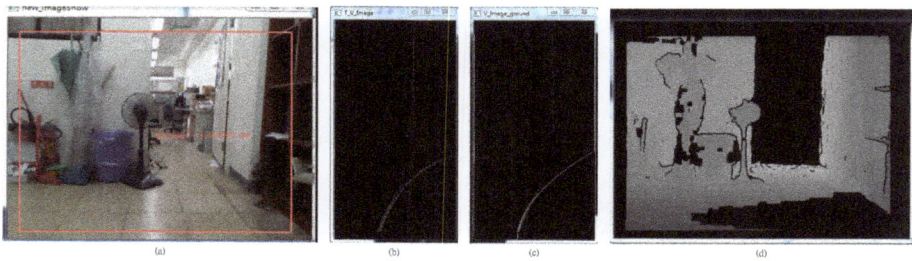

Figure 9. No offset. (**a**) Real scene; (**b**) No LSM Curve; (**c**) LSM Curve without offset; (**d**) Depth map.

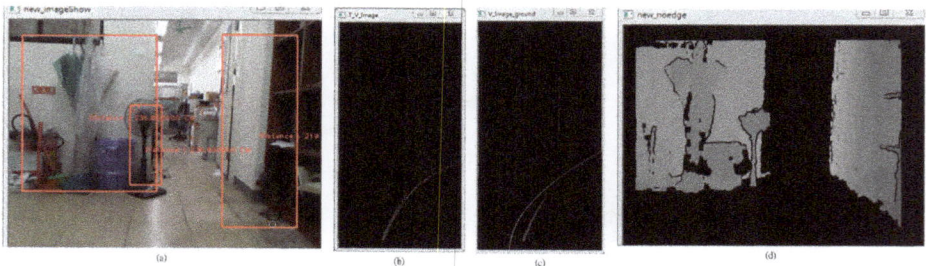

Figure 10. Offset value = 20. (**a**) Real scene; (**b**) No LSM Curve; (**c**) LSM Curve without offset; (**d**) Depth map.

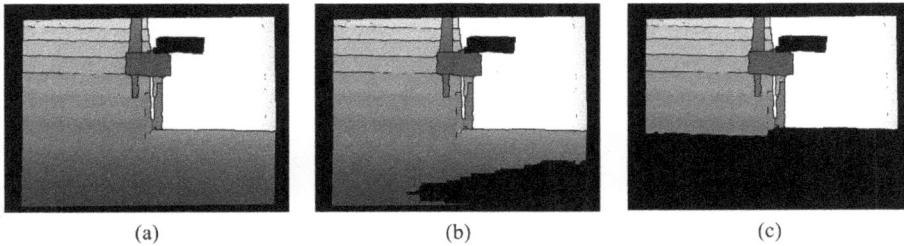

Figure 11. The result of the offset. (**a**) Original depth map; (**b**) Result image before offset; (**c**) Result image after offset.

2.4. Removal of the Edge

In the depth map, the depth represents the distance between the objects and the sensor. The variation in depth demonstrates whether the obstacles are the same. Variations in depth are usually not too significant for a specific object. If there are different objects, the relationship between the distances causes a significant variation in the depth. In this paper, in order to clarify the characteristics of different objects, the strong edge is removed. There are many edge detection methods, such as Roberts, Prewitt, Sobel, Laplace and Canny. In this paper, a function to detect the edge uses the following Equation (4):

$$P(x,y) = \begin{cases} 0 & , if \sum_{x_n,y_n \in S_n} |P(x_n,y_n) - P(x,y)| \geq TH2 \\ unchange & , others \end{cases} \tag{4}$$

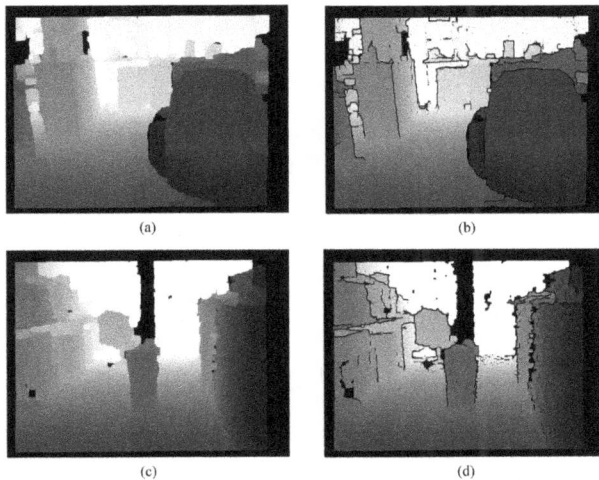

Figure 12. Removal of the edge. (**a**) Noiseless image; (**b**) Processing result; (**c**) Noiseless image; (**d**) Processing result.

The processing result is shown in Figure 12. Here, $P(\cdot)$ represents the pixel value of the coordinates (x,y) and $TH2$ represents the threshold. If $P(x_n,y_n)$ is $P(x,y)'s$ neighboring pixel and S_n is a set of $P(x,y)'s$ neighboring pixels and the image is traversed using Equation (4), then the edges in the image can be detected. When all of the edges in the depth map are found, objects can be isolated, so segmentation is accurate.

2.5. The Detection of Descending Stairs

In this section, a method to search and record points that exhibit significant variation from the noiseless image is proposed. In this study, the pixel values are larger than the setting threshold (50) and are defined as significant variation. The ground height threshold (*TH3*) is then used to filter out possible points, as shown in Figure 13a. These depth values of vertical adjacent point are very difference. After filtering, they become a group of points. We call these points "possible points". In depth map, the Hough Transform technique transforms the possible points into edge line of descending stairs. The Hough Transform technique then transforms the filtered points into a horizontal line, as shown in Figure 13b.

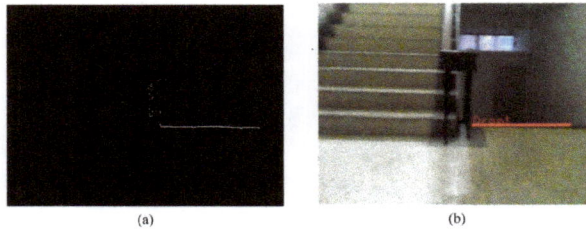

(a) (b)

Figure 13. The results for the detection of descending stairs. (**a**) Suspicious points of downstairs depth map; (**b**) The results of the Hough transform.

2.6. Removal of the Ground

If connected component labeling or other labeling methods are directly used to label tags, it is difficult to separate the obstacles from the ground, because the junctions between the ground and the obstacles have the same depth value. Therefore, the information for the ground must be removed. RANSAC plane fitting [35,37] is used to determine the ground plane in the 3D space. Because the sensor cannot be fixed, the calculation of the ground information requires an iterative approach. In order to improve the speed of the system, [38] and the following information are used to filter out the ground: (1) The ground is usually relatively flat and (2) Using the information on depth, the gray value varies from large to small (from far to near). (3) Only the large areas of the ground are required, so Equation (5) is used. Using these features, the planes of interest meet three conditions. The regions and the sizes of the different planes of interest are determined and then the ground plane is removed using Equation (5), which has a large area. The processing result is shown in Figure 14. These separated objects are label as different color in Figure 15. The least squares method (LSM) in a quadratic polynomial is used to approximate the ground curves and to determine the ground height threshold in the V-disparity:

$$Ground(x,y) = \begin{cases} 0 & , P(x,y) - P(x,y-n) \geq 2 \\ & \wedge \bigcap_{i=0}^{n-1} P(x,y-i) - P(x,y-i-1) \geq 0 \\ & \wedge P(x,y) > TH1 \\ unchanged & , others \end{cases} \quad (5)$$

where $P(\cdot)$ and $Ground(\cdot)$ represent the pixel value of the coordinates (x,y), n determines the range ($n = 10$) and $TH1$ represents the threshold ($TH1 = 35$). These characteristics of ground plane in depth map must meet the following three points: (1) Depth values of horizontally adjacent points of ground plane are almost the same; (2) Depth values of vertically adjacent points of ground plane are gradient; (3) Depth values must be greater than $TH1$.

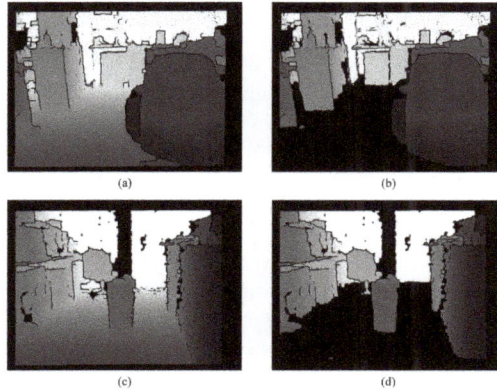

Figure 14. Removal of the ground. (**a**) Edge removed image 1; (**b**) Processing result of (**a**); (**c**) Edge removed image 2; (**d**) Processing result of (**c**).

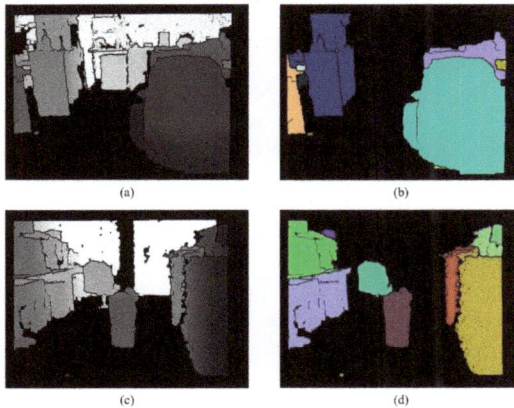

Figure 15. Labeling. (**a**) Ground removed image; (**b**) Labeling result; (**c**) Ground removed image; (**d**) Labeling result.

2.7. Labeling

The reason of using the labeling is easy to observe the experiment. After observations, we can stop this function, and then the performance is better. The Connected Component Method (CCM) and the region growth method [13,41] are the most common methods of labeling. The connected component method is used for a 2-D binary image. It scans an image, pixel-by-pixel (from top to bottom and left to right), in order to identify connected pixel regions, *i.e.*, regions of adjacent pixels, that share the same set of intensity values. CCM can be either 4-Connected Component or 8-Connected Component for two dimensions. The Connected Component Method can be a 6-connected neighborhood, an 18-connected neighborhood, or a 26-connected neighborhood for three dimensions. The disadvantage of the connected component method is that it is time-consuming.

A Region Growth Algorithm (RGA) is a simple, region-based image segmentation method. RGA is suitable for a gradient image. A Seeded Region Growth Method (SRG) [42] is a type of RGA. SRG is rapid, robust and allows free tuning of a parameter. SRG is faster than CCM, but it allows over-segmentation there is a problem with the initial positions of seeds. We briefly conclude the advantages and disadvantages of region growing. The advantages of region growing are as follows: (1) Region growing methods can correctly separate the regions that have the same properties we define; (2) Region growing methods can provide the original images, which have clear edges the good segmentation results; (3) The concept is simple. We only need a small numbers of seed point to represent the property we want, then grow the region; (4) We can determine the seed point and the criteria we want to make; (5) We can choose the multiple criteria at the same time; (6) It performs well with respect to noise. The Disadvantage of region growing as following: Noise or variation of intensity may result in holes or over-segmentation. We proposed system could solve this disadvantage of region-growing techniques.

The sensing range of Kinect is 0.8 to 4.0 m. When the range is greater than the maximum distance, it cannot determine the distance, so the distant information must be removed. In order to measure distances accurately, the distance information for less than 3 m is retained.

Different tags are then placed on different objects. The general labeling methods use eight connected component labeling and region growth, but tag harmonization for connected component labeling requires much iteration, because of the complex shape of the connected area:

$$
S(i,j) = \begin{cases} (i,j), \; if \; [\; (P(i-1,\,j-1)=0) \\ \quad \wedge (P(i,j-1)=0) \\ \quad \wedge (P(i+1,j-1)=0) \\ \quad \wedge (P(i-1,j)=0) \\ \quad \wedge (P(i,j) \neq 0)] \\ not\; seed,\; others \end{cases} \tag{6}
$$

Equation (6) is 8-connnected of image processing. According to neighbor state of $P(i,j)$, to determine $P(i,j)$ belongs to which seed (classification). Here, $S(i.j)$ represents the seed coordinate and $P(i,j)$ represents the pixel value at the coordinate (i,j).

In order to increase the efficiency of the system, Connected Component Region Growth is used. Traditional region growth initially sprinkles some seeds in the image. If the distribution of the sprinkled seeds is not appropriate, the growth results are imperfect, so the choice of the initial position of the seeds is improved in the proposed system. Information about object edges is used. Because the previous step removes the edge information for an object, each object is isolated by black color. Equation (6) and the mask for the initial seed are used to select the coordinates of initial seeds, as shown in Figure 16. These coordinates are then used to execute region growth. This ensures that each object has an initial seed and that any growth is not been repeated. Therefore, a system to reduce the amount of computation is proposed. The processing result is shown in Figure 17.

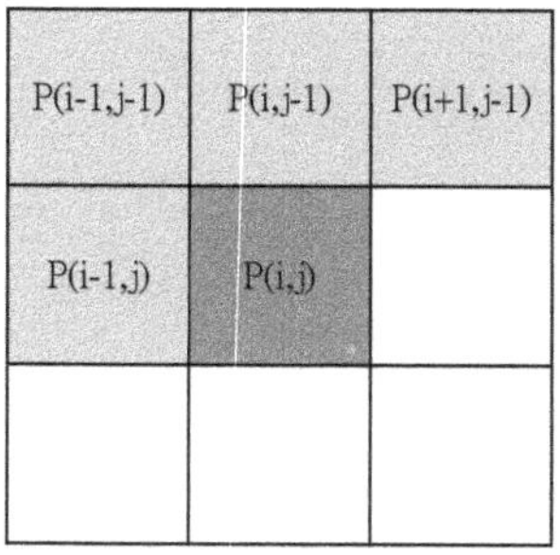

P(i-1,j-1)	P(i,j-1)	P(i+1,j-1)
P(i-1,j)	P(i,j)	

Figure 16. The mask for the initial seed.

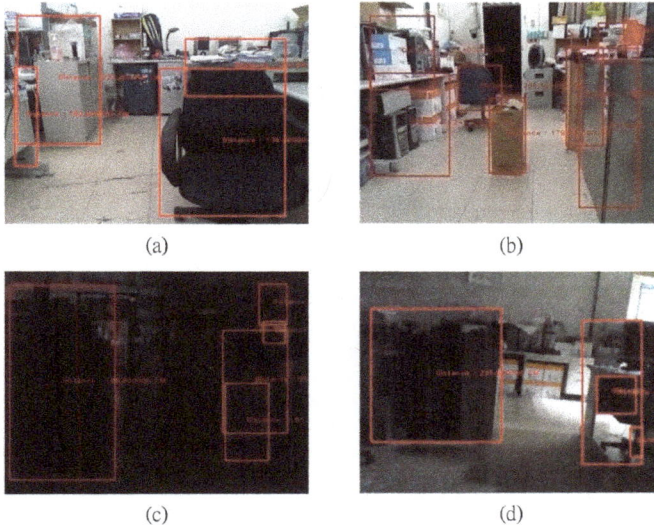

Figure 17. The results for obstacle detection. (**a**) Bright indoor; (**b**) Bright indoor; (**c**) Low-light indoor; (**d**) Low-light indoor.

2.8. The Detection of Rising Stairs

The system then analyzes each of the tagged objects individually, to determine whether the object is rising stairs because of a change in depth. The rising stairs depth value has a hierarchical characteristic, from top to bottom and from large to small. When the obstacle fulfills these characteristics, it is determined to be rising stairs. The detection results are shown in Figure 18.

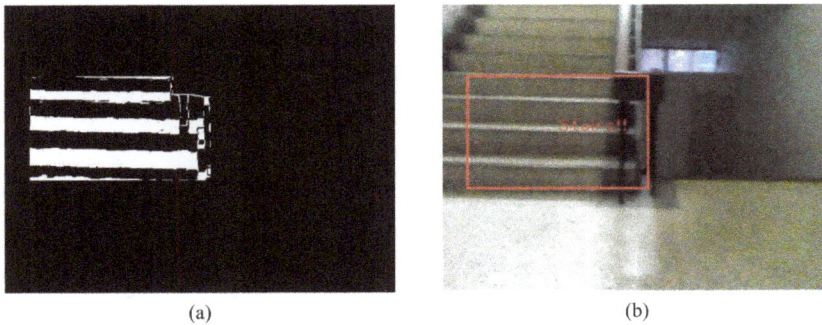

Figure 18. The detection of rising stairs. (**a**) Satisfied conditions of a suspicious plane; (**b**) Upstairs detection image.

2.9. The Labeling of Objects and Informing the User

This system labels objects with rectangle. It shows the information about detected objects on the image and the distance of the obstacle or the staircase. The results are shown in Figure 19.

Finally, the system uses Text-To-Speech (TTS) software [43]. When the obstacle is in front of the user, the system vocally informs the user of the distance to the obstacle and the obstacle category. When the system detects stairs, it gives the direction and the distance to the stairs to the user to ensure the user's safety. This vocal alarm is very short and focuses on concise information about the closest obstacle.

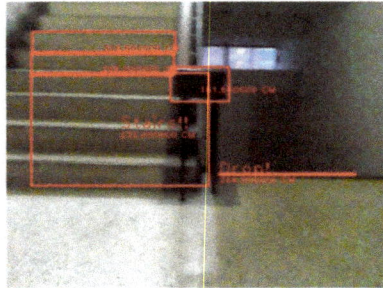

Figure 19. The result of labeling.

3. Experimental Results

A Microsoft Kinect sensor is a tool that captures images, as shown in Figure 5 and Table 1. The experimental platform is Windows 7. The programming language is Visual C++ 2010 with Opens 2.3, running on a notebook with an Intel(R) Core(TM) i5-3210M CPU@2.5GHz 8G 64 bits. The image resolution is 640 × 480 and the depth map capture rate is 30 frames per second. The sensing range is 0.8 to 4.0 m.

A Kinect sensor uses structured light methods to give an accurate depth map of a scene. Both the video and depth sensor cameras in the Kinect sensor have a 640 × 480-pixel resolution and run at 30 FPS (frames per second). There are two cameras and an IR projector. One camera is for color video and the other one with the IR Projector is for the depth map. Currently, there are two categories of SDK for Kinect: Open NI and Microsoft Kinect for Windows SDK.

Kinect configuration height and distance accuracy are related. If possible, the Kinect sensor keeps horizontally that experiment results are better. The Kinect sensor configuration is as shown in Figure 20. Our Kinect sensor is totally fixed on a helmet or chest and waist.

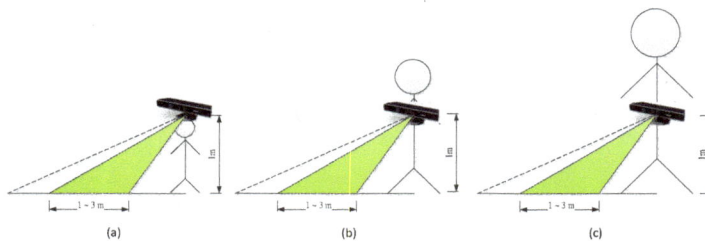

Figure 20. The Schematic diagram of the Kinect configuration for different stature person. (**a**) A short person; (**b**) A medium stature person; (**c**) A tall person.

Infrared rays are easily affected by sunlight [44]. The Kinect sensor depends on emitted infrared rays to generate a depth map, so the Kinect sensor has some hardware limitations. The Kinect sensor is easily affected by sunlight, so it can only be used for environments that lack sunlight, such as a night scene, a cloudy day or indoors. It is worthy of note that the Kinect sensor is not totally useless outdoors, but it cannot be used in sunny environments.

In this section, all of the experiment images are random images taken from the experiment. The experiments are divided into two different environments: simple and complicated. A simple environment does not include stairs and a complicated environment has stairs. Both environments are situated indoors and outdoors, with sufficient and insufficient light. The experiments use different brightness values for the indoor and outdoor environments and for with stairs and without stairs. Figure 17a,b shows the results for a bright indoor environment. Figure 17c,d shows the results for a low-light indoor environment. When obstacles are in front of the user, the system vocally informs the user of the distance to the obstacle.

3.1. System Testing in a Simple Environment

This section details the success rate for obstacle detection in a simple environment without stairs. In this study, an object that affects the path of a user is defined as an obstacle. If an obstacle is labeled, the detection is successful. If not, there is a failure to detect.

3.2. An Indoor Environment under Sufficient Light

The detection success rate and the failure rate are shown in Table 1. As shown in Figure 21, indoor ground is flatter than outdoor ground so the projection distribution of the ground in V-disparity is more concentrated. The success rate is excellent when the ground in the depth map is removed using the ground height threshold in the V-disparity. There are some failures due to the material nature of objects, such as a large expanse of transparent glass or smooth metal.

Table 1. The success rate and the failure rate for the detection of obstacles.

	Frame Amount (Total 2265 frames)	**Percentage (%)**
Success	2201	97.17%
Failure	64	2.83%

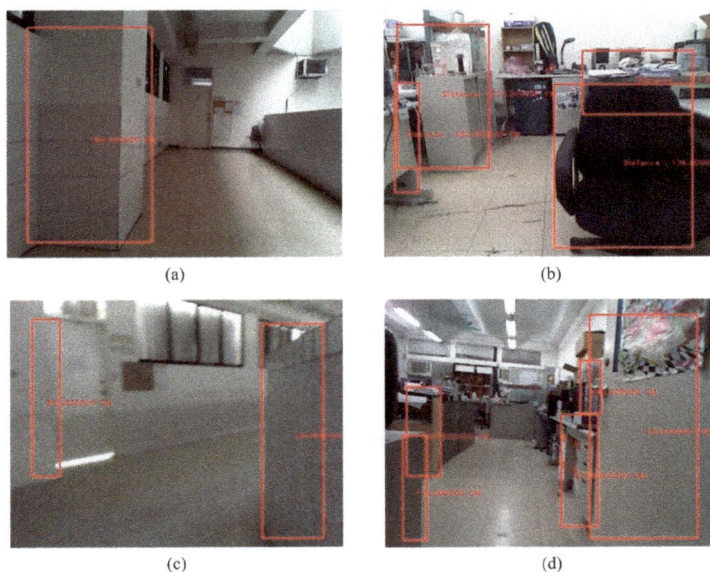

(a)   (b)

(c)   (d)

Figure 21. The detection of an obstacle indoors under sufficient light. (**a**) Corridor 1; (**b**) Laboratory 1; (**c**) Corridor 2; (**d**) Laboratory 2.

3.3. An Indoor Environment under Insufficient Light

The detection success rate and the failure rate for obstacle detection are shown in Table 2. As shown in Figure 22, the depth information is not affected by illumination because it is obtained from the Kinect sensor. Indoor ground is flatter than outdoor ground so the projection distribution of the ground in V-disparity is more concentrated. The success rate is excellent when the ground in depth map is removed using the ground height threshold in the V-disparity. The nature of the material of an object in the scene influences the success rate, for example, glass or metal.

Table 2. The success rate and the failure rate for obstacle detection.

	Frame Amount (Total 213 frames)	Percentage (%)
Success	206	96.71%
Failure	7	3.29%

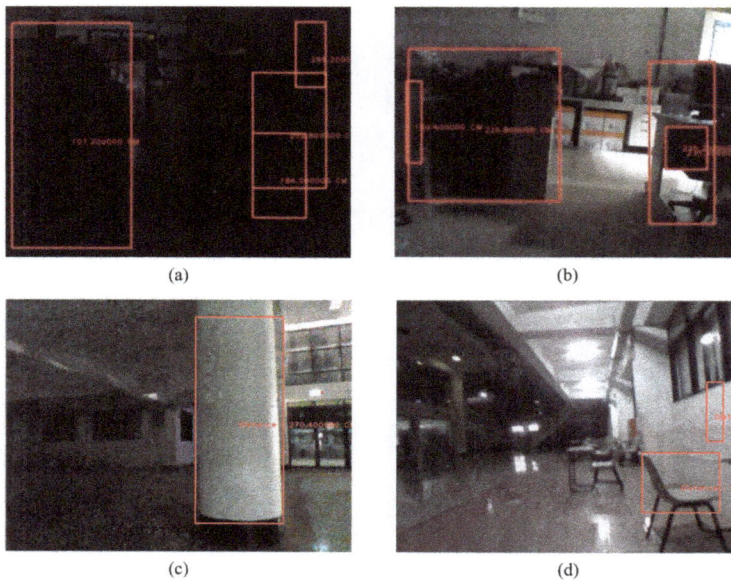

Figure 22. The detection of an indoor obstacle under insufficient light. (**a**) Laboratory 1; (**b**) Laboratory 2; (**c**) Lobby; (**d**) Corridor.

3.4. System Testing in a Complicated ENVIRONMENT

If the test environment contains stairs, it is defined as a complicated environment. The basic structure of the stairs is shown in Figure 23. This study focuses on rising and descending stair structures. If the system identifies the obstacles and the stairs accurately, it is a successful detection. If not, then it is a failure.

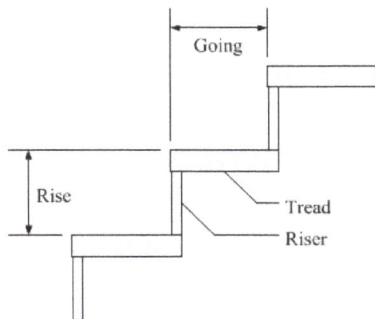

Figure 23. The structure of the stair.

3.5. An Indoor Environment under Sufficient Light

The success rate and the failure rate for detection are shown in Table 3. The types of stairs are simpler in the indoor environment, so there is no problem with detection. Figure 24 shows that if the most of the stair structures are not obscured by person or objects, it is successfully detected. The experimental results show that as long as most of the stair is not occluded, it is successfully detected.

Table 3. The success rate and the failure rate for obstacle detection.

	Frame Amount (Total 262 frames)	Percentage (%)
Success	245	93.5%
Failure	17	6.5%

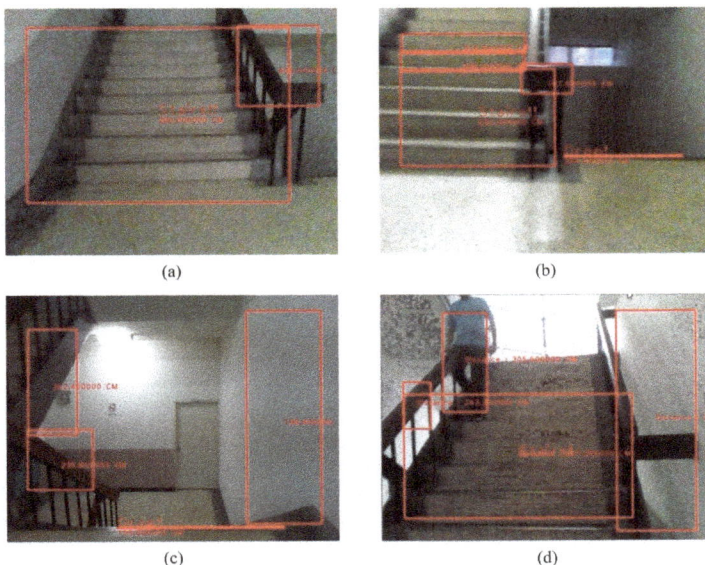

(a)

(b)

(c)

(d)

Figure 24. The detection of an obstacle indoors under sufficient light. (**a**) Rising stairs; (**b**) Rising and descending stairs; (**c**) Obstacle and descending stairs; (**d**) Obstacle and descending stairs.

3.6. An Indoor Environment under Insufficient Light

The success rate and the failure rate for obstacle detection are shown in Table 4. The success rate and failure rate for detection of descending stairs are shown in Table 5. To improve the accuracy and the capturing of images, the system uses a Kinect sensor, so that stairs can be easily detected, even in dimly lit environments as shown in Figure 25.

Table 4. The success rate and the failure rate for obstacle detection.

	Frame Amount (Total 104 frames)	Percentage (%)
Success	96	92.3%
Failure	8	7.7%

Table 5. The success rate and failure rate for detection of descending stairs.

	Frame Amount (Total 592 frames)	Percentage (%)
Success	498	84.12%
Failure	94	15.88%

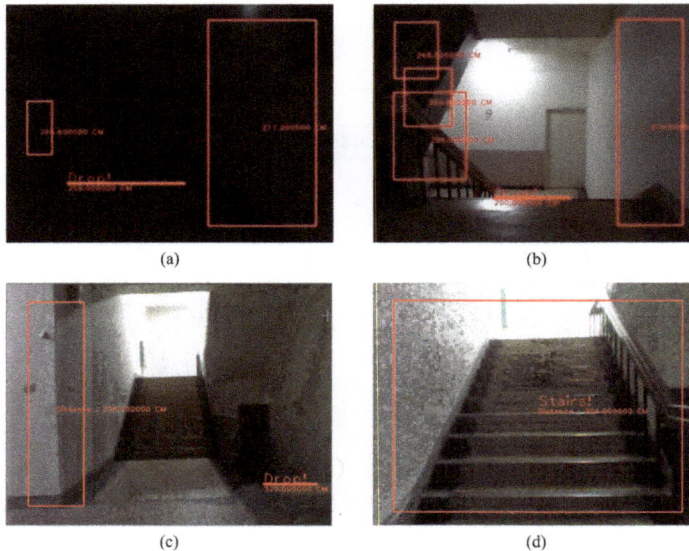

Figure 25. The detection of an obstacle indoors under insufficient light. (**a**) Descending stairs 1; (**b**) Descending stairs 2; (**c**) Rising stairs 1; (**d**) Rising stairs 2.

3.7. The Confusion Matrix for Experiment Results

The indoor experimental data is expressed using a confusion matrix, as shown in Table 6. If there is a large size break in the depth map, the obstacle is not detected. When the remaining part in depth map is calculated, it is so small as to be negligible. When rising stairs are to be detected, because there are broken parts in the image depth, some blocks are mistaken for obstacles. In an indoor environment there are fewer false assessments because the ground is uniform. The probability of a false assessment is greater in an outdoor environment because the ground is diverse, such as where there is a rough surface. The detection rate for an indoor obstacle reaches 97.40%.

Table 6. The confusion matrix for the indoor experiment results.

Confusion Matrix		Actual Output				
		Obstacle	Upstairs	Downstairs	Barrier Free	Recognition Rate
Expected output	Obstacle	1660	0	0	24	98.57%
	Upstairs	30	382	0	0	92.72%
	Downstairs	0	0	248	13	95.02%
	Barrier free	8	0	0	524	98.50%
Accuracy rate		2814/2889				97.40%

3.8. The Detection of Static and Dynamic Obstacles

Our system detects static and dynamic obstacles simultaneously as shown in Figure 24d. Figure 24a–c shows static obstacle detection. As illustrated in Figure 26, this testing is for dynamic obstacle detection. The scenario is that one man walks from the left to the right in the scene.

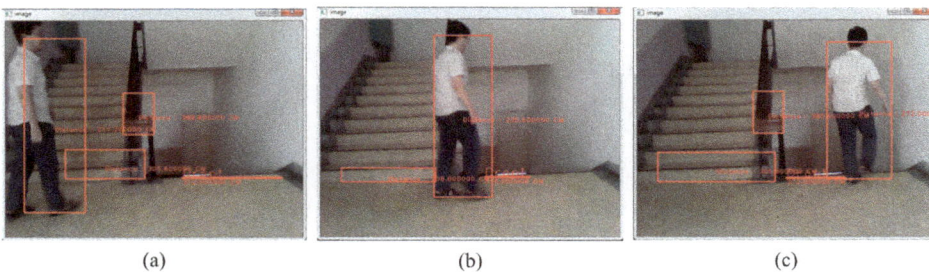

(a)   (b)   (c)

Figure 26. The detection of static and dynamic obstacles. (**a**) Walking people walks from the left side; (**b**) Walking people at the middle; (**c**) Walking people walks to the right side.

3.9. The Evaluation of the System by Blind and Blindfolded Participants

Three blind university students (as shown in Figure 27a,b) and thirty-eight blindfolded university students were used to evaluate the system. The system is not meant to take the place of a cane or a guide dog but to improve perception using a depth sensor-based sound system. A traditional cane, which is the standard navigation tool for the blind, is difficult to replace because a cane is cheap, light and can be folded.

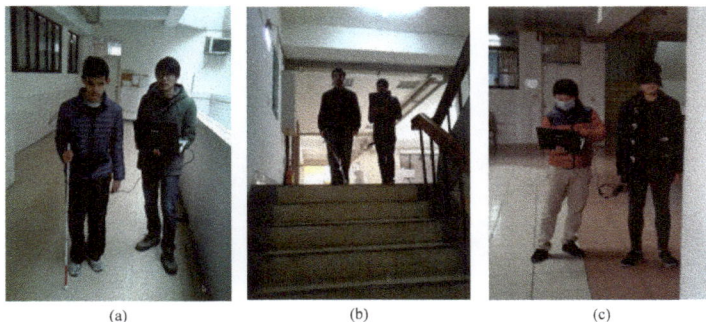

(a)   (b)   (c)

Figure 27. Blind and blindfolded participants. (**a**) Blind participant 1; (**b**) Blind participant 2; (**c**) Blind-folded participant.

These experiments use a control experiment. There is an experimental group and a control group. The experimental environment (as shown in Figure 28) includes rising stairs, descending stairs, static obstacles and dynamic obstacles along a specific path. The participants consisted of three blind junior students (Blind Participants: BP) and thirty-eight junior students (Blindfolded Participants: BFP). The best and worst experimental results were removed. The distribution of the experimental data is shown in Figure 29. Figure 30 shows that experimental results when only the proposed system is used are similar to the experimental results when only a cane is used. However, using the system and a cane together gives significantly improved experimental results that are closer experimental results of normal people.

We calculate the *p*-value for the cane and proposed system with cane as shown in Table 7. The calculating result of *p*-value is 0.001508556 (two-tail). In general, the significance level is 0.05 or 0.01. In our case, the two-tailed *p*-value suggests rejecting the null hypothesis of no difference. The *p*-value is less than 0.5 or 0.01, so the result is significant improvement.

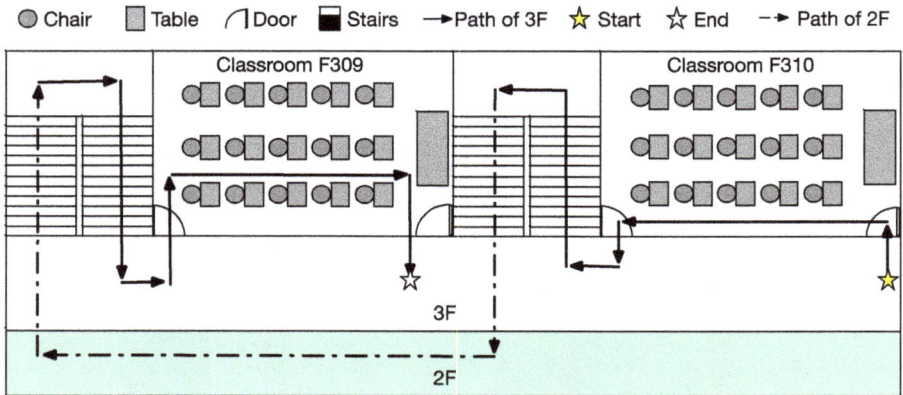

Figure 28. The experimental environment.

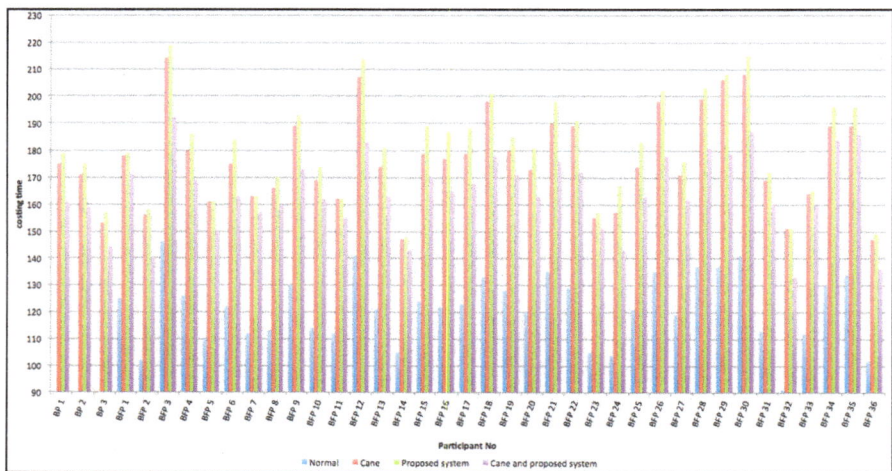

Figure 29. The statistical data of experiment.

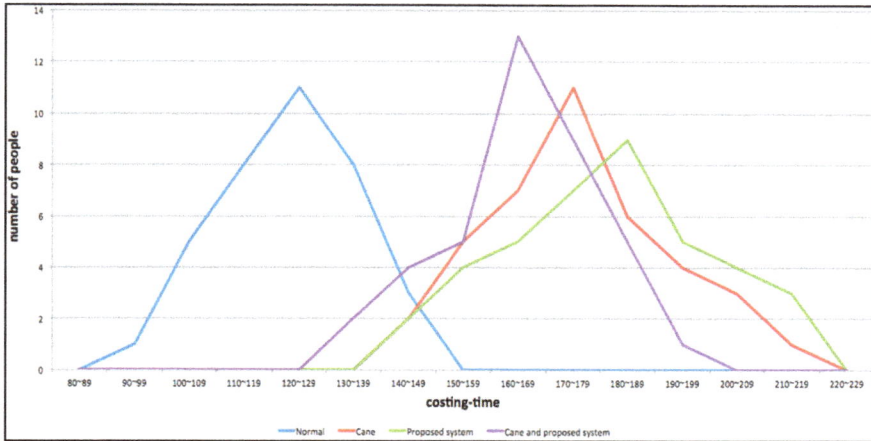

Figure 30. The distribution of the experimental data.

Table 7. *t*-Test: Paired Two Sample for Means.

	Cane	Cane and Proposed System
Mean	176.4615385	164.3846154
Variance	310.097166	213.5587045
Observations	39	39
Pearson Correlation	0	
Hypothesized Mean Difference	74	
t Stat	3.295836168	
P (T ≤ t) one-tail	0.000754278	
T Critical: one-tail	1.665706893	
P (T ≤ t) two-tail	0.001508556	
T Critical: two-tail	1.992543495	

4. Conclusions

This paper proposes an obstacle detection method that uses depth information. Because the depth information is obtained using an infrared sensor, the depth information is not affected by the degree of illumination. The proposed system is effective in detecting obstacles in a low light environment. The system addresses the problem of over-segmentation by removing the edge and eliminating the problem of the initial seed position for the region growth method, using CCM. It can also detect static and dynamic obstacles. These experimental results show that when only the proposed system is used similar to the experimental results when only a cane is used. However, using the system and a cane together gives significantly improved experimental results that are closer experimental results of normal people. The system is simple, robust and efficient.

Three thresholds are used: $TH1 = 35$ for the removal of the ground plane, $TH2 = 15$ for the removal of the obstacle edge and $TH3 = 50$ for the detection of descending stairs. The detection rate for an indoor obstacle is as high as 97.40%. The experimental results show that the proposed system is very robust, efficient and convenient in an indoor environment. The system can also detect rising stairs and descending stairs and ensures that visually impaired people have the environmental information that is required to avoid danger.

The system vocally informs the user of the distance of an obstacle and the category of the obstacle. This voice alarm is very short and focuses on the most concise information about the closest obstacle. The TTS voice is not a natural voice so it has a robotic sound. In the future, the system will be improved to support multiple languages. Image processing performance of our proposed system for ROI or fully image is different, but they are small and almost the same. The most of calculations are based on Kinect. To detect object in fully image is easier than in ROI. Our system detects complete object, not just a part.

Acknowledgments: This work is supported in part by the Ministry of Science and Technology of the Republic of China under grant number, MOST 103-2410-H-032-052. This support is gratefully acknowledged. The authors wish to thank participants in the experiments and the reviewers for their valuable comments, which have improved this paper considerably.

Author Contributions: Hsieh-Chang Huang and Ching-Tang Hsieh conceived and designed the experiments; Ching-Tang Hsieh defined the research line. Hsieh-Chang Huang and Cheng-Hsiang Yeh performed the experiments; Hsieh-Chang Huang and Ching-Tang Hsieh analyzed the data; Hsieh-Chang Huang wrote the paper.

Conflicts of Interest: The authors declare no conflict of interest.

References

1. International Agency for Prevention of Blindness. Available online: http://www.iapb.org/ (accessed on 5 January 2015).
2. Dakopoulos, D.; Bourbakis, N.G. Wearable Obstacle Avoidance Electronic Travel Aids for Blind: A Survey. *IEEE Trans. Syst. Man Cybern. Part C Appl. Rev.* **2010**, *40*, 25–35. [CrossRef]
3. Ma, G.; Dwivedi, M.; Li, R.; Sun, C.; Kummert, A. A Real-Time Rear View Camera Based Obstacle Detection. In Proceedings of the 12th IEEE International Conference on Intelligent Transportation Systems, St. Louis, MO, USA, 4–7 October 2009; pp. 1–6.
4. Zhang, Y.; Hong, C.; Weyrich, N. A single camera based rear obstacle detection system. In Proceedings of the 2011 IEEE Intelligent Vehicles Symposium, Baden-Baden, Germany, 5–9 June 2011; pp. 485–490.
5. Chen, L.; Guo, B.L.; Sun, W. Obstacle Detection System for Visually Impaired People Based on Stereo Vision. In Proceedings of the 2010 Fourth International Conference on Genetic and Evolutionary Computing, Shenzhen, China, 13–15 December 2010; pp. 723–726.
6. Ying, J.; Song, Y. Obstacle Detection of a Novel Travel Aid for Visual Impaired People. In Proceedings of the 2012 4th International Conference on Intelligent Human-Machine Systems and Cybernetics (IHMSC), Nanchang, Jiangxi, China, 26–27 August 2012; pp. 362–364.
7. Guerrero, L.A.; Vasquez, F.; Ochoa, S.F. An Indoor Navigation System for the Visually Impaired. *Sensors* **2012**, *12*, 8236–8258. [CrossRef] [PubMed]
8. Lin, Q.; Han, Y. A Context-Aware-Based Audio Guidance System for Blind People Using a Multimodal Profile Model. *Sensors* **2014**, *14*, 18670–18700. [CrossRef] [PubMed]
9. Zöllner, M.; Huber, S.; Jetter, H.C.; Reiterer, H. NAVI—A Proof-of-Concept of a Mobile Navigational Aid for Visually Impaired Based on the Microsoft Kinect. In Proceedings of 13th IFIP TC 13 International Conference, Lisbon, Portugal, 5–9 September 2011; pp. 584–587.
10. Filipe, V.; Fernandes, F.; Fernandes, H.; Sousa, A.; Paredes, H.; Barroso, J. Blind Navigation Support System Based on Microsoft Kinect. In Proceedings of the 4th International Conference on Software Development for Enhancing Accessibility and Fighting Info-Exclusion (DSAI 2012), Douro Region, Portugal, 19–22 July 2012; pp. 94–101.
11. Takizawa, H.; Yamaguchi, S.; Aoyagi, M.; Ezaki, N.; Mizuno, S.; Cane, K. An Assistive System for the Visually Impaired Based on the Concept of Object Recognition Aid. *Pers. Ubiquitous Comput.* **2015**, *19*, 955–965. [CrossRef]
12. Zhang, M.; Liu, P.; Zhao, X.; Zhao, X.; Zhang, Y. An Obstacle Detection Algorithm Based on U-V Disparity Map Analysis. In Proceedings of the 2010 IEEE International Conference on Information Theory and Information Security (ICITIS), Beijing, China, 17–19 December 2010; pp. 763–766.
13. Gao, Y.; Ai, X.; Rarity, J.; Dahnoun, N. Obstacle Detection with 3D Camera Using U-V Disparity. In Proceedings of the 2011 7th International Workshop on Systems, Signal Processing and their Applications (WOSSPA), Tipaza, Algeria, 9–11 May 2011; pp. 239–242.
14. Choi, J.; Kim, D.; Yoo, H.; Sohn, K. Rear Obstacle Detection System Based on Depth from Kinect. In Proceedings of the 2012 15th International IEEE Conference on Intelligent Transportation Systems (ITSC), Anchorage, AK, USA, 16–19 September 2012; pp. 98–101.

15. Sales, D.O.; Correa, D.; Osório, F.S.; Wolf, D.F. 3D Vision-Based Autonomous Navigation System Using ANN and Kinect Sensor. In Proceedings of the 13th International Conference, EANN 2012, London, UK, 20–23 September 2012; Volume 311, pp. 305–314.
16. Wang, S.; Pan, H.; Zhang, C.; Tian, Y. RGB-D Image-based Detection of Stairs, Pedestrian Crosswalks and Traffic Signs. *J. Vis. Commun. Image Represent.* **2014**, *25*, 263–272. [CrossRef]
17. Rodríguez, A.; Bergasa, L.M.; Alcantarilla, P.F.; Yebes, J.; Cela, A. Obstacle Avoidance System for Assisting Visually Impaired People. In Proceedings of the IEEE Intelligent Vehicles Symposium Workshops, Madrid, Spain, 3 June 2012; pp. 1–6.
18. Kim, D.; Kim, K.; Lee, S. Stereo Camera Based Virtual Cane System with Identifiable Distance Tactile Feedback for the Blind. *Sensors* **2014**, *14*, 10412–10431. [CrossRef] [PubMed]
19. Rodríguez, A.; Yebes, J.J.; Alcantarilla, P.F.; Bergasa, L.M.; Almazán, J.; Cela, A. Assisting the Visually Impaired: Obstacle Detection and Warning System by Acoustic Feedback. *Sensors* **2012**, *12*, 17476–17496. [CrossRef] [PubMed]
20. Saeid, F.; Hajar, M.D.; Payman, M. An Advanced Stereo Vision Based Obstacle Detection with a Robust Shadow Removal Technique. *World Acad. Sci. Eng. Technol.* **2010**, *4*, 935–940.
21. Aladren, A.; Lopez-Nicolas, G.; Puig, L.; Guerrero, J.J. Navigation Assistance for the Visually Impaired Using RGB-D Sensor With Range Expansion. *IEEE Syst. J.* **2014**, 1–11. [CrossRef]
22. Hub, A.; Hartter, T.; Ertl, T. Interactive tracking of movable objects for the blind on the basis of environment models and perception-oriented object recognition methods. In Proceedings of the 8th International ACM SIGACCESS Conference on Computers and Accessibility (Assets'06), Portland, OR, USA, 23–25 October 2006; pp. 111–118.
23. Skulimowski, P.; Strumiłło, P. Obstacle Localization in 3d Scenes from Stereoscopic Sequences. In Proceedings of the 15th European Signal Processing Conference (EUSIPCO 2007), Poznan, Poland, 3–7 September 2007; pp. 2095–2099.
24. Hsieh, C.-T.; Lai, W.-M.; Yeh, C.-H.; Huang, H.-C. An Obstacle Detection System Using Depth Information and Region Growing for Blind. *Res. Notes Inf. Sci. (RNIS)* **2013**, *14*, 465–470.
25. Leap motion. Available online: https://www.leapmotion.com (accessed on 5 January 2015).
26. Hokuyo. Available online: http://www.acroname.com/products/index_Hokuyo.html (accessed on 5 January 2015).
27. Intel® RealSense™ Integrated 3D Camera. Available online: https://software.intel.com/en-us/realsense/home (accessed on 5 January 2015).
28. Duda, R.O.; Hart, P.E. Use of the Hough Transformation to Detect Lines and Curves in Pictures. *Commun. ACM* **1972**, *15*, 11–15. [CrossRef]
29. McAndrew, A. *Introduction to Digital Image Processing with Matlab*; Asia Edition; Cengage Learning: Taipei, Taiwan, 2010; pp. 267–302.
30. Brock, M. Kristensson Supporting blind navigation using depth sensing and sonification. In Proceedings of the 2013 ACM International Joint Conference on Pervasive and Ubiquitous Computing (Ubicomp 2013), Zurich, Switzerland, 8–12 September 2013; pp. 255–258.
31. Edwards, N.; Rosenthal, J.; Moberly, D.; Lindsay, J.; Blair, K.; Krishna, S.; McDaniel, T.; Panchanathan, S. A pragmatic approach to the design and implementation of a vibrotactile belt and its applications. In Proceedings of the IEEE International Workshop on Haptic Audio Visual Environments and Games, 2009 (HAVE 2009), Lecco, Italy, 7–8 November 2009; pp. 13–18.
32. vOICe's Glasses for the Blind. Available online: http://www.artificialvision.com (accessed on 5 January 2015).
33. Mann, S.; Huang, J.; Janzen, R. Blind Navigation with a Wearable Range Camera and Vibrotactile Helmet. In Proceedings of the 19th ACM International Conference on Multimedia, Scottsdale, AZ, USA, 28 November–1 December 2011; ACM: New York, NY, USA, 2011; pp. 1325–1328.
34. Lee, C.H.; Su, Y.C.; Chen, L.G. An intelligent depth-based obstacle detection system for visually-impaired aid applications. In Proceedings of the 2012 13th International Workshop on Image Analysis for Multimedia Interactive Services (WIAMIS), Dublin, Ireland, 23–25 May 2012; pp. 1–4.
35. Zheng, C.; Green, R. *Feature Recognition and Obstacle Detection for Drive Assistance in Indoor Environments*; University of Canterbury: Christchurch, New Zealand, 2011.

36. Bhowmick, A.; Prakash, S.; Bhagat, R.; Prasad, V.; Hazarika, S.M. IntelliNavi: Navigation for Blind Based on Kinect and Machine Learning. *Multi-Discip. Trends Artif. Intell. Lect. Notes Comput. Sci.* **2014**, *8875*, 172–183.

37. Zheng, C. Richard Green Vision-based autonomous navigation in indoor environments. In Proceedings of the IEEE 25th International Conference of Image and Vision Computing New Zealand (IVCNZ), Queenstown, New Zealand, 8–9 November 2010; pp. 1–7.

38. Suzuki, S.; Abe, K. Topological structural analysis of digitized binary images by border following. *Comput. Vis. Graph. Image Process.* **1985**, *30*, 32–46. [CrossRef]

39. Soquet, N.; Aubert, D.; Hautiere, N. Road segmentation supervised by an extended v-disparity algorithm for autonomous navigation. In Proceedings of the 2007 IEEE Intelligent Vehicles Symposium, Istanbul, Turkey, 13–15 June 2007; pp. 160–165.

40. Hu, Z.; Lamosa, F. *Keiichi Uchimura a Complete U-V-Disparity Study for Stereovision Based 3D Driving Environment Analysis*; Kumamoto University: Kumamoto, Japan, 2005; pp. 204–211.

41. Region Growth. Available online: http://www.ijctee.org/files/VOLUME2ISSUE1/IJCTEE_0212_18.pdf (accessed on 5 January 2015).

42. Adams, R.; Bischof, L. Seeded region growing. *IEEE Trans. Pattern Anal. Mach. Intell.* **1994**, *16*, 641–647. [CrossRef]

43. TTS. Available online: http://msdn.microsoft.com/en-us/library/ms723627(v=vs.85).aspx (accessed on 5 January 2015).

44. Yu, H.; Zhu, J.; Wang, Y.; Jia, W.; Sun, M.; Tang, Y. Obstacle Classification and 3D Measurement in Unstructured Environments Based on ToF Cameras. *Sensors* **2014**, *14*, 10753–10782. [CrossRef] [PubMed]

sensors

MDPI

Article

Target Detection over the Diurnal Cycle Using a Multispectral Infrared Sensor

Huijie Zhao, Zheng Ji, Na Li *, Jianrong Gu and Yansong Li

School of Instrumentation Science & Opto-Electronics Engineering, Beihang University, 37 Xueyuan Road, Haidian District, Beijing 100191, China; hjzhao@buaa.edu.cn (H.Z.); jizhengss1988@buaa.edu.cn (Z.J.); karon@buaa.edu.cn (J.G.); lysbuaa@buaa.edu.cn (Y.L.)
* Correspondence: lina_17@buaa.edu.cn; Tel.: +86-10-8231-5884

Academic Editor: Gonzalo Pajares Martinsanz
Received: 14 September 2016; Accepted: 26 December 2016; Published: 29 December 2016

Abstract: When detecting a target over the diurnal cycle, a conventional infrared thermal sensor might lose the target due to the thermal crossover, which could happen at any time throughout the day when the infrared image contrast between target and background in a scene is indistinguishable due to the temperature variation. In this paper, the benefits of using a multispectral-based infrared sensor over the diurnal cycle have been shown. Firstly, a brief theoretical analysis on how the thermal crossover influences a conventional thermal sensor, within the conditions where the thermal crossover would happen and why the mid-infrared (3~5 μm) multispectral technology is effective, is presented. Furthermore, the effectiveness of this technology is also described and we describe how the prototype design and multispectral technology is employed to help solve the thermal crossover detection problem. Thirdly, several targets are set up outside and imaged in the field experiment over a 24-h period. The experimental results show that the multispectral infrared imaging system can enhance the contrast of the detected images and effectively solve the failure of the conventional infrared sensor during the diurnal cycle, which is of great significance for infrared surveillance applications.

Keywords: infrared sensor; multispectral; diurnal cycle; thermal crossover

1. Introduction

Infrared imaging detection systems are becoming more prevalent in numerous fields, including remote sensing [1], medical monitoring [2], military surveillance [3], and scientific research [4,5]. These systems offer major advantages over visual detection systems, such as their continuous day and night imaging capabilities, especially for target detection and acquisition [6].

When targets are aimed to be detected over the diurnal cycle using a conventional mid-infrared (3~5 μm) sensor, the results are generally affected by thermal crossover, where the infrared image contrast from the target and the background is difficult to discriminate from each other as the target would have integrated with the background and the radiation difference between the target and background was too low to be sensed by the infrared thermal sensor. Moreover, this could cause the targets to be blended into the background, lowering the detection accuracy, and even make the thermal sensor lose the target. In addition, the thermal crossover may also occur at any point in the day, because of solar loading, clouds, rain and fog. Therefore, it is critical to solve this problem for the conventional mid-infrared thermal sensor, especially for the infrared surveillance system.

In the last few decades, research has focused on how to solve the problem of infrared detection during thermal crossover periods and the thermal polarization technique, which is proposed as a method to enhance conventional thermal imaging, has been employed. Felton et al. [7–9] compared the crossover periods for mid-and long-wave infrared polarimetric and conventional thermal imagery. The mid-infrared (3~5 μm) imaging polarimeter they used was based on a division-of-aperture (DoA) lens technology developed by Polaris Sensor Technologies, which employed a 2×2 array of mini-lenses followed by four linear polarizers at different orientations, forming four identical images of the scene on four quadrants of the sensor focal plane array. The long-wave infrared (8~12 μm) polarimeter they used was a microbolometer-based rotating retarder imaging polarimeter developed by Polaris Sensor Technologies, which could capture up to 12 images sequentially in time with each image at a different orientation. Their experimental results showed that the polarimetric technology could be used as a method to enhance the conventional infrared image contrast between the targets and background during thermal crossover periods. However, their infrared image contrast improvement was not direct but resulted from the calculation of Stokes vector formula, which might not be suitable for the systems that requires high-performance of real-time processing. Still, as their work mainly focused on polarimetric detection experiments, what the pictures are when the target integrated with the background during the diurnal cycle and the theoretical analysis on how the thermal crossover influences the conventional thermal sensor and why the polarimetric technology could be used to solve the thermal crossover detection problem was also not mentioned. Based on Felton's research, Wilson et al. [10,11] used a single pixel scanning passive millimeter-wave polarimetric sensor, operating at a frequency of 77 GHz with a noise equivalent temperature difference (NETD) of 0.5 K, to measure the infrared image contrast during thermal crossover periods. As the passive millimeter-wave sensor is designed with capabilities to measure two linear polarization states simultaneously, it breaks the limitation that many of millimeter wave (mmW) sensors are only able to detect a single linear polarization state and improve the detection accuracy. Additionally, Retief et al. [12] studied the prediction method of thermal crossover based on imaging measurements under different weather conditions over the diurnal cycle. They used a series of infrared background objects images as the basis to establish the heat balance model and, on this basis, to predict when the thermal crossover may occur. In addition to the thermal polarization technique, the infrared multispectral technology is also considered as an important approach to solve the thermal crossover detection problem. The prior studies [13–17] on infrared multispectral technology mainly showed the potential benefits of infrared multispectral processing for clutter-limited ground target detection. However, due to constraints on the spectral resolution, band coverage, and radiometric sensitivity of existing sensors at that time, accurate measurement data and the real experimental image data were not available. Despite this, these studies firstly made the infrared multispectral technology a potential method for target/background identification. Furthermore, Schwartz and Eismann et al. [18–20] conducted a series of multispectral field measurements at Redstone Arsenal using a Bomem-developed high-sensitivity infrared Fourier Transform Spectrometer, which operates in the IR region (3–12 μm) with 8 cm^{-1} spectral resolution and noise equivalent spectral radiance (NESR, in nW/cm^{2}sr·cm^{-1} units) 7.5@3.8 μm, to enhance the capabilities of passive infrared surveillance. With the instrument, the data of several test panels, military vehicles and vegetated backgrounds at different times and under various environmental condition were obtained, their analysis of the experimental results statistically showed that the thermal sensor could detect the target hidden in vegetated and desert backgrounds with the use of multispectral techniques. As their work mainly focused on post-collection data analyses of infrared hyperspectral measurements and multispectral target detection algorithms, the design of the instrument, the real experimental image data and how the multispectral technology could be employed as an effective supplementary method for the conventional mid-infrared broadband thermal detection over the diurnal cycle was not mentioned. Nevertheless, their research results showed the potential and capacity of multispectral processing to detect low-contrast ground targets by providing valid estimates of targets to the background spectral contrast.

Overall, from the abovementioned research results, although the polarization technique was an effective solution to thermal crossover detection, there were still some disadvantages. Firstly, the improvement of the infrared image contrast resulted from the calculation of Stokes vector formula, which means that the contrast enhancement is not direct. Secondly, the time division imaging or simultaneous imaging technique are usually used in polarization detection, which would increase image processing time or the system size and weight. In addition, the environmental factors could affect the polarimetric contrast. Potential sources include vehicles, buildings, trees, clouds, water vapor, etc., which are not necessarily visible within the scene but still illuminate the objects in the field of view of the detector could be a reduction in the magnitude of polarimetric signature of a target. Compared with the polarization technique, as the target's infrared spectrum signature only differs with materials, one or some characteristic wavelengths could be enough to reflect the difference between the target and background without any redundant calculation. Thus, it would be faster and more direct to distinguish the target from the background in a complex environment with the multispectral technology if the characteristic wavelengths were acquired in advance according to prior knowledge. In this paper, our goal is to discuss how the multispectral technology could be employed to solve the problem of thermal crossover, design a fast, compact and light infrared multispectral prototype with the known characteristic wavelengths according to the prior knowledge and conclude that multispectral technology is capable of enhancing conventional thermal imaging.

Overview of Thermal Detection over the Diurnal Cycle

Thermal crossover is defined as a natural phenomenon that normally occurs twice daily, but may occur at any time throughout the day when temperature conditions are such that there is a loss of contrast between two adjacent objects on the infrared sensor. Figure 1 pictorially shows a schematic of an infrared system measuring the target radiance L_t and the background radiance L_{bg}. The infrared system can be any conventional infrared sensor or camera and located at any arbitrary orientation. The target can be any typical common objects, such as vehicles, and the background can be any natural or artificial objects, such as grass, tree, or road. To simplify, without considering the scattering, the total received radiance at the infrared system can be expressed by two components:

$$\begin{cases} L_{bg}(\lambda, \theta_v, \theta_s, \varphi) = L_{bg}^r(\lambda, \theta_v, \theta_s, \varphi) + L_{bg}^e(\lambda, \theta_v, \theta_s, \varphi, T) \\ L_t(\lambda, \theta_v, \theta_s, \varphi) = L_t^r(\lambda, \theta_v, \theta_s, \varphi) + L_t^e(\lambda, \theta_v, \theta_s, \varphi, T) \end{cases} \tag{1}$$

where L_{bg}^e and L_t^e are the emissive radiance (the radiant flux emitted by a surface, per unit solid angle, per unit projected area, per wavelength) of the background and target, L_{bg}^r and L_t^r are the reflection of the solar irradiance on the background and the target, λ is the wavelength of light, θ_v is the viewing zenith angle of the detection system, θ_s is the solar zenith angle, and φ is the azimuth angle between θ_v and θ_s, T is the temperature. As $DN = a \cdot L + b$, the DN difference between the targets and background objects (represented by C) can be expressed as [21]:

$$C = \left| DN_t - DN_{bg} \right| = \left| a(L_t - L_{bg}) \right| = a \left| (L_t^e - L_{bg}^e) + (L_t^r - L_{bg}^r) \right| \tag{2}$$

Furthermore, assuming that the reflectivity of the target and the background objects are ρ_t and ρ_{bg}, respectively, if ignoring the scattering and transmittance, the target and background objects' absorptivity would be $\alpha_t = 1 - \rho_t$ and $\alpha_{bg} = 1 - \rho_{bg}$. As the vast majority of objects in nature produce diffuse reflection, the reflectivity ρ_t and ρ_{bg} should be replaced by the Bidirectional Reflectance Distribution Function (BRDF, a function which defines the spectral and spatial reflection characteristic of a surface and is the ratio of reflected radiance to incident irradiance at a particular wavelength [22]) to represent the anisotropic properties of solar radiation effects on the reflectivity of objects. Therefore, Equation (2) can be rewritten as:

$$C = a \cdot \left| \begin{array}{l} \int_{\lambda_1}^{\lambda_2} \left[BRDF_t(\lambda, \theta_s, \theta_v, \varphi) - BRDF_{bg}(\lambda, \theta_s, \theta_v, \varphi) \right] L_s(\lambda) d\lambda \\ + \int_{\lambda_1}^{\lambda_2} \left([1 - BRDF_t(\lambda, \theta_s, \theta_v, \varphi)] L_t^e(\lambda, T) - \left[1 - BRDF_{bg}(\lambda, \theta_s, \theta_v, \varphi) \right] L_{bg}^e(\lambda, T) \right) d\lambda \end{array} \right| \tag{3}$$

where $L_s(\lambda)$ is the solar radiation and $\lambda_1 \sim \lambda_2$ is the working wavelength range of the infrared thermal sensor. In the case that θ_v, θ_s and λ are constant, BRDF only differs with the object's material.

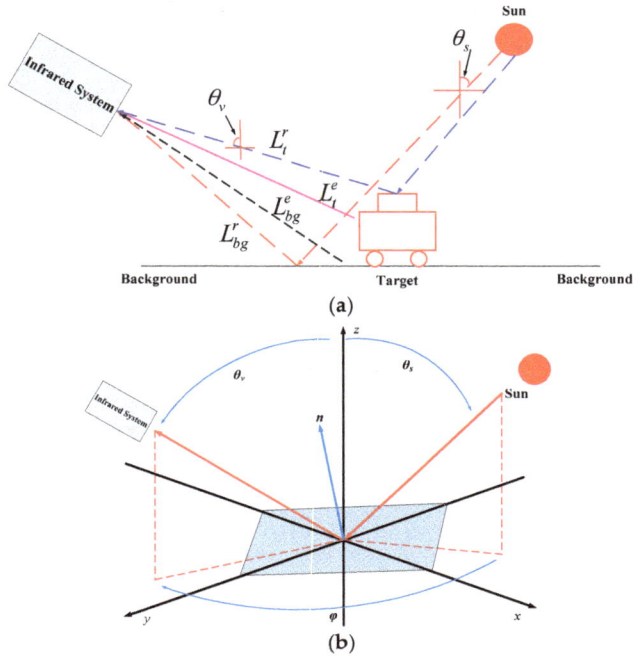

Figure 1. (**a**) Schematic of the infrared system measuring the radiance L_t and the background radiance L_{bg}. (**b**) Description of θ_v, θ_s and φ.

As can be seen from Equation (3), in general, the thermal crossover over the diurnal cycle would occur when C between the target and the background is zero or below the threshold value required to execute a specific task by the conventional infrared thermal sensor. Specifically, we divide one day, 24 h, into five time zones, as shown in Figure 2.

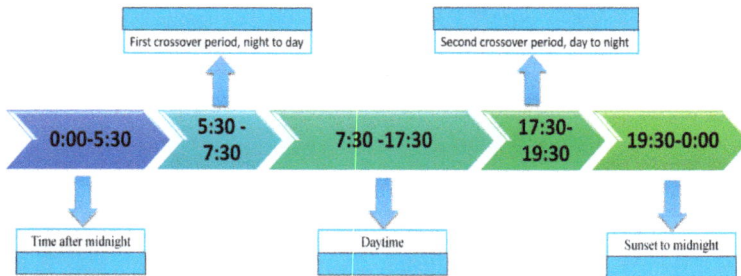

Figure 2. Five time zones from over one day.

Time after midnight:

$$C = a \left| \int_{\lambda_1}^{\lambda_2} \left([1 - BRDF_t(\lambda, \theta_s, \theta_v, \varphi)] L_t^e(\lambda, T) - \left[1 - BRDF_{bg}(\lambda, \theta_s, \theta_v, \varphi)\right] L_{bg}^e(\lambda, T) \right) d\lambda \right| \quad (4)$$

provided that the target and background have different material; in this case, the thermal crossover would not occur as C would not be zero.

First crossover period: in case of a sunny day, for the target with lower thermal inertia, such as metal, thermal crossover would occur. In this case, the multispectral exploration technique can be used to find the emissivity difference between the target and the background in $\Delta\lambda$ to enhance the contrast C. For the target with higher thermal inertia, such as water, thermal crossover might not occur.

Daytime: The circumstance is more complicated as $L_s(\lambda)$ would have an effect on thermal crossover. No matter whether the target has lower or higher thermal inertia, thermal crossover may occur at any time, depending both on temperature differences and environmental factors, such as rain, and fog. In this case, the multispectral exploration technology can still be used to find the emissivity difference in $\Delta\lambda$ to enhance the contrast C if thermal crossover occurs.

Second crossover period: similar to the "first crossover period", for the target with lower thermal inertia the thermal crossover would occur and the multispectral exploration technology can be used to solve the problem of infrared detection during thermal crossover periods.

Sunset to midnight: similar to the "time after midnight", provided that the target and background have different materials C would not be zero and thermal crossover would not occur.

2. Materials and Methods

2.1. Why the Infrared Multispectral Technology Works

From the abovementioned discussion and Equation (3), it can be seen that it is the combined impact of temperature difference, emissivity difference between the targets and background objects, and reflected solar radiation that leads to the occurrence of thermal crossover. To simplify the problem analysis, the single factor analysis of temperature and emissivity was specified in the following two cases.

In the first case, we assume that the targets and background objects have the same emissivity and use $RRD(\lambda, T)$ to represent the relative thermal radiation differences between the targets and background objects, which is shown as Equation (5).

$$RRD(\lambda, T) = \frac{\frac{1}{\pi}\int_{\lambda_1}^{\lambda_2}\left(\alpha_t L_t^e(\lambda, T_1) - \alpha_{bg}L_{bg}^e(\lambda, T_2)\right)d\lambda}{\frac{1}{\pi}\int_{\lambda_1}^{\lambda_2}\alpha_{bg}L_{bg}^e(\lambda, T_2)d\lambda} \tag{5}$$

Figure 3 shows the graphed outputs of Equation (5), provided that the ambient temperature was 300 K and the temperature difference between the targets and the background objects changes within ± 5 K. As can be seen from Figure 3, between the 3.7 μm–4.8 μm region, which is also the typical working wavelength range for a commercial infrared detector, $RRD(\lambda, T)$ changes within -20%–25% In addition, with the decrease of wavelength, the curve $RRD(\lambda, T)$ becomes steeper and would be more sensitive to the changes in temperature. Particularly, the calculation of Equation (5) in the whole 3.7 μm–4.8 μm region was also made (not shown in Figure 3) and $RRD(\lambda, T)$ changes within a smaller region, -15%–15%, which points out that, to a certain degree, for the traditional infrared broadband thermal sensor, compared to the one with several narrow wavebands, the thermal crossover would be more likely to happen and affect thermal detection for a longer time under the same conditions.

In the second case, we assume that the targets are grey plate and steel plate, and background objects are road and sand, respectively, both of them have the same temperature, 300 K. With the emissivity data obtained from the IR module using the software Sensors, the calculation results of Equation (4) is shown as Figure 4. As can be seen from Figure 4, in the 3.7 μm–4.8 μm region, the $RRD(\lambda, T)$ curve changes from 65% to 900%. Compared with $RRD(\lambda, T)$ in the first case, obviously, the change of $RRD(\lambda, T)$ caused by emissivity presents a greater volatility and wider range than that caused by temperature in Figure 3, which, in other words, indicates that the emissivity difference under characterized bands between the targets and background objects could be utilized to solve the detection problem during the thermal crossover periods.

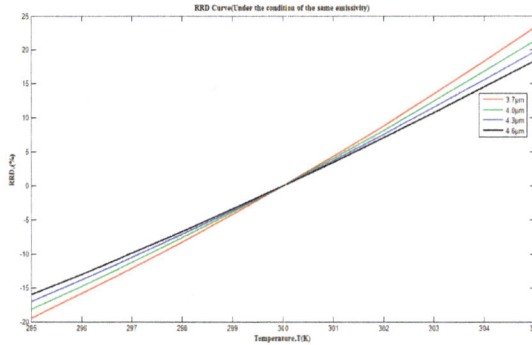

Figure 3. Relative thermal radiation differences Curve under the condition of the same emissivity in 3.7 μm–4.8 μm region.

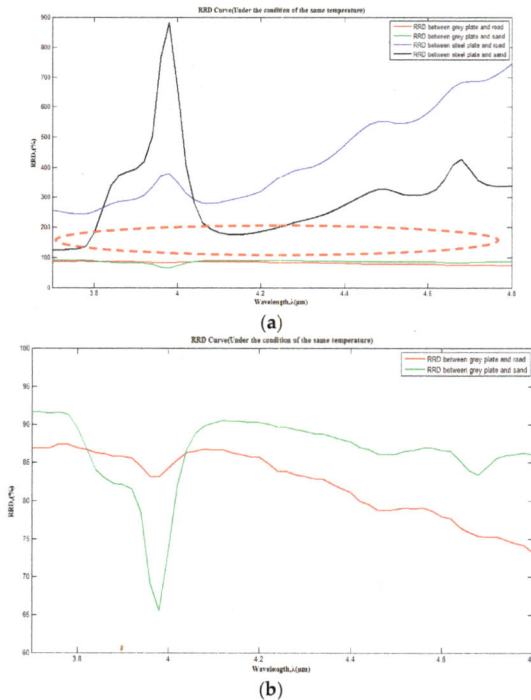

Figure 4. RRD curve under the condition of the same temperature in the 3.7 μm–4.8 μm region. (a) RRD curve among the targets grey plate, steel plate, and the background objects road and sand; (b) RRD curve among the target grey plate and the background objects road and sand.

2.2. Design of the Multispectral Infrared Imaging System

In order to verify the effectiveness of the multispectral infrared technology in solving the thermal crossover detection problem. A multispectral infrared imaging system prototype was designed and employed to conduct field experiment. The prototype consists of the infrared optical system, which is composed of the front infrared optical system and rear infrared optical system, a filter wheel with five band-pass filters, and a mid-infrared detector, as shown in Figure 5. The infrared camera lens has a focal length 100 mm. The mid-infrared detector is a France Sofradir Ltd. Model Mars 320 × 256 detector operating in region of 3.7–4.8 μm with a 5.5° × 4.4° field of view and up to 100 fps; this detector has a geometrical resolution of 0.3 mrad and a minimum detectable temperature difference between pixels of 0.03 °C and NETD of 9 mK. The five band-pass filters are produced by Sweden Spectrogon Ltd. with central wavelengths of 3700 nm, 3800 nm, 4120 nm, 4420 nm, and 4720 nm, respectively, and mounted on the filter wheel, which is driven by a stepper motor. In addition, the filter wheel reserves a hole without any filters so that the image comparison between the traditional broadband infrared image and narrowband infrared multispectral images can be conducted. Additionally, the cold reflection impact on the image has been considered and reduced to the minimum. The mid-infrared detector, filter wheel, and data acquisition and storage are controlled by a PC. The laboratory prototype is shown in Figure 6.

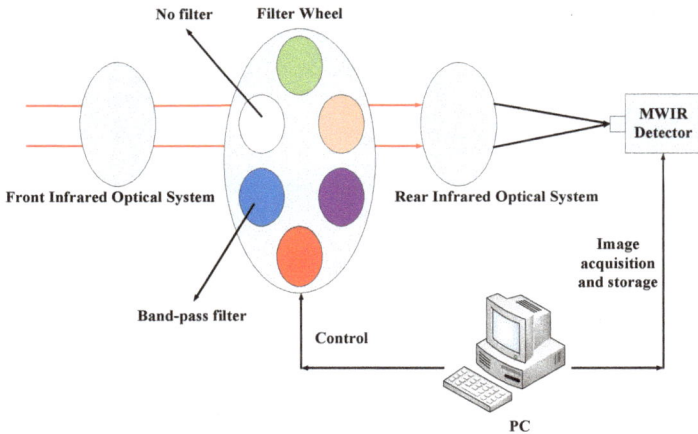

Figure 5. Schematic of multispectral infrared imaging system.

Figure 6. Prototype of multispectral infrared imaging system.

For the polarization technique used by Felton in [7–9], the MWIR imaging polarimeter employed division-of-aperture lens technology and the infrared image contrast improvement resulted from the calculation of Stokes vector formula. Thus, its image processing time, the system size and the system weight were longer and larger, the three parameters were 87 fps, the system size was 420L × 90W × 210H and weight was 4.99 kg. However, for our designed multispectral imaging system prototype, one or some characteristic wavelengths would be enough to reflect the difference between target and background without any redundant sensors and calculation if the characteristic wavelengths were acquired in advance according to the prior knowledge. Thus the prototype is faster, more compact, lighter and less costly. The image processing time of our prototype was up to 100 fps, the system size was 200L × 180W × 135H and was 3 kg, making it more suitable for practical applications. The specific specifications of the two sensors are listed in Table 1.

Table 1. Specifications of imaging polarimeter and multispectral infrared thermal sensor.

Parameter	Value of Felton's MWIR Imaging Polarimeter	Value of Our MWIR Imaging Spectrometer
Technological type	Polarization technique with four polarizers	Multispectral technique with five filters
FOV (°)	5.5	5.5 × 4.4
Focal Length (mm)	100	100
F/#	2.3	2
Total FPA pixels	Four FPA arrays 640 × 512 (single 220 × 220)	Single FPA 320 × 256
Pixel size (μm)	24 × 24	30 × 30
Max Frame Rate (fps)	87	up to 100
Sensor Dimensions (mm)	420L × 90W × 210H	200L × 180W × 135H
Sensor weight (kg)	4.99	3
Central wavelength (nm)	-	3700 nm, 3800 nm, 4120 nm, 4420 nm, 4720 nm
FWHM (nm)	-	80~100 nm
Sensitivity	10^{-7} W/cm^{2}sr	9 mK

Prior to the field experiment, the infrared multispectral imaging system is radiometrically calibrated in the field laboratory through a calibration procedure developed by EOI Ltd. so that the imaging capabilities of each wavelength can be assessed. The imaging capabilities measurement results are summarized in Table 2.

Table 2. Noise equivalent temperature difference (NETD) Measurement Results.

Wavelength	NETD (Background 298 K, f/2, Integration Time 6 ms)
3700 nm	476.6 mK
3800 nm	556.2 mK
4120 nm	533.9 mK
4420 nm	513.6 mK
4720 nm	490.1 mK

When conducting reconnaissance or surveillance tasks with the infrared multispectral imaging system, the blank hole without any filters is initially rotated to the optical axis of the system and the imaging system is just a conventional broadband infrared sensor under the initial state. With the change of time and weather conditions around the observed area, thermal crossover may occur. Once the observed target is hidden in the background caused by thermal crossover, the PC will rotate the filter wheel and control the detector to acquire the images under different multispectral wavebands to find the emissivity difference in the narrow wavebands between the targets and background objects and solve the thermal crossover problem. Afterwards, in order to highlight the multispectral information of the image and conform to the human eye's visual acquity at the same time, the narrow-band multispectral images and infrared broadband thermal image are blended to enhance the target recognition and solve the thermal crossover problem by using an HSV fusion algorithm [23]. HSV is one of the color systems that is used to pick a color from the color palette (H is hue, S is saturation and V is Value) and it is closer to people's experience and perception of color, compared with RGB.

2.3. Consideration of Experiment Design

To test the validity of the prototype, the field experiment was performed at the New Main Building at Beihang University. The infrared multispectral sensor was situated on the eighth floor of Tower B of the New Main Building (approximately 40 m) looking out of the window in the direction towards the target site, which was at approximately 100 m in distance. This open area was selected for the purpose of long-period image acquisition. The targets consisted of three different plates, galvanized sheet, steel sheet, and a wooden plate, and the natural backgrounds included grass, trees, and concrete road, which are shown in Figure 7a,b.

(a)

(b)

Figure 7. (a) Visible image of the target site consisting of three different plates and natural background; (b) The visible image of the test scene obtained on 15 March 2016.

The wooden plate was selected to make comparative experiments in order to prove the thermal radiation difference between the object with high inertia and one with low inertia over the diurnal cycle. The galvanized sheet and steel sheet were selected to demonstrate the thermal radiation difference under their characteristic wavebands and confirm the effectiveness of the multispectral technology. As some environmental parameters, like ambient temperature, relative humidity, and solar irradiance, may affect the experiment result, the experiment date was selected in advance according to the weather forecast, and the environmental parameters with a large ambient temperature difference, fewer clouds during the testing period, and relatively stable humidity, are advantageous to the experiment. The environmental data was collected on 15 March 2016 and the image data and environmental parameters were acquired continuously between 00:00 on 15 March 2016 and 23:59 on 15 March 2016 with a speed of half a minute per image and half a minute per measurement, respectively. The sunrise and sunset on 15 March 2016 occurred at roughly 06:25 and 18:21, respectively.

3. Results and Discussion

According to Figure 2, in which the 24 h day was divided into five time zones, the experimental results were demonstrated similarly. The contrast ratio between the DN values of the target and the background $C' = DN_{target}/DN_{bg}$ can be employed to reflect the contrast change of the regions of interest (ROI). At the time, after midnight, due to the different materials and temperature among the target galvanized sheet, steel sheet, and wooden plate, and the background road, C' did not approach 1 and thermal crossover did not occur, as shown in Figure 8a. At the first crossover period, for the wooden plate with higher thermal inertia, C' was 0.837 and thermal crossover did not occur, while for the galvanized sheet and steel sheet with lower thermal inertia, C' was approximately 1 (the exact number was 0.962 and 1.025, respectively) and thermal crossover did occur, as shown in Figure 8b. During the daytime, as $L_s(\lambda)$ had an effect on thermal crossover, the temperature difference among the background and the galvanized sheet and steel sheet increased gradually, C' was significantly greater than 1 and thermal crossover did not occur unless there was a rapid change in the weather conditions, as shown in Figure 8c. At the second crossover period, for the galvanized sheet and steel sheet with lower thermal inertia, C' was 1.011 and 1.045, respectively, and thermal crossover did occur again while, for the wooden plate, C' was 0.924 and the thermal crossover still did not occur, as shown in Figure 8d. From sunset to midnight, for each target C' was far less than 1 and thermal crossover did not occur, as shown in Figure 8e. Additionally, the average grey value of the image was larger than Figure 8a at the time after midnight because of the higher temperature. The contrast values among the three targets and the background in Figure 8 are listed in Table 3.

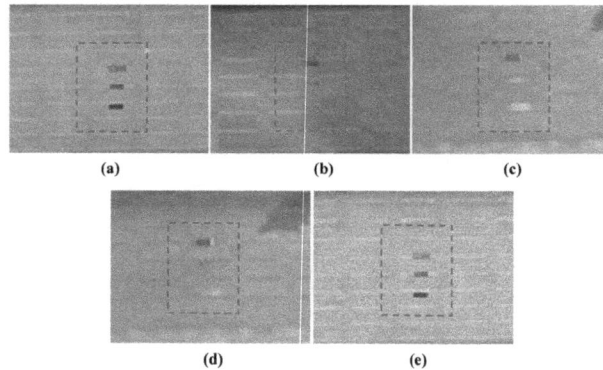

Figure 8. Infrared image obtained in each time zone. (**a**) Infrared image obtained at 03:00; (**b**) infrared image obtained at 06:50; (**c**) infrared image obtained at 12:30; (**d**) infrared image obtained at 18:05; and (**e**) infrared image obtained at 21:00.

Table 3. Contrast among the three targets and background in Figure 8.

Figure Number	Wooden Plate	Galvanized Sheet	Steel Sheet
a	0.822	0.784	0.632
b	0.837	0.962	1.025
c	0.825	1.098	1.151
d	0.810	1.011	1.045
e	0.807	0.786	0.612

In order to clarify the effectiveness of the multispectral technology to solve the thermal crossover problem, the multispectral images under central wavelengths of 4120 nm, 4420 nm, and 4720 nm at the first crossover period were obtained, and the results are presented in the form of image contrast plots, calculated using $C' = DN_{target}/DN_{bg}$. Included with each of these plots are the corresponding environmental data, as shown in Figure 9. Specifically, Figure 9b,c clearly show that the contrast curve of the galvanized sheet and steel sheet varied more significantly than the wooden plate during the 24-h test period and the contrast of the galvanized sheet and steel sheet was close to 1 during two diurnal cycles, while the contrast of the wooden plate fluctuated between 1.0 and 1.16 throughout the experiment time, proving the existence of thermal crossover for the objects with lower thermal inertia once again. Figure 9d showed the multispectral images, which were obtained at the same period with Figure 9b. As can be seen from Figure 9d, the multispectral technology was used to find the emissivity difference between the target and the background at 3700 nm, 3800 nm, 4120 nm, 4420 nm, and 4720 nm to enhance the contrast among the galvanized sheet, steel sheet, and road. Among the five wavebands the best contrast improvement was at 4720 nm, with 4420 nm following, which presented a consistent trend in accordance with Figure 4a and indicated the effectiveness of the multispectral technology in solving the thermal crossover problem. The contrast values among the three targets and the background under different wavebands in Figure 9d are listed in Table 4. Compared with the Figure 8, the image contrast enhancement in the target area is direct after employing the narrow band-pass filters.

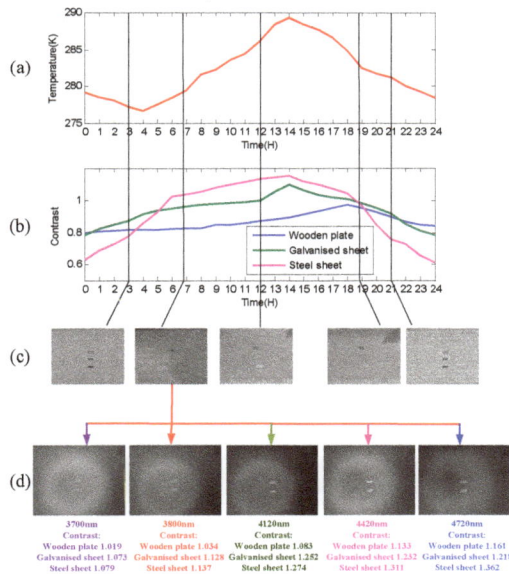

Figure 9. (**a**) Ambient temperature during the 24-h test; (**b**) contrast curve among the three targets and the background road; (**c**) images of the targets in the five time zones; and (**d**) images of the targets at 3700 nm, 3800 nm, 4120 nm, 4420 nm, and 4720 nm at the first diurnal cycle.

Figure 10 showed the pseudo-color image obtained by running the HSV image fusion algorithm described in Section 3 with Figures 8b and 9d. Combining the infrared broadband image with the infrared images under characteristic wavebands, the three targets were marked with different colors and presented clearly, by which the multispectral technology employed an effective supplementary method for the conventional mid-infrared broadband thermal sensor to solve the thermal crossover detection problem.

Further, it can be noted that the magnitude of contrast improvement is not as large as the calculation results in Figure 4a because of the solar radiation effect, stray radiation caused by band-pass filters, and the difference between the actual emissivity value and the real emissivity value. However, this does not influence our experimental conclusions that multispectral technology can be employed to solve the thermal crossover problem.

In order to further show the advantage of multispectral technology in solving the thermal crossover problem, the same field experiment with polarization technique by using $0°$, $45°$, $90°$, $135°$ four linear polarizers was also conducted and the four polarization state polarization images obtained at 07:00 were shown in Figure 11.

Table 4. Contrast among the three targets and the background in Figure 9.

Wavebands	Wooden Plate	Galvanized Sheet	Steel Sheet
3700 nm	1.019	1.073	1.079
3800 nm	1.034	1.128	1.137
4120 nm	1.083	1.252	1.274
4420 nm	1.133	1.232	1.311
4720 nm	1.161	1.218	1.362

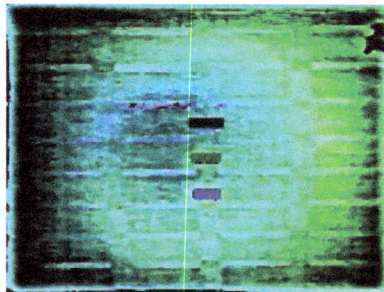

Figure 10. Pseudo-color image fused by Figures 7b and 8d.

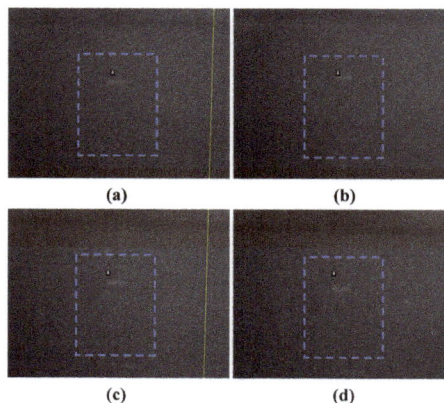

(a)   (b)

(c)   (d)

Figure 11. Infrared polarization images with four polarization states. (a) $0°$; (b) $45°$; (c) $90°$; (d) $135°$.

As can be seen from Figure 11, for the wooden plate, thermal crossover still did not occur, while for the galvanized sheet and steel sheet, the thermal crossover did occur, which was similar to the results with multispectral technology. The contrast values between the galvanized sheet, steel sheet and the background in Figure 11 are listed in Table 5.

Table 5. Contrast between galvanized sheet, steel sheet targets and background in Figure 11.

Figure Number	Galvanized Sheet	Steel Sheet
a	0.909	0.942
b	0.977	0.964
c	0.979	0.981
d	0.965	0.976

From Table 5, it can be found that, compared with the infrared multispectral images, without further image processing, the image contrast enhancement in the target area in the infrared polarization images with four polarization states were not obvious. Thus, the Stokes vectors, which completely characterized the polarization states of targets from the scene need to be calculated. The data products used in this experiment included S_0 and S_1 Stokes parameter images where S_0 is the horizontal ($0°$) plus the vertical ($90°$) components of polarization and the S_1 Stokes parameter is the horizontal minus the vertical components of polarization. The S_0 and S_1 Stokes parameter images are shown as Figure 12a,b respectively.

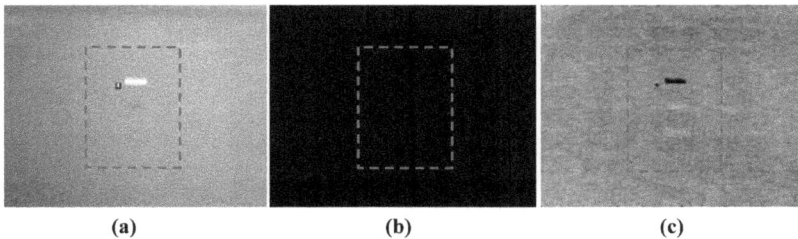

(a) (b) (c)

Figure 12. S_0 and S_1 Stokes parameter images (**a**) S_0 image; (**b**) S_1 image; (**c**) S_1 after contrast stretching.

The image contrast of the galvanized sheet and steel sheet in Figure 12a was 0.921 and 1.073, respectively, which showed some extent of improvement compared with Figure 11. However the DN difference between the galvanized sheet, steel sheet targets and background in Figure 12a were only 16 and 7. In Figure 12b, although the calculated image contrast, according to $C' = DN_{target}/DN_{bg}$, was improved, it was meaningless as the DN of targets and background were too low to be sensed by eyes. In fact, the DN difference between the galvanized sheet, steel sheet targets and background in Figure 12b were only 3 and 2, respectively. In order to show the targets in Figure 12b relatively clearly, the images were further processed with the contrast stretching algorithm, as shown in Figure 12c. Through the data processing procedure, it could be found that even with the Stokes parameter calculation, the difference between target and background had still not been improved significantly so that further image processing procedures were required. The main reason for the polarization detection experiment result was that the abundant geometry information contained in the background weakened the polarization characteristics differences. Because the polarization technology achieves distinction between target and background through the perception of their polarization characteristics differences, the background information might have an influence on the target detection. However, for the multispectral technology, as stated previously, compared with the polarization technique, as the target's infrared spectrum signature only differs with materials, it would be faster and more direct to distinguish the target from the background in the complex environment only if the characteristic wavelengths of the targets and backgrounds were acquired in advance.

4. Conclusions

As the thermal crossover has great influence on the infrared sensors working in a single wide range, it is significant to solve this problem for the conventional mid-infrared thermal sensor, especially for the infrared surveillance system. In this study, we analyze theoretically how the thermal crossover disables the conventional thermal sensor and under what conditions the thermal crossover would happen. Furthermore, based on the analysis, a fast, compact and light optical prototype based on infrared multispectral technology is designed with the known characteristic wavelengths according to the prior knowledge. Then the experimental process has been optimized and more image data is provided, especially regarding what the pictures are when the target integrated with the background during the diurnal cycle. Then, the whole process of employing the multispectral technology to solve the thermal crossover detection problem is clearly shown. In addition, a comparison experiment with polarization technique is also conducted to further show the advantage of multispectral technology.

The field experiment with multispectral technology was conducted over a 24-h period with the targets of galvanized sheet, steel sheet, and wooden plate, and the background road on a sunny day. The results showed that, for the galvanized sheet and steel sheet targets, the thermal crossover could affect a contrast for up to four hours at two diurnal cycles, jeopardizing the success of surveillance missions. For the wooden plate target, although the image contrast reduced over the diurnal cycle, it could still distinguish the targets from the background objects, which means that thermal crossover might not always occur, or even possibly not exist at all over the diurnal cycle for the objects with higher thermal inertia. Through employing the infrared narrow band-pass filters, thermal crossover in the first diurnal cycle was relieved as the contrast was upgraded to the levels such that the metal targets could be distinguished from the background objects. Furthermore, the experimental results provided us with the information about what the characterized bands between the targets and background objects were, which would be useful for system design in the future. In addition, the pseudo-colored image produced by multi-spectral image fusion method showed the effectiveness of the multispectral technology for contrast promotion of each target. Then, as a comparison, the same field experiment with polarization technique by using $0°$, $45°$, $90°$, $135°$ four linear polarizers was also conducted and the S_0 and S_1 Stokes parameter images showed that the image contrast showed some extent of improvement but no obvious improvement, as the background weakened the polarization characteristics differences.

While promising, the field experiment should just be considered as very preliminary practical application and the experimental results should also just be viewed as a proof-of-principle. Nevertheless, the conclusion that the multispectral technology can be employed to solve the thermal crossover problem is unambiguous. In future, it might be possible to further extend the range of applications for the conventional thermal infrared broadband sensor into the thermal crossover periods by exploiting the emissivity of infrared spectral signatures and fusing multispectral images from the perspective of mid-infrared thermal detecting system design. Research focusing on the characterized bands between different common targets and background objects and how the weather conditions influence the thermal crossover will be undertaken.

Acknowledgments: This work was supported by the National Natural Science Foundation of China (No. 61571029), the CAST Innovation Foundation of the China Academy of Space Technology, and the Changjiang Scholars and Innovative Research Team in University (No. IRT1203).

Author Contributions: Zheng Ji wrote the manuscript and was responsible for the research design, data collection, and analysis. Huijie Zhao, Na Li, Jianrong Gu and Yansong Li assisted in the methodology development and research design and participated in the writing of the manuscript and its revision.

Conflicts of Interest: The authors declare no conflict of interest.

Abbreviations

The following abbreviations are used in this manuscript:

mmW   Millimeter Wave
NESR   Noise Equivalent Spectral Radiance
NETD   Noise Equivalent Temperature Difference
DN   Digital Number
BRDF   Bidirectional Reflectance Distribution Function
RRD   Relative thermal radiation differences
IR   Infrared
PC   Personal Computer

References

1. Sidran, M. Broadband reflectance and emissivity of specular and rough water surfaces. *Appl. Opt.* **1981**, *20*, 3176–3183. [CrossRef] [PubMed]
2. Elsner, A.E.; Weber, A.; Cheney, M.C.; VanNasdale, D.A.; Miura, M. Imaging polarimetry in patients with neovascular age-related macular degeneration. *J. Opt. Soc. Am. A* **2007**, *24*, 1468–1480. [CrossRef]
3. Cooper, A.W.; Lentz, W.J.; Walker, P.L.; Chan, P.M. Infrared polarization measurements of ship signatures and background contrast. *Proc. SPIE* **1994**, *2223*, 300–309.
4. Jin, L.; Hamada, T.; Otani, Y.; Umeda, N. Measurement of characteristics of magnetic fluid by the Mueller matrix imaging polarimeter. *Opt. Eng.* **2004**, *43*, 181–185. [CrossRef]
5. Wijngaarden, R.J.; Heeck, K.; Welling, M.; Limburg, R.; Pannetier, M.; van Zetten, K.; Roorda, V.L.G.; Voorwinden, A.R. Fast imaging polarimeter for magneto-optical investigations. *Rev. Sci. Instrum.* **2001**, *72*, 2661–2664. [CrossRef]
6. Shaw, J.A. Degree of linear polarization in spectral radiances from water-viewing infrared radiometers. *Appl. Opt.* **1999**, *38*, 3157–3165. [CrossRef] [PubMed]
7. Felton, M.; Gurton, K.P.; Pezzaniti, J.L.; lt, D.B.C.; Roth, L.E. Measured comparison of the crossover periods for mid- and long-wave IR (MWIR and LWIR) polarimetric and conventional thermal imagery. *Opt. Express* **2010**, *18*, 15704–15713. [CrossRef] [PubMed]
8. Felton, M.; Gurton, K.P.; Pezzaniti, J.L.; Chenault, D.B.; Roth, L.E. *Comparison of the Inversion Periods for Mid-wave IR (MidIR) and Long-Wave IR (LWIR) Polarimetric and Conventional Thermal Imagery*; Army Research Laboratory: Adelphi, MD, USA, 2010; pp. 20783–21197.
9. Felton, M.; Gurton, K.P.; Roth, L.E.; Pezzaniti, J.L.; Chenault, D.B. Measured comparison of the inversion periods for polarimetric and conventional thermal long-wave IR (LWIR) imagery. *Proc. SPIE* **2009**, *7461*. [CrossRef]
10. Wilson, J.P.; Schuetz, C.A.; Harrity, C.E.; Kozacik, S.; Eng, D.L.K.; Prather, D.W. Measured comparison of contrast and crossover periods for passive millimeter-wave polarimetric imagery. *Opt. Express* **2013**, *21*, 12899–12907. [CrossRef] [PubMed]
11. Wilson, J.P.; Murakowski, M.; Schuetz, C.A.; Prather, D.W. Simulations of polarization dependent contrast during the diurnal heating cycle for passive millimeter-wave imagery. *Proc. SPIE* **2013**, *8873*. [CrossRef]
12. Retief, S.J.P.; Willers, C.J.; Wheeler, M.S. Prediction of thermal crossover based on imaging measurements over the diurnal cycle. *Proc. SPIE* **2003**, *5097*, 58–69.
13. Stotts, L.B.; Winter, E.M.; Hoff, L.E.; Reed, I.S. Clutter Rejection Using Multi-Spectral Processing. *Proc. SPIE* **1990**, *1305*. [CrossRef]
14. Winter, E.M. *Infrared Spectral Analysis*; Technical Research Associates Report No. TRA-90D-109; Air Force Research Laboratory: Dayton, OH, USA, 1990.
15. Stocker, A.D.; Reed, I.S.; Yu, X. Multi-Dimensional Signal Processing for Electro-Optical Target Detection. *Proc. SPIE* **1990**, *1305*, 218–231.
16. Stocker, A.D.; Yu, X.; Winter, E.M.; Hoff, L.E. Adaptive Detection of Sub-Pixel Targets Using Multi-Band Frame Sequences. *Proc. SPIE* **1991**, *1481*. [CrossRef]
17. Cederquist, J.N.; Johnson, R.O.; Reed, I.S. *Infrared Multispectral Imagery Program. Phase I: Model-Based Performance Predictions*; ERIM Final Report No. 232300-41-F to AF Wl/AARI-4, Contract No. F33615-90-C-1441; Environmental Research Institute of Michigan: Ann Arbor, MI, USA, 1993.

18. Eismann, M.T. Infrared Multispectral Target/Background Field Measurements. *Proc. SPIE* **1994**, *2235*, 135–147.
19. Stocker, A.D.; Seldin, A.O.H.; Cederquist, J.N.; Schwartz, C.R. Analysis of Infrared Multi-Spectral Target/Background Field Measurements. *Proc. SPIE* **1994**, *2235*, 148–161.
20. Schwartz, C.R.; Eismann, M.T.; Cederquist, J.N. Thermal multispectral detection of military vehicles in vegetated and desert backgrounds. *Proc. SPIE* **1996**, *2742*, 286–297.
21. Zhao, H.; Ji, Z.; Zhang, Y.; Sun, X.; Song, P.; Li, Y. Mid-infrared imaging system based on polarizers for detecting marine targets covered in sun glint. *Opt. Express* **2016**, *24*, 16396–16409. [CrossRef] [PubMed]
22. Nicodemus, F.E. Directional reflectance and emissivity of an opaque surface. *Appl. Opt.* **1965**, *4*, 767–775. [CrossRef]
23. Chen, W.; Wang, X.; Jin, W.; Li, F.; Cao, Y. Experiment of target detection based on medium infrared polarization imaging. *Infrared Laser Eng.* **2011**, *40*, 7–11.

sensors

MDPI

Article

Color Restoration of RGBN Multispectral Filter Array Sensor Images Based on Spectral Decomposition

Chulhee Park and Moon Gi Kang *

Department of Electrical and Electronic Engineering, Yonsei University, 50 Yonsei-ro, Seodaemun-gu, Seoul 03722, Korea; ascaron5@gmail.com
* Correspondence: mkang@yonsei.ac.kr; Tel.: +82-2-2123-4863

Academic Editor: Gonzalo Pajares Martinsanz
Received: 22 February 2016; Accepted: 13 May 2016; Published: 18 May 2016

Abstract: A multispectral filter array (MSFA) image sensor with red, green, blue and near-infrared (NIR) filters is useful for various imaging applications with the advantages that it obtains color information and NIR information simultaneously. Because the MSFA image sensor needs to acquire invisible band information, it is necessary to remove the IR cut-offfilter (IRCF). However, without the IRCF, the color of the image is desaturated by the interference of the additional NIR component of each RGB color channel. To overcome color degradation, a signal processing approach is required to restore natural color by removing the unwanted NIR contribution to the RGB color channels while the additional NIR information remains in the N channel. Thus, in this paper, we propose a color restoration method for an imaging system based on the MSFA image sensor with RGBN filters. To remove the unnecessary NIR component in each RGB color channel, spectral estimation and spectral decomposition are performed based on the spectral characteristics of the MSFA sensor. The proposed color restoration method estimates the spectral intensity in NIR band and recovers hue and color saturation by decomposing the visible band component and the NIR band component in each RGB color channel. The experimental results show that the proposed method effectively restores natural color and minimizes angular errors.

Keywords: color restoration; infrared cut-off filter removal; multispectral imaging; spectral estimation; spectral decomposition

1. Introduction

The near-infrared (NIR) is one of the regions closest in wavelength to the radiation detectable by the human eye. Unlike human eyes, sensors based on silicon (SiO_2) are sensitive to NIR up to 1100 nm, limited by the cut-off value of silicon. Due to the proximity of NIR to visible radiation, NIR images share many properties with visible images. However, surface reflection in the NIR bands is material dependent. For instance, most dyes and pigments used for material colorization are somewhat transparent to NIR. This means that the difference in the NIR intensities is not only due to the particular color of the material, but also to the absorption and reflectance of dyes. Therefore, the NIR intensity provides the useful information pertinent to material classes rather than the color of that object [1].

Recently, there have been several attempts to use NIR band information. In remote sensing applications [2,3], the multispectral images observed in a variety of spectrum bands have been used where both the visible and NIR bands are included. As each spectral band provided different kinds of information, the spectral bands were selectively used in the observation of the multispectral images. In surveillance cameras [4] and night vision cameras [5], the NIR band is used especially under low lighting conditions or invisible NIR lighting conditions. The NIR band is also used in biometric [6], face matching [7] and face recognition [8] applications, which have been studied based on the intrinsic reflectivity of the skin or eyes under NIR illumination. Since the reflection in NIR is material dependent, it is also used in material classification [1] and illuminant estimation [9]. NIR images can be used in image enhancement applications, such as image dehazing [10].

To develop an NIR image acquisition system, Kise *et al.* designed a three-band spectral imaging system composed of multiple cameras with a beam splitter [11]. This imaging system has been used to acquire multispectral images in user-selected spectral bands simultaneously by utilizing three interchangeable optical filters and various optical components. Similarly, Matsui *et al.* implemented a multispectral imaging system, where two infrared cut-off filter (IRCF)-removed cameras were used to capture the color and NIR images independently [12]. In this system, the IRCF-removed cameras were perpendicularly aligned, and the IRCF was used as a light splitter for the visible and NIR bands. By managing the shutter of two cameras with a single controller, each spectral band image pair was acquired, simultaneously. However, this imaging system requires a large space to attach two or more cameras and to perform the alignment process. Due to the lack of portability of these devices, multi camera-based imaging systems are not suitable for practical outdoor environments. C. Fredembach [13] suggests another approach in which an IRCF-removed single camera with multiple optical band pass filters can achieve smaller sizes than multi-camera systems. On the other hand, this imaging system requires too much time to change the optical filters. Because of this weakness, some artifacts, like motion blur and registration problems, can occur during the image acquisition process.

As an alternative approach, an IRCF-removed color filter array (CFA) image sensor, such as a Bayer image sensor without an IRCF, can be used [13]. By using a single digital camera without an IRCF, the spectral information of the visible bands and that of the NIR bands can be acquired at the same time. Figure 1 shows a conventional camera system approach with an IRCF and a spectral sensitivity of a complementary metal-oxide semiconductor (CMOS) imager integrated with traditional RGB Bayer filters. By removing the IRCF, the NIR contribution to the RGB channel can reach the CMOS imager. This additional NIR information can be used to allow for invisible monitoring in surveillance applications.

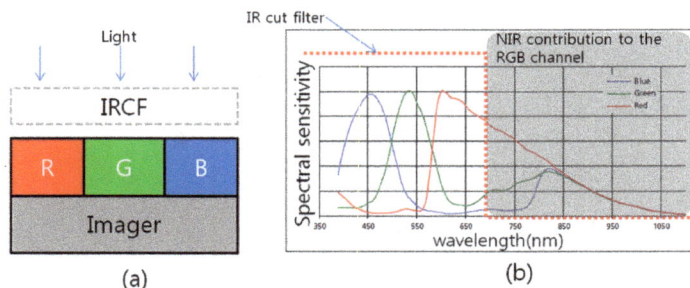

Figure 1. (**a**) Conventional camera system based on a color filter array (CFA) image sensor with the IR cut-off filter. (**b**) Spectral sensitivity of the camera system.

On the other hand, mixing color and NIR signals at the pixel level can result in extreme color desaturation if the illumination contains sufficient amounts of NIR. Although it may be possible to overcome the unwanted NIR contribution to the RGB color channel through the signal processing technique, it is hard to estimate the NIR spectral energy in each RGB color channel, because there is no way to detect the NIR band spectral characteristics.

As an improved system based on a single image sensor, an imaging system based on the multispectral filter array (MSFA), which simultaneously obtains visible and NIR band images, can be considered [14]. A pixel configuration of the RGB filters and another NIR pass filter, which transmits NIR light only, is shown in Figure 2. In the following descriptions, we refer to the four channels as RGBN channels, where RGB represents the red, green and blue channels and N represents the additional channel for the NIR band.

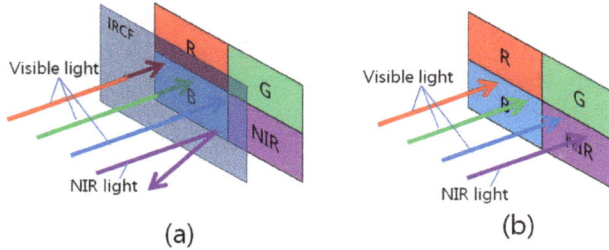

Figure 2. Infrared cut-off filter (IRCF): (**a**) typical imaging system using IRCF; (**b**) IRCF-removed imaging system.

Because these sensors based on the RGBN filter array need to acquire invisible range information, removing IRCF is necessary. Without IRCF, RGB and NIR signals can be obtained simultaneously. Because of this advantage, imaging systems based on MSFA sensors can be applied to a wide variety of applications. Under certain circumstances, especially low lighting conditions, this system can obtain wide spectral information simultaneously. Furthermore, by applying fusion technology that uses NIR band information, gaining additional sensitivity to colors that do not deviate considerably from the human visual system is possible [15].

However, without IRCF, the additional NIR component penetrates through the color filter to each R, G, B pixel. The unwanted NIR interference distorts the color information of each R, G, B color channel. Figure 3 is an example of an imaging system based on the MSFA image sensor. Many researchers studied the interpolation method, such as [15–17], to make a full resolution image in each RGBN channel. Since the input RGB signals contain NIR, natural RGB color information needs to be calculated by subtracting an NIR band component from the input RGB signals that have been deteriorated with NIR interference. During the process, the NIR channel information in the N pixel can be used to remove the unnecessary NIR contribution to the RGB channel. After restoring the color information of the RGB channel from the input signal received through the MSFA image sensors without an IRCF, a fusion method can be applied to generate the new blended images, which have not only natural color information, but also additional NIR spectral information. To take advantage of this benefit, it is necessary to restore natural color. As a result, the IRCF can be removed day and night with color restoration process.

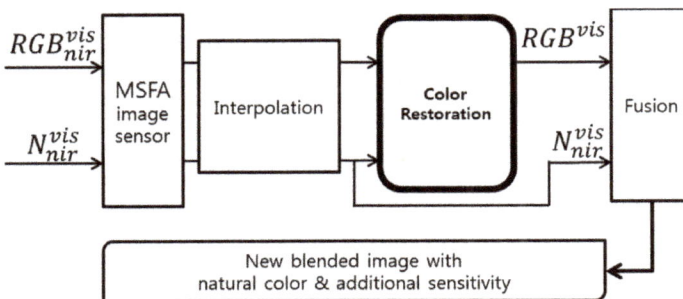

Figure 3. Example of a multispectral filter array (MSFA)-based imaging system.

In recent studies, researchers have proposed CFA for one-shot RGB and NIR capture in NIR imaging. However, the studies do not consider color restoration [16,17]. Although [18] addresses both crosstalk and demosaicing, it assumes crosstalk between the green and NIR channels only. Chen *et al.* proposed a color correction pipeline [15], which is able to apply only specific NIR illumination. The color correction method in [15] does not guarantee successful color correction results if the illumination spectrum is widely distributed in an NIR range. Furthermore, in [19], NIR restoration was proposed; however, the method does not consider crosstalk in a visible range in an NIR channel. The IR removal method proposed by Martinello *et al.* considers the crosstalk happening in the near IR range of 650 nm to 810 nm. This method assumes that the contribution from the wavelengths in the visible range ($\lambda < 650$ nm) to the IR channel can be ignored. On the contrary, the proposed color restoration method divides the visible and NIR bands to estimate the color correction matrix. In the visible band, the crosstalk in the N channel is estimated by using the linear regression of RGB channels by using the N channel decomposition matrix. By removing the estimated crosstalk in the N channel, the N channel information in the NIR band is obtained. The N channel information in the NIR band is used to estimate the NIR contribution to the RGB channels by using the RGB channel decomposition matrix. In this way, the proposed method copes with the different spectral responses of the visible and invisible bands, respectively. Furthermore, the proposed method considers the crosstalk happening in the near IR range from 650 nm to 1100 nm.

We proposed a brief idea to restore color information with an RGBN sensor [20]. However, since we focused only on color restoration under generally bright illumination environments, our previous work did not have good performance in low light conditions. In this paper, we proposes a color restoration method that removes the NIR component in each RGB color channel with an imaging system based on the IRCF-removed MSFA image sensor. To investigate the color restoration method for various illumination environments, we analyze the change of the chromaticity feature obtained by the additional NIR. In addition, the color restoration method for the low lighting condition based on the spectral energy distribution analysis is proposed. Since color degradation caused by IRCF removal is a huge limitation, the NIR contribution to each RGB color channel needs to be eliminated. To remove unwanted NIR components in each RGB channel, the color restoration model was subdivided into two parts of the spectral estimation and the spectral decomposition process.

The remainder of this paper is organized as follows: In Section 2, we discuss the problem that arises when a color image is acquired with the IRCF-removed MSFA image sensor. In Section 3, we analyze the color model of an IRCF-removed MSFA image sensor. In Section 4, we outline our proposed color restoration method with spectral estimation and spectral decomposition. In Section 5, we present our results and compare our solution to another state-of-the-art method. In Section 6, we provide a conclusion.

2. Color Degradation

To analyze the change of the chromaticity feature by the additional NIR, the RGB color space was converted to the HSI color space, as in [21]:

$$
\begin{aligned}
H &= \cos^{-1}\{\frac{\frac{1}{2}[(R-G)+(R-B)]}{[(R-G)^2+(R-B)(G-B)]^{1/2}}\} \\
S &= \frac{I-a}{I} \quad \text{where} \quad a = min[(R,G,B)] \\
I &= \frac{R+G+B}{3}
\end{aligned}
\tag{1}
$$

where $min[(\cdot)]$ represents the minimum value among three values. H, S and I represent the hue, saturation and intensity, respectively.

In Figure 4, the NIR band is divided into two sub-bands: we define these sub-bands as a chromatic NIR band (700 nm~800 nm) and an achromatic NIR band (800 nm~1100 nm), respectively. Figure 5 shows that the responses of the achromatic NIR bands are identical. To obtain achromatic NIR band information, we used an NIR band pass filter that passes a specific wavelength (800 nm∼1100 nm). The distribution of 96 color patch values in the Gretag color checker SG shows a linear response in the achromatic NIR band with respect to the NIR channel. Based on this, we define these responses as a constant at each pixel, such as $R_{nir(achr)} = G_{nir(achr)} = B_{nir(achr)} = \delta$. The $R_{nir(achr)}$, $G_{nir(achr)}$ and $B_{nir(achr)}$ represent the achromatic colors of the image sensor beyond an 800-nm wavelength in each channel. As a result, the RGB intensities at a pixel position are represented as:

$$
\begin{aligned}
R(i,j) &= R_{chr}(i,j) + \delta(i,j) \\
G(i,j) &= G_{chr}(i,j) + \delta(i,j) \\
B(i,j) &= B_{chr}(i,j) + \delta(i,j)
\end{aligned}
\tag{2}
$$

where $R_{chr}, G_{chr}, B_{chr}$ represent the chromatic colors of the image sensor under an 800-nm wavelength.

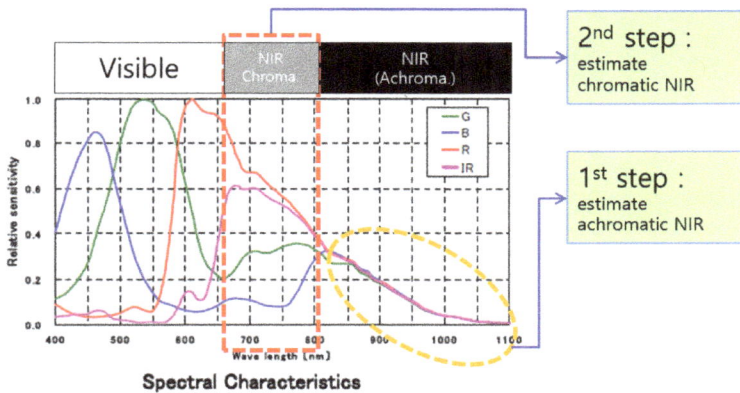

Figure 4. Spectral response of the MSFA image sensor.

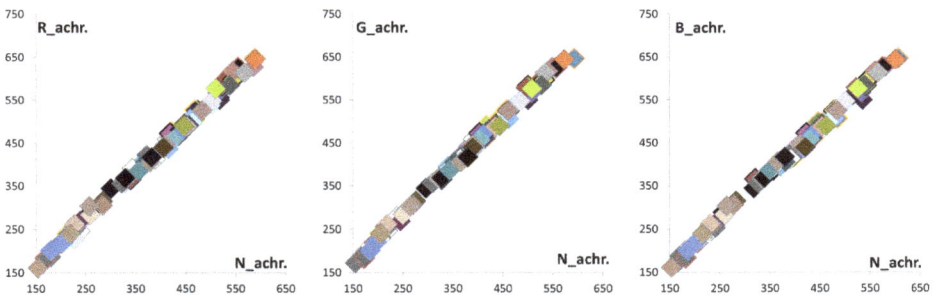

Figure 5. Correlation between the RGB channel and the N channel in the NIR band beyond 800 nm (**a**) $N_{nir(achr)}$ vs. $R_{nir(achr)}$ (**b**) $N_{nir(achr)}$ vs. $G_{nir(achr)}$ (**c**) $N_{nir(achr)}$ vs. $B_{nir(achr)}$.

With the RGB color values with offset δ, the intensity of the observed color is defined as follows:

$$I = \frac{[(R_{chr} + \delta) + (G_{chr} + \delta) + (G_{chr} + \delta)]}{3} \tag{3}$$
$$= I_{chr} + \delta$$

where $I_{chr} = (R_{chr} + G_{chr} + B_{chr})/3$ represents the intensity of the chromatic spectral band of the image sensor. The intensity of the IRCF-removed MSFA image sensor is changed by the amount of the offset value. The hue value in Equation (1) is redefined as:

$$H = \cos^{-1}\{\frac{\frac{1}{2}[(R - G) + (R - B)]}{[(R - G)^2 + (R - B)(G - B)]^{1/2}}\} \tag{4}$$
$$= \cos^{-1}\{\frac{\frac{1}{2}[(R_{chr} - G_{chr}) + (R_{chr} - B_{chr})]}{[(R_{chr} - G_{chr})^2 + (R_{chr} - B_{chr})(G_{chr} - B_{chr})]^{1/2}}\}$$

Because the achromatic offset value δ is removed during subtraction, an identical offset on the RGB channels could not change the hue value. Finally, the saturation value is described as:

$$S = \frac{I - a}{I} = \frac{I_{chr} - a_{chr}}{I} = \frac{I_{chr}}{I} \cdot S_{chr} \tag{5}$$

where $S_{chr} = (I_{chr} - a_{chr})/I_{chr}$ represents the saturation of the chromatic spectral band of the image sensor and $a_{chr} = min(R_{chr}, G_{chr}, B_{chr})$. Since the range of $\frac{I_{chr}}{I}$ is $0 \leq \frac{I_{chr}}{I} \leq 1$, the saturation of the image obtained by the IRCF-removed MSFA image sensor is degraded and becomes smaller than the image obtained by the chromatic spectral band of the image sensor.

Figure 6 describes how NIR affects the RGB color images. The illuminance was 200 lx, and the exposure time was 0.03 s. When objects are illuminated by an incandescent lamp, an image sensor with an IRCF obtains a yellowish hue due to the low color temperature of the illuminance. After performing a white balance technique from the grey color patch, a white-balanced color image was obtained as shown in Figure 6b. On the other hand, due to the additive NIR intensities included in the RGB channels, Figure 6c appears brighter than Figure 6a, and low color saturation was observed in Figure 6d.

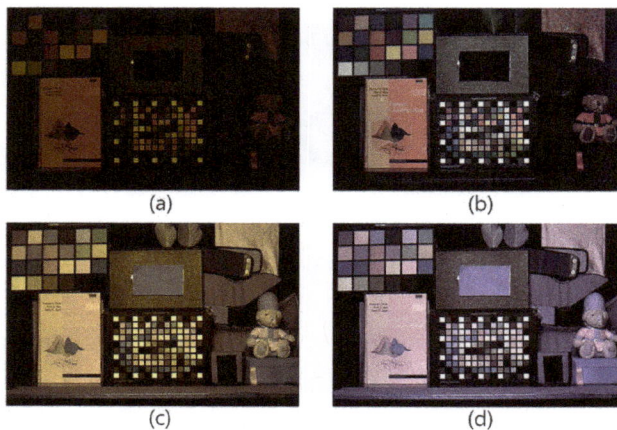

Figure 6. Color observation of the MSFA image sensor under incandescent light. (**a**) Image captured with IRCF; (**b**) (a) with white balance; (**c**) image captured with IRCF removal MSFA image sensor; (**d**) (c) with white balance.

To correct desaturated color from the input image acquired by the MSFA image sensor, several conventional methods can be considered, as described in [22,23]. A straightforward method is to train the matrix to reproduce a set of known reference colors. Given the observed color vector **Y** and the visible band color vector with canonical illuminance **X**, the color correction method is represented in a matrix form:

$$\mathbf{X} = \Phi^T \mathbf{Y} \tag{6}$$

where Φ is a matrix whose component corresponds to the ratio between the canonical and the current illuminance value of each channel. The illuminant color estimation was performed under unknown lighting conditions where pre-knowledge based approaches, such as gamut mapping [24] or the color correlation framework [25], were used.

However, color degradation caused by IRCF removal is not considered a multiplicative process, but an additive process. Applying a conventional color correction approach to the RGBN images yielded poor results, because it did not sufficiently remove the NIR contributions to the RGB channels. The higher the energy in the NIR band relative to that in the visible band, the higher the color errors caused by NIR contributions to the RGB signals. As a result, the conventional color correction method restored visible band color in a limited way. Although each color was obtained under the same illuminant conditions with and without an IRCF, respectively, the mixture of the exclusive NIR band intensity to the visible band intensity resulted in severe color distortion.

Figure 7 shows the result of the conventional color correction method for an MSFA image. In Figure 7c, the color correction matrix worked well for colors in the color chart with low reflectance in the NIR band. However, despite the fact that the colors of the black paper and velvet paper were the same in the visible band, the conventional color correction method could not restore the black color with high reflectance in the NIR band (such as fabric substance).

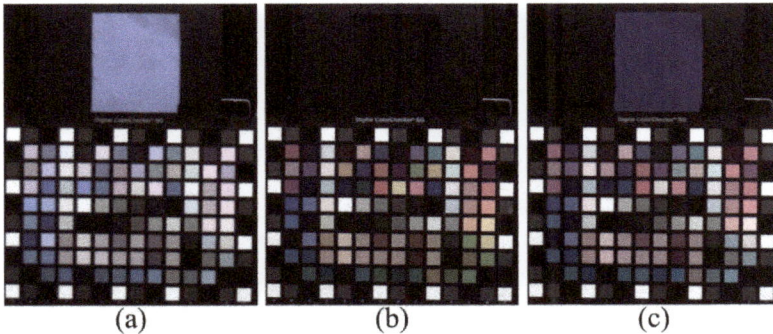

Figure 7. Example of the conventional color correction method for the MSFA image. (**a**) MSFA image without IRCF; (**b**) MSFA image with IRCF; (**c**) color correction result.

3. Color Model of an IRCF-Removed MSFA Image Sensor

A color image observed by a CMOS image sensor can be modeled as a spectral combination of three major components: illuminant spectra $E(\lambda)$, sensor function $R^{(k)}(\lambda)$ and the surface spectra $S(\lambda)$. The color image formation model in the visible band for channel k was defined as [26]:

$$C_{vis}^{(k)} = \int_{w_{vis}} E(\lambda) R^{(k)}(\lambda) S(\lambda) d\lambda \tag{7}$$

where w_{vis} represents the spectral range of the visible band between 400 nm and 700 nm. Since an IRCF-removed MSFA image sensor can acquire the additional NIR band spectral energy beyond a 700-nm wavelength, the range of these three major components in Equation (7) had to be expanded to the NIR band. The observed camera response for channel k when using the IRCF-removed MSFA image sensor is represented by the color image formation model $C_{MSFA}^{(k)}$ [19] from Equation (7):

$$
\begin{aligned}
C_{MSFA}^{(k)} &= \int_{w_{vis}+w_{nir}} E(\lambda)R^{(k)}(\lambda)S(\lambda)d\lambda \\
&= \int_{w_{vis}} E(\lambda)R^{(k)}(\lambda)S(\lambda)d\lambda + \int_{w_{nir}} E(\lambda)R^{(k)}(\lambda)S(\lambda)d\lambda \\
&= C_{vis}^{(k)} + C_{nir}^{(k)}
\end{aligned}
\tag{8}
$$

where w_{nir} represents the NIR band beyond 700 nm. $C_{vis}^{(k)}$ and $C_{nir}^{(k)}$ represent the camera response for channel k by using the IRCF-removed MSFA image sensor in the visible band and the NIR band, respectively. For an image sensor with RGBN filters, the intensities at each pixel position are represented as,

$$
\begin{aligned}
R(i,j) &= R_{vis}(i,j) + R_{nir}(i,j) \\
G(i,j) &= G_{vis}(i,j) + G_{nir}(i,j) \\
B(i,j) &= B_{vis}(i,j) + B_{nir}(i,j) \\
N(i,j) &= N_{vis}(i,j) + N_{nir}(i,j)
\end{aligned}
\tag{9}
$$

In Equation (9), each pixel contained additional NIR band information. Since this additional information can be helpful to increase the sensitivity of the sensor, this feature can be useful under low light condition. However, mixing color and NIR intensities can result in color degradation if the illumination contains high amounts of NIR. To restore the RGB channels corrupted by NIR band spectral energy, the additional NIR band components (R_{nir}, G_{nir}, B_{nir}) in the RGB channels have to be removed:

$$
\begin{aligned}
R_{vis} &= R - R_{nir} \\
G_{vis} &= G - G_{nir} \\
B_{vis} &= B - B_{nir} \\
N_{vis} &= N - N_{nir}
\end{aligned}
\tag{10}
$$

Since the spectral response function of the RGBN filter is not defined only in the NIR band, we used a signal processing approach to estimate the NIR band response. To decompose the spectral information of the RGBN channel, the unknown value N_{vis} or N_{nir} must be estimated. To cope with the different characteristics of the correlation in the visible band, as well as the NIR band, we set the correlation model in each sub-band, separately. In the visible band, the RGB channel filters show different peak spectral responses, while the N channel filter covered all spectral ranges without outstanding peaks. As a result, the N channel filter response function is modeled as a linear combination of the others:

$$
\begin{aligned}
N_{vis} &= \int_{w_{vis}} \omega_r(\lambda)E(\lambda)R^{(r)}(\lambda)S(\lambda)d\lambda \\
&+ \int_{w_{vis}} \omega_g(\lambda)E(\lambda)R^{(g)}(\lambda)S(\lambda)d\lambda \\
&+ \int_{w_{vis}} \omega_b(\lambda)E(\lambda)R^{(b)}(\lambda)S(\lambda)d\lambda
\end{aligned}
\tag{11}
$$

where $\omega_r(\lambda)$, $\omega_g(\lambda)$ and $\omega_b(\lambda)$ represent the coefficients that show cross-correlation in the visible band. Since the spectral response of the N channel in the visible band covers a wide spectral range without an outstanding peak, those coefficients are constrained to be constant in terms of the wavelength [27].

Using the constrained weights, the intensities of the N channel in the visible band are approximated as follows:

$$N_{vis}(i,j) \approx \omega_r \cdot R_{vis}(i,j) + \omega_g \cdot G_{vis}(i,j) + \omega_b \cdot B_{vis}(i,j) \tag{12}$$

where ω_r, ω_g and ω_b represent the visible band cross-correlation coefficients obtained by the linear transformation model:

$$\mathbf{N} = \mathbf{DC} \tag{13}$$

where $\mathbf{D}$ is a one by three matrix describing the mapping between the RGB to N channel values. The transformation $\mathbf{D}$ is obtained by solving the following minimization function:

$$\hat{\mathbf{D}}^T = argmin_{\mathbf{D}^T}||\mathbf{N} - \mathbf{D}^T\mathbf{C}||^2 \tag{14}$$

where $\mathbf{N}$ and $\mathbf{C}$ are matrices whose components are the NIR and the RGB components. Each cross-correlation coefficient could have been of any arbitrary form determined by the illuminance change and the spectral response of the sensor. As a result, the function ω depends not on the spectrum λ itself, but on the spectral response of the illuminance and the sensor. Figure 8 represents the comparison between the optical filtered N channel image in visible band and estimated N channel image in the visible band by using Equation (12).

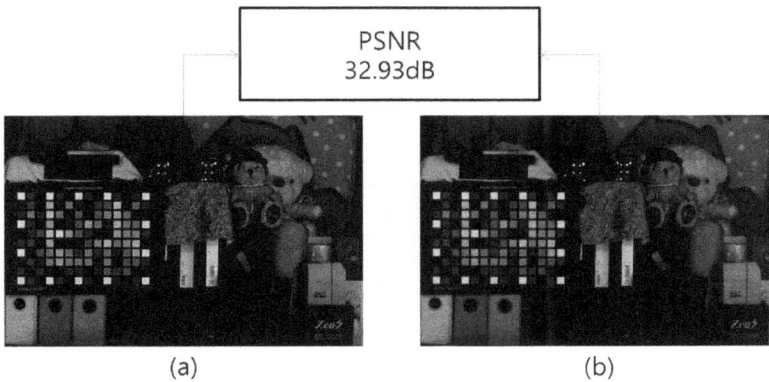

Figure 8. Comparison between (**a**) the optical filtered N channel image and (**b**) the estimated N channel image in visible bands.

In the NIR band, the cross-correlation is derived more intuitively, since the RGBN filters are all pass filters where the filter responses are highly correlated in the NIR spectral range. Since there is an energy difference between the two spectral ranges in the N filter response, the cross-correlation coefficients in Equation (12) have to be modified. To cope with the different energy ratios in the visible and the NIR bands, the response of the N channel in the NIR band is:

$$N_{nir}(i,j) \approx \beta_{v,n} \cdot (\omega_r \cdot R_{nir}(i,j) + \omega_g \cdot G_{nir}(i,j) + \omega_b \cdot B_{nir}(i,j)) \tag{15}$$

where $\beta_{v,n}$ is the inter-spectral correlation coefficient that considers the visible band to the NIR band energy balance. Figure 9 represents the comparison between the optical filtered N channel image in the NIR band and the estimated N channel image in the NIR band by using Equation (15).

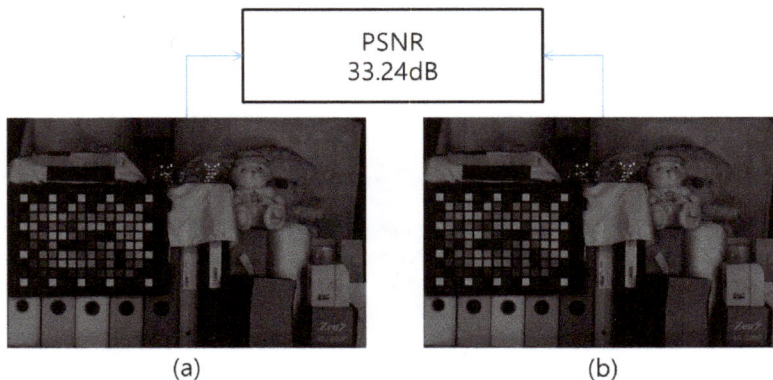

Figure 9. Comparison between (**a**) the optical filtered N channel image and (**b**) the estimated N channel image in NIR bands.

4. Proposed Methods

The purpose of the proposed method is to restore the original color in the visible bands from the mixed wide band signal. However, the color restoration in the spectral domain is an underdetermined problem, as described in Equation (9). Since MSFA image sensors have additional pixels whose intensity was represented in Equation (9), we redefined this underdetermined problem with eight unknown spectral values.

From Equation (8), the observed intensity vectors of the multispectral images are represented as $\mathbf{C}(i,j) = [R(i,j), G(i,j), B(i,j), N(i,j)]^T$. To focus on the color restoration at each pixel position, we assumed that the spatially-subsampled MSFA image was already interpolated. As a result, there are four different intensities at each RGBN pixel position.

In Figure 4, the spectral response of each channel is described with the corresponding RGB and N values. The energy of the NIR band is obtained by the RGB color filters, as well as the N filter. Similarly, a large amount of the energy in the visible band is obtained by the N channel. By considering the observed multispectral intensity vector $\mathbf{C}$, the spectral correlation between the channels in the visible band and the NIR band resulted in a mixture of exclusive responses in each channel, as represented in Equation (9).

From the sub-spectral band intensity mixture model, the color restoration problem is defined to find the unknown visible band intensity values $R_{vis}, G_{vis}, B_{vis}$ from the observed intensity values R, G, B and N, which contained the unknown NIR band intensity values and the unknown visible intensity values.

4.1. Color Restoration Based on Spectral Decomposition

When we spectrally decompose the N channel to the visible and NIR bands, the given N channel is represented by the RGB channel intensities in the visible and NIR bands from Equations (12) and (15):

$$
\begin{aligned}
N &= N_{vis} + N_{nir} \\
&= \omega_r \cdot (R_{vis} + \beta_{v,n} \cdot R_{nir}) + \omega_g \cdot (G_{vis} \\
&\quad + \beta_{v,n} \cdot G_{nir}) + \omega_b \cdot (B_{vis} + \beta_{v,n} \cdot B_{nir})
\end{aligned}
\tag{16}
$$

In Equation (16), the observed N channel is described with unknown RGB values in the visible bands and the NIR bands. Therefore, the decomposed N channel is obtained indirectly from Equation (16). Corresponding to the spectral response of the N channel, we define the artificial N channel $\hat{N}$ made by using the observed RGB channels and the visible band cross-correlation coefficients in Equation (12):

$$
\begin{aligned}
\hat{N} &= \omega_r \cdot R + \omega_g \cdot G + \omega_b \cdot B \\
&= \omega_r \cdot (R_{vis} + R_{nir}) + \omega_g \cdot (G_{vis} + G_{nir}) \\
&\quad + \omega_b \cdot (B_{vis} + B_{nir})
\end{aligned}
\tag{17}
$$

Since the visible band cross-correlation coefficients are designed to fit the N channel in the visible band, the estimated $\hat{N}$ value resembles the N channel filter responses in the visible band, but not in the NIR band. By using the energy difference between N and $\hat{N}$ in the NIR band, the observed N channel is decomposed into the two bands by subtracting the original N channel in Equation (16) and the artificial N channel $\hat{N}$ in Equation (17):

$$
\begin{aligned}
N - \hat{N} &= \omega_r \cdot (\beta_{v,n} - 1) \cdot R + \omega_g \cdot (\beta_{v,n} - 1) \cdot G \\
&\quad + \omega_b \cdot (\beta_{v,n} - 1) \cdot B \\
&= (\beta_{v,n} - 1) \cdot (\omega_r \cdot R_{nir} + \omega_g \cdot G_{nir} + \omega_b \cdot B_{nir}) \\
&= \frac{\beta_{v,n} - 1}{\beta_{v,n}} \cdot \hat{N}_{nir} \\
&= K \cdot \hat{N}_{nir}
\end{aligned}
\tag{18}
$$

where $K = \frac{\beta_{v,n} - 1}{\beta_{v,n}}$ is a scaling factor and $\hat{N}_{nir}$ represents the artificial N channel in the NIR band from Equation (15). Based on Equation (18), we decompose the spectral response of the N channel into two different channels, the visible band and the NIR band. The N channel information in the NIR band is recovered from the N channel that contained the energy of the entire spectrum of the MSFA image sensor. As a result, the decomposed N channel intensities in the NIR band and the RGB channel intensities in the NIR band are estimated from the result of Equation (18).

Figure 10 shows the relationship of the RGB channel intensities and the N channel intensity of 96 color patches of the Gretag color checker SG in the NIR band. As described in Figure 10, they are asymptotically linear in the NIR band. From this linear correlation, the decomposed RGB channel in the NIR band is defined as follows:

$$
\begin{aligned}
\hat{R}_{nir} &= \alpha_r \cdot \hat{N}_{nir} \\
\hat{G}_{nir} &= \alpha_g \cdot \hat{N}_{nir} \\
\hat{B}_{nir} &= \alpha_b \cdot \hat{N}_{nir}
\end{aligned}
\tag{19}
$$

where α_r, α_g and α_b represent the coefficients of the linear correlations between the RGB channels and the N channel in the NIR band. From the equation, the intensities of the RGB channel in the NIR band are estimated, and this color restoration model was processed with a single matrix transformation of:

$$
(\hat{R}_{vis}, \hat{G}_{vis}, \hat{B}_{vis})^T = \mathbf{M} \cdot (R, G, B, N)^T
\tag{20}
$$

where $\mathbf{M}$ is:

$$
\mathbf{M} = \mathbf{E} + \frac{1}{K} \mathbf{AW}
\tag{21}
$$

where **W** is the N channel decomposition matrix, **A** is the RGB channel decomposition matrix and **E** is a 3×4 matrix of zeros with ones along the leading diagonal. The N channel decomposition matrix **W** is defined as:

$$\mathbf{W} = \begin{pmatrix} \omega_r & \omega_g & \omega_b & -1 \\ \omega_r & \omega_g & \omega_b & -1 \\ \omega_r & \omega_g & \omega_b & -1 \\ \omega_r & \omega_g & \omega_b & -1 \end{pmatrix} \tag{22}$$

and the RGB channel decomposition matrix is defined as:

$$\mathbf{A} = \begin{pmatrix} \alpha_r & 0 & 0 & 0 \\ 0 & \alpha_g & 0 & 0 \\ 0 & 0 & \alpha_b & 0 \end{pmatrix} \tag{23}$$

Based on Equation (21), the unified matrix **M** is:

$$\begin{pmatrix} \hat{R}_{vis} \\ \hat{G}_{vis} \\ \hat{B}_{vis} \end{pmatrix} = \begin{pmatrix} \frac{\alpha_r \cdot \omega_r + K}{K} & \frac{\alpha_r \cdot \omega_g}{K} & \frac{\alpha_r \cdot \omega_b}{K} & -\frac{\alpha_r}{K} \\ \frac{\alpha_g \cdot \omega_r}{K} & \frac{\alpha_g \cdot \omega_g + K}{K} & \frac{\alpha_g \cdot \omega_b}{K} & -\frac{\alpha_g}{K} \\ \frac{\alpha_b \cdot \omega_r}{K} & \frac{\alpha_b \cdot \omega_g}{K} & \frac{\alpha_b \cdot \omega_b + K}{K} & -\frac{\alpha_b}{K} \end{pmatrix} \begin{pmatrix} R \\ G \\ B \\ N \end{pmatrix} \tag{24}$$

where $K = \frac{\beta_{v,n} - 1}{\beta_{v,n}}$ is a scaling factor in Equation (18), ω_r, ω_g, ω_b are the coefficients for the linear combination in Equation (11) and α_r, α_g and α_b are the coefficients that represent the linear correlation between the RGB channels and the N channel in the NIR band in Equation (19). Because Equation (24) is a combination of cascaded linear decomposition matrices **W** and **A**, the proposed color correction matrix is more flexible than the simple 3×4 linear color correction model. Further, because the sensor response function over the entire band is nonlinear, color correction error is inevitable when the linear color correction method is employed. Moreover, there is an energy difference between the visible and NIR bands. The spectral response of the local spectral band can be approximated to a linear model. On the basis of linear model approximation of each local spectral band, the proposed method separates the visible and NIR bands to estimate the color correction matrix and, thereby, obtain a more accurate estimation of the NIR interference in each RGB channel. Using **W**, the proposed method decomposes the N channel to the visible and NIR bands and uses the NIR band information obtained from **W** to estimate the NIR contribution in the RGB channels. The correlation between the RGB and N channels in the NIR band is estimated using **A**. Because the proposed method separates the visible and NIR bands to estimate the color correction matrix (CCM), it is possible to estimate the correlation between RGB and NIR in various illumination environments.

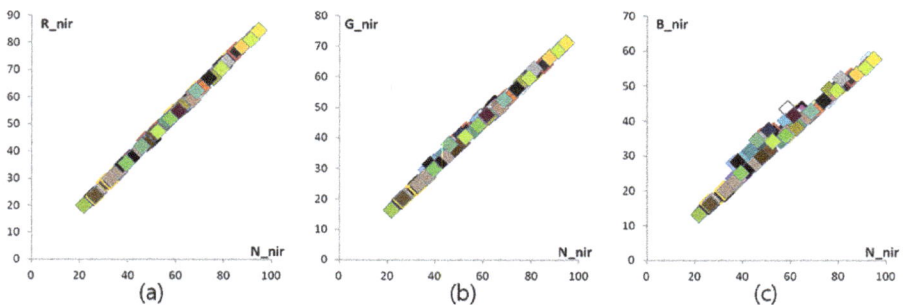

Figure 10. RGBN channel correlation in the NIR band: (**a**) N_{nir} *vs.* R_{nir}; (**b**) N_{nir} *vs.* G_{nir}; (**c**) N_{nir} *vs.* B_{nir}.

Figure 11 shows the experimental results obtained under an incandescent lamp with 300 lx illumination. Because the incandescent lamp emits an amount of spectral energy in the NIR band, we selected this lamp to show the advantage of the proposed method. By comparing Figure 11b and Figure 11c, the level of restoration of the overall colors of each color patch can be ascertained. In Figure 11a, which is the target optical filtered image, it can be seen that some color patches are slightly different. To investigate the color restoration accuracy, we calculated angular error. Table 1 shows the average angular error. From Table 1, it is clear that the proposed method restores color better than the linear 3 × 4 color correction method.

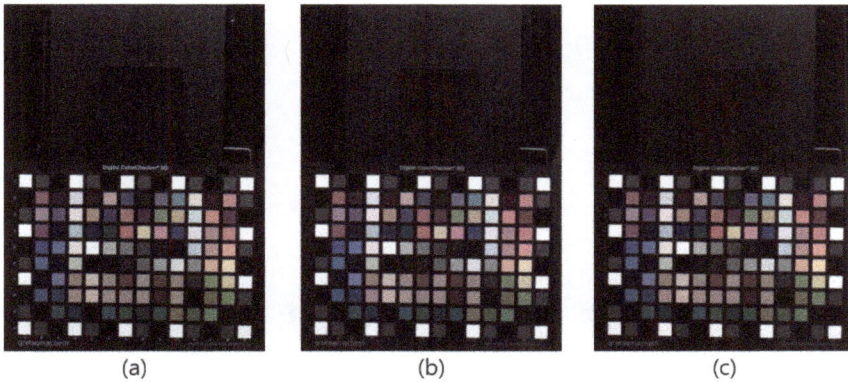

(a)   (b)   (c)

Figure 11. Experimental results under an incandescent lamp (300 lx). (**a**) Optical filtered visible band image; (**b**) 3 × 4 color correction method; (**c**) proposed method.

Table 1. Average angular error ($\times 10^{-2}$). CCM, color correction matrix.

	3 × 4 CCM	Proposed Method
Incandescent (300 lx)	5.12	4.17

4.2. Low Light Conditions

Because of the additional NIR band information, an IRCF-removed MSFA image sensor has advantages in low visible light conditions. From the perspective of color restoration, however, there is no advantage, since the unnecessary NIR interference to the RGB color channel does not have any visible band color information. Figure 12 represents the spectral energy distribution of an incandescent lamp with a variety of illuminance values. The correlated color temperature of the lamp is 3000 K. As illuminance decreased, the overall intensities of spectral energy decreased, too. In addition, the energy ratio between the visible band and the NIR band varied as the illuminance decreased.

Table 2 shows that decreasing illuminance increases the portion of the NIR band spectral energy under incandescent light. The numbers in Columns 2 and 3 represent the summation of the spectrum values in Figure 12. This implies that 60% of the unwanted NIR contributions in each RGB channel must be removed to obtain a natural color image under an incandescent lamp with 10 lx. Because the NIR contribution is greater than the color information in each RGB channel, it is important to estimate the NIR band spectral information precisely to prevent false color generation.

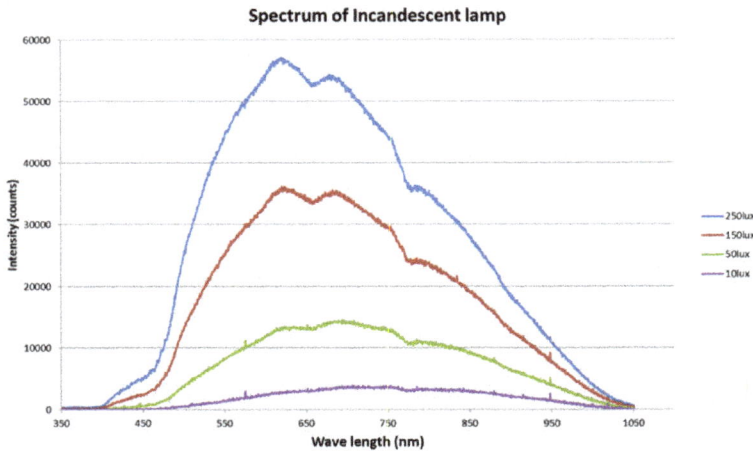

Figure 12. Spectrum of an incandescent lamp under various kinds of illumination (3000 K).

Table 2. Relationships between illuminance and the portion of the NIR band spectral energy.

Illuminance	Visible Band	NIR Band	Portion of the NIR Band (%)
250 lx	57,978.6	45,337.7	**43.8**
150 lx	26,833.2	23,587.7	**46.7**
50 lx	8045.7	8847.2	**52.3**
10 lx	1347.9	2042.7	**60.2**

4.3. Two-Step Color Restoration

In general lighting situations, the proposed color restoration method based on Equation (24) can decompose the NIR contribution in each RGB channel. However, as mentioned in Section 4.2, the spectral energy distribution changed under low lighting conditions. Furthermore, the ratio between the visible band and the NIR band changed. Therefore, the estimation of the N channel in the NIR band is more important under low lighting conditions. The color restoration model in Equation (24) is based on the assumption that the spectral response of the MSFA sensor in the NIR band correlated with the spectral linearity between the RGB and N channels. However, in the 700 nm to 800 nm spectral range, there was a lack of linear correlation between the channels, except for between the R and N channels. If the spectral energy distribution of the light source shows strong energy between this nonlinear range, the spectral decomposition error of the result will increase. Because the visible band information is smaller than the NIR band under low lighting conditions, the spectral decomposition error can produce a false color result.

To overcome this spectral nonlinearity problem, we used a two-step color restoration method that divides the spectral range into two parts and removes the NIR band information sequentially. Figure 4 represents the two-step color restoration process. In the first step, the intensities of the RGB channel in the NIR band with a spectral wavelength range greater than 800 nm were decomposed using the B channel. In Figure 13, the ratio between the B channel and the N channel of 96 color patches of the Gretag color checker SG is represented. Since the visible band information of the B channel is quite small under low lighting conditions, there is a strong correlation between the B channel and the N channel whose wavelength is above 800 nm, as described in Figure 13.

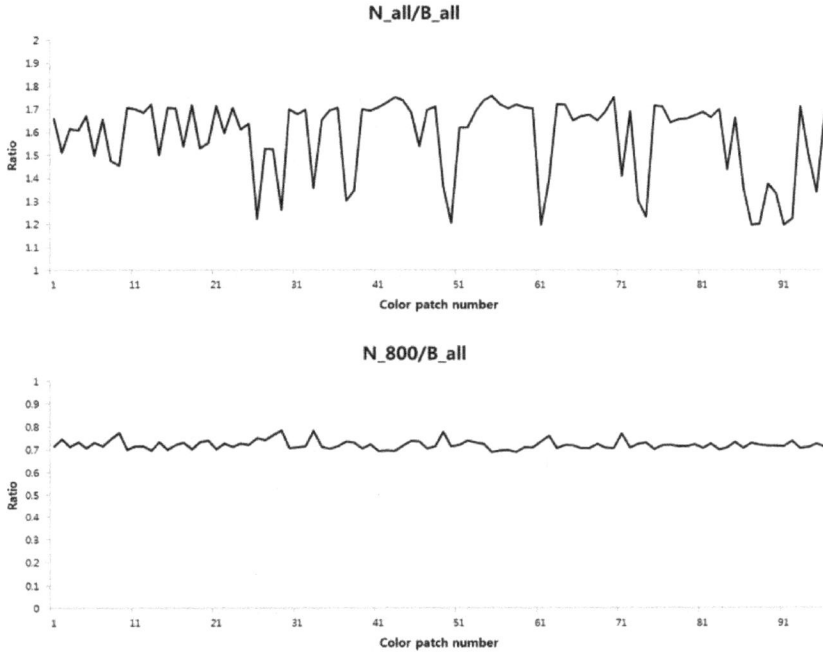

Figure 13. Relationship between the B channel and the N channel (incandescent lamp, 1 lx): the ratio between the B channel and the N channel in a wide spectral range (**Top**); the ratio between the B channel beyond 800 nm and the N channel in a wide spectral range (**Bottom**).

The N channel whose wavelength is beyond 800 nm was approximated from the B channel as follows:

$$\hat{N}_{nir}^{800} = \gamma \cdot B \tag{25}$$

where γ is the correlation coefficient between the B channel and the N channel above 800 nm. Figure 14 represents the result of Equation (25). Figure 14a is the image obtained with the optical filter, and Figure 14b is the result of the proposed method after the first step of color restoration. By comparing (a) to (b), the overall colors of the entire image were similar.

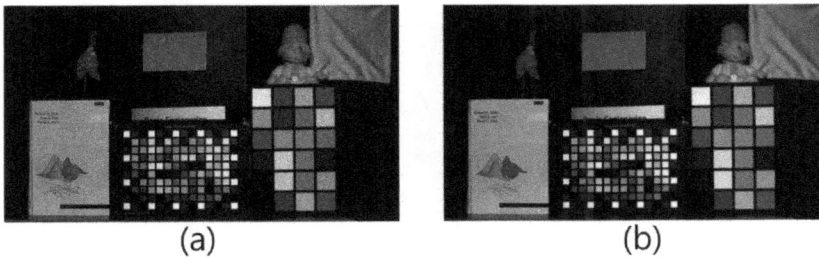

Figure 14. Result of the achromatic NIR band (above 800 nm) component removal (incandescent 5 lx). (**a**) Optical filtered image; (**b**) first step of the proposed method.

After the first step, the remaining NIR intensities in the RGB channel were removed through the spectral decomposition method as proposed in Equation (24). Based on Equation (20), the two-step color restoration model can be processed with a matrix equation as follows:

$$(\hat{R}_{vis}, \hat{G}_{vis}, \hat{B}_{vis})^T = \mathbf{M} \cdot (R, G, B, N)^T - \mathbf{P} \tag{26}$$

where **P** is defined as:

$$\mathbf{P} = (\gamma_r, \gamma_g, \gamma_b)^T \cdot B \tag{27}$$

The γ_r, γ_g and γ_b values represent the correlation coefficient between the B channel and the N channel whose wavelength was above 800 nm. The proposed two-step color restoration method was applied to estimate the NIR component of the image obtained under particular illumination situations, such as low light conditions, especially the illuminance of an incandescent lamp under 5 lx. In this paper, we use the proposed method with a two-step color restoration with Equation (26) when the illuminance of the light source is under 5 lx.

From Section 2, the achromatic NIR component δ did not affect the hue and saturation value of the images. The achromatic NIR component is not an important part of restoring the color component. Therefore, we estimated the spectral information of the chromatic NIR band precisely after removing the achromatic NIR component δ.

Figure 15 represents the result of the proposed method under an incandescent lamp with 5 lx.

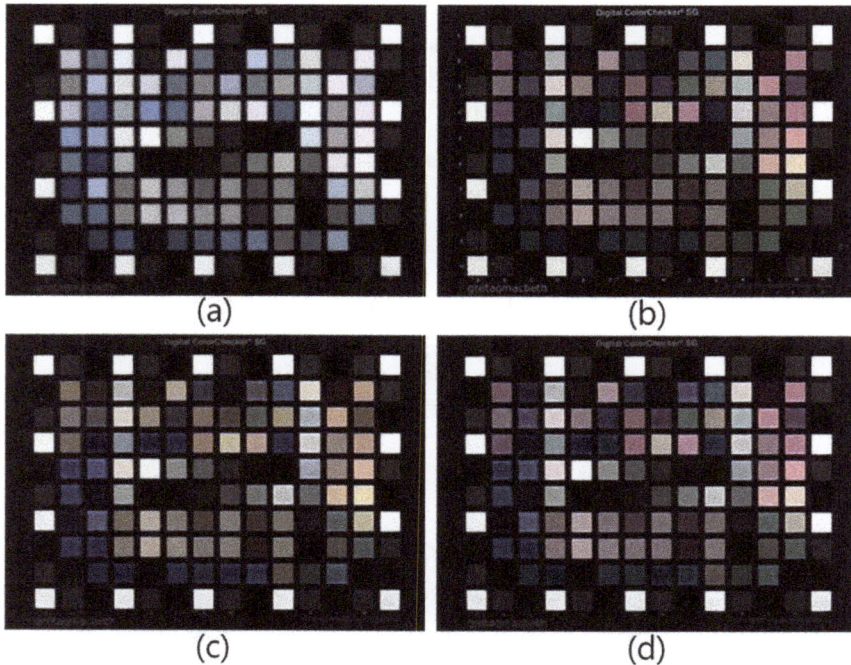

Figure 15. Comparison between proposed methods (incandescent, 5 lx). (**a**) Multi-spectral image; (**b**) optical filtered image; (**c**) proposed method without two-step color restoration; (**d**) proposed method with two-step color restoration.

Figure 15a is the input image, the color of which is desaturated by additional NIR, and Figure 15b is the optical filtered visible band image. Figure 15c is the result obtained using the proposed method in

Equation (24) as given in Section 4.1, and Figure 15d is the result that was obtained using the two-step color restoration described in Section 4.3. By comparing Figure 15c to Figure 15d, the overall color of Figure 15c is yellow-shifted, especially in red color patches. Since the spectral energy distribution changed under low lighting conditions, the unified color restoration model **M** in Equation (24) was limited in explaining the complicated nonlinear transformation. After removing the achromatic NIR band information, the only concern was the chromatic NIR band used to restore the color information. Since the unified color restoration model **M** handled the chromatic NIR band information, the color was successfully restored as represented in Figure 15d.

5. Experimental Results

The proposed color restoration method was tested with images captured under different standard illuminations: sunlight, incandescent lamp, sodium lamp and fluorescent lamp. Since the spectrum of these light sources was spread over a wide range, we used these lights as the target illuminance values as represented by Figure 16.

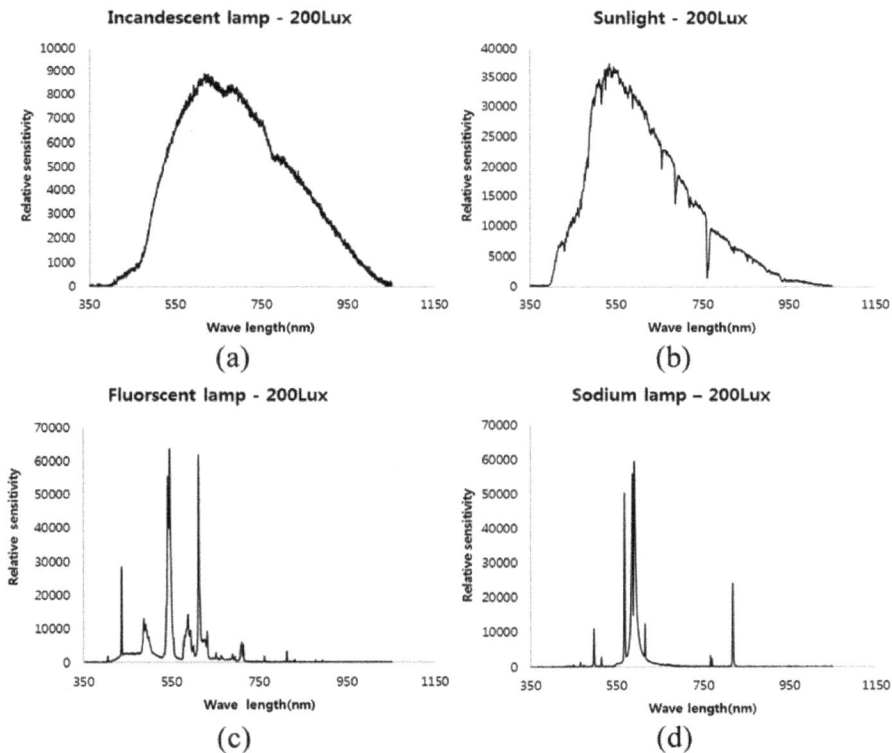

Figure 16. Spectral distribution of a variety of light sources. (a) Incandescent lamp (3000 K); (b) sunlight (6500 K); (c) fluorescent lamp (5000 K); (d) sodium lamp (2700 BK).

As the training set for the correlation coefficients, we used 96 standard colors of the Gretag color checker SG. Because the color samples were distributed widely, these colors were used for the training set. The input multispectral image was obtained by a camera system with an RGBN image sensor without IRCF, and we used a target visible band image with an IRCF as a reference image. The

96 patches were manually segmented, and we used the average RGB of each patch. The resulting average RGB values in the input image and the reference image were used to derive a set of color restoration models in Equation (24). We also measured the XYZ of each of the 96 patches using a spectrophotometer. If an illuminance value was less than 5 lx, we used an additional optical filter that passes wavelengths beyond 800 nm to derive a set of color restoration models in Equation (27). After setting a color restoration model, the proposed method was applied to an input multispectral image without IRCF. As mentioned in Section 4.3, we used two-step color restoration when the illuminance of the light source was darker than 5 lx. In our experiment, we measured the illumination level using an illuminometer. In practical situations, the light sensor commonly used to turn on the flash light or changing to night shot mode must be installed to measure the luminance level of the illumination. The light sensor performs the simple role of determining whether the luminance level corresponds to dark or bright. When the illuminance of the light source is brighter than 5 lx, we used the color restoration model in Equation (24).

As an error criterion, the angular error was calculated. Considering the Z color sample entities in the training set, the angular error for the z-th color was defined as:

$$\theta_z = \cos^{-1}\left(\frac{\mathbf{m}_z \cdot \mathbf{p}_z}{|\mathbf{m}_z||\mathbf{p}_z|}\right) \tag{28}$$

where θ_z is the angular error between the target color vector $\mathbf{m}_z$ and the color restoration result $\mathbf{p}_z$. '·' represents the inner product of two vectors, and $|\mathbf{m}|$ represents the magnitude of the vector $\mathbf{m}$. In addition, we measured the color difference ΔE of each color sample in the CIELAB color space defined by:

$$\Delta E_{ab}^* = [(\Delta L^*)^2 + (\Delta a^*)^2 + (\Delta b^*)^2]^{1/2} \tag{29}$$

We regarded the average of ΔE as the color correction error. To convert RGB to the CIELAB color space, the RGB signals were transformed to CIE tristimulus values by using a spectrophotometer with a standard illuminant, after which the CIELAB equation was applied [28]. The tristimulus values of the illuminant were A, F and D65 with respect to the incandescent lamp, fluorescent lamp and sunlight, respectively. We used a visible band image with IRCF as a reference image that was used to compare to the input image and the result image. As comparative methods for the proposed color restoration algorithm, we implemented the least squares-based color correction method [29] and the N-to-sRGB mapping color correction method based on root-polynomial mapping [30].

Figure 17 depicts the experimental results under a fluorescent lamp with 350-lx illumination. Since the fluorescent lamp did not emit NIR, the input image in Figure 17a and the optical filtered image in Figure 17b were almost similar. Our proposed method preserved the color of the input image (Figure 17f) and the other color correction methods (Figure 17c to Figure 17e) because of the absence of NIR color distortion in the input image.

Figure 18 shows the experimental result under sunlight, which has a wide range of spectral distribution and abundant visible band information. In this case, it was sufficient to restore color using the proposed method in Equation (24). Comparing Figure 18b and Figure 18c to Figure 18f, the resulting image of the proposed method restored the distorted color well, especially the materials with high reflectance in the NIR band. The root-polynomial mapping method in Figure 18e restored the overall colors of each color patch and black materials well. The comparison of Figure 18b,e shows that the saturation is slightly high. Since sunlight has plenty of spectral energy in visible bands, the root-polynomial mapping restores color information as well as the proposed method. To investigate color restoration accuracy, each method was compared in Tables 3 and 4.

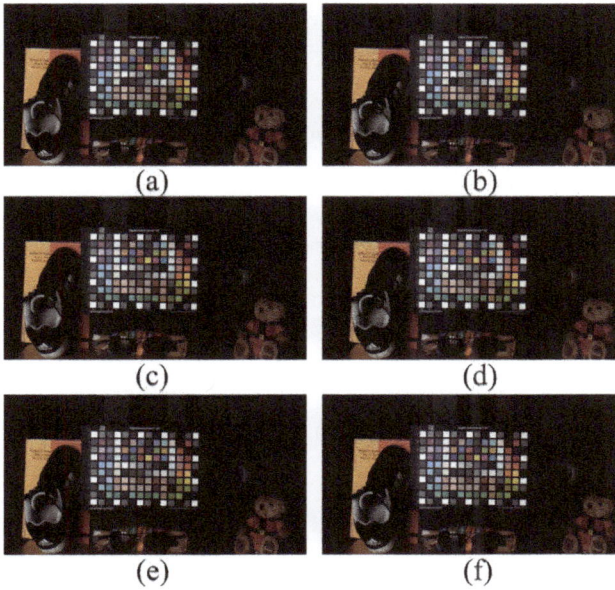

Figure 17. Experimental results under a fluorescent lamp (350 lx). (**a**) Input image; (**b**) optical filtered visible band image; (**c**) 3 × 3 CCM; (**d**) 3 × 4 CCM; (**e**) root-polynomial mapping; (**f**) proposed method.

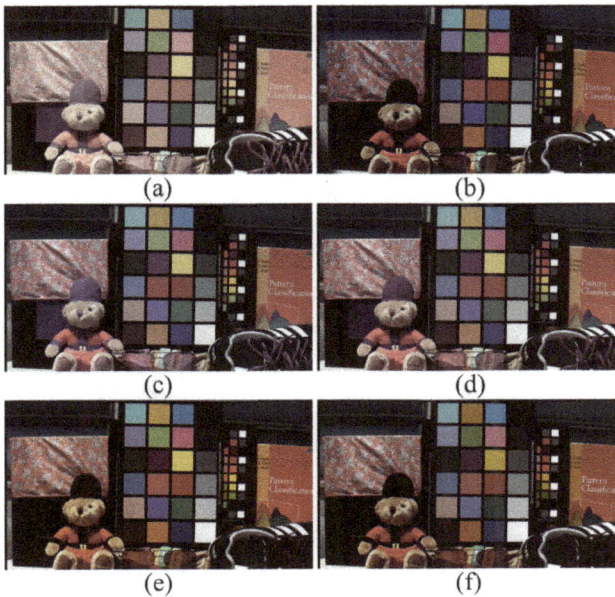

Figure 18. Experimental results under sunlight (400 lx) (**a**) Input image (**b**) optical filtered visible band image; (**c**) 3 × 3 CCM; (**d**) 3 × 4 CCM; (**e**) root-polynomial mapping; (**f**) proposed method.

Another set was tested under an incandescent lamp, which emits much spectral energy in the NIR band. Figure 19a represents the multispectral image obtained under the incandescent lamp. The color channels were white balanced without considering the color degradation caused by the additional NIR; therefore, the overall colors of the image show low saturation and blue hue over much of the NIR band. Figure 19c shows the result of the conventional color correction method. When comparing Figure 19c to Figure 19b, the overall colors of each color patch and object were close to the target image. The comparison of Figure 19d to Figure 19f shows that the overall colors of each color patch and object were close to the optical-filtered visible band image (Figure 19b). However, the color of the objects with high reflectance in the NIR band, such as fabric, leaf, and so on, was slightly different. This means that the accuracy of the NIR estimation was different. Figure 19f is much closer to the visible color in Figure 19b because the proposed method separates the visible and NIR bands to estimate the color correction matrix and, thereby, obtains a more accurate estimation of the NIR interference in each RGB channel. The black colors of the fabric patch in the upper side of the image, as well as the doll's cap and clothes were restored to their original colors successfully.

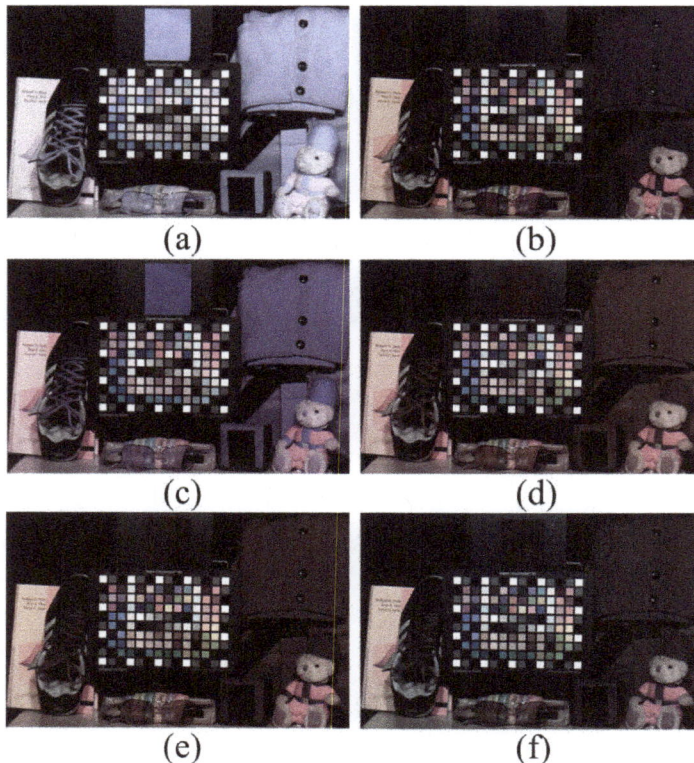

Figure 19. Experimental results under an incandescent lamp (200 lx). (**a**) Input image; (**b**) optical filtered visible band image; (**c**) 3×3 CCM; (**d**) 3×4 CCM; (**e**) root-polynomial mapping; (**f**) proposed method.

As discussed in Section 4.3, the proposed two-step color restoration method is useful under particular illumination. Figure 20 represents a comparison with and without two-step color restoration under an incandescent lamp at 1 lx. Since the visible band information was less than that of the NIR band in low lighting conditions, the spectral estimation error increased. As a result, Figure 20c shows a yellow image compared to Figure 20b. With the proposed two-step color restoration method, the color of the image was successfully restored, as shown in Figure 20d. Based on this result, we tested the proposed method under low lighting situations.

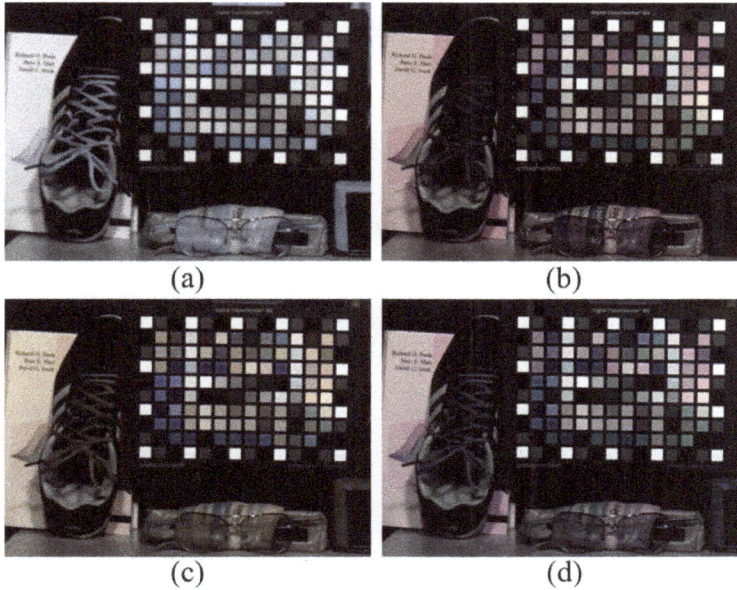

Figure 20. Two-step color restoration result comparison (1 lx). (**a**) Input image; (**b**) optical filtered visible band image; (**c**) proposed method without two-step color restoration; (**d**) proposed method with two-step color restoration.

Figure 21 represents the experimental results under an incandescent lamp at 1 lx. This illumination emits plenty of spectral energy in the NIR band. In Figure 16, the spectrum distribution of the incandescent lamp is spread evenly over a wide range. In low lighting conditions, the lack of visible band information makes the overall saturation of the images low. Figure 21c shows that the 3 × 3 CCM-based method could not restore the overall color of the input image (Figure 21a). By comparing Figure 21d to Figure 21f, the overall colors of each color patch and object were close to the optical-filtered visible band image (b). However, the colors of black materials were not restored correctly in Figure 21d,e. Since the spectral energy of the incandescent lamp under 550 nm and the MSFA sensor response in the blue channel were low, blue information is lacking in the black area. As a result, the blue intensity was boosted during the process of color constancy. Both root-polynomial mapping and our proposed color restoration method are based on least-square linear mapping; therefore, a large amount of NIR spectral energy in low-lighting condition (see Section 4.2) must be considered. Compared to Figure 21d,e, Figure 21f shows that the proposed method restored colors satisfactorily for both the patches and for materials with high NIR component.

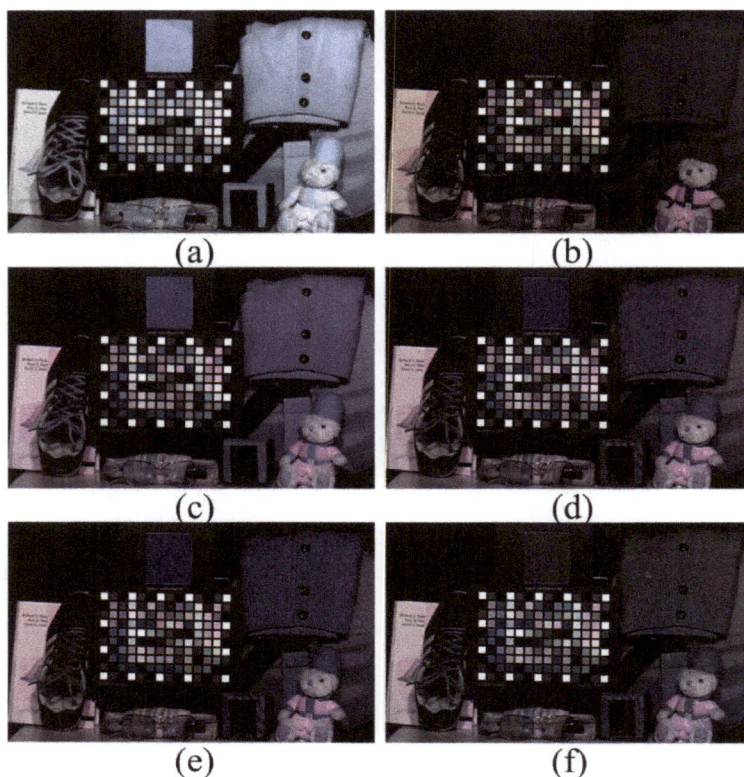

Figure 21. Experimental results under an incandescent lamp (1 lx) (**a**) Input image; (**b**) optical filtered visible band image; (**c**) 3 × 3 CCM; (**d**) 3 × 4 CCM; (**e**) root-polynomial mapping; (**f**) proposed method.

Figure 22 represents the experimental results under a sodium lamp at 1 lx. Figure 22c shows that the 3×3 CCM-based method could not restore the overall color of the input image (Figure 22a). The spectrum distribution of the sodium lamp is concentrated at a particular wavelength at 830 nm, as shown in Figure 16. In this case, the sensor spectral response of the local spectral band can be approximated to a linear model. For this reason, the experimental results in Figure 22d to Figure 22f show high restoration performance visually. To investigate the color restoration accuracy, each method was compared in Tables 3 and 4.

Table 3. Average angular error.

	Average Angular Error ($\times 10^{-2}$)				
	Input Image	**3 × 3 CCM**	**3 × 4 CCM**	**Root-Polynomial**	**Proposed**
fluorescent (350 lx)	0.77	0.80	0.77	0.78	0.77
sunlight (400 lx)	6.97	2.93	2.27	1.98	1.53
incandescent (200 lx)	28.73	7.79	5.31	5.05	4.53
incandescent (1 lx)	29.94	8.71	5.88	6.59	4.89
sodium (1 lx)	28.94	5.99	3.15	3.13	3.13

Table 4. Average color difference, ΔE.

	Average Color Difference ΔE				
	Input Image	**3 × 3 CCM**	**3 × 4 CCM**	**Root-Polynomial**	**Proposed**
fluorescent (350 lx)	0.98	1.12	1.06	1.04	1.04
sunlight (400 lx)	15.66	10.97	9.83	8.16	7.50
incandescent (200 lx)	20.32	8.62	4.98	4.55	4.19
incandescent (1 lx)	22.28	8.18	6.45	7.24	5.07

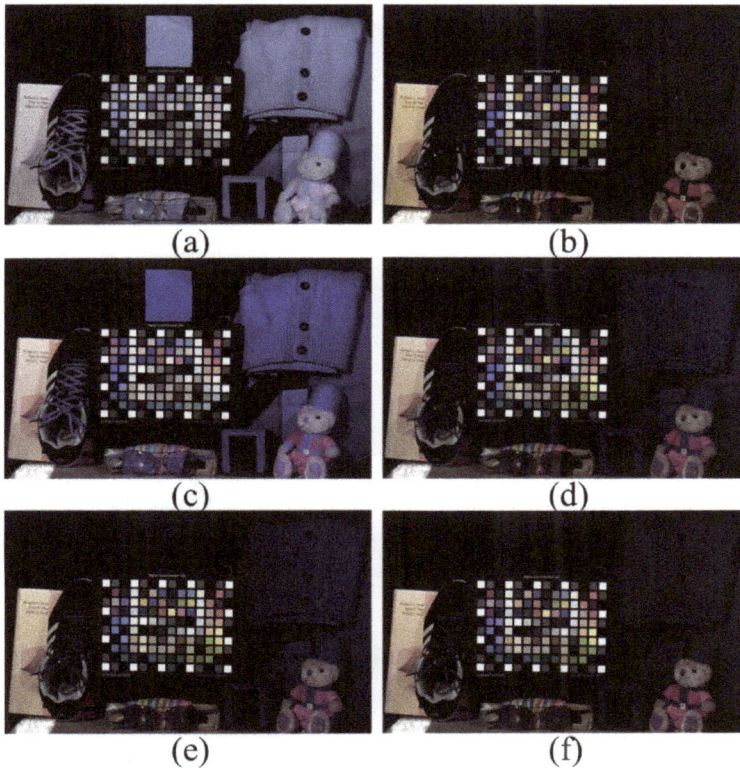

Figure 22. Experimental results under sodium lamp (1 lx) (**a**) Input image; (**b**) optical filtered visible band image; (**c**) 3 × 3 CCM; (**d**) 3 × 4 CCM; (**e**) root-polynomial mapping; (**f**) proposed method.

Tables 3 and 4 show the average angular error and the color difference with a variety of light sources. The performance of the proposed method was confirmed visually for materials with high reflectance in the NIR band. However, the performance of the proposed method for various colors in the color chart and substances had to be measured. Table 3 shows the amount of angular error, where our proposed method outperformed other methods. Since the color of the input image was severely distorted, the angular error between the input image and the optical filtered image was significantly high. After the application of color correction methods, the average angular errors were reduced, and the performance of the proposed method was better than that of the conventional methods. Similarly, the color difference in Table 4 shows that the color correction results obtained with the proposed method were better compared to the another methods.

In addition, to calculate the gain advantage provided with NIR information, we measured the intensities of the image obtained in various illuminations with or without IRCF. Figure 23 represents the sensitivity boosting provided by the NIR information. To measure the additional intensities, the image is divided into 16 sections. After that, the intensities are averaged in each section. As shown in Table 5 and Figure 23, the sensitivity was boosted by 10 dB without IRCF under an incandescent lamp. On the contrary, because the fluorescent lamp does not emit an NIR component, there is no gain advantage.

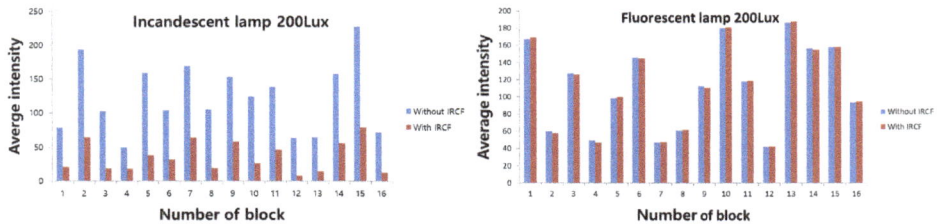

Figure 23. Sensitivity boosting provided by the NIR information.

Table 5. Average intensity value with or without IRCF in various illuminations.

Illumination	With IRCF	Without IRCF	Sensitivity Gain (dB)
Incandescent	35.6	122.3	10.71 dB
Fluorescent	112.7	112.5	0.01 dB

6. Conclusions

In this paper, a color restoration algorithm for an IRCF-removed MSFA image sensor in low light conditions was proposed. In the proposed method, the color degradation caused by the spectral composition of the visible and NIR band information was mainly considered. For the spectrally-degraded color information with RGB channels, the spectral estimation and spectral decomposition method were proposed to remove additional NIR band spectral information. Based on the channel estimation when considering the nonlinearity of the spectral response function of the MSFA sensor in low light conditions, the channel approximation using the B channel is for two-step color restoration. Based on the filter correlation, the inter-channel correlation on the visible and NIR band were assumed, respectively. When the N channel was decomposed into visible and NIR band information, the RGB channel in the visible band was finally restored with spectral decomposition. The experimental results show that the proposed method effectively restored the visible color from the color-degraded images caused by IRCF removal.

Acknowledgments: This research was supported by the Basic Science Research Program through the National Research Foundation of Korea (NRF) funded by the Ministry of Science, ICT and Future Planning (No. 2015R1A2A1A14000912).

Author Contributions: Chulhee Park conducted the experiments and wrote the manuscript under the supervision of Moon Gi Kang.

Conflicts of Interest: The authors declare no conflict of interest.

References

1. Salamati, N.; Fredembach, C.; Süsstrunk, S. Material classification using color and NIR images. In Proceedings of the 17th Color and Imaging Conference, Albuquerque, NX, USA, 9–13 November 2009; Volume 2009, pp. 216–222.

2. Pohl, C.; Van Genderen, J.L. Review article multisensor image fusion in remote sensing: concepts, methods and applications. *Int. J. Remote Sens.* **1998**, *19*, 823–854.

3. Choi, J.; Yu, K.; Kim, Y. A new adaptive component-substitution-based satellite image fusion by using partial replacement. *IEEE Trans. Geosci. Remote Sens.* **2011**, *49*, 295–309.

4. Hao, X.; Chen, H.; Yao, C.; Yang, N.; Bi, H.; Wang, C. A near-infrared imaging method for capturing the interior of a vehicle through windshield. In Proceedings of the 2010 IEEE Southwest Symposium on Image Analysis & Interpretation (SSIAI), Austin, TX, USA, 23–25 May 2010; pp. 109–112.

5. Hertel, D.; Marechal, H.; Tefera, D.A.; Fan, W.; Hicks, R. A low-cost VIS-NIR true color night vision video system based on a wide dynamic range CMOS imager. In Proceedings of the 2009 IEEE Intelligent Vehicles Symposium, Xi'an, China, 3–5 June 2009; pp. 273–278.

6. Kumar, A.; Prathyusha, K.V. Personal authentication using hand vein triangulation and knuckle shape. *IEEE Trans. Image Process.* **2009**, *18*, 2127–2136.

7. Yi, D.; Liu, R.; Chu, R.; Lei, Z.; Li, S. Face Matching Between Near Infrared and Visible Light Images. *Adv. Biometr.* **2007**, *4642*, 523–530.

8. Li, S.Z.; Chu, S.R.; Liao, S.; Zhang, L. Illumination invariant face recognition using near-infrared images. *IEEE Trans. Pattern Anal. Mach. Intell.* **2007**, *29*, 627–639.

9. Fredembach, C.; Susstrunk, S. Illuminant estimation and detection using near-infrared. *Proc. SPIE* **2009**, *7250*, 72500E.

10. Schaul, L.; Fredembach, C.; Süsstrunk, S. Color image dehazing using the near-infrared. In Proceedings of the 16th IEEE International Conference on Image Processing (ICIP), Cairo, Egypt, 7–10 November 2009; No. LCAV-CONF-2009-026.

11. Kise, M.; Park, B.; Heitschmidt, G.W.; Lawrence, K.C.; Windham, W.R. Multispectral imaging system with interchangeable filter design. *Comput. Electron. Agric.* **2010**, *72*, 61–68.

12. Matsui, S.; Okabe, T.; Shimano, M.; Sato, Y. Image Enhancement of Low-Light Scenes with Near-Infrared Flash Images. In Proceedings of the 9th Asian Conference on Computer Vision (ACCV 2009), Xi'an, China, 23–27 September 2009; pp. 213–223.

13. Fredembach, C.; Süsstrunk, S. Colouring the near-infrared. In Proceedings of the 16th Color and Imaging Conference (CIC 2008), Portland, OR, USA, 10–14 November 2008; Volume 2008, pp. 176–182.

14. Koyama, S.; Inaba, Y.; Kasano, M.; Murata, T. A day and night vision MOS imager with robust photonic-crystal-based RGB-and-IR. *IEEE Trans. Electron Dev.* **2008**, *55*, 754–759.

15. Chen, Z.; Wang, X.; Liang, R. RGB-NIR multispectral camera. *Opt. Expr.* **2014**, *22*, 4985–4994.

16. Lu, Y.M.; Fredembach, C.; Vetterli, M.; Süsstrunk, S. Designing color filter arrays for the joint capture of visible and near-infrared images. In Proceedings of the 16th IEEE International Conference on Image Processing (ICIP), Cairo, Egypt, 7–10 November 2009; pp. 3797–3800.

17. Kiku, D.; Monno, Y.; Tanaka, M.; Okutomi, M. Simultaneous capturing of RGB and additional band images using hybrid color filter array. *Proc. SPIE* **2014**, *9023*, doi:10.1117/12.2039396.

18. Sadeghipoor, Z.; Lu, Y.M.; Susstrunk, S. A novel compressive sensing approach to simultaneously acquire color and near-infrared images on a single sensor. In Proceedings of the 2013 IEEE International Conference on Acoustics, Speech and Signal Processing (ICASSP), Vancouver, BC, Canada, 26–31 May 2013; pp. 1646–1650.

19. Martinello, M.; Wajs, A.; Quan, S.; Lee, H.; Lim, C.; Woo, T.; Lee, W.; Kim, S.S.; Lee, D. Dual Aperture Photography: Image and Depth from a Mobile Camera. In Proceedings of the 2015 IEEE International Conference on Computational Photography (ICCP), Houston, TX, USA, 24–26 April 2015; pp. 1–10.

20. Park, C.H.; Oh, H.M.; Kang, M.G. Color restoration for infrared cutoff filter removed RGB multispectral filter array image sensor. In Proceedings of the 2015 International Conference on Computer Vision Theory and Applications (VISAPP 2015), Berlin, Germany, 11–14 March 2015; pp. 30–37.

21. Kong, F.; Peng, Y. Color image watermarking algorithm based on HSI color space. In Proceedings of the 2nd International Conference on Industrial and Information Systems (IIS), Dalian, China, 10–11 July 2010; Volume 2, pp. 464–467.

22. Funt, B.V.; Lewis, B.C. Diagonal versus affine transformations for color correction. *JOSA A* **2000**, *17*, 2108–2112.

23. Reinhard, E.; Ashikhmin, M.; Gooch, B.; Shirley, P. Color transfer between images. *IEEE Comput. Graph. Appl.* **2001**, *21*, 34–41.

24. Finlayson, G.; Hordley, S. Improving gamut mapping color constancy. *IEEE Trans. Image Process.* **2000**, *9*, 1774–1783.

25. Finlayson, G.D.; Hordley, S.D.; Hubel, P.M. Color by correlation: A simple, unifying framework for color constancy. *IEEE Trans. Pattern Anal. Mach. Intell.* **2001**, *23*, 1209–1221.

26. Barnard, K.; Cardei, V.; Funt, B. A comparison of computational color constancy algorithms. I: Methodology and experiments with synthesized data. *IEEE Trans. Image Process.* **2002**, *11*, 972–984.

27. Park, J.; Kang, M. Spatially adaptive multi-resolution multispectral image fusion. *Int. J. Remote Sens.* **2004**, *25*, 5491–5508.

28. Kang, H.R. *Computational Color Technology*; Spie Press: Bellingham, WA, USA, 2006.

29. Brainard, D.H.; Freeman, W.T. Bayesian color constancy. *JOSA A* **1997**, *14*, 1393–1411.

30. Monno, Y.; Tanaka, M.; Okutomi, M. N-to-SRGB Mapping for Single-Sensor Multispectral Imaging. In Proceedings of the 2015 IEEE International Conference on Computer Vision Workshop (ICCVW), Santiago, Chile, 7–13 December 2015; pp. 33–40.

MDPI

Article

Penetration Depth Measurement of Near-Infrared Hyperspectral Imaging Light for Milk Powder

Min Huang [1,2], Moon S. Kim [2,*], Kuanglin Chao [2], Jianwei Qin [2], Changyeun Mo [3], Carlos Esquerre [4,5], Stephen Delwiche [5] and Qibing Zhu [1]

[1] Key Laboratory of Advanced Process Control for Light Industry, Ministry of Education, Jiangnan University, Wuxi 214122, China; huangmzqb@163.com (M.H.); zhuqib@163.com (Q.Z.)

[2] Environmental Microbial and Food Safety Laboratory, Agricultural Research Service, USDA, Beltsville, MD 20705, USA; kevin.chao@ars.usda.gov (K.C.); jianwei.qin@ars.usda.gov (J.Q.)

[3] National Institute of Agricultural Science, Rural Development Administration, 310 Nongsaengmyeong-ro, Wansan-gu, Jueonju-si, Jeollabuk-do 54875, Korea; cymoh100@korea.kr

[4] School of Biosystems and Food Engineering, University College Dublin, Dublin 4, Ireland; carlos.esquerre@ucd.ie

[5] Food Quality, Agricultural Research Service, USDA, Beltsville, MD 20705, USA; stephen.delwiche@ars.usda.gov

* Correspondence: moon.kim@ars.usda.gov; Tel.: +1-301-504-8462; Fax: +1-301-504-9466

Academic Editor: Gonzalo Pajares Martinsanz

Received: 1 September 2015; Accepted: 22 March 2016; Published: 25 March 2016

Abstract: The increasingly common application of the near-infrared (NIR) hyperspectral imaging technique to the analysis of food powders has led to the need for optical characterization of samples. This study was aimed at exploring the feasibility of quantifying penetration depth of NIR hyperspectral imaging light for milk powder. Hyperspectral NIR reflectance images were collected for eight different milk powder products that included five brands of non-fat milk powder and three brands of whole milk powder. For each milk powder, five different powder depths ranging from 1 mm–5 mm were prepared on the top of a base layer of melamine, to test spectral-based detection of the melamine through the milk. A relationship was established between the NIR reflectance spectra (937.5–1653.7 nm) and the penetration depth was investigated by means of the partial least squares-discriminant analysis (PLS-DA) technique to classify pixels as being milk-only or a mixture of milk and melamine. With increasing milk depth, classification model accuracy was gradually decreased. The results from the 1-mm, 2-mm and 3-mm models showed that the average classification accuracy of the validation set for milk-melamine samples was reduced from 99.86% down to 94.93% as the milk depth increased from 1 mm–3 mm. As the milk depth increased to 4 mm and 5 mm, model performance deteriorated further to accuracies as low as 81.83% and 58.26%, respectively. The results suggest that a 2-mm sample depth is recommended for the screening/evaluation of milk powders using an online NIR hyperspectral imaging system similar to that used in this study.

Keywords: penetration depth; hyperspectral imaging; milk powder; PLS-DA

1. Introduction

Milk, both a nutritious food in itself and a functional ingredient in other food products, is a complex fluid consisting of fats, proteins, minerals, vitamins, enzymes, carbohydrates and water. However, fluid milk is difficult to transport and store. Therefore, milk powders are produced using drying technologies to turn fluid milk into dry milk powder. Nonfat milk and whole milk are the two most common milk powders and contribute nutritionally to many food formulations, including reconstituted milk, dairy products, baked goods, confectionery, processed meat products, nutritional beverages and prepared ready-to-eat foods. As an important food ingredient for human and animal food, milk powder safety is a worldwide concern. In recent years, incidents of milk powder adulteration by melamine (2,4,6-triamino-1,3,5-triazine) to boost apparent protein content caused illnesses and resulted in wide recognition of melamine contamination as a food safety problem. Traditional methods of melamine detection in foods involve analytical techniques, such as mass spectrometry and high performance liquid chromatography, are time consuming, expensive and require complicated sample preparation procedures [1,2]. Visible (VIS) and near-infrared (NIR) spectroscopy have been studied as non-destructive methods to detect melamine in milk by several research groups [3,4]; however, spectroscopic assessments with relatively small point-source measurements cannot provide information on the spatial distribution of melamine particles within a food sample.

As one of the most promising tools for non-destructive real-time evaluation of food quality and safety in numerous applications, hyperspectral imaging combines features of both imaging and spectroscopy, such that it is capable of not only directly assessing the presence of different components simultaneously, but also locating the spatial distribution of those components in the products under examination [5]. Hyperspectral imaging may be operated in either reflectance or transmittance modes, although reflectance is more commonly used. Generally, transmittance mode is used for thinly-prepared samples, allowing light to pass through the samples [6–9], while diffuse reflectance is used for thicker samples in hyperspectral imaging measurements of whole or larger portions of foods, such as apples [10], peach [11], mushrooms [12], cucumbers [13] and chickens [14]. Hyperspectral reflectance imaging has been used to detect defects and contaminants and to evaluate quality attributes for fruits, vegetables, meats and dairy products. Implemented with automated image processing and analysis algorithms, hyperspectral imaging has been demonstrated for effective real-time assessment of the quality and safety attributes of poultry [14]. For reflectance imaging, the NIR light must sufficiently penetrate the food material in order for the intensity of the remitted radiation, as a function of wavelength (or frequency), to have been influenced by the chemical nature of the absorbing compound.

Light penetration depth is defined as the depth in a sample material at which incident light is reduced by 99% and will vary with the status of sample, the type of sample and the detection wavelength [15]. Hyperspectral reflectance imaging is usually operated in the VIS-NIR (400–1000 nm) or NIR (1000–1700 nm) range. Limited research exists for the investigation of the penetration depth in the VIS and NIR ranges. Lammertyn *et al.* [16] reported light penetration depth in apples to be up to 4 mm in the 700–900-nm range and between 2 and 3 mm in the 900–1900-nm range. Qin and Lu [15] found that the light penetration depth in fruit tissue varies depending on the type of fruit, ranging from 7.1 mm for plums to 65.2 mm for zucchini. Most studies reporting light penetration depths were conducted on fruits. Further research would provide not only helpful references for thickness determination, but also valuable insight into appropriate sensing configurations, especially for milk powder products [17,18]. Fu *et al.* [18] coupled the NIR hyperspectral imaging technique (990–1700 nm) with spectral similarity analyses to detect melamine mixed into samples of dry milk powder. Imaging allowed visualization of the distribution of melamine particles in the milk mixture samples that were prepared at melamine concentrations ranging from 0.02%–1.00% and presented for imaging in plastic Petri dishes. However, it was not examined exactly how many millimeters the NIR hyperspectral imaging light could penetrate into the milk powder samples. The objective of this study was to determine the penetration depth of NIR hyperspectral imaging light in milk powder for the wavelengths between 937.5 nm and 1653.7 nm.

2. Experimental Section

2.1. Sample Preparation

Eight different milk powder products were purchased from commercial retailers, including 5 brands of nonfat milk powders ('valley (N)', Organic Valley, La Farge, WI, USA; 'nestle (N)', Nestle, Solon, OH, USA; 'hoosier (N)', Hoosier Hill Farm, Fort Wayne, IN, USA; 'now (N)', Now Real Food,

Blooming, IL, USA; 'bob (N)', Bob's Red Mill, Pheasant Court Milwaukie, OR, USA) and three brands of whole milk powders ('hoosier (W)', Hoosier Hill Farm, Fort Wayne, IN, USA; 'nestle (W)', Nestle, Glendale, CA, USA; 'peak (W)', Peak, Friesland Campina, P. Stuyvesantwet 1, AC Leeuwarden, Holland). Melamine powder (99% purity) was obtained from Sigma-Aldrich Company (M2659, St. Louis, MO, USA). For detecting the light penetration depth of the hyperspectral imaging system, different thicknesses (1 mm, 2 mm, 3 mm, 4 mm and 5 mm) of pure milk powders were layered on the top of pure melamine. These milk-melamine samples were prepared in black electroplated aluminum plates precisely machined with square wells (30-mm width and height, 10-mm depth). Each sample was first prepared with a leveled layer of pure melamine in the bottom of the well, followed by a leveled layer of pure milk powder on top of the melamine. The combined thicknesses of the melamine and milk powder layers completely filled the 10-mm depth of each well, with the milk depth ranging between 1 and 5 mm (*i.e.*, 1 mm-thick milk layer over 9 mm of melamine, 2 mm-thick milk layer over 8 mm of melamine, and so on). Figure 1 illustrates the preparation of a sample containing 3 mm of milk over 7 mm of melamine. For each kind of milk powder, three samples were prepared at each of the five milk depths, as well as three samples of pure dry milk (10 mm of milk with no melamine underneath) and one sample of pure melamine. With 19 plates prepared for each of the milk powder products, a total of 152 samples were measured for the eight milk products used in this study.

3 mm Milk

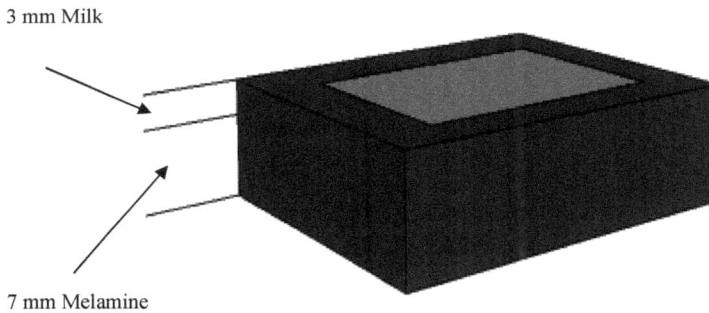

7 mm Melamine

Figure 1. Milk-melamine sample holder (e.g., 3 mm-thick milk and 7 mm-thick melamine layers). The light grey area shows the sample surface (30 mm × 30 mm).

2.2. Instrument and Experiment

As shown in the schematic in Figure 2 and described in detail in Kim *et al.* [19], the line-scan NIR hyperspectral imaging system consists of an InGaAs focal-plane-array (FPA) camera with 320 × 256 pixels (Xenics, Model Xeva-1.7-320, Leuven, Belgium), an imaging spectrograph (SWIR Hyperspec, Headwall Photonics, MA, USA) and a 25-mm zoom lens (Optec, Model OB-SWIR25/2, Parabiago, Italy), as well as a computer for controlling the camera and acquiring images, two 150-W DC light sources with fiber optic bundles (Dolan Jenner, Model DC-950, Boxborough, MA, USA) and a motorized uniaxial stage (Velmex, Model XN10-0180-M02-21, Bloomfield, NY, USA). The camera array sensor consists of 150 usable pixels in the spectral dimension over a wavelength range of 937.5–1653.7 nm, with an average wavelength spacing of 4.8 nm. Except for the two 150-W quartz tungsten halogen light sources, the system is entirely housed with an aluminum-framed enclosure. The light is conveyed via the low-OH fiber optic bundles, with each bundle terminating in a 250 mm-long line of fibers encased in a machined aluminum head. The angle of incidence from the line lights is 30° (from surface normal), and the distance between line lights and sample surface is 20 mm. For each light source, a mechanical iris was used to allow approximately 75% of 150-W light intensity to arrive at the sample surface. The motorized stage moved the samples incrementally across the linear field of view for step-by-step acquisition of line-scan images.

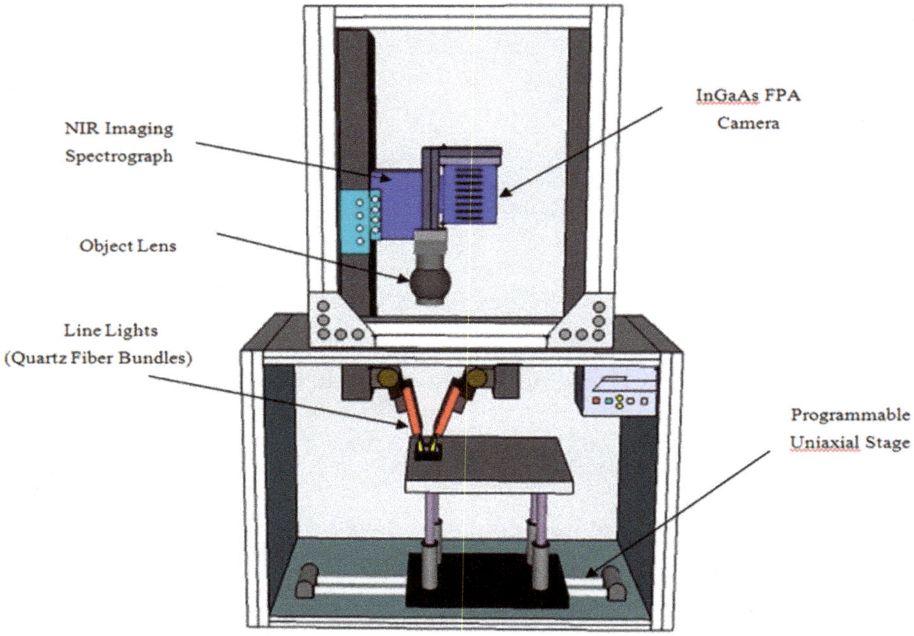

Figure 2. Schematic of the NIR hyperspectral imaging system used to acquire reflectance images of milk powders. FPA, focal-plane-array.

In this study, square imaging pixels were achieved by setting the incremental step size to 0.1 mm to match the pixel-to-pixel distance along the imaging line, which included 320 pixels. The camera exposure time was set at 2.5 ms, and a total of 360 scans were acquired in 1 min. The camera digitized raw energy readings in 14-bit resolution. Finally, each sample's spectral data were stored as a 16-bit hyperspectral image cube of dimensions $320 \times 360 \times 150$ containing spatial and spectral data. The hypercube is a three-dimensional image, which represents a 2D spatial image with x-axis and y-axis coordinate information and z-axis spectral information.

2.3. Hyperspectral Data Analysis

2.3.1. Image Preprocessing

In this study, dark current and white reference images were collected to correct the raw reflectance images for wavelength-dependent system responses and heterogeneous dark current in the FPA camera. The dark current image was captured while the lens was covered by a lens cap. An image of an illuminated 99% diffuse reflectance standard (Spectralon^{TM}, SRT-99-120, Labsphere, NH, USA) was acquired for use as the white reference image. These images were used to calculate the relative reflectance image of each sample, which was calculated by dividing the difference in energy readings between sample and dark current by the difference in energy readings between the white reference and dark current.

2.3.2. Spectral Preprocessing

The region of interest (ROI) for each sample was composed of a square region of 10,000 pixels (100×100 pixels around the image center), selected within the 30 mm $\times$ 30 mm sample area (shown in Figure 1) so as to include only milk powder areas and exclude background (plate) regions for further analyses. Spectral preprocessing techniques were used prior to develop a calibration model in the quest for improving the subsequent classification model. Common preprocessing methods for NIR spectra were used, including Standard Normal Variate (SNV), Multiplicative Scatter Correction (MSC), Extended MSC (EMSC), Normalized (NORMALIZED), the common (base 10) Logarithm (LOG10), Savitzky–Golay Smooth (SMOOTH) and the Savitzky–Golay 1st (1ST) and the Savitzky–Golay 2nd (2ND) derivative [20].

2.3.3. Development of the Classification Model

After spectral preprocessing, partial least squares discriminant analysis (PLS-DA) was used to classify each pixel as belonging to either the pure milk class or the milk-melamine mixture class. PLS-DA, an extension of PLS modeling, aims to find the variables and directions in a multivariate space that discriminate the known classes in the calibration set. PLS components are computed under the constraint of the maximization of covariance between inputs and outputs. Therefore, it can provide a set of orthogonal factors that have the best predictive power from the combinations of different methods with an increased number of variables [21–23].

Prior to model development, the triplicate samples (three samples imaged at each depth of milk over melamine and three samples for each pure milk powder product) were divided into two groups, with two samples (comprising 20,000 ROI pixels) assigned to a model development dataset and one sample (comprising 10,000 ROI pixels) assigned to a validation dataset for use in evaluating model performance. Model performance was compared based on classification accuracies, *i.e.*, the percentage of correctly-classified pixels over the total number of pixels. To better assess the performance of the classification models, calibrations and validations for each depth were run three times [24]. For each of the 5 different milk-melamine preparations, a separate PLS-DA model (1-mm model, 2-mm model, 3-mm model, 4-mm model and 5-mm model) was developed. The number of components chosen for the PLS-DA models was determined by a contiguous block cross-validation method, in which each block contained the samples from one milk powder product.

Image processing, selection of the square ROI, the spectral preprocessing operation and model development were performed in MATLAB (R2007b, MathWorks, Natick, MA, USA) equipped with the PLS Toolbox (v. 7.5, Eigenvector Research, Manson, WA, USA).

3. Results and Discussion

3.1. Hyperspectral Spectra

Figure 3 shows representative mean absorption spectra (calculated from 10,000-pixel ROI) of pure melamine and the eight pure milk products (including five nonfat and three whole) obtained from the LOG10 preprocessing technique. Significant differences, absorption peaks related to the first and second N-H functional group, can be clearly observed between melamine and milk spectra. The mean melamine spectrum had peaks near 1523.9 and 1490.3 nm, corresponding to the first overtone of N-H symmetric and anti-symmetric stretching vibration, respectively. The second overtone of N-H stretching vibration is located near 985.6–1033.7 nm (centered at 1009.6 nm). The most significant spectral difference between melamine and milk occurred near 1466.3 nm, which is attributed to aromatic amine structures [3] and showed the highest absorption in the melamine spectrum. For the nonfat milk spectra, the spectral patterns of the five different brands are similar, and another similar spectral pattern was observed for the three brands of whole milk from Figure 3. The absorbance of most nonfat milk spectra is lower than that of the whole milk spectra. For the visual difference between the nonfat and whole milk pattern, whole milk spectra have an evident absorption peak around 1211.5 nm, which is due to the second overtone of C-H stretching vibration constituted by saturated fat structures [25].

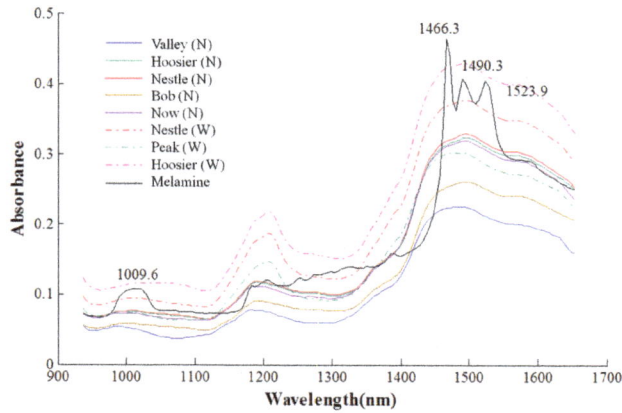

Figure 3. Representative mean spectra of pure nonfat (N) and whole (W) milk powders and pure melamine, each calculated from a 10,000-pixel ROI.

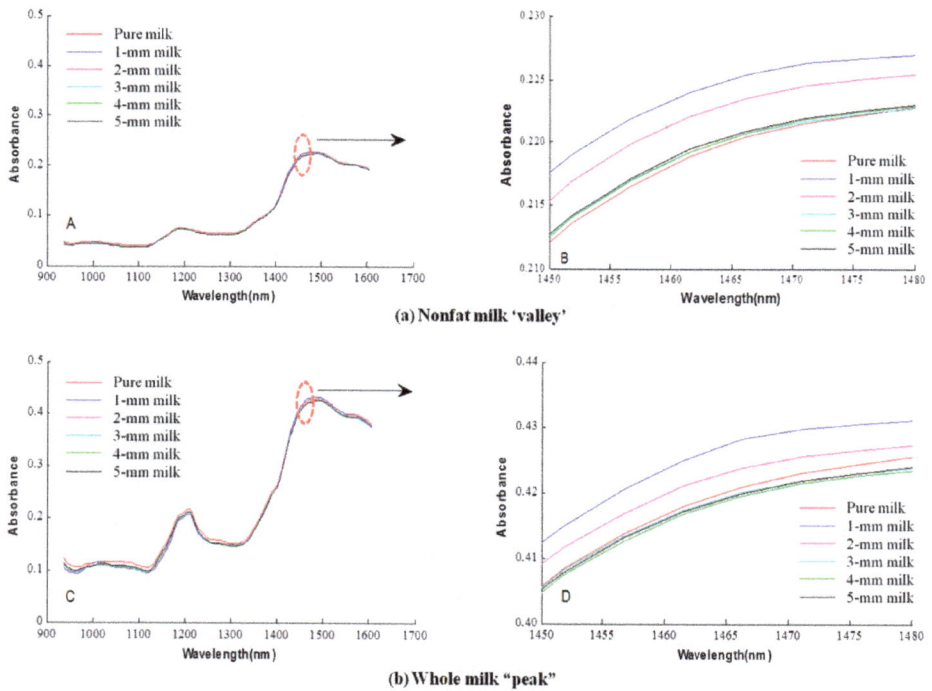

Figure 4. Mean ROI spectra of samples prepared using (**a**) 'valley' (N) nonfat milk and (**b**) 'peak' (W) whole milk, including pure milk samples and milk-melamine samples with milk depths from 1–5 mm (thickness). Plots A and C show the full spectra, while Plots B and D show the enlarged view of the mean spectra near the 1466.3-nm melamine peak.

Figure 4 shows representative mean spectra for one nonfat milk and one whole milk, valley (nonfat (N)) and peak (whole (W)), including milk-melamine samples for milk depths of 1–5 mm (thickness) and for pure milk. Among the eight milk products used in this study, 'peak' (W) had the strongest absorbance and 'valley' (N) had the weakest absorbance. The mean spectra of milk-melamine samples are very similar to the mean spectrum of pure milk; all exhibit no obvious melamine absorption features. This observation suggests that individual pixel-based spectral evaluations (instead of averaged spectra across spatial image areas) may allow better detection of melamine to measure the penetration depth of milk [18]. In the region of the 1466.3-nm melamine absorption peak, the mean absorbance notably decreased as the milk depth increased from 1 mm–3 mm. For 3 mm–5 mm, the mean spectra were nearly the same as that for pure milk. The same trends were observed for both nonfat milk 'valley' and whole milk 'peak'.

3.2. Discriminant Models for Milk Depth Classification

The plots in Figure 5 compare the validation set classification results (average of three runs) for the PLS-DA models coupled with specific spectral preprocessing algorithms, for milk-melamine at each of the five different milk depths (1-mm–5-mm models). As shown in Figure 5, the different preprocessing algorithms can have a great impact on model accuracy. Compared to the other spectra preprocessing algorithms, the 2ND derivative consistently resulted in low classification accuracy. The reason may be that it has a more prominent spectra shoulder after derivative transformation, but it affects the component number selection of PLS-DA. The models coupled with the SNV and MSC algorithm gave better, more robust performance than the NORMALIZED model (mean zero, unit variance), which is due to reducing the scattering influence from particle size.

Since the classification accuracies of the calibration set were better than those of the validation sets, only the classification results of the validation set for milk and milk-melamine samples are presented. Table 1 shows the classification results of the validation set based on the PLS-DA model coupled with the SNV preprocessing algorithm for the eight milk powders. The data show the classification accuracy notably decreasing as the milk depth increases from 1 mm–5 mm. For the 1-mm model, 99.65%–100% of milk samples (pixels) and 99.06%–100.00% of milk-melamine samples were correctly classified, for overall accuracies of 99.93% and 99.86% for milk and milk-melamine samples, respectively, across the eight milk powders. The 2-mm model's highest accuracy was the same as that for the 1-mm model, while its lowest accuracies were lower at 95.53% for milk and 96.4% for milk-melamine samples. The average misclassification rates were 1.39% for milk and 1.58% for milk-melamine samples across all eight milk powders. Although the 3-mm model achieved an average accuracy 95.54% for milk and 94.93% for milk-melamine samples, the classification accuracies of nonfat milk 'hoosier' and 'nestle' were lower than 90%, which is not suitable for melamine detection at lower concentrations. As the milk depth increased to 4 mm and 5 mm, the nearly identical spectra for milk and for milk-melamine (shown in Figure 3) resulted in deteriorated model performance. The average misclassification rate of the 4-mm model was greater than 10%, while about 20% of the samples were misclassified by the 5-mm model for some of the milk powders. This means that the 4-mm and 5-mm models were invalid for classification.

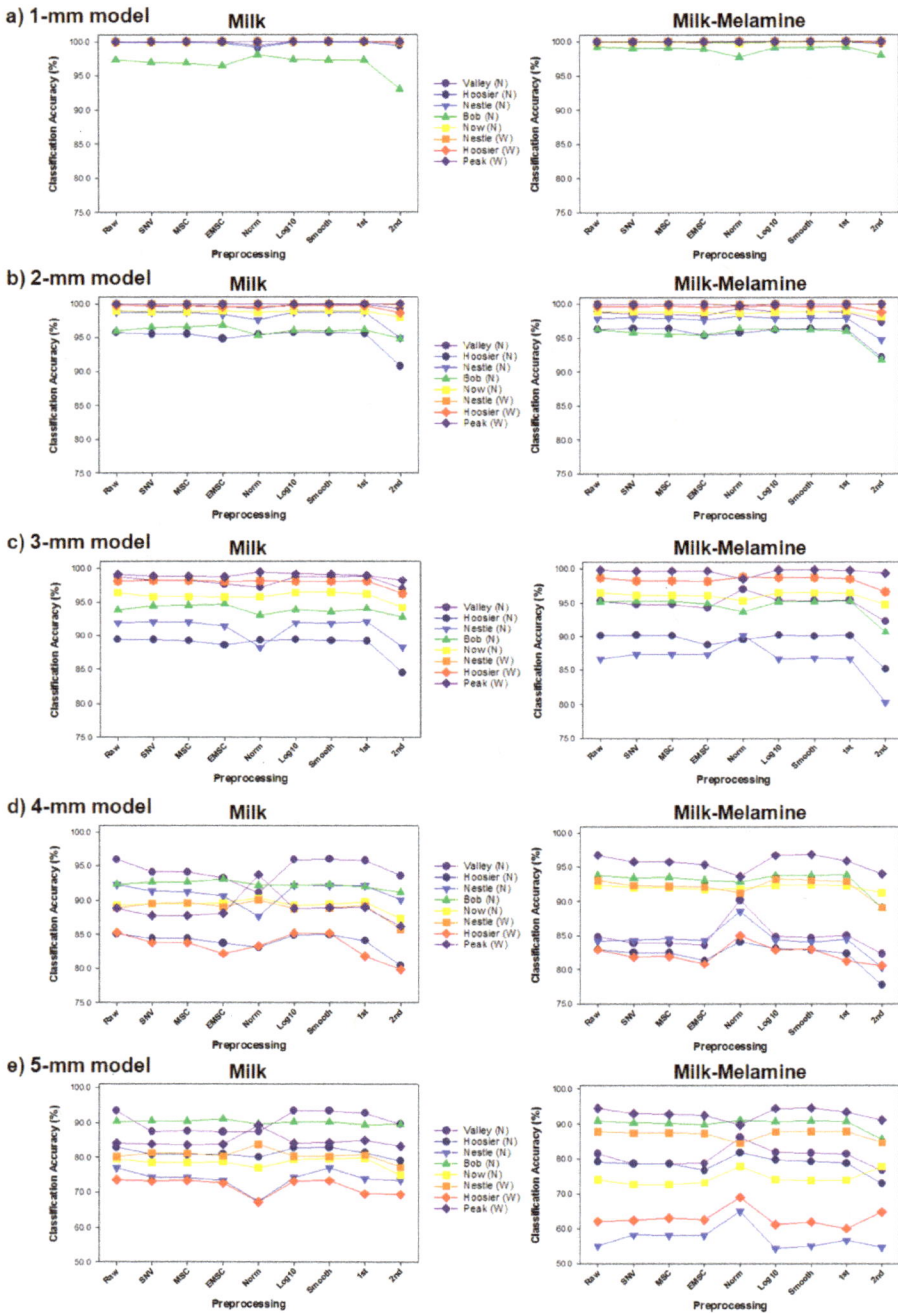

Figure 5. Classification comparison of classification results for PLS-DA models coupled with specific spectral preprocessing algorithms for milk-melamine samples at milk depths from 1 mm to 5 mm (**a–e**).

Table 1. Classification results of the validation set for milk and milk-melamine samples from a 1-mm–5-mm depth using PLS-DA coupled with the Standard Normal Variate (SNV) spectra preprocessing algorithm.

		Classification (%)				
	Depth	1 mm	2 mm	3 mm	4 mm	5 mm
	valley (N)	99.98	99.66	98.20	94.13	87.41
	hoosier (N)	99.86	95.53	89.37	84.43	80.91
	nestle (N)	100.00	98.72	92.08	91.39	74.26
	bob (N)	99.65	96.51	94.49	92.70	90.52
Milk	now (N)	99.97	98.80	95.86	89.56	78.50
	nestle (W)	100.00	99.98	98.19	89.57	81.29
	hoosier (W)	99.98	99.73	97.29	83.77	73.19
	peak (W)	99.99	99.98	98.86	87.81	83.85
	Average	99.93	98.61	95.54	89.17	81.24
	valley (N)	99.91	98.56	94.81	83.94	78.60
	hoosier (N)	99.96	96.43	90.21	82.47	78.62
	nestle (N)	100.00	97.97	87.34	84.35	58.26
Milk-melamine	bob (N)	99.06	95.82	95.29	93.42	90.30
	now (N)	99.94	98.89	96.21	92.03	72.73
	nestle (W)	100.00	100.00	98.27	92.32	87.42
	hoosier (W)	99.98	99.70	97.66	81.83	62.43
	peak (W)	99.99	99.99	99.70	95.76	92.92
	Average	99.86	98.42	94.93	88.26	77.66

Figure 6 shows the classification results for two brands of nonfat and whole milk using the PLS-DA model coupled with the SNV spectra preprocessing algorithm. For the same brand milk powder, the classification results of whole milk were slightly higher than those of nonfat milk for the 1-mm–3-mm valid models.

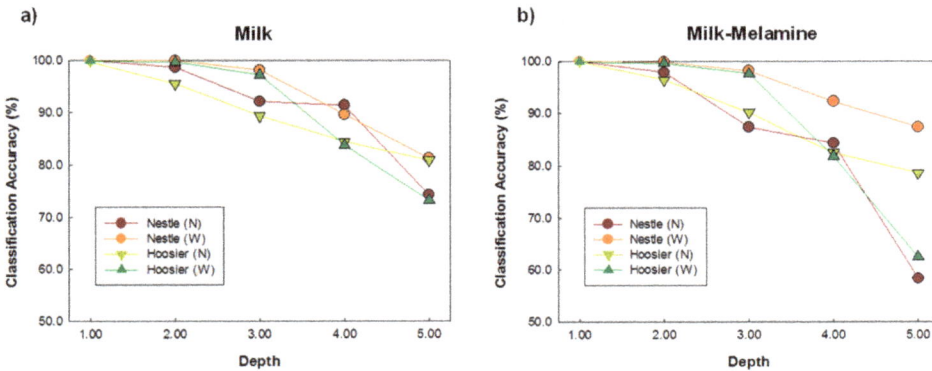

Figure 6. Comparison of the classification results for the same brand nonfat and whole milk using the PLS-DA model coupled with the SNV spectral preprocessing algorithm for (**a**) milk and (**b**) milk-melamine.

4. Conclusions

In this study, the penetration depth of near-infrared hyperspectral imaging light (937.5–1653.7 nm) was investigated for milk powders. Five different depths of milk powder, from 1 mm–5 mm, were investigated for the detection of melamine underneath the milk powder. Classification models were developed using the PLS-DA technique for milk and milk-melamine samples prepared using five brands of non-fat milk powder and three brands of whole milk powder. The classification results showed that the classification accuracy gradually decreased as the milk depth increased. For the 2-mm models, the classification accuracies were higher than 95% for both milk and milk-melamine samples for all of the milk powders under investigation. It can be concluded that the use of a 2-mm milk powder depth can be recommended for applying NIR hyperspectral imaging for the detection of contaminants in milk powder. In addition, the method described can also be potentially applied to other food powders for penetration depth measurement of NIR hyperspectral imaging system.

Acknowledgments: This work was partially supported by grants from the Agenda Programs (PJ009399) and (PJ012216), Rural Development Administration, Korea, the National Natural Science Foundation of China (Grant # 61271384 and 61275155), and the 111 (B12018) and Qing Lan Projects of China. The authors would like to thank Diane Chan of Environmental Microbial and Food Safety Laboratory, Agricultural Research Service, USDA, for reviewing the manuscript.

Author Contributions: M.H., M.S.K, K.C., J.Q., C.M and S.D. conceived of and designed the experiments. M.H. and M.S.K. performed the experiments. M.H. and C.E. analyzed the data. M.H. and Q.Z. wrote the paper. M.S.K. and S.D. edited the paper.

Conflicts of Interest: The authors declare no conflict of interest.

References

1. Filigenzi, M.S.; Puschner, B.; Aston, L.S.; Poppenga, R.H. Diagnostic determination of melamine and related compounds in kidney tissue by liquid chromatography/tandem mass spectrometry. *J. Agric. Food. Chem.* **2008**, *56*, 7593–7599. [CrossRef] [PubMed]
2. Xu, X.; Ren, Y.; Zhu, Y.; Cai, Z.; Han, J.; Huang, B.; Zhu, Y. Direct determination of melamine in dairy products by gas chromatography/mass spectrometry with coupled column separation. *Anal. Chim. Acta* **2009**, *650*, 39–43. [CrossRef] [PubMed]
3. Mauer, L.J.; Chernyshova, A.A.; Hiatt, A.; Deering, A.; Davis, R. Melamine detection in infant formula powder using near- and mid infrared spectroscopy. *J. Agric. Food. Chem.* **2009**, *57*, 3974–3980. [CrossRef] [PubMed]
4. Balabin, R.M.; Smirnov, S.V. Melamine detection by mid-and near-infrared (MIR/NIR) spectroscopy: A quick and sensitive method for dairy products analysis including liquid milk, infant formula, and milk powder. *Talanta* **2011**, *85*, 562–568. [CrossRef] [PubMed]
5. Wu, D.; Sun, D. Advanced applications of hyperspectral imaging technology for food quality and safety analysis and assessment: A review—Part I: Fundamentals. *Innovative Food Sci. Emerg. Technol.* **2013**, *19*, 1–14. [CrossRef]
6. Cogdill, R.; Hurburgh, C.; Rippke, G. Single-kernel maize analysis by near-infrared hyperspectral imaging. *Trans. ASAE* **2004**, *47*, 311–320. [CrossRef]
7. Qin, J.; Lu, R. Detection of pits in tart cherries by hyperspectral transmission imaging. *Trans. ASAE* **2005**, *48*, 1963–1970. [CrossRef]
8. Coelho, P.A.; Soto, M.E.; Torres, S.N.; Sbarbaro, D.G.; Pezoa, J.E. Hyperspectral transmittance imaging of the shell-free cooked clam *Mulinia edulis* for parasite detection. *J. Food Eng.* **2013**, *117*, 408–416. [CrossRef]
9. Huang, M.; Wan, X.; Zhang, M.; Zhu, Q. Detection of insect-damaged vegetable soybeans using hyperspectral transmittance image. *J. Food Eng.* **2013**, *116*, 45–49. [CrossRef]
10. Huang, M.; Lu, R. Apple mealiness detection using hyperspectral scattering technique. *Postharvest Biol. Technol.* **2010**, *58*, 168–175. [CrossRef]
11. Lu, R.; Peng, Y. Hyperspectral scattering for assessing peach fruit firmness. *Biosystems Eng.* **2006**, *93*, 161–171. [CrossRef]
12. Gowen, A.A.; Taghizadeh, M.; O'donnell, C.P. Identification of mushrooms subjected to freeze damage using hyperspectral imaging. *J. Food Eng.* **2009**, *93*, 7–12. [CrossRef]
13. Ariana, D.P.; Lu, R.; Guyer, D.E. Hyperspectral reflectance imaging for detection of bruises on pickling cucumbers. *Comput. Electron. Agric.* **2006**, *53*, 60–70. [CrossRef]
14. Chao, K.; Yang, C.; Kim, M.S.; Chan, D. High throughput spectral imaging system for wholesomeness inspection of chicken. *Appl. Eng. Agric.* **2008**, *24*, 475–485. [CrossRef]
15. Qin, J.; Lu, R. Measurement of the optical properties of fruits and vegetables using spatially resolved hyperspectral diffuse reflectance imaging technique. *Postharvest Biol. Technol.* **2008**, *49*, 355–365. [CrossRef]

16. Lammertyn, J.; Peirs, A.; de Baerdemaeker, J.; Nicolaï, B. Light penetration properties of NIR radiation in fruit with respect to non-destructive quality assessment. *Postharvest Biol. Technol.* **2000**, *18*, 121–132. [CrossRef]

17. Fernández Pierna, J.A.; Vincke, D.; Dardenne, P.; Yang, Z.; Han, L.; Baeten, V. Line scan hyperspectral imaging spectroscopy for the early detection of melamine and cyanuric acid in feed. *J. Near Infrared Spectrosc.* **2014**, *22*, 103–112. [CrossRef]

18. Fu, X.; Kim, M.S.; Chao, K.; Qin, J.; Lim, J.; Lee, H.; Garrido-Varo, A.; Pérez-Marín, D.; Ying, Y. Detection of melamine in milk powders based on NIR hyperspectral imaging and spectral similarity analysis. *J. Food Eng.* **2014**, *124*, 97–104. [CrossRef]

19. Kim, M.S.; Chen, Y.R.; Mehl, P.M. Hyperspectral reflectance and fluorescence imaging system for food quality and safety. *Trans. ASAE* **2001**, *44*, 721–729.

20. Rinnan, A.; van den Berg, F.; Engelsen, S.B. Review of the most common pre-processing techniques for near-infrared spectra. *Trends Anal. Chem.* **2009**, *28*, 1201–1222. [CrossRef]

21. Barker, M.; Rayens, W. Partial least squares for discrimination. *J. Chemom.* **2003**, *17*, 166–173. [CrossRef]

22. Westerhuis, J.A.; Hoefsloot, H.C.J.; Smit, S.; Vis, D.J.; Smilde, A.K.; van Velzen, E.J.J.; van Duijnhoven, J.P.M.; van Dorsten, F.A. Assessment of PLSDA cross validation. *Metabolomics* **2008**, *4*, 81–89. [CrossRef]

23. Huang, M.; Ma, Y.; Li, Y.; Zhu, Q.; Huang, G.; Bu, P. Hyperspectral image-based feature integration for insect-damaged hawthorn detection. *Anal. Methods* **2014**, *6*, 7793–7800. [CrossRef]

24. Lu, R.; Huang, M.; Qin, J. Analysis of hyperspectral scattering characteristics for predicting apple fruit firmness and soluble solids content. In Proceedings of the International Conference on SPIE 7315, Sensing for Agriculture and Food Quality and Safety, Orlando, FL, USA, 13 April 2009.

25. Nagarajan, R.; Singh, P.; Mehrotra, R. Direct determination of moisture in powder milk using near infrared spectroscopy. *J. Autom. Methods Manag. Chem.* **2006**, *2006*, 51342. [CrossRef] [PubMed]

sensors

MDPI

Technical Note

Forward-Looking Infrared Cameras for Micrometeorological Applications within Vineyards

Marwan Katurji * and Peyman Zawar-Reza

Center for Atmopsheric Research, University of Canterbury, Christchurch 8140, New Zealand; peyman.zawar-reza@canterbury.ac.nz
* Correspondence: marwan.katurji@canterbury.ac.nz; Tel.: +64-3-364-2987 (ext. 3088)

Academic Editor: Gonzalo Pajares Martinsanz
Received: 30 June 2016; Accepted: 13 September 2016; Published: 18 September 2016

Abstract: We apply the principles of atmospheric surface layer dynamics within a vineyard canopy to demonstrate the use of forward-looking infrared cameras measuring surface brightness temperature (spectrum bandwidth of 7.5 to 14 µm) at a relatively high temporal rate of 10 s. The temporal surface brightness signal over a few hours of the stable nighttime boundary layer, intermittently interrupted by periods of turbulent heat flux surges, was shown to be related to the observed meteorological measurements by an in situ eddy-covariance system, and reflected the above-canopy wind variability. The infrared raster images were collected and the resultant self-organized spatial cluster provided the meteorological context when compared to in situ data. The spatial brightness temperature pattern was explained in terms of the presence or absence of nighttime cloud cover and down-welling of long-wave radiation and the canopy turbulent heat flux. Time sequential thermography as demonstrated in this research provides positive evidence behind the application of thermal infrared cameras in the domain of micrometeorology, and to enhance our spatial understanding of turbulent eddy interactions with the surface.

Keywords: time sequential thermography; micrometeorology; self-organizing maps; surface energy balance; turbulence; microclimate; infrared camera

1. Introduction

There is great interest in using near-target remote sensing techniques such as time-sequential thermography (TST) in precision agriculture, ecology [1] and phenomics [2]. Thermography techniques have to address the thermal condition of the object of interest and the thermal and humidity conditions of the intervening atmosphere. Near-surface atmospheric temperature is influenced by synoptic weather patterns and their interaction with local topography at the smaller scale, which together determines the nature of the air turbulence that envelops the plant and controls the rate of water vapor and heat exchanges. On the other hand, plants are more than passive objects and employ stomata to sense the surrounding environment and respond rapidly to abiotic stresses, such as the air temperature. Their response is typically through stomatal conductance to water vapor and/or transpiration, which are critical physiological controls. The plant's surface temperature, or its brightness temperature as sensed by a thermal infrared camera, is the result of the interaction of the air temperature and the plant's physiological response. Thus, to understand the plant's microclimate through thermography (or the environment that embodies the plant to a few orders of magnitude in spatial scale relative to the plant's volume), it is important to understand the brightness temperature signal (measured by a infrared camera) as a function of near-surface meteorological parameters controlling the energy exchanges happening across the plant-environment envelope.

Land surface temperatures are influenced by surface energy balance [3] especially when horizontal advection processes are negligible. Surface temperature varies as a consequence of partitioning of net-all wave radiation (Q^*, or the balance between solar and infrared radiation input and output to the surface) into the subsurface conduction of heat (Q_G) and changes in sensible (Q_H) and latent heat exchange (Q_E) with the overlying atmosphere. On short time scales (less than an hour), radiative input is relatively constant, unless clouds interfere or overlying plant canopy causes rapid changes (flickering) in solar radiation [4]. Higher frequency (seconds to minutes) surface temperature fluctuations are a response to the turbulent sensible and latent heat fluxes. Turbulence, caused by eddy motion, is expected to control temperature fluctuations on the same length and time scales as the atmospheric eddy motions.

The brightness temperatures of objects within the surface layer were not typically considered in the atmospheric community as a proxy for near-surface turbulence, but as infrared cameras become cheaper and are able to record data at high spatial and temporal resolutions, it is now feasible to study turbulence through the acquisition of brightness temperature. One of the earlier studies to investigate the coupling between coherent turbulent structures and surface temperature over an agricultural field (maize canopy) employed a directional infrared thermometer (sampling at 10 Hz) in identifying ramp structures in the surface temperature signal of the canopy with significant correlation with fluctuations in the air temperature above the canopy [5]. Coherent structures were identified as temperature ramps in the surface and air temperature time series, with the magnitude of surface temperature ramps being significantly smaller than the air temperature ramps. Surface temperature ramps are caused by turbulent eddies mixing warmer (or cooler) air with cool (warm) air from aloft. A similar study was conducted over grass [3] and also found direct relationships between surface brightness temperatures and independently measured surface-layer turbulence parameters.

Application of time-sequential thermography (TST) to calculate urban sensible heat fluxes (from a building) was first demonstrated by Hoyano et al. [6], and was further developed conceptually by Voogt [7] as a method for viewing the "footprint" of the coherent flow structures, and it was later emphasized by Christen et al. [8] that brightness temperature fluctuations are largely controlled by atmospheric turbulence while the level of fluctuation becomes modulated by surface properties, especially its thermal admittance. The application of TST to detect large temperature fluctuations in the unstable surface layer to understand the turbulence structure has shown great promise in field experiments [9], and was successful in deriving surface wind velocities over simple grass areas [10] via the principle of turbulent eddy interaction with surface brightness temperatures.

As forward-looking infrared cameras become more affordable, TST will become an attractive method to measure the energy and moisture exchanges between the surface and overlying atmosphere. This research utilizes spatial brightness temperature data from infrared cameras looking onto a vineyard canopy. The canopy is also instrumented with an eddy covariance system measuring in situ turbulent and radiation fluxes and near-ground thermistor-based temperature sensors. The brightness temperature fluctuations (sampled sequentially over a nighttime period at a high frequency) are then used to interpret the spatial variability of the turbulent nature of the site using a combination of in situ metrological measurements and a pattern recognition algorithm (or self-organizing maps, SOM) applied to the acquired brightness temperature data. The SOM approach allows for clustering self-similar images into groups that could then be analyzed according to their unique meteorological context. This research highlights the significance and relevance of the methodology in terms of relating the brightness temperature variability to atmospheric turbulence, which also highlights the local meteorology. This approach is not only limited to vineyard applications and could be applied and assessed over various other crop types or surfaces.

2. Methods

2.1. Study Site and Instrumentation

The experimental site is located in Marlborough, situated in the South Island of New Zealand, renowned for abundance of vineyards and a major wine-producing region. The experimental site was chosen to be at the Lions Back vineyard in Seddon (41°41′51.5″S, 174°05′13.6″E) (Figure 1). This vineyard is ideal for the purpose of this experiment as it is easy to access and a nearby-elevated escarpment provides an ideal platform for placement of the infrared cameras overlooking most of the vineyard. Within the field of view of the cameras we have placed an eddy covariance system containing a 3D sonic anemometer, water vapor analyzer, measurements of all-wave radiation, and near surface air and soil temperature loggers in the same measurement area (Table 1). All measurements were controlled through data logging devices and/or monitored manually throughout the sampling period. The meteorological variables measured by the eddy covariance system (surface energy balance measurement system), climate station (standard meteorological parameters), and the near surface temperature logger (thermistor-based temperature sensing) were taken from 17 May 2014 12 a.m. up until 18 May 2014 9 p.m. While the thermography data was collected during the evening of the 17 May 2014 between 5 p.m. and 9 p.m. local standard time, which is also highlighted with a black rectangular box in Figure 2a.

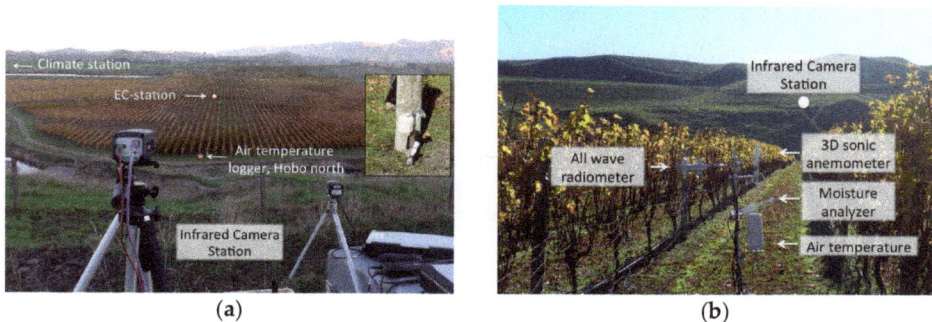

| (a) | (b) |

Figure 1. (**a**) The vineyard site covering most of the field of view of the two longwave infrared cameras shown on tripods in the foreground. The near-ground Hobo temperature logger is shown in the small figure inset; (**b**) A close up on the eddy covariance system (or EC-station in left panel) placed in the center of the camera's field of view.

Table 1. Instrumentation specifications and measured variables.

Instrument	Description	Measured Variable	Range and Accuracy	Sampling Frequency
Campbell Scientific CSAT3	Three-dimensional ultrasonic anemometer	Cartesian components of velocity and sonic temperature (u, v, w, Ts)	± 65 m·s^{-1} ± 0.08 m·s^{-1} for u, v ± 0.04 m·s^{-1} for w, and -30 to 50 °C ± 0.01 °C for Ts	20 Hz
LICOR-7500	Open path infrared H$_2$O analyzer (situated 30 cm below the sonic anemometer)	Specific humidity	0 to 60 parts per trillion (ppt) ± 0.6 ppt	20 Hz
Kipp and Zonen CNR1	Net radiometer	Incident and reflected long- and short-wave radiation components	$\pm 10\%$ over 24 h	1 Hz
Vaisala HMP45C	Temperature and relative humidity probe	Air temperature, soil surface temperature, and relative humidity	-40 to $+60$ °C ± 0.3 at 0 °C 0 to 90% $\pm 2\%$	1 Hz
HOBO U23	Radiation shielded temperature sensor	Air temperature at 35 cm above ground level or AGL	-40 to $+70$ °C ± 0.2 °C	0.1 Hz
FLIR A644sc	Uncooled infrared camera	Brightness temperature on a raster of 640×480 pixels	-40 to 150 °C ± 2 °C Spectral range 7.5 to 14 µm Thermal sensitivity 30 mK	50 Hz
Optris Pi 640	Uncooled infrared camera	Brightness temperature on raster of 640×480 pixels	-20 to 900 °C ± 2°C Spectral range 7.5 to 13 µm Thermal sensitivity 75 mK	0.1 Hz

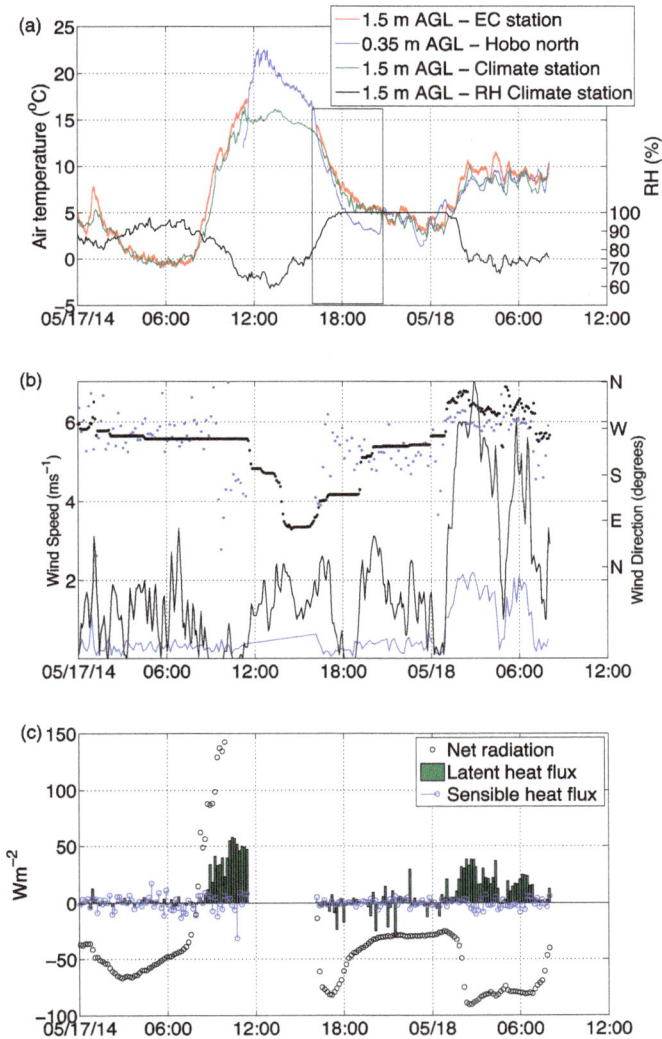

Figure 2. (**a**) Air temperature and relative humidity from within and above the vineyard canopy. The climate station data was collected from a low escarpment around 5 m above the canopy's horizon. The black box shows the time period of the operation of the infrared cameras; (**b**) Wind speed (line) and direction (dotted) from within (blue) and above (black) the canopy; (**c**) The surface radiation and turbulent energy budget from the eddy covariance system inside the canopy.

2.2. Self-Organizing Maps, SOMs

The SOMs algorithm for the pattern recognition used in this analysis is SOM_PAK, found at Helsinki University of Technology website [11]. SOM iterates through the input dataset while matching each input to the SOM node that is closest in terms of its Euclidean distance, and then adjusts the node and its neighbors to incorporate the input data. A learning rate parameter controls the rate at which the SOM absorbs the information from the input data while a neighborhood radius determines which other nodes, are affected by the input data. The learning rate decreases to zero and the neighborhood radius decreases to one as the algorithm iterates through the dataset. Several matrix sizes (representing the number of maps or patterns) were tested, and for each matrix size the number of iterations and learning rate function types were adjusted, with the aim of reducing the quantization error, or the mean Euclidean distance between the input data and the SOM; bigger matrices generally exhibit lower errors. When the change in the error is minimal, the process can be considered complete. Criterion referred to as the "Sammon Map" is used to inspect whether each node of the SOM has more in common with its neighboring nodes, than non-neighboring. If not, the SOM algorithm was rerun with adjusted parameters. For examples on using SOM for various sources of meteorological data see [12–18].

The brightness temperatures (180 × 180 pixels) extracted as a spatial subset from the total infrared camera image (640 × 480 pixels) were used as input data for the SOM algorithm. First the pixel-wise data was normalized so that all pixel variables have a variance of 1, this allows for a more effective way in extracting patterns without biasing regions towards extreme values. Before executing the SOM algorithm the infrared brightness temperature perturbations (Tb') were calculated based on the deviation of every sample (at 10 s interval) from the 10 min average. The de-trended Tb' were then related to the turbulence and/or radiation forcing measured by the eddy covariance unit. The number of nodes (or pattern groups) was chosen to be a 3 × 3 matrix arrangement after testing with several other arrangements and an optimization between the size of the matrix and the detail of output was reached. After constructing the SOM, the non-clustered data was then matched with its most representative node (or spatial pattern), and then a number count was found to compare the relative population of each node with the original or non-clustered data. As a result, all 9 nodes had best matching units between 60 and 120, which suggests that the nodes were relatively well populated.

3. Results

3.1. Micrometeorological Context

The experiment extends between midnight of 17 May 2014 up until the early evening of 18 May 2014. During this period the region was synoptically quiescent, which limited surface wind speeds to less than 3 m·s^{-1}, and the diurnal temperature ranged between just below freezing level and 20 °C (Figure 2). The air temperatures measured from within the vineyard canopy (EC station), near the surface air (Hobo north) and from a nearby climate station (Figure 2a) all show similar diurnal temperature variations within the experimental domain. The first morning period (17 May 12 a.m. to 8 a.m.) was colder by 5 to 10 °C than the following morning period (18 May 12 a.m. to 8 a.m.), mainly due to the more stable atmosphere maintained by weak surface wind speeds and the surface radiation cooling process. The following early morning wind speeds measured at 6 m above ground level (AGL) at the nearby climate station site increased up to 6 m·s^{-1}, causing higher levels of turbulence within the canopy as registered by the increase in the sonic wind speed (Figure 2b) and the increase in the latent heat flux during the period between 12 a.m. and 8 a.m.) on 18 May. The relative humidity during this period (70%) was also lower than the night before (90%, Figure 2b), and the wind direction was from the northwest sector which blows relatively dryer air from the elevated mountainous regions.

3.2. Brightness and Air Temperature Relationship

In this section we aim to present a direct comparison between the measured brightness temperatures as seen by both of the long-wave infrared cameras and in situ air temperatures measured from within the canopy and from the surface at the northern edge of the canopy. The method we have used relies on averaging the brightness temperature over a 25 m^2 area and a larger field of view area (see boxes (1), (2) and (3) in Figure 3a). Figure 3a shows a snapshot from the Optris camera of the entire field of view taken at 7:28 p.m. on 17 May 2014 local standard time. The color scale in Figure 3a represents the brightness temperature and a clear depiction of the warmer vegetated canopy (around 7 °C) rows running north to south and a cooler grass surface between the canopy rows with a brightness temperature of around 4 °C. Figure 3b is a derived image that represents the brightness temperature perturbation calculated by de-trending each sampled pixel from the temporal mean over a 10 min period. This statistical quantity represents a perturbation value that clearly shows pixels and regions that are either warmer (positive) or cooler (negative) than their neighbors. The resulting image highlights clouds in the upper sky section of the image that were invisible in Figure 3a, and warmer structures over the canopy. Figure 3c presents a time series of the area-averaged brightness temperature measured by the FLIR and Optris cameras over regions (2) and (3). A time series of the air temperature as measured by the EC-station and the HOBO temperature logger is also added to the figure to relate the brightness temperature to the air temperature as a function of the height of the air temperature measurement and location with respect to the vegetated canopy.

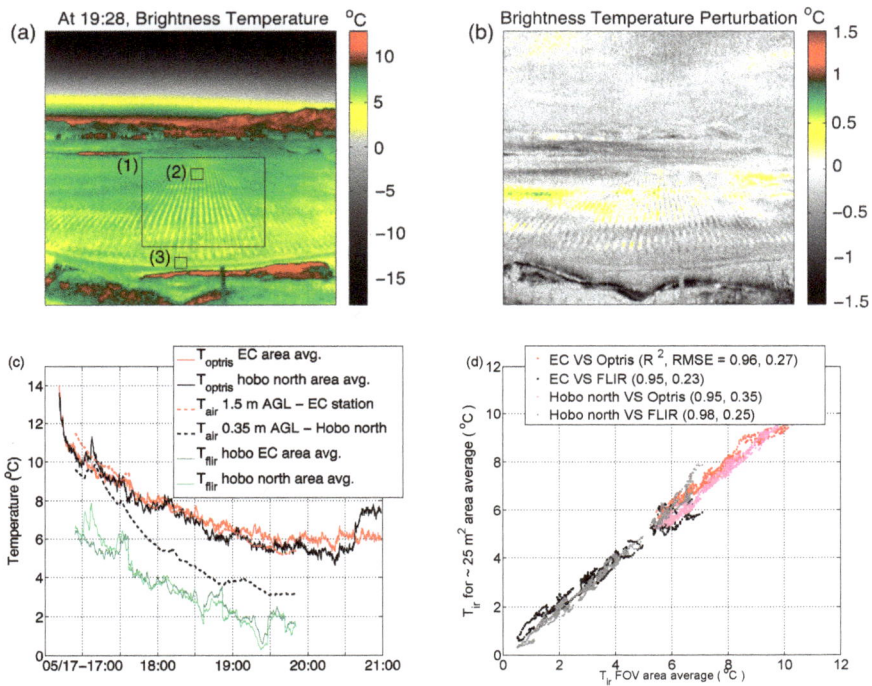

Figure 3. (**a**) A sample snapshot in time of the brightness temperature measured by the Optris camera. Regions (1), (2) and (3) represent the areas from which the mean was calculated from some of the analysis; (**b**) The same snapshot in time as in (**a**) but for the derived perturbation brightness temperature calculated from the deviation of each of the 10 s samples from the 10 min mean. Positive values show areas of increasing temperature in time; (**c**) Time series comparison of brightness temperature from two cameras, air temperature from the eddy covariance station and near-surface Hobo temperature logger; (**d**) Scatter plot of brightness temperature and in situ air temperatures.

3.3. Brightness Temperature and Turbulent Heat Flux

The horizontal and vertical kinematic heat flux components were calculated from the covariance of the horizontal (U, V) and vertical (W) velocity components and the sonic temperature recorded by the sonic anemometer. In Figure 4a the kinematic heat flux is sampled at 10 s periods from collected data at 20 Hz; the results show turbulent horizontal and vertical heat advection over the few hours of the evening when the brightness temperature was sampled via the infrared cameras. Figure 4a shows an initial period of moderate to little turbulent heat flux (5 p.m. to 6:30 p.m.), with an increase of heat flux over the rest of the evening. This result is also supported by an increase in above-canopy wind speeds after 6 p.m. as depicted by the climate station wind speed data in Figure 2b. Figure 4b shows the area-averaged (area (2) in Figure 3a) de-trended brightness temperature from the FLIR and Optris cameras around the EC-station, and the corresponding de-trended air temperatures form the EC-station in the green line.

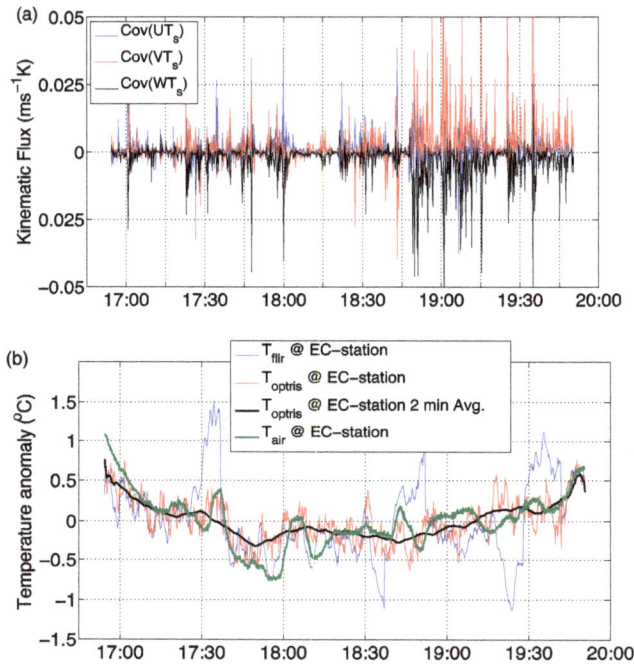

Figure 4. (a) Turbulent kinematic heat flux in the three Cartesian directions (U for east-west, V for north-south, and W in the vertical). Statistically, the heat flux was calculated based on the covariance of the velocity and sonic temperature; (b) Time series of brightness and air temperature at the eddy covariance area-averaged site.

3.4. Self-Organizing Maps (SOMs) of Brightness Temperature

In this section we relate the spatial pattern of the brightness temperature to the meteorological conditions observed at the center of the image via the eddy covariance system. An unsupervised pattern recognition algorithm (SOM: see descriptions in the Methods section) was used to cluster the brightness temperature field into nine different nodes. This method allows us to interpret the brightness temperature field within the right meteorological context when other parameters (such as data from the eddy covariance system) are composited as a function of individual clusters. Figure 5 is the resulting SOM of all of the nearly 1000 images that were taken at a 10 s sampling interval from the Optris infrared camera; the sky brightness temperatures (appearing as a white mask at the top half of the nodes) were not used in the clustering but used as a derived composite variable for further analysis. The variables, which appear in parentheses in Figure 5 and are illustrated in Equations (1) and (2), represent the (A) mean sky brightness temperature perturbation, and the (B) ratio of the mean sky brightness temperature perturbation to the mean of the absolute sum of the three kinematic turbulence heat flux components derived earlier for Figure 4a.

$$A \ (per \ node) = \frac{\sum_{i=1}^{n} \overline{Tb'\_sky}}{n} \tag{1}$$

where n is the number of images per node.

$\overline{Tb'\_sky}$ is the spatial average of the de-trended sky brightness temperature $Tb'\_sky = Tb\_sky - \overline{Tb\_sky}$, where Tb_sky is the instantaneous pixel-based value and $\overline{Tb\_sky}$ is the 10 min average.

$$B \ (per \ node) = \frac{A \ (per \ node)}{|cov(UTs)| + |cov(VTs)| + |cov(WTs)|} \tag{2}$$

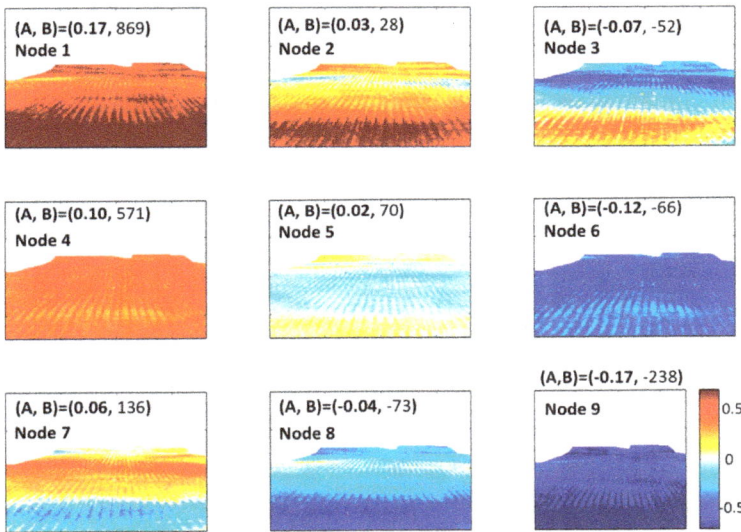

Figure 5. SOM nodes derived from the perturbation brightness temperature. Positive or red values or colors indicate an increase in Tir and negative or blue values or colors indicate a decrease in Tir or surface cooling. The numbers (A, B) on the top of each node are derived from the best matching units of that specific node and are perturbations of (A = average sky brightness perturbation temperature, B = average of the ratio of A by the absolute sum of the three kinematic turbulent heat flux components that were used in Figure 4a).

4. Discussion

The results comparing the brightness temperature measured by the Optris infrared camera and in situ air temperatures show a very good match for the canopy height air temperature measurement and a warm bias for the air temperature measurement at the near-ground level (Figure 3c). The FLIR camera results show a systematic cold bias, with larger temperature oscillations when compared with the Optris, which could be explained by the automatic focusing method employed by the FLIR camera, which tends to periodically auto-sharpen the image, but also could be explained by the need for a camera calibration. Figure 3d shows a correlation diagram between the brightness temperature and the in situ air temperature measurement between the smaller area-averaged regions (1) and (2) and the larger field of view region (3). The results show a good linear correlation and a low value of root mean square error; they also show the cold bias offset previously revealed by the FLIR camera. This results also shows that the one-to-one relationship between the brightness temperature and the in situ air temperature is still preserved while spatially up-scaling the image over a homogenous terrain.

The brightness temperature signal of the Optris camera when compared to the direct measurements of turbulent heat flux (shown in red in Figure 4b) follows the air temperature trend and responds to the cooling and warming period suggested by the heat flux advection. The brightness temperature oscillation range also scales to the ranges shown by the air temperature record. The FLIR brightness temperature trend (shown in blue in Figure 4b) generally follows the initial cooling and then warming cycle but tends to overestimate the range with around five relatively large peaks. These peaks are linked to the automatic focusing of this specific infrared camera. Between 5 p.m. and 6:30 p.m. the brightness and air temperature trend did not exhibit either a positive or negative trend in comparison with the negative (cooling) or warming (positive) trends outside this period. This period also reflects a period of quiescence and little to no turbulent heat flux as shown in Figure 4a.

The unsupervised clustering carried out by the SOM technique in Figure 5 was successful in distinguishing nine clusters that have a meteorological context when compared to in situ measurements. The color in Figure 5 represents the brightness temperature perturbation (red being a warming trend and blue a cooling trend). The SOM shows distinct features that are relatively different among nodes. For example, node 1 shows a field-wide warming trend, while node 9 a field-wide cooling trend. Nodes 4 and 6 show the same extremes but with lower magnitudes, while nodes 3 and 7 show a north to south cooling or warming gradient and appear to represent an opposite brightness temperature gradient. In order to link the SOM patterns to a meteorological context we have composited two different variables (A and B in Figure 5 and Equations (1) and (2)) from the un-clustered data behind the construction of each of the node patterns. The mean sky brightness temperature perturbations, or A (varying between −0.17 and 0.17), correlate well with the brightness temperature trends. Positive values of quantity A (such as in nodes 1 and 4 for example) indicate a warming sky brightness temperature that relates to nocturnal cloud cover, which reradiates long-wave radiation back onto the surface, creating a homogenous spatial brightness temperature positive trend. The opposite applies for clear sky conditions (for example nodes 6 and 9). This result does not apply for nodes 2, 3, 5, 7, and 8 which show an order of magnitude lower mean sky brightness temperature perturbation and higher values of turbulent heat flux, which clearly creates localized and distinct warming and cooling trends within the brightness temperature spatial pattern. The particular patterns shown by nodes 3 and 7 are intriguing, as one could hypothesize that these opposite patterns are related to turbulent advection or mixing events that are happening within this local topographic catchment, especially as these patterns only exist during high turbulent heat flux periods (see quantity B for these nodes in comparison to quantity A for the other nodes).

5. Conclusions

We have demonstrated the use of forward-looking infrared cameras measuring the surface brightness temperature over a vineyard in the spectrum bandwidth of 7.5 to 14 μm at a relatively high temporal rate of 10 s for the application of vineyard-scale micrometeorology. Our results show that this technique, when applied for interpreting the micrometeorology as a function of cloud cover and within-canopy turbulence, could become a useful tool for up-scaling point measurements to spatially wide footprints. The temporal surface brightness signal over a few hours of the stable nighttime boundary layer intermittently interrupted by periods of turbulent heat advection was shown to be related to the atmospheric surface-layer dynamics observed by the eddy-covariance measurements, and reflects the temporal evolution of above-canopy wind variability.

The analysis also introduced the SOM of the spatio-temporal brightness temperature data to reduce the dimensionality of this large dataset, but more importantly to highlight the physical dynamics of nighttime surface brightness temperature over a complex canopy measured by an infrared camera. The resultant spatial clusters were self-organized and compared to the meteorological context they reflected, and the spatial brightness temperature pattern was explained in terms of the presence or absence of nighttime cloud cover and down-welling of long-wave radiation and the canopy turbulence heat flux. Time sequential thermography as demonstrated in this research provides positive evidence behind the application of thermal infrared cameras in the domain of micrometeorology. The results of this experiment could then be used in accordance with the surface renewal theory (which assumes that surface-atmospheric turbulence exchanges are driven by ramp-like structures within the temperature time series), which will eventually allow for a spatial pixel-based derivation of sensible and latent heat flux which are essential for the canopy's water balance during daytime periods [19–22].

There are a couple of limitations to this study that need to be considered when it is applied for more complex terrain. The first limitation comes from the potential effect of air temperature and humidity fluctuations along the camera's line of sight on the interpretation of surface brightness fluctuations. This effect is usually addressed by simple one-dimensional radiative transfer modeling, which delineates the role of infrared signal attenuation. Both of these effects have been previously found to be less than 10% (for atmospheric temperature) and less than 3% (for atmospheric humidity) [8] for an urban setting and results may vary for other applications. The other limitation is the variable image pixel resolution as a function of depth. So the pixels furthest away from the camera have a different pixel resolution than pixels close to the camera. This could be only fixed with ortho-rectification when a high-resolution digital elevation map (DEM) for the site is available. A DEM was not available for this study, and given that the focus of this study was not to study atmospheric turbulence as a function of length scale, we considered that not affecting our major conclusions in this study.

Acknowledgments: The authors would like to thank Stuart Powell and Andrew Hammond for facilitating access and providing field support during the Lions Back vineyard experiment. The authors would also like to thank the anonymous reviewers for their efforts in the review process of this article. This research was supported by the Ministry of Business and Innovation, New Zealand, under research contract number UOWX1401.

Author Contributions: M.K. and P.Z. conceived the research idea, designed and performed the field experiment; M.K. analyzed the data and wrote the paper with contributions from P.Z.

Conflicts of Interest: The authors declare no conflict of interest.

References

1. Costa, J.M.; Grant, O.M.; Chaves, M.M. Thermography to explore plant–environment interactions. *J. Exp. Bot.* **2013**, *64*, 3937–3949. [CrossRef] [PubMed]
2. Araus, J.L.; Cairns, J.E. Field high-throughput phenotyping: The new crop breeding frontier. *Trends Plant Sci.* **2014**, *19*, 52–61. [CrossRef] [PubMed]
3. Katul, G.G.; Schieldge, J.; Hsieh, C.I. Skin temperature perturbations induced by surface layer turbulence above a grass surface. *Water Resour. Res.* **1998**, *34*, 1265–1274.
4. Chazdon, R.L. Sunflecks and their importance to forest understorey plants. *Adv. Ecol. Res.* **1988**, *18*, 1–63. [CrossRef]
5. Paw, U.K.T.; Brunet, Y.; Collineau, S.; Shaw, R.H.; Maitani, T.; Qiu, J.; Hipps, L. On coherent structures in turbulence above and within agricultural plant canopies. *Agric. For. Meteorol.* **1992**, *61*, 55–68. [CrossRef]
6. Hoyano, A.; Asano, K.; Kanamaru, T. Analysis of the sensible heat flux from the exterior surface of buildings using time sequential thermography. *Atmos. Environ.* **1999**, *33*, 3941–3951. [CrossRef]
7. Voogt, J.A. Assessment of an urban sensor view model for thermal anisotropy. *Remote Sens. Environ.* **2008**, *112*, 482–495. [CrossRef]
8. Christen, A.; Meier, F.; Scherer, D. High-frequency fluctuations of surface temperatures in an urban environment. *Theor. Appl. Climatol.* **2012**, *108*, 301–324. [CrossRef]
9. Garai, A.; Kleissl, J. Interaction between coherent structures and surface temperature and its effect on ground heat flux in an unstably stratified boundary layer. *J. Turbul.* **2013**, *14*, 1–23. [CrossRef]
10. Inagaki, A.; Kanda, M.; Onomura, S.; Kumemura, H. Thermal Image Velocimetry. *Bound.-Layer Meteor.* **2013**, *149*, 1–18. [CrossRef]
11. SOMPAK. Available online: http://www.cis.hut.fi/research/som-research/nnrc-programs.shtml (accessed on 16 September 2016).
12. Katurji, M.; Noonan, B.; Zawar-Reza, P.; Schulmann, T.; Sturman, A. Characteristics of the springtime alpine valley atmospheric boundary layer using self-organizing maps. *J. Appl. Meteorol. Climatol.* **2015**, *54*, 2077–2085. [CrossRef]

13. Sheridan, S.C.; Lee, C.C. The self-organizing map in synoptic climatological research. *Prog. Phys. Geogr.* **2011**, *35*, 109–119. [CrossRef]
14. Cassano, E.N.; Lynch, A.H.; Cassano, J.J.; Koslow, M.R. Classification of synoptic patterns in the western Arctic associated with extreme events at Barrow, Alaska, USA. *Clim. Res.* **2006**, *30*, 83–97. [CrossRef]
15. Cassano, J.J.; Uotila, P.; Lynch, A.H.; Cassano, E.N. Predicted changes in synoptic forcing of net precipitation in large Arctic river basins during the 21st century. *J. Geophys. Res.* **2007**, *112*, G04S49.
16. Cassano, E.N.; Cassano, J.J. Synoptic forcing of precipitation in the Mackenzie and Yukon River basins. *Int. J. Climatol.* **2010**, *30*, 658–674. [CrossRef]
17. Nigro, M.; Cassano, J.J.; Seefeldt, M.W. A weather pattern-based approach to evaluate the Antarctic Mesoscale Prediction System (AMPS) forecasts: Comparison to automatic weather station observations. *Weather Forecast.* **2011**, *26*, 184–198. [CrossRef]
18. Richardson, A.J.; Risien, C.; Shillington, F.A. Using self-organizing maps to identify patterns in satellite imagery. *Prog. Oceanogr.* **2003**, *59*, 223–239. [CrossRef]
19. Suvocarev, K.; Shapland, T.M.; Snyder, R.L.; Martínez-Cob, A. Surface renewal performance to independently estimate sensible and latent heat fluxes in heterogeneous crop surfaces. *J. Hydrol.* **2014**, *509*, 83–93. [CrossRef]
20. Paw, U.K.T.; Qiu, J.; Su, H.B.; Watanabe, T.; Brunet, Y. Surface renewal analysis: A new method to obtain scalar fluxes without velocity data. *Agric. For. Meteorol.* **1995**, *74*, 119–137.
21. Katul, G.; Hsieh, C.; Oren, R.; Ellsworth, D.; Philips, N. Latent and sensible heat flux predictions from a uniform pine forest using surface renewal and flux variance methods. *Bound.-Layer Meteor.* **1996**, *80*, 249–282.
22. Snyder, R.L.; Spano, D.; Pawu, K.T. Surface renewal analysis for sensible and latent heat flux density. *Bound.-Layer Meteor.* **1996**, *77*, 249–266. [CrossRef]

sensors

MDPI

Article

Test of the Practicality and Feasibility of EDoF-Empowered Image Sensors for Long-Range Biometrics

Sheng-Hsun Hsieh [1], Yung-Hui Li [2],* and Chung-Hao Tien [1]

[1] Department of Photonics, National Chiao Tung University, 1001 University Road, Hsinchu 30010, Taiwan; jack10313.eo00g@nctu.edu.tw (S.-H.H.); chtien@mail.nctu.edu.tw (C.-H.T.)

[2] Department of Computer Science & Information Engineering, National Central University, No. 300 Zhongda Road, Zhongli District, Taoyuan 32001, Taiwan

* Correspondence: yunghui@csie.ncu.edu.tw; Tel.: +886-3-422-7151 (ext. 35204)

Academic Editor: Gonzalo Pajares Martinsanz
Received: 28 September 2016; Accepted: 18 November 2016; Published: 25 November 2016

Abstract: For many practical applications of image sensors, how to extend the depth-of-field (DoF) is an important research topic; if successfully implemented, it could be beneficial in various applications, from photography to biometrics. In this work, we want to examine the feasibility and practicability of a well-known "extended DoF" (EDoF) technique, or "wavefront coding," by building real-time long-range iris recognition and performing large-scale iris recognition. The key to the success of long-range iris recognition includes long DoF and image quality invariance toward various object distance, which is strict and harsh enough to test the practicality and feasibility of EDoF-empowered image sensors. Besides image sensor modification, we also explored the possibility of varying enrollment/testing pairs. With 512 iris images from 32 Asian people as the database, 400-mm focal length and F/6.3 optics over 3 m working distance, our results prove that a sophisticated coding design scheme plus homogeneous enrollment/testing setups can effectively overcome the blurring caused by phase modulation and omit Wiener-based restoration. In our experiments, which are based on 3328 iris images in total, the EDoF factor can achieve a result 3.71 times better than the original system without a loss of recognition accuracy.

Keywords: wavefront coding; extended depth of field; iris recognition; biometrics

1. Introduction

Biometric recognition has been applied to many practical uses, including homeland security, e-commerce or other authentication management purposes. Basically, the personal attributes used for authentication were classified into two parts: (1) physiological attributes, such as DNA, facial features, retinal vasculature, fingerprint, hand geometry, iris texture and so on; and (2) individual behavior features, such as signature, keystroke, voice, and gait style [1]. Among these features, iris texture is one of the most attractive modalities because of its inherent distinctiveness, high stability over time and low risk of circumvention [2].

An iris recognition system consists of modules of the imaging optics unit, the image processing unit and the feature matching unit, as shown in Figure 1. The optical system, involving the camera and the irradiance, is used to capture a distant iris image with the highest fidelity possible. The captured images are subsequently processed through many steps. Firstly, the iris images are segmented by determining the centers and radii of the pupillary and limbic boundaries. A conventional segmentation method, such as an integro-differential operator [2–4] or Hough transform [4,5], can be applied. Then the iris images are normalized by transforming the coordinates from Cartesian to Polar accordingly. The prominent features of the iris texture are extracted using Gabor filters. Finally, the features are thresholded into binary codes (called iris codes) for the recognition algorithm [2–4]. Matching two iris codes using the bit-wise XOR operation generates a distance score. The distance score, Hamming Distance (HD), is employed to measure the distance between two iris codes. An appropriate threshold value of HD is determined so that a decision of acceptance or rejection can be made. For example, two iris images are said to be independent if their HD is above a certain threshold, which is about 0.33 according to Daugman's algorithm [6]. Otherwise they are assumed to be a match.

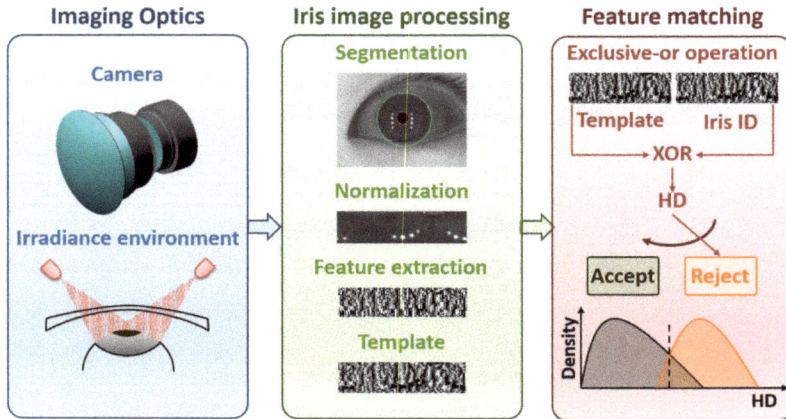

Figure 1. An iris recognition system is composed of the imaging optics unit, the iris image processing unit and the feature matching unit, respectively.

For the practical scenario in iris recognition, the acquisition volume, which is defined as the depth of field (DoF), should be large enough to preserve the high reliability and robustness of the system. Imaging optics with sufficient DoF while preserving satisfactory spatial resolution is highly desirable. The conventional approach to increase the DoF is to increase the F-number, which corresponds to using a smaller aperture or longer focal length. However, both scenarios have a side effect. A smaller aperture would lead to a poor optical throughput, and thus a low signal-to-noise ratio; a longer focal length would reduce the field-of-view (FoV), thereby adversely affecting the resolution of the system. Computational imaging proposed by Dowski and Cathey engineered the pupil function to resolve this dilemma in a successful way. After that, many studies applied the coded image for iris recognition and extended the acquisition volume without loss of recognition accuracy [7–11]. To our understanding, numerous previous studies were addressed by the simulation, where the phase mask is assumed to be on the pupil plane exactly. However, for most practical uses, the pupil plane is unreachable by end users because it is hidden inside a complex optomechanical layout. Meanwhile, imperfect irradiances such as glare reflection, non-uniform distribution and brightness level should be considered as well.

In this paper, we implemented the wavefront coded iris recognition system starting from the acquisition optics to the final score-matching stage. We experimentally compared the recognition performance with different enrollment/testing schemes. The results offer some insight for utilizing the wavefront coding image to provide maximal allowable DoF while maintaining the high recognition accuracy.

The remainder of this paper is organized as follows. Section 2 introduces the optical consideration and the corresponding terminologies. In Section 3, an extended depth of field (EDoF) system for iris recognition is implemented [12]. In Section 4, the experimental results are examined in terms of various figure of merits, including equal error rate (EER) and HD distributions. In Section 5, discussions on homogeneous or heterogeneous iris recognition are carried out to explore the performance difference between different setups. Section 6 concludes the paper.

2. Optical Consideration

2.1. Tradeoff between Resolution and Field-of-View

The major challenge of a prime lens for iris recognition lies in a constant acquisition volume (which can be expressed as resolution × FoV). The iris images need sufficient sampling resolution to ensure recognition performance. At the same time, the FoV should be wide enough to cover the entire ocular region and localize the facial landmarks. ISO/IEC 19794-6 suggests that the sampling rate across the iris region should exceed at least 150 pixels so as to contain sufficient features [13]. For an image sensor with a pixel size d, the minimum width of iris images D_1' is given by:

$$D_1' = 150 \times d, \tag{1}$$

With average width of an adult's iris $D_1 = 12$ mm [14], the magnification m of the camera can be obtained as:

$$m = \frac{D_1'}{D_1} \geq \frac{150 \times d}{12}, \tag{2}$$

The effective focal length f of a camera is related to the magnification m and object distance S_o [15]:

$$f = \frac{m}{1+m} S_o, \tag{3}$$

For the sensor pixel size $d = 8$ μm in our case, by Equations (2) and (3), the magnification $m \geq 0.1$ and focal length $f \geq 0.09\,S_o$, which defines the lower bound in terms of resolution. As shown in Figure 2, in case of $S_o = 3$ m working distance, the available focal length should be larger than 272 mm.

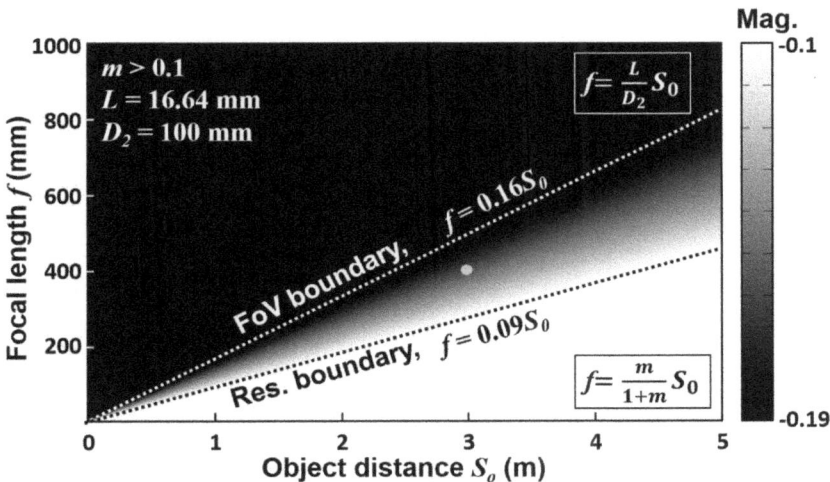

Figure 2. The focal length f is constrained by two boundaries for the resolution and the field of view. For object distance $S_o = 3$ m, the available focal length is in range of $f = 272$–499 mm. In this study, the focal length is set to 400 mm and illustrated as the orange point in this figure.

On the other hand, another boundary of focal length was defined by the FoV, which is given by:

$$\text{FoV} = 2tan^{-1}\left(\frac{D_2}{2S_0}\right) = 2tan^{-1}\left(\frac{L}{2S_i}\right), \tag{4}$$

where D_2 is the width of full ocular region. The ratio of D_2/S_o in object space is equivalent to the L/S_i is image space, where L is the full size of an image sensor and S_i is the image distance. For the distant imaging, the paraxial approximation holds that $S_i \sim f$, Equation (4) can be rewritten as:

$$f = \frac{L}{D_2}S_o, \tag{5}$$

Since the FoV should be large enough to encompass the entire ocular region, with typical size of an adult's ocular region $D_2 = 100$ mm and the available sensor diameter $L = 16.64$ mm, Equation (5) defines the upper boundary, $f = 0.16S_o$, as shown in Figure 2. For the object distance $S_o = 3$ m, the available focal length should be smaller than 499 mm accordingly. Taking both resolution and FoV into account, we employed a commercial telephoto lens, Sigma APO, with 400-mm effective focal length. Detailed specifications of the image sensor and lens set are listed in Table 1.

Table 1. Specification of image sensor and telephoto lens set.

MV1-D2080 IR Sensor		Sigma APO 150–500 mm	
Optical Format	23.5 mm	Field of View	5–16 degrees
Resolution	2080 × 2080	Minimum Distance	220 cm
Pixel Size	8 μm	Maximum Mag.	1:5.2
Dark current	0.65 fA/pixel	Caliber Diameter	86 mm

2.2. Depth of Field

Figure 3 illustrates the concept of DoF, which is marked as dotted zone on the left. When the subject is out of DoF, the point spread functions (PSFs) of the imaging system would increase by the path-length error. The most common merit to evaluate the defocus extent is the circle of confusion (CoC), which is defined as the largest blur PSFs indistinguishable from two distant point sources. For computational imaging with the aid of post-processing, currently there is no universal definition for CoC in optics. In our work, we defined CoC as the maximally allowable iris blurring with acceptable recognition performance [7]. Under the paraxial approximation, the imaging condition at the near and far limits of the DoF (D_N and D_F) can be described as:

$$\frac{1}{S_i} + \frac{1}{S_o} = \frac{1}{f}, \tag{6}$$

$$\frac{1}{S_i(1 + \frac{C}{P-C})} + \frac{1}{S_o - D_N} = \frac{1}{f}, \tag{7}$$

$$\frac{1}{S_i(1 + \frac{C}{P+C})} + \frac{1}{S_o + D_F} = \frac{1}{f}, \tag{8}$$

where P and C are the diameter of the pupil and CoC, respectively. According to Equations (6)–(8), the DoF of an imaging system can be obtained as:

$$D_N = \frac{CS_o(S_o - f)}{fP + C(S_o - f)}, \tag{9}$$

$$D_F = \frac{CS_o(S_o - f)}{fP - C(S_o - f)}, \tag{10}$$

$$DoF = D_N + D_F = \frac{2CS_o}{fP/(S_o - f) - C^2(S_o - f)/fP}, \tag{11}$$

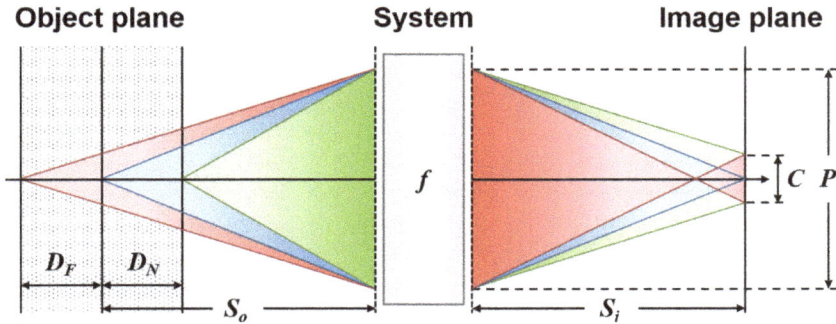

Figure 3. Schematic illustration of the DoF (dotted zone) in an imaging system. DoF is determined by four factors: circle of confusion C, exit pupil P, object distance S_o and focal length f, respectively.

In our case with P = 86 mm, f = 400 mm and C = 0.136 mm, the DoF is merely 60 mm, which is too shallow to be operated in a robust way. Motion blur inevitably occurs if users are allowed to move or walk, like the use case reported in [12]. One scheme is to increase the shutter speed or F-number by sacrificing the image brightness, which degrades the image quality with a low signal-to-noise ratio. In this study, we resort to the computational image, which cleverly enlarges the acquisition volume without any possible thermal hazard or glare reflection by strong irradiance.

2.3. Irradiation Condition

The performance of an iris recognition system depends greatly on captured image quality. Without cooperation of the subject, image quality is subject to many factors like the low contrast, the inconsistent illumination or the specular reflection. The low contrast is due to the reason that human iris has lower reflectance under visible light but higher reflectance under near infrared (NIR) light [16]. To overcome this issue, we equipped two LED illuminators (BE-IR80L, BlueEyes Technology, TW), which have 850 nm central wavelength and 50 nm full width at half maximum (FWHM). The average irradiance in continuous mode is about 13 mW/cm^2 to acquire enough information. We set the ISO value of the camera so that the iris image is dark when the NIR LEDs were in off state. Therefore, no NIR pass filter is needed. To take the specular reflection into consideration, we set the incident angle from the illuminators to 35 degrees. When the subject is at 70 cm from the illuminators, such geometry can avoid strong specular reflection even when the subjects wear glasses. An adequate irradiance setup can enhance the probability of the correctness of iris segmentation [17].

3. Method to Extend the Depth-of-Field

As the preceding discussion (Section 2.2) states, an iris image with insufficient DoF would inevitably cause motion blur and reduce recognition accuracy. Although decreasing the aperture size is the easiest way to alleviate the phase degradation which is quadratically proportional to the pupil size, smaller apertures will be accompanied by insufficient irradiance and low signal-to-noise ratio. In order to overcome this issue, we applied computational imaging techniques to extend DoF. Computational imaging integrating optics with post signal processing can keep the PSF more robust toward defocusing. Two issues are addressed about the coding scheme. One is to find the appropriate function of the phase mask on the basis of merit. The other is about the coding strength of the phase mask. In addition to coding strategy, decoding is a counterpart issue about the performance of recognition. The intermediate iris image with blurriness undergoes an approximately linear transformation, which ideally can be reversed by a linear reconstruction. A matched filter thus can fully restore the original iris texture. However, noise prevents the reconstruction from being perfect in practical case. In the second part of this section, we examine the decoding scheme coupled with matching algorithm. The following issues are addressed in this section: function of phase modulation with respect to optical transfer function (OTF) analysis; optimal coding strength, which is a tradeoff between increasing defocus insensitivity and loss of information; and the restoration process with optimal matched filter design parameters.

3.1. Wavefront Coding

Generally, there are two strategic approaches for phase modulation. One is to use a free-form phase plate, whose distribution is expressed as a polynomial expansion [18]:

$$\varphi(x,y) = \sum_{m=0}^{k} \left(\sum_{m=0}^{n} C_{nm} x^n y^{n-m} \right),$$ (12)

where C_{nm} are a set of coefficients that will be determined by the optimization algorithm to balance the factor of EDoF and zero nulls over a broad band range of DoF. Such free-form phase masks have circular symmetric OTFs. The major challenge for the free-form phase plate lies in its fabrication tolerance. More than 10 dominant coefficients in shape formulation would result in difficulties with fabrication [19]. Meanwhile, tilt or alignment error would drastically reduce the performance in an unexpected way. In contrast, the more popular scheme in phase coding is to use a separable function like the cubic phase form $P(x,y)$ in the rectangle coordinate [20]:

$$P(x,y) = exp \left[i\alpha(x^3 + y^3) \right],$$ (13)

where x and y are the normalized pupil coordinates. The phase coding strength, α, is determined by the numerical evaluation. In our study, we chose cubic phase form to be our candidate because the mask is easier to be fabricated and implemented in an iris recognition system.

We utilized the optical software Zemax™ to compromise the coding strength of the cubic phase mask and a quadratic defocus term $W_{02}(x^2 + y^2)$ [20]. Unlike the conventional approach which finds the coding strength based on diffraction-limited OTF in simulation, we set the PSF similarity as the merit function and find its mean-square-error (MSE) through the focus range. With different coding strengths, the PSF similarity and its derivative with respect to defocus provide insight into the optical layout that we could conduct in optical design. It should be noticed that the OTF of the coded system cannot cross zero, because the null point in the OTF will lead to permanent loss of information which cannot be restored by post-processing [20–22]. The worse situation is that when the strong coding is imposed, the negative value (contrast reversal) occurred. In order to keep the system within a safe margin, we allow OTF threshold at the Nyquist frequency to be larger than 0.169, which ensures most information is above the noise floor and thus well recoverable [23,24]. With an off-the-shelf telephoto lens system with focal length f = 400 mm (F/6.3), the maximum value of α is 42, which enables a three-fold DoF. The feasibility of EDoF was convincingly demonstrated in our simulation which coincides with the prior literatures [25–27].

3.2. Restoration Decoding Process

The coded PSFs are restored by the Wiener filter, which is one of the best known approaches to linear image restoration [28,29]. The Wiener filter expressed in Fourier domain (u, v are spatial frequencies in x and y direction, respectively) can be formulated as:

$$\hat{F}(u,v) = \left[\frac{1}{H(u,v)} \frac{|H(u,v)|^2}{|H(u,v)|^2 + S_\eta(u,v)/S_f(u,v)} \right] G(u,v),$$ (14)

where $H(u,v)$ is the coded transfer function, $G(u,v)$ is the intermediate iris image. The ratio $S_\eta(u,v)/S_f(u,v)$ is the noise-to-signal ratio (NSR) of the imaging system, where $S_\eta(u,v)$ and $S_f(u,v)$ is the power spectrum of the noise and the ideal image, respectively. Generally, the NSR of an imaging system is unknown, and it can only be obtained empirically. We conducted a preliminary test to fine-tune the NSR parameter (denoted as R) used in Wiener filtering.

Iris images of 64 subjects are collected and inversely filtered by Wiener filtering with different R values. Then those iris images are used as probe images to match with the iris images in gallery (iris images captured at on-focus position). In principle, Wiener filtering should not affect the inter-class iris matching scores since they are intrinsically different. For the purpose of decreasing computational complexity, we only consider the intra-class comparisons. By accumulating all HD of intra-class comparisons, we obtained the relation between the averaged HD and parameter R, as shown in Figure 4. Since HD indicates the distance of two iris images, by locating the minimum of the curve, we are able to estimate the optimal parameter R that leads to the best inverse filtering performance. The optimal R is found near 0.15 in our case.

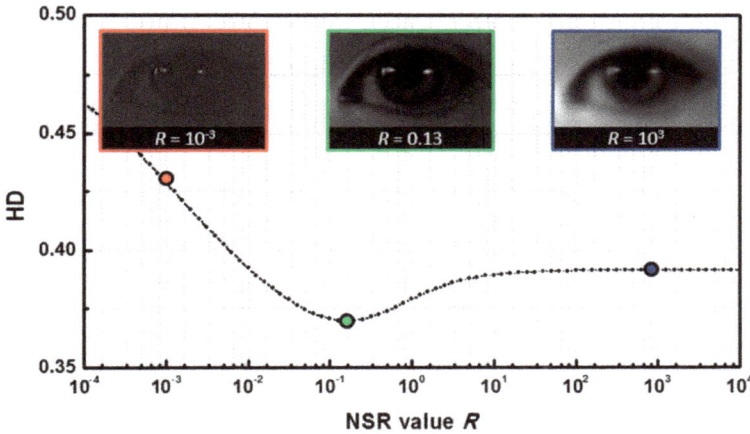

Figure 4. Hamming distance (HD) with different parametric estimation R in the Wiener filter, where the HD are averaged based on 64 intra-class comparisons. The optimal R is about 0.15.

4. Laboratory Experimentation

4.1. Optical Quality

We embedded a cubic phase mask with optimized coding strength $\alpha = 42$ into the off-the-shelf telephoto camera, as shown in Figure 5, where the phase mask was fabricated by the diamond turning process. We implemented wavefront coding by putting the cubic phase mask at the rear space of the system. The influence of mask displacement away from the focus has been examined by a series of testing in our past research work. Interested readers can refer to our previous research [30]. Figure 6 shows the PSFs across a range of object distances from −18 to 18 cm. It is apparent that the cubic mask helps to reduce the spreading of PSFs. When the FWHM of PSF is comparable to the size of CoC, the object defocus corresponds to −30 mm and +40 mm, which is very close to the theoretical prediction of 60 mm in Section 2.2.

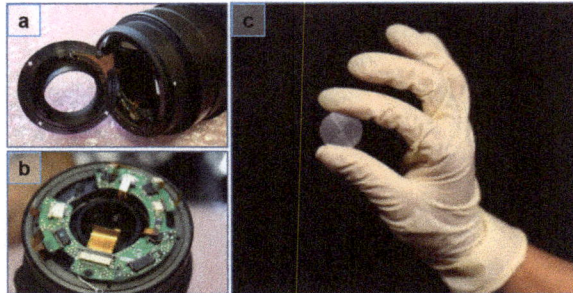

Figure 5. Cubic phase mask with optimized coding strength $\alpha = 42$ was imbedded into the off-the-shelf telephoto camera, where the mask was placed at the rear space of the system. (**a**) The off-the shelf telephoto lens; (**b**) the mask holder; (**c**) the cubic phase mask.

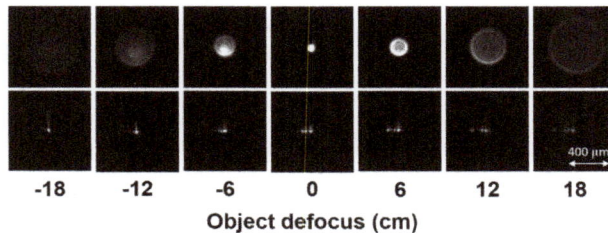

Object defocus (cm)

Figure 6. PSFs with different defocus position for top row: conventional, and bottom row: EDoF. Compared with conventional optics whose PSFs are quadratically broadened by defocus, EDoF enables PSFs to be more robust against the defocus.

4.2. Image Quality

Figure 7 shows a series of iris images captured with respect to different object distances. Compared with the conventional image (left column), wavefront coding with (middle column) or without (right column) Wiener filtering effectively kept the iris image insensitive to the defocus effect. From the right column of Figure 7, EDoF with Wiener filter restoration manifested the detailed texture of iris image. However, the fidelity of the restored image was deteriorated by the artificial and ringing effect, respectively. The artificial effect was due to the wavefront error caused by the cubic phase mask. In an imaging optics, it is obvious that slight rotation and displacement of the phase components in a real system would induce the changes of phase error with respect to the focus [30].

The second factor is the ringing effect caused by Wiener filtering itself. As a phenomenon already presented in the literature [31,32], when images were restored by either the inverse linear filtering or the Wiener filtering, there would be a certain amount of noticeable edge error. For the inverse linear filtering, coded transfer function occurring to be zeros at high frequency caused singularities. For the Wiener filtering, though the above problem was solved by replacing the singularities of the inverse filter at zeros, there also existed edge error, as discussed in [30]. An optimization process on parameter R of the Wiener filtering may adequately reduce the edge error to some degree, but it is virtually impossible to remove all of it. In addition, even when the R value has been fine-tuned, the Wiener filtering could still cause a smearing effect near the center of the restored frequency spectrum, resulting in a reduction of the image's resolution. Such a problem can be lessened or solved by further modification of the Wiener filtering, for example, using the method proposed in [32]. However, in order to restrict ourselves to focus on the main topic of this paper, we did not perform further analysis in this research direction. Detailed features were further examined on normalized iris image and iris code, as shown in Figure 8. The dissimilar iris code after Wiener filtering (bottom row) revealed that the restoration process was vulnerable to artificial noise and leads to increasing HDs.

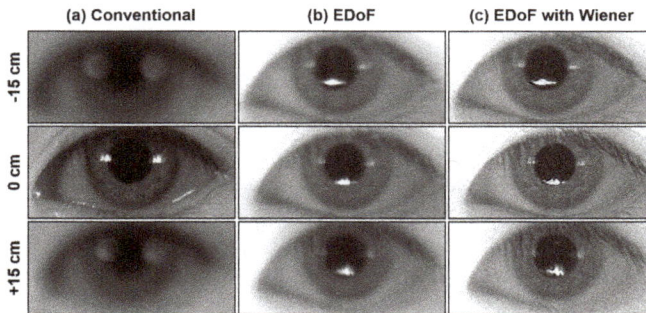

Figure 7. Iris image (640 × 480p) captured with different scenarios. From left to right: (**a**) conventional; (**b**) EDoF; and (**c**) EDoF with Wiener filtering. From top to bottom, the object distances are set to: (1) −15 cm; (2) on focus; and (3) +15 cm, respectively.

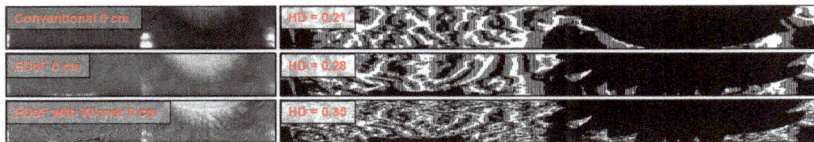

Figure 8. The intermediate images in stage of iris normalization and feature extraction (of images shown in Figure 7). The HD increased when wavefront coded image was used. The HD further increased when the wavefront coded image is restored using Wiener filtering.

4.3. Database

We collected iris images from 64 subjects (32 persons × 2 eyes) in National Chiao Tung University, Taiwan. Each subject stood at the on-focus position eight times, where the on-focus raw images and wavefront coded images were used as the different enrollment data. The total number of enrollment images was 512. Each subject stood at 11 defocus positions (from −15 to +15 cm, at 3-cm intervals) and was captured by both conventional and wavefront coded system. The total number of both conventional and wavefront coded probe images was 2816. The iris images were manually segmented and iris masks were also manually created. We used Libor Masek's iris recognition toolbox written in Matlab for iris feature extraction [33], which used 1D Log-Gabor filters for iris feature extraction. After the iris codes were extracted, we computed normalized HD as described in Section 1. Each testing iris code was compared with the enrollment data. After all the possible combination comparison finished, we plotted the HD distributions for evaluation.

4.4. EDoF Performance Evaluation Method

Because we aimed to extend the DoF without compromising the iris recognition performance, the extension factor of DoF was defined as the longest object distance that can be achieved under the same error rate (i.e., accuracy invariance). Four error rates were used to examine the recognition performance: (1) false acceptance rate (FAR): the probability of falsely accepting an impostor as an authentic sample; (2) false rejection rate (FRR): the probability of falsely rejecting an authentic sample as an impostor sample; (3) equal error rate (EER): the value when FAR is equivalent to FRR; and (4) sensitivity index (SI): a measure to describe the separability between scores of authentic and impostor distributions. Sensitivity index is represented as follows.

$$\text{SI} = \left(m_{im.}^2 - m_{au.}^2\right) / \sqrt{\left(\sigma_{im.}^2 - \sigma_{au.}^2\right)/2} \tag{15}$$

where m and σ are the mean and standard deviation, respectively. Afterward, we chose EER as the evaluation metric with different enrollment/testing schemes because EER is less sensitive to outliers.

4.5. Different Enrollment/Testing Schemes

The conventional scheme of an iris recognition system is to use the clear (on-focus) iris raw image as the enrollment data. During the testing stage, wavefront coded iris images were used as testing data. Such a scheme can be called heterogeneous matching. In this study, we would like to design a series of enrollment/testing pairs to test the feasibility of the combination of homogeneous and heterogeneous matching. A total of six approaches were carried out to inspect two issues: (1) whether the DoF of the iris recognition system can be effectively increased by employing the wavefront coded images as the enrollment data; and (2) among the various approaches, which (homogeneous or heterogeneous matching) approach is the best to balance the recognition accuracy and extended the DoF.

4.5.1. Approach 1: Raw/Raw Pair

Approach 1 is the conventional imaging, where both enrollment and testing are not coded. The gallery images (i.e., enrollment) were iris images captured at on-focus position, while the probe images (i.e., testing) were iris images captured with various object distances. It can be considered a practice of "homogeneous matching". Figure 9a shows the HD distribution when the subject stood on focus, where SI was 4.3, FRR was 4.7% when FAR was 0.1%, and EER was 1.0%. Blue and red bars represent the distribution of the HD of the authentic and impostor matching results, respectively. Such results were reasonable based on the small number of test subjects. For example, in the result of the Multiple Biometric Grand Challenge (MBGC) 2009 version 2, it reports that the best four groups had FRR ranging from roughly 10% when FAR was set to 0.1% [34]. Such great results were computed based on a dataset which consists of 4789 right iris and 4792 left iris images of 136 subjects. With much less iris data, our recognition rate outperformed theirs. In this way, the quality of our iris recognizer was assured, since our recognition results were comparable to the best four groups from the MBGC 2009 version 2.

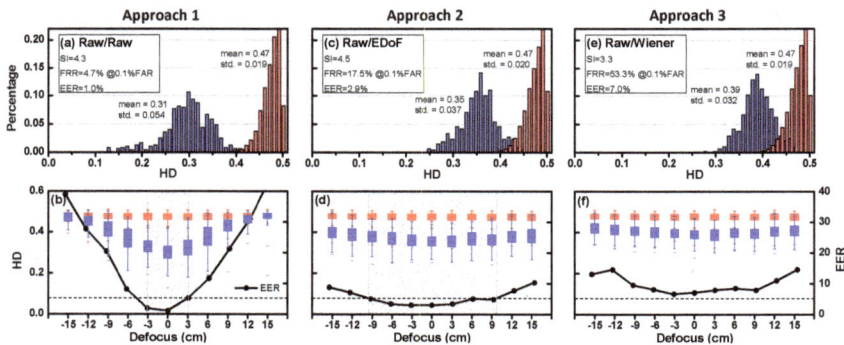

Figure 9. Experimental results, where the iris database (gallery images) is enrolled by the conventional optics, with testing images (probe images) captured by different schemes. **Top** row: HD Histogram distribution at on-focus: (**a**) conventional (Approach 1); (**c**) EDoF (Approach 2); and (**e**) EDoF with Wiener filtering (Approach 3). **Bottom** row: HD distribution (shown as the multiple boxplots colored in **red** and **blue**) and EER (shown as the **black** dotted curve) with different defocus: (**b**) Approach 1; (**d**) Approach 2; and (**f**) Approach 3. When EER was set to 5.2% as the baseline, Approach 2 extends the depth of field about 3.07-fold, whereas no improvement by Wiener filtering (Approach 3).

Figure 9b shows HD versus defocus. The boxplot represents the first quartile to third quartile of the data, while the five error bars from the top to the bottom represent the maximum, 99%, median, 1 percent and minimum values of the data. The HD of authentic matching rapidly increased as the subject was out of DoF, whereas the impostor matching was kept at a high value. The increasing authentic HD was due to the quality heterogeneity of iris images with defocus effect. The trend of EER was in close agreement with the theoretical prediction in Section 2.2. If we set ±30-mm as the DoF, the EER = 5.2% was defined as the baseline for the following comparison.

4.5.2. Approach 2: Raw/EDoF Pair

Approach 2 was the case where the wavefront coded images were captured as probe images. The gallery images were the same as Approach 1. It can be considered = a practice of "heterogeneous matching". Figure 9c shows the authentic and impostor HD histogram when the subjects were on focus. The SI was 4.5, FRR was 17.5% when FAR was 0.1%, and EER was 2.9%. As expected, the performance of recognition at the best focus was poorer than the conventional one due to a prior phase modulation.

Figure 9d shows the authentic and imposter HD with defocus. Compared to conventional imaging (Approach 1), EERs were increasing less rapidly with respect to increasing defocus. With the same merit in terms of EER was 5.2%, the DoF by wavefront coding was extended by a factor of 3.07. Such results validated the feasibility of DoF theory in Section 2.2, where the extended factor could be higher if the lower F-number optics were used.

4.5.3. Approach 3: Raw/Wiener Pair

In this approach, we aimed to examine the performance of restoration of coded iris images (probe images). The gallery images were the same as Approach 1, and the probe images were coded iris images after Wiener filtering. It can be considered another practice of "heterogeneous matching". Figure 9e shows the authentic and impostor HD histograms. The SI was reduced to 3.3, FRR was 53.3% when FAR was 0.1%, and EER was 7.0%. The large amount of overlapping would prevent the practical use of the system in most recognition requests.

Figure 9f shows that high EER over the capture zone causing the system to be highly unstable. Some of the literature claimed that a perfect digital filter had the capacity to restore the coded iris image over an extended DoF without adversely affecting the recognition accuracy. Unfortunately, in this study, no improvement was observed compared to the conventional iris imaging system (Approach 1) as well as wavefront coding image without the restoration (Approach 2).

4.5.4. Approach 4: EDoF/EDoF Pair

Iris recognition relies heavily on the correct feature matching of the iris codes between enrollment and test images. The failure of Approach 3 inspired us to investigate the performance of recognition with a new EDoF enrollment. In this approach, both the gallery images and the probe images were iris images acquired by a wavefront coded image without any restoration process. The difference between them lies in that gallery images were captured in focus, while the probe images were captured at various defocus positions. It can be considered another practice of "homogeneous matching".

Figure 10a shows the authentic and impostor HD histogram on focus. The SI was 4.0, FRR was 23.0% when FAR was 0.1%, and EER was 3.0%. Figure 10b shows authentic and impostor HD versus defocus range. With the same baseline (EER = 5.2%), the DoF was extended by a factor of 3.71. Surprisingly, the extended factor was higher than Approach 2. Compared with different types of the gallery and the probe images (i.e., heterogeneous pair) in Approach 2, the recognition rate was improved by the same types of gallery and the probe images (i.e., homogeneous pair).

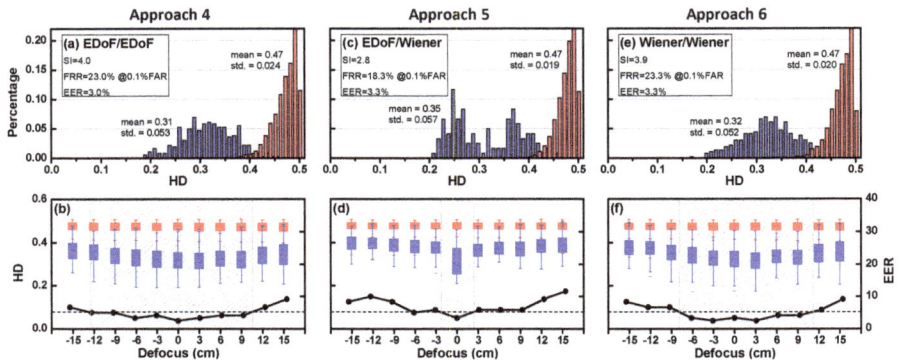

Figure 10. Experimental results, where the iris database (gallery images) is enrolled by the EDoF image (Approach 4 and 5); and EDoF+Wiener image (Approach 6), respectively; For different enrollment/testing pairs, the homogeneous imaging pairs (Approaches 4 and 6) is superior to heterogeneous one (Approach 5) in terms of recognition rate. **Top** row: HD Histogram distribution at on-focus: (**a**) EDoF/EDoF (Approach 4); (**c**) EDoF/Wiener (Approach 5); and (**e**) Wiener/Wiener (Approach 6). **Bottom** row: HD distribution (shown as the multiple boxplots colored in **red** and **blue**) and EER (shown as the **black** dotted curve) with different defocus: (**b**) Approach 4; (**d**) Approach 5; and (**f**) Approach 6. Compared to the conventional optics with EER was set to 5.2%, the DoF was extended by a factor of 3.71 (Approach 4) and 3.10 (Approach 6), respectively.

4.5.5. Approach 5: EDoF/Wiener Pair

For the sake of completeness of this study, we also examined the recognition performance through the Wiener filtering with EDoF enrollment. In this approach, the gallery images were the same as in Approach 4, while the probe images were coded iris images with restoration by Wiener filtering. It can be considered as another practice of "heterogeneous matching". Figure 10c shows the authentic and impostor HD histogram. The SI was 2.8, FRR was 18.3% when FAR was 0.1%, and EER was 3.3%. High EERs showed that the iris codes were dramatically changed by the noticeable artifacts from Wiener filtering. Therefore, Approach 5 was not suggested for the purpose of iris recognition.

4.5.6. Approach 6: Wiener/Wiener Pair

In the last scheme, both the gallery and the probe images were coded iris images restored by Wiener filtering. It can be considered another practice of "homogeneous matching". We aimed to check whether the homogeneous acquisition scheme (EDoF with Wiener filtering) in both enrollment and testing would alleviate the side effects caused by the Wiener filtering. Figure 10e shows the authentic and impostor HD histogram. The SI was 3.9, FRR was 23.3% when FAR was 0.1%, and EER was 3.3%. Figure 10f shows authentic and impostor HD versus defocus range.

For wavefront coded image, the performance of homogeneous pairs (Approaches 4 and 6) were better that of heterogeneous pairs (Approach 5). Compared to Approach 4, the global distribution of the HD in Approach 6 showed a decreasing trend, as could be observed for both authentic and imposter HD distributions. Such property reveals that after performing image restoration using the Wiener filtering, the induced artifacts would make images of different classes look more similar to each other, causing a decreased HD for inter-class comparison. For the purpose of biometric identification, such a phenomenon was not desirable and should be avoided.

5. Discussion

In this paper, six enrollment/testing system configurations were carried out for the iris recognition system. These configurations can be divided into homogeneous pairs (Approaches 1, 4 and 6) and heterogeneous pairs (Approaches 2, 3 and 5). Based on the statistical results summarized in Table 2, the DoF of the wavefront coding image was significantly extended. Taking the best scheme (Approach 4), the factor was 3.71 based on the criteria EER = 5.2%. For optimal system in operation, we suggest using the homogeneous optics (Approach 1, 4 or 6) to achieve a more satisfactory recognition rate. However, if we consider the power of the EDoF capability as one of the core objective functions from the experimental results, Approach 4 is the best approach with a 3.71 EDoF factor.

Table 2. The results of six experimental approaches.

Test Enrollment	Conventional	EDoF	EdoF with Wiener
Conventional	DoF = 6 cm SI = 4.3 FRR = 4.7% EER = 1.0%	DoF = 18.4 cm SI = 4.5 FRR = 17.5% EER = 2.9%	DoF = 0 cm SI = 3.3 FRR = 53.3% EER = 7.0%
EdoF		DoF = 22.2 cm SI = 4.0 FRR = 23.0% EER = 3.3%	DoF = 4.5 cm SI = 2.8 FRR = 18.3% EER = 3.3%
EdoF with Wiener			DoF = 18.6 cm SI = 3.9 FRR = 23.3% EER = 3.3%

One lesson we learned from the experiment is how to design an imaging system for the purpose of pattern recognition. In order to achieve the highest recognition rate, one should make sure to put into the gallery set those images which were processed in exactly the same procedure as the test images. Otherwise, the heterogeneity caused by the hardware mismatch would degrade the accuracy. From the perspective of pattern recognition theory, it is better that the gallery image set involves the largest possible amount of variations which could possibly be observed in the probe image set. In such conditions, pattern recognition or machine learning algorithms could estimate the density of the image sample distribution correctly. During the testing stage the learned decision boundary can be robust enough to achieve higher recognition rate.

Another question we can ask ourselves is, given the aforementioned principle, why is the performance of Approach 4 better than 6? As discussed in Section 4.2, using Wiener filtering for image restoration may introduce additional artifacts, which may degrade the image and modify the detailed textural structures in iris images. Iris recognition relies heavily on the correct pattern matching of the iris code between training and test images. If the detailed textural components of an iris image are changed by some unpredictable factor, the iris code changes dramatically. That is the reason why the recognition performance of Approach 6 is worse than 4. Such experimental results also coincide with the practice proposed in [11], which shows that such methodology is supported by two independent research groups.

The comparison between the proposed method in the best scheme and existing works is summarized in Table 3 [7–11]. Compared to other works, the desired distance is set to 300 cm for a long range iris recognition system. As it is strict and harsh enough to test the practicality and feasibility of the EDoF-empowered image sensors, the database is abundantly captured. Due to the difficulty of long range image acquisition, the optics, sensor and wavefront coding technique are systematically designed and integrated into our laboratory. Finally, the EDoF factor reached 3.71 times that of the original system without loss of recognition accuracy.

Table 3. Comparison between existing works and this study.

	Proposal	Gracht [7]	Narayanswamy [8]	Smith [9]	Barwick [10]	Boddeti [11]
Scheme	Experiment	Experiment	Simulation	Simulation	Simulation	Simulation
Database	Laboratory	Laboratory	Laboratory	ICE	UPOL	ICE
	3328 images 64 classes	- one class	44 images two classes	150 images 50 classes	168 images 56 classes	1061 images 61 classes
Distance	300 cm	50 cm	55 cm	50 cm	55 cm	-
Optics	f = 400 mm F/6.3 λ = 850 nm	f = 57 mm F/8 λ = 830 nm	f = 50 mm F/3.5 λ = 780 nm	f = 53 mm F/2 λ = 760 nm	f = 50 mm F/2.85 λ = 768 nm	-
Sensor	2080 × 2080 8 μm	1300 × 1300 6.7 μm	1024 × 768 -	- 5.134 μm	- 3 μm	-
Wavefront coding	Cubic	Cubic	Cubic	Cubic	Cubic-pentic	Cubic
	α = 42	α = 11	α = 156	α = 30	(−16, 71, −265, 370, 267)	α = 60
Restoration	without	with	with	without	without	without
Merit function	Accuracy invariant	HD = 0.32	Iris score [1] set to 0.3	HD = 0.33	SI = 5	Error bars of the authentic and impostor scores do not overlap
Extended factor	3.71	over 2	over 3.3	2.8	2.2	4.8

[1] Iris score: using exclusive-NOR operator for bit comparison. The values 1 and 0 represents the match and mismatch bit pairs, respectively.

6. Conclusions

In this paper, we examine a number of EDoF approaches for the purpose of a distant iris recognition system. Unlike prior studies that mostly addressed this in a simulation, we experimentally overhauled the entire computational imaging flow via an EDoF imagery and verified the ultimate performance with different homo- and hetero-enrollment/testing image pairs.

On the basis of experimental results, the DoF of the wavefront coding system is significantly increased in comparison with the conventional imaging. Taking the best scheme (Approach 4) as the benchmark, the EDoF factor was 3.71 under the constraint EER = 5.2%. For optimal system configurations of testing and enrollment image sets, we suggest using the homogeneous pair (Approaches 1, 4 and 6) to achieve a more satisfactory recognition rate.

The EDoF function via pupil engineering is validated based on the assumption that the pupil mask should be in place of the pupil or equivalent in the imaging system. For practical use, different positions of the phase mask would lead to diverse coding effects with respect to field of view. As a result, the fidelity of the restored image is difficult to keep constant within a wide acquisition volume. To keep the uniform phase coding satisfying the linear shift invariant, the position of the aperture stop in the system layout should be further examined in future work.

Acknowledgments: This study was accomplished with the support of the National Science Council in Taiwan under contract No. MOST 105-2221-E-009-009-088, MOST 105-2622-E-009-006-CC2 and MOST 105-2221-E-008-111.

Author Contributions: S.-H.H. designed the experimental procedure and performed the experiments. Y.-H.L. implemented code for iris recognition. C.-H.T. supervised the project. All authors contributed to writing the paper.

Conflicts of Interest: The authors declare no conflict of interest.

References

1. AJain, K.; Ross, A.; Prabhakar, S. An introduction to biometric recognition. *IEEE Trans. Circuits Syst. Video Technol.* **2004**, *14*, 4–20.

2. Daugman, J.G. High confidence visual recognition of persons by a test of statistical independence. *IEEE Trans. Pattern Anal. Mach. Intell.* **1993**, *15*, 1148–1161. [CrossRef]

3. Daugman, J.G. The importance of being random: Statistical principles of iris recognition. *Pattern Recognit.* **2003**, *36*, 279–291. [CrossRef]

4. Bowyer, K.; Hollingsworth, K.; Flynn, P.J. Image understanding for iris biometrics: A survey. *Comput. Vis. Image Underst.* **2008**, *110*, 181–307. [CrossRef]

5. Camus, T.A.; Wildes, R.P. Reliable and Fast Eye Finding in Close-up Images. In Proceedings of the 16th International Conference on Pattern Recognition, Quebec City, QC, Canada, 11–15 August 2002; Volume 1, pp. 389–394.

6. Daugman, J.G. How iris recognition works. *IEEE Trans. Circuits Syst. Video Technol.* **2004**, *14*, 21–30. [CrossRef]

7. Van der Gracht, J.; Pauca, V.P.; Setty, H.; Narayanswamy, R.; Plemmons, R.J.; Prasad, S.; Torgersen, T. Iris Recognition with Enhanced Depth-of-Field Image Acquisition. In *Visual Information Processing XIII*; Society of Photo Optical: Bellingham, WA, USA, 2004; Volume 5438, pp. 120–129.

8. Narayanswamy, R.; Johnson, G.E.; Silveira, P.E.X.; Wach, H.B. Extending the imaging volume for biometric iris recognition. *Appl. Opt.* **2005**, *44*, 701–712. [CrossRef] [PubMed]

9. Smith, K.N.; Pauca, V.P.; Ross, A.; Torgersen, T.; King, M.C. Extended Evaluation of Simulated Wavefront Coding Technology in Iris Recognition. In Proceedings of the First IEEE International Conference on Biometrics: Theory, Applications, and Systems, Washington, DC, USA, 27–29 September 2007; pp. 259–265.

10. Barwick, D.S. Increasing the information acquisition volume in iris recognition systems. *Appl. Opt.* **2008**, *47*, 4684–4691. [CrossRef] [PubMed]

11. Boddeti, V.N.; Kumar, B.V.K.V. Extended-depth-of-field iris recognition using unrestored wavefront-coded imagery. *IEEE Trans. Syst. Man Cybern. Part A Syst. Hum.* **2010**, *40*, 495–508. [CrossRef]

12. Matey, J.R.; Naroditsky, O.; Hanna, K.; Kolczynski, R.; Lolacono, D.J.; Mangru, S.; Tinker, M.; Zappia, T.M.; Zhao, W.Y. Iris on the move: Acquisition of images for iris recognition in less constrained environments. *Proc. IEEE* **2006**, *94*, 1936–1947. [CrossRef]

13. ISO/IEC 19794–6:2005. Information Technology—Biometric Data Interchange Formats—Part 6: Iris Image Data. Available online: http://www.iso.org/iso/catalogue_detail.htm?csnumber=38750 (accessed on 23 November 2016).

14. Forrester, J.V.; Dick, A.D.; McMenamin, P.G.; Lee, W. *The Eye: Basic Sciences in Practice*, 2nd ed.; Saunders: London, UK, 2001.

15. Smith, W.J. *Modern Optical Engineering: The Design of Optical System*, 4th ed.; McGraw-Hill: New York, NY, USA, 2008.

16. Boyce, C.; Ross, A.; Monaco, M.; Hornak, L.; Li, X. Multispectral Iris Analysis: A Preliminary Study51. In Proceedings of the Conference on Computer Vision and Pattern Recognition Workshop, CVPRW'06, New York, NY, USA, 17–22 June 2006.

17. Lian, Z. Miniming Specular Reflections of the Iris Image Acquisition System. Master's Thesis, National Chiao Tung University, Hsinchu, Taiwan, 2014.

18. Stamnes, J.J. *Waves in Focal Regions: Propagation: Diffraction, and Focusing of Light, Sounds, and Water Waves*; CRC Press: Boca Raton, FL, USA, 1986.

19. Takahashi, Y.; Komatsu, S. Optimized free-form phase mask for extension of depth of field in wavefront-coded imaging. *Opt. Lett.* **2008**, *33*, 1515–1517. [CrossRef] [PubMed]

20. Dowski, E.R.; Cathey, W.T. Extended depth of field through wave-front coding. *Appl. Opt.* **1995**, *34*, 1859–1866. [CrossRef] [PubMed]

21. Ojeda-Castaneda, J.; Tepichin, E.; Diaz, A. Arbitrarily high focal depth with quasioptimum real and positive transmittance apodizer. *Appl. Opt.* **1989**, *28*, 2666–2670. [CrossRef] [PubMed]

22. Bagheri, S.; Silveira, P.E.X.; Barbastathis, G. Signal-to-noise-ratio limit to the depth-of-field extension for imaging systems with an arbitrary pupil function. *J. Opt. Soc. Am. A-Opt. Image Sci. Vis.* **2009**, *26*, 895–908. [CrossRef] [PubMed]

23. Sherif, S.S.; Dowski, E.R.; Cathey, W.T. A Logarithmic Phase Filter to Extend the Depth of Field of Incoherent Hybrid Imaging Systems. *Int. Symp. Opt. Sci. Technol.* **2001**, *4471*, 272–280.

24. Kubala, K.; Dowski, E.R.; Kobus, J.; Brown, R. Design and Optimization of Aberration and Error Invariant Space Telescope Systems. In Proceedings of the SPIE 40th Annual Meeting Optical Science and Technology, Denver, CO, USA, 2–6 August 2004; Volume 5524, pp. 54–65.

25. Chen, Y.-L.; Hsieh, S.-H.; Hung, K.-E.; Yang, S.-W.; Li, Y.-H.; Tien, C.-H. Extended Depth of Field System for Long Distance Iris Acquisition. In Proceedings of the Novel Optical Systems Design and Optimization XV, San Diego, CA, USA, 12–16 August 2012.

26. Hsieh, S.-H.; Yang, S.-W.; Li, Y.-H.; Tien, C.-H. Long Distance Iris Recognition System. In Proceedings of the International Conference 2012 Optics and Photonics Taiwan, Taipei, Taiwan, 6–8 December 2012.

27. Hsieh, S.-H.; Yang, H.-W.; Huang, S.-H.; Li, Y.-H.; Tien, C.-H. Biometric iris image acquisition system with wavefront coding technology. In Proceedings of the International Symposium on Photoelectronic Detection and Imaging 2013: Infrared Imaging and Applications, Beijing, China, 25–27 June 2013; p. 890730.

28. Gonzalez, R.C.; Woods, R.E. *Digital Image Processing*, 3rd ed.; Prentice Hall: Upper Saddle River, NJ, USA, 2008.

29. Gonzalez, R.C.; Woods, R.E.; Eddins, S.L. *Digital Image Processing Using MATLAB*, 2nd ed.; Gatesmark Pub. S.I.: Knoxville, TN, USA, 2009.

30. Hsieh, S.-H.; Lian, Z.-H.; Chang, C.-M.; Tien, C.-H. The Influence of Phase Mask Position upon EDoF System. In Proceedings of the Novel Optical Systems Design and Optimization XVI, San Diego, CA, USA, 25 August 2013.

31. Tan, K.-C.; Lim, H.; Tan, B.T.G. Windowing techniques for image-restoration. *Graph. Models Image Process.* **1991**, *53*, 491–500. [CrossRef]

32. Aghdasi, F.; Ward, R.K. Reduction of boundary artifacts in image restoration. *IEEE Trans. Image Process.* **1996**, *5*, 611–618. [CrossRef] [PubMed]

33. Masek, L.; Kovesi, P. *Matlab Source Code for a Biometric Identification System Based on Iris Patterns*; The School of Computer Science and Software Engineering, The University of Western Australia: Crawley, Australia, 2003; Volume 2.

34. Philips, J. *Portal Challenge Problem Multiple Biometric Grand Challenge Preliminary Results of Version 2*; MBGC 3rd Workshop; National Institute of Standards and Technology: Gaithersburg, MD, USA, 2009.

Article

Nonintrusive Finger-Vein Recognition System Using NIR Image Sensor and Accuracy Analyses According to Various Factors

Tuyen Danh Pham, Young Ho Park, Dat Tien Nguyen, Seung Yong Kwon and Kang Ryoung Park *

Division of Electronics and Electrical Engineering, Dongguk University, 26 Pil-dong 3-ga, Jung-gu, Seoul 100-715, Korea; phamdanhtuyen@gmail.com (T.D.P.); fdsarew@dongguk.edu (Y.H.P.); nguyentiendat@dongguk.edu (D.T.N.); sbaru07@dgu.edu (S.Y.K.)
* Correspondence: parkgr@dongguk.edu; Tel.: +82-10-3111-7022; Fax: +82-2-2277-8735

Academic Editor: Gonzalo Pajares Martinsanz
Received: 1 May 2015; Accepted: 9 July 2015; Published: 13 July 2015

Abstract: Biometrics is a technology that enables an individual person to be identified based on human physiological and behavioral characteristics. Among biometrics technologies, face recognition has been widely used because of its advantages in terms of convenience and non-contact operation. However, its performance is affected by factors such as variation in the illumination, facial expression, and head pose. Therefore, fingerprint and iris recognitions are preferred alternatives. However, the performance of the former can be adversely affected by the skin condition, including scarring and dryness. In addition, the latter has the disadvantages of high cost, large system size, and inconvenience to the user, who has to align their eyes with the iris camera. In an attempt to overcome these problems, finger-vein recognition has been vigorously researched, but an analysis of its accuracies according to various factors has not received much attention. Therefore, we propose a nonintrusive finger-vein recognition system using a near infrared (NIR) image sensor and analyze its accuracies considering various factors. The experimental results obtained with three databases showed that our system can be operated in real applications with high accuracy; and the dissimilarity of the finger-veins of different people is larger than that of the finger types and hands.

Keywords: nonintrusive finger-vein capturing device using NIR image sensor; misalignment of finger-vein image; multiple images for enrollment; score-level fusion

1. Introduction

Recent developments have led to the widespread use of biometric technologies, such as face, fingerprint, vein, iris, and voice recognition, in a variety of applications in access control, financial transactions on mobile devices, and automatic teller machines (ATMs) [1–4]. Among them, finger-vein recognition has been highlighted because it can overcome several drawbacks of other biometric technologies, such as the effect of sweat, skin distortions, and scars in fingerprint recognition, or the effect of poses and illumination changes in face recognition. Moreover, a finger-vein recognition system is cost effective in comparison, and offers high accuracy together with the advantages of fake detection and a bio-cryptography system [5]. Finger-vein recognition uses the vascular patterns inside human fingers to uniquely identify individuals. Vein imaging technology relies on the use of near infrared (NIR) illuminators at a wavelength longer than about 750 nm, because the deoxyhemoglobin in veins absorbs light in this range [6,7]. Previous work on finger-vein recognition include research aimed at enhancing vein image quality, increasing recognition accuracy by various feature extraction methods, considering finger veins as a factor for individual recognition in multimodal systems, as well as detecting fake finger veins. The research on finger-vein image enhancement, which is based on a software algorithm, can be classified into restoration-based and non-restoration-based methods [7,8]. The restoration-based methods proposed by Yang *et al.* [9–11] were able to produce enhanced finger-vein images by considering the effect of the layered structure of skin and restored the images by using a point-spread function (PSF) model [10], and a biological optical model (BOM) [11]. In the non-restoration-based approaches, Gabor filtering was popularly used [6–8,12,13]. Yang *et al.* introduced an enhancement method that uses multi-channel even-symmetric Gabor filters with four directions to strengthen the vein information in different orientations [6]. A study by Park *et al.* [8] led to the proposal of an image enhancement method using an optimal Gabor filter based on the directions and thickness of the vein line. An adaptive version of the Gabor filter was used in the research of Cho *et al.* [12] to enhance the distinctiveness of the finger-vein region in the original image. The Gabor filter was also used in combination with a Retinex filter, by using fuzzy rules in the method proposed by Shin *et al.* [7]. Zhang *et al.* proposed gray-level grouping (GLG) for the enhancement of image contrast, and a circular Gabor filter (CGF) for the enhancement of finger-vein images [13].

Pi *et al.* proposed a quality improvement method based on edge-preserving and elliptical high-pass filters capable of maintaining the edges and removing blur [14]. In addition, Yu *et al.* proposed a fuzzy-based multi-threshold algorithm considering the characteristics of the vein patterns and skin region [15].

Work has also been conducted on extracting and combining various features from finger-vein images to increase the quality of the recognition results [16–19]. In [16], they used both the global feature of the moment-invariants method and Gabor filter-based local features. In the method proposed by Lu *et al.* [17], eight-channel Gabor features were extracted and analyzed prior to application to score-level fusion to obtain the final matching score. Qian *et al.* [18] proposed a finger-vein recognition algorithm based on the fusion of score level moment invariants by the weighted-average method. In [19], Yang *et al.* proposed a binary feature for finger-vein matching, termed personalized best bit map (PBBM), which was extracted based on the consistent bits in local binary pattern (LBP) codes. Finger-vein recognition was also considered as a sub-system in multimodal biometric systems [20–23] along with other individual recognition methods to compensate for the drawbacks of each of the recognition methods. The results of finger-vein and fingerprint recognitions were matched and combined by using various methods, such as decision level fusion of "AND" or "OR" rules as in [20], a support vector machine (SVM) as in [21], or score level fusion as in [22]. He *et al.* [23] proposed a multimodal biometric system by considering the three biometric characteristics of fingerprint, face, and finger-vein, and evaluated the performance of the system with the use of sum rule-based and SVM-based score level fusion. The research on finger-vein recognition has also taken counterfeit vein information into account, as in [24,25]. In the anti-spoofing system for vein identification in [24], live fingers were detected by continuously capturing successive heart-rate-based images and then examining the details in the series of images. Nguyen *et al.* [25] proposed an image-analysis method for fake finger-vein detection based on Fourier transform and wavelet transforms.

A number of research efforts on finger-vein recognition have considered the quality of the preprocessed images, as well as the effectiveness of the matching features. However, the evaluation of the discriminant factors on finger-vein information, such as the differences between people, left and right hands, and the type of finger, has not received much attention. In our research, we propose a nonintrusive finger-vein capturing device.

Table 1. Comparison of the proposed method with previous methods.

Category	Methods	Strengths	Weaknesses
Accuracy evaluation without considering the various factors of people, hands, finger types, and the number of images	EER or ROC curve-based evaluation of finger-vein recognition with the assumption that the veins from different hands or finger types are different classes without comparing the dissimilarity of finger-vein among people, hands, and finger types [7–9,11,16–23,25]	New methods for enhancing finger-vein images with feature extraction or score fusions for enhancing the recognition accuracy are proposed	Assuming the veins from different hands or finger types are different classes without any theoretical or experimental ground
Accuracy evaluation according to people, hands, finger types, and the number of images	The dissimilarity of finger-veins among people, hands, and finger types are quantitatively evaluated (Proposed method)	Providing the experimental ground for the dissimilarity of finger-veins among people, hands, and finger types	Not providing the experimental ground for the dissimilarity of palm-veins or hand dorsal veins among people and hands

Our research is novel in the following three ways compared to previous work.

· We propose a nonintrusive finger-vein capturing device using a small-sized web-camera and NIR light-emitting diodes. To reduce the misalignment of captured images while ensuring minimal user inconvenience, two guiding bars for positioning the fingertip and side of the finger were attached to the device.

· The accuracies of recognition were compared by assuming that images from the same person, the same hand, and the same finger types form the same classes. Based on the receiver operational characteristic curve, equal error rate, authentic and imposter matching distributions, and d-prime value, the dissimilarity of finger-veins among people, hands, and finger types are quantitatively evaluated.

· The accuracies of recognition are compared according to the number of finger-vein images combined by score-level fusion for recognition, and the number of images for enrollment.

Table 1 presents a comparison of the proposed method with previous methods.

The remainder of this paper is organized as follows. Section 2, explains the details of the proposed method and Section 3, shows the experimental results and discussions. Finally, the conclusions and opportunities for future work are given in Section 4.

2. Finger-Vein Recognition and Evaluation Method

2.1. Overview of the Finger-Vein Recognition System

An overview of the proposed method is shown in Figure 1. Because the input finger-vein image consists of two parts, *i.e.*, the finger region containing the finger-vein information and the background region, the method to detect the finger region is first applied in order to remove the background, which contains unnecessary information. In the next step, based on the detected upper and lower finger boundaries detected in the previous step, the segmented finger region is stretched into a rectangular form in the normalization step. The processing time is reduced by obtaining a sub-sample of the stretched finger-vein image to reduce the size of the image. Before the recognition features are extracted, the quality of the finger-vein image is enhanced by using Gabor filtering, subsequent to which the preprocessed image is applied to the feature extraction step using the local binary pattern (LBP) method. In the next step, the hamming distance (HD) is calculated to determine the matching distance between the extracted code features of the input finger-vein image and the enrolled image. The input finger-vein image is then classified as either being genuine or being that of an imposter by using the enrolled data based on the matching distance.

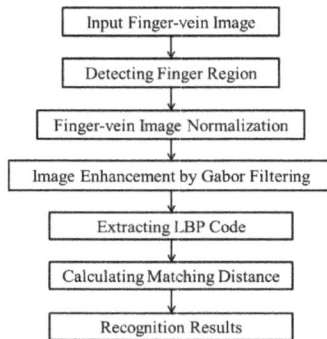

Figure 1. Flowchart of the experimental procedure of our research.

2.2. Finger Region Detection and Normalization

As shown in Figure 2, a captured finger-vein image consists of the background surrounding the finger region, which contains the vein pattern, which is used for recognition purposes, and which has higher gray levels than the background. The background is removed from the captured image by detecting the four boundaries of the finger region consisting of the left and right boundaries in the horizontal direction, and upper and lower boundaries in the vertical direction, based on previous research [7]. In the images from the three databases, the left and right finger region boundaries are restricted by the size of the hole in the device for capturing the finger-vein image. Detailed explanations of the three databases and the device are provided in Section 3. As such, the values of X_L and X_R, which determine the left and right boundaries, as shown in Figure 2, are experimentally defined for the three databases. In the case of the good-quality database with 640 × 480 pixel images, the values of X_L and X_R are 180 and 480 pixels, respectively. For the mid-quality database with the same image size, the values of X_L and X_R are 220 and 470 pixels, respectively. The third (open) database, which consists of images with a size of 320 × 240 pixels, X_L and X_R are 20 and 268 pixels, respectively.

The 1st (Figure 2a) and 2nd database (Figure 2c) are collected by our lab-made devices (see Section 3). In our devices, each person puts his or her finger on the hole of the upper-part of device, and the size of the hole in the device for capturing the finger-vein image is fixed and limited in order to remove the effect by the environmental light into the captured image. Therefore, the part of the finger area can be acquired in the image, and the positions of left and right finger boundaries are restricted and same in all the captured images as shown in Figure 2a,c. Therefore, in order to enhance the processing speed of segmenting the finger area from the image, we use the pre-determined X_L and X_R values as the horizontal (X) position of the left and right boundary of the finger area, respectively.

In case of the 3rd database (Figure 2e), although the whole finger area can be acquired in the image, the left and right-most areas of finger are so dark (caused by the insufficient illumination of NIR light) that these areas are difficult to be used for finger-vein recognition. Therefore, we use the part of finger area by removing these left and right-most areas, based on pre-determined X_L and X_R values. The positions of the left and right boundaries can be automatically segmented with the 3rd database, but these positions can be different from each other among images, according to the performance of the segmentation algorithm of the finger area. The main goal of our research is not focused on the segmentation algorithm but on comparing the accuracies of recognition by assuming that images from the same person, the same hand, and the same finger types form the same classes. In addition, another goal is to compare the accuracies of recognition according to the number of finger-vein images combined by score-level fusion for recognition, and the number of images for enrollment. Therefore, we use the part of finger area by removing these left and right-most areas, based on pre-determined X_L and X_R values.

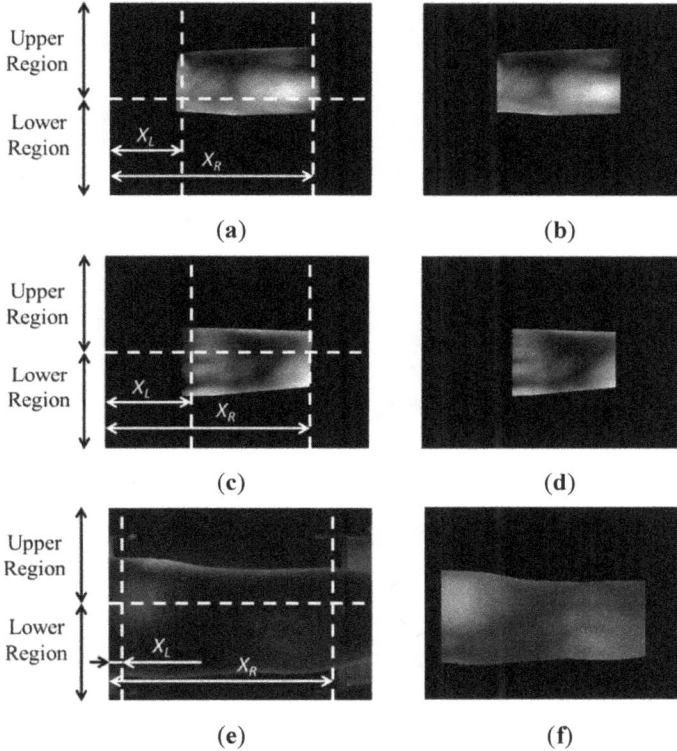

Figure 2. Examples of input finger-vein images and finger-region detection results obtained with images from the three databases: Original images from the (**a**) good-quality; (**c**) mid-quality; and (**e**) open databases with their corresponding finger-region detection results shown in (**b,d,f**), respectively.

Then, two masks of 4×20 pixels, which are shown in Figure 3, were used to detect the upper and lower boundaries of the finger region. Because the gray level of the background region is lower than that of the finger region, as shown in Figure 2, the value that was calculated by using the masks in Figure 3 is maximized at the position of the finger boundary. Examples of the finger region detection results are given in Figure 2b,d,f.

Based on the detected finger boundaries, the finger-vein image is normalized to the size of 150×60 pixels by using a linear stretching method, and it is then sub-sampled to produce a 50×20 pixel image to enhance the processing speed [7]. This is done by averaging the gray values in each 3×3 pixel block of the 150×60 pixel image. Figure 4 shows examples of the normalization results of the images in Figure 2.

(a)

(b)

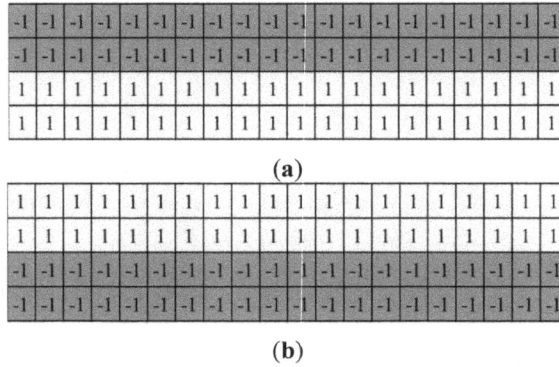

Figure 3. Masks for detecting finger-region boundaries in the vertical direction: Masks for detecting (**a**) the upper boundary; and (**b**) the lower boundary of the finger region.

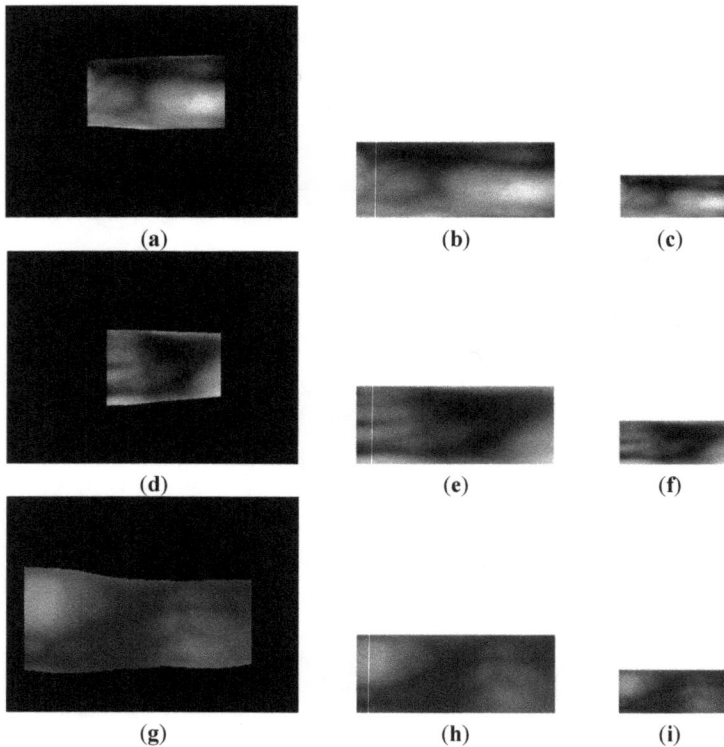

Figure 4. Linear stretching and sub-sampled results of finger-vein images from the three databases: Detected finger-region image from the (**a**) good-quality; (**d**) mid-quality; and (**g**) open databases, with their corresponding 150 × 60 pixel stretched images shown in (**b,e,h**), respectively, and their corresponding 50 × 20 pixel sub-sampled images shown in (**c,f,i**), respectively.

2.3. Four-Directional Gabor Filtering

Gabor filtering has been popularly used in finger-vein recognition for enhancing the image quality [6–8]. In this research, we apply a four-directional Gabor filter to the 50 × 20 pixel sub-sampled image prior to extracting the finger-vein code to enhance the distinctiveness of the vein image. Gabor filtering of the sub-sampled image could also be helpful to reduce the processing time compared to that of the original finger-vein image [7,8]. A two-dimensional Gabor filter can be represented as follows [6–8]:

$$G(x,y) = \frac{1}{2\pi\sigma_x\sigma_y} \exp\left\{ -\frac{1}{2}\left(\frac{x_\theta^2}{\sigma_x^2} + \frac{y_\theta^2}{\sigma_y^2} \right) \right\} \exp(j2\pi f_0 x_\theta) \tag{1}$$

with

$$\begin{bmatrix} x_\theta \\ y_\theta \end{bmatrix} = \begin{bmatrix} \cos\theta & \sin\theta \\ -\sin\theta & \cos\theta \end{bmatrix} \begin{bmatrix} x \\ y \end{bmatrix}$$

where $j = \sqrt{-1}$, θ is the direction, and f_0 is the central frequency of the Gabor kernel. The two coordinates (x, y) are rotated to x_θ and y_θ, respectively, and on each coordinate, the spatial envelopes of the Gaussian function are represented by σ_x and σ_y, respectively. By eliminating the imaginary part of the Gabor filter, the real part, namely the even-symmetric Gabor filter, is used in this research because of the effectiveness with which it processes time. An even-symmetric Gabor filter is represented as Equation (2) as follows [6–8]:

$$G_k^E(x,y) = \frac{1}{2\pi\sigma_x\sigma_y} \exp\left\{ -\frac{1}{2}\left(\frac{x_{\theta_k}^2}{\sigma_x^2} + \frac{y_{\theta_k}^2}{\sigma_y^2} \right) \right\} \cos(2\pi f_k x_{\theta_k}) \tag{2}$$

with

$$\theta_k = k\pi/4; \ k = 1, 2, 3, 4$$

where k is the index of the directional channel, and θ_k and f_k represent the orientation and spatial frequency of the kth channel, respectively. Based on previous research [6], the optimal parameters of f_k, σ_x, and σ_y, are determined to be 0.2, 2.38, and 2.38, respectively, for the four channels in the 0°, 45°, 90°, and 135° directions of the Gabor filter applied to the sub-sampled image of 50 × 20 pixels. A convolution operation is applied to an input finger-vein image with the Gabor filter of the four channels to obtain the filtered image in the form of four separated convolution result images. These images are then combined by selecting, at each pixel position, the pixel with the lowest gray-level value among the four pixels of the four result images to be the final result of Gabor filtering, because, generally, the vein line is darker than the skin region [7]. Figure 5 provides example results of four-directional Gabor filtering on the sub-sampled images in Figure 4c,f,i.

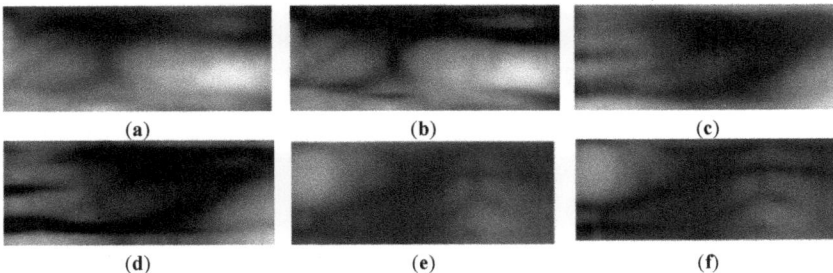

Figure 5. Gabor filtering results of the 50 × 20 pixel sub-sampled images from the three databases: Sub-sampled image from the (**a**) good-quality; (**c**) mid-quality; and (**e**) open databases with their respective Gabor filtered images shown in (**b**,**d**,**f**).

2.4. Finger-Vein Code Extraction Using LBP and Matching

The binary codes are extracted from the quality enhanced finger-vein image by using the LBP method, which was selected based on its high performance [7]. This method encodes the difference between the gray level of each central pixel (I_C) and that of its neighboring pixels (I_N) to the binary values of 0 or 1, as described by Equation (3) and illustrated in Figure 6. For each pixel position in a 50×20 pixel image, an 8-bit code string is extracted. Consequently, for each finger-vein image, a 6912-bit binary code (8 bits $\times$ 48 columns $\times$ 18 rows) is generated by the LBP operator.

$$LBP(x_C, y_C) = \sum_{k==0}^{7} S(I_N - I_C) \cdot 2^k$$
$$S(t) = \begin{cases} 1 & if \ \ i \geq 0 \\ 0 & otherwise \end{cases} \tag{3}$$

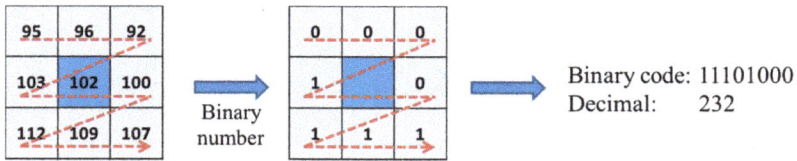

Figure 6. LBP operator.

The matching distance is calculated by using the HD between the enrolled and input LBP binary codes. In this research, we used the normalized version of the Hamming distance on all of the 6912 bits of each finger-vein image as the following Equation (4) [7]:

$$HD = \frac{VCE \oplus VCI}{N} \tag{4}$$

where *VCE* and *VCI* are the binary codes extracted from the enrolled and input images, respectively, $\oplus$ is the Boolean exclusive OR (XOR) operator, and *N* is the total number of bits (6912).

3. Experimental Results

3.1. Proposed Finger-Vein Capturing Device

Figure 7 depicts our finger-vein capturing device. This device consists of six NIR light-emitting diodes (LEDs) operating at a wavelength of 850 nm and a webcam (Logitech Webcam C600) [26]. Alignment of the input finger-vein image in the capturing process was achieved by attaching two bars to the device to guide the positioning of the fingertip and the side of the finger. This was done to ensure a high similarity between images acquired from the same finger of an individual and thus, increase the matching accuracy. By adding the guiding bars, our device is able to acquire finger-vein images for each person non-intrusively. This enabled us to create a good-quality finger-vein database with enhanced alignment of the finger position, and to use the database for the following experiments.

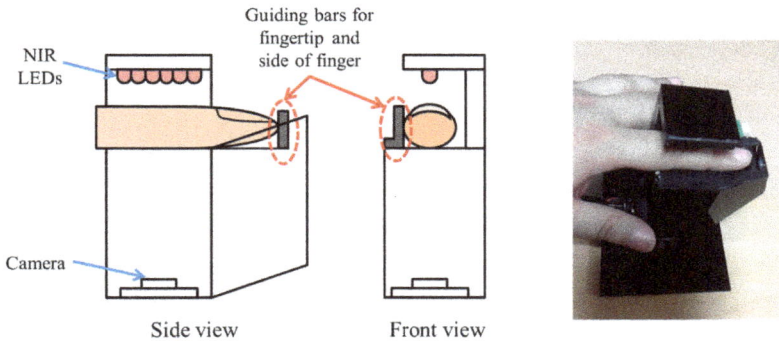

Figure 7. Proposed finger-vein capturing device used to build a good-quality database.

3.2. Performance Evaluation on Three Databases

For this research, we used three different finger-vein databases to evaluate the factors that affect the matching accuracy. The first database was created by collecting finger-vein data from 20 people using the device proposed in Section 3.1. [27]. For each person, the vein images of the index, middle, and ring fingers on both the left and right hands were captured 10 times with an image resolution of 640 × 480 pixels. The total number of images in our database was 1200 (20 people × 2 hands × 3 fingers × 10 images). Because the finger alignment and image quality of the images in this database were strictly assured, it was considered a good-quality database.

In addition, we used two other databases, the first of which was constructed by selecting the vein images of six fingers among the images of 10 fingers in the database I (which were collected by the finger-vein capturing device without the guiding bar in previous research [7]). The device, which was used for collecting the database I [7], is shown in Figure 8. Because the guiding bar was absent, the misalignment among trial images of each finger in this database is relatively high; therefore, this was considered mid-quality database. In detail, each people puts his or her finger on the hole of the upper-part of device, and the size of the hole in the device for capturing the finger-vein image is fixed and limited in order to remove the effect by the environmental light into the captured image. Therefore, the part of finger area can be acquired in the image as shown in Figure 9a,b. Consequently, it is often the case that some part of the finger area (which is seen in the enrolled image) is not seen in the recognized image, which cannot be compensated by preprocessing step and can reduce the accuracy of recognition. In order to solve this problem, we propose a new device including two guiding bars for fingertip and side of finger as shown in Figure 7, which can make the consistent finger area be acquired by our device with reduced misalignment. However, no guiding bar is used in the other device of Figure 8, which is used for collecting the 2nd database. Therefore, we call the 1st and 2nd databases collected by the devices of Figures 7 and 8 as good-quality and mid-quality databases, respectively.

The number of images in the mid-quality database is 1980 (33 people × 2 hands × 3 fingers × 10 trials), and each image has the same size as that of the images in the good-quality database, *i.e.*, 640 × 480 pixels [27].

The last database used in this study is an open finger-vein database (SDUMLA-HMT Finger-vein database) [28], which comprises 3816 images, with a size of 320 × 240 pixels, from 106 people, including six fingers from each person and six trials for each finger. Example images of different trials of one individual (same finger) from each database are given in Figure 9. It can be seen in Figure 9 that the degree of misalignment among the trials of each finger from the mid-quality and open databases is larger than that from the good-quality database.

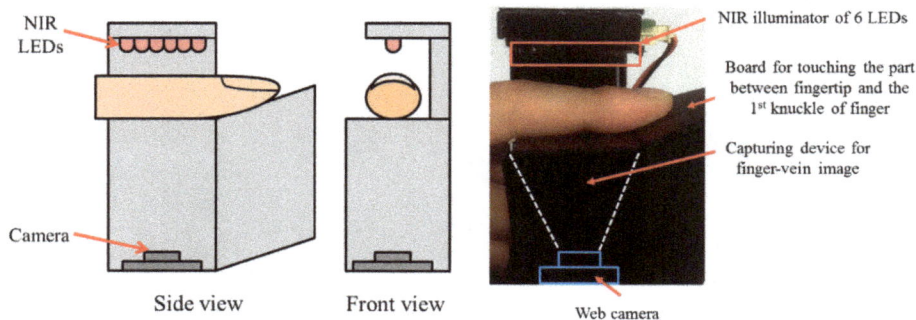

Figure 8. Device for capturing finger-vein images for the second (mid-quality) database.

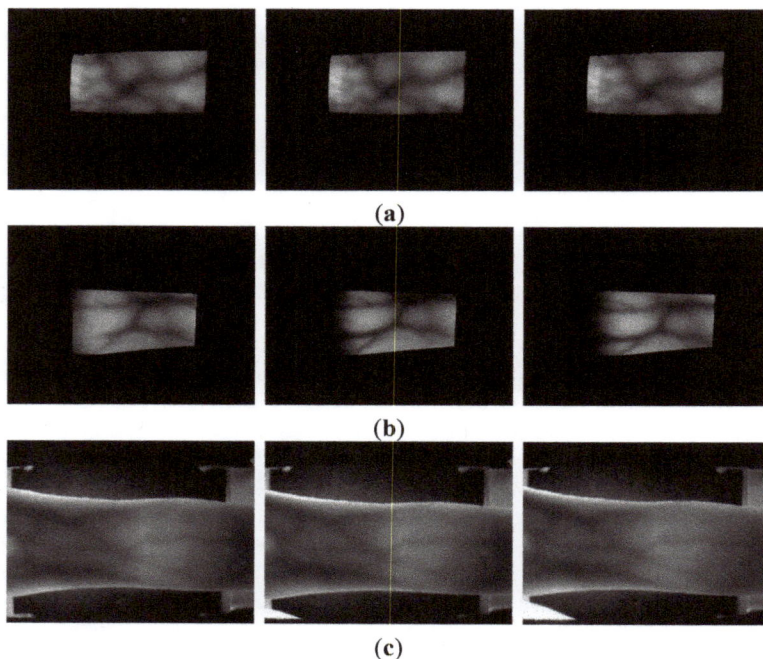

Figure 9. Input images of different trials from the same finger of one individual from each database: (**a**) good-quality; (**b**) mid-quality; and (**c**) open database.

The accuracies of the finger-vein recognition method were evaluated by performing authentic and imposter matching tests. In our experiments, the images in each finger-vein database could be classified in various ways to allow the discrimination factors to be evaluated. In each experiment, authentic matching tests were used to calculate the pairwise matching distances between the images selected from the same class, whereas for the imposter matching tests, the distances between the pairs of images from different classes were calculated. Assuming that for a particular database, we classify finger-vein images into M classes and each class has N images, then the number of authentic and imposter matching tests denoted by A and I are determined by the following Equations (5) and (6), respectively.

$$A = {}_NC_2 \times M \tag{5}$$

$$I = N \times {}_MC_2 \times N \tag{6}$$

where ${}_NC_2 = N(N-1)/2$ is the number of two-combinations from a set of N elements.

By applying and adjusting the threshold on the matching Hamming distance, we calculated the false acceptance rates (FARs), false rejection rates (FRRs), and the EER. FAR refers to the error rates of imposter matching cases, which are misclassified into authentic classes, whereas FRR refers to the error rates of misclassified authentic testing cases into imposter classes. EER is the error rate when the difference between FAR and FRR is minimized. In addition, we measured the d-prime (d') value, which represents the classifying ability between authentic and imposter matching distributions as the following Equation (7) [3].

$$d' = \frac{\mu_A - \mu_I}{\sqrt{\frac{\sigma_A^2 + \sigma_I^2}{2}}} \tag{7}$$

where μ_A and μ_I represent the mean values of the authentic and imposter matching distributions, respectively, and σ_A and σ_I denote the standard deviations of authentic and imposter matching distributions, respectively. A higher d-prime value indicates a larger separation between the authentic and imposter matching distributions, which corresponds to a lower error of recognition, in case that the distributions of authentic and imposter matching scores are similar to Gaussian shape, respectively.

We conducted the following experiments to evaluate the various factors that affect the results of the finger-vein recognition system.

First, we considered each finger of each person to form a different class. This method is used by conventional finger-vein recognition systems to evaluate the recognition accuracy [7–9,11,16–23,25]. Consequently, for the good-quality, mid-quality, and open databases, the number of classes were 120 (20 people × 6 fingers), 198 (33 people × 6 fingers), and 636 (106 people × 6 fingers), respectively. As this class definition method includes the dissimilarity information of fingers, hands, and people in the finger-vein database, we considered this as the 1st experiment (classified by fingers, hands, and people).

In the 2nd experiment, we classified the finger-vein images based on people (classified by people), by assuming that the images of all the fingers on both hands from the same person formed the same class. As a result, in the 2nd experiment, the number of classes in each database equaled the number of users, which was 20, 33, and 106 for the good-quality, mid-quality, and open databases, respectively.

In the 3rd experiment, we assumed that the finger-vein images of all the fingers on the left hands of all the people belong to the same class, and those on the right hands of all the people form another class. Thus, there were two classes based on different hand sides in this experiment (classified by hands).

In the 4th experiment, we evaluated the dissimilarities of finger types by assuming that the images from the index fingers, middle fingers, and ring fingers on both hands of all the people belong to three different classes. This assumption is referred to as (classified by fingers). The organization of these experiments is summarized in Figure 10. The numbers of authentic and imposter matching tests in the experiments on the three databases are determined by Equations (5) and (6), and are shown in Table 2.

Table 2. Number of matching tests (authentic and imposter) for the experiments on the three finger-vein databases (M is the number of classes in each experiment and N is the number of images belonging to one class. Authentic and Imposter refer to the numbers of authentic and imposter matching tests, respectively).

Experiments Databases		1st Experiment	2nd Experiment	3rd Experiment	4th Experiment
		Classified by Fingers, Hands and People	Classified by People	Classified by Hands	Classified by Fingers
Good-quality Database	N/M	10/120	60/20	600/2	400/3
	Authentic	5400	35,400	359,400	239,400
	Imposter	714,000	684,000	360,000	480,000
Mid-quality Database	N/M	10/198	60/33	990/2	660/3
	Authentic	8910	58,410	979,110	652,410
	Imposter	1,950,300	1,900,800	980,100	1,306,800
Open Database	N/M	6/636	36/106	1908/2	1272/3
	Authentic	9540	66,780	3,638,556	2,425,068
	Imposter	7,269,480	7,212,240	3,640,464	4,853,952

Table 3 shows the comparative results of the four experiments defined in Table 2 and Figure 10 for the three databases. In the 1st experiment, in which finger-vein images were classified by fingers, hands, and people, the lowest EER (0.474%) was obtained for the good-quality database. This is due to the fact that this database was captured by the proposed capturing device, which uses a guiding bar to reduce the misalignment among input finger-vein images. In the case of the open database, the authors did not apply any guiding bar for alignment in the image-capturing device [29]. As a result, the EER obtained from this database was the highest (8.096%) because of the misalignment of captured fingers. The results of the first experiment also indicate that the matching accuracies from images in the good-quality database were the highest, followed by those in the mid-quality database, whereas the worst matching accuracies were obtained for the open database, in terms of EERs (0.474%, 2.393%, and 8.096%, respectively). These results correspond to the level of misalignment in each finger-vein database. The resulting plots of the ROC curves and matching distance distributions obtained from the experiments classified by fingers, hands, and people for the three databases are shown in Figures 11 and 12.

Table 3. Comparative results of the four experiments for the three databases.

Experiments		Good-Quality Database		Mid-Quality Database		Open Database	
		EER (%)	d-Prime	EER (%)	d-Prime	EER (%)	d-Prime
1st Experiment	Classified by Fingers, Hands, and People	0.474	5.805	2.393	4.022	8.096	2.727
2nd Experiment	Classified by People	40.223	0.695	39.280	0.759	36.095	0.791
3rd Experiment	Classified by Hands	48.427	0.136	49.320	0.072	49.137	0.039
4th Experiment	Classified by Fingers	45.434	0.277	45.506	0.267	47.299	0.147

< Finger descriptions >
 Lr – Left Ring, Lm – Left Middle, Li – Left Index;
 Rr – Right Ring, Rm – Right Middle, Ri – Right Index;

- 1st experiment (Classified by **Fingers**, **Hands**, and **People**):
 Class1 = {1Lr};
 Class2 = {1Lm};
 Class3 = {1Li};
 ...

- 2nd experiment (Classified by **People**):
 Class1 (person1) = {1Lr, 1Lm, 1Li, 1Ri, 1Rm, 1Rr};
 Class2 (person2) = {2Lr, 2Lm, 2Li, 2Ri, 2Rm, 2Rr};
 ...

- 3rd experiment (Classified by **Hands**):
 Class1 (left hand) = {1Lr, 1Lm, 1Li, 2Lr, 2Lm, 2Li, ...};
 Class2 (right hand) = {1Ri, 1Rm, 1Rr, 2Ri, 2Rm, 2Rr, ...};

- 4th experiment (Classified by **Fingers**):
 Class1 (index) = {1Li, 1Ri, 2Li, 2Ri, ...};
 Class2 (middle) = {1Lm, 1Rm, 2Lm, 2Rm, ...};
 Class3 (ring) = {1Lr, 1Rr, 2Lr, 2Rr, ...};

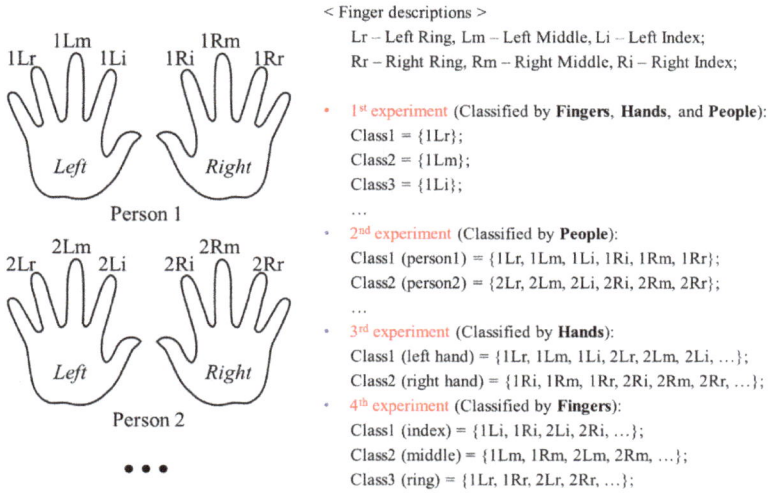

Figure 10. Organization of experiments for finger-vein database.

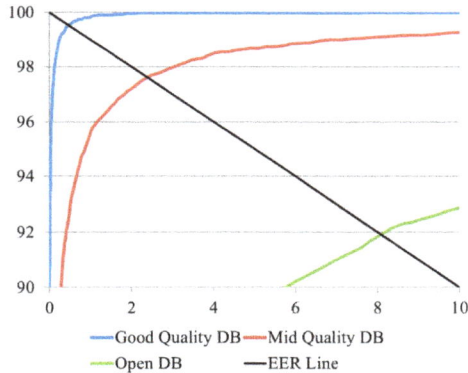

Figure 11. ROC curves of the 1st experiment on the three databases (DBs).

In the 2nd experiment, we classified the finger-vein images from the three databases based on people. In this way, the finger-vein images from the same person were considered as belonging to the same class; hence, the finger-vein images in different classes indicated the dissimilarities between different people. Likewise, the 3rd and 4th experiments on the three databases, considered images from the same hand side (*i.e.*, either the left or the right hand), and images from the same finger type (*i.e.*, the index, middle, or ring fingers) of all the people to be from the same classes, respectively. A comparison of the results of the three experiments (2nd, 3rd and 4th) on each database by considering the finger-vein dissimilarity between people, hands, and fingers, enabled us to evaluate the effect of each of these factors on the accuracy of the finger-vein recognition system.

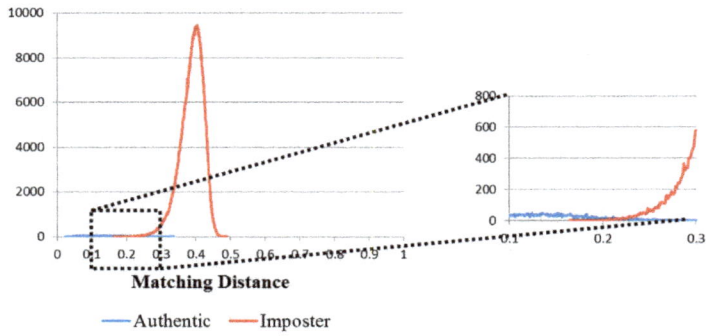

(a)

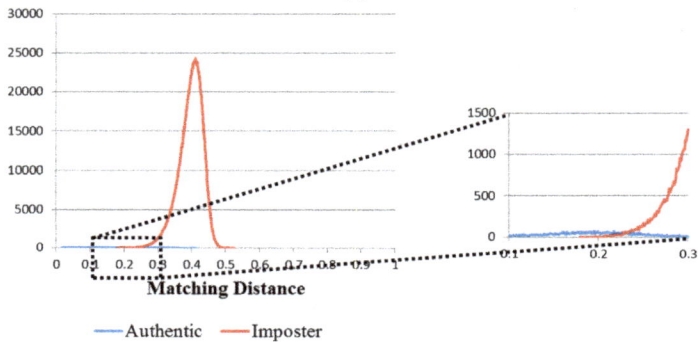

(b)

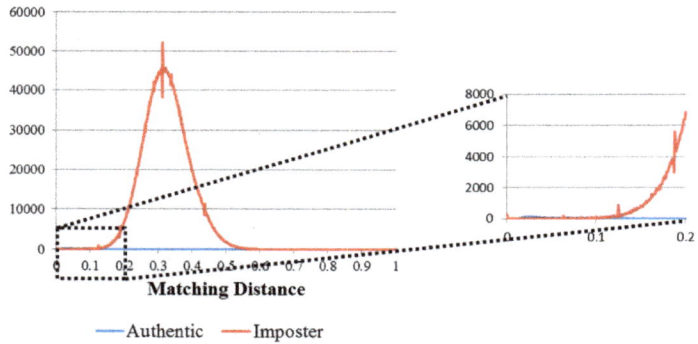

(c)

Figure 12. Matching distance distribution of authentic and imposter matching tests in the 1st experiment on the three databases: (**a**) good-quality; (**b**) mid-quality; and (**c**) open database.

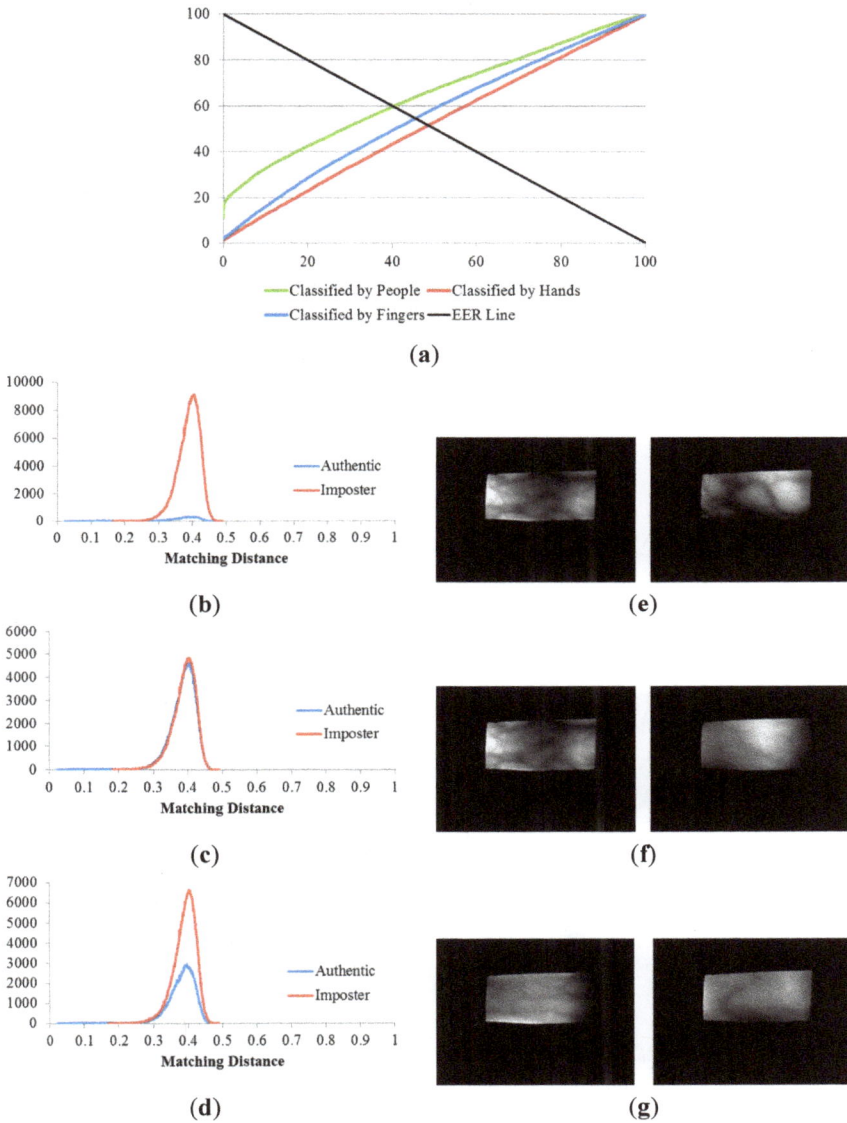

Figure 13. Results of the 2nd, 3rd and 4th experiments on the good-quality database: (**a**) ROC curves of the results of the three experiments; matching distribution of (**b**) the experiment classified by people (2nd experiment); (**c**) the experiment classified by hands (3rd experiment); and (**d**) the experiment classified by fingers (4th experiment), each shown with its corresponding false rejection error case: (**e**) images of the ring and index fingers on left hand of the same person; (**f**) images of the ring and index fingers on left hands of two different people; and (**g**) images of the middle fingers on left and right hands of two different people.

(a)

(b)

(c)

(d)

(e)

(f)

(g)

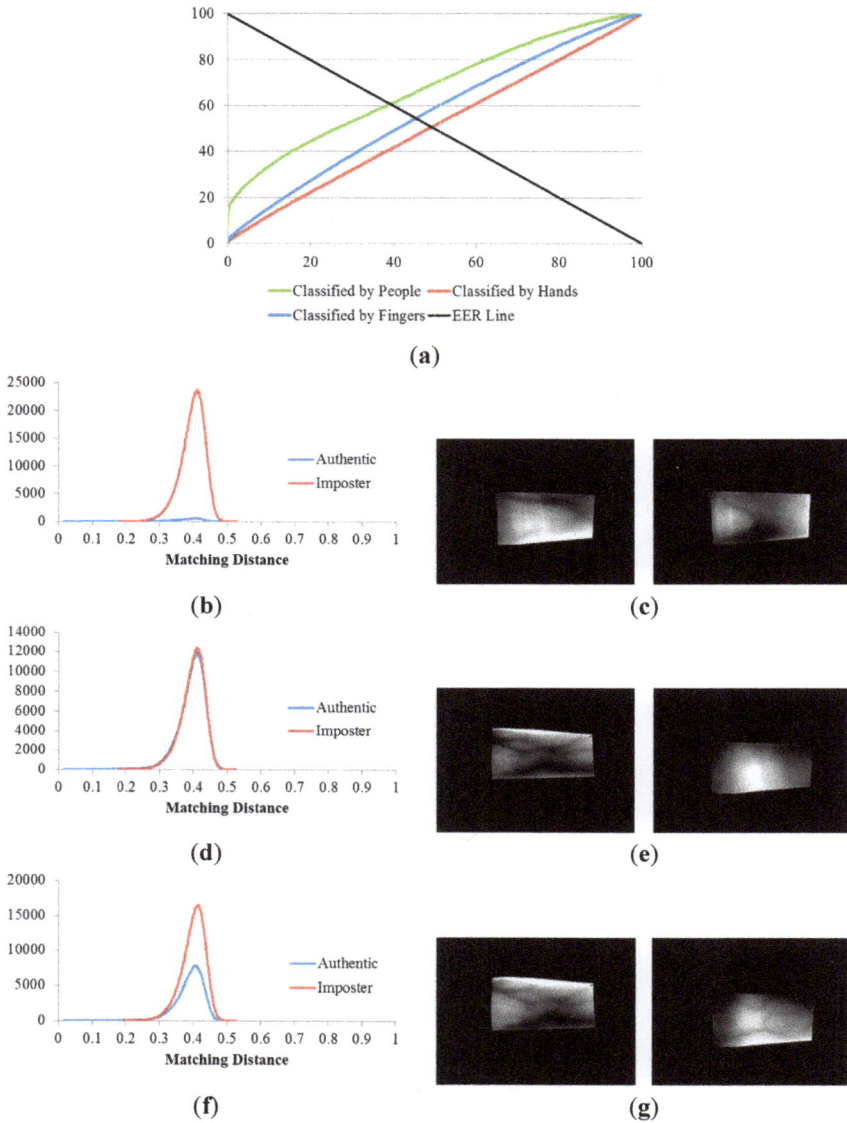

Figure 14. Results of the 2nd, 3rd, and 4th experiments on the mid-quality database: (**a**) ROC curves of the results of the three experiments; matching distribution of (**b**) the experiment classified by people (2nd experiment); (**d**) the experiment classified by hands (3rd experiment); and (**f**) the experiment classified by fingers (4th experiment), each shown with its corresponding false rejection error case: (**c**) images of the right ring and left middle fingers of the same person; (**e**) images of the ring and index fingers on right hands of two different people; and (**g**) images of the ring fingers on right hands of two different people.

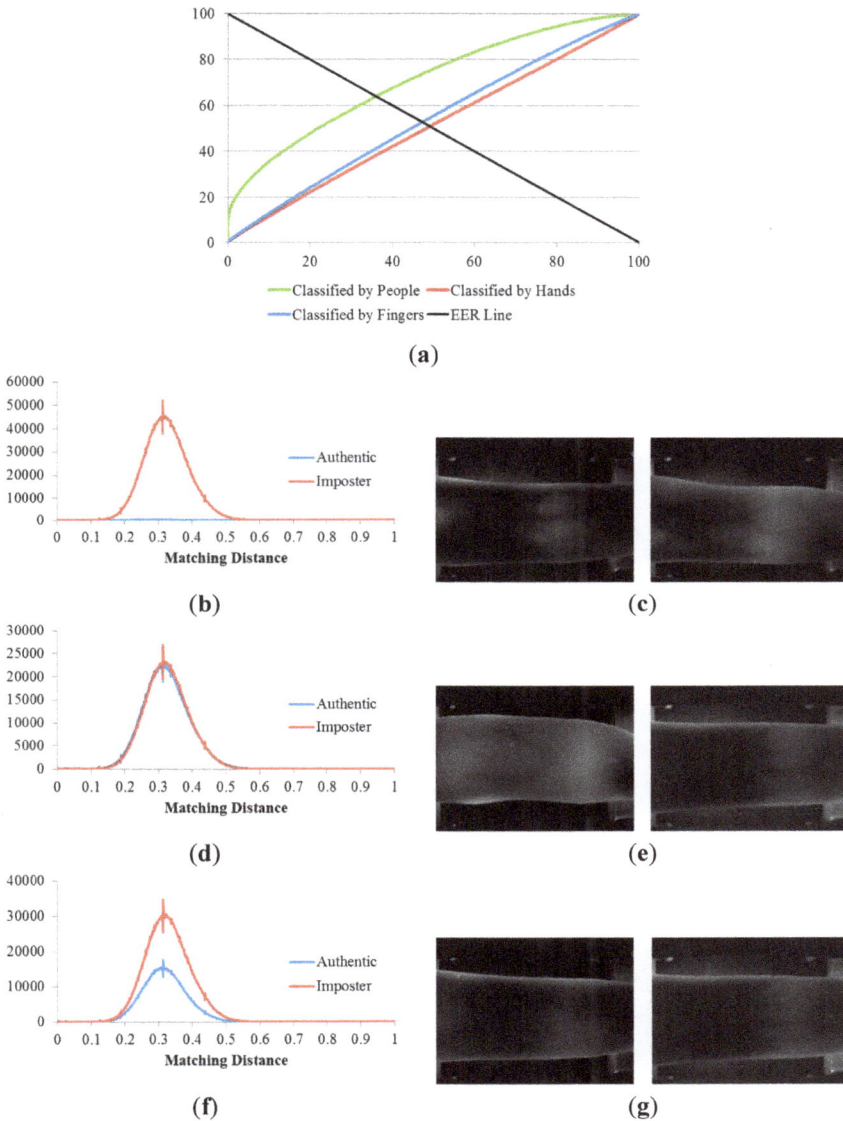

Figure 15. Results of the 2nd, 3rd, and 4th experiments on the open database: (**a**) ROC curves of the results of the three experiments; matching distribution of (**b**) the experiment classified by people (2nd experiment); (**d**) the experiment classified by hands (3rd experiment); and (**f**) the experiment classified by fingers (4th experiment); each shown with its corresponding false rejection error case: (**c**) images of the right middle and right ring fingers of the same person; (**e**) images of the index and middle fingers on left hands of two different people; and (**g**) images of the middle fingers on two hands of two different people.

Table 3 indicates that, when the three databases are compared, the lowest EERs (the highest d-prime value) were obtained from the experiment classified by people (the 2nd experiment), the second lowest EERs (the second highest d-prime value) were obtained from the experiment classified by fingers (the 4th experiment), and that classified by hands (3rd experiment) produced the highest EERs (the lowest d-prime value). This sequence was consistent for all three of the databases.

Consequently, we are able to conclude that, finger-vein dissimilarity increases in the order people, fingers, and hands, respectively. In other words, the discrimination between finger-vein images from different people is larger than that between the different finger types (index, middle, and ring fingers) and that between hands from different sides (left or right hands).

The plots of the ROC curves and matching distributions of authentic and imposter tests obtained from the three experiments (the 2nd, 3rd, and 4th experiments) as well as the error cases for the good-quality, mid-quality, and open databases are shown in Figures 13–15, respectively. In the 2nd experiment (classified by people), the cases for which a false rejection was obtained were for different fingers from the same person. The false rejection cases of the 3rd experiment (classified by hands) were the matching pair of vein images of fingers from the same hand side, but belonging to different people or captured from different fingers. Similarly, the false rejections of the 4th experiment (classified by fingers) are cases in which images were captured from the same finger types (*i.e.*, index, middle, or ring fingers) but belonged to different people or hand sides.

3.3. Experimental Results Using Multiple Images for Enrollment

In this experiment, we used a number of input finger-vein images for enrollment instead of using only one image as was done previously [7–9,11,16–23,25]. The method involving the enrollment of finger-vein data using the average of multiple finger-vein images is as follows. After the input images were captured for enrollment, they were processed and normalized by the methods described in Sections 2.2 and 2.3. From the image consisting of 50×20 pixels, obtained as a result of sub-sampling, we obtained the average image from which we extracted the LBP code which was then enrolled into the system. We applied this method by using either three or five enrollment finger-vein images to compare the matching accuracies with the conventional method, which only uses one image for enrollment. Examples of the average images generated from the 50×20 pixel images are shown in Figure 16. The experiments were conducted on the good-quality database as demonstrated in Figure 9a.

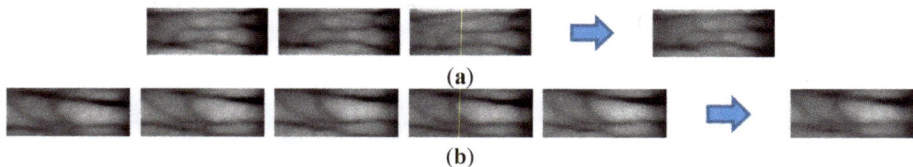

Figure 16. Normalized finger-vein images and their average images when: (**a**) three images; and (**b**) five images were used for enrollment.

When three images were used for enrollment, these were selected from the 10 images of each finger of the same user. Then, we extracted the finger-vein code from the average of these three images, and used the data extracted from the remaining seven images of the same finger to perform authentic matching tests. For the imposter matching tests, we used the images of the other fingers to perform matching with the average image generated for the enrolled finger. This experimental method is illustrated in Figure 17.

Assuming that the images from different fingers, hands, and people belong to different classes, the good-quality database contained 120 classes in total, as shown in Table 2. When three images were used for enrollment, the number of authentic matching tests was 100,800 ($_{10}C_3 \times 7$ (the number of remaining images in the same class) $\times$ 120 (the number of classes)), whereas the number of imposter matching tests was 17,136,000 ($_{10}C_3 \times 10$ (the number of images in other classes) $\times$ 119 (the number of other classes) $\times$ 120 (the number of classes from which images for enrollment were selected)).

When five images were used for enrollment, the number of authentic matches was 151,200 ($_{10}C_5 \times 5$ (the number of remaining images in the same class) $\times$ 120 (the number of classes)), whereas that of imposter matches was 35,985,600 ($_{10}C_5 \times 10$ (the number of images in other classes) $\times$ 119 (the number of other classes) $\times$ 120 (the number of classes from which images for enrollment were selected)).

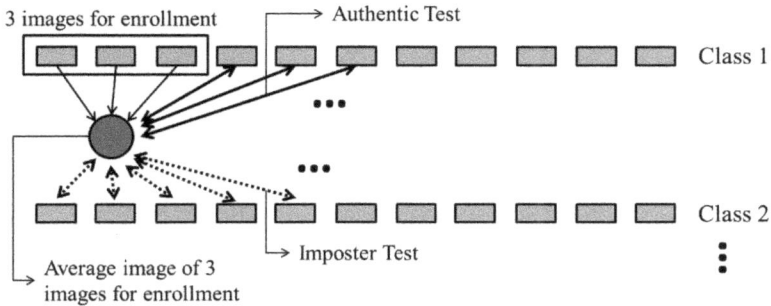

Figure 17. Experimental method when three images were used for finger-vein enrollment.

The experimental results of the methods in which three and five images were used for enrollment, are compared with those obtained by the conventional method (using one image for enrollment) in Table 4, where it can be seen that the matching accuracy was enhanced by increasing the number of enrollment images, in terms of low EER and high d-prime values. The ROC curves and the distribution plots of authentic and imposter tests corresponding to the results in Table 4 are shown in Figures 18 and 19, respectively, and can be explained as follows. In the finger-vein database, matching errors were mostly caused by misalignment at the time when the input finger-vein images were initially recorded, which subsequently resulted in translation errors in the normalized images of 50 $\times$ 20 pixels. The use of image averaging reduced the translation errors within the normalized images and increased the similarities between the enrolled and the matched finger-vein data. Table 5 shows examples of error cases resulting in false rejection when the enrolled images were compared with the test image in the same class, listed according to the number of images used for enrollment.

Table 4. Comparative results when multiple images were used for enrollment.

Number of Images for Enrollment	EER (%)	d-Prime
1	0.474	5.805
3	0.454	6.406
5	0.362	6.633

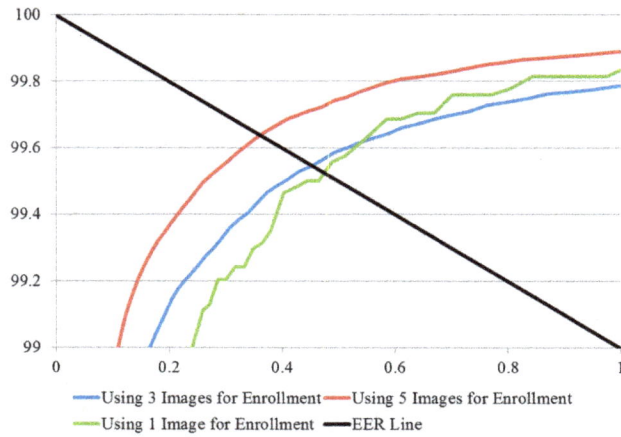

Figure 18. ROC curves of using multiple images for enrollment methods.

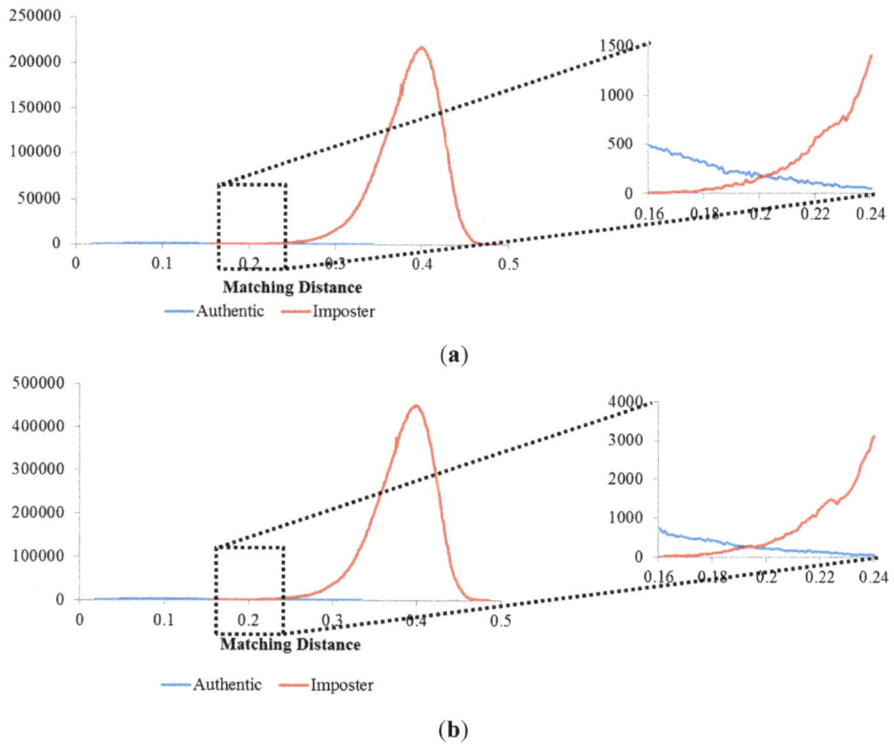

(a)

(b)

Figure 19. Matching distance distributions of authentic and imposter tests using (a) three images; and (b) five images for enrollment methods.

Table 5. False rejection cases: Comparison of the detected input finger-vein images with the enrolled images.

Number of Images for Enrollment	Enrolled Images	Average Image for Enrollment	Input Image
1		N/A	
		N/A	
3			
5			

From Table 5, we can see that, when one image was used for finger-vein enrollment, false rejection was mostly caused by translational errors between images of the same finger. These errors can either occur during translations in the horizontal direction of the image (the first row of Table 5) or in the vertical direction of the image (the second row of Table 5).

The reason why false rejections occur when either three or five images are used for enrollment is as follows. The misalignment between finger-vein images selected for enrollment resulted in blurred vein lines and the appearance of artifacts in the average images that were generated. Consequently, this led to high matching distances between the enrolled finger-vein data and test data, and these cases were misclassified into the imposter matching class.

3.4. Experimental Results Using Score-level Fusion Methods with Multiple Input Images

In the final set of experiments, we evaluated the matching accuracies and classifying ability of the system by using score-level fusion methods with multiple input images. This involved the application of SUM and PRODUCT rules, of which the formulas are expressed by Equations (8) and (9), to combine either three or five matching scores, which were then used to classify images as being either authentic or those of an imposter.

$$SUM\ rule: \qquad d_S = \sum_{i=1}^{N} d_i \tag{8}$$

$$PRODUCT\ rule: \quad d_P = \prod_{i=1}^{N} d_i \tag{9}$$

where d_i is the original matching score between the ith input finger-vein image and the one that was enrolled, and d_S and d_P are the resulting matching scores obtained by using the SUM and PRODUCT rules, respectively. N (3 or 5) is the number of scores to be combined.

The experiments were conducted on the good-quality database of Figure 9a as follows. From the 10 finger-vein images of each finger of an individual person in the database, we selected either three or five images as the authentic test images and used the remaining seven or five images as the enrolled images, respectively. For the imposter tests, we considered each of the 10 images of the other fingers in the database as the enrolled finger-vein image. For each enrolled image, we calculated the matching scores with the test images, combined these scores using the SUM and PRODUCT rules, and used the fused scores for final decisions. The use of this experimental method produced the same number of authentic and imposter-matching test results as for the experiments in which multiple images were used for enrollment in Section 3.3. That is because the number of images used for enrollment in the previous experiments (Section 3.3) and the number of scores used for score-level fusion in these experiments were the same, *i.e.*, three and five. Therefore, for each of the rules, SUM and PRODUCT, when three scores were used for fusion purposes, the numbers of authentic and imposter tests were 100,800 and 17,136,000, respectively, whereas the use of the five-score-level fusion method produced 151,200 authentic matches and 35,985,600 imposter matches.

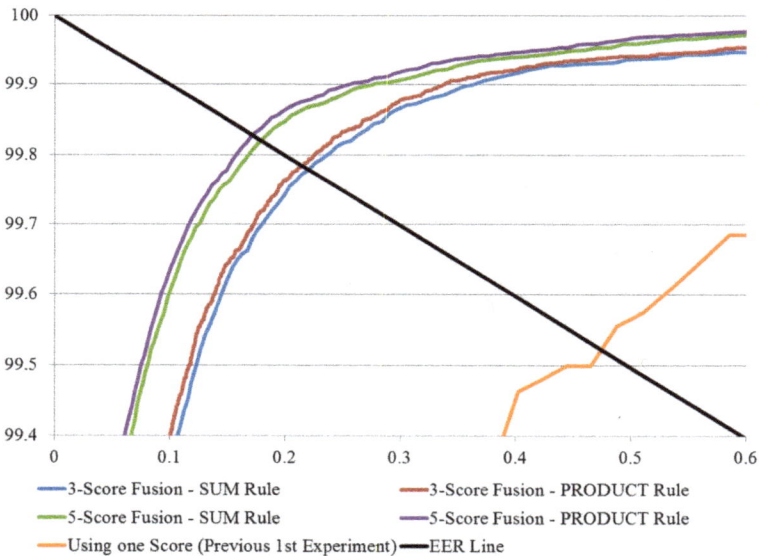

Figure 20. ROC curves obtained from the experiments with or without score-level fusion (the 1st experiment of Table 3) on good-quality database.

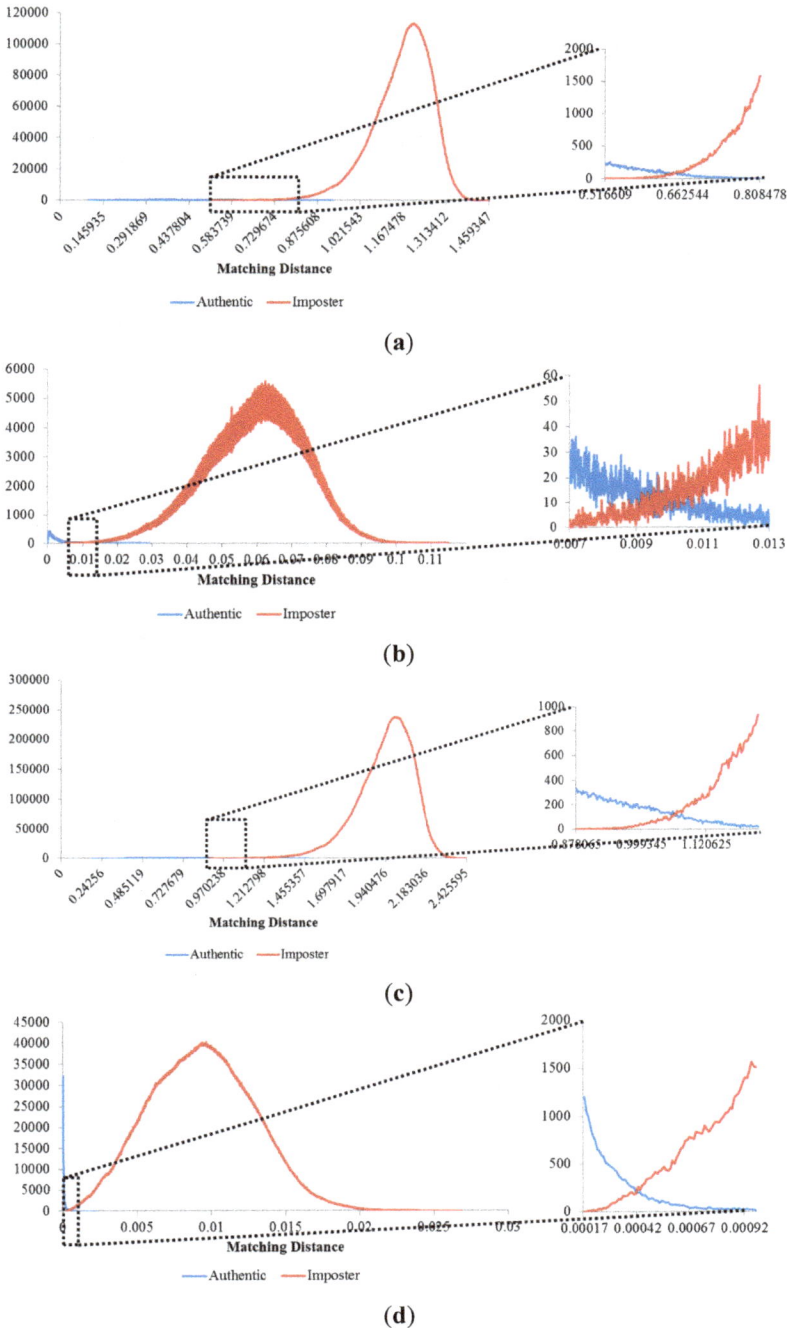

Figure 21. Matching distance distributions of authentic and imposter tests using score-level fusion methods: Three-score fusion method (**a**) using the SUM rule; and (**b**) using the PRODUCT rule; five-score fusion method (**c**) using the SUM rule; and (**d**) using the PRODUCT rule.

The experimental results of the score-level fusion methods were compared with the previous experiment in which one score was used for finger-vein recognition (the 1st experiment performed on the good-quality database of Table 3 in Section 3.2) and shown in Table 6. The ROC curves and the distribution plots of the authentic and imposter matching scores obtained from these experiments are shown in Figures 20 and 21.

Table 6. Comparisons of the matching accuracies of the score-level fusion methods and the previous matching method in which one finger-vein image was used.

Number of Score	Score-Level Fusion Rule	EER (%)	d-Prime
1	N/A	0.474	5.805
3	SUM	0.220	6.892
	PRODUCT	0.215	5.573
5	SUM	0.180	7.183
	PRODUCT	0.172	3.755

It can be seen from Table 6 that the score-level fusion methods enhanced the matching accuracies in that they resulted in low EER values and that the best results (the EER of 0.172%) were obtained in case that five matching scores were fused with the PRODUCT rule. In the case of using the same number of fused scores, the PRODUCT rule produced a lower EER value compared to the SUM rule. However the d-prime value of PRODUCT rule was lower than that of SUM rule. In general, the case of lower EER produces that of the higher d-prime value only in the case that the authentic and imposter distributions are similar to Gaussian shape, respectively. However, the authentic distributions obtained by PRODUCT rule of Figure 21b,d are different from the Gaussian shape, which causes the d-prime value not to correctly reflect the accuracy of finger-vein recognition. Therefore, the d-prime value of PRODUCT rule was inconsistently lower than that of SUM rule in Table 6.

3.5. Discussions

Regarding the issue of using average images for feature extraction as shown in Figures 16 and 17, the method of selecting one enrolled image (whose finger-vein code shows the minimum distances compared to the codes of other enrolled images) has been most widely used (1st method). However, the finger-vein code of one image among three or five enrolled images for enrollment is selected by this method, which cannot fully compensate for the differences among three or five enrolled images. Therefore, we adopt the method of using average image for enrollment as shown in Figures 16 and 17 (2nd method). To prove this, we compared the accuracy of finger-vein recognition by this 1st method with that by the 2nd method. The EER (d-prime) by the 1st method with three and five images for enrollment are 0.468% (6.128) and 0.412% (6.597), respectively. By comparing the EER (d-prime) by the 2nd method as shown in the 3rd and 4th rows of Table 4, we confirm that our 2nd method using average image for enrollment outperforms the 1st method.

The method of simply averaging the images for enrollment can be sensitive to image alignment and detailed features can be lost in the average image. In order to solve this problem, in our research, the misalignment among the images was firstly compensated by template matching before obtaining the average image. For example in Table 5, in the case that the number of images for enrollment is 3, the horizontal and vertical movements of the second enrolled image based on the first one are measured by template matching with the first enrolled image. If the measured horizontal and vertical movements of the second enrolled image are −2 and −1 pixels, respectively, for example, the compensated image is obtained by moving the original second enrolled image by +2 and +1 pixels, respectively, in the horizontal and vertical directions. From this, we can obtain the (compensated) second enrolled image where the misalignment based on the first enrolled image is minimized. Same procedure is iterated with the third enrolled image. From this procedure, we can obtain three (compensated) enrolled images where the misalignment between each other is minimized, and these three images are averaged for obtaining one enrolled image. Therefore, we can solve the problem that the average image is sensitive to image alignment and detailed features can be lost in the average image.

The total number of images in the good-quality database was 1200 (20 people × 2 hands × 3 fingers × 10 images), and that in the mid-quality database is 1980 (33 people × 2 hands × 3 fingers × 10 images). In order to obtain the meaningful conclusions and prove our conclusion irrespective of kinds of database, we also include the third open database for experiments. The total number of images in the open database was 3816 (106 people × 2 hands × 3 fingers × 6 images). Consequently, a total of 6996 images were used for our experiments, and we obtained the conclusion through a great deal of authentic and imposter matching, as shown in Table 2.

The original LBP used in our method can be more sensitive to noise than the uniform LBP. Therefore, in our method, the sub-sampled image of 50 × 20 pixel is used for feature extraction by LBP as shown in Figure 4c,f,i, which can reduce the noise in the image for feature extraction. In addition, the two cases of LBP codes in Figure 22c are assigned as same decimal code of 1 by the uniform LBP although they are actually different LBP code (00000001 (left case) and 00010000 (right case)), which can reduce the dissimilarity between two different patterns of finger-vein image. Therefore, we use the original LBP method in our research.

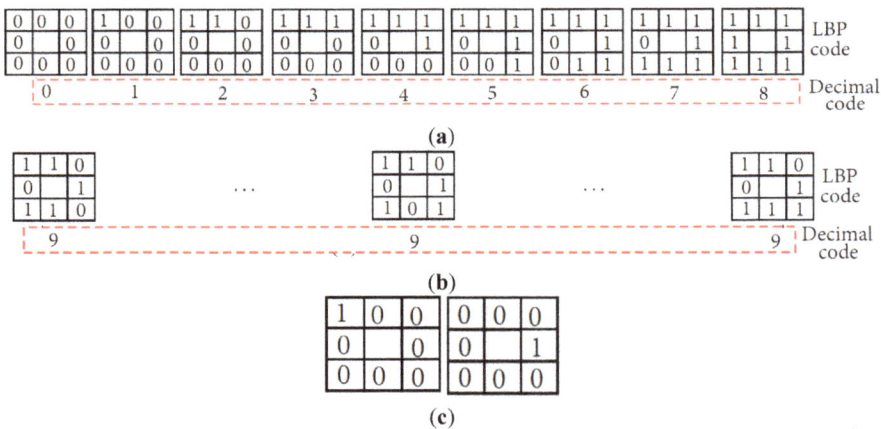

Figure 22. Example of uniform and nonuniform patterns and their assigned decimal codes by uniform LBP, respectively: (**a**) uniform patterns; (**b**) nonuniform patterns; (**c**) two cases of decimal code of 1 by uniform LBP.

We compared the accuracies by our original LBP and those by uniform LBP. The EER (d-prime) by uniform LBP with 1st, 2nd, and 3rd databases are 0.925% (5.122), 4.513% (3.443), and 12.489% (2.114). The EERs by uniform LBP are larger than those by original LBP of the 1st row of Table 3. In addition, the d-prime values by uniform LBP are smaller than those by original LBP of the 1st row of Table 3. From that, we can confirm that the performance by our original LBP is better than that by uniform LBP.

4. Conclusions

This paper proposed a new finger-vein capturing device that relies on accurate finger positioning to reduce misalignment when vein images are captured. This device was used to capture images to construct a database composed of good-quality finger-vein images, which were compared to the images in the mid-quality database (which was used in previous research) and an open database. The images in the good-quality database produced lower matching EER and higher d-prime values. Based on the comparative experimental results considering finger-vein dissimilarities between people, hands, and fingers in the three databases, we evaluated the factors that affect the accuracy of finger-vein recognition and concluded that finger-vein dissimilarity decreases for people, fingers, and hands in that order. We also proposed a method based on the use of multiple images to generate an image for finger-vein enrollment, instead of using one image as done previously. For our final set of experiments, we proposed a recognition method using score-level fusion obtained by using SUM and PRODUCT rules. The experimental results obtained for images from the database captured by our device, showed that the use of multiple enrollment images and score-level fusion could enhance the matching accuracies by reducing the EER. For future work, we plan to evaluate the various factors determining the accuracies of hand vein or palm vein recognition systems. In addition, we would also consider evaluating the effect of race, age, and gender on the accuracy of vein recognition.

Acknowledgments: This research was supported by the Basic Science Research Program through the National Research Foundation of Korea (NRF) funded by the Ministry of Education (NRF-2012R1A1A2038666).

Author Contributions: Tuyen Danh Pham and Kang Ryoung Park designed the overall system for finger-vein recognition. In addition, they wrote and revised the paper. Young Ho Park, Dat Tien Nguyen, and Seung Yong Kwon implemented the algorithm of score-level fusion, and helped with the collecting database with the analyses of experimental results.

Conflicts of Interest: The authors declare no conflict of interest.

References

1. Prabhakar, S.; Pankanti, S.; Jain, A.K. Biometric recognition: Security and privacy concerns. *IEEE Secur. Priv.* **2003**, *1*, 33–42. [CrossRef]
2. Turk, M.A.; Pentland, A.P. Face Recognition Using Eigenfaces. In Proceedings of the IEEE Conference on Computer Vision and Pattern Recognition, Maui, HI, USA, 3–6 June 1991; pp. 586–591.
3. Daugman, J. How iris recognition works. *IEEE Trans. Circuits Syst. Video Technol.* **2004**, *14*, 21–30. [CrossRef]
4. Jain, A.K.; Ross, A.; Prabhakar, S. An introduction to biometric recognition. *IEEE Trans. Circuits Syst. Video Technol.* **2004**, *14*, 4–20. [CrossRef]
5. Yang, W.; Hu, J.; Wang, S. A finger-vein based cancellable bio-cryptosystem. *Lect. Notes Comput. Sci.* **2013**, *7873*, 784–790.
6. Yang, J.F.; Yang, J.L. Multi-Channel Gabor Filter Design for Finger-Vein Image Enhancement. In Proceedings of the Fifth International Conference on Image and Graphics, Xi'an, China, 20–23 September 2009; pp. 87–91.
7. Shin, K.Y.; Park, Y.H.; Nguyen, D.T.; Park, K.R. Finger-vein image enhancement using a fuzzy-based fusion method with Gabor and Retinex filtering. *Sensors* **2014**, *14*, 3095–3129. [CrossRef] [PubMed]
8. Park, Y.H.; Park, K.R. Image quality enhancement using the direction and thickness of vein lines for finger-vein recognition. *Int. J. Adv. Robot. Syst.* **2012**, *9*, 1–10.
9. Yang, J.; Shi, Y. Finger-vein ROI localization and vein ridge enhancement. *Pattern Recognit. Lett.* **2012**, *33*, 1569–1579. [CrossRef]
10. Yang, J.; Wang, J. Finger-Vein Image Restoration Considering Skin Layer Structure. In Proceedings of the International Conference on Hand-Based Biometrics, Hong Kong, China, 17–18 November 2011; pp. 1–5.
11. Yang, J.; Zhang, B.; Shi, Y. Scattering removal for finger-vein image restoration. *Sensors* **2012**, *12*, 3627–3640. [CrossRef]
12. Cho, S.R.; Park, Y.H.; Nam, G.P.; Shin, K.Y.; Lee, H.C.; Park, K.R.; Kim, S.M.; Kim, H.C. Enhancement of finger-vein image by vein line tracking and adaptive gabor filtering for finger-vein recognition. *Appl. Mech. Mater.* **2011**, *145*, 219–223. [CrossRef]
13. Zhang, J.; Yang, J. Finger-Vein Image Enhancement Based on Combination of Gray-Level Grouping and Circular Gabor Filter. In Proceedings of the International Conference on Information Engineering and Computer Science, Wuhan, China, 19–20 December 2009; pp. 1–4.
14. Pi, W.; Shin, J.; Park, D. An Effective Quality Improvement Approach for Low Quality Finger Vein Image. In Proceedings of the International Conference on Electronics and Information Engineering, Kyoto, Japan, 1–3 August 2010; pp. V1-424–V1-427.

15. Yu, C.B.; Zhang, D.M.; Li, H.B.; Zhang, F.F. Finger-Vein Image Enhancement Based on Multi-Threshold Fuzzy Algorithm. In Proceedings of the International Congress on Image and Signal Processing, Tianjin, China, 17–19 October 2009; pp. 1–3.

16. Yang, J.; Zhang, X. Feature-level Fusion of Global and Local Features for Finger-Vein Recognition. In Proceedings of IEEE 10th International Conference on Signal Processing, Beijing, China, 24–28 October 2010; pp. 1702–1705.

17. Lu, Y.; Yoon, S.; Park, D.S. Finger vein recognition based on matching score-level fusion of gabor features. *J. Korean. Inst. Commun. Inf. Sci.* **2013**, *38A*, 174–182. [CrossRef]

18. Qian, X.; Guo, S.; Li, X.; Zhong, F.; Shao, X. Finger-vein Recognition Based on the Score Level Moment Invariants Fusion. In Proceedings of International Conference on Computational Intelligence and Software Engineering, Wuhan, China, 11–13 December 2009; pp. 1–4.

19. Yang, G.; Xi, X.; Yin, Y. Finger vein recognition based on a personalized best bit map. *Sensors* **2012**, *12*, 1738–1757. [CrossRef] [PubMed]

20. Park, Y.H.; Tien, D.N.; Lee, H.C.; Park, K.R.; Lee, E.C.; Kim, S.M.; Kim, H.C. A Multimodal Biometric Recognition of Touched Fingerprint and Finger-Vein. In Proceedings of International Conference on Multimedia and Signal Processing, Guillin, China, 14–15 May 2011; pp. 247–250.

21. Nguyen, D.T.; Park, Y.H.; Lee, H.C.; Shin, K.Y.; Kang, B.J.; Park, K.R. Combining touched fingerprint and finger-vein of a finger, and its usability evaluation. *Adv. Sci. Lett.* **2012**, *5*, 85–95. [CrossRef]

22. Cui, F.; Yang, G. Score level fusion of fingerprint and finger vein recognition. *J. Comput. Inf. Syst.* **2011**, *7*, 5723–5731.

23. He, M.; Horng, S.J.; Fan, P.; Run, R.S.; Chen, R.J.; Lai, J.L.; Khan, M.K.; Sentosa, K.O. Performance evaluation of score level fusion in multimodal biometric systems. *Pattern Recognit.* **2010**, *43*, 1789–1800. [CrossRef]

24. Qin, B.; Pan, J.-F.; Cao, G.-Z.; Du, G.-G. The Anti-spoofing Study of Vein Identification System. In Proceedings of International Conference on Computational Intelligence and Security, Beijing, China, 11–14 December 2009; pp. 357–360.

25. Nguyen, D.T.; Park, Y.H.; Shin, K.Y.; Kwon, S.Y.; Lee, H.C.; Park, K.R. Fake finger-vein image detection based on fourier and wavelet transforms. *Digit. Signal Process.* **2013**, *23*, 1401–1413. [CrossRef]

26. Webcam C600. Available online: http://www.logitech.com/en-us/support/5869 (accessed on 15 April 2015).

27. Dongguk Finger-Vein Database. Available online: http://dm.dgu.edu/link.html (accessed on 4 June 2015).

28. SDUMLA-HMT Finger Vein Image Database. Available online: http://mla.sdu.edu.cn/sdumla-hmt.html (accessed on 15 April 2015).

29. Yin, Y.; Liu, L.; Sun, X. SDUMLA-HMT: A multimodal biometric database. *Lect. Notes Comp. Sci.* **2011**, *7098*, 260–268.

sensors

[MDPI]

Article

Full-Field Optical Coherence Tomography Using Galvo Filter-Based Wavelength Swept Laser

Muhammad Faizan Shirazi [1], Pilun Kim [2], Mansik Jeon [1,*] and Jeehyun Kim [1,2]

[1] School of Electronics Engineering, College of IT Engineering, Kyungpook National University,
80 Daehak-ro, Bukgu, 41566 Daegu, Korea; faizanshirazi110@gmail.com (M.F.S.); jeehk@knu.ac.kr (J.K.)
[2] Oz-Tec Co. Ltd., Office 901, IT Convergence Industrial Building, 47 Gyeongdae-ro, 17-gil, Bukgu,
41566 Daegu, Korea; pukim@oz-tec.com
* Correspondence: msjeon@knu.ac.kr; Tel./Fax: +82-53-950-7221

Academic Editor: Gonzalo Pajares Martinsanz
Received: 30 September 2016; Accepted: 15 November 2016; Published: 17 November 2016

Abstract: We report a wavelength swept laser-based full-field optical coherence tomography for measuring the surfaces and thicknesses of refractive and reflective samples. The system consists of a galvo filter–based wavelength swept laser and a simple Michelson interferometer. Combinations of the reflective and refractive samples are used to demonstrate the performance of the system. By synchronizing the camera with the source, the cross-sectional information of the samples can be seen after each sweep of the swept source. This system can be effective for the thickness measurement of optical thin films as well as for the depth investigation of samples in industrial applications. A resolution target with a glass cover slip and a step height standard target are imaged, showing the cross-sectional and topographical information of the samples.

Keywords: OCT; wavelength swept laser; full-field OCT; large area scanning; galvo filter

1. Introduction

With the rapid development of technology, the precision requirement in semiconductor chip technology, printed electronics, and various component industries has been increased owing to miniaturization. Additionally, the complexity of products has been increased using micro-processing technology. As a result, surface profile measurement has become an important issue in industrial inspection processes to reduce the defect rate and to fulfill the consumers' demand. Many nondestructive three-dimensional (3D) surface profiling techniques have been developed to monitor product quality and to improve process efficiency [1–4]. Furthermore, different optical inspection methods have been proposed for 3D surface profile imaging with cost-effective, nondestructive, noncontact, high-speed, and high-accuracy solutions [5–9]. The amplitude variation is recorded by an optical interferometer with axial scanning methods utilizing the mechanical movement [6–8]. However, the susceptibility to hysteresis of the mechanical movement in imaging systems makes it unsuitable for profile measurement.

Optical coherence tomography (OCT) is a noninvasive and nondestructive technique for obtaining high-resolution tomographic images of microstructures [10]. OCT utilizes a Michelson interferometer with a low-coherence light source to obtain the cross-sectional depth information of an object. Using this method, three-dimensional images are obtained either in the time or frequency domain. Therefore, OCT is classified into two categories, the time domain and frequency domain [11]. The time domain (TD) OCT is relatively slow and has a low signal-to-noise ratio compared to the frequency domain (FD). The depth information obtained in the TD-OCT is because of the mechanical movement of the reference arm. In the frequency domain, the Fourier transform of the fringe patterns reveals the depth information of the sample in the FD-OCT [12,13]. In the frequency domain, the Fourier transform of the fringe patterns reveals the depth information of the sample. FD-OCT is further classified as spectral domain (SD) OCT and swept source (SS) OCT. In the SD-OCT, a broadband source is used for the illumination of the sample and reference, whereas the interference fringe is detected by a spectrometer. In the SS-OCT, a broadband swept source is employed to obtain the spectral fringes using a photodetector. Further, with the advancement in OCT, parallel acquisition of the interference fringe is possible by using an area camera in a full-field configuration; as a result, fast acquisition of the volume data is possible in the time domain as well as in the frequency domain [14,15]. The frequency domain full-field setup employs a wavelength swept laser for acquiring the interference signal from the sample and the reference arms without mechanical scanning, compared to the time domain. The frame rate in the frequency domain is a function of the sweep rate of the source and the camera acquisition speed.

OCT has extended its applications not only in biomedical imaging but also in agriculture, where it is used for the early detection of diseases in seeds and plants by investigating morphological abnormalities [16,17]. Numerous industrial applications of the OCT such as optical thin film, touch-screen panel assessment, liquid crystal display (LCD), and light-emitting diode (LED) inspections have been reported [18–21]. Similarly, various techniques have been employed in the OCT for increasing the image resolution, penetration depth, sensitivity, and field of view [22–24]. In the parallel area's acquisition scheme, various samples have been investigated using a swept source with different filtering mechanisms and optical architecture. These samples include finger print detection, silicon integrated circuits, composite material reconstruction, micro-lens array measurement, in vivo human retina, ex vivo porcine eye, and tree shrew retina [25–28]. Similarly, wavelength scanning interferometry has been demonstrated for the profile measurement of transparent films and industrial objects [29–32]. In several applications, the thickness of a thin film including the measurements of both the top and bottom surfaces is critical. Similarly, in the semiconductor industry, a micrometer analysis includes the solder mask connection, through silicon via (TSV), and multilayer silicon wafer investigation.

In order to develop the wavelength swept laser, different filtering mechanisms have been used which includes the Fabry-Perot tunable filter, acousto-optic tunable filter, fiber Bragg grating, polygon mirror scanning filter, etc. Wavelength swept laser-based full-field OCT systems have several advantages compared to the other systems. One of the advantages is the parallel acquisition of signals using an area detector to get the area image of the sample. As a result, more power can be spread across the sample. The other advantage is that it does not need mechanical scanning in the sample arm to acquire a volume image. Because of the absence of x-y-z scanning, the full-field system is robust with high phase stability. The depth information is obtained by sweeping the wavelength in SS-OCT; the wavelength sweeping is done by the filtering mechanism incorporated in the source.

In this study, a simple, high-power galvo filter–based wavelength swept laser is integrated with a full-field OCT system for obtaining the tomographical as well as topographical features of the samples. The custom-built high-power wavelength swept laser is used to demonstrate the performance of the full-field OCT system for area scanning of samples. This study will contribute to the application of full-field OCT systems in the semiconductor industry for defect identification and investigation of samples.

2. Materials and Methods

2.1. Wavelength Swept Laser

Figure 1a shows the schematic diagram of a galvo filter–based wavelength swept laser. The wavelength swept laser is composed of a fiber coupled semiconductor optical amplifier (SOA), two isolators, a 50:50 optical coupler, an optical circulator for the formation of a unidirectional ring cavity, two polarization controllers, and a galvo filter–based external cavity filter for selection of the lasing wavelength. The semiconductor optical amplifier (SOA-372, Superlum, Cork, Ireland) with maximum gain at a center wavelength of 848 nm is utilized as the broadband gain medium.

All the components are connected in a ring cavity configuration through optical fibers. The SOAs have a high gain coefficient, robustness, and quick response characteristics in the gain medium for wavelength amplification.

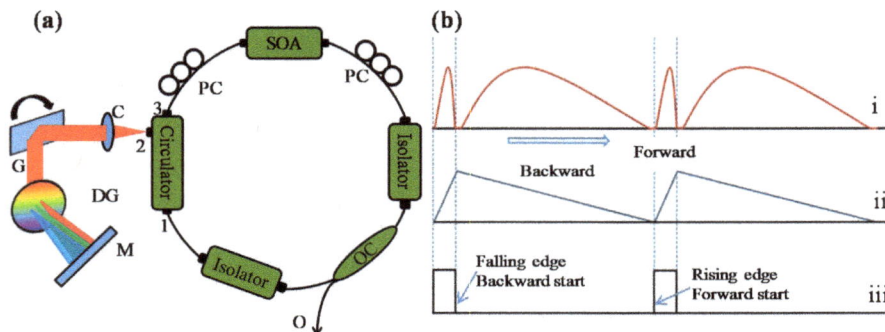

Figure 1. Schematic and timing diagram of the wavelength swept laser. (**a**) Schematic diagram of a galvo filter–based wavelength swept laser. C: collimator, DG: diffraction grating, G: galvo scanner, M: mirror, O: final output, OC: optical coupler, PC: polarization controller, SOA: semiconductor optical amplifier. (**b**) Timing of the main signals of the wavelength swept laser (i) scanning spectrum, (ii) waveform input to the galvo scanner driving board, and (iii) trigger signal at a 1 Hz frequency and a 10% duty cycle.

The SOA has a broad amplified spontaneous emission (ASE) spectrum with 6 mW of optical power at a current of 145 mA. One side of the SOA is connected to the 50:50 optical fiber coupler through an isolator and the other side is connected to port-3 of the optical circulator. One of the output terminals of the coupler is connected to port-1 of the optical circulator through the other isolator, while the other output provides the wavelength swept laser spectrum. In this wavelength swept laser setup an external cavity is utilized as the band pass filter for selecting the wavelength. The wavelength selection filter is connected to port-2 of the optical circulator. The components used for wavelength selection in the external cavity configuration are a transmission type diffraction grating with 1800 lines per millimeter from Wasatch Photonics Inc. (Logan, UT, USA), a galvo scanner (GS001, Thorlabs, Newton, NJ, USA) and a mirror. The optical alignment of the filter is as follows: The output light from port-2 of the optical circulator is collimated and is incident on the center of the galvo scanning mirror after which it is reflected and passed through the diffraction grating, where it is dispersed into discrete wavelengths. The grating is aligned for providing the angular diffraction as a function of the input angle and wavelength with a maximum light intensity. As the grating's grove direction is vertical, the incident light on the grating is diffracted horizontally.

The spectrally distributed light reflects back from the mirror placed at a centimeter distance from diffraction grating. This light follows the same return path and is coupled back to the optical circulator after passing through the grating, reflecting from the galvo mirror and is then incident on the collimator. Note that only the component of light that is normal to the mirror is coupled back to the fiber-based ring cavity. The galvo scanner is repeatedly scanned to produce the lasing output that results in an amplified optical signal at the output of the SOA. The isolators in the ring cavity ensure a unidirectional flow and protect the source by preventing reflections back from the application system to the cavity, to avoid catastrophic damages to the SOA. The sweeping output as a function of the galvo scanner frequency can be obtained from the output terminal of the optical coupler. The spectral line width of the feedback light is 0.05 nm with instantaneous coherence length of ~12 mm. The ring cavity length of the wavelength swept laser is approximately 2 m. This cavity length is acceptable for low sweep rate.

Figure 1b shows the input and output signal timing diagrams during the working of the wavelength swept laser. Three different signals are depicted: (i) the scanning spectrum of the wavelength swept laser, achieved when a (ii) triangular input signal is applied to the scan drive and simultaneously a (iii) trigger signal is generated. The arrow indicates the directions of the signals. The trigger signal has a 10% duty cycle that corresponds to the ramp-up duration of the triangular signal. The trigger signal can be synchronized to the application system for starting and stopping the data acquisition. In a single cycle of the input triangular signal to the galvo scanner drive, two sweeps are generated, as shown in (i). The forward and backward in (ii) indicate the voltage increment and decrement applied to the galvo scanner. In this wavelength swept laser the sweeping frequency, duty cycle, and the output power can be controlled by a customized electronic circuit with software control. This laser has a broad sweeping range with a 3 dB bandwidth of 48 nm as shown in Figure 2. Detailed information regarding this wavelength swept laser has been provided in a previous research [33]. The output profile of this source is assumed almost Gaussian. Therefore, the coherence length (l_c) of the wavelength swept laser can be calculated using the following equation:

$$lc = \frac{4\ln(2)}{\pi}\frac{\lambda_0^2}{\Delta\lambda} \tag{1}$$

From the above description of the source, we know that $\Delta\lambda$ = 48 nm and λ_0 = 848 nm; therefore, substituting these values in Equation (1) will give us a coherence length of 13.184 μm. The axial resolution of the OCT system is half of l_c; therefore, it is approximately 6.59 μm.

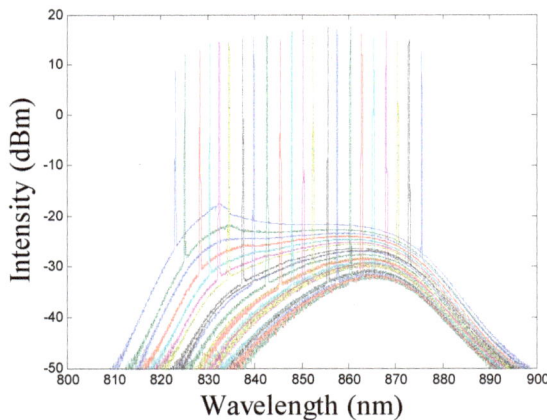

Figure 2. Swept source spectral bandwidth profile.

2.2. Full-Field OCT System

The optical architecture of a wavelength swept laser-based full-field OCT system is shown in Figure 3. A Michelson interferometer setup is used to obtain an interference fringe signal by reflections from the sample and reference arms. The light from the wavelength swept laser is collimated using a metal-coated reflective parabolic mirror collimator (RC12APC-P01, Thorlabs, USA). The collimated light then passes through the beam splitter and it is divided into two paths. A portion of the light travel towards the sample and the remaining light is incident on the reference mirror. A variable neutral density filter is placed in the reference arm to decrease the intensity of the light in order to control the optical power in the reference arm. A continuously variable ND filter (NDC-50C-2M, Thorlabs, USA) is rotated to adjust the reflection from the reference arm in proportion to the back scattering from the sample. Data are acquired by a frame grabber through two camera link cables. A customized electronic control board is used to control the sweep frequency, SOA current, galvo signal, and the trigger signal.

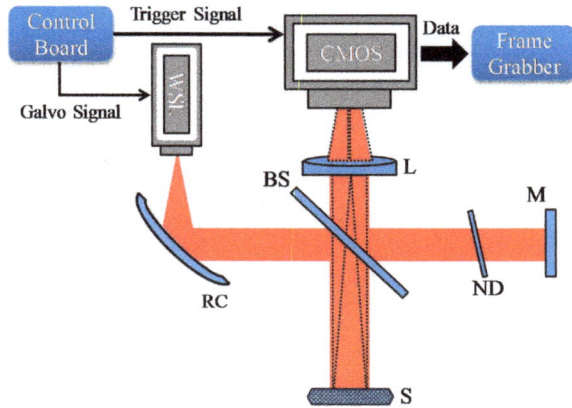

Figure 3. Schematic diagram of a full-field OCT system. BS: pellicle beam splitter, CMOS: complementary metal oxide semiconductor, L: lens, M: mirror, ND: neutral density filter, RC: reflective collimator, S: sample, WSL: wavelength swept laser.

An achromatic lens with a 30 mm focal length is utilized as the imaging lens for acquiring the raw fringe data from the sample and reference arms. The near infrared 8-bit area camera utilized here is the Basler ace (acA2000-340kmNIR, Basler, Ahrensburg, Germany) with a pixel resolution of 2048 × 1088 and maximum frame rate of 340 fps at full resolution. In order to avoid memory errors owing to large data and frame rate reduction problems, the region of interest is reduced to 500 × 500 pixels. The lateral resolution depends upon the numerical aperture of the lens and the pixel size of the camera. The axial resolution is a function of the spectral bandwidth of the source. The measured axial and lateral resolution of the system in current setup is ~10 μm and ~15 μm, respectively. The penetration depth depends on the wavelength and properties of sample. The sampling frequency of the camera also impacts the depth (axial) range. At sampling frequency of 340 fps, the wavelength sampling interval is ~0.15 nm which in turns gives the maximum imaging depth of ~2 mm. The system sensitivity drops off to 10 dB at 400 μm. The system has signal to noise ratio of 75 dB.

In the Fourier domain, the depth information can be obtained from the frequency components of the interference pattern after signal processing. If $S(x,y;k)$ is the spectral density of the wavelength swept laser, I_R and I_s are the intensities of the reflected light from the reference and sample arms, respectively; hence, the total intensity detected at the camera is given as:

$$I(x,y;k) = S(x,y;k) I_R + S(x,y;k) \iint_{-\infty}^{\infty} \sqrt{I_S(z) I_S(z')} \exp\left(i\left\{k\left[(z-z') + \theta(z) - \theta(z')\right]\right\}\right) dz dz'$$
$$+ 2S(x,y;k) \sqrt{I_R} \int_{-\infty}^{\infty} I_S(z) \cos\left[kz + \theta(z)\right] dz \tag{2}$$

where k is the wave number, z is the depth, and $\theta(z)$ is the phase shift. The first two terms in Equation (2) are dc components; the first is generated by the intensity reflection from the reference mirror and the second is owing to the mutual interference between different layers of the sample. The third component is the result of the interference between the reference and sample signals. The cross-sectional depth information, z, of the sample can be retrieved from the third term after performing fast Fourier transform from the k to the z domain. Due to the non-linearity of the galvo filter, the interference signals are first resampled and linearized before Fourier transformation. The wavelength swept laser has phase stability with a standard deviation of ±0.01 radian.

2.3. Data Processing Steps

For capturing images, a camera is synchronized with the wavelength swept laser using a trigger signal. The frame grabber utilized in this system is the PCIe-1433 (National Instruments, Austin, TX, USA) and a camera link full configuration is used to run the camera at a maximum frame rate. Figure 4 shows the signal processing mechanism employed for obtaining the OCT images in a wavelength swept laser based full-field OCT system. The mechanism is initiated by the acquisition of raw area images from the camera in the wavelength scanning direction of the wavelength swept laser. The camera starts the acquisition with the falling edge of the trigger and continues up to the rising edge of the trigger. All the acquired frames contain the interference fringe information along with the number of frames. Fast Fourier transform (FFT) of this acquired signal will provide the depth information. The data is resampled and linearized before FFT to avoid wave number non-linearization due to the non-linearity of galvo filter. In the case of the mirror, a single peak represents the mirror position as a function of the path difference between the two arms. In the case of complex samples, multiple peaks are obtained in the Fourier spectrum. By applying FFT for each interference signal, all the frames are processed and the resulting enface images at different depths can be seen (Figure 4). Similarly, by selective filtering of the peaks, the amplitude and phase information of a particular surface can be extracted with a micrometer to nanometer resolution. A LabVIEW program is coded to acquire the data and view the real-time cross-sectional image with the depth information. For enface OCT images, a program is coded to post-process the saved data. As mentioned earlier that the area image has dimension of 500×500 pixels with 340 frames per second. Therefore, the system has B-scan frame rate of 500 with each frame contains 500 A-scans and each A-scan contains 340 sample points.

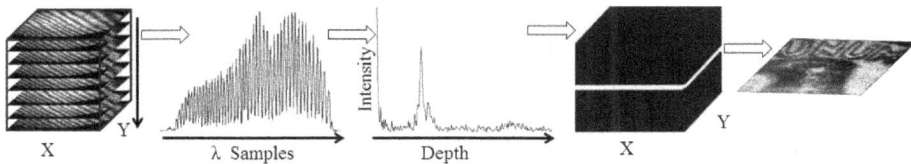

Figure 4. Signal processing steps (left to right) for acquiring the enface depth image. A stack of images is acquired by the camera along the wavelength axis, as indicated by the downward arrow. Next, a single interference signal is depicted with respect to the wavelength samples. A fast Fourier transform of this signal will provide the depth information. Hence, the respective enface depth image can be extracted from the 3D cube after processing all the acquired data.

3. Results and Discussion

Figure 5 shows the cross-sectional images using the proposed wavelength swept laser-based full-field OCT system. The wavelength swept laser is operated at 1 Hz with two sweep spectrums corresponding to a 10% forward sweep and a 90% backward sweep, as shown in (i) of Figure 1b. The camera starts the acquisition at the falling edge of the trigger signal connected to the frame grabber. The data acquired in the 90% backward sweep is processed in the 10% forward sweep duration and simultaneously displayed. Figure 5a,c show the raw area images of an Edmund optics target and an Edmund optics target with a 170-μm-thick cover slip, respectively. Similarly, Figure 5b,d are the cross-sectional images along the red-colored dashed line of the respective targets. Different fringe patterns appear in regions with and without the cover slip in Figure 5c. The fringe pattern is obtained as a result of the interference between the sample and reference arms, provided that the path difference is within the coherence length of the source. In Figure 5c, the fringe pattern in the right half is due to the interference between the reference arm and light reflected from the top of the cover slip and the Edmond optics target. The Edmund optics target has various lines per millimeter (l pmm), starting from five and reaching up to 200. In this experiment, the portions containing 25, 20, and 15 lines per millimeter are imaged with a field of view of 3 mm × 3 mm, corresponding to 500 × 500 camera pixels. In Figure 5b, the line separation can be seen clearly, corresponding to a lateral distance of 20 μm according to the target. The refractive index of the cover slip is 1.52 μm. The separation between the cover slip and the target can be clearly seen in Figure 5d, corresponding to an optical thickness of 112 μm. Owing to the change in the refractive index, there is a shift in the position of the surface of the resolution target, as can be seen at the intersection. The solid arrows indicate the fringe pattern, while the dashed arrows point to the background noise. Using this experiment, the tomographical thickness variation in the samples can be observed, and with calibration, an accurate measurement can be computed.

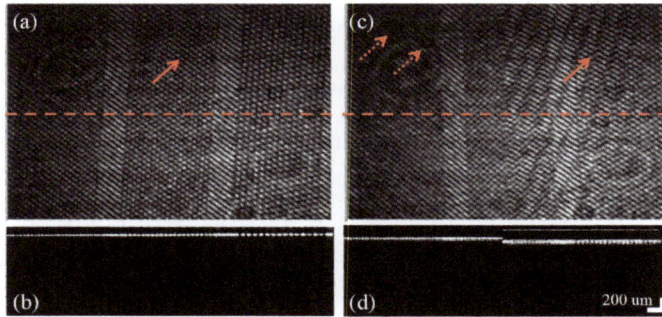

Figure 5. Tomographic images of the full-field OCT system. (**a**,**c**) Raw enface images of the Edmund optics target and the Edmund optics target with a 170-μm-thick cover slip, respectively. (**b**,**d**) Cross-sectional images, clearly distinguishing the line separations along the dashed lines in (**a**,**c**), respectively. The solid arrow in (**a**) shows the interference fringe because of the reference mirror and the resolution target, while in (**c**) the arrow indicates the fringe pattern owing to the cover slip, reference mirror, and resolution target. The dotted arrow shows the background noise on the sample.

The next experiment is performed for visualizing the topological information of a very large scale integration (VLSI) height standard target with a step height of 50 μm. Figure 6a shows a 3 mm × 3 mm portion of the standard mark on the VLSI target. The height variations are shown in Figure 6b,c along the vertical and horizontal cross-sections, respectively, of the red-colored dashed line in Figure 6a. These tomographical features can be seen in entirety in real time using the proposed system. For visualizing and detecting defects in industrial samples with refractive and reflective characteristics, the proposed system can be effective and can generate fault detection results with microscopic resolutions. Similarly, tomographical and topographical variation in samples can be precisely monitored using this system.

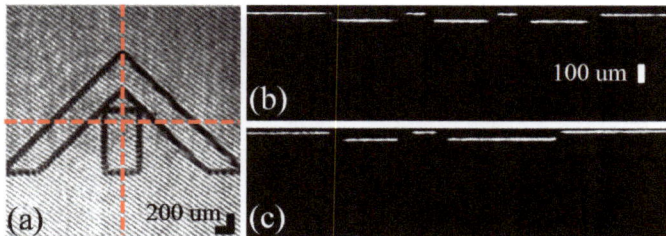

Figure 6. Tomographic images of the full-field OCT system; (**a**) Raw enface image of the VLSI 50 μm step height standard target; (**b**,**c**) Cross-sectional profiles along the horizontal and vertical red-colored dashed lines in (**a**), respectively.

Figure 7 shows the enface image of a portion of the VLSI target. The color-coded image in Figure 7a shows the topographical variations with an almost 50 μm step height with a 3 mm × 3 mm field of view. Figure 7b,c show the height variation along the horizontal and vertical white dashed line in Figure 7a, respectively. By utilizing this information, the topographical variations in industrial samples can be monitored with high precision. The system has height variation within the standard deviation of ±1 μm.

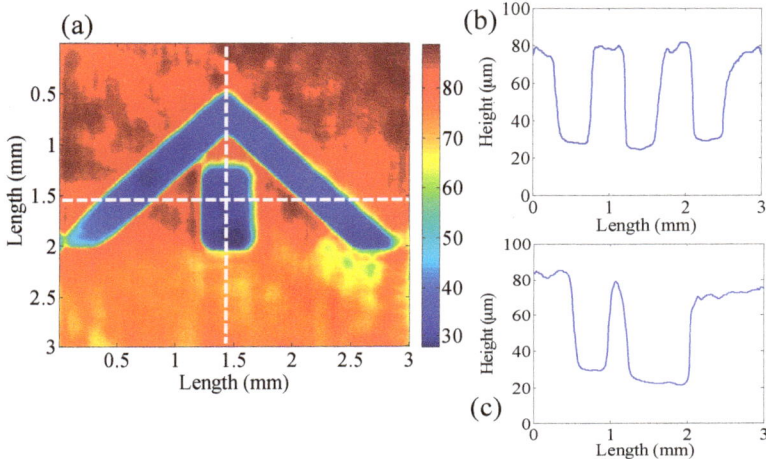

Figure 7. Tomographic image of the full-field OCT system; (**a**) Enface image of the VLSI 50 μm step height standard target; (**b**,**c**) Cross-sectional profiles along the horizontal and vertical white-colored dashed lines in (**a**), respectively.

In future works, the performance of this system can be improved by utilizing the 4*f* configuration in the detection part of the camera. By increasing the sweep duration of the wavelength swept laser, more images can be acquired to improve the imaging depth of the system. The further improvement of the system can be done by increasing the sweep rate of the source with a high speed camera. The spatial interference of light reflected from the sample affects the transverse resolution of the system. Therefore, the transverse resolution of the system can be improved by destroying the spatial coherence of the illumination beam by utilizing the diffuser or mode mixer.

4. Conclusions

We have demonstrated the tomographical and topographical features of samples using the wavelength swept laser-based full-field optical coherence tomography system. The Edmund optics target with a transparent cover slip indicates the lateral resolution and thickness of the cover slip. The optical arrangement is realized by using a high-power wavelength swept laser. The simple proposed optical system can play an important role in the inspection of industrial samples. The system has a comparatively large field of view, is easy to align with non-mechanical scanning components, and has a compact optical setup. In future works, this system will be used for measuring the varying thicknesses of optical thin films including both the top and bottom surfaces. Solder mask connection tests, the connectivity of through silicon vias, and multilayer silicon wafer investigations in the semiconductor industry are the areas in which the system performance can be further evaluated.

Acknowledgments: This study was supported by the BK21 Plus project funded by the Ministry of Education, Korea (21A20131600011). This research was also financially supported by the "Over regional linked 3D convergence industry promotion program (R0004589)" through the Ministry of Trade, Industry & Energy (MOTIE) and the Korea Institute for Advancement of Technology (KIAT).

Author Contributions: M.J. and P.K. designed the study. M.F.S. performed experiments, interpreted the data, and drafted the manuscript. J.K. critically revised and finalized the intellectual contents of manuscript.

Conflicts of Interest: The authors declare no conflict of interest.

References

1. Wang, T.J.; Sze, T. The image moment method for the limited range CT image reconstruction and pattern recognition. *Pattern Recognit.* **2001**, *34*, 2145–2154. [CrossRef]
2. Marapane, S.B.; Trivedi, M.M. Region-based stereo analysis for robotic applications. *IEEE Trans. Syst. Man Cybern.* **1989**, *19*, 1447–1464. [CrossRef]
3. Svetkoff, D.J.; Kilgus, D.B.; Boman, R.C. 3D Line-Scan Imaging with Orthogonal Symmetry. In Proceedings of the Dimensional and Unconventional Imaging for Industrial Inspection and Metrology, Philadelphia, PA, USA, 22 October 1995; pp. 177–188.
4. Stein, N.; Frohn, H. Laser Light Stripe Measurements Assure Correct Piston Assembly. In Proceedings of the Computer Vision for Industry, Munich, Germany, 21 June 1993; pp. 302–307.
5. Smith, M.L.; Smith, G.; Hill, T. Gradient space analysis of surface defects using a photometric stereo derived bump map. *Image Vision Comput.* **1999**, *17*, 321–332. [CrossRef]
6. Creath, K. Step height measurement using two-wavelength phase-shifting interferometry. *Appl. Opt.* **1987**, *26*, 2810–2816. [CrossRef] [PubMed]
7. Lai, C.-C.; Hsu, I.J. Surface profilometry with composite interferometer. *Opt. Express* **2007**, *15*, 13949–13956. [CrossRef] [PubMed]
8. Harasaki, A.; Schmit, J.; Wyant, J.C. Improved vertical-scanning interferometry. *Appl. Opt.* **2000**, *39*, 2107–2115. [CrossRef] [PubMed]
9. Katafuchi, N.; Sano, M.; Ohara, S.; Okudaira, M. A method for inspecting industrial parts surfaces based on an optics model. *Machine Vision Appl.* **2000**, *12*, 170–176. [CrossRef]
10. Huang, D.; Swanson, E.A.; Lin, C.P.; Schuman, J.S.; Stinson, W.G.; Chang, W.; Hee, M.R.; Flotte, T.; Gregory, K.; Puliafito, C.A.; et al. Optical coherence tomography. *Science* **1991**, *254*, 1178–1181. [CrossRef] [PubMed]
11. Tomlins, P.H.; Wang, R.K. Theory, developments and applications of optical coherence tomography. *J. Phys. D Appl. Phys.* **2005**, *38*, 2519. [CrossRef]
12. Chinn, S.R.; Swanson, E.A.; Fujimoto, J.G. Optical coherence tomography using a frequency-tunable optical source. *Opt. Lett.* **1997**, *22*, 340–342. [CrossRef] [PubMed]
13. Lee, S.-W.; Jeong, H.-W.; Kim, B.-M.; Ahn, Y.-C.; Jung, W.; Chen, Z. Optimization for axial resolution, depth range, and sensitivity of spectral domain optical coherence tomography at 1.3 μm. *J. Korean Phys. Soc.* **2009**, *55*, 2354–2360. [PubMed]
14. Považay, B.; Unterhuber, A.; Hermann, B.; Sattmann, H.; Arthaber, H.; Drexler, W. Full-field time-encoded frequency-domain optical coherence tomography. *Opt. Express* **2006**, *14*, 7661–7669. [CrossRef] [PubMed]
15. Dubois, A.; Boccara, A.C. Full-field optical coherence tomography. In *Optical Coherence Tomography*; Drexler, W., Fujimoto, J., Eds.; Springer: Berlin/Heidelberg, Germany, 2008; pp. 565–591.
16. Lee, C.; Lee, S.-Y.; Kim, J.-Y.; Jung, H.-Y.; Kim, J. Optical sensing method for screening disease in melon seeds by using optical coherence tomography. *Sensors* **2011**, *11*, 9467–9477. [CrossRef] [PubMed]
17. Ravichandran, N.K.; Wijesinghe, R.E.; Shirazi, M.F.; Park, K.; Lee, S.-Y.; Jung, H.-Y.; Jeon, M.; Kim, J. In vivo monitoring on growth and spread of gray leaf spot disease in capsicum annuum leaf using spectral domain optical coherence tomography. *J. Spectrosc.* **2016**, *2016*, 6. [CrossRef]
18. Cho, N.H.; Park, K.; Kim, J.-Y.; Jung, Y.; Kim, J. Quantitative assessment of touch-screen panel by nondestructive inspection with three-dimensional real-time display optical coherence tomography. *Opt. Lasers Eng.* **2015**, *68*, 50–57. [CrossRef]
19. Cho, N.H.; Jung, U.; Kim, S.; Kim, J. Non-destructive inspection methods for leds using real-time displaying optical coherence tomography. *Sensors* **2012**, *12*, 10395–10406. [CrossRef] [PubMed]
20. Kim, S.-H.; Kim, J.-H.; Kang, S.-W. Nondestructive defect inspection for lcds using optical coherence tomography. *Displays* **2011**, *32*, 325–329. [CrossRef]
21. Shirazi, M.; Park, K.; Wijesinghe, R.; Jeong, H.; Han, S.; Kim, P.; Jeon, M.; Kim, J. Fast industrial inspection of optical thin film using optical coherence tomography. *Sensors* **2016**, *16*, 1598. [CrossRef] [PubMed]

22. Jeon, M.; Kim, J.; Jung, U.; Lee, C.; Jung, W.; Boppart, S.A. Full-range k-domain linearization in spectral-domain optical coherence tomography. *Appl. Opt.* **2011**, *50*, 1158–1163. [CrossRef] [PubMed]

23. Shirazi, M.F.; Cho, N.H.; Jung, W.; Kim, J. Lateral resolution enhancement using programmable phase modulator in optical coherence tomography. *Bio-Med. Mater. Eng.* **2015**, *26* (Suppl. S1), S1465–S1471. [CrossRef] [PubMed]

24. Lee, J.; Kim, K.; Wijesinghe, R.E.; Jeon, D.; Lee, S.H.; Jeon, M.; Jang, J.H. Decalcification using ethylenediaminetetraacetic acid for clear microstructure imaging of cochlea through optical coherence tomography. *J. Biomed. Opt.* **2016**, *21*, 081204. [CrossRef] [PubMed]

25. Satish Kumar, D.; Dalip Singh, M.; Arun, A.; Chandra, S. Simultaneous topography and tomography of latent fingerprints using full-field swept-source optical coherence tomography. *J. Opt. A Pure Appl. Opt.* **2008**, *10*, 015307.

26. Anna, T.; Shakher, C.; Mehta, D.S. Three-dimensional shape measurement of micro-lens arrays using full-field swept-source optical coherence tomography. *Opt. Lasers Eng.* **2010**, *48*, 1145–1151. [CrossRef]

27. Fergusson, J.; Považay, B.; Hofer, B.; Drexler, W. In Vitro Retinal Imaging with Full Field Swept Source Optical Coherence Tomography. In Proceedings of the Optical Coherence Tomography and Coherence Domain Optical Methods in Biomedicine XIV, San Francisco, CA, USA, 23 January 2010.

28. Bonin, T.; Franke, G.; Hagen-Eggert, M.; Koch, P.; Hüttmann, G. In vivo fourier-domain full-field OCT of the human retina with 1.5 million a-lines/s. *Opt. Lett.* **2010**, *35*, 3432–3434. [CrossRef] [PubMed]

29. Lee, S.H.; Kim, M.Y.; Ser, J.I.; Park, J. Asymmetric polarization-based frequency scanning interferometer. *Opt. Express* **2015**, *23*, 7333–7344. [CrossRef] [PubMed]

30. Lee, H.; Cho, S.-W.; Kim, G.; Jeong, M.; Won, Y.; Kim, C.-S. Parallel imaging of 3d surface profile with space-division multiplexing. *Sensors* **2016**, *16*, 129. [CrossRef] [PubMed]

31. Kim, H.J.; Cho, J.; Noh, Y.O.; Oh, M.C.; Chen, Z.; Kim, C.S. Three-dimensional surface phase imaging based on integrated thermo-optic swept laser. *Meas. Sci. Technol.* **2014**, *25*, 035201. [CrossRef]

32. Kitagawa, K. Surface and thickness profile measurement of a transparent film by three-wavelength vertical scanning interferometry. *Opt. Lett.* **2014**, *39*, 4172–4175. [CrossRef] [PubMed]

33. Shirazi, M.F.; Jeon, M.; Kim, J. 850 nm centered wavelength-swept laser based on a wavelength selection galvo filter. *Chin. Opt. Lett.* **2016**, *14*, 011401. [CrossRef]

![sensors logo] *sensors*

MDPI

Article

A Selective Change Driven System for High-Speed Motion Analysis

Jose A. Boluda *, Fernando Pardo and Francisco Vegara

Departament d'Informàtica, Escola Tècnica Superior d'Enginyeria, Universitat de València,
Avd. de la Universidad, s/n, 46100 Burjassot, Spain; Fernando.Pardo@uv.es (F.P.); Francisco.Vegara@uv.es (F.V.)
* Correspondence: Jose.A.Boluda@uv.es; Tel.: +34-963-543-944; Fax: +34-963-544-768

Academic Editor: Gonzalo Pajares Martinsanz
Received: 29 July 2016; Accepted: 3 November 2016; Published: 8 November 2016

Abstract: Vision-based sensing algorithms are computationally-demanding tasks due to the large amount of data acquired and processed. Visual sensors deliver much information, even if data are redundant, and do not give any additional information. A Selective Change Driven (SCD) sensing system is based on a sensor that delivers, ordered by the magnitude of its change, only those pixels that have changed most since the last read-out. This allows the information stream to be adjusted to the computation capabilities. Following this strategy, a new SCD processing architecture for high-speed motion analysis, based on processing pixels instead of full frames, has been developed and implemented into a Field Programmable Gate-Array (FPGA). The programmable device controls the data stream, delivering a new object distance calculation for every new pixel. The acquisition, processing and delivery of a new object distance takes just 1.7 µs. Obtaining a similar result using a conventional frame-based camera would require a device working at roughly 500 Kfps, which is far from being practical or even feasible. This system, built with the recently-developed 64 × 64 CMOS SCD sensor, shows the potential of the SCD approach when combined with a hardware processing system.

Keywords: CMOS image sensor; event-based vision; high-speed visual acquisition; data-flow architecture; FPGA system; laser scanning

---

1. Introduction

Most common artificial vision systems are based on full-frame image processing [1]. The representation of a scene in an instant t as a still image is the typical source of data to extract visual information. Conventional video systems are based on the sequential acquisition and processing of full-frame images. Independently of whether there have been changes in the scene or not, all of the pixels are acquired, stored and processed, which is not efficient in terms of resources if there are no relevant changes. Moreover, this sequential nature makes it more difficult to reduce the control loop delay in real-time applications.

Nature, with evolution being the key point, has developed one of the most perfect machines any engineer could possibly conceive of: living beings. Engineering has drawn on nature as a source of inspiration to solve many problems, particularly in the field of sensing [2,3]. Biological vision systems do not follow the policy of capturing and sending sequences of full frame images at a fixed rate. The idea of a snapshot sequence is not present in biological systems. Visual systems in living beings are based on different types of photoreceptors, which respond to light stimuli, sending information asynchronously to the upper levels of cognitive systems [4].

A Selective Change Driven (SCD) vision sensor delivers only the pixels that have changed most since the last read-out, ordered by the magnitude of their change [5]. Therefore, an SCD sensor only delivers information that is not redundant; as a consequence, there is an efficient use of time and energy, contrary to conventional vision sensors. Additionally, since this information is ordered and delivered synchronously according to the absolute magnitude of its change, the most significant changes (that are related to higher light intensity variations) will be processed first. This pixel prioritization, based on change ordering, can be relevant for some real-time applications that must be accomplished with time restrictions, because it could not be possible to process all of the events delivered by the sensor. The SCD policy ensures that in the case of time constraints, the most relevant changes will be processed.

1.1. Event-Based Sensors

Most sensors that have been inspired by this biological approach are based on the Address Event Representation (AER) model [6,7]. In the AER model, pixels operate as individual processing units and fire themselves according to their spatial or temporal change of illumination level. Moreover, event-based sensors can also be classified taking into account how they transform light into an electrical signal. There are integration-based sensors and continuous conversion-based sensors. Light integration is based on a capacitor that stores a charge, which is proportional to both illumination intensity and integration time. Sensors designed with integration photoreceptors offer better image quality. As a drawback, these sensors lose part of their event-based philosophy, because the integration time degrades the fast event-driven response speed. Instead, continuous conversion-based sensors offer a faster response to the stimuli, better mimicking the visual system of living beings.

Many event-based sensors have been developed up to the present. Several of them are especially significant. For instance, [8] has only eleven transistors per pixel and can work in three event triggering modes: illumination level, spatial contrast and temporal contrast. However, this sensor has a low temporal resolution since it is based on fixed time integration. Similarly, the sensors described in [9,10] have good signal quality, since they are based on integration to a fixed voltage, but with a lower time resolution than the events. Worth mentioning is the so-called Dynamic Vision Sensor (DVS) where each pixel autonomously computes the normalized time derivative from the sensed light and provides an output event with its coordinate when this amount exceeds a programmed contrast. DVS cameras offer contrast coding under wide illumination variation and microsecond latency response [11,12]. Hence, it is possible to track fast motion without special lighting conditions

All of the above-mentioned sensors have their particular advantages in certain circumstances. The recently-developed 64 × 64 SCD sensor [13] presents, in our view, a good trade-off, taking advantage of the data reduction that event-driven sensors have, while keeping a synchronous interface that delivers information when the processing system is capable of processing it. Moreover, the feature of reading-out events ordered by the magnitude of their change contributes to the implementation of systems that can work even without full data processing. The fact of having this computing-oriented interface makes the SCD sensor a good candidate to be easily integrated into an embedded processing system. Additionally, the SCD sensor is the only one that combines illumination level and temporal contrast in continuous time, thus providing high-speed operation in both event and illumination response. The SCD sensor will be described in depth in Section 2.2. There is an earlier SCD sensor, with a 32 × 32 resolution based on an integration photoreceptor, which gave a resolution time of 500 µs [14]. However, the current 64 × 64 SCD sensor takes advantage of a continuous conversion cell, allowing higher working speeds.

1.2. Event-Based Systems

The next natural step after event-based sensors is the development of vision systems based on that philosophy. However, a problem arises because there is an inherent contradiction when trying to mimic complex mammal vision systems with a traditional computing system. The human brain is a massive parallel system with nearly 100 billion neurons [15]; its good performance relies on the huge amount of connections and on its parallel functioning ability. However, a traditional computing system is sequential in nature. It is true that this drawback can be partially overcome with parallel architectures, but their performance is still very far away from the vision systems of real living beings.

Neuromorphic systems [16,17] appear as implementations in VLSI (Very Large Scale of Integration) circuits of sensor and neural systems, whose architecture and design are based on neurobiology. These systems try to mimic neuro-biological structures present in the nervous system, and AER fits perfectly into this strategy [18].

There are some examples of neuromorphic systems implemented in full custom chips and some others that use FPGAs as processing elements. In [18], a 32 × 32 convolution chip with a 155-ns event latency and a theoretical throughput of 20 mega events per second is presented. The low latency between the input and output streams in a neuromorphic system they term pseudosimultaneity. In this paper, several experiments with both dynamic and recorded AER stimuli are shown, although the highest speed experiments are performed with simulated data. In [19], another convolution module is presented, with a similar speed performance, but in this case with 64 × 64 pixels. It has been designed to allow many of them to be assembled to build modular and hierarchical Convolutional Neural Networks (ConvNets). Similarly, in [20], a neuromorphic system is implemented mixing the DVS event-driven sensor chip together with event-driven convolution module arrays implemented on FPGAs. Experimental results in this paper are the implementation of Gabor filters and 3D stereo reconstruction systems. More recently, a fully-digital implementation of a spiking convolutional event-driven core that can be implemented in FPGAs [21] has been presented. This system uses a DVS sensor, an FPGA and two USB AER mini boards that send AER spikes through a USB connection to a computer. This system is capable of updating 128 synaptic connections in 12 ns, this being an improvement with respect to previously reported FPGA convolutional event-driven cores.

Some processing systems based on AER sensors show the desired pseudosimultaneity, to reduce the control loop delay to its minimum, achieving in this way real-time performance. Most of them are complex systems with a high quantity of resources, which means they cannot be used in embedded systems. Neuromorphic systems try to implement image processing algorithms mimicking living beings' neural systems as a guideline. In our view, the asynchronous nature of AER sensors makes the subsequent processing system difficult. Sometimes, the processing stages must deal with an explosion of events that are hard to process. In our opinion, event-based systems tailored to embedded systems should have a traditional synchronous interface rather than a neuromorphic approach, which in the end results in a more resource-heavy, less feasible system.

Many AER systems are usually based on FPGAs, often on several boards, due to the complexity of dealing with asynchronous events. Conversely, our approach, based on SCD vision, uses very few resources. We have developed a high-speed event-based motion tracking system, with just the sensor, a medium-size FPGA and some support components. This characteristic means that the system can be integrated into an autonomous platform or any system with limited resources.

1.3. Laser Scanning

High speed object detection and scene mapping is an extensively-investigated topic in computer vision. It is useful, for instance, in autonomous vehicle navigation, where it is important to detect obstacles located in the direction the vehicle is traveling [22]. One of the sensing methods used to measure depth without physical contact is laser-based 3D scanning. It is common for laser-based scanners to generate a huge amount of data, for instance in the case of Laser Imaging Detection and Ranging systems (LiDAR) [23]. There are several solutions to measure distances with laser-based scanners: photogrammetry, interferometry or ToF (Time of Flight). The most common technique is known as active triangulation. This method is relatively easy to implement, giving good results for measuring distances in the range of millimeters to several meters. These scanners mainly consist of a laser line generator and a camera that records the pattern projected onto the surface to be measured. Active triangulation has several error sources that limit its resolution, such as any other measurement technique [24]. Many contributions have been published trying to overcome the drawbacks of this technique. For instance, in [25], some solutions are proposed to increase the accuracy of measurements through sub-pixel resolution. Moreover, the basic configuration sometimes varies in order to minimize the occlusion problem due to laser/receiver distance [26].

Nowadays, additional developments have produced small commercial sensors that are employed in robotic applications. For instance, the SICK LMS 200 sensor (Sick AG, Waldkirch, Germany) is based on the ToF measurement principle, and the Hokuyo URG-04LX scanner (Hokuyo Automatic Co. ltd., Osaka, Japan) uses amplitude modulated laser light. By measuring the phase shift between the emitted light wave and its reflection, it computes the target distance. In [27], there is an in-depth comparison of both sensors. It is shown how, in the case of the Hokuyo sensor, the accuracy is strongly dependent on the target surface properties. ToF techniques have been employed in other commercial scanners, such as the MESA SR4000 scanner sensor (Mesa Imaging AG, Zurich, Switzerland). It has a resolution of 176 × 144 pixels and 50 fps. Similarly, PMD's CamCube 3.0 (PMD Technologies AG, Siegen, Germany) also uses ToF and achieves 40 fps for a resolution of 200 × 200 pixels or 30 fps for 160 × 120 pixels. An analysis of the state-of-the art in the field of lock-in ToF cameras can be seen in [28]. The performance of ToF range cameras has been improved over the last few years; error sources are minor, and higher resolution and frame rates can be achieved. Despite these improvements, ToF cameras cannot yet achieve the depth accuracy obtained by classical triangulation systems. As a final example, Microsoft's Kinect is being used nowadays for depth mapping. In [29], there is an in-depth analysis of this sensing system. Similarly, high-speed cameras, devices capable of frame rates in excess of 250 frames per second (typically over 1000), are being used nowadays intensively in many applications [30,31].

The volume of data involved that needs to be processed in real time, together with the power and size restrictions inherent in an embedded system suggest the use of an SCD vision system to address these problems. It has been already proven, with the previous 32 × 32 SCD sensor based on an integration cell, that this approach can be employed to detect the frequency of a rotating movement at very high speeds [32]. In this study, movement detection is done off-line by software in a PC (with the Fourier transform), due to the relatively long delay of the data in the order of tenths of µs. With these experiments, it has been shown that the SCD approach has been able to detect frequencies up to 240 rps, frequencies that cannot be detected with an equivalent classical sensor.

In this paper, we present a real-time high-speed working system, developed with the new 64 × 64 SCD sensor based on a conversion cell. The system is able to track movements in real time. This tracking is done by means of computing an object's distance. This distance computation is made in real time for each new pixel that arrives, instead of off-line, as reported in previous experiments. As will be shown in Section 3, it has been possible to track a rotating movement with a delay of 1.7 µs. This speed is above one order of magnitude faster than the speed of previous systems. The proposed system can track arbitrary movements in real time, thanks to the use and combination of a new SCD sensor, a new pipelined processing architecture and the use of hardware based on a portable FPGA board. The following sections describe the system in detail.

2. System Description

Our original motivation was to build a real-time working system as a proof-of-concept of SCD vision. This sensor combines the advantages of event-based sensors, though it can be used in real embedded systems because of its easily integrated synchronous interface. A good application of SCD sensing is object-distance detection for autonomous vehicle navigation. With a few resources, the proposed system should be able to detect objects moving at very high speeds, something that a conventional vision system would not be capable of. Relative movement between the camera and the detected object is necessary. Either the vehicle with the detection system or the obstacle if the vehicle is stationary must be moving to generate a stream of changes. This condition could appear to be a serious restriction, but in fact, it is not. If the vehicle is moving, objects in the moving direction can be detected. If the vehicle is stationary, approaching objects can be detected. Only if there is no relative movement between the vehicle and the obstacle would the object not be detected. In this case, this would not cause any problems.

Figure 1 shows the system conception. In order to push the system to the limit instead of a real, slow moving vehicle, we have tested the system with the fastest mechanical system available in our lab: a high-speed rotating tool.

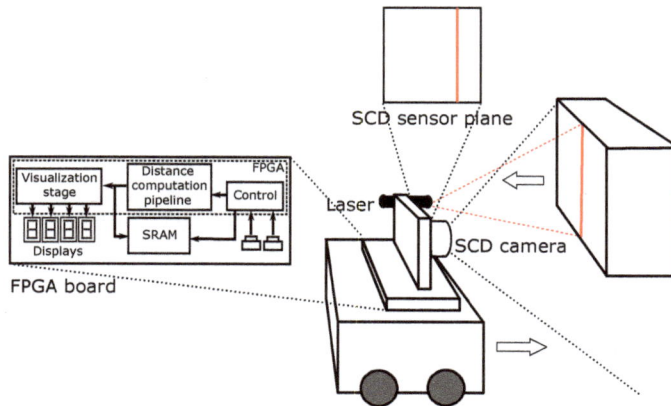

Figure 1. Motion detection system.

The system, as Figure 1 shows, is based on active laser triangulation. The detection system must be set in front of the moving vehicle. There is a laser that projects a line in front of the camera. In the case of Figure 1, the laser is projected onto a moving object. The laser line is captured by the SCD camera, placed at a known distance from the laser. The position of the laser in the sensor image gives the distance between the camera and the surface, as explained in Section 2.1. In fact, for a fixed y position (row), the x column will give the horizontal distance to the obstacle. As long as the vehicle moves, the laser image will change when there are distance variations. In any case, the SCD sensor will only deliver pixels that have changed. This characteristic dramatically reduces the amount of data to be processed. Of course, it is possible for each row to obtain a different column value. The line can be projected onto an irregular surface, so the line image will give a different column for each different distance, providing an exact depth map. This task is not complicated because as there are only 64 columns and each column position is bi-univocally related to a distance, then it is possible to implement a Look-Up Table (LUT) to instantaneously calculate each pixel distance. This distance profile can be sent to the vehicle control, which decides what to do. Instead of a LUT, because we have assumed that the object has a regular surface, we have decided to compute the average distance of the surface profile. This calculation is an example of how a system can take advantage of the SCD architecture. Each new pixel arrival updates its contribution to the average distance computation, being obtained a new distance value with each pixel. The same algorithm using a classical approach would have required a complete image acquisition. Afterwards, it would have been necessary to binarize the image taking into account only the pixels illuminated by the laser. Then, the average column value of these pixels would have been employed to compute the distance. In our system, each new pixel produces a new distance value in the fastest possible way.

2.1. Laser Triangulation System

As already mentioned in Section 1.3, active triangulation is an easy, widely-used technique for non-contact distance measurement. There are some differences in the method implementation, which depends on the final goal; accuracy, range, etc. Some well-known configurations can be seen in [29] or in [33]. Equations in those papers give accurate values when all of the system variables are known. After all, if extremely high accuracy is not a key factor in the experiment, as in the case of our SCD proof-of-concept, it is very common to use simplified formulae. Figure 2 depicts a simplified pin-hole representation of triangle equivalences shown in those papers. The basic principle of the method consists of projecting a pattern of light (usually a laser line) on the surface to be measured. Afterwards, the pattern image is captured in the sensor plane. Because the laser line is so narrow and the sensor has only 64 columns, it is almost guaranteed that in the range of operation, the laser line will be in one column, or two, while it changes from one column to the neighboring column.

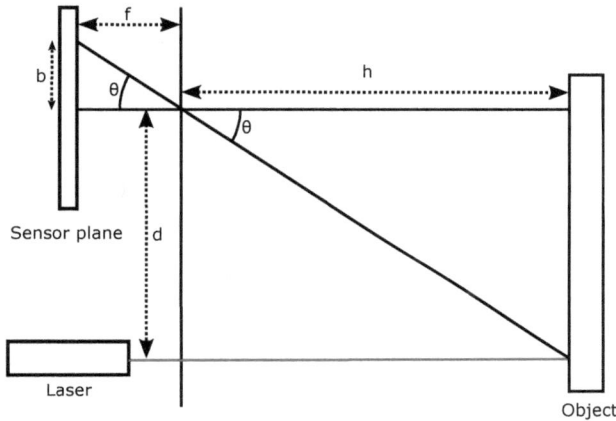

Figure 2. Simple pin-hole triangulation scheme.

From Figure 2, it is possible to infer:

$$h = \frac{d}{tan(\theta)} \qquad (1)$$

Equation (1) gives the value of the distance h between the laser-camera system and the surface to be measured. It can be obtained from d, the known laser-camera gap, and the angle θ. This angle can be obtained as a function of the shift in the image plane (in pixels).

In a linear model, the angle θ can be expressed as:

$$\theta = x \cdot \omega + \phi \qquad (2)$$

where x is the distance in pixels from the computed pixel to the image center, ω represents the radians per pixel and ϕ is a useful parameter for alignment error compensation. Hence, Equation (1) can be rewritten as:

$$h = \frac{d}{tan(x \cdot \omega + \phi)} \qquad (3)$$

It is possible to easily obtain x from the sensor data stream (in fact, it is the column value), but ω and ϕ must be obtained through the calibration process; this is shown in Section 3.1. The system range is adjusted by computing these parameters. Of course, higher polynomial fitting would give more accurate values of the angle θ as a function of column y. This higher precision has a major drawback: higher Programmable Logic Device (PLD) complexity and, consequently, a higher delay, since all of the computations must be made by hardware in the FPGA. Nevertheless, a coefficient of determination R^2 of 0.9984 has been achieved with the linear adjustment, as is shown in Section 3.1. This coefficient of determination corresponds to a percent of standard deviation of 96%.

2.2. SCD Sensor

The first version of an SCD sensor has been designed [34]. Although it is just a 32 × 32 sensor based on an integration cell, it has shown its utility in resource-limited systems [35]. Recently, a 64 × 64 SCD sensor has been designed based on a conversion cell, which will allow it to take advantage of higher working speeds [13].

The basic SCD sensor cell scheme can be seen in Figure 3. This sensor has an array of 64 × 64 pixels. Each pixel can detect whether it has experienced the largest change in illumination since the last time it has been read-out. Any pixel can detect if it is the winner because there is a Winner-Take-All (WTA) circuit that decides which is the pixel with the greatest change. The winner selection has two stages: the first one consists of an analog WTA that selects the set of pixels that have changed most. This set usually already has one single pixel, but in the case of several potential pixels, the second stage digitally selects one of them. All this winner selection takes place in less than 1 μs.

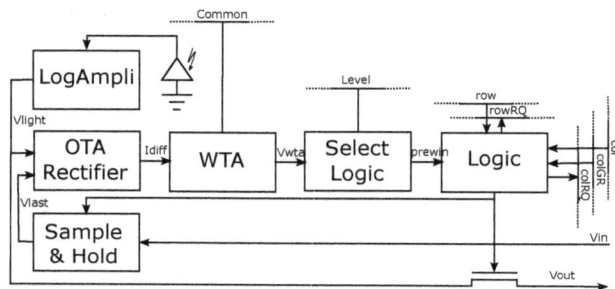

Figure 3. Basic SCD cell.

A photodiode transforms incident light into current in each cell. This current generates a logarithmic dependent voltage with a dependency through a log-amplifier configuration, based on a weak inversion transistor negative feedback amplifier and a source follower. The signal *Vlight* is the logarithmic voltage of the incident light. There is a sample and hold circuit, which stores the last read-out value (*Vlast*) in a capacitor. All pixels compare the difference between their last read-out value *Vlast* and the present incident light *Vlight*. This absolute difference is calculated using an Operational Transconductance Amplifier and rectifier (OTA rectifier) that transforms it into the current *Idiff*. All of the *Idiff* currents are compared through the WTA circuit. The *Common* line, shared by all WTA circuits in the sensor array, allows any pixel to generate the *Vwta* signal. The WTA is designed to pull-down the *Vwta* of the winner pixel and pull-up the *Vwta* of the rest. Because there are 64 × 64 = 4096 competitors, it is possible to have more than one pixel signaled as a winner. All of these pixels have their *prewin* signal asserted, and all of them will enter a second-stage competition to select just a single winner. The logic block allows only one of the columns of the attempting winners, setting the *colGR* of the selected column. Each pixel detects this *colGR* setting its row request *rowRQ*, because again, there could be more than one pre-winner pixel cell in the selected column. Immediately, an arbitration circuit decides a single row winner, giving a final winner pixel. This winner will not be sent out until the sensor receives an external clock signal. This signal latches the column and row winners, so the winner will set its *win* signal when both *col* and *row* are set. The sample and hold circuit, triggered by the *win* signal, charges the capacitor to *Vlight*. Consequently, since there is now no difference, the pixel loses the present competition and un-sets the *prewin* signal. The *col* and *row* signals remain unchanged until a new clock signal is received.

The chip, which can be seen in Figure 4, has been designed with 0.18 μm 6M1P (6-Metal 1-Polysilicon) MIM (Metal-Insulator-Metal) CMOS technology. A single pixel uses 41 transistors and occupies 30 ×30 μm² with a fill factor of 4%. Additionally, the sensor has the feature of working as a conventional camera. The sensor has an input signal, *SCDena*, which selects whether the camera works following the SCD function (if set to one) or whether it works as a conventional camera. In the latter, the pixel address must be supplied in order to obtain the corresponding illumination value as a random access memory. This characteristic is useful for system calibration, as is shown in Section 3.1.

Figure 4. Details of the SCD sensor in the camera.

2.3. SCD Camera

One of the main advantages of the SCD sensor, compared to other event-based sensors, is that it offers a simple interface. The sensor always works as the slave of a processing unit to which it communicates in a synchronous way. A SAM4S Xplained pro micro-board (Atmel Corporation, San Jose, CA, USA) has been used to implement the SCD camera. The camera offers a USB interface to a computer and digitalizes the analogue illumination level value obtained by the sensor. Both functions have not been used in the system, since the pixel stream control has been carried out directly in the FPGA, and the illumination value is not being used. Nevertheless, the camera has been kept in the system because the sensor needs nine polarization analogue values. The camera also adapts the voltage levels between the sensor (1.8 V) and the FPGA (3.3 V). Figure 4 shows the sensor and part of the camera. Similar to most AER systems, only the event address has been taken into account. This has been done in this way in order to reduce the 2-µs conversion time needed for the analogue to digital conversion.

The sensor always sends the pixel that has changed the most based on an external request. Figure 5 shows the timing schema for this. Initially, the competition signal *Comp* must be asserted. After that, the *Ck* signal must be set to one and then set to zero again to generate a pulse while *Comp* is asserted. Then, *Comp* can be released. A few nanoseconds later, the column and row of the pixel appear in the sensor bus. Exact signal timing has been tested with FPGA system clock multiples ($t = 20$ ns period), the event generation being stable with the timing shown in Figure 5. A Finite State Machine (FSM) in the FPGA is in charge of generating these signals. As a conclusion, it is possible to generate a new event each 120 ns in our system, which would be the highest temporal resolution of the sensor without the illumination level.

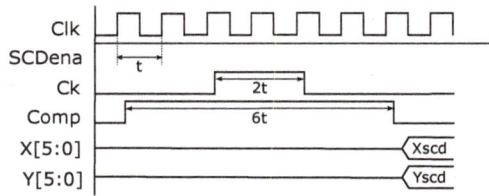

Figure 5. Sensor protocol and timing.

2.4. High-Speed Computation Pipeline

Our assumption is that due to laser brightness, the most significant changes will mostly occur in the laser line. Equation (3) gives the distance of the point where the laser is being projected depending on x, the column position of the laser line in the sensor. In our demonstration, we have assumed that there is an almost constant distance where the laser line is being projected. Of course, different values of h could be obtained for each different x. In this case, and because there are only 64 possible y values, a simple LUT with the 64 possible pre-computed depth values would solve the problem. However, our system is not just a pre-computed LUT. Ideally, all of the lines will have the same column, but in a real situation, there would be different values for the columns corresponding to the 64 rows. Therefore, our system computes the average column value, or x, as an average distance from the object.

Figure 6 shows a scheme of the computation pipeline. There is a column of 64 registers, one for each row. When an event arrives with its (x, y) coordinates, the y-th row updates its column value x. Each row in the registers column stores the present laser position in the image, and it is updated as soon as there is a change. All column registers are added with a tree of carry-lookahead adders. Due to the reduced number of bits involved, these kinds of adders are usually fast enough. The limitation would appear if the adder delay exceeded the clock period; taking into account that the FPGA clock has a period of 20 ns, simulations proved that this restriction is not going to be exceeded. As will be shown later in Section 2.6, the maximum system clock after synthesis is 97.84 MHz, which means that the delays in the carry chain of the adders are below 10.3 ns. This is the expected result since these adders have 12 bits [36].

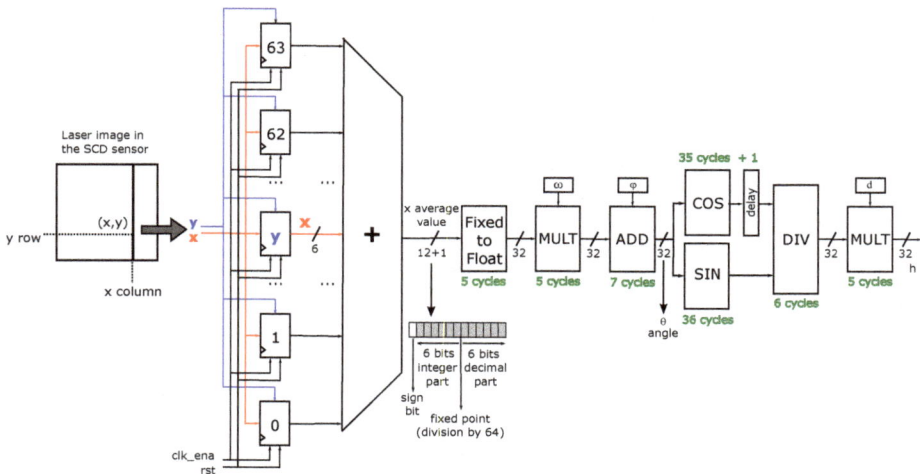

Figure 6. Computation stage.

The maximum result of the addition of 64 registers of six bits fits in a 12-bit register, but an additional bit has been added to the left with a zero value. This has been done to guarantee that the result is interpreted as positive in the next stage. The average value must be computed by dividing the sum by 64, something that is easily done just by moving the point six bits to the left. This operation converts a natural number into a real fixed point number, as Figure 6 shows.

Once the average column value has been computed, Equation (3) must be applied in order to obtain the depth related to that column displacement. To do so, the Altera Library of Parameterized Modules (LPM) has been used since the FPGA employed is the Cyclone II 2C35 from Altera (San Jose, CA, USA). The sequence of mathematical operations that can be inferred from Equation (3) will be done by hardware, as Figure 6 shows.

First of all, the average column value must be converted from fixed-point representation to IEEE 754 floating-point representation, in order to serve as input to the subsequent modules. Secondly, the θ angle is calculated by first multiplying x by ω and then adding the ϕ parameter. Next, the cosine and sine of this angle are computed. Because the cosine calculus is one cycle shorter than the sine, a one cycle delay is necessary to equalize the delay in both paths. Afterwards, the division of both magnitudes is performed in the subsequent LPM module, this result being finally multiplied by d.

The latency of each LPM module can be seen in Figure 6. Adding these values, the latency of the system is 64 cycles. This is not a bad result, taking into account the complexity of the operations performed. The system has been successfully compiled in the target FPGA obtaining a clock frequency of nearly 100 MHz, which gives a latency of 0.6 µs. Nevertheless, it has been decided to utilize the 50-MHz system clock generated by the board, so the initial latency is 1.3 µs. After this, the system can give a new result each clock cycle, that is each 20 ns or even each 12.5 ns with a 100-MHz clock. In any case, the bottleneck of the system is not the computation pipeline, but the sensor data stream. An FSM needs six cycles to obtain each new pixel event, as shown in Figure 5.

All of the LPM modules have been slightly tuned to reduce their bus widths and the number of pipeline stages. Moreover, the synthesis parameters have been fine-tuned towards maximizing speed instead of reducing area.

2.5. Display Stage and Memory Access

Our system is focused on taking advantage, through specialized hardware, of the high-speed resolution time achieved with the SCD sensor. This system is beyond any other previous system using an SCD sensor. Once the distance to the moving object has been computed, it can be used to activate some actuators or just help to make decisions in an autonomous navigation vehicle. In our case, we implemented a simple display system to see, in real time, the computed distance. This has been useful to gauge the accuracy of the system at low speeds. Nevertheless, in order to prove the correctness of the computed data in high-speed experiments, the implemented control FSM stores distance data in an external SRAM (Static Random Access Memory), which is in the FPGA board. Afterwards, data can be extracted from the SRAM and plotted.

Figure 7 shows a schema of the system with the display stage, the SRAM and the control FSM. In order to visualize distance data in real time, distance is converted to a fixed point representation using an LPM module and then to BCD (Binary-coded decimal) and seven-segment format through a custom combinatorial module. The involved magnitudes are estimated to be from millimeters to roughly two hundred centimeters. The distance between the laser and the camera d has been expressed in centimeters, so h will also be expressed in centimeters. This has been done with the intention of representing these figures in the display. The theoretical range of distances achieved with the triangulation system would be, as later shown in Section 3.1, in the order of 250 cm. To express these figures, eight bits are necessary for the integer part. Consequently, the decimals appearing as a result will be millimeters. Unfortunately, the low sensor resolution makes it difficult to detect the laser when it is in the range of more than roughly 100 cm.

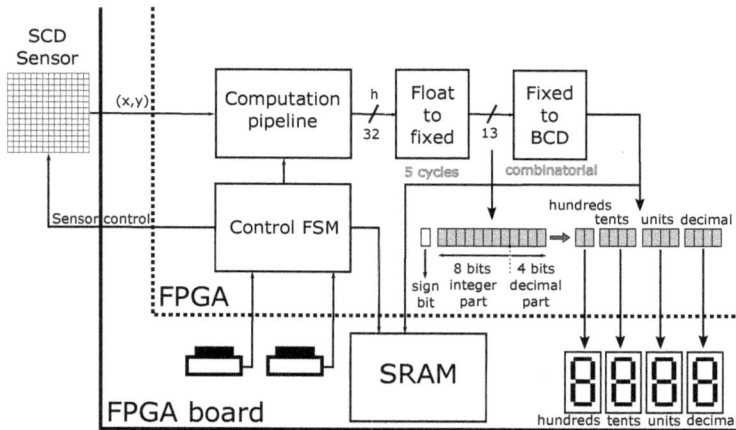

Figure 7. Display and memory stages.

2.6. Synthesis Details

The FPGA board used is the Altera DE2 board. This board contains a Cyclone II FPGA device, with:

- 33,216 logic elements
- 105 M4K RAM (Random Access Memory) blocks
- 483,840 total RAM bits
- 35 embedded multipliers
- 4 PLLs (Phase-Locked Loops)
- 475 user I/O pins
- FineLine BGA (Ball Grid Array) 672-pin package.

The synthesis results shown in Table 1 justify the use of this device. There is still room to add some improvements, although all of the multipliers have already been used.

Table 1. Synthesis results.

Parameter	Value
System clock	97.84 MHz
Total logic elements	16,064 (48%)
Total registers	8285 (25%)
Total memory bits	4608 (1%)
Embedded Multiplier 9-bit elements	70 (100%)

The FPGA board has many interfaces and additional hardware, although only a few buttons, four seven-segment displays and the SRAM have been used to store the results in the experimentation process. The SRAM is a 512-Kbytes chip, organized as a 256 K × 16, with a 10-ns access time.

3. Experimentation

Most smart sensors are developed in CMOS technology because it is possible to include some processing or "smart" capabilities added to the sensing part. Unfortunately, CMOS technology offers worse image quality compared to older CCD technologies. CMOS vision sensors have some noise problems that can be overcome with some additional strategies. In the case of our system, the illumination level has been discarded, and only the events have been taken into account.

There is a source of noise that must be faced in most event-based sensors, the Random Telegraph Signal (RTS). This effect causes some pixels (hotspots) to behave randomly, registering large changes when they should not and, consequently, firing wrong events [37]. This source of noise is related to the presence of traps in the transistor channel and affects a really low quantity of photosensors. The practical effect of RTS noise in a CMOS imager is that it invalidates a few pixels. Each sensor must be characterized to find out which pixels must be marked as faulty. In the case of the 64 × 64 SCD sensor used in our system, only 28 pixels were discarded, giving a ratio of 0.7%. The control FSM incorporates a module that identifies the received pixel, and it is not sent to the pipeline if it is in the faulty pixel list.

The experiment was focused on proving the high-speed capabilities of the system, but beforehand, some low-speed experiments were performed in order to characterize the sensor.

3.1. Calibration

Equation (1) expresses the relationship between the object distance h and the angle θ. In Equation (2), a linear relationship between the angle variation and the column value has been supposed. Finally, Equation (3) gives the object distance as a function of the column number x. Since the SCD sensor can work in conventional mode, it has been possible to obtain the column number x of the laser in the image plane for several h values. The value of d has been fixed to 7.5 cm, and the angle $\theta = arctan(d/h)$ has been represented.

Figure 8 shows the least squares adjustment of *theta* as a function of x. This adjustment gives a coefficient of determination $R^2 = 0.9984$, which justifies, in our view, the linear model for θ. The linear equation obtained was $\theta = \omega \cdot x + \phi = 0.0065 \cdot x - 0.1012$. These values give a theoretical distance range between 23 cm, as the nearest detectable point, and roughly 250 cm as the farthest measurable distance. Unfortunately, due to the sensor resolution, it has not been possible to keep this range of distances with a single lens. The system has been proven to work effectively in a range up to roughly 100 cm.

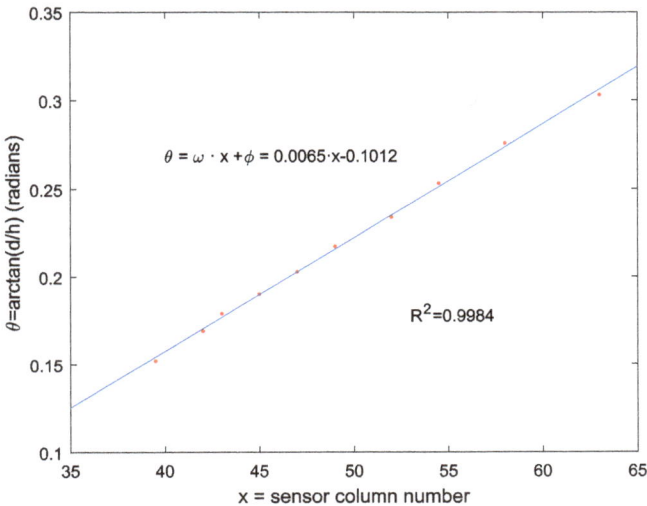

Figure 8. Least squares adjustment of sensor parameters.

Three supply sources are needed for the system: one 9-V supply for the FPGA board, a USB 5-V connection to power the camera and another 5 V to power the laser, which has also been taken from a USB connection. A portable battery with these supply sources has been employed, making a fully-portable system. Everything was set on the top of a wheeled cart, moving it in a controlled environment. Figure 9 shows the upper part of the system and the laser projected onto a wall. With the sensor parameters already computed, some low-speed controlled experiments were performed to verify that the system was working correctly.

During start up, the system may give wrong distance values. This happens because initially, all of the registers do not store the right column values. This is going to happen until all of the registers that store the row value for each column are updated, as Figure 6 shows. It is necessary to update the row of 64 pixels where the laser is projected, but this happens very quickly because these changing events are going to be the first events to be sent out and processed. This is guaranteed by the SCD policy of first sending the pixels that have changed most.

Figure 9. System on the upper part of a wheeled cart in the approaching wall experiment.

Several low-speed experiments, with different surfaces, have been carried out to test the accuracy of the system at low speed. All of the experiments have been repeated until two standard deviations in confidence have been obtained. The first experiment has been to move the wheeled cart towards a white plaster wall to measure the distance. The experiment needed to be repeated five times to achieve the required degree of certainty. Figure 10a shows the results of the average distance versus time (a new event each 328 µs), with a Root Mean Square Error (RMSE) of 2 mm.

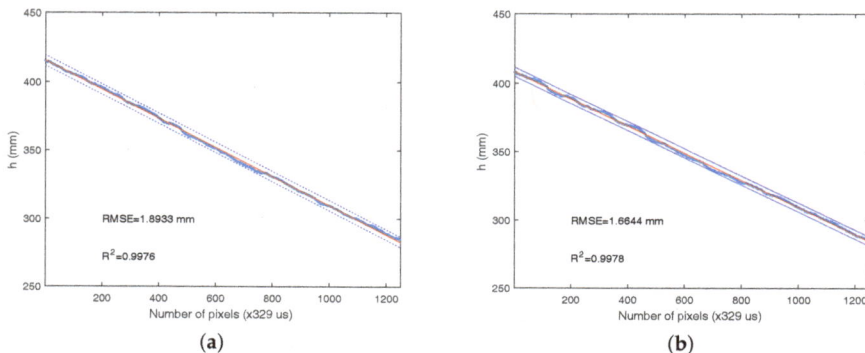

Figure 10. (**a**) Approaching a white plaster wall experiment; (**b**) approaching a blue painted wooden door experiment.

The second experiment has been carried out with a blue painted wooden door. Figure 10b shows a better result in terms of less error and a better coefficient of determination. In fact, this second experiment has been repeated only three times, since the required level of certainty has been achieved with three measurements. Additionally, a third experiment with a brown paperboard sheet fixed to a wall has been performed, as Figure 11a shows, with slightly better results.

Better results could be expected with the white surface, since that color is clearer, and it offers greater laser reflection; but this expectation has been proven to be wrong. The brighter color has been shown to be worse for laser line definition in the sensor plane. Nevertheless, experiments done with flatter surfaces and matte colors have worked better. Figure 11b shows the inverse experiment of moving the wheeled cart away from the wall with the brown paperboard; this has been repeated three times.

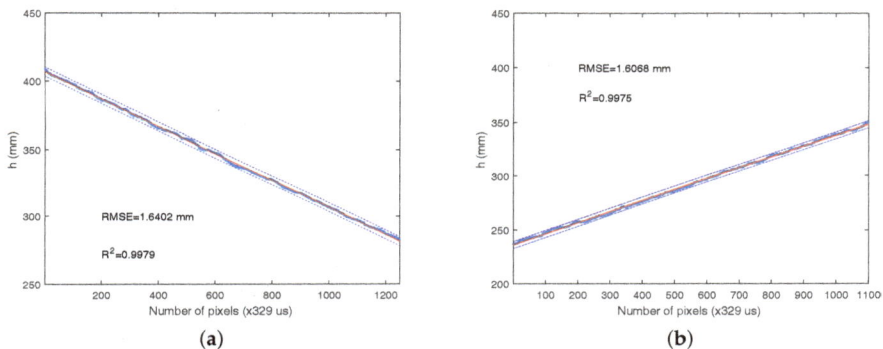

Figure 11. (**a**) Approaching a brown paperboard experiment; (**b**) moving away from a brown paperboard experiment.

3.2. High-Speed Experiments

The system challenge is to test its movement detection capabilities at high speed. This is really one of the most important facets of any event-based vision system. The delay loop in a control system can be reduced to the minimum possible, because only pixels that have changed are acquired and processed. In our view, the SCD system has the additional advantage that it offers a constant event rate controlled by the processing system, and it also has a very simple interface. This fact allows very high-speed movements in an embedded system to be analyzed, with very few resources. To do so, the fastest tool that we could find was used: a rotating tool with a theoretical maximum speed of 33,000 rpm with no load.

Figure 12 shows three pictures of the experimental setup. A plastic stick is fixed to the rotating tool, with the laser beam illuminating the stick in a perpendicular manner. When the stick is fully vertical (Figure 12a), it offers the maximum surface to the laser beam, being fully detected and giving the minimum distance to the system. Conversely, when the stick is fully horizontal (Figure 12b), it offers the minimum surface to the laser beam contributing, in this case, mostly the background to the average distance.

The display is not going to be useful in this case, because the figures change so quickly. Instead, the recorded file in the SRAM is going to be especially useful. Moreover, a theoretical model of what is going to be measured as the average distance in the experiment has been developed. Figure 13 shows a parametrized scheme of the experimental setup.

The stick is rotating at a constant speed of α radians per second. There are some assumptions in the model to simplify it: only the laser projection is considered to generate a change in the image, and the laser beam fully occupies one column and only one column. The stick width and the rotating tool axis are not taken into consideration. The upper and lower part of the image are fixed in a symmetrical scenario. The image column received, that is the lateral view of Figure 13, is discretized into 64 rows, so the model adds the contribution of each row and computes the average value, exactly as the proposed system does.

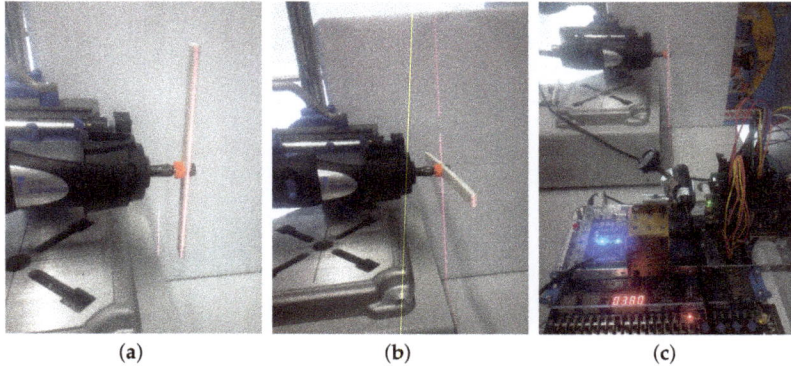

Figure 12. Rotating stick experiment. (**a**) Stick side fully illuminated by the laser (minimum distance to the camera); (**b**) Background mostly illuminated by the laser (maximum distance to the camera); (**c**) Overall experiment set-up view with the rotating stick.

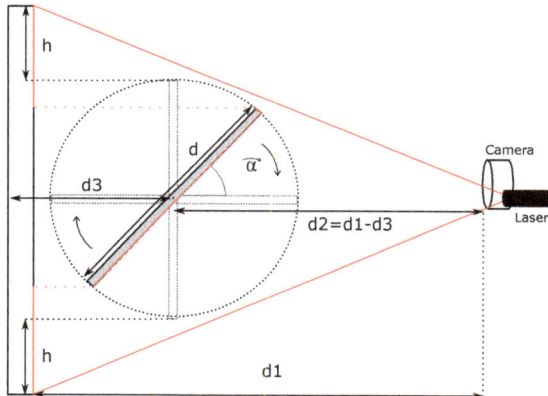

Figure 13. Parametrized model of the rotating stick experiment.

The first experiment performed was done at the lowest speed of the rotating tool. Figure 14 shows the distance position obtained experimentally (red points). In this experiment, an event has been received each 1.7 µs. Experimental data show that the tool is rotating at roughly 12,400 rpm. There is a high coincidence between the theoretical model (blue line) and the experimental result. The coincidence is not absolute, because the model is quite limited, but the periodic movement is perfectly detected. The magnitude of the involved distances is quite coincident, as well.

The experiment shows that a pixel can be acquired and processed each 1.7 µs. A frame rate of 588 Kfps would be needed to achieve this temporal resolution with a conventional camera. A frame-based camera working at that speed is almost impractical, apart from the extremely high required data bandwidth or processing power. It is true that in a conventional system, we would obtain full frames each acquisition time, but there is no way of acquiring, storing and processing such a rate of frames in an embedded system. The SCD system has been able to track a very high-speed movement with very low resources. An equivalent traditional system is not capable of tracking the movement of an object with so much detail at such a speed.

More experiments were performed with the rotating tool, changing some parameters and, more importantly, increasing the tool speed. Figure 15a shows the experimental results with the tool rotating at 21,000 rpm. Again, the periodic movement can be reconstructed, and there is some correspondence with the theoretical mode; for instance, some of the wave peaks are quite sharp as predicted by the model. At this rotation speed, some uncontrolled vibrations appear, which makes it difficult for the laser to point to the side of the stick.

Finally, Figure 15b shows the stick distance at the maximum speed achieved with the rotating tool. This speed is roughly 26,000 rpm, less than the theoretical speed of 33,000 reported in the tool data sheet. We suppose that the tool's maximum speed must be without any load or just with the tool drills. The stick's inertia and air resistance are probably why the theoretical maximum speed could not be achieved. In this experiment, the periodicity of the movement is also perfectly detected, and the sharp form of the curves corresponds to the theoretical model. Again, vibrations appear in the system, which make it difficult to obtain more accurate results.

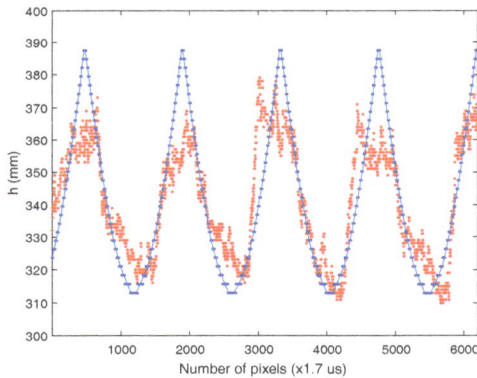

Figure 14. Rotating stick at 12,400 rpm. Red points: experimental data. Blue line: predicted data by the model.

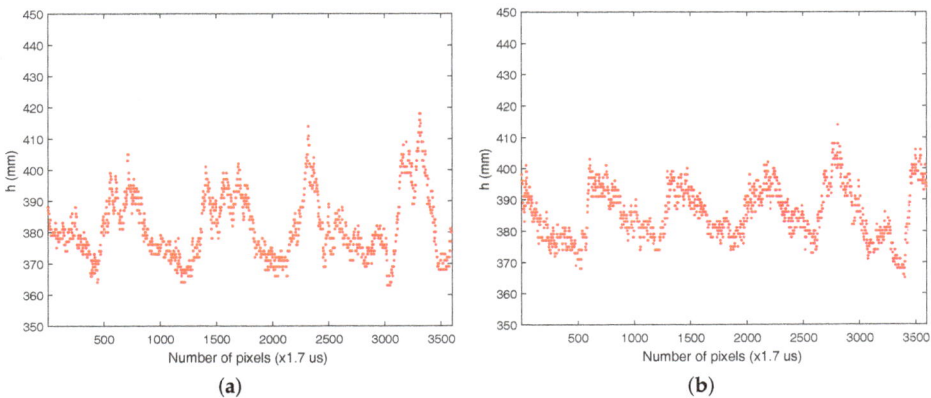

(a)

(b)

Figure 15. (a) Rotating stick at 21,000 rpm; (b) rotating stick at 26,000 rpm.

4. Conclusions

The combination of SCD sensing, which is a case of event-based sensing, together with full custom hardware processing gives a temporal resolution of 1.7 μs, which is about three orders of magnitude better than that offered by frame-based systems, with even fewer processing requirements. With this temporal resolution, it is possible to track movement at very high speeds, which would be almost impossible with a classical full frame approach. Theoretically, it would be necessary to have a camera working at more than 500,000 fps in order to achieve a similar temporal resolution. Moreover, the hardware needed to acquire, store and process such a massive data stream would be beyond any practical application and would be prohibitive for an embedded system. The system is fully portable and has a simple interface. This new system is also one order of magnitude faster than previous realizations of similar laser scan systems based on event-based sensors.

The implemented algorithm in the FPGA can ask for a new event every six cycles = 120 ns. The processing pipeline can deliver a new distance data each cycle, with a latency of 64 cycles. Since data are being stored in the SRAM and in order to avoid contention, a sequential process of acquiring, computing and storing data has been implemented. This is why the delay has been set to 1.7 μs. The system could work with a temporal resolution of 120 ns, which has been proven to be the minimum sensor event access time (without the illumination value). Sometimes, this time will be greater since some events must be discarded because they are coming from faulty pixels. The processing stage would not be the throughput bottleneck, although as in any pipelined system, it would include some latency. In any case, this latency is several orders of magnitude lower than the latency in a classical full-frame vision system.

Additionally, the SCD approach offers an easy interface, with a processing system that can control the stream data and adjust them to its computation capabilities. The implemented system is also a proof-of-concept of how event-based systems, and particularly selective change-driven vision, can be implemented in real high-speed applications.

Some system limitations must be noted: Object detection with this approach only works when there is movement perpendicular to the sensor plane under the laser line. Furthermore, this first prototype has a relatively low measurement range (distances up to 100 cm) because of the optics and sensor resolution. Additionally, 64×64 pixels can be considered a low resolution for some practical applications. Finally, system vibrations and some CMOS sources of noise may affect the distance measurements.

Acknowledgments: This work has been supported by the Spanish Ministry of Economy and Competitiveness (MINECO) and the EU regional development funds (FEDER) project: TEC2015-66947-R.

Author Contributions: J.A.B. conceived of the system, designed the FPGA architecture, conceived of and performed the experiments and wrote the paper. F.P. designed the SCD sensor and camera and revised the paper. F.V. designed the laser subsystem and performed the experiments.

Conflicts of Interest: The authors declare no conflict of interest. The founding sponsors had no role in the design of the study; in the collection, analyses or interpretation of data; in the writing of the manuscript; nor in the decision to publish the results.

Abbreviations

The following abbreviations are used in this manuscript:

AER	Address Event Representation
CK	Clock sensor signal
CMOS	Complementary Metal Oxide Semiconductor
Comp	Competition sensor signal
ConvNets	Convolutional Neural Networks
DVS	Dynamic Vision Sensor
FPGA	Field Programmable Gate Array
fps	frames per second

Sensors **2016**, *16*, 1875

FSM	Finite State Machine
Kfps	Kilo frames per second
LiDAR	Light Detection and Ranging or Laser Imaging Detection and Ranging
LPM	Library of Parameterized Modules
LUT	Look-Up Table
PLD	Programmable Logic Device
RMSE	Root Mean Square Error
RTS	Random Telegraph Signal
SCD	Selective Change-Driven
SRAM	Static Random Access Memory
ToF	Time of Flight
VLSI	Very Large Scale of Integration
WTA	Winner-Take-All

References

1. Cha, Y.J.; You, K.; Choi, W. Vision-Based Detection of Loosened Bolts using the Hough Transform and Support Vector Machines. *Autom. Constr.* **2016**, *71*, 181–188.
2. Vincent, J.F.V. Biomimetics—A Review. *Proc. Inst. Mech. Eng. Part H* **2009**, *223*, 919–939.
3. Antonietti, A.; Casellato, C.; Garrido, J.A.; Luque, N.R.; Naveros, F.; Ros, E.; D'Angelo, E.; Pedrocchi, A. Spiking Neural Network with Distributed Plasticity Reproduces Cerebellar Learning in Eye Blink Conditioning Paradigms. *IEEE Trans. Biomed. Eng.* **2016**, *63*, 210–219.
4. Gollisch, T.; Meister, M. Rapid Neural Coding in the Retina with Relative Spike Latencies. *Science* **2008**, *319*, 1108–1111.
5. Pardo, F.; Benavent, X.; Boluda, J.A.; Vegara, F. Selective Change-Driven Image Processing for High-Speed Motion estimation. In Proceedings of the 13th International Conference on Systems, Signals and Image Processing (IWSSIP), Budapest, Hungary, 21–23 September 2006; pp. 163–166.
6. Mahowald, M. VLSI Analogs of Neural Visual Processing: A Synthesis of Form and Function. Ph.D. Thesis, Computer Science Divivision, California Institute of Technology, Pasadena, CA, USA, 1992.
7. Vanarse, A.; Osseiran, A.; Rassau, A. A Review of Current Neuromorphic Approaches for Vision, Auditory, and Olfactory Sensors. *Front. Neurosci.* **2016**, *10*, 115.
8. Kim, D.; Culurciello, E. Tri-Mode Smart Vision Sensor With 11-Transistors/Pixel for Wireless Sensor Networks. *IEEE Sens. J.* **2013**, *13*, 2102–2108.
9. Posch, C.; Matolin, D.; Wohlgenannt, R. A QVGA 143 dB Dynamic Range Frame-Free PWM Image Sensor with Lossless Pixel-Level Video Compression and Time-Domain CDS. *IEEE J. Solid State Circuits* **2011**, *46*, 259–275.
10. Brandli, C.; Berner, R.; Yang, M.; Liu, S.C.; Delbruck, T. A 240×180 130 dB 3 µs Latency Global Shutter Spatiotemporal Vision Sensor. *IEEE J. Solid State Circuits* **2014**, *49*, 2333–2341.
11. Serrano-Gotarredona, T.; Linares-Barranco, B. A 128×128 1.5% Contrast Sensitivity 0.9% FPN 3 µs Latency 4 mW Asynchronous Frame-Free Dynamic Vision Sensor Using Transimpedance Preamplifiers. *IEEE J. Solid State Circuits* **2013**, *48*, 827–838.
12. Lichtsteiner, P.; Posch, C.; Delbruck, T. A 128×128 dB 15 µs Latency Asynchronous Temporal Contrast Vision Sensor. *IEEE J. Solid State Circuits* **2008**, *43*, 566–576.
13. Pardo, F.; Boluda, J.A.; Vegara, F. Selective Change Driven Vision Sensor with Continuous-Time Logarithmic Photoreceptor and Winner-Take-All Circuit for Pixel Selection. *IEEE J. Solid State Circuits* **2015**, *50*, 786–798.
14. Zuccarello, P.; Pardo, F.; de la Plaza, A.; Boluda, J.A. 32×32 Winner-Take-All matrix with single winner selection. *Electron. Lett.* **2010**, *46*, 333–335.
15. Herculano-Houzel, S. The Human Brain in Numbers: A Linearly Scaled-up Primate Brain. *Front. Hum. Neurosci.* **2009**, *3*, 31.
16. van Schaik, A.; Delbruck, T.; Hasler, J. *Neuromorphic Engineering Systems and Applications*; Frontiers in Neuroscience, Frontiers Media: Lausanne, Switzerland, 2015.
17. Liu, S.C.; Delbruck, T.; Indiveri, G.; Whatley, A.; Douglas, R. *Event-Based Neuromorphic Systems*; John Wiley & Sons Ltd.: Chichester, UK, 2015.

18. Ramos, C.Z. Modular and Scalable Implementation of AER Neuromorphic Systems. Ph.D. Thesis, Universidad de Sevilla, Sevilla, Spain, 2011.

19. Camunas-Mesa, L.; Zamarreno-Ramos, C.; Linares-Barranco, A.; Acosta-Jimenez, A.J.; Serrano-Gotarredona, T.; Linares-Barranco, B. An Event-Driven Multi-Kernel Convolution Processor Module for Event-Driven Vision Sensors. *IEEE J. Solid State Circuits* **2012**, *47*, 504–517.

20. Camunas-Mesa, L.A.; Serrano-Gotarredona, T.; Linares-Barranco, B. Event-Driven Sensing and Processing for High-Speed Robotic Vision. In Proceedings of the IEEE Biomedical Circuits and Systems Conference (BioCAS), Lausanne, Switzerland, 22–24 October 2014; pp. 516–519.

21. Yousefzadeh, A.; Serrano-Gotarredona, T.; Linares-Barranco, B. Fast Pipeline 128 ×128 pixel Spiking Convolution Core for Event-Driven Vision Processing in FPGAs. In Proceedings of the First IEEE International Conference on Event-based Control, Communication, and Signal Processing (EBCCSP), Krakow, Poland, 17–19 June 2015; pp. 1–8.

22. Budzan, S.; Kasprzyk, J. Fusion of 3D Laser Scanner and Depth Images for Obstacle Recognition in Mobile Applications. *Opt. Laser Eng.* **2016**, *77*, 230–240.

23. Guana, H.; Libc, J.; Caoa, S.; Yud, Y. Use of Mobile LiDAR in Road Information Inventory: A Review. *Int. J. Image Data Fusion* **2016**, *7*, 219–242.

24. Clarke, T.; Grattan, K.; Lindsey, N. Laser-based Triangularion Techniques in Optical Inspection of Industrial Structures. *Proc. SPIE* **1990**, *1332*, 474–486.

25. Khademi, S.; Darudi, A.; Abbasi, Z. A Sub Pixel Resolution Method. *World Acad. Sci. Eng. Technol.* **2010**, *70*, 578–581.

26. Peiravi, A.; Taabbodi, B. A Reliable 3D Laser Triangulation-based Scanner with a New Simple but Accurate Procedure for Finding Scanner Parameters. *J. Am. Sci.* **2010**, *6*, 80–85.

27. Kneip, L.; Tache, F.; Caprari, G.; Siegwart, R. Characterization of the Compact Hokuyo URG-04LX 2D Laser Range Scanner. In Proceedings of the IEEE International Conference on Robotics and Automation (ICRA), Kobe, Japan, 12–17 May 2009; pp. 2522–2529.

28. Foix, S.; Alenya, G.; Torras, C. Lock-in Time-of-Flight (ToF) Cameras: A Survey. *IEEE Sens. J.* **2011**, *11*, 1917–1926.

29. Khoshelham, K.; Elberink, S.O. Accuracy and Resolution of Kinect Depth Data for Indoor Mapping Applications. *Sensors* **2012**, *12*, 1437–1454.

30. Chen, J.G.; Wadhwa, N.; Cha, Y.J.; Durand, F.; Freeman, W.T.; Buyukozturk, O. Modal Identification of Simple Structures with High-Speed Video using Motion Magnification. *J. Sound Vib.* **2015**, *345*, 58–71.

31. Cha, Y.J.; Chen, J.G.; Buyukozturk, O. Motion Magnification Based Damage Detection Using High Speed Video. In Proceedings of the 10th International Workshop On Structural Health Monitoring (IWSHM), Stanford, CA, USA, 1–3 September 2015.

32. Vegara, F.; Zuccarello, P.; Boluda, J.A.; Pardo, F. Taking Advantage of Selective Change Driven Processing for 3D Scanning. *Sensors* **2013**, *13*, 13143–13162.

33. Acosta, D.; Garcia, O.; Aponte, J. LaserTriangulation for Shape Acquisition in a 3D Scanner Plus Scanner. In Proceedings of the Electronics, Robotics and Automotive Mechanics Conference (CERMA), Cuernavaca, Mexico, 26–29 September 2006; pp. 14–19.

34. Zuccarello, P.; Pardo, F.; de la Plaza, A.; Boluda, J.A. A 32 × 32 Pixels Vision Sensor for Selective Change Driven Readout Strategy. In Proceedings of the 36th European Solid State Circuits Conference (ESSCIRC), Sevilla, Spain, 14–16 September 2010.

35. Pardo, F.; Zuccarello, P.; Boluda, J.A.; Vegara, F. Advantages of Selective Change Driven Vision for Resource-Limited Systems. *IEEE Trans. Circuits Syst. Video* **2011**, *21*, 1415–1423.

36. Kiran, R.; Nampally, S. Analyzing the Performance of Carry Tree Adders Based on FPGA's. *Int. J. Electron. Signals Syst.* **2012**, *2*, 54–58.

37. Pardo, F.; Boluda, J.A.; Vegara, F. Random Telegraph Signal Transients in Active Logarithmic Continuous-Time Vision Sensors. *Solid State Electron.* **2015**, *114*, 111–114.

Article

Geometric Calibration and Validation of Kompsat-3A AEISS-A Camera

Doocheon Seo [1], Jaehong Oh [2,*], Changno Lee [3], Donghan Lee [1] and Haejin Choi [1]

[1] Korea Aerospace Research Institute, Daejeon 34133, Korea; dcivil@kari.re.kr (D.S.); dhlee@kari.re.kr (D.L.); hjchoi@kari.re.kr (H.C.)
[2] Department of Civil Engineering, Chonnam National University, Gwangju 61186, Korea
[3] Department of Civil Engineering, Seoul National University of Science and Technology, Seoul 01811, Korea; changno@seoultech.ac.kr
* Correspondence: ojh@jnu.ac.kr; Tel.: +82-62-530-1654

Academic Editor: Gonzalo Pajares Martinsanz
Received: 26 September 2016; Accepted: 20 October 2016; Published: 24 October 2016

Abstract: Kompsat-3A, which was launched on 25 March 2015, is a sister spacecraft of the Kompsat-3 developed by the Korea Aerospace Research Institute (KARI). Kompsat-3A's AEISS-A (Advanced Electronic Image Scanning System-A) camera is similar to Kompsat-3's AEISS but it was designed to provide PAN (Panchromatic) resolution of 0.55 m, MS (multispectral) resolution of 2.20 m, and TIR (thermal infrared) at 5.5 m resolution. In this paper we present the geometric calibration and validation work of Kompsat-3A that was completed last year. A set of images over the test sites was taken for two months and was utilized for the work. The workflow includes the boresight calibration, CCDs (charge-coupled devices) alignment and focal length determination, the merge of two CCD lines, and the band-to-band registration. Then, the positional accuracies without any GCPs (ground control points) were validated for hundreds of test sites across the world using various image acquisition modes. In addition, we checked the planimetric accuracy by bundle adjustments with GCPs.

Keywords: Kompsat-3A; AEISS-A; calibration; validation

1. Introduction

Kompsat-3A, which was launched on 25 March 2015, is a sister spacecraft of the Kompsat-3 developed by the Korea Aerospace Research Institute (KARI). Kompsat-3A's AEISS-A (Advanced Electronic Image Scanning System-A) camera is similar to Kompsat-3's AEISS but it was designed to provide PAN (Panchromatic) resolution of 0.55 m, MS (multispectral) resolution of 2.20 m, and TIR (thermal infrared) at 5.5 m resolution as presented in Table 1. The altitude of Kompsat-3A is 528 km—which is lower than that of Kompsat-3 (685 km)—for better spatial resolution, sacrificing the swath width.

In-orbit geometric calibration and validation of high-resolution Earth-observation satellites is important because various impulses and vibrations during the launch process may have affected the satellite's payload [1–8]. Therefore, the in-orbit geometric calibration process determines the focal length, the distortions of lenses, CCD (charge-coupled device) alignments, and other geometric distortions. For this purpose, bundle adjustments are carried out utilizing GCPs (ground control points) at the reference sites that give accurate location information. This leads to the elimination of a series of systematic errors and reduction of correlation between the interior and exterior orientation parameters to improve the geometric accuracy. Then the validation process is conducted to check and ensure the mapping accuracy.

Table 1. Kompsat-3A specifications.

	PAN (Panchromatic)	MS (Multispectral)
Spectral Bands	450–900 μm	Blue: 450–520 μm Green: 520–600 μm Red: 630–690 μm NIR (Near infra-red): 760–900 μm
GSD (Ground Sample Distance)	0.55 m at nadir	2.2 m at nadir
Focal Length	8.6 m	8.6 m
Swath Width at Nadir	12 km	12 km
Data Quantization	14 bit	14 bit
CCD (Charge-Coupled Device) Detector	Array of 24,000 pixels (2 × 12,000)	Arrays of 4 (RGB and IR) × 6,000 pixels (2 × 3,000)
Pixel Pitch	8.75 μm	35 μm

The geometric calibration and validation of the Kompsat-3A AEISS-A camera have been completed [9]. The geometric calibration consisted of two phases. At phase I, AOCS (attitude and orbit control subsystem) in-orbit calibration was performed with the satellite's position and attitude data, which are estimated through time synchronization of GPS, AOCS, and payloads. Phase II included the calibration of CCD alignments and the focal length, the CCD overlap area correction, and band-to-band alignments. This was followed by the validation of positional accuracy.

In this paper, we introduce the Kompsat-3A AEISS-A camera and the physical sensor model that incorporates the interior and exterior orientation parameters. Based on the rigorous sensor model, the geometric calibration method of Kompsat-3A will be explained. This includes the boresight calibration, and the calibration of the CCD alignments and the focal length. Then we present the results of the geometric calibration and validation including not only the aforementioned sensor calibrations but the merge of sub-images and the band-to-band registrations. Finally, the positional accuracy after the calibration is presented.

2. Kompsat-3A AEISS-A Camera

2.1. AEISS-A Sensor

Figure 1a shows the configuration of Kompsat-3A AEISS-A camera. Blue, PAN1, PAN2, TIR, green, red, and near-infrared (NIR) channels are aligned in a unifocal camera. Figure 1b depicts the design of PAN, MS, and TIR (written IR in the figure). The gaps between the sensors in the focal plane correspond to the differences of the projection centers. The telescope uses a Korsch combination with three aspheric mirrors and two folding mirrors, using an aperture diameter of 80 cm. This design was chosen because of its simplicity and compact size, allowing it to fit within the small spacecraft platform. Also, the camera was designed to minimize the aberration.

Figure 2 presents the detailed configuration of the panchromatic CCD-lines [10]. A single CCD-line consists of 12,080 pixels with 20 dark pixels on each side and the overlapping area is 100 pixels in the center. The pixel size is 8.75 micron. Each CCD-line produces a single subimage with overlapping areas, and the two produced subimages must be merged together into a single image that is 24,020 pixels of image width.

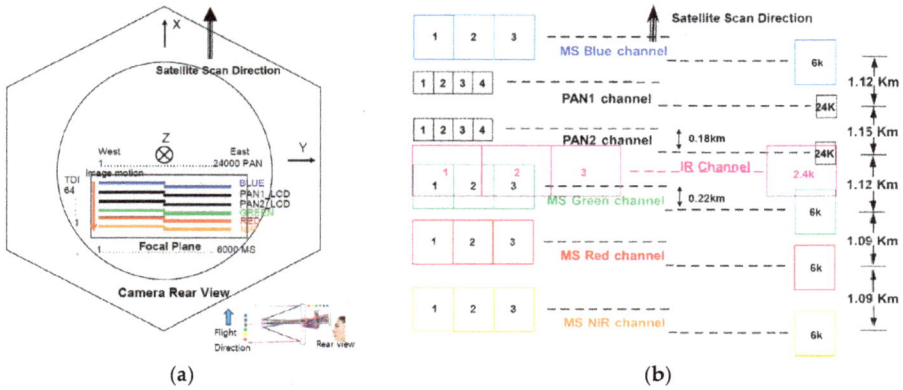

Figure 1. Kompsat-3A AEISS-A (Advanced Electronic Image Scanning System-A) sensor configuration. (a) Camera rear view; (b) CCD array configurations.

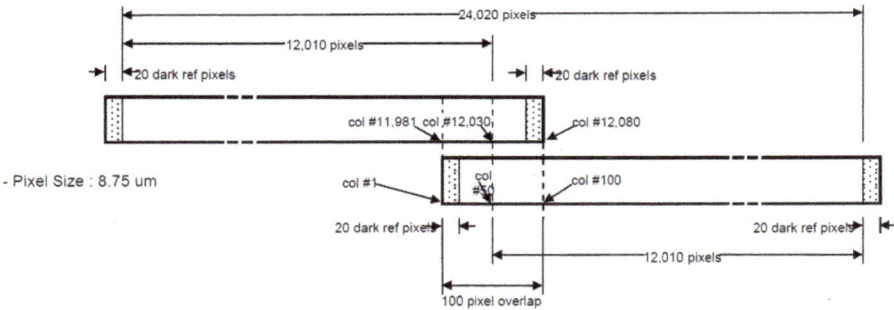

Figure 2. Kompsat-3A AEISS-A panchromatic CCD-lines configuration with an overlapping zone (the scan direction is upward).

2.2. Physical Sensor Modeling

The physical sensor model of Kompsat-3A is in a nonlinear form of projection from a given ground point in an Earth-centered Earth-fixed (ECEF) coordinate frame to a point in the body coordinate frame as shown in Equations (1) and (2). We call this the forward model. The exterior orientation parameters (EOPs) can be interpolated from the ephemeris data given an instant time.

$$
\begin{bmatrix} U \\ V \\ W \end{bmatrix} = M_{Bore}^{T} M_{Orbit}^{Body} M_{ECEF}^{Orbit} \begin{bmatrix} X - X_S \\ Y - Y_S \\ Z - Z_S \end{bmatrix}
\tag{1}
$$

$$
\begin{bmatrix} x_b \\ y_b \\ z_b \end{bmatrix} = \lambda \begin{bmatrix} U \\ V \\ W \end{bmatrix}
\tag{2}
$$

where $\begin{bmatrix} X & Y & Z \end{bmatrix}^{T}$ is the ground point in the ECEF coordinate frame, $\begin{bmatrix} X_S & Y_S & Z_S \end{bmatrix}^{T}$ is the satellite position in the ECEF coordinate frame, M_{ECEF}^{ECI} is the time-dependent rotation matrix from the ECEF coordinate frame to the inertial coordinate frame, M_{ECI}^{Body} is the time-dependent rotation matrix from the inertial coordinate frame to the body coordinates frame, M_{Bore} is the boresight rotation matrix,

x_b, y_b, z_b are the coordinates in the body coordinate frame (x_b is the flight direction, z_b is the direction to the Earth, and y_b completes the right-handed coordinate system), and λ is the scale factor.

The position and the rotation of the satellite at time t can be computed using Equation (3). A scan time t corresponding to an image line is used for the computation of the position and the rotation using the Lagrange interpolation of 8 neighboring ephemeris data.

$$\vec{P}(t) = \sum_{j=1}^{8} \vec{P}(t_j) \times \prod_{\substack{i=1 \\ i \neq j}}^{8} (t - t_i) / \prod_{\substack{i=1 \\ i \neq j}}^{8} (t_j - t_i) \tag{3}$$

where $\vec{P}(t)$ is the position and the rotation of the satellite at time t.

The relationship between the sensor coordinate frame and the body coordinates frame is presented in Figure 3 and Equation (4). In, x_s, y_s show the sensor coordinate frame and z_s completes the right-handed coordinate system.

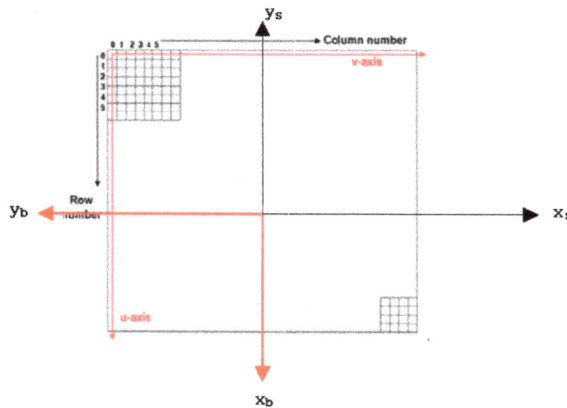

Figure 3. The relationship between the sensor coordinate frame and the body coordinate frame.

$$\begin{bmatrix} -y_s \\ -x_s \\ f \end{bmatrix} = M_b^s \begin{bmatrix} x_b \\ y_b \\ z_b \end{bmatrix}, \quad M_b^s = \begin{bmatrix} 0 & -1 & 0 \\ -1 & 0 & 0 \\ 0 & 0 & -1 \end{bmatrix} \tag{4}$$

where M_b^s is the transformation matrix from the body coordinate frame to the sensor coordinate frame, and f is the focal length.

The sensor coordinates can be converted from the image coordinates using CCD-line alignment information parameters as shown in Equation (5). An individual CCD-line requires unique alignment parameters. The CCD alignment equation was determined based on the precise calibration performed before the launch. The second-order equation showed 0.01% difference compared to the reference, satisfying the requirement of distortion 0.2%.

$$\begin{aligned} x_s &= a_{0i} + a_{1i} \times (c - c_{0i}) + a_{2i} \times (c - c_{0i})^2 \\ y_s &= b_{0i} + b_{1i} \times (c - c_{0i}) + b_{2i} \times (c - c_{0i})^2 \end{aligned} \tag{5}$$

where i is the CCD chip index, a_0, b_0 are the x, y coordinates of the first pixel in the CCD chip, a_1, a_2 are related to the pixel size (a_2 is the nonlinearity part), b_1, b_2 are the alignment parameters of the non-straight line CCD chip, and c is the column (sample) coordinate in pixels (c_0 is for the first column of the CCD chip).

2.3. Sensor Geometric Calibrations

During the calibration process, the focal length, the boresight angles, and CCD alignment parameters are estimated. Removing the scale factor in Equation (2), observation equations can be established as Equation (6).

$$F_x = x_b - f\frac{U}{W} = 0$$
$$F_y = y_b - f\frac{V}{W} = 0 \tag{6}$$

The first step of the AOCS absolute calibration is to perform the boresight calibration between the star tracker and the other payloads using GCPs (ground control points) located at calibration sites. To this end, the boresight rotation matrix M_{Bore} must be estimated.

$$M_{Bore} = M_{Y_b} M_{P_b} M_{R_b} \tag{7}$$

$$M_{Y_b} = \begin{bmatrix} cY_b & sY_b & 0 \\ -sY_b & cY_b & 0 \\ 0 & 0 & 1 \end{bmatrix}, M_{P_b} = \begin{bmatrix} cP_b & 0 & -sP_b \\ 0 & 1 & 0 \\ sP_b & 0 & cP_b \end{bmatrix}, M_{R_b} = \begin{bmatrix} 1 & 0 & 0 \\ 0 & cR_b & sR_b \\ 0 & -sR_b & cR_b \end{bmatrix} \tag{8}$$

$$cY_b = \cos(yaw_B), sY_b = \sin(yaw_B)$$
$$cP_b = \cos(pitch_B), sP_b = \sin(pitch_B)$$
$$cR_b = \cos(roll_B), sR_b = \sin(roll_B) \tag{9}$$

The partial derivatives with respect to the boresight angles can be expressed as Equations (10) and (11), which show the case of the boresight roll angle. The same analogy is applied to the other angle cases, such as the pitch and yaw.

$$\frac{\partial F_x}{\partial R_b} = -\frac{f}{W}\left(\frac{\partial U}{\partial R_b} - \frac{U}{W}\frac{\partial W}{\partial R_b}\right)$$
$$\frac{\partial F_y}{\partial R_b} = -\frac{f}{W}\left(\frac{\partial V}{\partial R_b} - \frac{V}{W}\frac{\partial W}{\partial R_b}\right) \tag{10}$$

$$\frac{\partial}{\partial R_b}\begin{bmatrix} U \\ V \\ W \end{bmatrix} = \begin{bmatrix} 0 & 0 & 0 \\ 0 & -sR_b & -cR_b \\ 0 & cR_b & -sR_b \end{bmatrix} M_{P_b}^T M_{Y_b}^T M_{Orbit}^{Body} M_{ECEF}^{Orbit} \begin{bmatrix} X - X_S \\ Y - Y_S \\ Z - Z_S \end{bmatrix} \tag{11}$$

Secondly, the calibration of the focal length is simply carried out by deriving the partial derivatives with respect to the focal length as Equation (12).

$$\frac{\partial F_x}{\partial f} = -\frac{U}{W}, \frac{\partial F_y}{\partial f} = -\frac{V}{W} \tag{12}$$

Finally, the calibration for the CCD alignment parameters can also be carried out by computing partial derivatives as shown in Equation (13). Note that Kompsat-3A AEISS-A sensor consists of several CCD lines and they are calibrated all together. Note that i is the CCD chip index.

$$\begin{bmatrix} \frac{\partial F_{x_i}}{\partial a_{0i}} & \frac{\partial F_{x_i}}{\partial a_{1i}} & \frac{\partial F_{x_i}}{\partial a_{2i}} & \frac{\partial F_{x_i}}{\partial b_{0i}} & \frac{\partial F_{x_i}}{\partial b_{1i}} & \frac{\partial F_{x_i}}{\partial b_{2i}} \\ \frac{\partial F_{y_i}}{\partial a_{0i}} & \frac{\partial F_{y_i}}{\partial a_{1i}} & \frac{\partial F_{y_i}}{\partial a_{2i}} & \frac{\partial F_{y_i}}{\partial b_{0i}} & \frac{\partial F_{y_i}}{\partial b_{1i}} & \frac{\partial F_{y_i}}{\partial b_{2i}} \end{bmatrix} = \begin{bmatrix} 1 & (c - c_{0i}) & (c - c_{0i})^2 & 0 & 0 & 0 \\ 0 & 0 & 0 & 1 & (c - c_{0i}) & (c - c_{0i})^2 \end{bmatrix} \tag{13}$$

The partial derivatives with respect to the focal length, the boresight angles, and CCD alignment parameters, as well as EOPs (exterior orientation parameters) are used to form a design matrix A for the linearized observation equation in Equation (14) and iteratively solved using the least square adjustment as Equation (15). Note that the calibrations of boresight angles, focal length, and CCD alignment parameters are carried out sequentially to avoid large correlations among the parameters.

Note that single iterative least squares can make the normal matrix not invertible. Therefore, other systems utilized a step-by-step approach for the geometric calibration [11].

$$A\zeta = y \tag{14}$$

$$\hat{\zeta} = \left(A^T P A\right)^{-1} A^T P y \tag{15}$$

3. Geometric Calibration and Validation

3.1. Workflow

The geometric calibration consists of AOCS in-orbit calibration (boresight calibration), the focal length calibration, the calibration of CCD alignments, the CCD overlap area correction, and band-to-band registration, as shown in Figure 4. The details on the AOCS absolute calibration, focal length, and the CCD alignments were presented in the previous section. For the CCD overlap area correction and band-to-band registration we applied the same merging process used for Kompsat-3 data [10]. This method generates a grid of tie points over the subimages based on the physical sensor model and uses them for similarity transformation with the compensation of ephemeris and terrain variation.

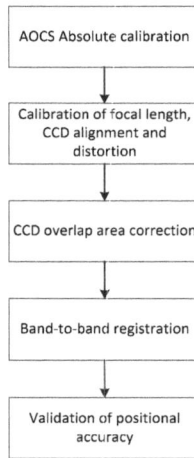

Figure 4. Summarized geometric calibration procedure of Kompsat-3A.

3.2. Calibration Sites

We classified test sites into two categories according to their positional accuracies. Level 0 sites are located in several sites over Mongolia and Korea where about 150~180 circle targets of 3 m diameter were established. The coordinates of the targets were acquired by GNSS surveys and they showed the positional accuracy of 5 cm in RMSE (root mean square error). They were used for the CCD alignment, the focal length calibration, AOCS absolute calibration, and validation of mapping accuracy. Average 9~20 points were used for each scene.

Level 1 sites are distributed at 82 locations across the world as shown in Figure 5. Locations such as road intersection were global navigation satellite system (GNSS)-surveyed with about 70 cm accuracy in RMSE. Level 1 reference data were used for AOCS absolute calibration and validation of positional accuracy.

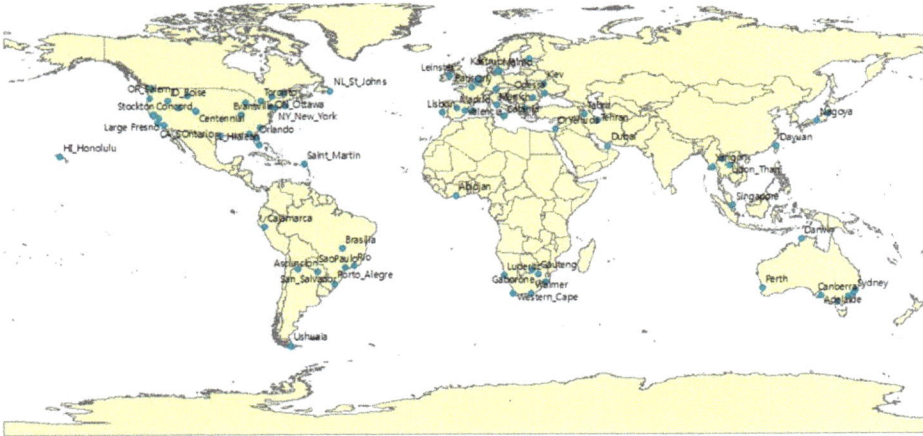

Figure 5. Level 1 site locations.

3.3. AOCS Absolute Calibration

First we conducted the AOCS absolute calibration. The accuracy of the AOCS absolute calibration highly depends on the accuracy of GCPs, and its reliability can also be affected by the temperature characteristics of the star tracker. Note that the accuracy of the star tracker used (Sodern SED36) is 1 arcsec for the cross-boresight and 6 arcsec for the boresight axis. This necessitated using the GCPs from different calibration sites of the southern and northern hemispheres to carry out the boresight calibration. We used five image strips over level 0 sites and six strips over the level 1 sites with difference roll angles ranging −25.5°~27.9° for the calibration as shown in Table 2.

The result of the calibration, which is the rotation matrix M_{Bore} in Equation (1), was used to update the system. The horizontal accuracy of the check points was estimated to 2.9 km (CE90) before the calibration, but the error was reduced to 13.6 m (CE90) after the system update, as shown in Figure 6.

Table 2. Attitude angles of Kompsat-3 test data used for the AOCS (attitude and orbit control subsystem) absolute calibration.

Strip	Roll (°)	Pitch (°)	Yaw (°)
1	−15.6	−0.96	2.85
2	0.56	0.20	2.78
3	12.52	0.63	2.09
4	27.91	0.71	1.65
5	0.05	0.17	3.04
6	6.41	0.23	3.33
7	−25.48	−0.99	2.42
8	−9.24	−0.30	2.89
9	27.58	1.16	1.90
10	15.09	0.80	2.31
11	11.88	0.67	2.38

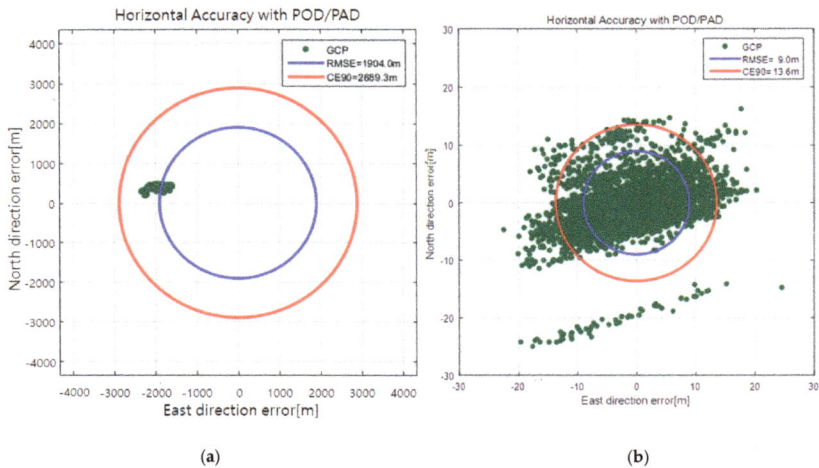

(a) (b)

Figure 6. Horizontal accuracy in the ground before (**a**) and after the AOCS calibration (**b**).

3.4. Calibration of Focal Length and CCD Alignments

Focal length and CCD alignments are determined before the satellite launch, but the information may change due to the large acceleration during launch. Therefore, in-orbit calibrations should be carried out for better geometric accuracy.

For the in-orbit calibration of focal length and CCD alignment, we utilized 29 images acquired over level 0 calibration sites. The roll and pitch angle ranges are $-29.1°\sim+30.3°$ and $-1.1°\sim+1.2°$, respectively. The focal length of the camera was determined to 8.56181 m and the determined alignment parameters for each CCD sensor are presented in Table 3. Precisions of the calibration for each CCD line are less than one pixel and less than half-pixels for PAN and the others, respectively. Note that the across-track is the direction along the CCD lines and the along-track is opposite to the flight direction as shown in Equation (5). a_1 and a_2, linearity, and nonlinearity parameters of the pixel size are determined to be slightly larger than 1.0 and 0.0, respectively, meaning the pixel size is not exactly regular. Also, small but non-zero b_1, b_2 values indicate non-straightness of the CCD lines. Thus we plotted the PAN#1 and BLUE#1 CCD lines to see the patterns (Figure 7). PAN #1 shows that the non-regularity of the pixel size accumulates up to about 40 pixels in the across-track alignment. The along-track direction plot shows that the CCD is almost straight with less than half pixels of discrepancy. In case of BLUE#1, the non-regularity of the pixel is accumulated up to about 10 pixels and the non-straightness is about a quarter pixels.

Table 3. Determined alignment parameters for each CCD.

Detector	Across-Track (LOD)			Along-Track (LOS)			RMSE
	a_0	a_1	a_2	b_0	b_1	b_2	[pixels]
PAN#1	−12053.14	1.00142	1.26346×10^{-8}	0.41	−0.00006	1.88813×10^{-9}	1.00
PAN#2	−12053.07	1.00142	1.25988×10^{-8}	−340.02	−0.00003	1.96001×10^{-9}	0.86
Red#1	−3013.24	1.00126	3.50715×10^{-8}	1556.21	0.00023	-1.25847×10^{-8}	0.33
Red#2	−3013.16	1.00119	4.51743×10^{-8}	1454.06	0.00026	-1.34342×10^{-8}	0.33
Green#1	−3014.48	1.00137	3.73793×10^{-8}	1058.21	0.00024	-2.70587×10^{-8}	0.32
Green#2	−3014.41	1.00130	4.71767×10^{-8}	956.04	0.00029	-2.85802×10^{-8}	0.32
Blue#1	−3013.74	1.00160	3.68874×10^{-8}	−496.87	0.00004	5.24967×10^{-10}	0.32
Blue#2	−3013.68	1.00152	4.80332×10^{-8}	−599.04	0.00006	1.53648×10^{-9}	0.28
NIR#1	−3014.57	1.00130	2.56360×10^{-8}	2052.60	0.00022	-1.60160×10^{-8}	0.35
NIR#2	−3014.50	1.00125	3.30289×10^{-8}	1950.17	0.00037	-2.97782×10^{-8}	0.31

Figure 8 is the plotted horizontal accuracy of GCPs before and after the in-orbit CCD alignment calibration. We can clearly observe the accuracy improvement from 12.5 m (CE90) in Figure 8a to 8.0 m (CE90) in Figure 8b.

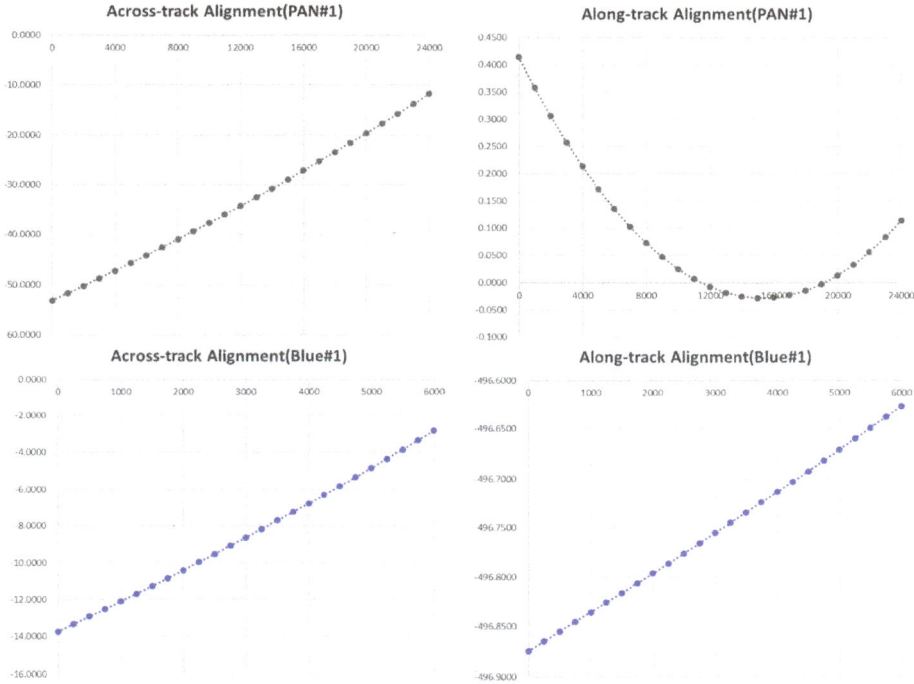

Figure 7. CCD alignment plots for PAN#1 and BLUE#1.

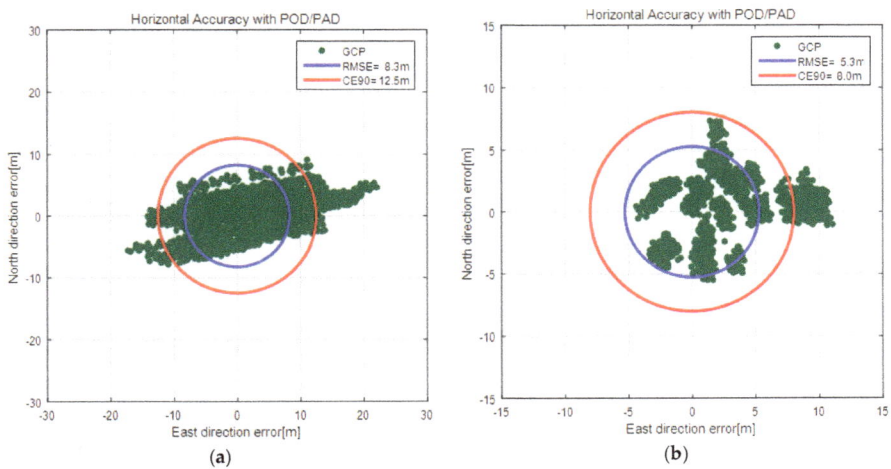

Figure 8. Horizontal accuracy before (**a**) and after the CCD alignment calibration (**b**).

3.5. Merge of Subimages and Band-to-Band Registration

As described in Figure 2, individual CCD lines of Kompsat-3A produce overlapping subimages. These subimages should be merged side-by-side for a larger swath width, but the process is not simple because the sensor alignment, ephemeris effects, and terrain elevations should be considered each time. We applied an automated approach using virtual tie points to estimate the shift and similarity transformation, as well as to compute compensations according to the satellite's attitude differences and terrain elevations due to the gap between the CCD lines [10]. Figure 9 presents an example image before and after the merging. We can hardly identify the discrepancy by applying the process.

We tested the quality of the merging results to plot the estimated discrepancies for various acquisition modes of Kompsat-3, as presented in Figure 10. The results showed that the discrepancy between the sub-images from the PAN sensors was estimated to 0.25 pixels (CE90). In addition, the merge quality is not affected by the acquisition modes.

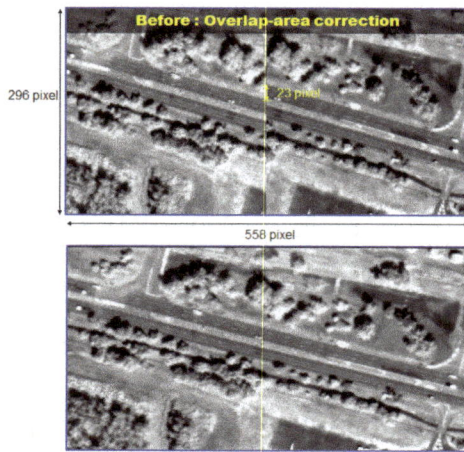

Figure 9. Comparison between before and after the merge of subimages.

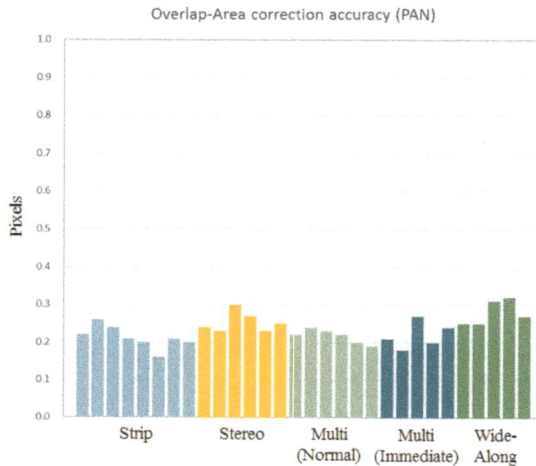

Figure 10. Discrepancy between subimages after the merge for various image acquisition modes.

Following the subimage merge, we continue the band-to-band registration process. We utilized a total 30 image strips of various acquisition modes, such as 9 single strips, 6 stereos, 6 standard multi, 5 immediate multi, and 5 wide-along modes. In addition, we used SRTM (shuttle radar topography mission) V2 for the elevation information. Table 4 shows that the accuracy of the registration between PAN and MS sensor is lower than a half pixel in RMSE. Figure 11 shows examples of the band-to-band registration showing the negligible saturation of multispectral colors after the process.

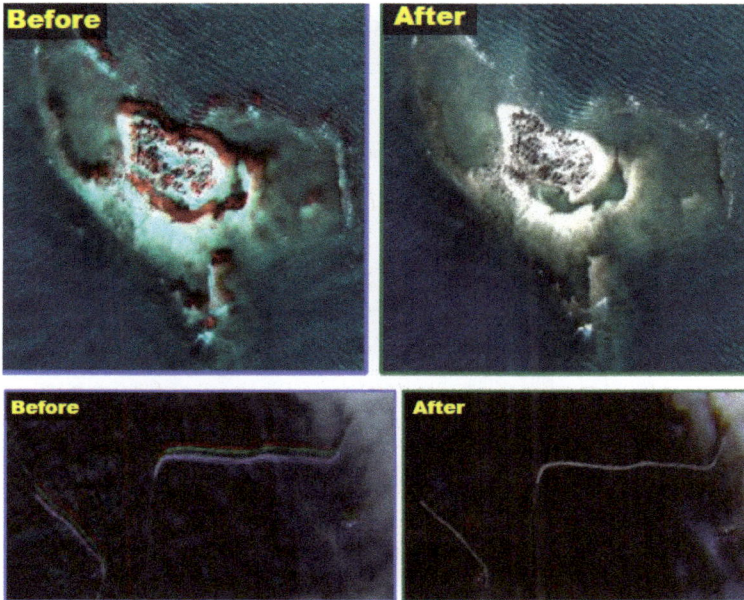

Figure 11. Examples of the band-to-band registration results.

3.6. Validation of Positional Accuracy

Completing the geometric calibration, we validated the positional accuracy. To this end we used 325 sets of test images across the world. The validation data were categorized for several acquisition modes which Kompsat-3A is capable of, as shown in Table 5.

Figure 12 presents the horizontal accuracy for various image acquisition modes in RMSE and CE90. The strip mode showed the best accuracy among the acquisition modes with 8.9 m (RMSE) and 13.5 m (CE90). The one-pass stereo mode showed about 1 m larger error range than the strip mode. In the cases of multi (normal) and wide-along modes, the accuracy decreased to 13~14 m (RMSE) and 20~21 m (CE90).

Next, we validated the potential mapping accuracy using GCPs. A total of 16 image strips with −21.5°~+29.8° of roll angle range were used for the bundle adjustment. Note that these data were not used for the calibration. Each image was adjusted with 8~9 GCPs and 32~148 check points were used to validate the accuracy. The resultant errors were 0.91 (0.5 m) and 1.39 pixels (0.8 m) in RSME and CE90, respectively, as shown in Figure 13.

Table 4. Band-to-band registration accuracy.

Imaging Type	CalVal_ID	Cloud Level	Scene Center					Average Height (m)	Red-to-PAN		Green-to-PAN		Blue-to-PAN		NIR-to-PAN	
			Latitude	Longitude	Roll	Pitch	Yaw		No	RMSE	No	RMSE	No	RMSE	No	RMSE
Strip	Geo_006639	B	037.7359	−097.1630	30.0	01.4	02.1	416	370	0.23	508	0.22	367	0.27	590	0.22
Strip	Geo_007744	A	−033.4062	−070.5701	−29.8	−01.4	02.2	574	1325	0.19	1307	0.24	1133	0.20	1254	0.19
Strip	GKJ_006331	A	035.8358	126.9800	−29.1	−01.1	02.3	30	1145	0.24	1122	0.24	1035	0.26	1192	0.22
Strip	GSS_006331	A	036.8496	126.6604	−29.1	−01.1	02.2	80	1243	0.18	1284	0.19	1034	0.26	1267	0.25
Strip	GUB_006345	B	047.9453	107.0282	29.8	01.2	01.8	1309	796	0.33	855	0.30	782	0.25	780	0.33
Strip	Geo_006892	C	007.4304	125.9018	29.4	01.6	02.8	304	136	0.30	193	1.18	160	0.58	243	0.39
Strip	Geo_008337	C	045.0623	−093.1176	20.8	01.0	02.2	275	1024	2.81	1022	1.89	1047	0.86	813	4.75
Strip	Geo_005772	B	−012.4020	130.9393	06.7	00.3	03.6	25	869	0.34	873	0.30	721	0.24	909	0.35
Stereo	Geo_006477	B	022.5447	088.4104	−00.4	−30.0	03.2	08	1074	0.25	1263	0.26	1246	0.27	1223	0.28
Stereo	Geo_006478	B	022.5011	088.4200	−00.4	30.2	03.7	07	1076	0.21	1290	0.26	1204	0.26	1194	0.21
Stereo	Geo_008764	B	−035.9403	145.7189	−18.6	−29.4	−07.2	98	228	0.30	377	0.31	323	0.34	1014	0.21
Stereo	Geo_008765	B	−035.9067	145.7085	−18.2	27.6	12.5	102	641	0.36	836	0.34	452	0.41	1063	0.36
Stereo	Geo_009983	B	040.9669	−082.7330	17.5	−27.8	12.0	304	129	0.27	450	0.30	223	0.22	1296	0.24
Stereo	Geo_009984	B	040.9260	−082.7204	17.8	29.5	−07.0	308	240	0.26	542	0.25	401	0.28	1250	0.23
Multi(Normal)	Geo_007676	B	−019.2687	146.8494	28.7	01.4	02.7	26	1179	0.24	1220	0.15	1166	0.19	1192	0.37
Multi(Normal)	Geo_006201	B	−043.5198	172.6841	24.4	00.9	02.3	31	790	0.16	845	0.13	749	0.23	856	0.22
Multi(Normal)	Geo_010131	B	−024.9108	152.4551	28.0	01.3	02.6	24	630	0.34	842	0.22	716	0.25	790	0.44
Multi(Normal)	Geo_008877	B	053.6063	−113.4369	−06.5	−00.1	02.2	672	1124	0.19	1139	0.21	1165	0.25	1313	0.18
Multi(Normal)	Geo_009344	B	047.4359	019.3405	−01.1	00.1	02.5	131	650	0.32	704	0.31	577	0.36	878	0.35
Multi(Normal)	Geo_009502	B	−015.5594	−056.0030	28.1	01.4	02.8	190	1085	0.19	1117	0.09	1108	0.14	1062	0.27
Multi(Immediate)	Geo_002483	B	014.4249	033.6011	−27.5	−01.4	02.8	407	1167	0.20	1150	0.21	1154	0.17	1175	0.28
Multi(Immediate)	Geo_006662	B	−016.7051	−043.7942	02.3	00.0	03.6	672	1033	0.27	1064	0.27	988	0.26	996	0.18
Multi(Immediate)	Geo_007785	B	−027.1914	151.3482	28.3	01.3	02.6	341	767	1.01	822	0.72	866	0.55	848	1.05
Multi(Immediate)	Geo_008606	B	053.5697	−113.3945	26.7	01.0	01.7	667	1282	0.19	1147	0.16	1211	0.12	1157	0.15
Wide-Along	Geo_006826	B	040.7115	−076.6778	11.2	−28.9	08.8	230	136	0.22	366	0.29	190	0.25	672	0.27
Wide-Along	Geo_006827	B	040.7103	−076.5101	13.0	30.0	−04.3	280	178	0.36	474	0.14	311	0.12	594	0.38
Wide-Along	Geo_010468	B	044.3926	−100.2630	−12.2	−29.7	−04.0	492	369	0.37	628	0.29	528	0.33	1213	0.31
Wide-Along	Geo_010469	B	044.3308	−100.4222	−13.8	28.8	10.0	520	66	1.93	392	0.35	102	0.50	758	0.30
Wide-Along	Geo_009221	B	035.6860	−000.5169	28.1	−25.2	17.2	118	1303	0.24	1288	0.21	1220	0.23	1215	0.27

Table 5. Test data used for the positional accuracy validation.

Acquisition Mode	Number of Data Sets
Strip	94
Multi (Immediate/Normal)	63/63
Wide-Along	63
Along-Track Stereo	56

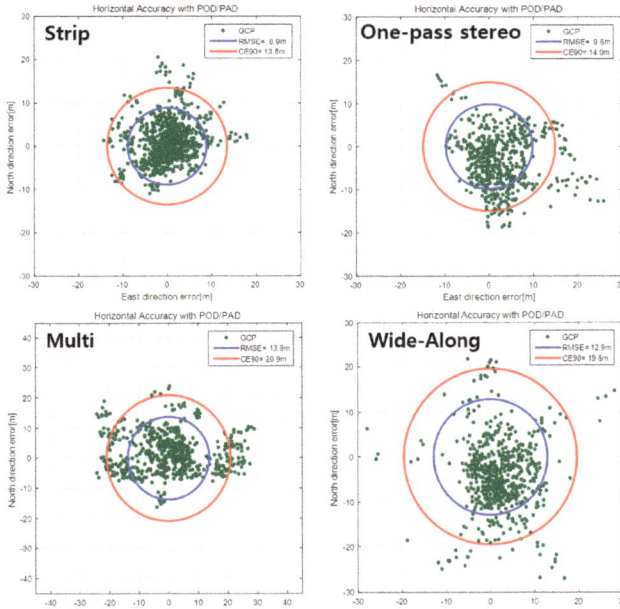

Figure 12. Horizontal accuracy for various acquisition modes without any GCPs.

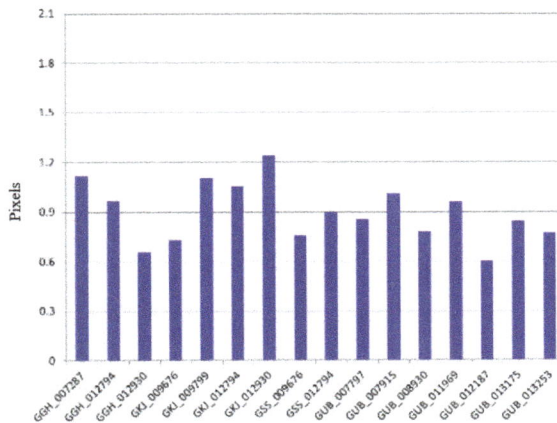

Figure 13. Mapping accuracy for 16 independent data.

4. Conclusions

We presented the geometric calibration and validation work of Kompsat-3A that was completed last year. A set of images over the test sites was taken for two months and was utilized for the work. The works include the AOCS absolute calibration, the calibration of the focal length and CCD alignments, the merge of CCD lines, the band-to-band registration, and, finally, by the validation of the positional accuracy.

The successful AOCS' calibration increased the horizontal accuracy from 2.9 km (CE90) to 13.6 m and the CCD alignment calibration determined the non-regular and nonlinear CCD distortions, improving the accuracy from 12.5 m (CE90) to 8.0 m. Based on the calibration results, we could successfully merge the subimages from each CCD line with a negligible discrepancy of 0.25 pixels (CE90). Finally, we validated the positional accuracy with completion of the geometric calibrations. Without any GCP, the popularly used image acquisition mode, the strip mode, showed 13.5 m of horizontal accuracy, though other modes showed slightly lower accuracies. When GCPs were used for the bundle adjustment, we could obtain less than 1 m of horizontal accuracy in CE90.

Acknowledgments: This study was supported by the Korea Aerospace Research Institute (FR16720W02).

Author Contributions: Doocheon Seo and Jaehong Oh carried out the experiment and wrote this paper. Changno Lee participated in the experiment. Donghan Lee and Haejin Choi were responsible for the data acquisition. All authors reviewed the manuscript.

Conflicts of Interest: The authors declare no conflict of interest.

References

1. Christian, J.A.; Benhacine, L.; Hikes, J.; D'Souza, C. Geometric Calibration of the Orion Optical Navigation Camera using Star Field Images. *J. Astronaut. Sci.* **2016**, *63*, 335–353. [CrossRef]
2. Jacobsen, K. High-Resolution Imaging Satellite Systems. In Proceedings of the 3D-Remote Sensing Workshop, Porto, Portugal, 6–11 June 2005.
3. Kornus, W.; Lehner, M.; Schroeder, M. Geometric In-flight Calibration by Block Adjustment Using MOMS−2P Imagery of the Three Intersecting Stereo-strips. In Proceedings of the ISPRS joint workshop "Sensors and Mapping from Space 1999", Hanover, Germany, 27–30 September 1999.
4. Oberst, J.; Brinkmann, B.; Giese, B. Geometric Calibration of the MICAS CCD Sensor for the DS1 (Deep Space One) Spacecraft: Laboratory vs. In-Flight Data Analysis. In Proceedings of the International Archives of Photogrammetry and Remote Sensing, Amsterdam, The Netherlands, 16–22 July 2000.
5. Schröder, S.; Maue, T.; Marquès, P.; Mottola, S.; Aye, K.; Sierks, H.; Keller, H. In-Flight Calibration of the Dawn Framing Camera. *ICARUS* **2013**, *226*, 1304–1317. [CrossRef]
6. Toutin, T.; Blondel, E.; Rother, K.; Mietke, S. In-flight Calibration of SPOT5 and Formosat−2. In Proceedings of the ISPRS Commission I Symposium on From Sensors to Imagery, Paris, France, 4–6 May 2006.
7. Kubik, P.; Lebegue, L.; Fourest, S.; Delvit, J.M.; de Lussy, F.; Greslou, D.; Blanchet, G. First In-flight Results of Pleiades 1A Innovative Methods for Optical Calibration. In Proceedings of the International Conference on Space Optical Systems and Applications, Ajaccio, France, 9–12 October 2012; pp. 1–9.
8. Chen, Y.; Xie, Z.; Qui, Z.; Zhang, Q.; Hu, Z. Calibration and validation of ZY−3 optical sensors. *IEEE Trans. Geosci. Remote Sens.* **2015**, *53*, 4616–4626. [CrossRef]
9. Seo, D.C.; Hong, G.B.; Jin, C.G. Kompsat-3A Direct Georeferencing Mode and Geometric Calibration/ validation. In Proceedings of the Asian Conference on Remote Sensing, Quezon City, Philippines, 19–23 October 2015.
10. Seo, D.C.; Lee, C.N.; Oh, J.H. Merge of sub-images from two PAN CCD lines of Kompsat-3 AEISS. *KSCE J. Civ. Eng.* **2016**, *20*, 863–872. [CrossRef]
11. Mulawa, D. On-orbit Geometric Calibration of the Orbview−3 High Resolution Imaging Satellite. *Int. Arch. Photogramm. Remote Sens. Spat. Inf. Sci.* **2004**, *35*, 1–6.

sensors

MDPI

Article

Design and Evaluation of a Scalable and Reconfigurable Multi-Platform System for Acoustic Imaging

Alberto Izquierdo [1,*], Juan José Villacorta [1], Lara del Val Puente [2] and Luis Suárez [3]

[1] Signal Theory and Communications Department, University of Valladolid, Valladolid 47011, Spain; juavil@tel.uva.es
[2] Mechanical Engineering Area, Industrial Engineering School, University of Valladolid, Valladolid 47011, Spain; lvalpue@eii.uva.es
[3] Civil Engineering Department, Superior Technical College, University of Burgos, Burgos 09001, Spain; luis.a.suarez.vivar@gmail.com
* Correspondence: alberto.izquierdo@tel.uva.es; Tel.: +34-983-185-801

Academic Editor: Gonzalo Pajares Martinsanz
Received: 3 May 2016; Accepted: 8 October 2016; Published: 11 October 2016

Abstract: This paper proposes a scalable and multi-platform framework for signal acquisition and processing, which allows for the generation of acoustic images using planar arrays of MEMS (Micro-Electro-Mechanical Systems) microphones with low development and deployment costs. Acoustic characterization of MEMS sensors was performed, and the beam pattern of a module, based on an 8 × 8 planar array and of several clusters of modules, was obtained. A flexible framework, formed by an FPGA, an embedded processor, a computer desktop, and a graphic processing unit, was defined. The processing times of the algorithms used to obtain the acoustic images, including signal processing and wideband beamforming via FFT, were evaluated in each subsystem of the framework. Based on this analysis, three frameworks are proposed, defined by the specific subsystems used and the algorithms shared. Finally, a set of acoustic images obtained from sound reflected from a person are presented as a case study in the field of biometric identification. These results reveal the feasibility of the proposed system.

Keywords: MEMS array; scalable system; multi-platform framework; acoustic imaging

1. Introduction

In recent years, techniques for obtaining acoustic images have been developed rapidly. At present, acoustic images are associated with a wide variety of applications [1], such as non-destructive testing of materials, medical imaging, underwater imaging, SONAR, geophysical exploration, etc. These techniques for obtaining acoustic images are based on the RADAR (RAdio Detection and Ranging) principles, which form an image of an object from the radio waves that have been reflected on it [2]. RADAR systems require high-cost hardware and their application with people and specific materials is difficult, due to their low reflectivity. These are the reasons why acoustic imaging techniques, also called SODAR (Sound Detection and Ranging) techniques, were developed. These SODAR techniques, mainly based on the use of arrays, represent a simple, low-cost alternative for obtaining "acoustic images" of an object.

An array is an arranged set of identical sensors, excited in a specific manner [3]. Microphone arrays are a particular case, used in applications such as speech processing, echo cancellation, localization, and sound sources separation [4]. By using beamforming techniques [5], the array beam pattern can be electronically steered to different spatial positions, allowing the discrimination of acoustic sources based on their position.

The authors of this paper have experience in the design [6] and development of acoustic imaging systems, based in acoustic arrays, used in many different fields, such as detection and tracking systems [7,8], Ambient Assisted Living [9], or biometric identification systems [10–12]. The arrays used were ULA (Uniform Linear Array), formed by acoustic sensors distributed uniformly along a line. These arrays are simple, but limited to one dimension (azimuth or elevation) to estimate the spatial position of the sound source. In order to obtain spatial information in two dimensions, it is necessary to work with planar arrays with sensors distributed on a surface.

Working with planar arrays leads to an increase in both system complexity and space required by the acoustic sensors and the associated hardware, since the extension from a 1D array of N element, to a 2D array of N × N elements increases the number of the required channels in an order of N^2. These factors are accentuated when the number of sensors to work with is large. In the design of a system for acquiring and processing signals from an acoustic array, it should be noted that costs and complexity are directly related to the number of channels/sensors of the system. A typical system to obtain acoustic images has four basic elements: sensors, signal conditioners, acquisition devices and signal processor. For the first three elements, system cost increases linearly with the number of channels, as each sensor needs a signal conditioner and an acquisition device.

Digital MEMS (Micro-Electro-Mechanical System) microphones include a microphone, a signal conditioner, and an acquisition device incorporated in the chip itself. For this reason, an acquisition and processing system based on MEMS microphone arrays is reduced to two basic elements: MEMS microphone and a processing system. This processing system is usually based on an FPGA (Field Programmable Gate Array), as it is a digital system with multiple I/O nodes. The integration of the microphone preamplifier and the ADC in a single chip significantly reduces costs, if compared with solutions based on analog microphones. This is one of the main reasons why MEMS microphones are used in acoustic arrays with a large number of channels. However, cost reduction is not the only advantage of working with MEMS microphones. This technology also reduces the space occupied by the system, which makes it feasible to build arrays with hundreds or even thousands of sensors.

Arrays of MEMS microphones are specially designed for acoustic source localization [13–15]; however, they are also used in other applications such as DOA (direction of arrival) estimation for vehicles in motion [16], speech processing [17,18], turbulence measurements [19], identifying geometric dimensions and internal defects of concrete structures [20], or acoustic imaging [21–23]. The system presented in this paper has two main characteristics/novelties: (i) it is scalable using several subarrays or modules, increasing the number of sensors used, and reconfigurable in the position and orientation of the modules; and (ii) its multi-platform framework is reconfigurable. These characteristics made our system flexible and suitable for many different applications. Many of the systems found on the literature show fixed solutions to particular problems, where the hardware is not scalable or the software is not reconfigurable as in the system that is proposed in this paper. The system shown in [22] is reconfigurable, i.e., in the position of its modules, but its acquisition framework, based on an USB port, limits the maximum number of sensors that could be added to the system. For its part, the system shown by Turqueti [16] is scalable; it allows multiple arrays to be aggregated, as the system detailed in this paper. However, our system has two advantages in comparison with Turqueti's system: (i) it can be a standalone system, due to the use of a myRIO platform (embedded hardware with a FPGA and an ARM processor) as an acquisition and data processing system; and (ii) it allows for emitting sounds, making it an active system.

This paper presents in detail the framework of a system that acquires and processes acoustic images. This system is based on a multi-platform framework, where each processing task can be interchanged between the different levels of the framework. Thus, the system can be adapted to different cost and mobility scenarios by means of its reconfigurable framework. The paper also shows the evaluation of processing times of the algorithms involved in obtaining an acoustic image, including signal processing and wideband beamforming via FFT, in each subsystem of the framework. The hardware of the system is based on modules of 8 × 8 square arrays of MEMS microphones, allowing the system to be scalable, by means of the cluster of several modules. This paper also shows the acoustic characterization of the MEMS microphones of the square array, as well as a comparative analysis of the theoretical and measured beam patterns of one of the array modules and of some clusters formed by several array modules.

Section 2 describes the hardware setup of the system and the implemented software algorithms, both defined on the basis of the requirements stated. Section 3 presents the planar array designed for the system: it introduces MEMS microphones technology and characterizes the frequency response of the microphones used, and it characterizes the array module acoustically, obtaining its beam pattern and also the beam pattern of some clusters of modules. A reconfigurable acquisition and processing framework is proposed in Section 4, and its performance is analyzed for several scenarios. Section 5 presents a case study using the system for biometric identification tasks. Finally, Section 6 contains the conclusions and future research lines.

2. System Description

In this section, the requirements for the implementation of the acquisition and processing system for a 2D array, based on MEMS sensors, are analyzed. Then, the hardware chosen for the implementation is defined, and the processing algorithms to obtain an acoustic image using beamforming techniques are explained.

2.1. Requirements

The size of an acquisition and processing system for an array depends on the number of sensors and the set of processing algorithms. Thus, it is necessary to use a very low-cost technology per channel, in order to build a viable high dimensional array. Using MEMS microphones allows for a cost reduction of two main elements of the system: sensors and acquisition systems. Digital MEMS microphones need only one digital input to be read, although the received digital data must be processed to obtain the waveform signal. Most acquisition systems are based on FPGAs, which have a large number of digital inputs, between 40 and 1400 depending on the model; so, one FPGA can acquire as much channels as the number of digital inputs it contains. Besides, FPGA processing capacity allows the system to carry out simple operations with the acquired signals without increasing costs. The processing capacity of the FPGA is not enough to obtain the final acoustic image so it is necessary that another processor with higher capacity be joined to the FPGA.

The system must be modular and scalable. Thus, for applications that require a large number of channels, it would only be necessary to add modules of lower dimension instead of designing a new system with higher capacity. In this way, reusing arrays and their processing systems reduces costs. Extra arrays, with their acquisition subsystem, can always be added in order to build higher dimensional systems. The array modules can increase the dimension of the array, but they can also form different kinds of module configurations in order to obtain specific beam patterns, improving the performance of the array, and thus of the complete system. The use of modular subsystems implies the use of a central unit that joins the data from all the modules and controls them.

Finally, the tools and programming languages to be used should also be defined. As this system has different processing platforms, one solution can be the use of a specific language for each platform, but the use of a common programming language on all platforms is desirable.

2.2. Hardware Setup

2.2.1. MEMS Array

The acoustic images acquisition system shown in this paper is based on a Uniform Planar Array (UPA) of MEMS microphones. This array, which has been entirely developed by the authors, is a square array of 64 (8 × 8) MEMS microphones that are spaced uniformly, every 2.125 cm, in a rectangular Printed Circuit Board (PCB), as shown in Figure 1. As can also be observed in Figure 1, the PCB where the MEMS microphones are placed has square gaps between the acoustic sensors, in order to make the array as light and portable as possible.

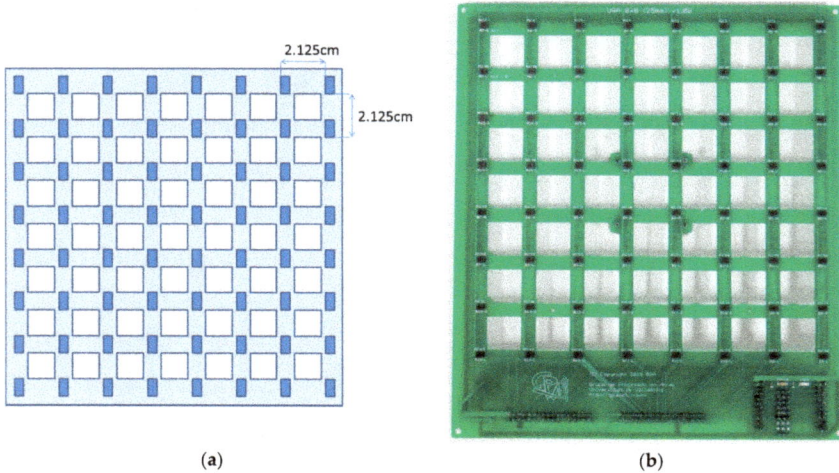

Figure 1. System microphone array: (**a**) sensor spacing; (**b**) system array.

This array was designed to work in an acoustic frequency range between 4 and 16 kHz. The 2.125 cm spacing corresponds to $\lambda/2$ for the 8 kHz frequency. This spacing allows a good resolution for low frequencies, while avoiding grating lobes for high frequencies in the angular exploration zone of interest [10].

For the implementation of the array, MP34DT01 microphones of STMicroelectronics were chosen. They are digital MEMS microphones with a PDM (Pulse Density Modulation) interface and with a one-bit digital signal output, obtained using a high sampling rate (1 MHz to 3.25 MHz) [24–26]. The main features of these microphones are: low-power, omnidirectional response, 63 dB SNR, high sensitivity (−26 dB FS) and an almost flat frequency response (±6 dB in the range of 20 Hz to 20 kHz).

2.2.2. Processing System

Taken into account the previous requirements, the hardware used to implement the system was selected. The design of a specific hardware was rejected due to the high cost of the design and the required time, so a search for a commercial solution was done.

MyRIO platform [27] has been selected as the base unit for this system. This platform belongs to the Reconfigurable Input-Output (RIO) family of devices from National Instruments that is oriented to sensors with nonstandard acquisition procedures, allowing low-level programming of the acquisition routines. Specifically, myRIO platform is an embedded hardware based on a Xilinx Zynq 7010 chip, which incorporates a FPGA and a dual-core ARM® Cortex™-A9 processor. The FPGA has 40 lines of digital input/output, 32 of which are used as the connection interface with the 64 MEMS microphones of the array, multiplexing two microphones in each I/O line; while the other eight lines are used to clock generation and synchronization. The ARM processor is equipped with 256 MB of DDR3 RAM, 512 MB of built-in storage space, USB Host port, and Wi-Fi interface. All this hardware is enclosed in a small box (136 × 89 × 25 mm) that costs about $1,000.

The embedded processor included in myRIO is capable of running all the software algorithms to generate acoustic images, so it can be used as a standalone array module formed by a myRIO connected to a MEMS array board as shown in Figure 2. The acoustic images can be stored in the internal storage of myRIO or in an external disk connected through the USB port.

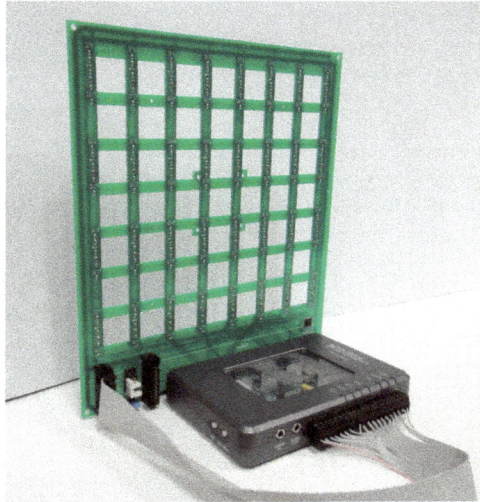

Figure 2. Array module with myRIO and MEMS array board.

Although myRIO can work as a standalone system, the lack of display means that it is usually controlled from a PC connected using a Wi-Fi interface. In a global hardware setup, as shown in Figure 3, the system includes a PC and one or more array modules. The PC performs three main functions:

- As user interface, the PC allows changing the system parameters and visualizing the acoustic images.
- As processing unit, the processors inside the PC could be used to execute the algorithms in order to obtain the acoustic images faster. Two processors are available in the PC: a general-purpose PC processor and a Graphical Processing Unit (GPU) included in the graphics card.
- As a control unit, a single PC can manage several myRIO platforms, each one associated to an array module. This feature allows clustering several modules for a proper operation of the system, which are synchronized between them using their I/O lines, where one myRIO is the master and the others are slaves.

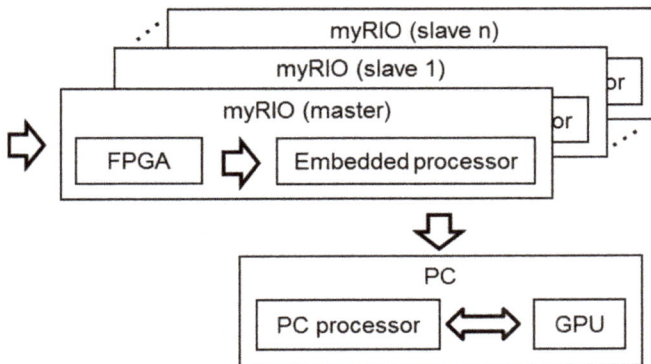

Figure 3. Global hardware setup.

2.3. Software Algorithms

The algorithms implemented in the system, shown in Figure 4, can be divided into three blocks: MEMS acquisition, signal processing, and image generation.

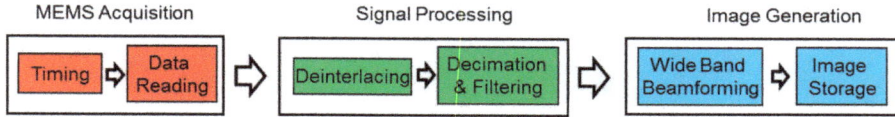

Figure 4. Software algorithms diagram.

The programming language used is LabVIEW 2015, along with its Real Time, FPGA, and GPU modules, which allows developing applications on different hardware platforms like those used in the system: FPGA, Embedded Processor (EP), PC, and GPU. In addition, most of the developed algorithms can run on any of the platforms without reprogramming.

In the acquisition block, each MEMS microphone with a PDM interface, which internally incorporates a one-bit sigma-delta converter with a sampling frequency of 2 MHz, performs signal acquisition. So, each acquired signal is coded with only one bit per sample. This block is implemented in the FPGA, generating a common clock signal for all MEMS, and reading simultaneously 64 sensors signals via the digital inputs of the FPGA. These signals are stored in 64-bit binary words, where each bit stores the signal of each MEMS. Thus, the size of the data is minimal and the transfer rate is high.

In the signal processing block two routines are implemented: (i) Deinterlacing: Through this process, 64 one-bit signals are extracted from each binary word and (ii) Decimate & Filtering: Applying downsampling techniques, based on decimation and filtering [25], 64 independent signals are obtained and the sampling frequency is reduced from 2 MHz to 50 kHz.

Finally, in the image generation block, based on wideband beamforming, a set of N × N steering directions are defined, and the beam former output are assessed for each of these steering directions. Wideband beamforming [3] computes the FFT of the MEMS signals $x_i[n]$; multiplies, element by element, each FFT $X_i[k]$ by a phase vector, that depends on the steering direction and the sensor position; and finally takes the sum of the FFT shifted in phase, as shown in Figure 5. The images generated are then displayed and stored in the system.

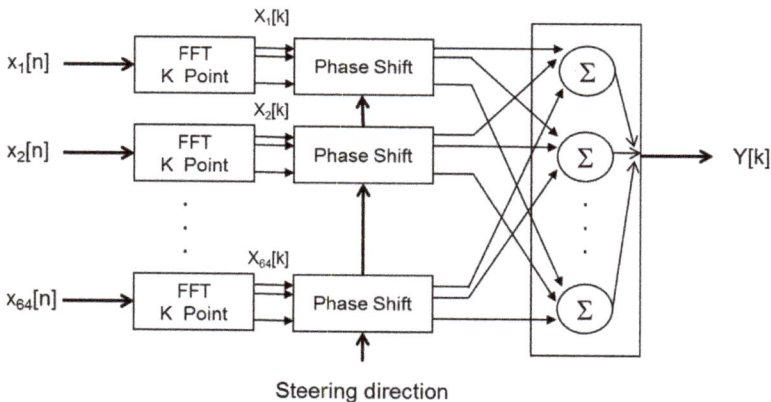

Figure 5. Wideband beamforming.

3. MEMS Array Description and Characterization

The acronym MEMS refers to mechanical systems with a dimension smaller than 1 mm [28] manufactured with tools and technology arising from the integrated circuits (ICs) field. These systems are mainly used for the miniaturization of mechanical sensors. Their small size makes interconnection with other discrete components more difficult. Therefore, when ordered, they are supplied as part of an encapsulated micro-mechanical system composed by a sensor, a signal conditioning circuit and an electric interface [29].

3.1. MEMS Microphones Characterization

An analysis of the frequency response of all MEMS microphones included in the array was performed. A sinusoidal 4 ms pulse, with a frequency changing between 2 and 18 kHz, was generated using a reference loudspeaker. Previously, the frequency response of the reference loudspeaker was calibrated using a measurement microphone (Behringer ECM 8000) and placing it in the same position as the array. Figure 6 shows the arrangement of the components used to perform this analysis. All measurements were performed in an anechoic chamber.

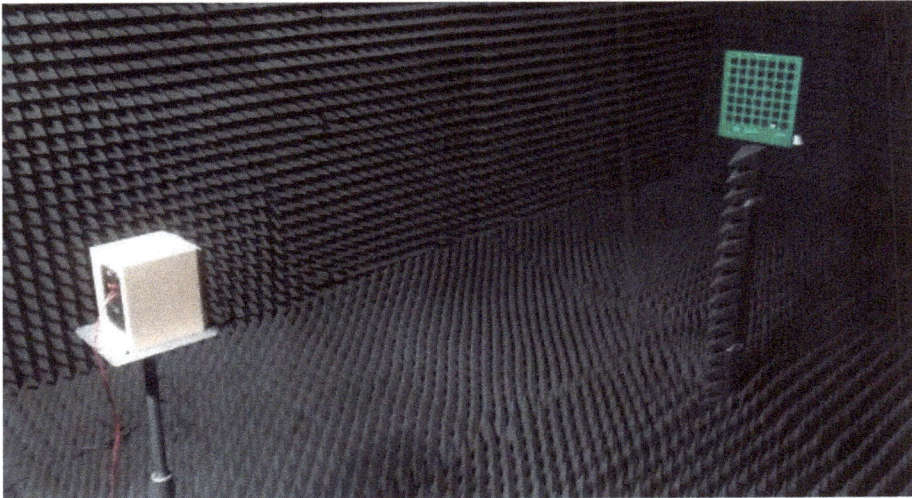

Figure 6. Arrangement of the components in the anechoic chamber.

The frequency response of each MEMS sensor was obtained and normalized according to the loudspeaker's response. Then, the average of the frequency responses was assessed. Figure 6 shows all the responses. It can be observed that the averaged frequency response is essentially flat, with a slight increase at high frequencies. This averaged response is bounded within a range of ± 3 dB. Figure 7 also shows that the frequency response of MEMS sensors varies in a range of ± 2 dB around the averaged value.

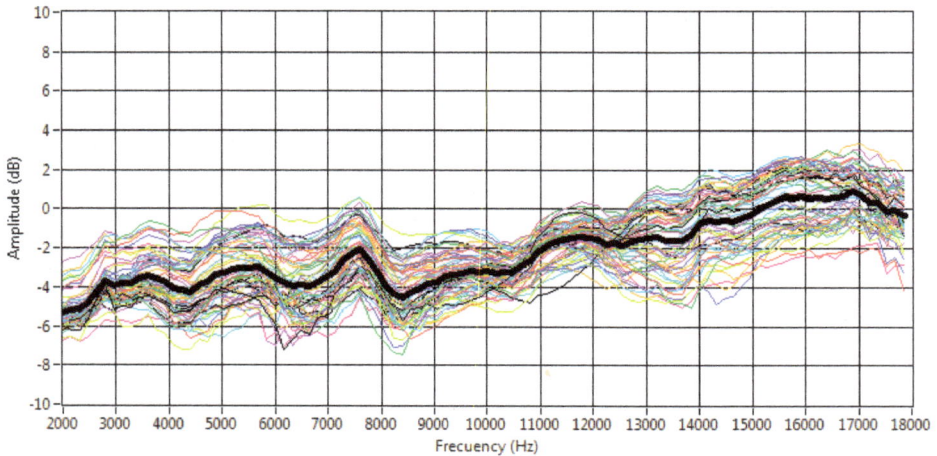

Figure 7. Frequency responses and averaged response of the MEMS microphones.

3.2. MEMS Array Characterization

3.2.1. Acoustic Characterization of an Array Module

Figure 8 shows some theoretical beam patterns of the UPA system, which are pointed towards several steering angles ([azimuth, elevation]): $[-15°, -15°]$, $[0°, 0°]$ and $[5°, 10°]$, for several working frequencies: 4 kHz, 8 kHz and 16 kHz. It can be observed that: (i) as the frequency increases, the array angular resolution also increases, because the mainlobe beamwidth is reduced; and (ii) the sidelobe level is constant –at -13 dB for all frequencies. Figure 8b,c show that, for high frequencies, there are not grating lobes in the angular exploration zone of interest.

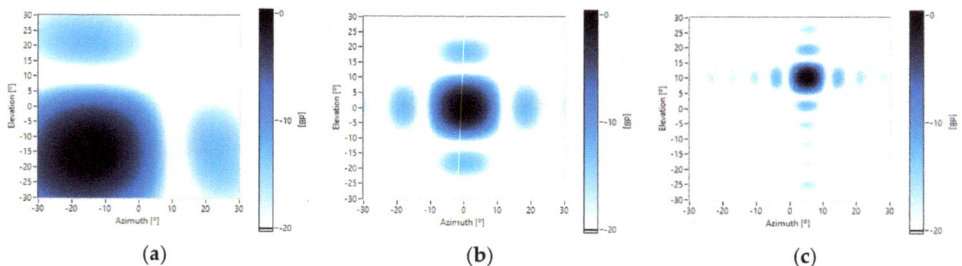

Figure 8. Theoretical beam patterns for different frequencies and pointing to different steering angles [azimuth, elevation]: (**a**) 4 kHz $[-15°, -15°]$; (**b**) 8 kHz and $[0°, 0°]$; and (**c**) 16 kHz and $[5°, 10°]$.

For the acoustic characterization of the MEMS array, a reference loudspeaker placed in different positions was employed to obtain its beam patterns. Beamforming was carried out with a wideband FFT algorithm, focused on the loudspeaker position. Figure 9 shows some of these beam patterns.

The measured beam patterns are very similar to the theoretical ones, which assume that the acoustic sensors are omnidirectional and paired in phases. Nevertheless, a more detailed analysis of the measured beam patterns shows: (i) there are more sidelobes with a level higher than -20 dB; and (ii) there is a very small displacement of the sidelobes, which are closer. These effects are because the gain of each microphone is slightly different for each frequency, as shown in Figure 7. This is the same effect as applying windowing techniques to the beamforming weight vector, which modifies the level and the position of the sidelobes. Thus, as the variations of the measured beam pattern, with respect to the theoretical one, are limited, it is not necessary to apply calibration techniques to the array.

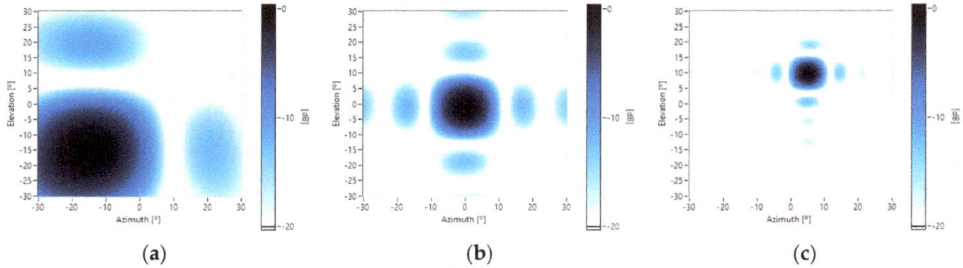

(a)   (b)   (c)

Figure 9. Measured beam patterns for different frequencies and pointing to different steering angles [azimuth, elevation]: (**a**) 4 kHz and $[-15°, -15°]$; (**b**) 8 kHz and $[0°, 0°]$; and (**c**) 16 kHz and $[5°, 10°]$.

3.2.2. Acoustic Characterization of Array Clusters

The proposed system, characterized by its modularity and scalability, can group together multiple modules with 64 sensors to build clusters with a very large number of sensors, where their geometry and spatial properties could be adapted to specific application requirements.

As an example, the acoustic characterizations of three clusters geometries are shown:

- A row cluster, to increase the directivity in one direction.
- A square cluster, to increase directivity in two orthogonal directions.
- A star cluster, to implement special radiation beam patterns.

Figure 10 shows the implemented cluster, the theoretical beam pattern, steered towards the broadside for 8 kHz, and the measured beam pattern.

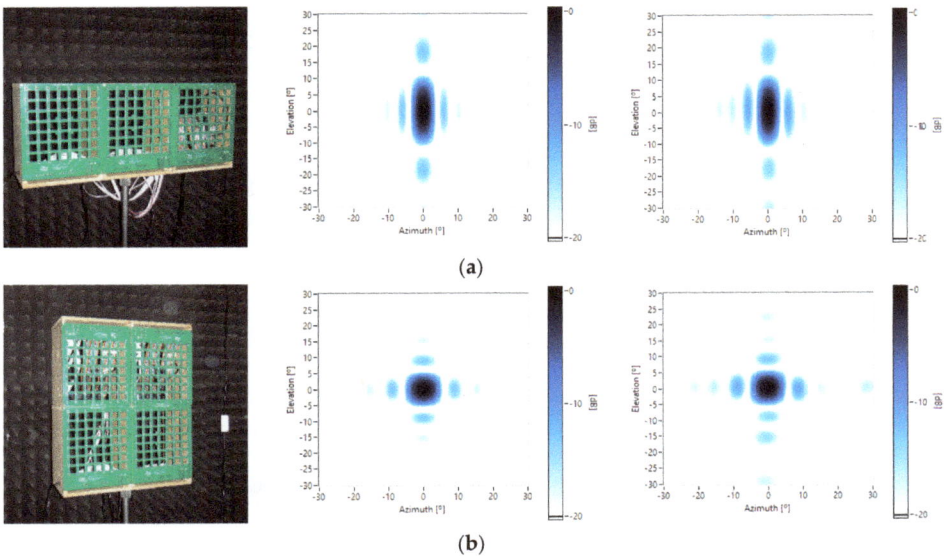

(a)

(b)

Figure 10. *Cont.*

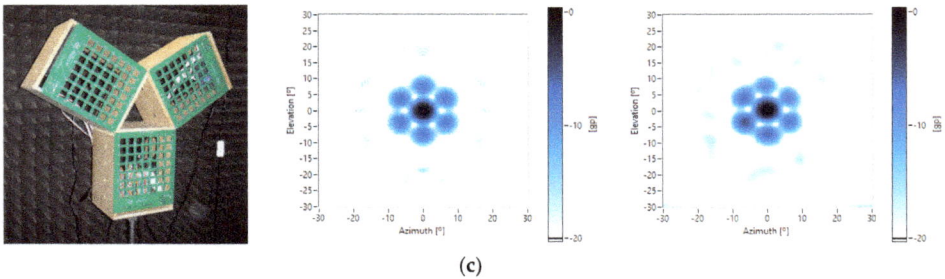

(c)

Figure 10. Cluster (col. 1), theoretical (col. 2) and measured beampatterns (col. 3) for 8 kHz, pointed to the broadside. Cluster geometries: row (**a**); square (**b**); and star (**c**).

The row cluster shows that the beamwidth in azimuth has been reduced by a factor of 3, increasing the angular resolution of the image in that direction. In the square cluster, the beamwidth in azimuth and elevation is halved. Finally, in the star cluster, a radial symmetrical pattern is achieved with a similar beamwidth in multiple directions.

4. Multiplatform Processing Framework

On the basis of the global hardware setup presented in Section 2, a multiplatform framework with four processing levels, each one implemented over a hardware platform, was defined. These processing levels are:

- Level 1 (L1) corresponds to the FPGA based on its capacity to carry out simple tasks of filtering and decimation. The parallelization degree is maximum and limited by the number of the FPGA resources (Look Up Tables, multipliers, DSP units, RAMs, etc.).
- Level 2 (L2) is based on an Embedded Processor (EP), such as an ARM processor that picks up the FPGA signals and carries out the first processing stages. It has limited memory as well as processing and storage capacity.
- Level 3 (L3) is based on a PC processor, such as an Intel Core i5/i7 with two to four cores. It is in charge of the main processing of the application, with medium cost and consumption. It has a high processing capacity, a great amount of memory (up to 64 Gb) and storage capacity based on a disk.
- Level 4 (L4) is formed by coprocessors, which can carry out massive FFT and lineal algebra operations, such as a Graphical Processing Unit (GPU), and they have from 200 to 1200 cores.

Processing time, parallelization degree and required memory must be analyzed for all the algorithms, needed to obtain an acoustic image, described in Section 2. The objective of this analysis is to determine the platforms/levels where these algorithms can be implemented, and the optimal distribution between the available algorithms and platforms.

The time required to transfer data between levels should also be taken into account. This transfer time, in many situations, can be similar to, or even greater than the algorithm processing time. Ideally, these transfers should be minimized, in order to process data on the same level and work with a one-way processing flow, i.e., the FPGA sends data to the EP, and it sends its data to the PC. When one level is used as a coprocessor, bidirectional flows are established, i.e., between the EP and the PC processor, through a TCP-IP interface, or between the PC processor and the GPU, using a PCIe interface.

4.1. Analysis Settings

In order to analyze the performance of the algorithms in each level, a work scheme, based on an active acoustic system, was defined. This system sends a multifrequency acoustic signal that reaches the person under test and then, the reflected signal is collected by the MEMS array. Finally, a multichannel signal is processed following the block diagram shown in Figure 11.

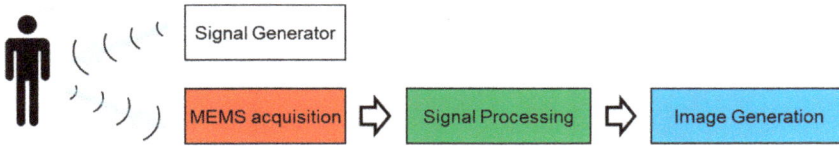

Figure 11. Processing block diagram.

There are hardware constrains that make the implementation of some algorithms in every processing level unfeasible, i.e., MEMS acquisition can only be carried out in the FPGA, or image storage cannot be executed either in the FPGA or in the GPU. Table 1 shows the main algorithms used and the processing levels where they could be implemented.

Table 1. Algorithms vs. processing level.

Algorithms	L1 FPGA	L2 EP	L3 PC	L4 GPU
MEMS acquisition	•			
Deinterlacing	•	•	•	
Decimation and filtering	•	•	•	
Wideband beamforming		•	•	•
Image storage		•	•	

The performance measurements have been carried out using the global hardware setup, with one array module, controlled by a PC. The selected PC is based on an i5 processor with four cores and 32 GB RAM, including a NVIDIA GTX 660 card with 960 cores. As algorithm parameters, an acquisition time of 30 ms, 256-point FFTs, and a grid of 40 × 40 steering directions have been assumed.

4.2. Performance Analysis

4.2.1. Signal Processing

The time required to implement each of the algorithms included in the signal processing on the different levels are presented in Table 2.

Table 2. Processing and transfer times [1] for the signal processing tasks.

Algorithms	L1 FPGA	L2 EP	L3 PC	L4 GPU
Processing time: Deinterlacing	0	270.3	22.4	-
Processing time: Decimation/Filtering	0	354.1	9.7	-
Transfer time: EP→PC			113.0	
Total		624.4	145.1	

[1] Time is expressed in ms.

Level 1 allows the implementation of all these algorithms using FPGA hardware resources, simultaneously with capture and signal processing without consuming additional processing time. Analyzing data from Levels 2 and 3, it can be observed that PC processing time is about 20 times lower than the time required by the EP. The times on Level 3 can be increased by transferring time from the EP to the PC for further processing. This transfer time was measured and its value is about 113 ms. Level 4, based on GPUs, was discarded because the algorithms required to perform deinterlacing and decimation/filtering are not available for this platform in LabVIEW.

4.2.2. Image Generation

Table 3 shows the required processing times related with wideband beamforming and transfer times between the PC and the GPU, for the generation of an acoustic image.

Table 3. Processing and transfer times [1] for wideband beamforming.

Algorithms	L1 FPGA	L2 EP	L3 PC	L4 GPU
Processing time: Wideband beamforming	-	257.3	18.5	25.8
Transfer time: PC→GPU→PC				7.6
Total		257.3	18.5	33.4

[1] Time is expressed in ms.

Level 1 is discarded due to the fact that (i) the FPGA included in myRIO does not have enough memory to generate and store images; and (ii) the implementation of the beamforming algorithms requires FPGAs with a large number of slices which makes it more costly. Processing times on Level 2 are the longest and the memory capacity of the EP is limited, therefore, this level is also discarded.

In order to compare Levels 3 and 4, transfer times between the PC and the GPU should be considered. The results in Table 3 show that it is preferable to perform the beamforming on the PC. These results might seem contradictory, since the processing power of the GPU is higher than the PC's. This is due to the following: (i) for 64 sensors, the GPU capacity of parallel processing is not totally used; and (ii) libraries included in LabVIEW-GPU are limited, which forces multiple data transfers between PC and GPU memories. The implementation of the beamforming algorithm in native code for the GPU and its subsequent invocation from LabVIEW would significantly reduce the overall processing time.

If the number of sensors of the system increases, by adding multiple modules of 64 sensors, the performed operations increase proportionately, taking advantage of the whole capacity of GPU parallel processing. In Table 4, the processing and transfer times versus the number of sensors is analyzed. If the number of sensors is larger than 128, GPU performance improves.

Table 4. Processing and transfer times [1] vs. number of sensors in the system.

| # Sensors | PC | GPU | | |
	Processing Time: Wideband BF [2]	Processing Time: Wideband BF	Transfer Time: PC↔GPU	Total
64	18.5	25.8	7.6	33.4
128	49.1	40.7	8.1	48.8
256	100.2	72.9	8.7	81.6
512	198.6	138.8	11.2	150.0
1024	394.3	267.8	12.5	280.3

[1] Time is expressed in ms; [2] BF: beamforming.

Image storage can be implemented in Levels 2 and 3 because embedded microprocessors and PCs have the capacity to store data. Embedded systems incorporate a low capacity internal disk and can use external high-capacity USB drives/disks. In the case of PCs, they include both high-capacity internal and external hard drives. As storage times are very dependent of the type and model of the used media, they were discarded for the analysis.

4.3. Framework Proposals

Depending on the level where the signal processing algorithm runs, there are three implementation options. In turn, for each of these possibilities, there are up to three options depending on where the wideband beamforming algorithm is implemented. Table 5 summarizes the different options to implement the acoustic imaging algorithms with their corresponding processing and transfer times for all feasible processing levels.

Table 5. Processing and transfer times [1] for acoustic imaging algorithms.

Algorithms	Processing Level							
Processing time: Signal processing	**L1: FPGA** 0			**L2: EP** 524.4			**L3: PC** 32.1	
	L2: EP	**L3: PC**	**L4: GPU**	**L2: EP**	**L3: PC**	**L4: GPU**	**L3: PC**	**L4: GPU**
Wideband BF [2]	257.3	18.5	25.8	257.3	18.5	25.8	18.5	25.8
Transfer time: EP→PC	-	113.0	113.0		113.0	113.0	113.0	113.0
PC→GPU	-	-	7.6			7.6		7.6
Total	257.3	131.5	146.4	881.7	755.9	770.8	163.6	178.5

[1] Time is expressed in ms; [2] BF: Beamforming.

The frameworks that are based on the implementation of the signal processing algorithms through an embedded processor or a PC (grey columns) are discarded, as the FPGA allows the parallel implementation of the capture and signal processing. Thus, only three framework proposals (white columns) have been considered, associating each one to a particular use:

(1) Embedded system: The whole processing takes place in the embedded processor/FPGA without a PC, as shown in Figure 12a. This framework is optimal for applications that require portable systems and where the processing speed is not critical.

(2) PC system: The processing is shared between the FPGA and the PC. The embedded processor is used to control and transfer data between the PC and FPGA, as shown in Figure 12b. This framework presents the lowest processing time using 64 sensors. It is optimal for systems that require a short response time and/or a small number of sensors.

(3) PC system with GPU: The algorithms are implemented in the same way as in the previous framework excluding beamforming, which is implemented on the GPU, as shown in Figure 12c. This framework improves processing time as the number of sensors increases. It is the most versatile framework and it is optimal for systems that require a large number of sensors.

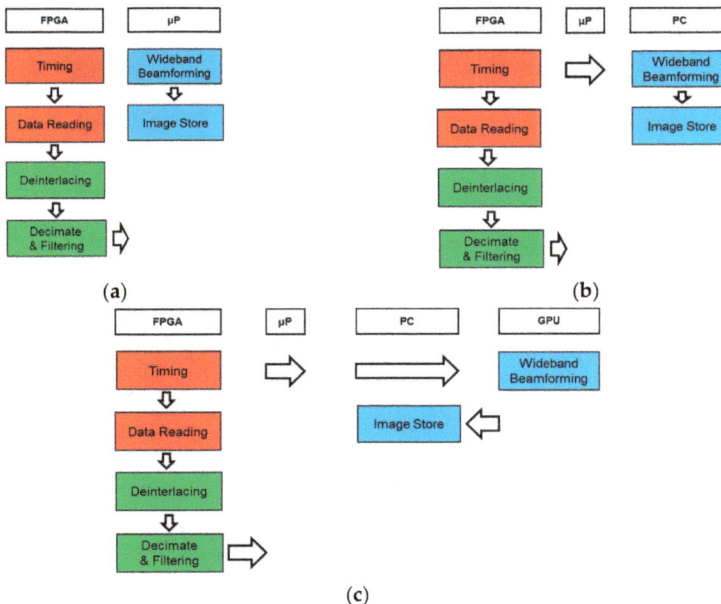

Figure 12. Embedded system (**a**); PC system (**b**); and PC system with GPU framework (**c**).

5. Case Study: Biometric Identification of People

The proposed system has multiple applications: localization and characterization of noise or vibration sources, spatial filtering and elimination of acoustic interference, etc. This case study is focused on biometric identification, as an extension of the previous system developed by the authors with a linear array of analog microphones [10]. This system could be used as an access control to enter in a medium-sized research department, where only several subjects have authorized access.

The biometric identification system is based on placing the person in front of the array and sending a multifrequency signal that is reflected on the person to be identified. The person and the system are placed inside an anechoic chamber in order to simplify the processing, but the chamber could be avoided if clutter removal techniques are used. The reflected signal is captured by the microphone array, obtaining several acoustic images for different ranges. The acoustic images are pre-processed to extract the information needed for further biometric identification. Figure 13 shows the image of a person under testing.

Figure 13. Example of a person under testing.

Figure 14 shows an example of some acoustic images obtained for a range of 2 m with a ±22° angle in the azimuth coordinate and ±15° angle in the elevation. It is observed that if the frequency increases, the spatial resolution improves and the main parts of the body could be discerned. The use of parameter extraction algorithms will improve the classification, as shown in the authors' previous work, using 1D microphone arrays [12].

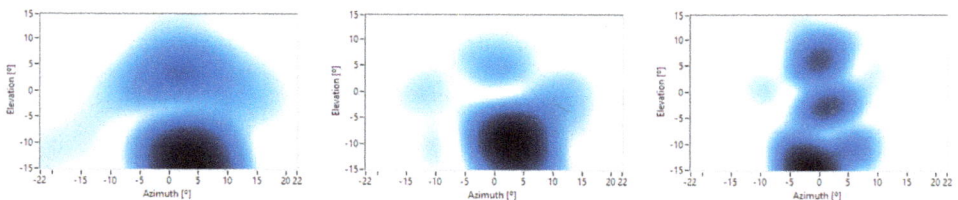

Figure 14. Acoustic images of a person with increasing frequencies, from left to right, at a constant range.

The range information can also be obtained from the captured images. Figure 15 shows images for a constant frequency and an elevation of 5°, 0° and −10°, with 0.5 m range intervals and a ±22° azimuth. Figure 15b shows that the torso is closer to the MEMS array than the arms. These results show that the designed system allows for the acquisition of 3D acoustic images.

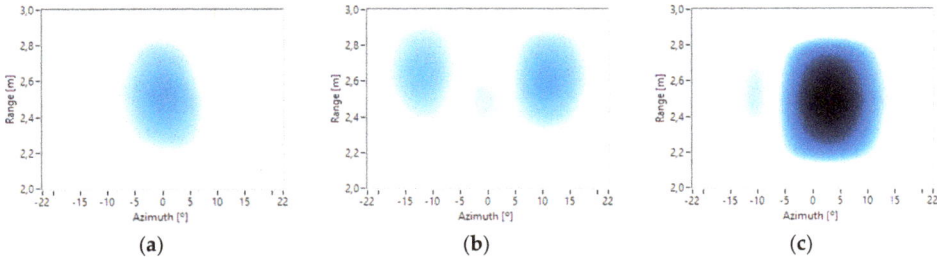

Figure 15. Acoustic images of a person at different elevations: (**a**) 5°; (**b**) 0°; and (**c**) −10° with a constant frequency.

6. Conclusions

A modular, scalable, multiplatform, and low-cost acquisition and processing system was designed to obtain acoustic 3D images. This system is based on a module with a myRIO platform and a planar array that consists of 64 MEMS microphones uniformly distributed on an 8 × 8 grid. The system can work with only one module or with a cluster of several of them, using the same PC unit. The system can be adapted to different cost and mobility scenarios, by means of its reconfigurable multi-platform framework, where each processing tasks can be interchanged between its different levels.

A digital MEMS microphone was selected as the acoustic sensor. This microphone allows the integration of a large number of sensors on a small sized board at low cost. An analysis of the frequency responses of these 64 MEMS microphones was carried out, obtaining: (i) a flat average response in the acoustic band, with a variation of ±3 dB; and (ii) dispersion between the responses lower than ±2 dB. The beampatterns of an array module and of different clusters of modules were also characterized, verifying that they indeed fit the theoretical models, so it is not necessary to calibrate the array.

Finally, a multiplatform framework and the necessary algorithms to obtain acoustic images from the data captured by each microphone were jointly defined. The processing time of the algorithms was evaluated on each platform. Based on these results, three different frameworks were defined for specific uses.

Thus, a versatile system for different applications, due to its modularity, scalability and reconfigurability, was designed to obtain acoustic images. Currently, the authors are using the system in the field of biometric identification, working on feature extraction and person identification from acoustic images.

Acknowledgments: This work has been funded by the Spanish research project SAM: TEC 2015-68170-R (MINECO/FEDER, UE).

Author Contributions: Juan José Villacorta and Lara Del Val Puente designed and implemented the software related with levels 0 and 1 of the framework. Alberto Izquierdo and Juan José Villacorta designed and implemented the software related with levels 2 and 3 of the framework. Luis Suárez and Juan José Villacorta designed and made the hardware prototypes. Lara Del Val Puente and Alberto Izquierdo conceived and performed the experiments, and analyzed the data. Alberto Izquierdo, Juan José Villacorta and Lara Del Val Puente wrote the paper.

Conflicts of Interest: The authors declare no conflict of interest.

Abbreviations

The following abbreviations are used in this manuscript:

MEMS	Micro-Electro-Mechanical Systems
FPGA	Field Programmable Gate Array
FFT	Fast Fourier Transform
ULA	Uniform Linear Array
ADC	Analog-Digital Converter
DOA	Direction of Arrival
IC	Integrated Circuit
SNR	Signal to Noise Ratio
PDM	Pulse Density Modulation
I2S	Integrated Interchip Sound
PCM	Pulse Code Modulation
UPA	Uniform Planar Array
PCB	Printed Circuit Board
GPU	Graphical Processing Unit
RIO	Reconfigurable I/O
PC	Personal Computer
OS	Operative System

References

1. Gan, W.S. *Acoustic Imaging: Techniques and Applications for Engineers*; John Wiley & Sons: New York, NY, USA, 2012.
2. Skolnik, M.I. *Introduction to RADAR Systems*, 3rd ed.; McGraw-Hill Education: New York, NY, USA, 2002.
3. Van Trees, H. *Optimum Array Processing: Part IV of Detection, Estimation and Modulation Theory*; John Wiley & Sons: New York, NY, USA, 2002.
4. Brandstein, M.; Ward, D. *Microphone Arrays*; Springer: New York, NY, USA, 2001.
5. Van Veen, B.D.; Buckley, K.M. Beamforming: A Versatile Approach to Spatial Filtering. *IEEE ASSP Mag.* **1988**, *5*, 4–24. [CrossRef]
6. Del Val, L.; Jiménez, M.; Izquierdo, A.; Villacorta, J. Optimisation of sensor positions in random linear arrays based on statistical relations between geometry and performance. *Appl. Acoust.* **2012**, *73*, 78–82. [CrossRef]
7. Duran, J.D.; Fuente, A.I.; Calvo, J.J.V. Multisensorial modular system of monitoring and tracking with information fusion techniques and neural networks. In Proceedings of the IEEE International Carnahan Conference on Security Technology, Madrid, Spain, 5–7 October 1999; pp. 59–66.
8. Izquierdo-Fuente, A.; Villacorta-Calvo, J.; Raboso-Mateos, M.; Martínez-Arribas, A.; Rodríguez-Merino, D.; del Val-Puente, L. A human classification system for a video-acoustic detection platform. In Proceedings of the International Carnahan Conference on Security Technology, Albuquerque, NM, USA, 12–15 October 2004; pp. 145–152.
9. Villacorta-Calvo, J.; Jiménez-Gómez, M.I.; del Val-Puente, L.; Izquierdo-Fuente, A. A Configurable Sensor Network Applied to Ambient Assisted Living. *Sensors* **2011**, *11*, 10724–10737. [CrossRef] [PubMed]
10. Izquierdo-Fuente, A.; del Val-Puente, L.; Jiménez-Gómez, M.I.; Villacorta-Calvo, J. Performance evaluation of a biometric system based on acoustic images. *Sensors* **2011**, *11*, 9499–9519. [CrossRef] [PubMed]
11. Izquierdo-Fuente, A.; del Val-Puente, L.; Villacorta-Calvo, J.; Raboso-Mateos, M. Optimization of a biometric system based on acoustic images. *Sci. World J.* **2014**, *2014*. [CrossRef] [PubMed]
12. Del Val, L.; Izquierdo, A.; Villacorta, J.; Raboso, M. Acoustic Biometric System Based on Preprocessing Techniques and Support Vector Machines. *Sensors* **2015**, *15*, 14241–14260. [CrossRef] [PubMed]
13. Tiete, J.; Domínguez, F.; da Silva, B.; Segers, L.; Steenhaut, K.; Touhafi, A. SoundCompass: A Distributed MEMS Microphone Array-Based Sensor for Sound Source Localization. *Sensors* **2014**, *14*, 1918–1949. [CrossRef] [PubMed]

14. Edstrand, A.; Bahr, C.; Williams, M.; Meloy, J.; Reagan, T.; Wetzel, D.; Sheplak, M.; Cattafesta, L. An Aeroacoustic Microelectromechanical Systems (MEMS) Phased Microphone Array. In Proceedings of the 21st AIAA Aerodynamic Decelerator Systems Technology Conference and Seminar, Dublin, Ireland, 23–26 May 2011.

15. Zhang, X.; Song, E.; Huang, J.; Liu, H.; Wang, Y.; Li, B.; Yuan, X. Acoustic Source Localization via Subspace Based Method Using Small Aperture MEMS Arrays. *J. Sens.* **2014**, *2014*. [CrossRef]

16. Zhang, X.; Huang, J.; Song, E.; Liu, H.; Li, B.; Yuan, X. Design of Small MEMS Microphone Array Systems for Direction Finding of Outdoors Moving Vehicles. *Sensors* **2014**, *14*, 4384–4398. [CrossRef] [PubMed]

17. Zwyssig, E.; Faubel, F.; Renals, S.; Lincoln, M. Recognition of overlapping speech using digital MEMS microphone arrays. In Proceedings of the IEEE International Conference on Acoustics, Speech and Signal Processing (ICASSP), Vancouver, BC, Canada, 26–31 May 2013; pp. 7068–7072.

18. Hafizovic, I.; Nilsen, C.; Kjølerbakken, M.; Jahr, V. Design and implementation of a MEMS microphone array system for real-time speech acquisition. *Appl. Acoust.* **2012**, *73*, 132–143. [CrossRef]

19. White, R.; Krause, J.; De Jong, R.; Holup, G.; Gallman, J.; Moeller, M. MEMS Microphone Array on a Chip for Turbulent Boundary Layer Measurements. In Proceedings of the 50th AIAA Aerospace Sciences Meeting including the New Horizons Forum and Aerospace Exposition, Nashville, TN, USA, 9–12 January 2012.

20. Groschup, R.; Grosse, C.U. MEMS Microphone Array Sensor for Aid-Coupled Impact Echo. *Sensors* **2015**, *15*, 14932–14945. [CrossRef] [PubMed]

21. Turqueti, M.; Saniie, Y.; Oruklu, E. Scalable Acoustic Imaging Platform Using MEMS Array. In Proceedings of the 2010 IEEE International Conference on Electro/Information Technology (EIT), Normal, IL, USA, 20–22 May 2010; pp. 1–4.

22. Vanwynsberghe, C.; Marchiano, R.; Ollivier, F.; Challande, P.; Moingeon, H.; Marchal, J. Design and implementation of a multi-octave-band audio camera for real time diagnosis. *Appl. Acoust.* **2015**, *89*, 281–287. [CrossRef]

23. Sorama: Sound Solutions. Available online: www.sorama.eu/Solution/measurements (accessed on 9 June 2016).

24. Scheeper, P.R.; van der Donk, A.G.H.; Olthuis, W.; Bergveld, P. A review of silicon microphones. *Sens. Actuators A Phys.* **1994**, *44*, 1–11. [CrossRef]

25. Park, S. Principles of Sigma-Delta Modulation for Analog-to-Digital Converters. Available online: http://www.numerix-dsp.com/appsnotes/APR8-sigma-delta.pdf (accessed on 2 May 2016).

26. Norsworthy, S.; Schreier, R.; Temes, G. *Delta-Sigma Data Converters: Theory, Design, and Simulation*; Wiley-IEEE Press: New York, NY, USA, 1996.

27. Acoustic System of Array Processing Based on High-Dimensional MEMS Sensors for Biometry and Analysis of Noise and Vibration. Available online: http://sine.ni.com/cs/app/doc/p/id/cs-16913 (accessed on 14 September 2016).

28. Hsieh, C.T.; Ting, J.-M.; Yang, C.; Chung, C.K. The introduction of MEMS packaging technology. In Proceedings of the 4th International Symposium on Electronic, Materials and Packaging, Kaohsiung, Taiwan, 4–6 December 2002; pp. 300–306.

29. Beeby, S.; Ensell, G.; Kraft, M.; White, N. *MEMS Mechanical Sensors*; Artech House Publishers: Norwood, MA, USA, 2004.

sensors

MDPI

Article

Underwater Imaging Using a 1 × 16 CMUT Linear Array

Rui Zhang [1,2], Wendong Zhang [1,2], Changde He [1,2], Yongmei Zhang [3], Jinlong Song [1,2] and Chenyang Xue [1,2,*]

[1] Key Laboratory of Instrumentation Science & Dynamic Measurement, North University of China, Ministry of Education, Taiyuan 030051, China; fly_zr@126.com (R.Z.); wdzhang@nuc.edu.cn (W.Z.); changde_henuc@163.com (C.H.); nucsong@163.com (J.S.)

[2] Science and Technology on Electronic Test and Measurement Laboratory, North University of China, Taiyuan 030051, China

[3] School of Computer Science, North China University of Technology, Beijing 100144, China; zhangym@ncut.edu.cn

* Correspondence: xuechenyang@nuc.edu.cn; Tel.: +86-351-3921-756

Academic Editor: Gonzalo Pajares Martinsanz
Received: 25 January 2016; Accepted: 25 February 2016; Published: 1 March 2016

Abstract: A 1 × 16 capacitive micro-machined ultrasonic transducer linear array was designed, fabricated, and tested for underwater imaging in the low frequency range. The linear array was fabricated using Si-SOI bonding techniques. Underwater transmission performance was tested in a water tank, and the array has a resonant frequency of 700 kHz, with pressure amplitude 182 dB (μPa·m/V) at 1 m. The −3 dB main beam width of the designed dense linear array is approximately 5 degrees. Synthetic aperture focusing technique was applied to improve the resolution of reconstructed images, with promising results. Thus, the proposed array was shown to be suitable for underwater imaging applications.

Keywords: capacitive micro-machined ultrasonic transducer linear array; transmission performance; synthetic aperture focusing technique; underwater imaging

1. Introduction

Ultrasound imaging has played an important role in various areas, such as medical diagnosis, medical treatment, nondestructive testing, and ultrasound microscopy [1–3]. The ultrasonic transducer is the core component of ultrasound imaging, and currently piezoelectric micro-machined ultrasonic transducers (PMUTs) based on the piezoelectric effect are widely used [4,5]. However, PMUT performance in underwater and medical applications is limited by material properties and impedance matching issues [2,6,7]. Capacitive micro-machined ultrasonic transducers (CMUTs) have many advantages over conventional PMUTs, such as wide bandwidth, high mechanical-electrical conversion efficiency, and ease of integration with electronic circuits to enhance signal-to-noise ratio [2,8–12]. Furthermore, CMUT membranes have low mechanical impedance, which makes them match well with air and other fluid media, and are suitable for manufacturing in large arrays [2,6]. These characteristics promote CMUTs as the development direction for next generation ultrasonic transducers.

Much research has been conducted regarding CMUT structural design, fabrication methods, and implementations. However, most studies have considered high frequency CMUTs (≥3 MHz) for medical imaging applications, with few studies on underwater imaging applications, which require low frequency. Roh *et al.* used finite element models to design a 1dimensional (1D) CMUT array robust to crosstalk [13]. ChiaHung *et al.* [14] designed and fabricated an underwater CMUT using full surface micro-machining technique. The transducer operated underwater at approximately 2 MHz with a detection range of 273 mm. Cheng *et al.* [15] realized B-mode imaging of a metal wire phantom using a 21-element 1-D array with 3.8 MHz central frequency and fractional bandwidth 116% in water, which limited the detection range. Doody *et al.* [16] designed a CMUT-in-CMOS array, which achieved central frequency 3.5 MHz, fractional bandwidth 32%–44%, and pressure amplitude 181–184 dB (μPa·m/V) at 15 mm when operated in a water tank. In this work, we designed and fabricated a 1 × 16 CMUT linear array with resonance frequency 700 kHz, and pressure amplitude 182 dB (μPa·m/V) at 1 m for use in underwater imaging.

2. Structural Design

CMUT array elements are composed of multiple sensitive cells connected in parallel. Each cell is composed of electrodes, vibrating membrane, vacuum cavity, insulating layer, and silicon substrate, as shown cross-sectional in Figure 1. A single CMUT array element often cannot meet imaging requirements with its low lateral resolution, low transmission power, and poor directivity. Therefore, a CMUT array composed of *N* identical elements is used to improve the imaging resolution.

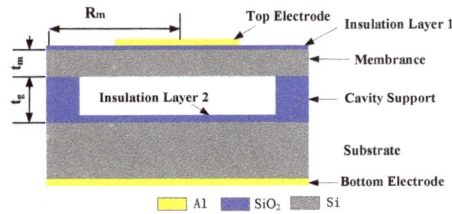

Figure 1. The structure of a cell.

The natural frequency is one of important performance parameters, whether operating in emission or receiving mode. When CMUTs work in the liquid, the resonant frequency of circular membrane f_r is [17,18]:

$$f_r = \frac{\dfrac{0.469 t_m}{R_m^2}\sqrt{\dfrac{E}{\rho\left(1-\sigma^2\right)}}}{\sqrt{1+0.67\dfrac{\rho_l R_m}{\rho t_m}}} \tag{1}$$

where E, ρ, ρ_l, σ, t_m, R_m represent the Young's modulus, the density of membrane, the density of liquid, the Poisson's ratio, the thickness of the membrane, and the radius of the membrane, respectively. High-resistivity silicon was chosen as the membrane material. From previous finite element software ANSYS analysis [19,20], following Equation (1), $t_m = 3.5$ μm and $R_m = 90$ μm.

The emission performance of ultrasonic transducers is closely linked with its structural parameters. From previous studies [21], several directivity functions were deduced by Huygens' Principle to guide the array structure design, and the resultant 1 × 16 CMUT structure is shown in Figure 2.

Si-SOI low temperature wafer bonding technology [22,23] was used to fabricate the CMUT linear array (see Figure 3). The silicon wafer was first thermally oxidized, then part-etched to form cavities ($t_g = 0.8$ μm) and an insulation layer (0.15 μm) (Figure 3a). The silicon wafer and SOI wafer were bonded using low temperature bonding (Figure 3b), and the silicon substrate and buried oxide layer of the SOI wafer was eliminated (Figure 3c) to form the silicon membrane ($t_m = 3.5$ μm). Finally, an isolation channel was formed using photolithography and dry etching, and low pressure chemical vapor deposition was used to form a 0.15 μm insulation layer to prevent conductive contact between the top electrodes and the vibration membrane, and evaporation methods were used to form the top electrodes with Al. To ensure fine conductive contact between the bottom electrode and silicon substrate, the other side of the Si wafer had phosphorus ions implanted and metal Al deposited (Figure 3d).

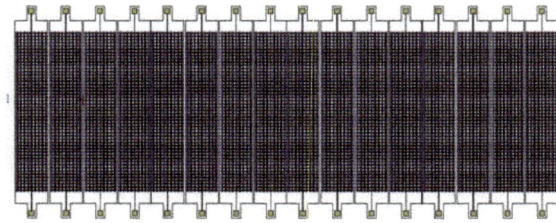

Figure 2. The resultant 1 × 16 CMUT array structure.

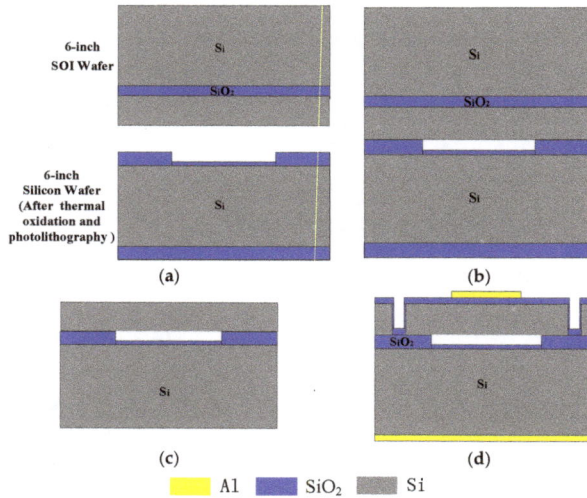

Figure 3. The main fabrication flow-charts. (**a**) part etched to form cavities; (**b**) bonding; (**c**) form membrane; (**d**) form isolation channel, insulation layer and electrodes.

3. Underwater Experimental

The CMUT resonant frequency was found as shown in Figure 4. The linear array was encapsulated for insulation from the water [21], and a precision impedance analyzer (Agilent 4294A, Agilent Technologies, Santa Clara, CA, United States) was used to measure the array impedance in water (underwater penetration 0.45 m, water temperature 13 °C). The CMUT resonant frequency = 700 kHz, and electric conductance = 222.35 mS. The CMUT transmission characteristics (transmitting voltage response and directivity) were analyzed over the frequency range of interest (100–1000 kHz).

The CMUT transmitting voltage response was measured as shown in Figure 5. A CMUT linear array (transmitter) and a standard hydrophone (receiver) were placed face to face, 1 m apart. The received electrical signal was displayed on an oscilloscope (Agilent 54624A), with direct current (DC) bias 20 V. The CMUT array was driven with a 5 cycle burst signal incorporating 100–1000 kHz and amplitude of 20 V_{pp}.

The transmitting voltage response can be expressed as [24]:

$$Sv = 20lg\frac{u_s \cdot l}{u_f} - M_o \tag{2}$$

where l is the distance between the two transducers, u_f is the applied driving voltage of the transmitting CMUT, u_s is the collected voltage standard hydrophone, and M_o is the receiving sensitivity of the hydrophone, the CMUT transmitting voltage response was obtained at different frequencies, as shown Figure 5b. The 1 × 16 CMUT linear array has resonant frequency 700 kHz, and pressure amplitude of 182 dB (μPa· m/V) at 1 m for underwater applications.

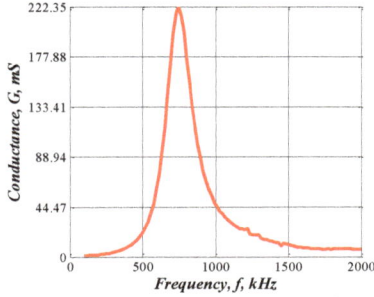

Figure 4. Resonant frequency.

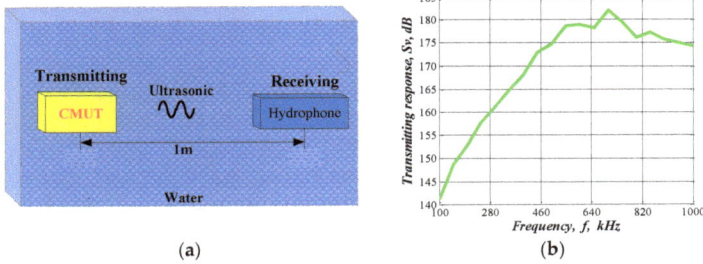

(a)   (b)

Figure 5. (**a**) Schematic diagram of the transmitting voltage response experiment; (**b**) Transmitting response.

The sector scanning experimental setup is shown in Figure 6. The operating conditions of the CMUT array were the same as previously, except it was now excited with a burst signal at 700 kHz for two cycles. The receiving and transmitting array was fixed on a precision rotary table and rotated to implement sector scanning for two-obstacle imaging.

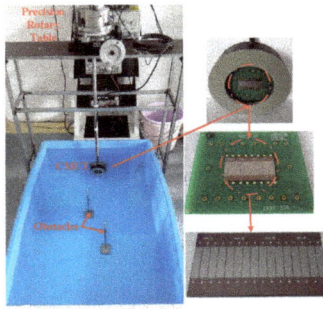

Figure 6. Sector scanning configuration.

Figure 7 shows the S-scan results and corresponding directivity diagrams for four transmitting conditions Figure 7b,d,f,h, where N is the number of array elements, d is the element spacing between array elements, and λ is the wavelength. The −3 dB main beam width of Figure 7b,d,f,h are approximately 21°, 6.4°, 8°, and 5°, respectively.

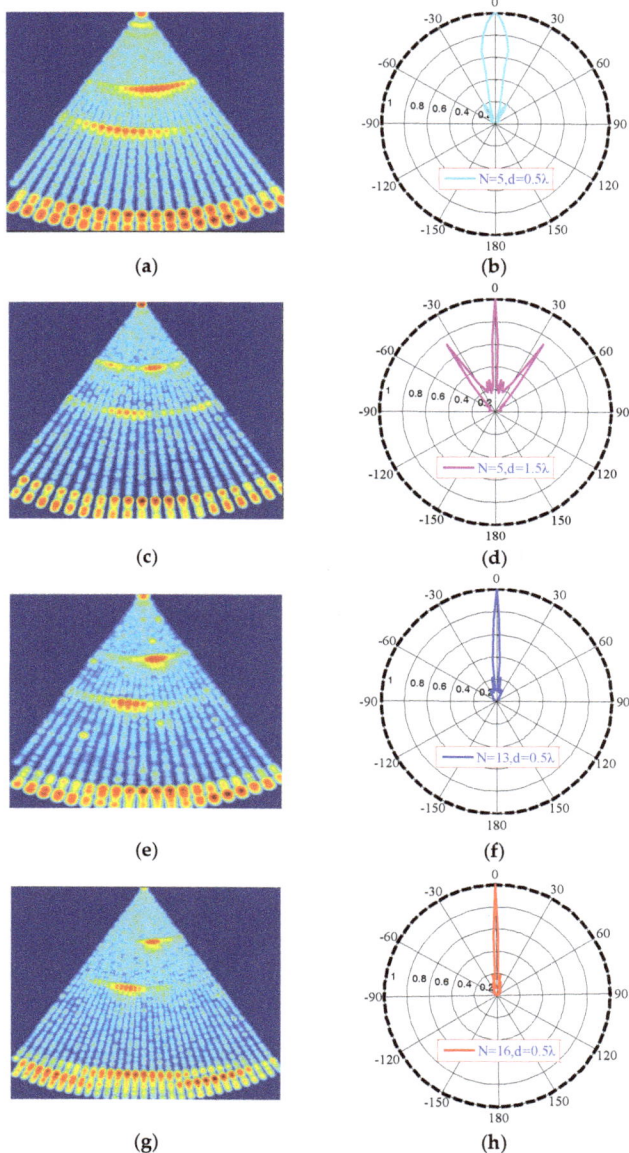

(a)

(b)

(c)

(d)

(e)

(f)

(g)

(h)

Figure 7. Sector scanning results: (**a**) N = 5, d = 0.5· λ; (**b**) directivity of diagram (a); (**c**) N = 5, d = 1.5· λ; (**d**) directivity of diagram (c); (**e**) N = 13, d = 0.5· λ; (**f**) directivity of diagram (e); (**g**) N = 16, d = 0.5· λ; (**h**) directivity of diagram (g).

(1) For the fixed array element number (Figure 7a,c, or Figure 7b,d), the main lobe becomes sharper as d increases. However, when $d > \lambda$ [19], strong grating lobes emerge, causing interference in the ultrasound image.

(2) For fixed array length (Figure 7c,e, or Figure 7d,f), the directivity of a dense array is better than that of a sparse array.

(3) For fixed element spacing (Figure 7a,e,g or Figure 7b,f,g), directivity improves as N increases. The wider main lobe also make severe interference.

Thus, $N = 16$, $d = 0.5\lambda$ was selected as the transmission mode for subsequent underwater imaging.

4. Imaging

To reduce the influence of side lobes and improve the lateral resolution of reconstructed CMUT underwater images, the synthetic aperture focusing technique (SAFT) [25–27], an optimization method involving B-scan implemented by delay-and-sum on the received A-scan signals, was applied [28,29].

Figure 8 shows the SAFT experimental setup. The operating conditions of the CMUT linear array were the same as for Section 3. The CMUT array was fixed on the electronic control guideway and moved horizontally to implement linear scanning for two-obstacle imaging. 16 elements were simultaneously excited. Figure 9 shows the SAFT reconstructed image (Figure 9a,b) is a significant improvement over traditional B-scan (Figure 9c,d), and artifacts caused by side lobes are effectively suppressed. Thus, the proposed array greatly reduces transmission issues without requiring phased transmission, thereby improving lateral resolution, which will be of great benefit in underwater imaging.

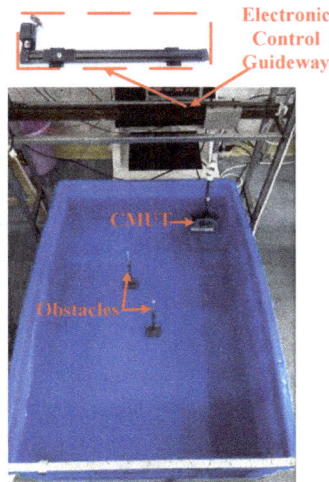

Figure 8. Linear scanning imaging configuration.

(a)

(b)

(c)

(d)

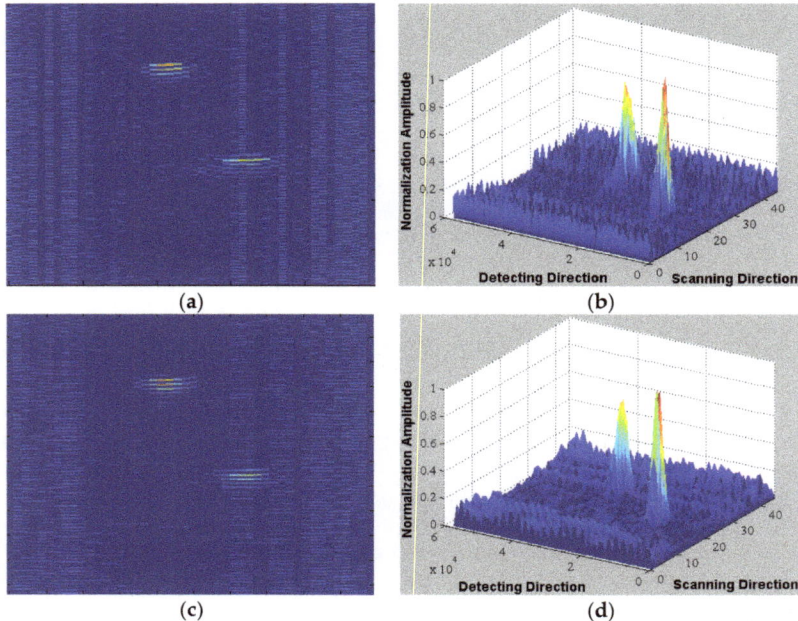

Figure 9. Reconstructed images (**a**) B-san reconstructed image; (**b**) 3-dimensional view of diagram (a); (**c**) SAFT reconstructed image; (**d**) 3 dimensional view of diagram (c).

5. Conclusions

A 1 × 16 CMUT array was designed and fabricated for underwater imaging in the low frequency range. The transmission performance of the array was analyzed in a water tank, showing transmission voltage response 182 dB (μPa· m/V) was achieved at 1 m underwater. Directivity and sector scanning were also analyzed to determine the optimal transmission mode for linear underwater imaging. Significant resolution improvement was obtained by applying SAFT to the received A-scan signals. Thus, the proposed CMUT array shows great benefit for underwater detection applications, such as obstacle avoidance, distance measuring, and imaging.

Acknowledgments: This work was supported by National Science Foundation for Distinguished Young Scholars of China under Grant 61525107, National Natural Science Foundation of China under Grant 61127008 and 61371143, and the "863" program of China under Grant 2013AA09A412.

Author Contributions: The author Rui Zhang designed the CMUT array structure, organized the experiments, performed the data collection and analysis, and drafted the manuscript. Chenyang Xue provided with the funding support. Wendong Zhang and Changde He designed the fabrication process. Jinglong Song completed the processing of the transducer. Yongmei Zhang performed provided the imaging algorithm guide.

Conflicts of Interest: The authors declare no conflict of interest.

References

1. Caronti, A.; Caliano, G.; Carotenuto, R.; Savoia, A.; Pappalardo, M.; Cianci, E.; Foglietti, V. Capacitive micromachined ultrasonic transducer (CMUT) arrays for medical imaging. *Microelectron. J.* **2006**, *37*, 770–777. [CrossRef]
2. Oralkan, O.; Ergun, A.S.; Johnson, J.A.; Karaman, M.; Demirci, U.; Kaviani, K.; Lee, T.H.; Khuri-Yakub, B.T. Capacitive micromachined ultrasonic transducers: Next-generation arrays for acoustic imaging? *IEEE Trans. Ultrason. Ferroelectr. Freq. Control* **2002**, *49*, 1596–1610. [CrossRef] [PubMed]

3.   Choe, J.W.; Oralkan, O.; Nikoozadeh, A.; Gencel, M.; Stephens, D.N.; O'Donnell, M.; Sahn, D.J.; Khuri-Yakub, B.T. Volumetric Real-Time Imaging Using a CMUT Ring Array. *IEEE Trans. Ultrason. Ferroelectr. Freq. Control* **2012**, *59*, 1201–1211. [CrossRef] [PubMed]

4.   Emadi, T.A.; Buchanan, D.A. A novel 6 × 6 element MEMS capacitive ultrasonic transducer with multiple moving membranes for high performance imaging applications. *Sens. Actuators A Phys.* **2015**, *222*, 309–313. [CrossRef]

5.   Chang, C.; Moini, A.; Nikoozadeh, A.; Sarioglu, A.F.; Apte, N.; Zhuang, X.; Khuri-Yakub, B.T. Singulation for imaging ring arrays of capacitive micromachined ultrasonic transducers. *J. Micromech. Microeng.* **2014**, *24*, 10700210. [CrossRef]

6.   Wang, H.; Wang, X.; He, C.; Xue, C. Design and Performance Analysis of Capacitive Micromachined Ultrasonic Transducer Linear Array. *Micromachines* **2014**, *5*, 420–431. [CrossRef]

7.   Ladabaum, I.; Jin, X.; Soh, H.T.; Atalar, A.; Khuri-Yakub, B.T. Surface micromachined capacitive ultrasonic transducers. *IEEE Trans. Ultrason. Ferroelectr. Freq. Control* **1998**, *45*, 678–690. [CrossRef] [PubMed]

8.   Emadi, T.A.; Buchanan, D.A. Multiple Moving Membrane CMUT with Enlarged Membrane Displacement and Low Pull-Down Voltage. *IEEE Electron Device Lett.* **2013**, *34*, 1578–1580. [CrossRef]

9.   Jeong, B.; Kim, D.; Hong, S. Performance and reliability of new CMUT design with improved efficiency. *Sens. Actuators A Phys.* **2013**, *199*, 325–333. [CrossRef]

10. Ergun, A.S.; Yaralioslu, G.G.; Khuri-Yakub, B.T. Capacitive micromachined ultrasonic Transducers: Theory and technology. *J. Aerosp. Eng.* **2003**, *16*, 76–84. [CrossRef]

11. Mills, D.M.; Smith, L.S. Real Time *in vivo* Imaging with Capacitive Micromachined Ultrasonic Transducer (CMUT) Linear Arrays. In Proceedings of the 2003 IEEE Ultrasonic Symposium, Honolulu, HA, USA, 5–8 October 2003; pp. 568–571.

12. Eccardt, P.C.; Niederer, K.; Fischer, B. Micromachined Transducers for Ultrasonic Applications. In Proceedings of the 1997 IEEE Ultrasonic Symposium, Toronto, ON, Canada, 5–8 October 1997.

13. Roh, Y.; Khuri-Yakub, B.T. Finite element analysis of underwater capacitor micromachined ultrasonic transducers. *IEEE Trans. Ultrason. Ferroelectr. Freq. Control* **2002**, *49*, 293–298. [CrossRef] [PubMed]

14. Liu, C.; Chen, P. Surface micromachined capacitive ultrasonic transducer for underwater imaging. *J. Chin. Inst. Eng.* **2007**, *30*, 447–458. [CrossRef]

15. Cheng, X.; Chen, J.; Li, C.; Liu, J.; Shen, L.; Li, P. A Miniature Capacitive Ultrasonic Imager Array. *IEEE Sens. J.* **2009**, *9*, 569–577. [CrossRef]

16. Doody, C.B.; Cheng, X.; Rich, C.A.; Lemmerhirt, D.F.; White, R.D. Modeling and Characterization of CMOS-Fabricated Capacitive Micromachined Ultrasound Transducers. *J. Microelectromech. Syst.* **2011**, *20*, 104–118. [CrossRef]

17. Peake, W.H.; Thurston, E.G. The lowest resonant frequency of a water-loaded circular plate. *Acoust. Soc. Am.* **1954**, *26*, 166–168. [CrossRef]

18. Wong, A.C. VHF Microelectromechanical Mixer-Filters. Ph.D. Thesis, The University of Michigan, Ann Arbor, MI, USA, 2001.

19. Shen, W.; Miao, J.; Xiong, J.; He, C.; Xue, C. Micro-electro-mechanical systems capacitive ultrasonic transducer with a higher electromechanical coupling coefficient. *IET Micro Nano Lett.* **2015**, *10*, 541–544.

20. Li, Y.P.; He, C.D.; Zhang, J.T.; Song, J.L.; Zhang, W.D.; Xue, C.Y. Design and analysis of Capacitive Micromachined Ultrasonic Transducer based on SU-8. *Key Eng. Mater.* **2014**, *645–646*, 577–582. [CrossRef]

21. Zhang, R.; Zhang, W.D.; He, C.D.; Song, J.L.; Mu, L.F.; Cui, J.; Zhang, Y.M.; Xue, C.Y. Design of Capacitive Micromachined Ultrasonic Transducer (CMUT) linear array for underwater imaging. *Sens. Rev.* **2015**, *36*, 77–85. [CrossRef]

22. Song, J.L.; Xue, C.Y.; He, C.D.; Zhang, R.; Mu, L.F.; Cui, J.; Miao, J.; Liu, Y.; Zhang, W.D. Capacitive Micromachined Ultrasonic Transducers (CMUTs) for Underwater Imaging Applications. *Sensors* **2015**, *15*, 23205–23217. [CrossRef] [PubMed]

23. Zhang, R.; Xue, C.Y.; He, C.D.; Zhang, Y.M.; Song, J.L.; Zhang, W.D. Design and performance analysis of capacitive micromachined ultrasonic transducer (CMUT) array for underwater imaging. *Microsyst. Technol.* **2015**. [CrossRef]

24. Zheng, S.J.; Yuan, W.J.; Miu, R.X.; Xue, Y.Q. *Underwater Acoustic Measurement Testing Technology*; Harbin Engineering University Press: Harbin, China, 1995; pp. 223–232.

25. Neild, A.; Hutchins, D.A.; Billson, D.R. Imaging using air-coupled polymer-membrane capacitive ultrasonic arrays. *Ultrasonics* **2004**, *42*, 859–864.
26. Guarneri, G.A.; Pipa, D.R.; Junior, F.N.; Valéria, L.; de Arruda, R.; Victor, M.; Zibetti, W. A Sparse Reconstruction Algorithm for Ultrasonic Images in Nondestructive Testing. *Sensors* **2015**, *15*, 9324–9343. [CrossRef] [PubMed]
27. Skjelvareid, M.H.; Olofsson, T.; Birkelund, Y.; Larsen, Y. Synthetic aperture focusing of ultrasonic data from multilayered media using an omega-K algorithm. *IEEE Trans. Ultrason. Ferroelectr. Freq. Control* **2011**, *58*, 1037–1048. [CrossRef] [PubMed]
28. Corl, P.D.; Grant, P.M.; Kino, G. A Digital Synthetic Focus Acoustic Imaging System for NDE. In Proceedings of the 1978 Ultrasonics Symposium, Cherry Hill, NJ, USA, 25–27 September 1978.
29. Frederick, J.; Seydel, J.; Fairchild, R. *Improved Ultrasonic Nondestructive Testing of Pressure Vessels*; Technical Report; Michigan University: Ann Arbor, MI, USA, 1976.

sensors

MDPI

Article

Ultraviolet Imaging with Low Cost Smartphone Sensors: Development and Application of a Raspberry Pi-Based UV Camera

Thomas C. Wilkes [1,*], **Andrew J. S. McGonigle** [1,2], **Tom D. Pering** [1], **Angus J. Taggart** [1], **Benjamin S. White** [3], **Robert G. Bryant** [1] and **Jon R. Willmott** [3]

[1] Department of Geography, The University of Sheffield, Winter Street, Sheffield S10 2TN, UK; a.mcgonigle@sheffield.ac.uk (A.J.S.M.); t.pering@sheffield.ac.uk (T.D.P.); a.taggart@sheffield.ac.uk (A.J.T.); r.g.bryant@sheffield.ac.uk (R.G.B.)

[2] Istituto Nazionale di Geofisica e Vulcanologia, Sezione di Palermo, via Ugo La Malfa 153, Palermo 90146, Italy

[3] Department of Electronic and Electrical Engineering, The University of Sheffield, Portobello Centre, Pitt Street, Sheffield S1 4ET, UK; ben.white@sheffield.ac.uk (B.S.W.); j.r.willmott@sheffield.ac.uk (J.R.W.)

* Correspondence: tcwilkes1@sheffield.ac.uk; Tel.: +44-7798-837-894

Academic Editor: Gonzalo Pajares Martinsanz
Received: 5 September 2016; Accepted: 3 October 2016; Published: 6 October 2016

Abstract: Here, we report, for what we believe to be the first time, on the modification of a low cost sensor, designed for the smartphone camera market, to develop an ultraviolet (UV) camera system. This was achieved via adaptation of Raspberry Pi cameras, which are based on back-illuminated complementary metal-oxide semiconductor (CMOS) sensors, and we demonstrated the utility of these devices for applications at wavelengths as low as 310 nm, by remotely sensing power station smokestack emissions in this spectral region. Given the very low cost of these units, $\approx$ USD 25, they are suitable for widespread proliferation in a variety of UV imaging applications, e.g., in atmospheric science, volcanology, forensics and surface smoothness measurements.

Keywords: UV camera; UV imaging; smartphone sensor technology; sulphur dioxide emissions; Raspberry Pi; low-cost camera

1. Introduction

Ultraviolet (UV) imaging has a wide variety of scientific, industrial and medical applications, for instance in forensics [1], industrial fault inspection [2], astronomy, monitoring skin conditions [3] and in remote sensing [4,5]. To date, scientific grade UV cameras, which have elevated quantum efficiencies in this spectral region, have been applied in this context. However, these systems are relatively expensive (typical unit costs thousands of dollars) and can be power intensive, since they may incorporate thermo-electric cooling. Although these units may provide high signal-to-noise ratios, a lower price point solution could expedite more widespread implementation of UV imaging.

Recently, considerable effort has been invested in developing low cost back-illuminated complementary metal-oxide semiconductor (CMOS) sensor technology. Previously, this sensor architecture was applied only within specialist, low light imaging arenas, e.g., in astronomy. In the last few years, however, the manufacturing costs of these devices have been reduced markedly, such that they now feature prominently in consumer electronic products, particularly smartphones. The key advantage of the back-illuminated sensor architecture, over the conventional CMOS configuration, is that the photo-diodes are placed in front of the metal wiring matrix layer of the sensor, thereby improving fill factor and substantially increasing optical throughput to the photoreceptors, particularly in the UV, where these detectors are photosensitive.

To date, however, application of these inexpensive sensors has predominantly been focused on visible imaging, due to the choice of fore-optics and the Bayer filter layer applied to the sensors, in order to generate red-green-blue (RGB) mosaics. Recent studies have highlighted that these smartphone cameras do have some UV sensitivity [6–8], even with the above optical arrangement. Here, we build on this work by modifying such a camera sensor, to maximize UV throughput to the sensor. To the best of our knowledge, this constitutes the first report of an imaging system specifically adapted for the UV, based on a back-illuminated CMOS device developed for the smartphone market. In particular, we developed a UV camera system, based on Raspberry Pi camera boards, and demonstrated its utility for UV imaging applications down to 310 nm, via a case study involving remote sensing of sulphur dioxide (SO_2) emissions from a power station smokestack.

2. UV Camera Development

This work was achieved using Raspberry Pi Camera Module v1.3 boards (referred to as PiCams hereafter), of cost $\approx$ USD 25 (Raspberry Pi Foundation). The PiCam is based on an Omnivision OV5647 back-illuminated CMOS sensor, developed primarily for the mobile phone market; the OV5647 is a 1/4" 5-megapixel (2592 × 1944 active array) backlit-CMOS unit, with 8-/10-bit RGB/RAW image output. PiCam boards were chosen here due to the ease of data acquisition and image processing using Raspberry Pi computer boards (Raspberry Pi Foundation), via the Python programming language. Due to their low expense and power consumption, Raspberry Pi computers are increasingly being used in a variety of sensing applications (e.g., [9,10]). Whilst this work is focused on one sensor type, preliminary tests with another such back-illuminated CMOS device, the Sony IMX219 (Sony Corporation, Minato, Tokyo, Japan), also evidence UV sensitivity.

The PiCam lens and filter, housed above the sensor within a plastic casing unit, both absorb the UV signal, and were therefore removed as a first step to developing the UV camera. This was achieved chemically using a Posistrip® EKC830™ (DuPont, Wilmington, DE, USA) bath. To improve UV sensitivity of the sensor, we then removed the microlens and Bayer filter layers, the latter of which masks the sensor in a mosaic of RGB colour filters (Figure 1), and attenuates much of the incident UV radiation. This process effectively turns the PiCam into a monochrome sensor.

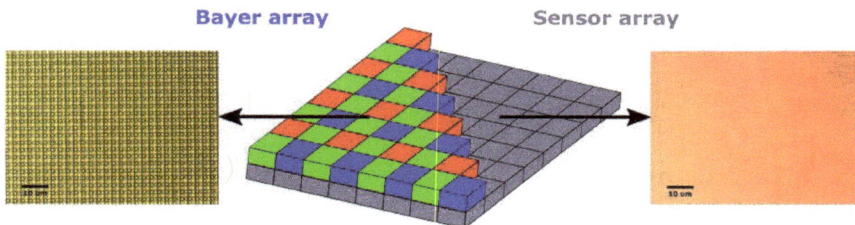

Figure 1. A schematic (**Centre**) of the Bayer filter array and its positioning on the photodetector array. Also shown are microscope images of the PiCam sensor pre- (**Left**) and post- (**Right**) Bayer removal process.

Whilst the filter can be removed by careful scratching (a range of tools, such as metal tweezers or a pointed wooden object, can be used for this process), a far more uniform finish is achievable chemically; we adopted the latter approach using a five step procedure. We first submerged the sensors in photoresist remover (Posistrip® EKC830™) and heated (70–100 °C) until the filter was entirely removed; this generally takes 10–30 minutes, depending on the level of applied heating and agitation. A second bath of this remover was then applied to optimise the cleaning procedure. Following this, the photoresist remover was washed from the sensor by a three stage cleaning procedure, using successive baths of n-butyl acetate, acetone and isopropyl alcohol. All the above chemicals were applied undiluted. For each of these steps, the sensor was submerged on the order of minutes, to ensure a thorough cleaning. Careful and rigorous application of this methodology can result in uniform and clean sensors which exhibit greatly enhanced UV sensitivity relative to non-de-Bayered units. For example, a clear-sky image taken with a partially de-Bayered unit, captured through a 310 nm filter and with a shutter speed of 400 ms, exhibits an increase of $\approx$ 600% in a de-Bayered region relative to the unmodified section of the sensor.

For image capture, it is necessary to retrieve RAW sensor data, rather than the standard JPEG image output, as the former excludes the de-Bayering algorithm and image processing, which results in a non-linear response from the sensor. This, and all the below acquisition and processing steps were achieved with in-house authored Python codes. Here the RAW images were saved to the camera, where they were stored in the metadata of the 8-bit JPEG image. These binary data are then extracted and saved as PNG images, to preserve the 10-bit RAW digital number (DN) format, in files of $\approx$ 6 MB. These images could then be straightforwardly processed and analysed.

A UV transmissive anti-reflection (AR) coated plano-convex quartz lens of 6 mm diameter and 9 mm focal length (Edmund Optics Ltd., Barrington, NJ, USA; $100), was then mounted to the fore of the PiCam, using an in-house designed, 3D printed lens holder, attached to the board using its pre-existing mount holes; this provided a field of view of $\approx$ 28°. This lens housing consists of two parts, enabling straightforward focusing via screw thread adjustment. As the housing covers the entire camera board, it was necessary to disable the board's light-emitting diode (LED), which, by default, is programmed to turn on during image acquisition (see Figure 2). By way of comparison: the all in cost of this camera configuration (e.g., PiCam, lens, filter, filter holder, chemical removal costs) including a UV bandpass filter is $\approx$ USD 200–300, depending on the filter, in comparison to a typical figure of at least ten times the upper value, for such a system based on currently available scientific grade UV cameras.

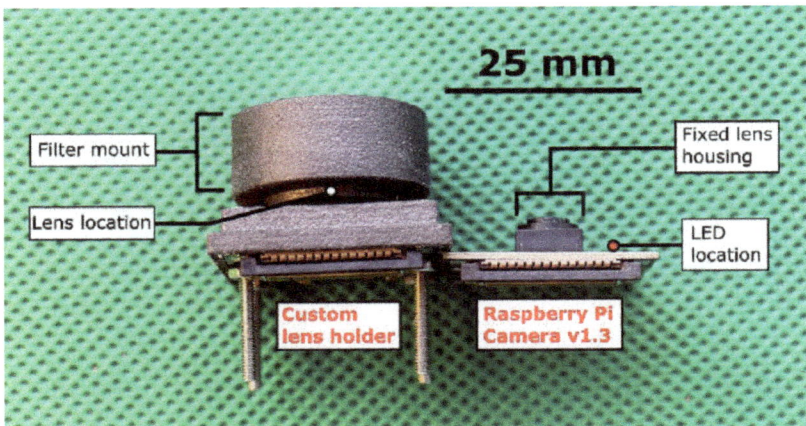

Figure 2. A profile image of the Raspberry Pi Camera Module (**Right**), and the modified system with custom built optics (**Left**). The custom design is bolted to the camera board using the pre-existing mount holes.

Following construction of the imaging system, the sensor linearity, which is important for quantitative applications, and UV sensitivity were tested. This was achieved by mounting a UV bandpass filter, of 12.5 mm diameter, centred on 310 nm and of 10 nm full width at half maximum (FWHM) (Edmund Optics Ltd.) to the fore of the camera lens; these filters have no other transmission features in the spectral sensitivity range of the CMOS detectors.

To test the UV sensor response, images of uniformly illuminated clear-sky were taken through the 310 nm filter at varying shutter speeds; these experiments were performed in Sheffield, UK, at approximately 11:00 local time (solar zenith angle of around 50°), during February 2016. At each shutter speed four images were captured and analog gain was fixed throughout, for consistency; the operating software does not allow gain specification, but it does enable this to be fixed, following stabilization of the system after the camera start-up process. By averaging pixel DNs from a 800 × 600 pixel region in the centre of each image, we observed a linear increase in average pixel DN relative to shutter speed, for the RAW 10-bit sensor data (Figure 3A). Furthermore, the RAW images demonstrated near saturation at shutter speeds well below 1 s, in a spectral region where there is very little skylight due to ozone absorption, indicating usable UV sensor sensitivity.

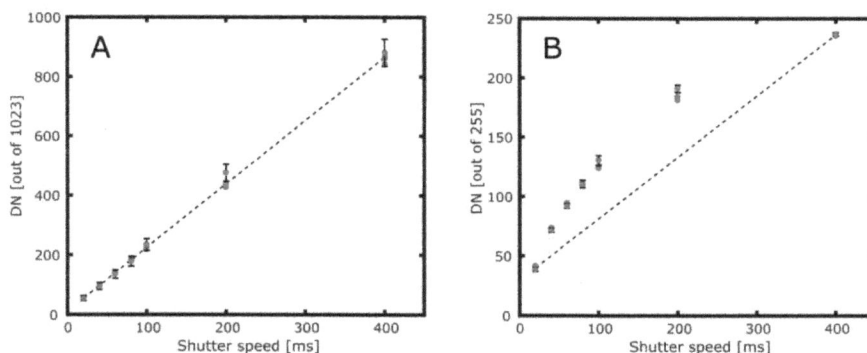

Figure 3. Plots of average pixel signal (in digital number; DN) vs. shutter speed (ms) for a cropped region (800 × 600 pixels) of clear-sky images taken at 310 nm: (**A**) the 10-bit RAW image output shows a linear increase in DN with respect to shutter speed; (**B**) the 8-bit standard output JPEG image shows a non-linear response in all three of the red-green-blue (RGB) channels (Red-channel DNs are plotted here) in line with gamma correction. An error bar is inserted on one data point per exposure time, indicating the standard deviation of the pixel intensities in the cropped region; the bar heights are approximately the same for all points of equivalent shutter speed, and just one bar is displayed for clarity.

3. Measurements of Power Station Sulphur Dioxide Emissions

Sulphur dioxide has strong UV absorption bands between 300 and 320 nm [11], which have been exploited in a range of atmospheric remote sensing measurements of this species, using differential optical absorption spectroscopy and UV imagery [12–14]. In particular, power stations release SO_2 to the atmosphere from their smokestacks, and remotely sensing these emissions is one means of ensuring regulatory compliance and constraining effects on the atmosphere [4,5,15,16].

In order to demonstrate proof of concept of the utility of the developed sensors for UV imaging applications, we deployed two UV PiCam units at Drax power station in the United Kingdom over a ≈ 15 min period on 15 August 2016. Bandpass filters, centred at 310 and 330 nm (10 nm FWHM; Edmund Optics Ltd.), were mounted to the fore of the co-aligned cameras, one on one unit, with the other filter on the other camera. Both devices simultaneously imaged the rising smokestack plume. Shutter speeds of ≈ 300 and 40 ms, respectively, were applied for the cameras, in view of the greater scattered skylight intensity at the latter wavelength, and images were acquired at 0.25 Hz. Camera-to-camera pixel mapping was achieved subsequently in software, using the smokestack as a reference. The acquisition and retrieval protocols followed those of Kantzas et al. [17], in particular by collecting dark images for each camera, which were subtracted from all acquired sky images. Clear sky images, taken adjacent to the plume, were also acquired, then normalised to generate a mask, which all acquired plume images were divided by; this eliminates vignetting as well as any sensor non-uniformity which may have resulted from the de-Bayering process. As SO_2 absorbs at 310 nm, but not at 330 nm, contrasting resulting images provides a means of constraining the spatial distribution of gas concentration across the field of view, whilst eliminating sources of extinction common to both wavelengths, e.g., due to aerosols. In particular, SO$_2$ apparent absorbance (*AA*) for each pixel was determined using the following relation [17]:

$$AA = -\log_{10}\left[\frac{IP_{310}}{IB_{310}} \middle/ \frac{IP_{330}}{IB_{330}}\right] \tag{1}$$

where *IP* is the in-plume intensity, and *IB* is the background intensity, taken as the average value in the clear sky adjacent to the rising gas plume; the subscripts specify the camera filter in question, for the dark subtracted, mask corrected imagery. Calibration of *AA* was then carried out using quartz cells containing known column amounts of SO$_2$ (in our case 0, 300 and 900 ppm.m), in particular measuring *AA* values for each cell when pointing at plume free sky, plotting concentration vs. these, and then multiplying all acquired plume image *AA* values by this gradient. The cell column amounts were verified using an Ocean Optics Inc. USB2000 spectrometer, using the VolcanoSO2 differential optical absorption spectroscopy code [18].

Figure 4C shows a typical calibrated SO2 image, after binning pixels to a 648 × 486 array to reduce noise, highlighting the ability of the cameras to clearly resolve the smokestack emissions. Furthermore, background image noise levels are remarkably low (typically ≈ 25 ppm.m standard deviation in a 648 × 486 binned image), in contrast to values of ≈ 30 ppm.m, previously reported from more expensive scientific grade camera systems in atmospheric SO$_2$ monitoring [5], indicating the potential of these low cost units in UV imaging applications.

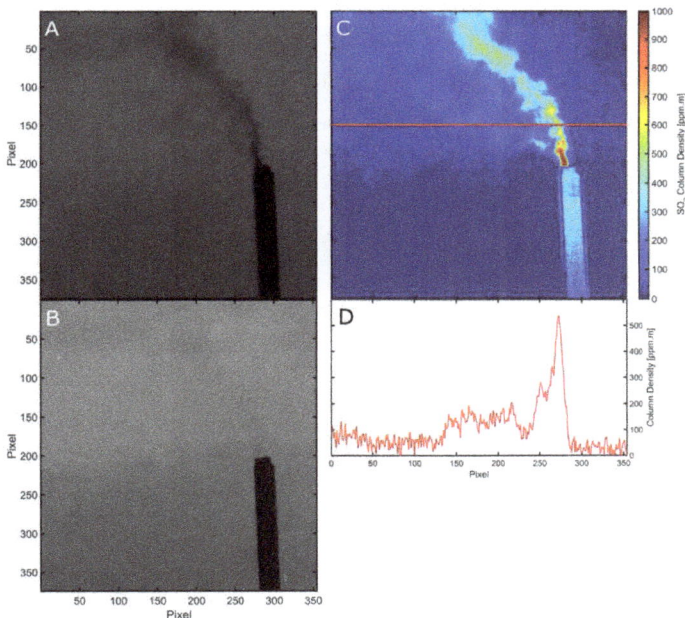

Figure 4. (**A**) A cropped image of the Drax smokestack taken at 310 nm with a shutter speed of 300 ms. The initial image pixels are binned to generate a pixel resolution of 648 × 486, to reduce noise; dark image subtraction and mask corrections have been applied. (**B**) As in (A) but at 330 nm with a shutter speed of 40 ms. (**C**) The resulting calibrated SO$_2$ image of Drax power station stack and plume showing the clear capacity of the system to resolve the plume emissions. (**D**) A cross-section of (C) showing gas concentrations along the row delineated by the red line. The background noise level can be clearly observed between pixels 300 to 350.

By integrating across the plume, perpendicular to the plume transport direction, integrated column amounts (ICAs) of SO_2 were obtained, and a time series of ICAs through successive images was produced. A plume speed of $\approx$4.9 m/s was calculated by generating two such ICA time series, determined for parallel plume cross sections, located at different distances from the source, and then cross correlating these. This plume speed was then multiplied by the ICA values to generate a SO_2 flux time series, as shown in Figure 5. The SO_2 flux was found to be relatively stable from the smokestack, with a mean emission of 0.44 kg/s; nevertheless, notable gas "puffs" are apparent, and can be seen both in the flux time series and SO_2 absorption image video (see Video S1 in the supplementary materials). The Drax Annual Reviews of Environmental Performance [19], provide the best available ancillary emissions data for comparison, stating annual SO_2 loadings ranging 24.5–35.1 kt/yr for the period 2008–2013, and thus, average annual fluxes ranging 0.78–1.1 kg/s. These data corroborate our observations, falling within an order of magnitude, given that Drax's operational output will fall somewhat below the mean annual value during the British summer.

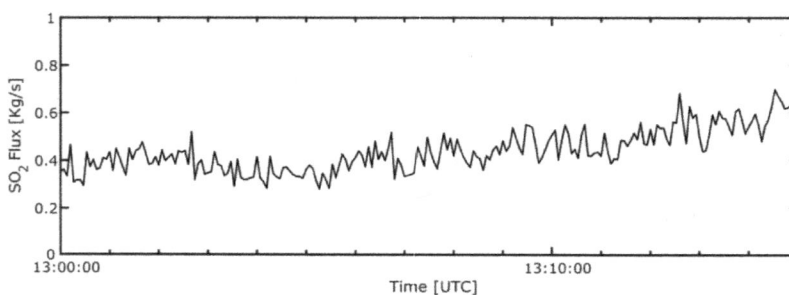

Figure 5. Time series of SO_2 flux from Drax power station for a 15 min acquisition period.

4. Discussion and Concluding Remarks

In this paper we have presented, for the first time, a relatively simple methodology for the development of low-cost UV cameras, based on inexpensive sensors developed for the smartphone market. We show that by modification of Raspberry Pi camera sensors, and rebuilding of the optical systems, camera boards as cheap as $\approx$ USD 25 can be adapted to applications down to wavelengths of at least 310 nm. The potential utility of such devices in UV imaging applications was illustrated via a case study, in which we used the cameras to perform ultraviolet remote sensing of SO_2 fluxes from a power station smokestack.

With this in mind, other possible areas in which this technology could be applied include, but are not limited to: ground-based mineral aerosol detection (331/360 nm), to mimic satellite based TOMS retrievals [20,21], monitoring skin conditions [3,22], forensics during crime scene investigation [1,23], fault detection in power systems [2], fault detection in vehicles, and astronomy. The low cost would be of particular benefit in application areas where budgets are limited, or arrays of these units are required. One example of such a scenario is in the field of volcanology, where UV cameras are used to image SO_2 emissions in volcanic plumes [4,24,25] as a means of investigating magmatic processes. As many volcanoes are located in developing countries, lower cost sensors could greatly expedite more comprehensive monitoring of global volcanic hazards, with allowance for sensor destruction in the event of explosions.

One potential current limitation to this system is its maximum achievable frame rate. The shutter speeds used here at the power station (300 ms at 310 nm) correspond to a theoretically achievable acquisition rate of >1 Hz. However, in reality obtainable frame rates are rather lower due to on chip processing. Preliminary tests suggest that framerates >0.5 Hz are achievable, however, over prolonged periods this resulted in sporadic dropping of frames, which disappeared altogether for acquisition at

Sensors **2016**, *16*, 1649

0.25 Hz. More work is required to identify the limiting factor in frame capture, and whether stable higher acquisition rates might be achievable with further software optimisation. For the majority of applications, however, we suggest that framerates of 0.25 Hz will be more than adequate.

Future work could focus on the spectral range of such units; the sensitivity of this system below 310 nm has not been quantified herein, and whilst the sensors do likely respond at deeper UV wavelengths, there may be a trade off in terms of signal to noise. Investigating the longer wavelength cut off of these monochrome devices could also be of interest, in establishing their use in high temperature thermal imaging applications. Other work may include consideration of thermal effects on sensor stability/noise, and the degree to which cooling/temperature stabilisation might improve signal to noise. Finally, whilst signal to noise levels appear promising here, particularly given the low sensor price point, further work is now also merited in comparing the system sensitivity and noise characteristics against more expensive scientific grade UV cameras.

Supplementary Materials: The following are available online at http://www.mdpi.com/1424-8220/16/10/1649/ s1, Video S1: Absorbance image sequence of Drax SO_2 emissions. Warmer colours indicate larger SO_2 column densities. Images were captured at 0.25 Hz, and this video sequence is speeded up to eight times real-time.

Acknowledgments: This work was funded by a Scholarship from the University of Sheffield, awarded to TCW. TCW would like to thank the Raspberry Pi community (found at https://www.raspberrypi.org/forums/), who provided great support whilst working with the PiCam module. We would also like to thank Matthew Davies for his help in selecting the object lens used herein. TDP acknowledges the support of a NERC studentship (NE/K500914/1) and AMcG acknowledges a Leverhulme Trust Research Fellowship (RF-2016-580).

Author Contributions: The paper was written by T.C.W and A.Mc.G. Development of the camera system and data processing/analysis was predominantly performed by T.C.W.; A.Mc.G. aided in initial camera tests, in the field work, and is the supervisor; T.D.P. provided the initial concept and aided in field work; A.T. aided in early-stage camera development and software development for camera control; B.W. devised and performed the methodology for removal of the Bayer filter from the camera sensor; R.B. is a co-supervisor; J.R.W. aided in the development of the camera optics.

Conflicts of Interest: The authors declare no conflict of interest. The founding sponsors had no role in the design of the study; in the collection, analyses, or interpretation of data; in the writing of the manuscript, and in the decision to publish the results.

References

1. Krauss, T.; Warlen, S. The Forensic Science Use of Reflective Ultraviolet Photography BT—The Forensic Science Use of Reflective Ultraviolet Photography. *J. Forensic Sci.* **1985**, *30*, 262–268. [CrossRef] [PubMed]
2. Chen, Z.; Wang, P.; Yu, B. Research of UV detection system based on embedded computer. World Automation Congress 2008. Available online: http://ieeexplore.ieee.org/stamp/stamp.jsp?tp=&arnumber=4699210&isnumber=4698939 (accessed on 26 August 2016).
3. Fulton, J.E. Utilizing the Ultraviolet (UV Detect) Camera to Enhance the Appearance of Photodamage and Other Skin Conditions. *Dermatol. Surg.* **1997**, *23*, 163–169. [CrossRef] [PubMed]
4. McElhoe, H.B.; Conner, W.D. Remote Measurement of Sulfur Dioxide Emissions Using an Ultraviolet Light Sensitive Video System. *JAPCA J. Air Pollut. Control Assoc.* **1986**, *36*, 42–47. [CrossRef]
5. Smekens, J.-F.; Burton, M.R.; Clarke, A.B. Validation of the SO_2 camera for high temporal and spatial resolution monitoring of SO_2 emissions. *J. Volcanol. Geoth. Res.* **2015**, *300*. [CrossRef]
6. Igoe, D.; Parisi, A.V.; Carter, B. Characterization of a smartphone camera's response to ultraviolet a radiation. *Photochem. Photobiol.* **2013**, *89*, 215–218. [CrossRef] [PubMed]
7. Igoe, D.; Parisi, A.V. Evaluation of a Smartphone Sensor to Broadband and Narrowband Ultraviolet A Radiation. *Instrum. Sci. Technol.* **2015**, *43*, 283–289. [CrossRef]
8. Igoe, D.; Parisi, A.V. Broadband direct UVA irradiance measurement for clear skies evaluated using a smartphone. *Radiat. Prot. Dosim.* **2015**, *167*, 485–489. [CrossRef] [PubMed]
9. Moure, D.; Torres, P.; Casas, B.; Toma, D.; Blanco, M.J.; Del Río, J.; Manuel, A. Use of Low-Cost Acquisition Systems with an Embedded Linux Device for Volcanic Monitoring. *Sensors* **2015**, *15*, 20436–20462. [CrossRef] [PubMed]
10. Schlobohm, J.; Pösch, A.; Reithmeier, E. A Raspberry Pi Based Portable Endoscopic 3D Measurement System. *Electronics* **2016**, *5*. [CrossRef]

11. Vandaele, A.C.; Simon, P.C.; Guilmot, J.M.; Carleer, M.; Colin, R. SO_2 absorption cross section measurements in the UV using a Fourier transform spectrometer. *J. Geophys. Res.* **1994**, *99*, 25599–25605. [CrossRef]

12. Mori, T.; Burton, M.R. The SO_2 camera: A simple, fast and cheap method for ground-based imaging of SO_2 in volcanic plumes. *Geophys. Res. Lett.* **2006**, *33*. [CrossRef]

13. Kantzas, E.P.; McGonigle, A.J.S. Ground Based Ultraviolet Remote Sensing of Volcanic Gas Plumes. *Sensors* **2008**, *8*, 1559–1574. [CrossRef]

14. Theys, N.; De Smedt, I.; van Gent, J.; Danckaert, T.; Wang, T.; Hendrick, F.; Stavrakou, T.; Bauduin, S.; Clarisse, L.; Li, C.; et al. Sulfur dioxide vertical column DOAS retrievals from the Ozone Monitoring Instrument: Global observations and comparison to ground-based and satellite data. *J. Geophys. Res. Atmos.* **2015**, *120*, 2470–2491. [CrossRef]

15. Millán, M.M. Remote sensing of air pollutants. A study of some atmospheric scattering effects. *Atmos. Environ.* **1980**, *14*, 1241–1253. [CrossRef]

16. McGonigle, A.J.S.; Thomson, C.L.; Tsanev, V.I.; Oppenheimer, C. A simple technique for measuring power station SO_2 and NO_2 emissions. *Atmos. Environ.* **2004**, *38*, 21–25. [CrossRef]

17. Kantzas, E.P.; McGonigle, A.J.S.; Tamburello, G.; Aiuppa, A.; Bryant, R.G. Protocols for UV camera volcanic SO_2 measurements. *J. Volcanol. Geoth. Res.* **2010**, *194*, 55–60. [CrossRef]

18. McGonigle, A.J.S. Measurement of volcanic SO_2 fluxes with differential optical absorption spectroscopy. *J. Volcanol. Geotherm. Res.* **2007**, *162*, 111–122. [CrossRef]

19. Annual review of Environmental Performance, Drax, 2013. Available online: http://www.drax.com/media/ 56551/Environmental-Performance-Review-2013.pdf (accessed on 26 August 2016).

20. Torres, O.; Bhartia, P.K.; Herman, J.R.; Ahmad, Z.; Gleason, J. Derivation of aerosol properties from satellite measurements of backscattered ultraviolet radiation: Theoretical basis. *J. Geophys. Res. Atmos.* **1998**, *103*, 17099–17110. [CrossRef]

21. Sinyuk, A.; Torres, O.; Dubovik, O. Combined use of satellite and surface observations to infer the imaginary part of refractive index of Saharan dust. *Geophys. Res. Lett.* **2003**, *30*. [CrossRef]

22. Mahler, H.M.; Kulik, J.A.; Harrell, J.; Correa, A.; Gibbons, F.X.; Gerrard, M. Effects of UV photographs, photoaging information, and use of sunless tanning lotion on sun protection behaviors. *Arch. Dermatol.* **2005**, *141*, 373–380. [CrossRef] [PubMed]

23. Yosef, N.; Almog, J.; Frank, A.; Springer, E.; Cantu, A. Short UV Luminescence for Forensic Applications: Design of a Real-Time Observation System for Detection of Latent Fingerprints and Body Fluids. *J. Forensic Sci.* **1998**, *43*, 299–304. [CrossRef] [PubMed]

24. Bluth, G.; Shannon, J.; Watson, I.M.; Prata, A.J.; Realmuto, V. Development of an ultra-violet digital camera for volcanic SO_2 imaging. *J. Volcanol. Geoth. Res.* **2007**, *161*, 47–56. [CrossRef]

25. Peters, N.; Hoffmann, A.; Barnie, T.; Herzog, M.; Oppenheimer, C. Use of motion estimation algorithms for improved flux measurements using SO_2 cameras. *J. Volcanol. Geoth. Res.* **2015**, *300*, 58–69. [CrossRef]

sensors

MDPI

Article

Design of a Sub-Picosecond Jitter with Adjustable-Range CMOS Delay-Locked Loop for High-Speed and Low-Power Applications

Bilal I. Abdulrazzaq [1,2,*], Omar J. Ibrahim [1], Shoji Kawahito [3], Roslina M. Sidek [1], Suhaidi Shafie [1], Nurul Amziah Md. Yunus [1], Lini Lee [4] and Izhal Abdul Halin [1]

[1] Department of Electrical and Electronic Engineering, Faculty of Engineering, Universiti Putra Malaysia, Serdang 43400, Selangor, Malaysia; omar.j.ibrahim@gmail.com (O.J.I.); roslinams@upm.edu.my (R.M.S.); suhaidi@upm.edu.my (S.S.); amziah@upm.edu.my (N.A.M.Y.); izhal@upm.edu.my (I.A.H.)
[2] Department of Electronic and Communications Engineering, Al-Nahrain University, Al-Jadriya Complex, Baghdad 10070, Iraq
[3] Imaging Devices Laboratory, Research Institute of Electronics, Shizuoka University, 3-5-1 Johoku, Nakaku, Hamamatsu, Shizuoka 432-8011, Japan; kawahito@idl.rie.shizuoka.ac.jp
[4] Faculty of Engineering, Multimedia University, Persiaran Multimedia, Cyberjaya 63100, Malaysia; linilee@mmu.edu.my
* Correspondence: bilal.i.abdulrazzaq@gmail.com; Tel.: +60-14-934-0794

Academic Editor: Gonzalo Pajares Martinsanz
Received: 28 July 2016; Accepted: 5 September 2016; Published: 28 September 2016

Abstract: A Delay-Locked Loop (DLL) with a modified charge pump circuit is proposed for generating high-resolution linear delay steps with sub-picosecond jitter performance and adjustable delay range. The small-signal model of the modified charge pump circuit is analyzed to bring forth the relationship between the DLL's internal control voltage and output time delay. Circuit post-layout simulation shows that a 0.97 ps delay step within a 69 ps delay range with 0.26 ps Root-Mean Square (RMS) jitter performance is achievable using a standard 0.13 µm Complementary Metal-Oxide Semiconductor (CMOS) process. The post-layout simulation results show that the power consumption of the proposed DLL architecture's circuit is 0.1 mW when the DLL is operated at 2 GHz.

Keywords: delay step; delay range; time jitter; Delay-Locked Loop (DLL); charge pump; Capacitor-Reset Circuit (CRC)

1. Introduction

Delay-Locked Loops (DLLs) with high-resolution delay steps are extensively used for time management of large systems [1]. For example, they are used in Fluorescence Lifetime Imaging Microscopy (FLIM) sensors where a light pulse is modulated with the capture window that is shifted in picosecond-order delay steps for a total range of tens of picoseconds [2]. Furthermore, high-resolution DLLs are used in the compensation for PVT variations and any delay mismatch that may be caused to signals during the operation of many high-frequency VLSI circuits [3]. For all of these applications, DLLs should generate an adequate amount of lock/delay range while maintaining the output jitter as low as possible. This is because there is a trade-off relation between delay range and the jitter performance [4]. In addition, the total delay fluctuations including jitter should be less than the delay resolution for optimum operation [5].

Since DLLs only adjust the phase (delay) of an input signal and not its frequency, DLLs suffer from limited delay range. Therefore, a considerable amount of new techniques has been developed to address this issue. For example, a technique employing a Digital-to-Analog Converter (DAC) with Parallel Variable Resistor (PVR) is used to realize high-resolution delay steps with a wide delay range by accurately controlling the Current-Controlled Delay Element (CCDE) of the DLL [1]. Another technique developed is the use of a dual-loop architecture which utilizes multiple delay lines [6]. The first "reference" loop generates a clock with quadrature phases. In the second "main" loop, these phases are delayed by four Voltage-Controlled Delay Lines (VCDLs) and then multiplexed to generate the output clock. A new technique based on cycle-controlled delay unit was proposed by [7] to enlarge the delay range by reusing the delay units in a cycle-like process without the need for cascading a large number of delay units. A DLL with a new voltage-controlled delay element based on body-controlled current source and body-feed technique was also developed to widen the delay range [8]. In this method, the Phase Detector (PD) is replaced by a Phase/Frequency Detector (PFD) with a start controller to achieve a sufficient locking range. Another new architecture was proposed by [9] which employs a mixed-mode Time-to-Digital Converter (TDC) for enabling a frequency-range selector. The frequency-range selector can generate digital control signals to switch the delay range of the multi-controlled delay cell in the VCDL and the current of the digitally-controlled charge pump.

However, the majority of the techniques mentioned above result in complex circuit architectures that lead to degraded jitter performance as well as increased area overhead, cost, and power consumption. Motivated by this research gap, this work proposes a new and simple technique using a Capacitor-Reset Circuit (CRC) to reset the loop filter capacitor for delay range extension and at the same time reducing the jitter performance into the sub-picosecond range. The capacitor-reset technique is widely used to reinitialize a control voltage to a fixed initial value and has been applied in many circuits such as pixels of image sensors and PLL circuits [10,11]. At this point, mathematical analysis confirmed by circuit simulation, our proposed technique is capable of generating a comparably wide delay range and picosecond-resolution delay steps with a sub-picosecond jitter performance. In addition, this architecture consumes a relatively small area and power compared with the available techniques reported in literature.

Table 1 below shows performance specifications of the most recent and relevant high-resolution DLL designs reported in the literature.

Table 1. Performance specifications of previously reported high-resolution DLLs.

Variable	[12]	[13]	[14]	[15]
CMOS technology	130 nm	55 nm	350 nm	65 nm
Supply voltage	1.5 V	1 V	3.5 V	1 V
Delay range	345 ps	128 ps	375 ps	161 ps
Delay resolution	4 ps	8.5 ps	7.5 ps	5.21 ps
No. of steps	63	15	7	31
Operating frequency range	1.5–2.5 GHz	200–850 MHz	N/A	3 MHz–1.8 GHz
RMS jitter	N/A	0.04 ps @ 850 MHz	7.5 ps @ 400 MHz	0.85 ps @ 1.8 GHz
Power consumption	30 mW @ 2.5 GHz	1.02 mW @ 850 MHz	N/A	9.5 mW @ 1.8 GHz
Active area	0.03 mm^2	0.007 mm^2	N/A	0.0153 mm^2

Table 1 summarizes the DLL's parameters that have a direct impact on the performance in terms of speed and power consumption. For example, information about achievable delay range, delay resolution, number of delay steps, operating frequency range, RMS jitter, and power consumption is provided in Table 1. It is worth noting that the finest delay step is 4 ps achieved by [12]. It also generates a comparably long delay range of approximately 345 ps.

The proposed design is explained in the subsequent section. The results and discussion are presented in Section 3. Finally, Section 4 summarizes and concludes this paper.

2. Materials and Methods

Our proposed circuit is shown in Figure 1a. It consists of a conventional VCDL, an Exclusive-OR (XOR) gate-based Phase Detector (PD), a Charge Pump (CP), and a modified Loop Filter (LF) with the addition of the CRC. It works by resetting the loop filter's capacitor by a pre-determined time constant before lock is achieved. The reset operation is performed by a reset signal, φ_R. By varying the pulse width of φ_R using a simple Pulse-Width Generator (PWG) circuit that will be illustrated at the end of this section, a change in the time constant τ_R of the modified loop filter is achieved. This results in a change in delay range of the DLL.

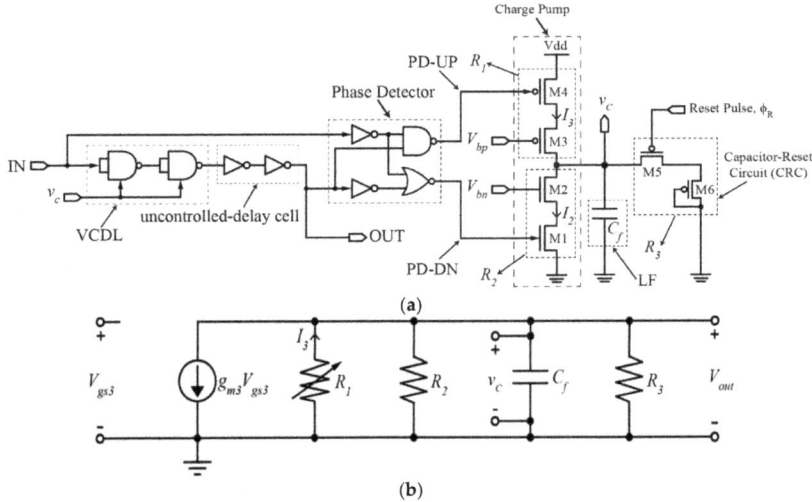

Figure 1. (**a**) Schematic of DLL with the CRC; and (**b**) small-signal model of DLL's charge pump with CRC.

Figure 1a shows the modification made to the loop filter of a DLL where M5 and M6 are used to create the CRC. M5 acts as a switch that resets the loop filter's capacitor, C_f. On the other hand, M6 is connected as a diode whose resistance together with the capacitance C_f of the loop filter's capacitor creates a time constant τ_R that controls the magnitude of v_c which is fed to the VCDL current to control bias current and propagation delay. This ultimately controls the delay step and lock/delay range. The aspect ratio of both pMOS transistors, M5 and M6, is 0.35 μm/0.13 μm.

The capacitor-reset operation at a pre-determined reset signal duration results in a varying charge/discharge rate of C_f when V_{bp} is changed as opposed to a DLL without the CRC. This results in changes in v_c settling time that controls the delay of the VCDL. To illustrate this operation, the charge pump's small-signal model shown in Figure 1b is used. v_c is expressed in Equations (1) and (2) during charging and discharging operations, respectively.

$$v_c(t) = v_{ch0}\left(1 - e^{-\frac{t}{\tau_R}}\right), \tag{1}$$

$$v_c(t) = v_{dis0}\left(e^{-\frac{t}{\tau_R}}\right). \tag{2}$$

where v_{ch0}, v_{dis0}, and τ_R are the initial voltage across C_f during charging (which is equal to $-V_{Tp}$), initial voltage across C_f during discharging (which is equal to maximum v_c), and the time constant of the loop filter, respectively. This time constant, τ_R, is written as:

$$\tau_R = R_3 C_f. \tag{3}$$

where R_3 is the equivalent resistance of the diode-connected transistor's (M6) output resistance in series with transistor M5's output resistance.

From Equations (1) and (2), it is obvious that the capacitor's voltage v_c is directly dependent on charging/discharging time, t. Equations (1) and (2) also implies that the charging/discharging time of C_f can be changed by changing the time constant τ_R of the CRC, which will in turn change v_c. This is achieved through changing the reset duration of the reset signal φ_R that is applied to the gate of M5 (see Figure 1a).

The small-signal model of the charge pump connected to the CRC shown in Figure 1b is also used to analyze how the DLL generates fine-linear delay steps within a selectable delay range. V_{bp} is varied in order to vary the delay steps. The series output resistance of transistors M4 and M3 is modeled as R_1 in Figure 1b. Likewise, R_2 in Figure 1b models the series output resistance of transistors M2 and M1 and R_3 models the output resistance of the diode-connected transistor M6 in series with transistor M5's output resistance when M5 is turned on. It should be mentioned that the aspect ratio of the nMOS transistors M1 and M2 is 0.6 μm/0.13 μm, and that of the pMOS transistors M3 and M4 is 1.2 μm/0.13 μm. The value of the capacitance C_f in Figure 1a,b is 0.63 fF.

R_1, R_2 and R_3 are given by Equations (4)–(6), respectively [16]:

$$R_1 \approx (g_{m3}r_{ds3})\, r_{ds4}. \tag{4}$$

$$R_2 \approx (g_{m2}r_{ds2})\, r_{ds1}. \tag{5}$$

$$R_3 \approx (g_{m5}r_{ds5})\, r_{ds6}. \tag{6}$$

When V_{bp} is varied, it is obvious that r_{ds3} and g_{m3} change accordingly, resulting in a change in R_1. Due to this, R_1 is written as:

$$\begin{aligned}R_1 &\approx (g_{m3} + \Delta g_{m3})\,(r_{ds3} + \Delta r_{ds3})\, r_{ds4},\\ R_1 &\approx g_{m3}r_{ds3}r_{ds4} + r_{ds4}\,(g_{m3}\Delta r_{ds3} + r_{ds3}\Delta g_{m3} + \Delta r_{ds3}\Delta g_{m3}).\end{aligned} \tag{7}$$

where Δr_{ds3} and Δg_{m3} are the changes in r_{ds3} and g_{m3}, respectively. Equation (7) represents the change in R_1 when $V_{bp} \neq 0$ and can also be written in the following form:

$$R_1 \approx R_{C_0} + R_P. \tag{8}$$

where R_{C0} is a constant corresponding to the term $(g_{m3}r_{ds3}r_{ds4})$ and R_P is a variable corresponding to the term $(r_{ds4}(g_{m3}\Delta r_{ds3} + r_{ds3}\Delta g_{m3} + \Delta r_{ds3}\,\Delta g_{m3}))$ in Equation (7). According to simulation results shown in Figure 2, when V_{bp} is varied from 1 V to 0.8935 V, Δr_{ds3} changes from 10.66 GΩ to 4.29 GΩ. Likewise, for the same V_{bp} range, the charge pump's charging current I_3 changes from 110.37 pA to 117.99 pA. However, when $V_{bp} = 0$, r_{ds3} and g_{m3} are at their minimum values. This implies that Δr_{ds3} and Δg_{m3} will have very small values which can be neglected compared with other R_1 cases in which $V_{bp} \neq 0$. Hence, Equation (8) can be rewritten as:

$$R_{1,0} \approx R_{C_0}. \tag{9}$$

where $R_{1,0}$ represents the case when $V_{bp} = 0$. Thus, Equation (8) can be rewritten as follows:

$$R_1 \approx R_{1,0} + R_P. \tag{10}$$

It should be mentioned that the non-monotonicity points in Figure 2b can be a consequence of the non-convergence problems. These problems can be caused during the simulation if the resistance of the transistor is very high or very low. This can be solved by adjusting either the simulator options or the transistor model parameters (Ron and/or transconductance g_m) [17]. However, no significant impact can be observed in the overall behavior and performance of the DLL circuit, as will be demonstrated in the delay steps linearity results explained in the next section. In addition, a linear regression has been employed and superimposed on the plot in Figure 2b regarding the charge pump charging current

I_3 versus the control voltage V_{bp}. It can be seen from Figure 2b that the Root-Mean Square Error (RMSE) of the linear regression plot is only 0.06, which indicates that the original plot of I_3 versus V_{bp} is almost linear.

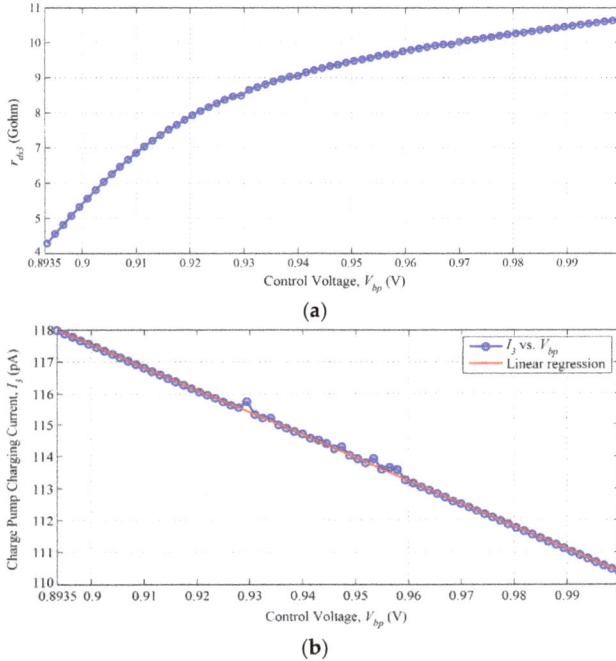

(a)

(b)

Figure 2. (**a**) r_{ds3} versus control voltage, V_{bp}; and (**b**) I_3 versus control voltage, V_{bp}.

According to [1], the relationship between the change in resistance and the change in current is expressed as:

$$\frac{R_1(t-1)}{R_1(t-1) + \Delta R_1(t)} \approx \frac{I_3(t-1) - \Delta I_3(t)}{I_3(t-1)}.$$

(11)

The time delay, t_d, of the VCDL is given as [18]:

$$t_d = T_{ref} - (K_{VCDL}v_c) + \Delta d.$$

(12)

where T_{ref}, K_{VCDL}, and Δd are the period time of the input clock signal, the gain of the VCDL, and the jitter caused by the VCDL, respectively. Equation (12) indicates that the voltage, v_c, across the capacitor determines the time delay, t_d. The voltage v_c can be written as [18]:

$$v_c(t) = \frac{1}{C_f}\int_0^{\Delta t} I_3(t) \times dt + v_c(0),$$

$$v_c(t) = \frac{1}{C_f}\int_0^{\Delta t} I_3(t) \times dt + (-V_{Tp}).$$

(13)

Substituting Equation (13) into Equation (12), t_d is written as:

$$t_d = T_{ref} - \left(K_{VCDL} \times \left(\frac{1}{C_f}\int_0^{\Delta t} I_3(t) \times dt + (-V_{Tp}) \right) \right) + \Delta d.$$

(14)

Substituting Equation (11) for I_3 into Equation (14) yields t_d in terms of the change in R_1 and I_3 and is written as:

$$t_d = T_{ref} - \left(K_{VCDL} \times \left(\frac{1}{C_f} \int_0^{\Delta t} \frac{(R_1(t-1)) \times (I_3(t-1))}{R_1(t-1) + \Delta R_1(t)} \times dt + (-V_{Tp}) \right) \right) + \Delta d. \quad (15)$$

t_d in Equation (15) represents the delay step. On the other hand, the delay range, t_{dr}, is defined as the difference between maximum and minimum delays and can be written as:

$$t_{dr} = t_{d(max)} - t_{d(min)}. \quad (16)$$

where $t_{d(max)}$ and $t_{d(min)}$ are the maximum and the minimum delays. On the other hand, the maximum and minimum delays are expressed by Equations (17) and (18), respectively:

$$t_{d(max)} = T_{ref} - \left(K_{VCDL} \times \left(\frac{1}{C_f} \int_0^{\Delta t\ max} \frac{(R_1(t-1)) \times (I_3(t-1))}{R_1(t-1) + \Delta R_{1(max)}(t)} \times dt + (-V_{Tp}) \right) \right) + \Delta d_{max}, \quad (17)$$

$$t_{d(min)} = T_{ref} - \left(K_{VCDL} \times \left(\frac{1}{C_f} \int_0^{\Delta t\ min} \frac{(R_1(t-1)) \times (I_3(t-1))}{R_1(t-1)} \times dt + (-V_{Tp}) \right) \right) + \Delta d_{min}. \quad (18)$$

In order to demonstrate the operation of the charge pump circuit without and with the proposed CRC technique, Figure 3 is considered. This figure is an illustration figure that illustrates the differences in the discharge rates for two extreme values of V_{bn} (1 V and 0 V). Figure 3a highlights the discharge rates for a charge pump without the proposed CRC and Figure 3b with the CRC. It is obvious from Figure 3b that the difference in discharge rates is significantly higher than that of the case in Figure 3a. To clarify this, according to the simulations, the discharge rates' difference for the case with the CRC technique is 2.49 mV/ps, while that for the case without CRC is only 0.2 mV/ps. The higher is the difference in the discharge rate, the bigger is the difference in v_c settling values according to Equations (1), (2), (15), (17) and (18). In relation to the simulation results, the discharge rate of v_c is faster when V_{bn} value is 1 V, causing the capacitance C_f to fully discharge faster and the discharging time to have a lower value compared to the case when V_{bn} is 0 V. The discharge rate is directly proportional to the discharge current I_2. Since the discharge rate is different, the settling voltage for v_c is also different, causing a change in the control voltage of the VCDL and resulting in a change in time delay of the DLL.

The charge pump circuit of a DLL suffers from amplifier noise charge injected from its amplifier into the loop filter capacitor, thus reducing this noise will result in better time jitter performance [19]. Figure 4 is used to illustrate error charge accumulation in charge pump's loop filter capacitor. Figure 4a shows a conventional charge pump where initially amplifier noise charge q_n is injected into C_f from the amplifier at the ON phase of the input signal. When the input signal goes low, C_f discharges but a small amount of residual noise charge q_{nr} is left in C_f. The next ON phase of the input signal injects new amplifier noise charge and it is added with the residual noise charge left from the previous discharge cycle. Therefore, for simplicity, the output voltage v_c of the charge pump, is given by:

$$v_c = \frac{q_n + q_{nr}}{C_f}. \quad (19)$$

The numerator of Equation (19) gives the total noise charge of a conventional charge pump. On the other hand, Figure 4b shows that noise charge is also transferred into the loop filter's capacitor. However, when the signal goes high, the CRC is activated and causes C_f to fully discharge. Only a small amount of reset charge q_r is injected into C_f from transistors M5 and M6 that make up the CRC (see Figure 1a). Thus, the output voltage v_c of the charge pump is given by:

$$v_c = \frac{q_r}{C_f} \qquad (20)$$

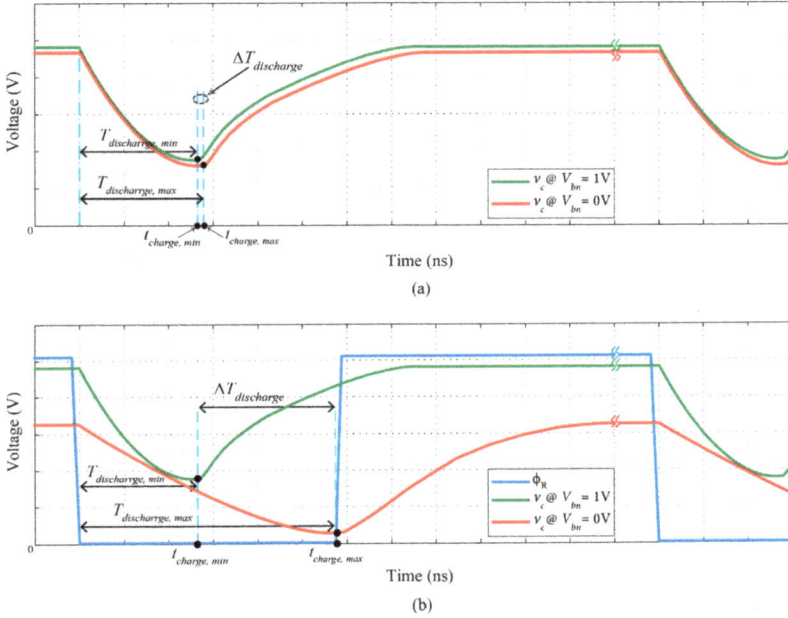

(a)

(b)

Figure 3. Difference in discharge rates for two cases: (**a**) without CRC; and (**b**) with CRC.

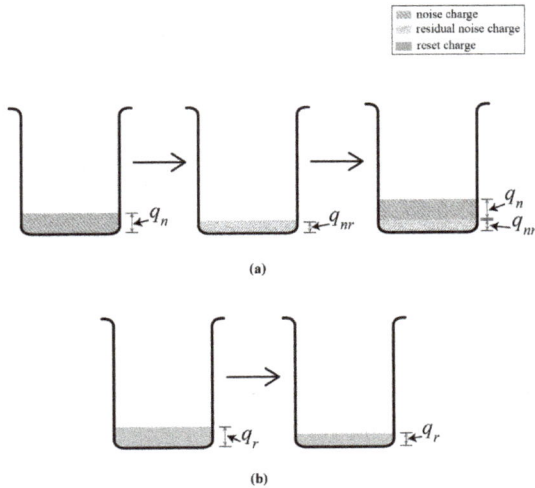

(a)

(b)

Figure 4. Error charge accumulation in loop filter's capacitor: (**a**) without CRC; and (**b**) with CRC.

We can also view q_r as noise charge since it is random in nature. However, q_r is much less than $q_n + q_{nr}$, thus the jitter of v_c is significantly reduced for a charge pump with the CRC. Moreover, this technique also produces a wider delay range since a larger and accurate level of v_c can be achieved when the ON phase period of the reset signal φ_R is made longer through the PWG circuit.

The PWG circuit, used to control the pulse width of the reset signal φ_R, is shown in Figure 5. This circuit is used to set the time constant of the loop filter in order to set the desired discharge rate of C_f (see Equations (1) and (2)). Once a desired delay range is acquired, the charge pump's charging current I_3 can be varied through the charge pump amplifier's bias voltage, V_{bp} (see Figure 1a), to allow small-linear changes in the DLL's output signal time delay.

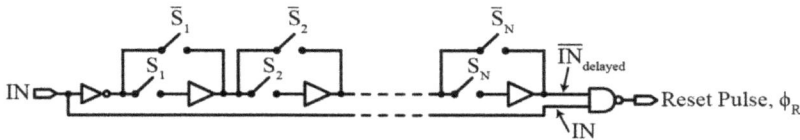

Figure 5. Simple Pulse-Width Generator (PWG) circuit.

It is also noted that the input of the PWG circuit shown in Figure 5 is fed from the input signal of the DLL itself in order to synchronize the discharge time of node v_c with the input pulse, as illustrated in the timing diagram shown in Figure 6.

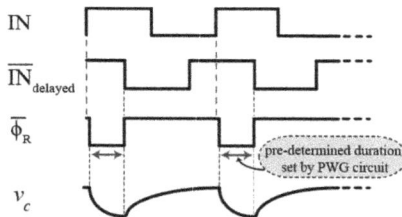

Figure 6. Timing diagram showing how the reset signal activation is synchronized with the input reference pulse.

3. Results and Discussion

The proposed DLL is simulated using a 0.13 μm CMOS process. The power supply voltage is 1.2 V. From post-layout simulations, the delay is controlled from zero to 69 ps by varying V_{bp} from 0.8935 V to 1 V in steps of 1.5 mV. In this simulation, parametric analysis was used to change V_{bp}; however, the value of V_{bp} can be controlled by a 10-bit Digital-to-Analog Converter (DAC). V_{bn} is fixed at 0.2 V.

Figure 7 shows the generated output time delay t_d as a function of the control voltage V_{bp} with respect to the delay steps linearity. It is clear from Figure 7 that the time delay increases linearly with the increase in V_{bp}. The sensitivity of the linear regression plot is approximately 644 (ps/V) with Root-Mean Square Error (RMSE) equals to 0.64. For an LSB of 0.97 ps, it can be seen in Figure 7a that the delay steps' *Differential Non-Linearity (DNL)* does not exceed 0.86. Moreover, the *DNL* values of the delay steps located between the 41st and 70th delay steps are all concentrated in the positive region. This has mainly caused the slight deviation observed between the linear regression and the simulated output delay steps shown in Figure 7a,b, and it has also resulted in the maximum 1.5 *Integral Non-Linearity (INL)* value at the end of the *INL* plot in Figure 7b. On the other hand, the *INL* values across the generated delay steps in Figure 7b are somewhat concentrated in the negative region. This indicates that the resolution of most of the delay steps is very close to one LSB of the output delay.

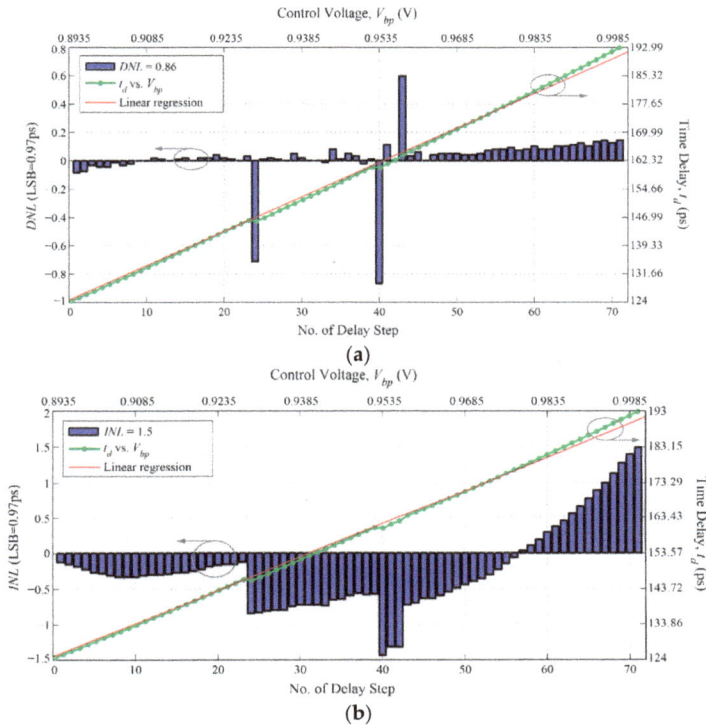

Figure 7. DLL's total time delay and control voltage ranges with respect to the delay steps linearity: (a) *DNL*; and (b) *INL*.

Figure 8 shows the simulated DLL's output signal, which is delayed by 0.97 ps as the minimum delay step and total of 69 ps as the maximum delay range, when operated at 2 GHz of the input reference signal. The lock-in time of the DLL is only 14 cycles.

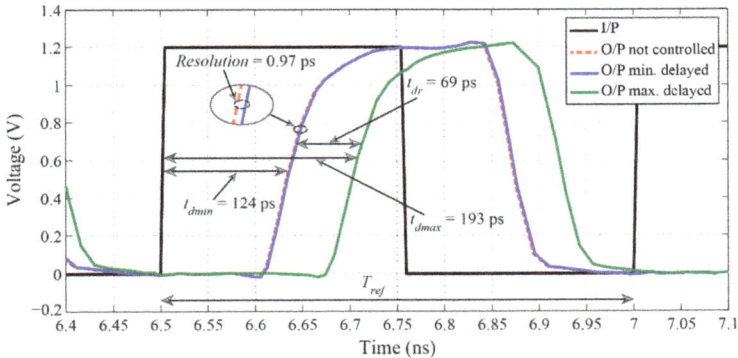

Figure 8. DLL's input and output signals when operated at 2 GHz of the input reference signal, where t_{dmin} and t_{dmax} correspond to the minimum and maximum times of the output time delay t_d and their values are 124 ps and 193 ps, respectively, the loop filter's time constant τ_R values are 2.32 μs at t_{dmin} and 14.22 μs at t_{dmax}, and the reference signal's period T_{ref} is 0.5 ns.

Figure 9 shows the simulated voltage v_c across the loop filter's capacitor with the reset signal φ_R.

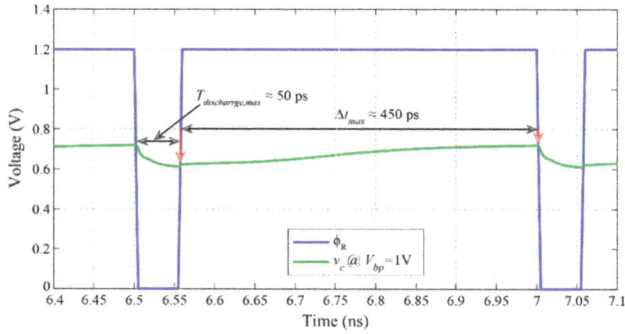

Figure 9. Voltage across the loop filter's capacitor v_c and reset signal φ_R at locked state.

It can be seen in Figure 9 that the duration of the reset signal φ_R is almost identical to the maximum discharge time, $T_{discharge,max}$, obtained when V_{bp} equals to 1 V. The waveforms plotted in Figure 9 have been obtained after the locked state has been achieved, i.e., after 14 cycles. Likewise, at locked state and when V_{bp} equals to 1 V, the input and output signals of the phase detector are presented in Figure 10. Figure 10a shows the input reference and output delayed signals which are fed to the two inputs of the phase detector. According to the phase difference between these two inputs, the phase detector generates phase difference information, signal "PD-UP" and signal "PD-DN" in Figure 10b, which is fed to the charge pump to keep the operation of the DLL in the locked state.

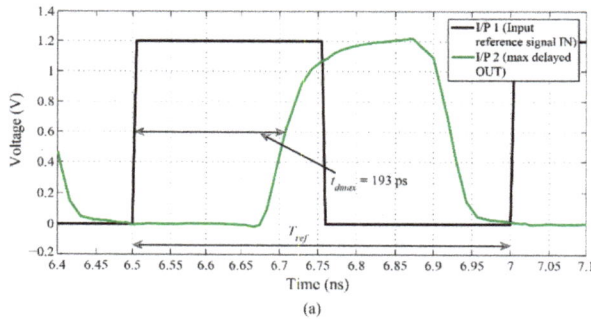

(a)

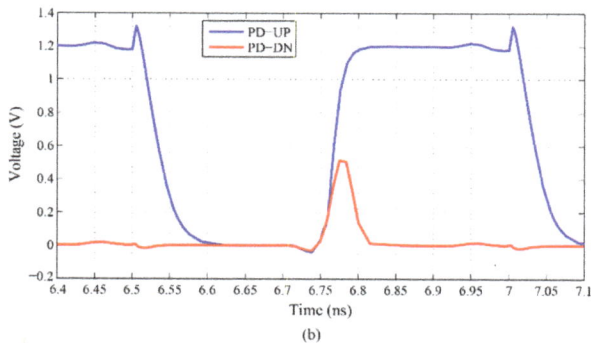

(b)

Figure 10. Input and output signals of the phase detector at locked state: (**a**) Inputs; and (**b**) Outputs.

For completeness, the PVT variations effects on the DLL's delay range have been simulated and analyzed, as shown in Figure 11. Since the maximum achievable delay range is 69 ps, it can be noted in Figure 11a that the process corner FF can degrade the delay range and the corner SS can mostly degrade the jitter through the extremely extended range. Nonetheless, extending or narrowing the pulse width of the reset pulse, φ_R, can solve these shortcomings. In Figure 11b, three temperature and voltage variations all located at 1.38 V for 0 °C, 27 °C, and 70 °C, which are all dark black-colored, can degrade the delay range by 12 ps, 15 ps, and 18 ps, respectively. Similarly, the small violations in the delay range with the other PVT variations can be compensated by extending the pulse width of φ_R without significantly degrading the total output jitter or the delay steps linearity.

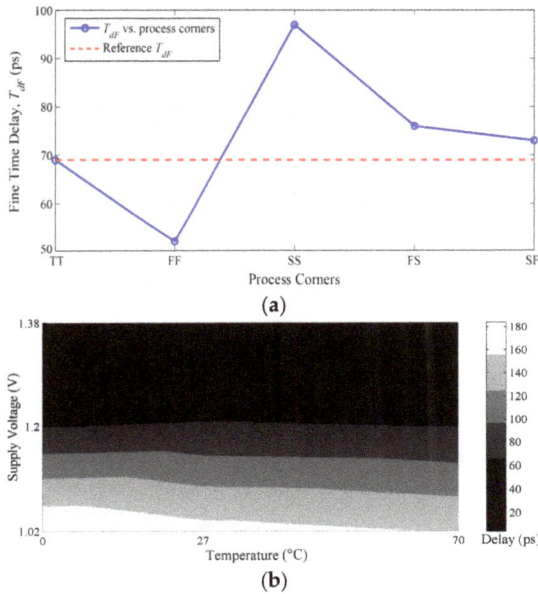

Figure 11. Maximum DLL's delay range versus: (**a**) process corners; and (**b**) temperature and supply voltage variations.

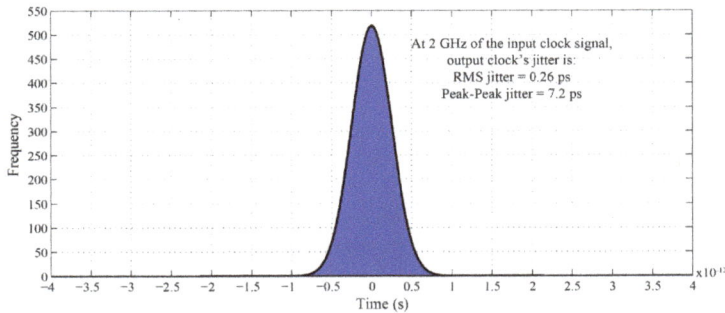

Figure 12. DLL output jitter histogram.

Simulation results on jitter show that the output jitter of the DLL is remarkably low. Figure 12 shows the simulated jitter when the DLL is operated at 2 GHz of the input reference signal.

The peak-to-peak and RMS values are 7.2 ps and 0.26 ps, respectively. As mentioned, this is attributed to the cycle-to-cycle reset operation of the charge pump's capacitor, which significantly minimizes the accumulated noise originated from the charge pump's amplifier. It is also worth mentioning that the low jitter is attributed to the use of only one NAND gate-based buffer in the VCDL circuit.

In addition, the effects of the PVT variations on the jitter performance have also been simulated and analyzed, as shown in Figure 13.

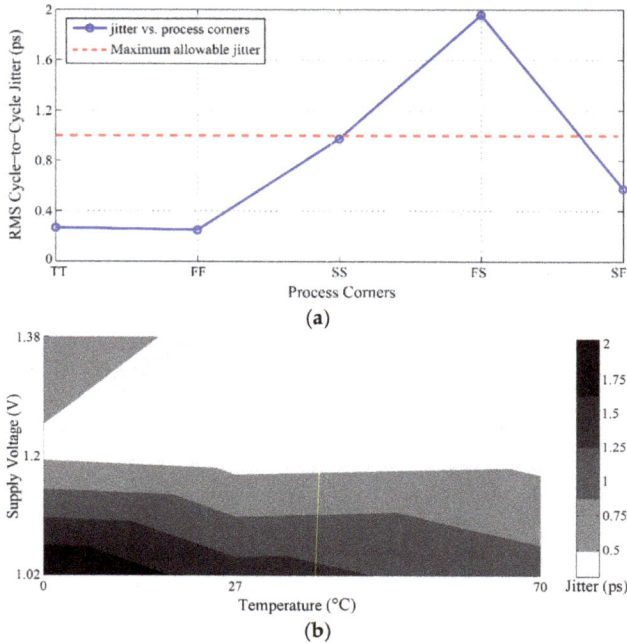

Figure 13. RMS cycle-to-cycle jitter versus: (a) process corners; and (b) temperature and supply voltage variations.

Since the desired value of the output jitter is in the sub-picosecond range, it can be noted in Figure 13a that only the process corner FS degrades the jitter. However, this shortcoming can be mitigated by optimizing the pulse width of the reset pulse, φ_R. In Figure 13b, only two temperature and supply voltage variations located at 1.02 V for 0 °C and 27 °C, which are dark black-colored and dark grey-colored, can degrade the output jitter to over 1.75 ps RMS and 1.3 ps RMS, respectively.

A summary of the performance specifications and results of the proposed work is presented in Table 2. The proposed CRC technique successfully achieves sub-picosecond-resolution delay step, a high number of delay steps within a specific range, sub-picosecond jitter performance, a wide operating frequency range, sub-milliwatt power consumption, and a small occupied active area for layout. The layout area is significantly minimized because the VCDL, which is followed by an uncontrolled inverter-based buffer as shown in Figure 1a, only uses a single NAND-based buffer. It is worth mentioning that the achieved delay range for the case without the proposed CRC technique is only 2 ps using the same transistor sizes and operating conditions as in the case with the CRC whose achieved delay range is 69 ps.

Table 2. Summary of performance specifications and results achieved by the proposed DLL design.

Variable	Value
CMOS technology	130 nm
Supply voltage	1.2 V
Delay range	69 ps
Delay resolution	0.97 ps
No. of steps	71
Operating frequency range	50 MHz–2 GHz
RMS jitter	0.26 ps @ 2 GHz
Power consumption	0.1 mW @ 2 GHz
Active area	0.001 mm^2

The layout of the proposed DLL circuit architecture is shown in Figure 14.

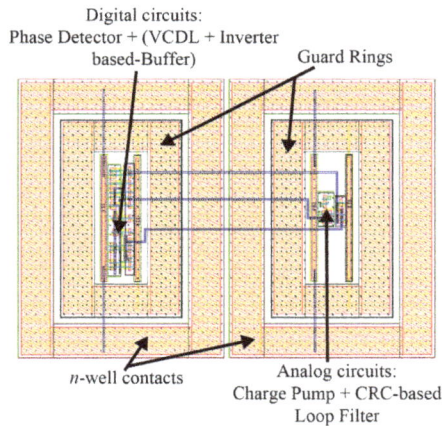

Figure 14. Layout of the proposed DLL.

It can be seen in Figure 14 that guard rings and n-well contacts have been used for the proposed DLL's layout in order to reduce the effects of the substrate and power noise. In addition, separation of the digital circuits from the analog circuits as well as utilizing separate VDD and GND lines for each of these circuits have been employed to further reduce the substrate noise effects.

In order to compare the performance of this work with other reported high-resolution DLL circuits, Table 3 is presented. In this table, the proposed work has been compared with the work reported by [12], which has been presented earlier in Table 1 and has shown to have almost the best performance compared with the other works in Table 1.

Table 3. Performance comparison of this work with a reported high-resolution DLL.

Variable	[12]	This Work
CMOS technology	130 nm	130 nm
Supply voltage	1.5 V	1.2 V
Delay range	345 ps	69 ps
Delay resolution	4 ps	0.97 ps
No. of steps	63	71
Operating frequency range	1.5–2.5 GHz	50 MHz–2 GHz
RMS jitter	N/A	0.26 ps @ 2 GHz
Power consumption	30 mW @ 2.5 GHz	0.1 mW @ 2 GHz
Active area	0.03 mm^2	0.001 mm^2

According to Table 3, the proposed lock-range extension technique in this work achieves higher-resolution delay step, higher number of delay steps within a specific range, better jitter performance, lower power consumption, and smaller occupied active area.

4. Conclusions

The proposed DLL architecture uses a CRC at the output of the DLL's charge pump in order to change the delay range and generate small steps with sub-picosecond jitter performance. Through simulation, the DLL maximum delay is 69 ps with 0.97 ps delay steps, while maintaining the total jitter at the output in the sub-picosecond range. In terms of circuit complexity, our proposed technique is much simpler when compared to others as only a reset circuit is added to the charge pump. This not only allows a smaller layout area, but also enhances the DLL's jitter performance and output range.

Acknowledgments: The authors would like to acknowledge Universiti Putra Malaysia (UPM) for contributing the facilities and funds under the research grant initiative GP-IPS/2014/9438713 for this work to be possible.

Author Contributions: Izhal Abdul Halin and Shoji Kawahito conceived and designed the experiments; Bilal I. Abdulrazzaq performed the experiments; Bilal I. Abdulrazzaq and Izhal Abdul Halin analyzed the data; Omar J. Ibrahim, Roslina M. Sidek, Suhaidi Shafie, Nurul Amziah Md. Yunus, and Lini Lee contributed to analysis tools; Bilal I. Abdulrazzaq and Izhal Abdul Halin wrote the paper.

Conflicts of Interest: The authors declare no conflict of interest.

References

1. Eto, S.; Akita, H.; Isobe, K.; Tsuchida, K.; Toda, H.; Seki, T. A 333 MHz, 20 mW, 18 ps resolution digital DLL using current-controlled delay with parallel variable resistor DAC (PVR-DAC). In Proceedings of the Second IEEE Asia Pacific Conference on ASICs (AP-ASIC 2000), Cheju, Korea, 28–30 August 2000; pp. 349–350.
2. Seo, M.W.; Kagawa, K.; Yasutomi, K.; Takasawa, T.; Kawata, Y.; Teranishi, N.; Li, Z.; Halin, I.A.; Kawahito, S. A 10.8 ps-time-resolution 256 × 512 image sensor with 2-Tap true-CDS lock-in pixels for fluorescence lifetime imaging. In Proceedings of the 2015 IEEE International Solid-State Circuits Conference (ISSCC), San Francisco, CA, USA, 22–26 February 2015; pp. 1–3.
3. Markovic, B.; Tisa, S.; Villa, F.A.; Tosi, A.; Zappa, F. A high-linearity, 17 ps precision time-to-digital converter based on a single-stage vernier delay loop fine interpolation. *IEEE Trans. Circuits Syst. I Regul. Pap.* **2013**, *60*, 557–569. [CrossRef]
4. Jaehyouk, C.; Kim, S.T.; Woonyun, K.; Kwan-Woo, K.; Kyutae, L.; Laskar, J. A low power and wide range programmable clock generator with a high multiplication factor. *IEEE Trans. Very Large Scale Integr. Syst.* **2011**, *19*, 701–705.
5. Zhang, R.; Kaneko, M. Robust and low-power digitally programmable delay element designs employing neuron-MOS mechanism. *ACM Trans. Des. Autom. Electron. Syst.* **2015**, *20*, 1–19. [CrossRef]
6. Jung, Y.J.; Lee, S.W.; Shim, D.; Kim, W.; Kim, C.; Cho, S.I. A dual-loop delay-locked loop using multiple voltage-controlled delay lines. *IEEE J. Solid State Circuits* **2001**, *36*, 784–791. [CrossRef]
7. Chang, H.H.; Liu, S.I. A wide-range and fast-locking all-digital cycle-controlled delay-locked loop. *IEEE J. Solid State Circuits* **2005**, *40*, 661–670. [CrossRef]
8. Lu, C.T.; Hsieh, H.H.; Lu, L.H. A 0.6 V low-power wide-range delay-locked loop in 0.18 μm CMOS. *IEEE Microw. Wirel. Compon. Lett.* **2009**, *19*, 662–664.
9. Cheng, K.H.; Lo, Y.L. A fast-lock wide-range delay-locked loop using frequency-range selector for multiphase clock generator. *IEEE Trans. Circuits Syst. II Express Briefs* **2007**, *54*, 561–565. [CrossRef]
10. El Gamal, A.; Eltoukhy, H. CMOS image sensors. *IEEE Circuits Devices Mag.* **2005**, *21*, 6–20. [CrossRef]
11. Maxim, A.; Scott, B.; Schneider, E.; Hagge, M.; Chacko, S.; Stiurca, D. Sample-reset loop filter architecture for process independent and ripple-pole-less low jitter CMOS charge-pump PLLs. In Proceedings of the 2001 IEEE International Symposium on Circuits and Systems (ISCAS 2001), Sydney, Australia, 6–9 May 2001; Volume 4, pp. 766–769.
12. Yang, R.J.; Liu, S.I. A 2.5 GHz all-digital delay-locked loop in 0.13 μm CMOS technology. *IEEE J. Solid State Circuits* **2007**, *42*, 2338–2347. [CrossRef]

13. Wang, J.-S.; Cheng, C.-Y.; Liu, J.-C.; Liu, Y.-C.; Wang, Y.-M. A duty-cycle-distortion-tolerant half-delay-line low-power fast-lock-in all-digital delay-locked loop. *IEEE J. Solid State Circuits* **2010**, *45*, 1036–1047. [CrossRef]
14. Klepacki, K.; Szplet, R.; Pelka, R. A 7.5 ps single-shot precision integrated time counter with segmented delay line. *Rev. Sci. Instrum.* **2014**, *85*, 034703. [CrossRef] [PubMed]
15. Wang, J.-S.; Cheng, C.-Y.; Chou, P.-Y.; Yang, T.-Y. A wide-range, low-power, all-digital delay-locked loop with cyclic half-delay-line architecture. *IEEE J. Solid State Circuits* **2015**, *50*, 2635–2644. [CrossRef]
16. Sedra, A.S.; Smith, K.C. Single-stage integrated circuit amplifiers. In *Microelectronic Circuit*, 5th ed.; Oxford University Press: New York, NY, USA, 2004; pp. 545–716.
17. Baker, R.J. *CMOS Circuit Design, Layout, and Simulation*; Wiley-IEEE Press: Hoboken, NJ, USA, 2010.
18. Van de Beek, R.C.H.; Klumperink, E.A.M.; Vaucher, C.S.; Nauta, B. Low-jitter clock multiplication: A comparison between PLLs and DLLs. *IEEE Trans. Circuits Syst. II Analog Digit. Signal Process.* **2002**, *49*, 555–566. [CrossRef]
19. Kim, B.K.; Im, D.; Choi, J.; Lee, K. A highly linear 1 GHz 1.3 dB NF CMOS low-noise amplifier with complementary transconductance linearization. *IEEE J. Solid State Circuits* **2014**, *49*, 1286–1302. [CrossRef]

sensors

MDPI

Article

A Low Power Digital Accumulation Technique for Digital-Domain CMOS TDI Image Sensor

Changwei Yu, Kaiming Nie, Jiangtao Xu and Jing Gao *

School of Electronic Information Engineering, Tianjin University, 92 Weijin Road, Nankai District, Tianjin 300072, China; yuchangwei@tju.edu.cn (C.Y.); niekaiming@tju.edu.cn (K.N.); xujiangtao@tju.edu.cn (J.X.)
* Correspondence: gaojing@tju.edu.cn; Tel.: +86-22-2789-0832

Academic Editor: Gonzalo Pajares Martinsanz
Received: 13 July 2016; Accepted: 19 September 2016; Published: 23 September 2016

Abstract: In this paper, an accumulation technique suitable for digital domain CMOS time delay integration (TDI) image sensors is proposed to reduce power consumption without degrading the rate of imaging. In terms of the slight variations of quantization codes among different pixel exposures towards the same object, the pixel array is divided into two groups: one is for coarse quantization of high bits only, and the other one is for fine quantization of low bits. Then, the complete quantization codes are composed of both results from the coarse-and-fine quantization. The equivalent operation comparably reduces the total required bit numbers of the quantization. In the 0.18 μm CMOS process, two versions of 16-stage digital domain CMOS TDI image sensor chains based on a 10-bit successive approximate register (SAR) analog-to-digital converter (ADC), with and without the proposed technique, are designed. The simulation results show that the average power consumption of slices of the two versions are 6.47×10^{-8} J/line and 7.4×10^{-8} J/line, respectively. Meanwhile, the linearity of the two versions are 99.74% and 99.99%, respectively.

Keywords: CMOS TDI image sensor; accumulation technique; coarse quantization; fine quantization

1. Introduction

Linear image sensors capture images by scanning the target scene. In low light or high speed scanning condition, the integration time of each pixel is too short to collect enough photons, which results in low signal-to-noise ratio (SNR). The time delay integration (TDI) images' sensor technology adopts more rows of pixels to scan the same scene, and the integrated signals are accumulated together, which extends the integration time equivalently. Thus, a better SNR can be achieved using a TDI image sensor. Compared with the regular linear image sensor, the accumulated signal and uncorrelated noise of a TDI image sensor are amplified by a factor N and square root of N respectively, which results in a square root of N times improvement of SNR, where N is the stage of accumulation. TDI image sensors are suitable in the situations where a line-scan system is required for high quality and low noise imaging even under low illuminations and high speed scanning [1,2]. The TDI technology could be easily implemented in a charge coupled device (CCD) [3] because it allows noiseless accumulation of signals in the charge domain. However, the CCD requires high supply voltage, which results in large power consumption. Additionally, due to the incompatibility with the CMOS process, the CCD could not integrate with other functions on the same chip with the sensor [4]. In recent years, TDI image sensors implemented with the CMOS process have been reported [5–14]. Compared with the CCD, CMOS TDI image sensors could successfully integrate various analog and digital circuits as an on-chip camera [4]. Such integration further reduces the power consumption and size [4].

There are typically three kinds of accumulation methods: charge-domain accumulation scheme [5–8], analog-domain accumulation scheme [9–11] and digital domain accumulation scheme [12–14] for CMOS TDI image sensors. In the charge-domain accumulation scheme, the noiseless accumulation enables good SNR. However, the charge transfer efficiency and full well capacity, which strongly depend on process, limit its momentum of development [6,7]. With respect to the analog-domain accumulation scheme, the generated signals from pixels are firstly accumulated by the analog accumulator, and then the accumulated result is quantized by column analog-to-digital converter (ADC) [9,10]. Because the maximum accumulated voltage depends on the supply voltage, the dynamic range of the analog-domain CMOS TDI image sensor is limited by the supply voltage [9,10]. Since the analog memory cell of the analog accumulator is implemented with capacitors, the accumulator area is considerably large [9,10]. In the digital domain accumulation scheme, the outputs of pixels are firstly quantized by column ADC, and then the corresponding digital results are added together by the digital accumulator [14]. The memory cell of the digital accumulator is implemented by digital registers, so the digital accumulator area is less than an analog accumulator with the same accumulation stages [14]. In order to make the most use of the quantization range of ADC, the output signals of pixels are usually amplified several times before being sent to the ADC, which results in better SNR than that of analog-domain accumulation [13,14]. However, at high speed scanning and high accumulation stages, a large amount of data generated by ADC bring in more power consumption. In this paper, a low power digital accumulation technique for a CMOS TDI image sensor is proposed to reduce power consumption without degrading the rate of imaging by means of downscaling the bit numbers of the quantization.

The remainder of this paper is organized as follows. In Section 2, we firstly describe the principle of the low power accumulation technique, the optimized implementation scheme is then discussed, and finally a behavioral simulation is completed with MATLAB (R2011b, MathWorks, Natick, MA, USA). Section 3 presents the circuit implementation scheme of the low power accumulation technique and the simulation results. A brief conclusion is drawn in Section 4.

2. Low Power Digital Accumulation Technique

2.1. Principle of the Low Power Digital Accumulation Technique

The accumulation scheme of the traditional N-stage digital domain CMOS TDI image sensor with L-bit column ADC, where $L = (M + K)$ is shown in Figure 1. During the exposure time, the exposed object moves at a constant speed relative to the pixel array. The line time (T_L) is defined as the time that the exposed object moves over exactly one pixel pitch, and the line rate is defined as $1/T_L$. The operation procedure is briefly described as follows. At the first T_L, the object is captured by pixel1. At the end of the integration of pixel1, the exposure result is sent to the column ADC and quantized into L-bit digital data. Then, the data is sent to the digital accumulator. At the second T_L, the same object is captured by pixel2 and its quantized data is also sent to the digital accumulator. In the same way, the object is exposed N times, and the corresponding quantized data are added in the digital accumulator. With the imaging flow above, the ADC converts $L \times N$ bits data at the N accumulation stage.

Ideally, the output results of pixels detecting the same object should be same. However, the noise may bring some variation in the low-bit codes of the quantized data. The influenced bit numbers depend on the SNR of output signals. The high bits of the quantized results remain the same among pixels detecting the same object. Therefore, there is no need to fully quantify every output signal of pixels. A low power digital accumulation technique for CMOS TDI image sensors is proposed by introducing coarse quantization and fine quantization, aiming at less data and lower power consumption. Based on the discussion above, when the output signals of pixels are quantized into L-bit codes by ADC, only the low K bits varies. The rest of the high bits is indicated by M and $L = (M + K)$. The high M bits of the quantized results of the same object are the same.

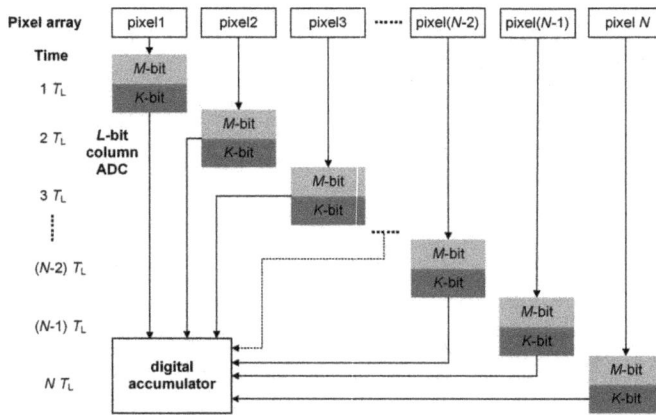

Figure 1. Traditional digital domain accumulation scheme.

The proposed low power digital domain accumulation scheme with *L*-bit column ADC is shown in Figure 2. The digital accumulator is divided into high *M*-bit accumulator and low *K*-bit accumulator. At the first T_L, the object is captured by pixel1. At the end of integration of pixel1, its analog result is sent to the column ADC and only coarsely quantized into high *M* bits, where the low power accumulation scheme differs from the traditional accumulation scheme. Then, the high *M*-bit codes are sent to the high *M*-bit accumulator. During the second T_L, the same object is captured by pixel2 and its analog result is sent to the column ADC for fine quantization. It only needs to be quantized into low *K* bits, as the high *M* bits have been already quantized in the first T_L. Then, the low *K*-bit codes are sent to the low *K*-bit accumulator. In the same way, the remaining pixels are all quantized with only low *K* bits and then transmitted to the low *K*-bit accumulator. Finally, the complete quantization result is obtained by combining high *M* bits with low *K* bits. To accomplish *N*-stage accumulation, the ADC needs to convert $(M + N \times K)$ bits data in the proposed low power accumulation scheme. Compared with the traditional accumulation scheme, the ADC saves $(N - 1) \times M$ bits conversion time with the low power accumulation scheme.

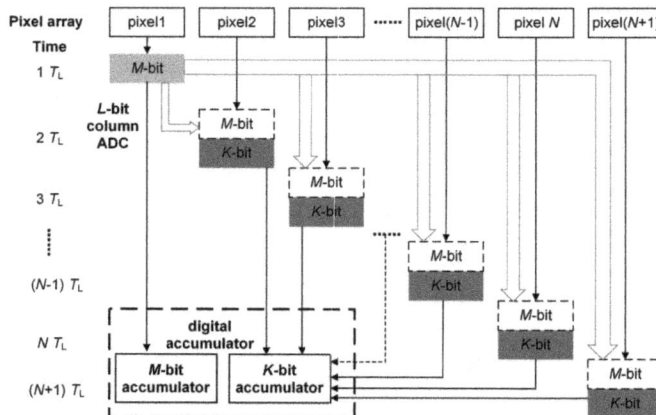

Figure 2. Low power digital domain accumulation scheme.

In order to increase the quantization precision of the high M bits, some lines of pixels are arranged to repeat the coarse quantization, and then the quantized high M-bit codes are averaged. Assuming that P lines of pixels are used to repeat the coarse quantization, the ADC of the low power digital accumulation scheme will save $(N - P) \times M$ bits conversion time compared with the traditional digital accumulation scheme for a complete N-stage accumulation. The low power digital domain CMOS TDI image sensor needs much less conversion time at the same conversion speed and accumulation stages, which could achieve lower power consumption than the traditional digital domain CMOS TDI image sensor. According to the operation procedure, successive approximate register (SAR) ADC is very suitable for the low power digital domain accumulation scheme.

2.2. Analysis and Optimization of the Coarse Quantization Bit Numbers and Coarse Quantization Times

To analyze and optimize the coarse quantization bit numbers and coarse quantization times of the low power accumulation scheme, a 16-stage digital domain CMOS TDI image sensor based on 10-bit column parallel SAR ADC is established. The total conversion time of a SAR ADC is divided into two parts: sampling time and quantization time. While the sampling time is negligible, it needs 11 clock cycles to complete one conversion for a 10-bit SAR ADC. In the low power accumulation scheme, the full 10-bit conversion is divided into a coarse one and a fine one. Then, the saved time of SAR ADC depends on the maximum consumption time between the coarse quantization and the fine quantization, and the saved time could be roughly calculated as

$$\text{Saved time} = \frac{11 - \max(K, M) - 1}{11} \times 100\%, \tag{1}$$

where M is the coarse quantization bit numbers and K is the fine quantization bit numbers. The relationship between the coarse quantization bit numbers and the saved time of the ADC conversion is illustrated in Figure 3. When the coarse quantization bit numbers are 5-bit, the saved time could reach a maximum value of 45.45%.

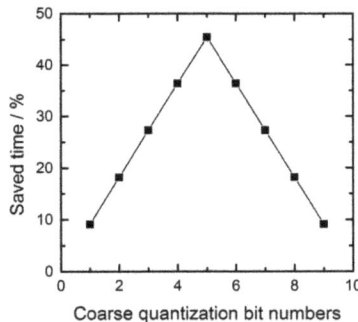

Figure 3. Relation curve between the saved time of SAR ADC and coarse quantization bit numbers.

Because the output signals of pixels are influenced by the pixel noise and readout circuit noise, the precision of the coarse quantization decreases with the increase of the coarse quantization bit numbers. Therefore, the precision of the coarse quantization should be taken into account considering the saved time. More than one line of pixels can be used to repeat the coarse quantization procedure to improve the precision. A behavioral simulation is implemented with MATLAB to explore the precision difference between the traditional accumulation scheme and the low power accumulation scheme that adopts different coarse quantization bit numbers and different lines of pixels for coarse quantization. The simulation result is shown in Figure 4. The difference between the results of the proposed scheme and the traditional digital accumulation scheme is indicated by probability that the quantization results of the proposed scheme can reach that of the traditional one. As shown in Figure 4, the probability decreases with the increase of the coarse quantization bit numbers and it is proportional to the lines of pixels for coarse quantization. Considering the quantization precision, chip area and saved time, 4-bit coarse quantization and four-line pixels for coarse quantization are selected for a 16-stage low power digital domain CMOS TDI image sensor with 10-bit SAR ADC. The analysis method is suitable for digital domain CMOS TDI image sensors with various accumulation stages.

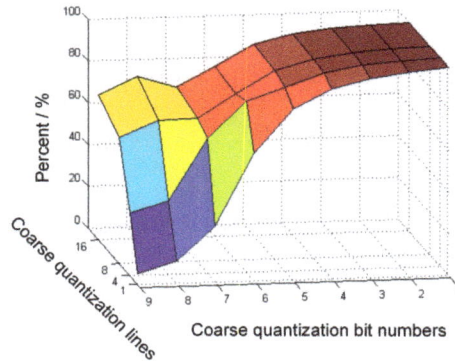

Figure 4. Comparing results between the low power accumulation scheme and the traditional accumulation scheme.

As the sampling time of SAR ADC is taken into consideration, a complete conversion needs $(11 + H)$ clock periods for a 10-bit SAR ADC in the actual design, where H is the number of clocks for sampling. If an N-stage low power digital domain CMOS TDI image sensor adopts the above optimized parameters, a complete N-stage accumulation consumes $(N + 4) \times (11 + H - 4)$ clock periods. With the same accumulation stages, a complete N-stage accumulation consumes $N \times (11 + H)$ clock periods using the traditional accumulation scheme. If the two kinds of image sensors work in the same conversion rate, the ratio of the line rate between them could be expressed as

$$\text{line rate ratio} = \frac{N \times (11 + H)}{(N + 4) \times (11 - 4 + H)}. \tag{2}$$

Since the power consumption is proportional to the line rate, the low power accumulation scheme could achieve lower power consumption than the traditional accumulation scheme with the same line rate. The saved power consumption increases with the increase of accumulation stages for the fixed coarse quantization bit numbers and coarse quantization times.

2.3. MATLAB Behavioral Simulation

In order to verify the efficiency of the low power accumulation technique, a behavioral simulation is completed with MATLAB to compare the imaging quality between the proposed and the traditional 16-stage digital domain accumulation schemes. The brief block diagram of the behavioral model is shown in Figure 5. First, a 256×256 greyscale image is obtained under the low illumination condition, and then the greyscale image is transferred to analog voltage array according to the performance of SAR ADC. The terms 'V_{ref}' and 'G' shown in Figure 5 indicate the quantization range and precision of SAR ADC, respectively. According to the voltage information, the Gaussian noise is mapped to the greyscale image. Then, the greyscale images with Gaussian noise are amplified by digital programmable gain amplifier (DPGA) and are handled with the two kinds of accumulation schemes. The greyscale image with auxiliary Gaussian noise is shown in Figure 6a and the image that is amplified by DPGA is shown in Figure 6b. Figure 6c,d illustrate the reconstructed images processed with the two kinds of accumulation schemes. There are no more differences between the two images from the visual effect.

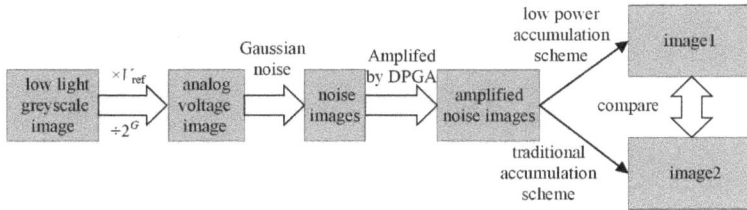

Figure 5. Block diagram of the MATLAB behavioral simulation.

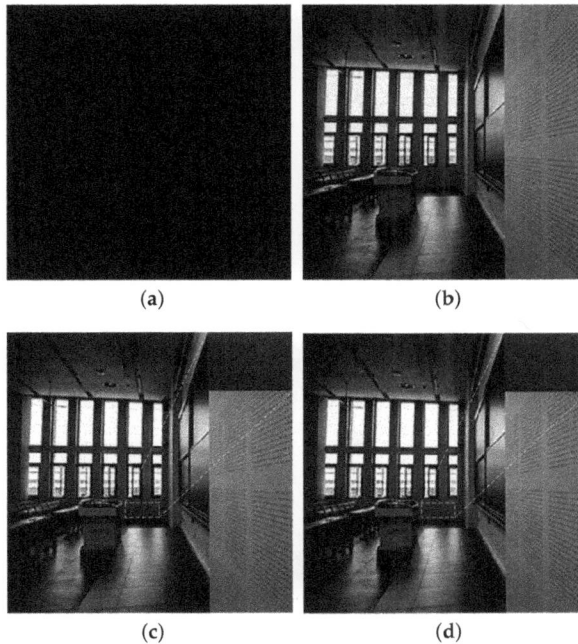

(a)

(b)

(c)

(d)

Figure 6. Greyscale images (**a**) Greyscale image with auxiliary noise; (**b**) Amplified image: (**c**) Reconstructed image processed by the traditional accumulation scheme; and (**d**) Reconstructed image processed by the low power accumulation scheme.

3. Circuit Implementation and Simulation Results

3.1. Circuit Implementation

In this work, two versions of 16-stage digital domain CMOS TDI image sensor chains, with and without the low power accumulation technique, are designed in the 0.18 μm CMOS process. The image sensor chain consists of one column pixel and one column readout circuit. According to the analysis in Section 2, 4-bit coarse quantization and four lines of pixels for coarse quantization are selected in the 16-stage low power version. The low power digital domain CMOS TDI image sensor chain is shown in

Figure 7. It consists of a pixel column, a DPGA, a 10-bit SAR ADC and a digital accumulator. The pixel column consists of a coarse quantization pixel column and a fine quantization pixel column.

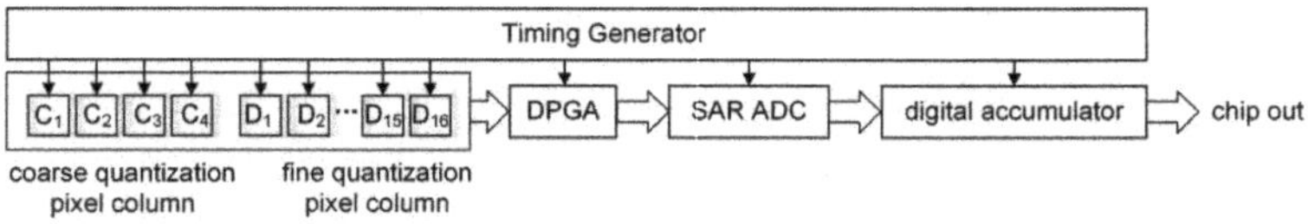

Figure 7. Block diagram of the 16-stage low power digital domain CMOS TDI image sensor chain.

3.1.1. Pixel and DPGA

As is shown in Figure 8, the pixel circuit consists of a normal 3T structure and a class AB output stage for driving the large load capacitor provided by the DPGA. The schematic of the DPGA is shown in Figure 9. It is composed of some switches, one group capacitor of $C_2 = C_2'$, one group programmable capacitor array of $C_{sp} = C_{sn}$, and a two-stage fully differential operational amplifier (OPA) [15] with hybrid cascade compensation technique [16]. The simulation results of the OPA are shown in Table 1 with a 3.3 V supply voltage and 5 pF capacitance load. The corners 'ss', 'tt' and 'ff' are defined as (NMOS: slow and PMOS: slow), (NMOS: typical and PMOS: typical) and (NMOS: fast and PMOS: fast).

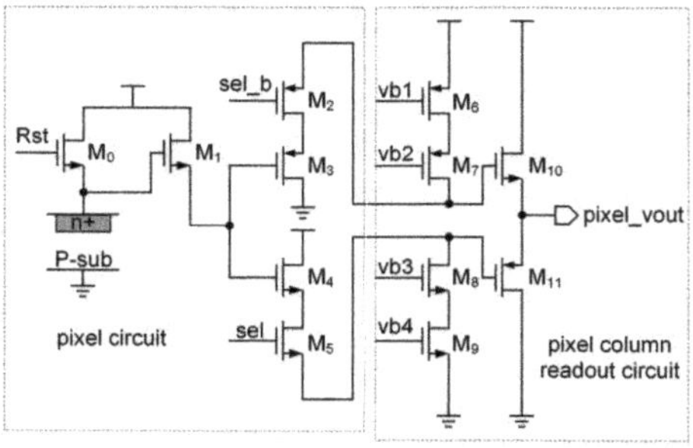

Figure 8. Schematic of the pixel circuit.

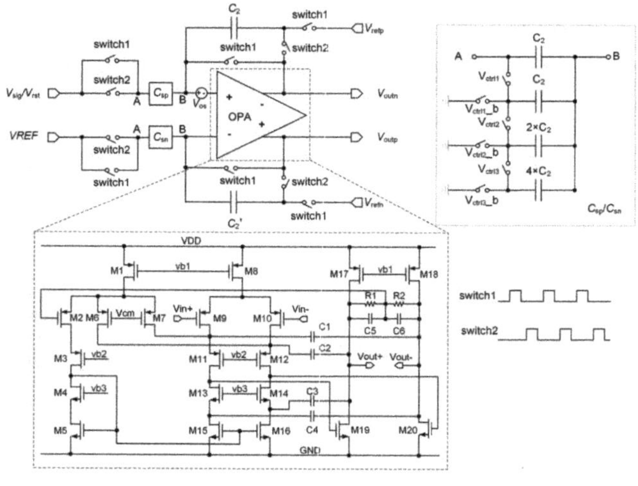

Figure 9. Schematic of the DPGA.

Table 1. Simulation results of the OPA.

Corner	Temperature	Supply Voltage of OPA	Output Voltage Swing of OPA	Unity Gain-Bandwidth	Gain	Phase Margin
tt	−40°C–80°C	2.97 V –3.63 V	0.85 V–2.45 V	135 MHz–167 MHz	105 dB–117 dB	68°–69°
ss	−40°C–80°C	2.97 V –3.63 V	0.85 V–2.45 V	118 MHz–161 MHz	107 dB–117 dB	67°–69°
ff	−40°C–80°C	2.97 V –3.63 V	0.85 V–2.45 V	124 MHz–173 MHz	97 dB–116 dB	68°–71°

3.1.2. SAR ADC

The SAR ADC uses a split-capacitor structure with redundancy and a digital correction technique [17]. The schematic of the SAR ADC is shown in Figure 10. It is composed of a digital-to-analog converter (DAC), a comparator and one SAR logic. The comparator of the SAR ADC uses three cascaded low-gain amplifiers as the preamplifier and a latch, applying a dynamic input offset cancellation technique [18]. Thus, the input offset is quite small [19]. Figure 11a shows the simulation results of the fast Fourier transform (FFT) spectrum of the SAR ADC with an input frequency close to 400 KHz at a 1.96 MS/s sampling rate. Figure 11b plots the simulation results of total harmonic distortion (THD), spurious free dynamic range (SFDR), SNR, signal to noise distortion ratio (SNDR) versus the input frequency at 1.96 MS/s sampling rate.

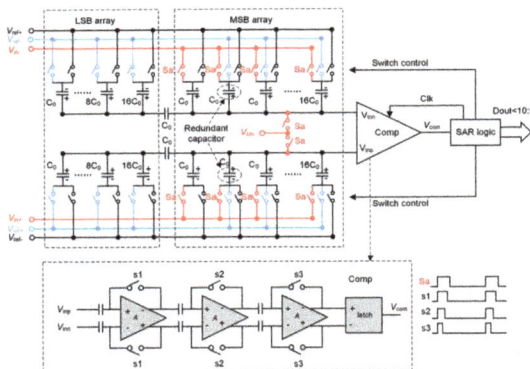

Figure 10. Schematic of the SAR ADC.

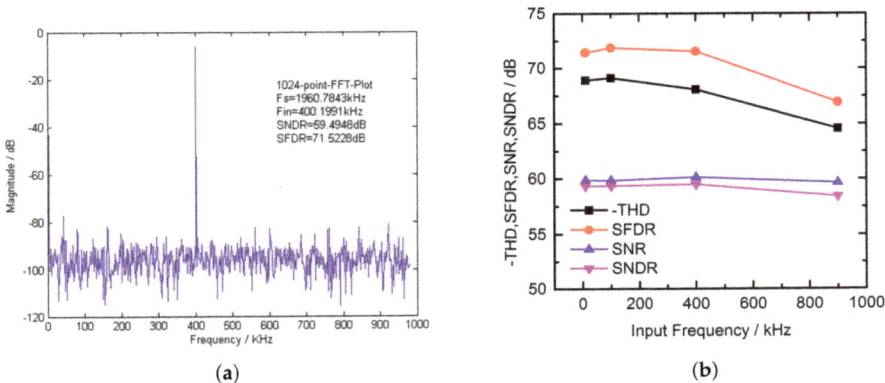

(a)   (b)

Figure 11. Simulation results of the SAR ADC (**a**) 1024-point FFT spectrum at 1.96 MS/s of SAR ADC; and (**b**) dynamic performance of SAR ADC versus input frequency.

3.1.3. Digital Accumulator

The block diagram of the digital accumulator is shown in Figure 12. It consists of a 10-bit digital adder, a multiplexer, a memory array A, a memory array B, a combination adder, a divider, a round divider, three one-way buses (bus1 to bus3), and three two-way buses (bus4 to bus6). The memory array A is used to store the coarse quantization results, and the memory array B is used to store the fine quantization results. The combination adder is used to combine separate quantization results, and the round divider is used to calculate the round average value of the accumulated high-bit codes. As the discussion above, the coarse quantization pixel column consists of four lines (C_1–C_4) of pixels and the fine quantization pixel column is composed of 16 lines (D_1–D_{16}) of pixels. Based on the oversampling exposure method [20], $2 \times (4 + 6 + 1)$ memories are needed to store the results of coarse quantization and fine quantization.

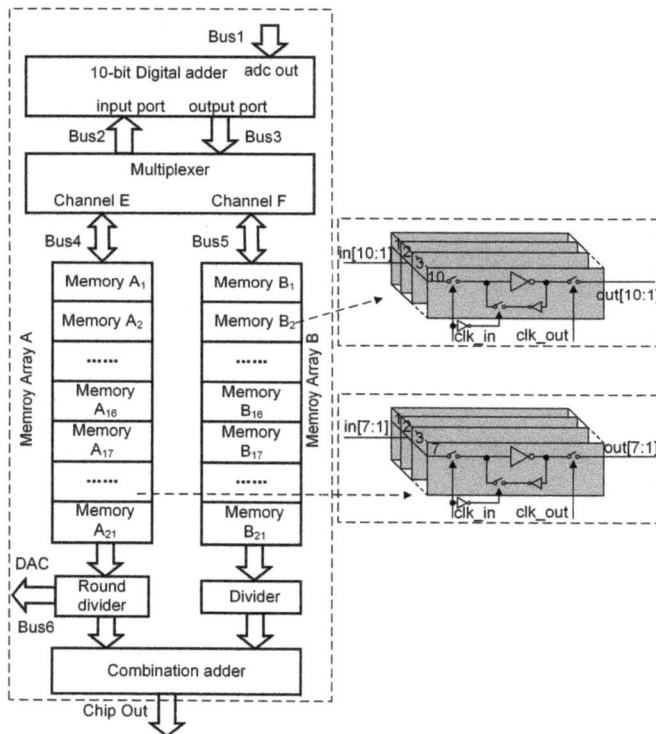

Figure 12. Block diagram of the digital accumulator.

The detailed operation procedure of the digital accumulator is described as follows. Because the SAR ADC uses a redundancy method, the actual bit numbers of coarse quantization increase to 5 bits. When the image sensor starts to work, the four lines coarse quantization pixels (C_1–C_4) perform in advance to capture the object four times. The output result of the C_{1st} line of pixels undergoes the coarse quantization by SAR ADC, and then the quantized high 5-bit data are stored in the memory A_{16}. After that, the output result from the C_{2nd} line of pixels also completes the coarse quantization through SAR ADC, and the quantized digital codes are transferred to the digital adder via bus1. The data stored in the memory A_{16} is also delivered to the digital adder through bus4, multiplexer and bus2. Then, the output of the digital adder is sent back to the memory A_{16} through bus3, multiplexer and bus4.

With the same technique, the signals output by four lines of coarse quantization pixels are quantized and accumulated. Then, the accumulated result is stored in the memory A_{16}.

During the fine quantization, after the D_{1st} line of pixels finishes the exposure of the same object, the result is sampled by SAR ADC. In the meantime, the round average value of the accumulated data stored in the memory A_{16} is sent back to the DAC of SAR ADC through bus6. Based on the coarse quantization results, the SAR ADC performs the remaining successive approximation to the analog input signal, which is defined as the fine quantization. The quantized digital codes are stored in the memory B_{16}. In a similar manner, the signal stored in the memory B_{16} is accumulated 16 times. Then, the round average value of the data stored in the memory A_{16} is delivered to an input port of the combination adder. Meanwhile, the data stored in the memory B_{16} is divided by 16, and the quotient is also delivered to the other port of the combination adder. Finally, the output of the combination adder is served as the complete quantization result of the scanned object.

3.1.4. Noise Analysis of the Low Power Readout Circuit

The noise of the low power digital domain readout circuit is mainly contributed by the pixel circuit, the DPGA and the SAR ADC. There are two important intrinsic noise sources: thermal noise and flicker noise [21]. The flicker noise of the pixel circuit and DPGA are reduced to a negligible level by using theinut input–offset storing technique [22]. Because the pixel uses a 3T structure, *KTC* noise could not be eliminated. Benefitting from the DPGA, the total equivalent input noise of DPGA and SAR ADC is divided by the amplified magnitude of the DPGA. Therefore, we mainly concentrate on the effect of the thermal noise contributed by the source follower. Assuming that the source follower is equivalent to a unity-gain amplifier and the total equivalent output power spectral density of the noise of the source follower could be expressed as

$$
\begin{aligned}
V_{nSF\_out}(f)^2 &= 4kT\gamma \times \left(\frac{1}{g_{m1}} + \frac{1}{g_{m3}} + \frac{1}{g_{m4}} + \frac{1}{g_{m10}} + \frac{1}{g_{m11}}\right) \\
&+ \frac{4kT\gamma}{g_{m3}^2} \times (g_{m2} + g_{m6} + g_{m7}) + \frac{4kT\gamma}{g_{m4}^2} \times (g_{m5} + g_{m8} + g_{m9}),
\end{aligned}
\tag{3}
$$

where k is the Boltzmann constant, T is the absolute temperature, and γ is a coefficient of process. Since the source follower could be equivalent to a single-pole system, the equivalent bandwidth of noise could be expressed as

$$
BW_{nSF\_equ} = \frac{1}{2} \times \frac{g_{m1}}{2 \times (C_{bus} + C_s)},
\tag{4}
$$

where C_{bus} and C_s are the total node parasitic capacitance and load capacitance of the pixel readout circuit. Thus, the total equivalent output noise of the source follower could be expressed as

$$
\begin{aligned}
\overline{V_{nSF\_out}}^2 &= \int_1^{BW_{nSF\_equ}} V_{nSF\_out}(f)^2 df = \frac{g_{m1} \times \xi_1}{(C_{bus} + C_s)} \\
\xi_1 &= kT\gamma \times \left(\frac{1}{g_{m1}} + \frac{1}{g_{m3}} + \frac{1}{g_{m4}} + \frac{1}{g_{m10}} + \frac{1}{g_{m11}} + \frac{g_{m2}}{g_{m3}^2} + \frac{g_{m6}}{g_{m3}^2} + \frac{g_{m7}}{g_{m3}^2} + \frac{g_{m5}}{g_{m4}^2} + \frac{g_{m8}}{g_{m4}^2} + \frac{g_{m9}}{g_{m4}^2}\right).
\end{aligned}
\tag{5}
$$

The noise optimization of the source follower is completed by properly increasing g_{m3-4}, g_{m10-11} and reducing g_{m2}, g_{m5-9}. Since the output signals of pixels are amplified H times, the total equivalent input noise contributed by the readout circuit after N times accumulation could be expressed as

$$
\overline{V_{Ntot}} = \sqrt{N \times \left(\frac{kT}{C_{PD}} + \overline{V_{nSF\_out}}^2\right) + \frac{N}{H^2} \times \left(\overline{V_{nDPGA\_out}}^2 + \overline{V_{nADC\_out}}^2\right)},
\tag{6}
$$

where C_{PD} is the photo-diode node capacitance, $\overline{V_{nDPGA\_out}}^2$ is the noise energy contributed by the DPGA, and $\overline{V_{nADC\_out}}^2$ is the noise energy contributed by the SAR ADC.

3.2. Linearity Simulation Results and Power Consumption Analysis

According to the non-linearity measured method of CMOS image sensors [23], the linearity is defined as

$$\text{Linearity} = (1 - \frac{\Delta V_{max}}{V_{max}}) \times 100\%,$$ (7)

where the ΔV_{max} is the maximum difference between the simulated and fitted transmission curves and the V_{max} is the full scale. The linearity simulation results of the two versions are shown in Figure 13. The linearity of the image sensor chains are 99.74% with the low power accumulation scheme and 99.99% with the traditional accumulation scheme.

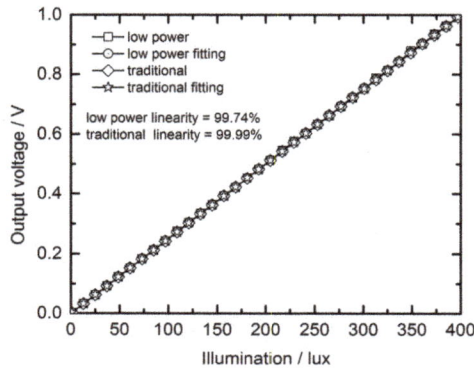

Figure 13. Linearity simulation results of the two versions.

The total power consumption of the image sensor chain is contributed by the DPGA, the SAR ADC, the accumulator, the pixel circuit and other circuits. The other circuits include a brief current generation circuit and some digital buffers. The distributions of the power consumptions of the two versions are shown in Figure 14. The comparisons of some main parameters of the two kinds of image sensor chains with our prior works are shown in Table 2. Compared with our prior works and the traditional accumulation scheme, the proposed low power digital accumulation technique can achieve lower average power consumption of each slice. The average power consumption of each slice is defined as

$$\text{average power consumption of slice} = \frac{\text{total power consumption}}{(\text{horizontal resolution}) \times (\text{line rate})}.$$ (8)

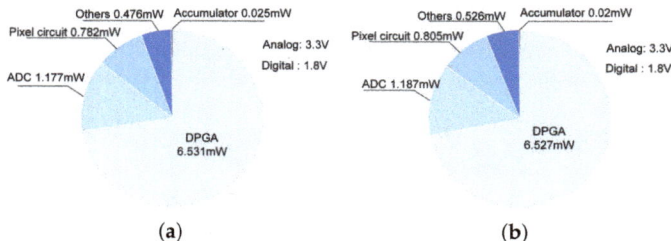

(a)   (b)

Figure 14. Power consumption distributions of the two kinds of image sensor chains (a) low power accumulation scheme; and (b) traditional accumulation scheme.

Table 2. Comparisons of some main parameters of the two kinds of image sensor chains with prior works. (H-Horizontal, V-Vertical).

Parameter	This Work		[10]	[14]
	Low Power	Traditional		
Technology	0.18 μm CMOS	0.18 μm CMOS	0.18 μm CMOS	0.18 μm CMOS
Array size	1 (H)×20 (V)	1 (H) ×16 (V)	1024 (H)×128 (V)	1024(H)×128 (V)
Maximum stage	16	16	128	128
Maximum line rate	138888 lines/s	122549 lines/s	3875 lines/s	3875 lines/s
Total power consumption	8.991 mW	9.065 mW	500 mW	290 mW
Average power consumption of slice	6.47×10^{-8} J/line	7.4×10^{-8} J/line	12.6×10^{-8} J/line	7.3×10^{-8} J/line

4. Conclusions

In this paper, a low power digital accumulation technique for a digital domain CMOS TDI image sensor is proposed to reduce power consumption without degrading the rate of imaging. The low power accumulation scheme is composed of coarse quantization and fine quantization. The optimized bit numbers and times of coarse quantization are analyzed. With the 0.18 μm CMOS process, two versions of 16-stage digital domain CMOS TDI image sensor chains, with and without the proposed technique, were designed. The simulation results show that the average power consumption of each slice are 6.47×10^{-8} J/line and 7.4×10^{-8} J/line, respectively. The linearity of the image sensor chains are 99.74% with the low power accumulation scheme and 99.99% with the traditional accumulation scheme.

Acknowledgments: This work is supported by the National Natural Science Foundation of China (61504091, 61404090).

Author Contributions: Changwei Yu and Kaiming Nie conceived the ideas and innovations; Changwei Yu and Kaiming Nie performed the simulations; Jiangtao Xu and Jing Gao provided supervision and guidance in this work; and Changwei Yu wrote the paper.

Conflicts of Interest: The authors declare no conflict of interest.

References

1. Farrier, M.G.; Dyck, R.H. A large area TDI image sensor for low light level imaging. *IEEE J. Solid-State Circuits* **1980**, *15*, 753–758.
2. Barbe, D.F. Time delay and integration image sensors. In *Solid State Imaging*; Jespers, P.G., van de Wiele, F., White, M.H., Eds.; Noordhoff: Leyden, MA, USA, 1976; pp. 659–671.
3. Wong, H.S.; Yao, Y.L.; Schlig, E.S. TDI charge-coupled devices: Design and applications. *IBM J. Res. Dev.* **1992**, *36*, 83–106.
4. Gamal, A.E.; Eltoukhy, H. CMOS image sensors. *IEEE Circuits Devices Mag.* **2005**, *21*, 6–20.
5. Mayer, F.; Bugnet, H.; Pesenti, S.; Guicherd, C.; Gili, B.; Bell, R.; Monte, B.D.; Ligozat, T. First Measurements of True Charge Transfer TDI (Time Delay Integration) Using a standard CMOS technology. In Proceedings of the International Conference on Space Optics, Ajaccio, France, 9–12 October 2012.
6. Ercan, A.; Haspeslagh, L.; Munck, K.D.; Minoglou, K.; Lauwers, A.; Moor, P.D. Prototype TDI sensors in embedded CCD in CMOS technology. In Proceedings of the International Image Sensor Workshop, Snowbird Resort, UT, USA, 12–16 June 2013.
7. Moor, P.D.; Robbelein, J.; Haspeslagh, L.; Boulenc, P.; Ercan, A.; Minoglou, K.; Lauwers, A.; Munk, K.D.; Rosmeulen, M. Enhanced time delay integration imaging using embedded CCD in CMOS technology. In Processings of the 2014 IEEE Electron Devices Meeting (IEDM), San Franciscom, CA, USA, 15–17 December 2014.
8. Mayer, F.; Pesenti, S.; Barbier, F.; Bugnet, H.; Endicott, F.; Ligozat, T. CMOS Charge Transfer TDI with Front Side Enhanced Quantum Efficiency. In Proceedings of the International Image Sensor Workshop, Vaals, The Netherlands, 8–11 June 2015.
9. Nie, K.; Yao, S.; Xu, J.; Gao, J. Thirty two-stage CMOS TDI image sensor with on-chip analog accumulator. *IEEE Trans. Very Large Scale Integr. (VLSI) Syst.* **2014**, *22*, 951–956.

Sensors **2016**, *16*, 1572

10. Nie, K.; Yao, S.; Xu, J.; Gao, J.; Xia, Y. A 128-stage analog accumulator for CMOS TDI image sensor. *IEEE Trans. Circuits Syst. I.* **2014**, *61*, 1952–1961.

11. Xia, Y.; Nie, K.; Xu, J.; Yao, S. A two-step analog accumulator for CMOS TDI image sensor with temporal undersampling exposure method. *IEEE Trans. Very Large Scale Integr. (VLSI) Syst.* **2016**, *34*, 1104–1117.

12. Lepage, G.; Dantès, D.; Diels, W. CMOS long linear array for space application. In Proceedings of the International Society for Optical Engineering, San Jose, CA, USA, 17–19 January 2014.

13. Materne, A. CNES developments of key detection technologies to prepare next generation focal planes for high resolution Earth observation. In Proceedings of the International Society for Optical Engineering, Amsterdam, The Netherlands, 22–25 September 2014.

14. Nie, K.; Xu, J.; Gao, Z. A 128-Stage CMOS TDI image sensor with on-chip digital accumulator. *IEEE Sens. J.* **2016**, *16*, 1319–1324.

15. Banu, M.; Khoury, J.M.; Tsividis, Y. Fully differential operational amplifiers with accurate output balancing. *IEEE J. Solid-State Circuits* **1988**, *23*, 71410–1414.

16. Yavari, M.; Shoaei, O.; Svelto, F. Hybrid cascode compensation for two-stage CMOS operational amplifiers. In Proceedings of the IEEE International Symposium on Circuits and Systems, Kobe, Japan, 23–26 May 2005.

17. Chang, A.H.; Lee, H.S.; Boning, D. A 12 b 50 Ms/s 2.1 mW SAR ADC with redundancy and digital background calibration. In Proceedings of thr IEEE 39th European Solid State Circuits Conference, Bucharest, Romanian, 16–20 September 2013.

18. Doernberg, J.; Gray, P.R.; Hodges, D.A. A 10-bit 5-Msample/s CMOS two-step flash ADC. *IEEE J. Solid-State Circuits* **1989**, *24*, 241–249.

19. Lyu, T.; Yao, S.; Nie, K.; Xu, J. A 12-Bit High-Speed Column-Parallel Two-Step Single-Slope Analog-to-Digital Converter (ADC) for CMOS Image Sensors. *Sensors* **2014**, *13*, 21603–21625.

20. Lepage, G.; Bogaerts, J.; Meynants, G. Time-delay-integration architectures in CMOS image sensor. *IEEE Trans. Electron Devices* **2009**, *56*, 2525–2533.

21. Schreier, R.; Silva, J.; Steensgaard, J.; Temes, G.C. Design-oriented estimation of thermal noise in switch-capacitor circuits. *IEEE Trans. Circuits Syst. I.* **2005**, *52*, 2358–2368.

22. Razavi, B. *Design of Analog CMOS Integrated Circuits*, 1st ed.; McGraw-Hill: New York, NY, USA, 2000; pp. 471–477.

23. Theuwissen, A. How to Measure Non-Linearity. Available online: http://harvestimaging.com/blog/?p=1125 (accessed on 31 October 2014).

Article

A 75-ps Gated CMOS Image Sensor with Low Parasitic Light Sensitivity

Fan Zhang and Hanben Niu *

Key Laboratory of Optoelectronic Devices and Systems of Education Ministry and Guangdong Province,
College of Optoelectronic Engineering, Shenzhen University, Shenzhen 518060, China;
zhangfan3@email.szu.edu.cn
* Correspondence: hbniu@szu.edu.cn; Tel.: +86-755-2653-8579 (ext. 859)

Academic Editor: Gonzalo Pajares Martinsanz
Received: 3 April 2016; Accepted: 22 June 2016; Published: 29 June 2016

Abstract: In this study, a 40×48 pixel global shutter complementary metal-oxide-semiconductor (CMOS) image sensor with an adjustable shutter time as low as 75 ps was implemented using a 0.5-µm mixed-signal CMOS process. The implementation consisted of a continuous contact ring around each p+/n-well photodiode in the pixel array in order to apply sufficient light shielding. The parasitic light sensitivity of the in-pixel storage node was measured to be $1/8.5 \times 10^7$ when illuminated by a 405-nm diode laser and $1/1.4 \times 10^4$ when illuminated by a 650-nm diode laser. The pixel pitch was 24 µm, the size of the square p+/n-well photodiode in each pixel was 7 µm per side, the measured random readout noise was 217 e$^-$ rms, and the measured dynamic range of the pixel of the designed chip was 5500:1. The type of gated CMOS image sensor (CIS) that is proposed here can be used in ultra-fast framing cameras to observe non-repeatable fast-evolving phenomena.

Keywords: CMOS image sensor (CIS); gated imager; snapshot imager; ultra-fast global shutter; framing camera; low parasitic light sensitivity; high shutter efficiency

1. Introduction

Fast gated or global shutter cameras with shutter time at a level of tens of picoseconds are widely used in the observation of fast-evolving phenomena, including repeatable and non-repeatable processes. Traditionally, micro-channel plate (MCP)-based gating cameras are used in range imaging systems (time-of-flight depth cameras) and wide-field fluorescence-lifetime imaging microscopy to observe repeatable fast evolving phenomena. In recent years, a large number of solid-state devices have been developed for such applications [1–5].

Currently, MCP-based gated cameras are almost the only type of receive-only 2D imaging device used in applications that require the observation of non-repeatable fast-evolving phenomena, with a time resolution as little as approximately 35 ps [6]. Such applications include plasma expansion dynamics research, charged particle accelerator diagnosis, optical time-of-flight measurements of fast moving objects, and high-resolution photo-acoustic imaging. However, some other successful efforts have also been presented for these purposes that use streak cameras [7] or that rely upon light-absorption-induced modulation of the optical refractive index of a semiconductor sensor medium [8].

A pulse-dilation enhanced gated optical imager can achieve a time resolution of approximately 5 ps [9–11], which is an overwhelmingly high speed for receive-only 2D imaging. The drawback of such a device is that it is bulky in size, sensitive to magnetic fields, and relatively low in spatial resolution.

Very fast global shutter complementary metal-oxide-semiconductor (CMOS) readout test chips with on-chip photodiodes have also been implemented [12]. Tests to measure the minimum exposure time when using on-chip photodiodes have achieved results of approximately 200 ps [13], but without any reports on the successful implementation of the sensor chip it requires, their use in practical applications is still limited. To the best of the authors' knowledge, no reports have been given on the parasitic light sensitivity of the global shutter CMOS readout chip while using the on-chip photodiodes.

If the parasitic light sensitivity does not meet requirements, artifacts will be captured within an image from bright moving objects or light spots after exposure and before the readout [14].

In this paper, the authors present the detailed design, test methods, and results of a low parasitic light sensitivity 40 × 48 pixel gated CMOS image sensor that is sensitive to visible and near ultraviolet light with a shutter time as low as 75 ps, which is manufactured using a 0.5-µm 2-poly 3-metal polycide mixed-signal CMOS process. The type of CMOS image sensor proposed in this paper can be used in ultra-fast framing cameras to observe single-shot fast-evolving phenomena.

2. Pixel Circuit Design

In order to obtain the desirable fast photo response, p+/n-well photodiodes are used in the pixel array. Semiconductor processes and device simulations have shown that a small-sized p+/n-well photodiode manufactured using this process has an impulse response time that is shorter than approximately 5 ps, for visible light or near ultraviolet, at a bias of 5 V. Figure 1a shows the simulated photocurrent response of the small-sized p+/n-well photodiode at a 5-V bias after illumination by 1 ps short light pulses with 0.2 pJ of energy. The software used in the device simulation was Silvaco Atlas, and Figure 1b shows the structure and the doping profile of the photodiode used in the device simulation. A wavelength of 558 nm was used in the simulation, which is the maximum emission of a bright ultra-fast scintillator (n-$C_6H_{13}NH_3$)$_2$$PbI_4$, which is in a natural multiple quantum well (MQW) structure and has a decay component of 390 ps (30%) at room temperature [15]. X-ray or electron sensitivity can be achieved by coupling the proposed image sensor to this type of bright ultra-fast scintillator screen by microscopy.

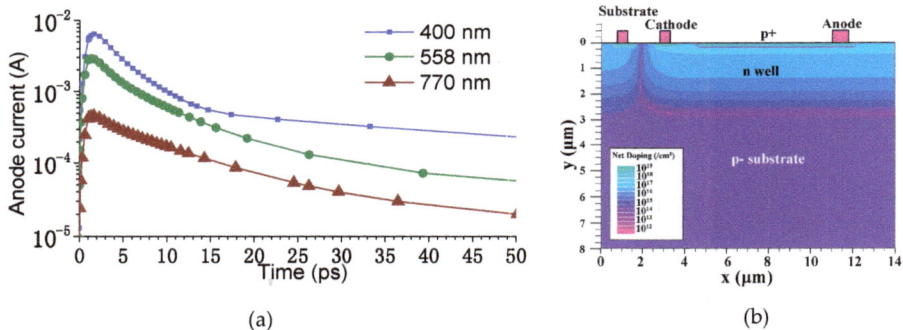

Figure 1. (**a**) The simulated transient response of the p+/n-well photodiode at three different wavelengths; (**b**) The structure of the photodiode used in the transient simulation.

The circuit diagram for a single pixel is shown in Figure 2a. In the design, V_{reset}, V_{start}, and V_{end} are set to the same value as V_{DD}, and V_{select} is set to ground when waiting for the trigger signal. Thus, only the transistors M1 and M2 are turned on. Once triggered, V_{start} and subsequently V_{end} are pulled down to ground in the sequence, within a time interval that is slightly shorter than the exposure time. During exposure, when M1 is turned off and M2 is still on, part of the photo-induced charge is stored on the polysilicon-insulator-polysilicon (PIP) capacitor C1. After exposure, all five transistors of the pixel are turned off. After approximately 14 nanoseconds, V_{reset} is pulled down to ground, thus pulling up the anode of the photodiode to V_{DD} and forming the final signal voltage on the gate of M3 for the read-out.

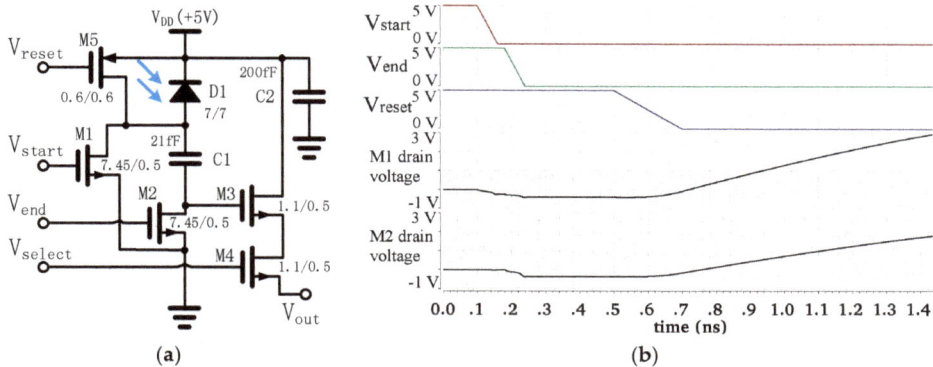

Figure 2. (a) The pixel circuit schematic; (b) The timing chart of the pixel and the simulated results of the drain voltage of M1 and M2.

A timing chart for V_{start}, V_{end}, V_{reset}, and the simulated results of the drain voltages of M1 and M2 is shown in Figure 2b. In the simulation, a 6.25-fF capacitor representing the parasitic capacitance was added between the bottom plate of C1 and the ground. Figure 2b shows that the M2 drain voltage drops to around -0.396 V after V_{end} has been pulled down to ground. This value is acceptable as the simulated leakage current of the 0.396-V forward-biased p− substrate/n+ drain diode of M2 is only 216 pA at room temperature. The relatively large parasitic capacitance between the bottom plate of C1 and the grounded p− substrate is the key to keeping the M2 drain voltage from dropping deeper, while it also restricts the sensitivity of the proposed image sensor.

In more recently developed global shutter CMOS image sensors, a photodiode substrate and an in-pixel storage node substrate is interconnected by microbumps to achieve excellent parasitic light sensitivity [14]. In an ultra-fast gated CMOS image sensor, this type of strategy is not the best choice as the parasitic capacitance of the microbump interconnections is too large. Instead, sufficient light shield, and some shield for the carriers, is applied to a single-chip CMOS image sensor to achieve a low enough level of parasitic light sensitivity. In the proposed CMOS image sensor, the entire area in the pixel array, with the exception of the photodiodes, is shielded by the top metal layer in order to achieve high shutter efficiency. Furthermore, for each pixel, a continuous contact ring is included in the design, which is in contact with the n+ active area within the photodiode n-well and surrounding the photodiode p+ active area, in order to achieve superior light shielding efficiency. The anode of the photodiode (the p+ area) is led out by metalized polysilicon through a small opening on the contact ring. There are also continuous via rings between the metal layers (1,2) and (2,3) surrounding the photodiode without any openings. Although using continuous contact rings or via rings in the circuit violates the topological design rule from the foundry, the proposed design works well. It is based on the 0.5-μm CMOS process without any changes to the default process parameters. The 0.5-μm CMOS process that is used to fabricate the proposed chip does not include any chemical mechanical planarization (CMP) processing.

The layout of a couple of pixels in the pixel array is shown in Figure 3a, and a cross-sectional diagram of the photodiode in the pixel is shown in Figure 3b. Transistors M1 and M2 of the pixels in the even and odd columns share the same active areas. Thus, the drain of transistor M2 is far away from the nearest contact opening, which helps provide sufficient light shielding to the drain of transistor M2. This approach also simplifies the layout of the vertical clock tree in the pixel array.

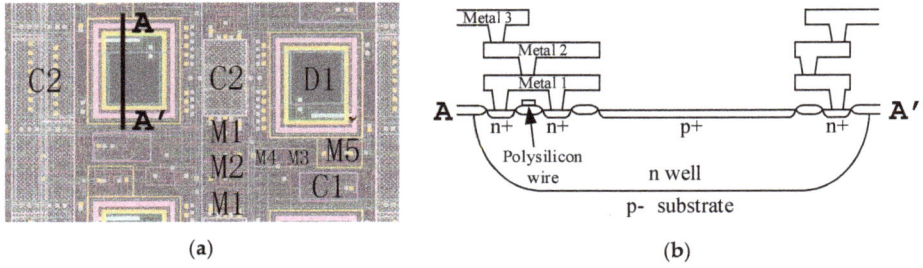

(a)   (b)

Figure 3. (**a**) The layout of two pixels; (**b**) Cross-sectional diagram of the p+/n-well photodiode in a pixel.

In order to minimize the drain capacitance, both gates of transistors M1 and M2 are configured in a square annular structure. This also provides some shield for the photoelectrons for the drain of M2 and increases the shutter efficiency. The authors also used several depletion NMOS capacitors with an approximate total value of 200 fF in each pixel for power-decoupling purposes.

3. Limitations of Shortest Shutter Time

Figure 4a shows a simplified circuit model of a pixel before exposure at the moment when transistors M1 and M2 are both on. C_d includes the capacitance from the photodiode D1 and the transistors M1 and M5. C_1 represents the capacitance of the sampling capacitor C_1 in Figure 2a. R_1 and R_2 represent the on-state resistance of the transistors M1 and M2, respectively. It can be assumed that the pulse current source I_d emits a short enough current pulse with a total charge of Q_p before exposure. The time interval between the current pulse and the start of exposure is t_1, and Q_1 is the charge on capacitor C_1 after the shutter has remained in the "open" state, as shown in Figure 4b, for sufficient time. For simplicity, let $C = C_d = C_1$, and then Q_1 can be expressed as

$$Q_1 = \frac{(2R_1 - R_2)\sinh(\tau t_1) + 2R_1 R_2 C\tau \cdot \cosh(\tau t_1)}{4R_1 R_2 C\tau} \cdot e^{-\frac{2R_1 + R_2}{2R_1 R_2 C} t_1} \cdot Q_p \tag{1}$$

$$\tau = \frac{\sqrt{4R_1{}^2 + R_2{}^2}}{2R_1 R_2 C} \tag{2}$$

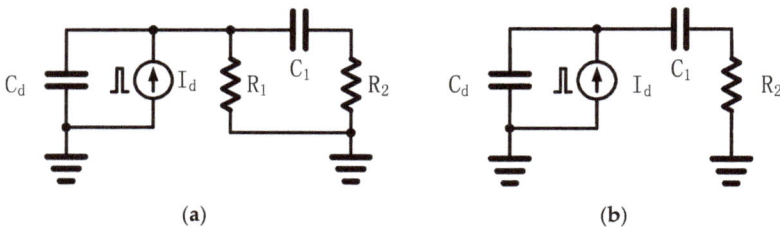

(a)   (b)

Figure 4. Simplified circuit model of a pixel. (**a**) Before exposure. (**b**) During exposure.

Q_1 will decline as t_1 increases. In the proposed design, $R_1 = R_2 = 1413\ \Omega$, and we assume $C = 21$ fF. Let t_{half1} be the value of t_1 when Q_1 has decreased to half of its maximum value. For the values given above, $t_{half1} = 31.5$ ps.

Figure 4b shows the simplified circuit model of a pixel during exposure, when transistor M1 is off and M2 is still on. It can be assumed that the current source I_d emits a short enough current pulse with a total charge Q_p during the shutter "open" state. If t_2 denotes the time interval after the current pulse and Q_2 denotes the charge on capacitor C_1 at time t_2, Q_2 can be expressed as

$$Q_2 = \frac{C_1}{C_1 + C_d}\left[1 - e^{-\frac{C_1 + C_d}{R_2 C_1 C_d}t_2}\right]Q_p \tag{3}$$

Q_2 will increase after the current pulse and eventually reaches a maximum value. Let t_{half2} be the time after the current pulse when Q_2 reaches half of its maximum value; t_{half2} can then be expressed as

$$t_{half2} = \ln(2)\frac{R_2 C_1 C_d}{C_1 + C_d} \tag{4}$$

In the proposed design, C_1 = 21 fF, R_2 = 1413 Ω, and the simulated value of C_d is 27 fF with a 5-V power supply, giving t_{half2} = 11.7 ps. The shortest shutter time of the proposed design should therefore be longer than $t_{\text{half1}} + t_{\text{half2}}$, i.e., 43.2 ps.

4. Sensor Chip Architecture

Figure 5a shows a micrograph for the designed image sensor, and Figure 5b illustrates the circuit architecture of the sensor. The exposure control signals V_{start} and V_{end} can be configured to be directly controlled by an external digital input, or alternatively the exposure process can be triggered using an external digital signal. When the exposure process is triggered by an external signal, the time between the falling edge of V_{start} and V_{end} signal (roughly the exposure time) is controlled by a voltage-controlled delayer, which is located in the exposure clock control circuits at the bottom of the chip.

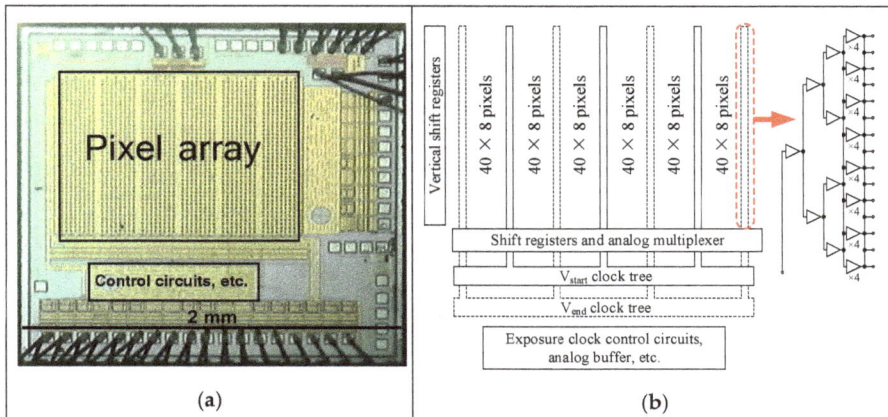

Figure 5. (a) Micrograph of the designed image sensor. (b) Structural diagram of the designed image sensor; simplified schematic of a single vertical component in the clock trees.

The exposure control signals V_{start} and V_{end} are firstly distributed across the horizontal components [4,5], then across the vertical components of the clock trees, and finally to the pixels. The vertical components of the clock trees are placed in the pixel array by pruning one row of pixels after every eight rows, as shown in Figure 5b. The three even and the four odd vertical components of the clock trees belong to the V_{start} and the V_{end} signals, respectively. A simplified schematic of the single vertical components of the clock trees is shown in the right part of Figure 5b. These clock trees consist of fast falling-edge digital buffers [12] and distributed power decoupling capacitors. This type of design of clock trees can be easily extended to large-format gated CMOS image sensors.

Since balanced clock trees with fast falling-edge digital buffers are used in the exposure control signal distribution in both the horizontal and vertical directions, and the output of the final nodes of all vertical components of the V_{start} clock tree are connected together as shown in the right part of Figure 5b, as are the horizontal components and the V_{end} clock tree, the exposure signal skew should be relatively small compared with the shortest shutter time of the small designed image sensor.

The image signal from the pixels is first multiplexed by an analog multiplexer to a voltage shifter, and it is then buffered by an on-chip analog buffer and eventually drives an off-chip analog-to-digital (A/D) converter.

5. Test Methods and Results

A test board connected to a PCI digital data acquisition board was used to test the designed chip. A 12-bit A/D converter chip operating at a 5-V input range was used on the test board. The highest speed achievable by the digital data acquisition board when operating bi-directionally is 10 M samples per second. This speed limits the A/D converter clock frequency and the sampling rate to a maximum of 5 M samples per second.

Figure 6 shows the measured photo response curve and the photo response non-uniformity (PRNU) between pixels of the proposed image sensor. The photo response curve and the PRNU were obtained by varying the exposure time, while keeping a constant uniform illumination by a blue LED. The measured PRNU for the selected area at half of the saturated voltage for all columns, odd columns only and even columns only was 1.39%, 1.42%, and 1.28%, respectively. This is normal and the difference in the pixel layouts for the odd and even columns show no significant influence on the PRNU.

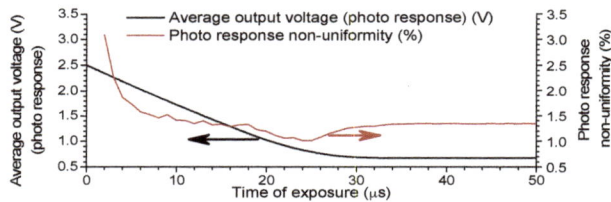

Figure 6. The measured photo response curve and the photo response non-uniformity of the designed image sensor.

To measure small signal responsivity or the charge-to-voltage gain of the designed image sensor, the central area of the pixel array was illuminated by a defocused 405-nm wavelength continuous-wave (CW) diode laser spot, as shown in Figure 7a. The difference between the measured total supply current of the chip when the 405-nm laser was on and off was the measured total photocurrent. During such measurements, the pixel array was in a state of waiting for the trigger signal, and V_{select} of all pixels was set to ground, so that the output analog buffer remained in the same state. The measured small signal responsivity of the designed chip was 1.47 $\mu V/e^-$. The linear range of the output signal was from 2.5 V to about 0.7 V, so the full capacity of the pixel was around 1,200,000 e^-. The measured random readout noise of the output signal was 475 μV rms. The quantization noise of a 12-bit readout with a 5-V full range is 352 μV rms. Thus, the random readout noise of the designed chip was 319 μV rms, which is equivalent to 217 photoelectrons generated by the photodiode. Therefore, the dynamic range of the designed chip was about 5500:1.

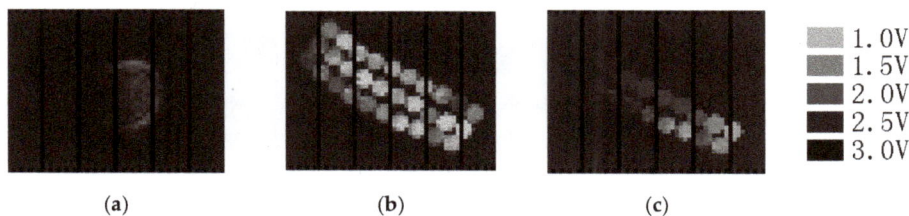

Figure 7. Images captured by the designed image sensor. (**a**) Image of the defocused laser spot used in the small signal responsivity measurement; (**b**) Image of the fiber bundle at a 17 ns shutter time. (**c**) Image of the fiber bundle at a 30 ps simulated shutter control signal delay.

The measured parasitic light sensitivity of the in-pixel storage node was very low when illuminated by a continuous-wave diode laser with a peak wavelength of 405 nm. The parasitic light sensitivity was measured by comparing the following two images. For the first image, the shutter time was set to approximately 300 ps. The exposure to the 405-nm diode laser lasted 0.5 s after the shutter was closed, and the captured image was then read out. The second image was taken by setting the shutter time to 100 ns and read out immediately after the shutter was closed. The dark image taken with the laser off was subtracted from both images to eliminate the output signal bias and fixed pattern noise. The resulting two images were then used to calculate the parasitic light sensitivity. The final measured parasitic light sensitivity when illuminated by a 405-nm diode laser was $1/8.5 \times 10^7$.

The parasitic light sensitivity when illuminated by a 650-nm continuous-wave diode laser was measured using a similar method, and the measured value was $1/1.4 \times 10^4$.

The measured leakage signal in a dark environment after the global shutter was closed was 0.7 V/s. According to the simulation, this leakage signal value is equivalent to a leakage current of approximately 22 fA on the storage node in the pixel.

The shortest shutter time (fastest shutter speed, best temporal resolution) of the designed chip was measured using a frequency-doubled 400-nm wavelength Ti:sapphire laser system with a 130-fs pulse width. The 400-nm laser flash was used to uniformly illuminate a fiber cable. The cable was composed of 30 silica fibers of different lengths [11]. The difference in length between adjacent fibers in the fiber cable was 2.0 mm. The output port of the fibers was imaged on the image sensor using the lens. During the shutter time measurement, the image sensor was triggered by a biased p-i-n photodiode outside the chip. Figure 7b shows the image that was obtained at a 1-V exposure time control voltage, which corresponds to a shutter time of 17 ns, whereas Figure 7c was obtained at a 4-V exposal time control voltage, which corresponds to a 30 ps simulated shutter control signal delay. The two images were used to obtain a normalized exposure curve, as shown in Figure 8. The measured shortest shutter time of this camera was less than 75 ps.

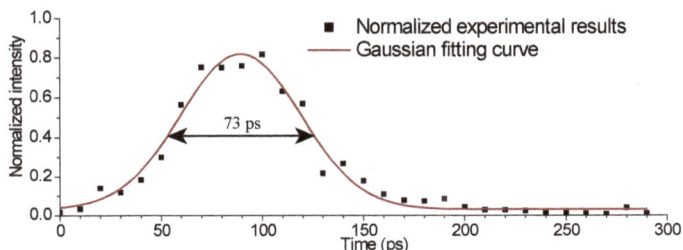

Figure 8. Exposure curve of the designed image sensor.

The characteristics and measurement results of the designed image sensor and a comparison with prior works are summarized in Table 1.

Table 1. Comparison with previous ultra-fast gated CMOS image sensors.

Reference	JSSC 2008 [12] SPIE 2012 [13]	SPIE 2012 [13]	JSSC 2016 [3]	This Work
Design aim	Test readout chip for ultra-fast gated X-ray imager	Readout chip for ultra-fast gated X-ray imager	Fluorescence lifetime imaging	Test ultra-fast gated imager for visible light
Supply voltage	1.8 V	1.8 V	–	5 V
Process	0.18-μm CMOS	0.18-μm CMOS	0.11-μm CIS CMOS	0.5-μm CMOS
Chip size	3 mm × 3 mm	> 15 × 15 mm	7.0 mm × 9.3 mm	2 mm × 2 mm
Resolution	64 × 64 pixels	512 × 512 pixels	256 × 512 pixels	40 × 48 pixels
Pixel pitch	30 μm	30 μm	11.2 μm × 5.6 μm	24 μm
Photodiode aperture area	–	–	~10.5 μm^2	6.9 μm × 6.9 μm
Power consumption	125 mW	–	540 mW	50 mW
Fixed pattern noise (rms)	9 mV	–	0.12 e$^-$ (vertical)	23.3 mV
Random readout noise	115 e$^-$	–	1.75 e$^-$	475 μV
Random readout noise (rms, with quantization noise subtracted)	–	–	–	217 e$^-$ (319 μV)
PRNU	–	–	–	1.4%
Full capacity	310, 000 e$^-$	–	2,700 e$^-$	1,200,000 e$^-$
Small signal responsivity	11 μV/e$^-$	–	85 μV/e$^-$	1.47 μV/e$^-$
Output swing	0.8 V	–	0.3 V	1.8 V
Leakage signal (global shutter closed)	< 125 fA	–	–	22 fA (0.7 V/s)
Parasitic light sensitivity	–	–	1/16.7 (472 nm)	1/8.5 × 10^7(405 nm) 1/1.4 × 10^4(650 nm)
Shortest shutter time	200 ps	250 ps	180 ps (374 nm)	75 ps

6. Discussion

A measured leakage signal of 0.7 V/s in a dark environment is too large compared to the readout time of a large-format imager when there are 5-M samples being read out per second. Therefore, either the readout speed needs to be increased, or the leakage current needs to be lowered for an imager with much more pixels. Methods such as cooling or improving the pixel circuit design can be used to lower the leakage current.

The measured minimum shutter time of 75 ps is much larger than the calculated value of 43.2 ps, and the exposure curve shown in Figure 8 seems to be symmetric. This is as expected, since the shutter time is limited mainly by the fall time of the exposure control signals V_{start} and V_{end} driving the gates of M1 and M2, and not by the intrinsic minimum shutter time of the pixel circuit.

The parasitic light sensitivity that is measured when a 650-nm diode laser is used for illumination is much higher than that obtained using a 405-nm diode laser. This is due to the fact that the absorption depths of light at 405 nm and 650 nm in intrinsic silicon is approximately 0.12 μm and 3.56 μm, respectively [16]. Therefore, much more photoelectrons are generated in the p− substrate under the photodiode when it is illuminated by the 650-nm light, and some of these photoelectrons drift to the n+ drain of the transistor M2, although the p-well of transistor M2 provides some shield to the photoelectrons generated in the p− substrate [17]. Therefore, placing the p-well of transistor M2 in a deep n-well isolated area may provide considerable improvement to the shutter efficiency.

Since the exposure signal skew is relatively small compared with the shortest shutter time in the small designed image sensor, and there is a lack of pixels with skew test circuits [12] in the pixel array, it is hard to measure the exact exposure signal skew. Precise measurement may be possible in the future using an ultra-fast gated CMOS image sensor based on a similar design, but with a much larger imaging area.

7. Conclusions

For this paper, a 40 × 48-pixel ultra-fast global shutter CMOS image sensor was designed and manufactured using a 0.5-µm mixed-signal CMOS process. The measured parasitic light sensitivity for a 405-nm diode laser was $1/8.5 \times 10^7$, which is comparable to MCP-based gated cameras and is low enough for most applications. The measured shutter time can be as short as 75 ps, and the measured dynamic range of the pixel of the designed chip was 5500:1, which is no worse than MCP-based picosecond framing cameras that are currently used [18]. The authors are confident that further significant improvements can be made to the proposed design's temporal resolution through the combined use of more advanced CMOS processes such as advanced silicon-on-insulator (SOI) CMOS technologies and by overdriving the gates of M1 and M2 immediately before and during the exposure process.

Acknowledgments: This work was supported in part by the National Natural Science Foundation of China (Grant No. 61227802). The authors would like to thank Jinyuan Liu of our college for providing the Ti:sapphire laser system and guidance during the temporal resolution measurements.

Author Contributions: Hanben Niu conceived the ideas and supervised the entire work, especially the idea that an ultra-fast gated CMOS image sensor for visible light with very low parasitic light sensitivity would be very useful. Fan Zhang performed the simulations and deductions, designed the circuits, drew the layouts, performed the tests, and wrote the paper.

Conflicts of Interest: The authors declare no conflict of interest.

References

1. Field, R.M.; Realov, S.; Shepard, K.L. A 100 fps, time-correlated single-photon-counting-based fluorescence-lifetime imager in 130 nm CMOS. *IEEE J. Solid-State Circuit* **2014**, *49*, 867–880. [CrossRef]
2. Bamji, C.S.; O'Connor, P.; Elkhatib, T.; Mehta, S.; Thompson, B.; Prather, L.A.; Snow, D.; Akkaya, O.A.; Daniel, A.; Payne, A.D.; et al. A 0.13 µm CMOS system-on-chip for a 512 × 424 time-of-flight image sensor with multi-frequency photo-demodulation up to 130 MHz and 2 GS/s ADC. *IEEE J. Solid-State Circuit* **2015**, *50*, 303–319. [CrossRef]
3. Seo, M.W.; Kagawa, K.; Yasutomi, K.; Kawata, Y.; Teranishi, N.; Li, Z.; Halin, I.A.; Kawahito, S. A 10 ps Time-resolution CMOS image sensor with two-tap true-CDS lock-in pixels for fluorescence lifetime imaging. *IEEE J. Solid-State Circuit* **2016**, *51*, 141–154.
4. Maruyama, Y.; Blacksberg, J.; Charbon, E. A 1024 × 8, 700-ps Time-gated SPAD line sensor for planetary surface exploration with laser raman spectroscopy and LIBS. *IEEE J. Solid-State Circuit* **2014**, *49*, 179–189. [CrossRef]
5. Burri, S.; Maruyama, Y.; Michalet, X.; Regazzoni, F.; Bruschini, C.; Charbon, E. Architecture and applications of a high resolution gated SPAD image sensor. *Opt. Express* **2014**, *22*, 17573–17589. [CrossRef] [PubMed]
6. Bell, P.M.; Kilkenny, J.D.; Hanks, R.L.; Landen, O.L. Measurements with a 35-psec gate time microchannel plate camera. *Proc. SPIE* **1991**, *1346*, 456–464.
7. Gao, L.; Liang, J.; Li, C.; Wang, L.V. Single-shot compressed ultrafast photography at one hundred billion frames per second. *Nature* **2014**, *516*, 74–77. [CrossRef] [PubMed]
8. Vernon, S.; Lowry, M.; Baker, K.; Bennett, C.; Celeste, J.; Cerjan, C.; Haynes, S.; Hernandez, V.J.; Hsing, W.W.; LaCaille, G.A.; et al. X-ray bang-time and fusion reaction history at picosecond resolution using RadOptic detection. *Rev. Sci. Instrum.* **2012**, *83*, 10D307. [CrossRef] [PubMed]
9. Hilsabeck, T.; Hares, J.; Kilkenny, J.; Bell, P.; Dymoke-Bradshaw, A.; Koch, J.A.; Celliers, P.M.; Bradley, D.K.; McCarville, T.; Pivovaroff, M.; et al. Pulse-dilation enhanced gated optical imager with 5 ps resolution (invited). *Rev. Sci. Instrum.* **2010**, *81*, 10E317. [CrossRef] [PubMed]
10. Nagel, S.; Hilsabeck, T.; Bell, P.; Bradley, D.; Ayers, M.; Barrios, M.A.; Felker, B.; Smith, R.F.; Collins, G.W.; Jones, O.S.; et al. Dilation x-ray imager a new/faster gated x-ray imager for the NIF. *Rev. Sci. Instrum.* **2012**, *83*, 10E116. [CrossRef] [PubMed]
11. Bai, Y.; Long, J.; Liu, J.; Cai, H.; Niu, L.; Zhang, D.; Ma, X.; Liu, D.; Yang, Q.; Niu, H. Demonstration of 11-ps exposure time of a framing camera using pulse-dilation technology and a magnetic lens. *Opt. Eng.* **2015**, *54*, 124103. [CrossRef]

12. Berger, R.; Rathman, D.D.; Tyrrell, B.M.; Kohler, E.; Rose, M.K.; Murphy, R.A.; Perry, T.S.; Robey, H.F.; Weber, F.A.; Craig, D.M.; et al. A 64 × 64-pixel CMOS test chip for the development of large-format ultra-high-speed snapshot imagers. *IEEE J. Solid-State Circuit* **2008**, *43*, 1940–1950. [CrossRef]

13. Teruya, A.T.; Vernon, S.P.; Moody, J.D.; Hsing, W.W.; Brown, C.G.; Griffin, M.; Mead, A.S.; Tran, V. Performance of a 512 × 512 gated CMOS imager with a 250 ps exposure time. *Proc. SPIE* **2012**, *8505*, 85050F.

14. Kondo, T.; Takazawa, N.; Takemoto, Y.; Tsukimura, M.; Saito, H.; Kato, H.; Aoki, J.; Kobayashi, K.; Suzuki, S.; Gomi, Y.; et al. 3-D-stacked 16-mpixel global shutter CMOS image sensor using reliable in-pixel four million microbump interconnections with 7.6-μm pitch. *IEEE Trans. Electron. Device* **2016**, *63*, 128–137. [CrossRef]

15. Shibuya, K.; Koshimizu, M.; Murakami, H.; Muroya, Y.; Katsumura, Y.; Asai, K. Development of ultra-fast semiconducting scintillators using quantum confinement effect. *Jpn. J. Appl. Phys.* **2004**, *43*, 1333–1336. [CrossRef]

16. Green, M.A.; Keevers, M.J. Optical properties of intrinsic silicon at 300 K. *Prog. Photovolt. Res. Appl.* **1995**, *3*, 189–192. [CrossRef]

17. Yasutomi, K.; Itoh, S.; Kawahito, S. A two-stage charge transfer active pixel CMOS image sensor with low-noise global shuttering and a dual-shuttering mode. *IEEE Trans. Electron. Device* **2011**, *58*, 740–747. [CrossRef]

18. Landen, O.L.; Bell, P.M.; Oertel, J.A.; Satariano, J.J.; Bradley, D.K. Gain uniformity, linearity, saturation, and depletion in gated microchannel-plate x-ray framing cameras. *Proc. SPIE* **1993**, *2002*, 2–13.

sensors

MDPI

Article

A Fast Multiple Sampling Method for Low-Noise CMOS Image Sensors With Column-Parallel 12-bit SAR ADCs

Min-Kyu Kim, Seong-Kwan Hong and Oh-Kyong Kwon *

Department of Electronics and Computer Engineering, Hanyang University, Seoul 133-791, Korea;
gimmingyu@hanyang.ac.kr (M.-K.K.); seongkhong@hanyang.ac.kr (S.-K.H.)
* Correspondence: okwon@hanyang.ac.kr; Tel.: +82-2-2220-0359; Fax: +82-2-2297-2231

Academic Editor: Gonzalo Pajares Martinsanz
Received: 17 November 2015; Accepted: 22 December 2015; Published: 26 December 2015

Abstract: This paper presents a fast multiple sampling method for low-noise CMOS image sensor (CIS) applications with column-parallel successive approximation register analog-to-digital converters (SAR ADCs). The 12-bit SAR ADC using the proposed multiple sampling method decreases the A/D conversion time by repeatedly converting a pixel output to 4-bit after the first 12-bit A/D conversion, reducing noise of the CIS by one over the square root of the number of samplings. The area of the 12-bit SAR ADC is reduced by using a 10-bit capacitor digital-to-analog converter (DAC) with four scaled reference voltages. In addition, a simple up/down counter-based digital processing logic is proposed to perform complex calculations for multiple sampling and digital correlated double sampling. To verify the proposed multiple sampling method, a 256 × 128 pixel array CIS with 12-bit SAR ADCs was fabricated using 0.18 µm CMOS process. The measurement results shows that the proposed multiple sampling method reduces each A/D conversion time from 1.2 µs to 0.45 µs and random noise from 848.3 µV to 270.4 µV, achieving a dynamic range of 68.1 dB and an SNR of 39.2 dB.

Keywords: successive approximation register ADC; column parallel readout; CMOS image sensor

1. Introduction

Recently, noise performance of CMOS image sensors (CISs) has become an important factor for images captured under low light conditions. The CIS typically uses a programmable gain amplifier (PGA) and a multiple sampling method to suppress noise caused by the pixel and readout circuit [1–8]. The PGA amplifies the pixel output and reduces noise with respect to the gain of the PGA. The readout circuit using the multiple sampling method repeatedly samples the pixel output and then averages the sampled pixel outputs to reduce noise by one over the square root of the number of samplings [7,8].

Several multiple sampling methods, such as the correlated multiple sampling (CMS), digital correlated multiple sampling (DCMS), and pseudo-multiple sampling (PMS) methods, have been studied for CISs with single-slope analog-to-digital converters (SS ADCs) [4–6]. The CMS method repeatedly integrates and averages the pixel output in the analog domain, but requires a power-consuming amplifier [4]. The DCMS method repeatedly converts the pixel output to a digital signal and averages the A/D conversion results in the digital domain. However, the total A/D conversion time increases in proportion to the number of samplings [5]. An alternative solution to DCMS, the PMS method which uses multiple ramp signals with different offsets is reported [6]. However, it requires an accurate ramp generator to control the offset of multiple ramp signals.

Several ADCs using the multiple sampling method have been researched to achieve low noise and overcome the drawbacks of SS ADC, which makes achieving short conversion time and high resolution difficult. A ΔΣ ADC easily achieves low noise by repeating sampling and integrating operations, but requires many clocks and a complex decimation filter [9]. An extended counting

ADC (EC ADC) achieves short conversion time by sequentially converting the pixel output to the upper bit by using $\Delta\Sigma$ ADC and the lower bit by using cyclic ADC. However, an operational amplifier in EC ADC increases power consumption [10,11]. A successive approximation register ADC (SAR ADC) consumes less power due to its simple structure [3,12–15] and reduces noise by using the PMS method [14]. However, since the SAR ADC repeats the operation of the 1st A/D conversion, long A/D conversion time is required.

This paper proposes a fast multiple sampling (FMS) method for CIS with SAR ADC to achieve short conversion times and low noise. A 12-bit SAR ADC using the proposed FMS method repeatedly converts the pixel output to the lower 4-bit among the 12-bit output. Therefore, the required number of bit conversion steps is reduced to one-third. The 12-bit SAR ADC in the readout channel employs a 10-bit capacitor DAC with four scaled reference voltages to reduce the area. In addition, a simple digital processing logic consisting of a 3-input MUX and toggle flip-flop (T-F/F) is proposed to perform the complex calculations for multiple sampling and digital correlated double sampling (DCDS). In Section 2, the architecture of the developed CIS is described, along with the operating principle of the proposed FMS method. Section 3 presents the circuit implementations of the SAR ADC and the digital processing logic in the readout channel. In Section 4, the experimental results of the developed CIS are analyzed and compared with prior works. Finally, conclusions are given in Section 5.

2. CIS Architecture

2.1. Block Diagram

Figure 1 shows the block diagram of the developed CIS employing the proposed FMS method. The pixel array is composed of 256 × 128 pixels with a pixel size of 4.4 μm × 4.4 μm. Each readout channel, consisting of a PGA, a 12-bit SAR ADC, a digital processing logic, and a column decoder, has a pitch of 17.6 μm and converts the four column outputs of the pixel array to digital signals.

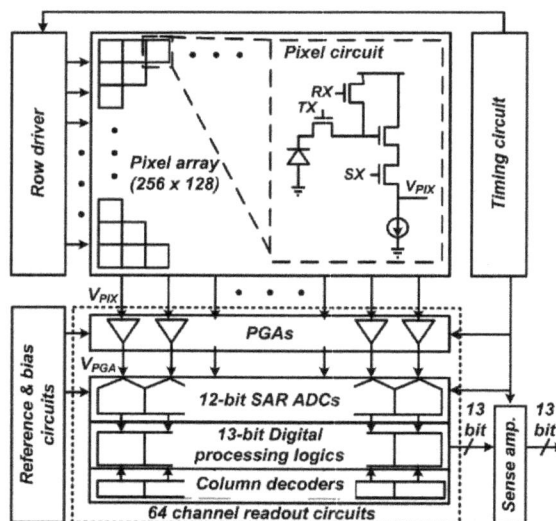

Figure 1. Block diagram of the developed CIS with a schematic of pixel circuit.

The PGA amplifies the pixel output, V_{PIX}, and then the 12-bit SAR ADC using the proposed FMS method repeatedly converts the PGA output, V_{PGA}, to a digital signal. The digital processing logic simultaneously performs the calculations for multiple sampling and DCDS. The output of the digital processing logic selected by the column decoder is transferred to the sense amplifier. To reduce the area and power consumption, four images obtained from each column output are externally combined to form an entire image. The timing circuit generates control signals for the row driver, readout circuit, and sense amplifier, while the bias and reference circuits provide the bias and reference voltages, respectively, for the PGA and 12-bit SAR ADC.

Figure 2 shows the operating sequence of the developed CIS in a row line time. The pixels selected by SX sequentially generates a pixel reset voltage, V_{PIX_RST}, and a photo-induced signal voltage, V_{PIX_SIG}, according to the control signals, RX and TX, respectively. The PGA with a gain of G generates a PGA reset voltage, V_{PGA_RST}, and an amplified pixel signal voltage, $V_{PGA\_RST} + G \times (V_{PIX\_RST} - V_{PIX\_SIG})$, according to the pixel output. The SAR ADC using the proposed FMS method has the maximum number of samplings of 17. At the first A/D conversion, the 12-bit SAR ADC converts V_{PGA} to 12-bit, and then repeatedly converted V_{PGA} to the lower 4-bit. The digital processing logic combines the sequential outputs of the comparator to obtain the A/D conversion result, and then subtracts the A/D conversion result of V_{PGA_RST} from that of $V_{PGA\_RST} + G \times (V_{PIX\_RST} - V_{PIX\_SIG})$. Finally, the digital processing logic output becomes the A/D conversion result of $G \times (V_{PIX\_RST} - V_{PIX\_SIG})$ [16,17].

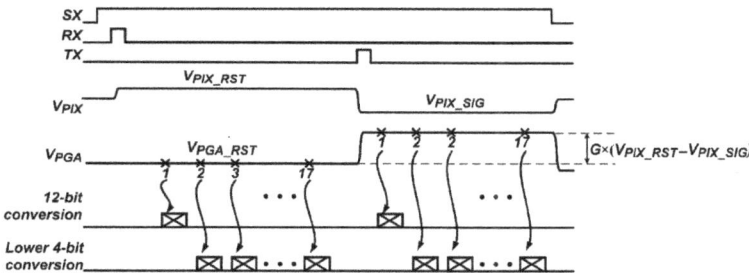

Figure 2. Operating sequence in a row line time.

2.2. Operating Principle of the Proposed FMS Method

Figure 3 shows the signal flow diagram of the proposed FMS method. The proposed FMS method operates in four steps: (1) An N-bit ADC converts the input voltage, V_{IN}, to an N-bit digital signal, D_{1st_N-bit}, for the first A/D conversion, where V_{IN} is a constant for CIS applications; (2) An N-bit DAC converts D_{1st_N-bit} to an analog voltage, V_{1st}, and then the error voltage, V_{ERR}, is obtained by subtracting V_{1st} from V_{IN}. With no quantization errors and noise, V_{1st} is equal to V_{IN}, and V_{ERR} becomes GND. (3) An M-bit ADC converts V_{ERR} to the lower M-bit among the N-bit output, where the range of V_{ERR} is determined by the quantization error and noise. The proposed FMS method repeats the second and third steps $(L - 1)$ times, where L is the number of samplings. Therefore, the lower M-bit conversion results, D_{Kth_M-bit}'s, are repeatedly obtained from the second to Lth A/D conversion, where D_{Kth_M-bit} corresponds to the Kth A/D conversion result. (4) To obtain the final A/D conversion result, D_{FIN_N-bit}, the digital processing logic adds D_{1st_N-bit} to the average value of D_{Kth_M-bit}'s.

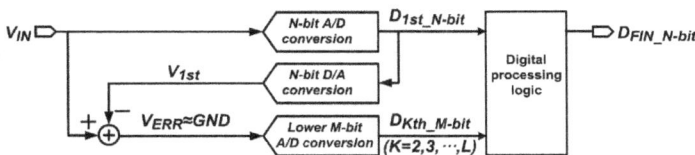

Figure 3. Signal flow diagram of the proposed FMS method.

Figure 4 shows the block diagram of the N-bit SAR ADC employing the proposed FMS method, without the use of an additional analog circuit for the N-bit D/A conversion in the second step and lower M-bit conversion in the third step. The N-bit SAR ADC consists of an N-bit capacitor DAC, a SAR logic, and a comparator. The reference voltages, $+V_{REF}$, $-V_{REF}$, and GND are used for the SAR ADC.

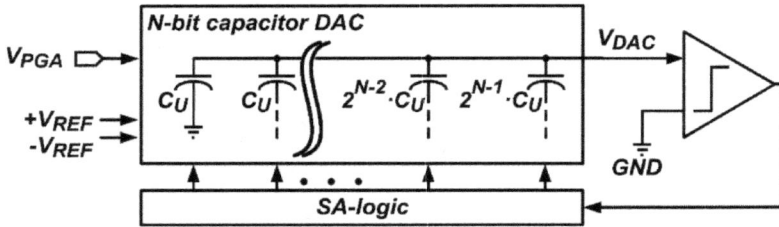

Figure 4. Block diagram of the N-bit SAR ADC.

In the first step of the proposed FMS method, the N-bit capacitor DAC samples V_{IN} by connecting the top and bottom plates of all capacitors to V_{IN} and GND, respectively, and then its output, V_{DAC}, became V_{IN}. The comparator compares V_{DAC} with GND to obtain the most significant bit (MSB), and the SAR logic connects the largest capacitor, $2^{N-1} \cdot C_U$, to $+V_{REF}$ or $-V_{REF}$. This operation is repeated until the least significant bit (LSB) is obtained. Then, the smallest capacitor, C_U, is connected to $+V_{REF}$ or $-V_{REF}$. After the first A/D conversion by the N-bit SAR ADC, V_{DAC} is expressed as:

$$V_{DAC} = V_{IN} + V_{1st\_NOISE} - \sum_{i=1}^{N} \left(D_{1st\_N-bit}[i] \times \frac{V_{REF}}{2^i} \right) \approx GND \qquad (1)$$

where $D_{1st\_N-bit}[i]$ corresponding to the ith bit of the first A/D conversion result has a value of "1" or "-1", and V_{1st_NOISE} is an input-referred noise which includes the sampling and circuit noises at the first A/D conversion.

Figure 5a shows an example of the 4-bit capacitor DAC when D_{1st_N-bit} is "1001". The capacitors, $8 \cdot C_U$, $4 \cdot C_U$, $2 \cdot C_U$, and C_U, are connected to $-V_{REF}$, $+V_{REF}$, $+V_{REF}$, and $-V_{REF}$, respectively, and V_{DAC}, which is equal to $V_{IN} + V_{1st\_NOISE} - 3/16 \cdot V_{REF}$, converges to GND. In the second step, V_{IN} is sampled again in the 4-bit capacitor DAC, and the capacitors, $8 \cdot C_U$, $4 \cdot C_U$, $2 \cdot C_U$, and C_U, are connected to $+V_{REF}$, $-V_{REF}$, $-V_{REF}$, and $+V_{REF}$, respectively, as shown in Figure 5b, which are inversely connected as compared with Figure 5a. Afterwards, all capacitors are connected to GND as shown in Figure 5c, and V_{DAC} becomes $V_{IN} + V_{2nd\_NOISE} - 3/16 \cdot V_{REF}$ which is equal to V_{ERR}, where V_{2nd_NOISE} is an input-referred noise at the second A/D conversion. From the second to Lth A/D conversions, the N-bit capacitor DAC repeats the second step, and V_{DAC} after the second step is expressed as:

$$V_{DAC} = V_{IN} + V_{Kth\_NOISE} - \sum_{i=1}^{N} \left(D_{1st\_N-bit}[i] \times \frac{V_{REF}}{2^i} \right) = V_{Kth\_ERR} \qquad (2)$$

where V_{Kth_NOISE} and V_{Kth_ERR} are the input-referred noise and error voltage at the Kth A/D conversion, respectively. Using Equations (1) and (2), V_{Kth_ERR} can be simplified as:

$$V_{Kth\_ERR} \approx GND - V_{1st\_NOISE} + V_{Kth\_NOISE} \qquad (3)$$

V_{Kth_ERR} has a Gaussian distribution around $GND - V_{1st\_NOISE}$ and its minimum and maximum voltages are determined by V_{Kth_NOISE}.

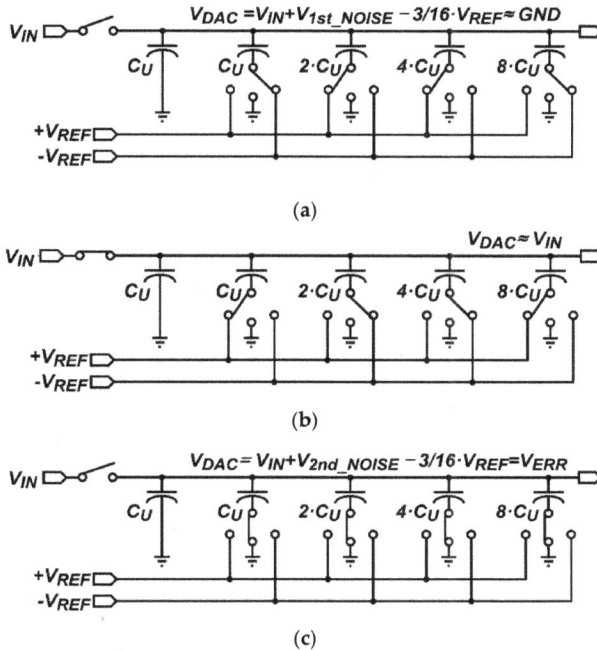

Figure 5. Schematics of the 4-bit capacitor DAC when the first A/D conversion result is "1001": (**a**) when converting the 4-bit; (**b**) when sampling V_{IN}; and (**c**) when obtaining V_{ERR}.

In the third step, since the range of V_{Kth_ERR} is relatively small compared with that of V_{IN}, V_{Kth_ERR} can be converted to a digital signal using only lower capacitors in the N-bit capacitor DAC. After the N-bit SAR ADC converts V_{Kth_ERR} to the lower M-bit among the N-bit output, V_{DAC} is expressed as:

$$V_{DAC} = V_{Kth\_ERR} - \sum_{i=1}^{M} \left(D_{Kth\_M-bit}[i] \times \frac{V_{REF}}{2^{N-M+i}} \right) \approx GND \qquad (4)$$

where $D_{Kth\_M-bit}[i]$ corresponding to the ith bit of the Kth A/D conversion result has a value of "1" or "−1". From the second to Lth A/D conversions, the SAR ADC repeats the second and third steps. After completing the total A/D conversion, the analog voltage corresponding to an average value of the lower M-bits, $AVR[V(D_{Kth\_M-bit})]$, is expressed as:

$$AVR\left[V\left(D_{Kth\_M-bit}\right)\right] = \frac{1}{L-1} \times \sum_{k=2}^{L} \sum_{i=1}^{M} \left(D_{Kth\_M-bit}[i] \times \frac{V_{REF}}{2^{N-M+i}} \right) \approx \frac{1}{L-1} \times \sum_{k=2}^{L} (V_{Kth\_ERR} - GND)$$

$$\approx \frac{1}{L-1} \times \sum_{k=2}^{L} (-V_{1st\_NOISE} + V_{Kth\_NOISE}) \approx -V_{1st\_NOISE} + \frac{1}{L-1} \times \sum_{k=2}^{L} V_{Kth\_NOISE} \qquad (5)$$

where $(L-1)$ is the repeated number of the second and third steps at the number of samplings of L. The final A/D conversion result, D_{FIN_N-bit}, is obtained by adding D_{1st_N-bit} and the average value of the lower M-bits, while the analog voltage corresponding to D_{FIN_N-bit}, $V(D_{FIN\_N-bit})$, can be expressed as:

$$
\begin{aligned}
V\left(D_{FIN\_N-bit}\right) &= \sum_{i=1}^{N}\left(D_{FIN\_N-bit}[i] \times \frac{V_{REF}}{2^i}\right) \\
&= \sum_{i=1}^{N}\left(D_{1st\_N-bit}[i] \times \frac{V_{REF}}{2^i}\right) + \frac{1}{L-1} \times \sum_{k=2}^{L}\sum_{i=1}^{M}\left(D_{Kth\_M-bit}[i] \times \frac{V_{REF}}{2^{N-M+i}}\right) \\
&\approx V_{IN} - GND + \frac{1}{L-1} \times \sum_{k=2}^{L} V_{Kth\_NOISE}
\end{aligned} \tag{6}
$$

where $D_{FIN\_N-bit}[i]$ corresponding to the ith bit of the final A/D conversion result has a value of "1" or "−1". Since the effect of V_{Kth_NOISE} on $V(D_{FIN\_N-bit})$ decreases by averaging V_{Kth_NOISE}, $V(D_{FIN\_N-bit})$ converges to $V_{IN} - GND$ as L increases.

3. Circuit Implementation

3.1. Design of 12-bit SAR ADC Using the Proposed FMS Method

Figure 6 shows a schematic of the 12-bit SAR ADC which converts the PGA output, V_{PGA}, to a digital signal. The PGA in [2,3], which has gains of ×1, ×2, and ×4, is used for the developed CIS. To reduce the area, the SAR ADC uses a 10-bit capacitor DAC, which has a split capacitor structure with an attenuation capacitor, C_{ATT}, instead of a 12-bit capacitor DAC. To obtain the additional lower 2-bit, four scaled reference voltages, +1/2· V_{REF}, +1/4· V_{REF}, −1/4· V_{REF}, and −1/2· V_{REF}, are used in the 10-bit capacitor DAC. The reference voltages, +V_{REF}, −V_{REF}, and GND, are generated via an off-chip DAC, whereas the scaled reference voltages are generated by using an internal R-string. All reference voltages are provided to the 12-bit SAR ADC via on-chip reference buffers. At the rising edge of the control signal, EN_CMP, the clocked comparator compares two outputs of the preamplifier which amplifies the difference between V_{DAC} and GND. The SAR ADC converts V_{PGA} to 12-bit at the first conversion, and then repeatedly converts V_{PGA} to the lower 4-bit. 12 latches and four latches operate as SAR logic at the 12-bit and lower 4-bit conversions, respectively. The SAR logic sequentially stores the output of the clocked comparator and selects the reference voltages connected to the capacitors in the 10-bit capacitor DAC.

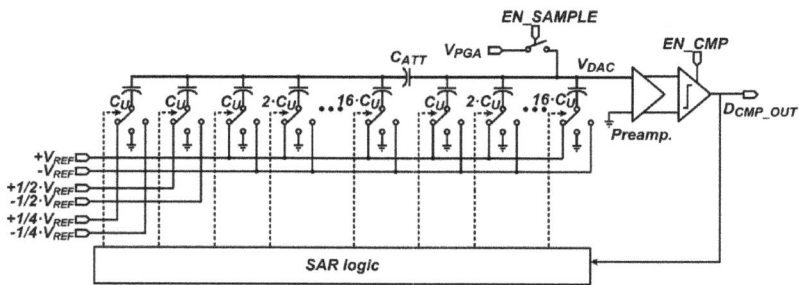

Figure 6. Schematic of the 12-bit SAR ADC.

Figure 7 shows the maximum switching energies of the capacitor DAC according to the bit conversion step of the 12-bit conversion. Since charges in the capacitor DAC are provided from the reference voltages, the switching energy determines the settling time for each bit conversion step [12], and it is calculated using equations in [15]. When the capacitor DAC is reset and samples V_{PGA}, all the capacitors are simultaneously charged to V_{PGA} and the largest switching energy is required in the bit conversion steps. The maximum switching energy decreases with an increase of the bit conversion step with an exception of the second bit, sixth bit, and seventh bit conversion steps, due to the split capacitor structure. Considering the driving capability of the PGA and the decrease in switching energy, the 12-bit SAR ADC is designed to take 200 ns to sample V_{PGA}, 100 ns to convert each upper 8-bit, 50 ns to convert each lower 4-bit, and 100 ns to obtain V_{Kth_ERR}. Therefore, each A/D conversion time becomes 1.2 µs for the first conversion, and then is reduced to 0.45 µs for the second to seventeenth conversion.

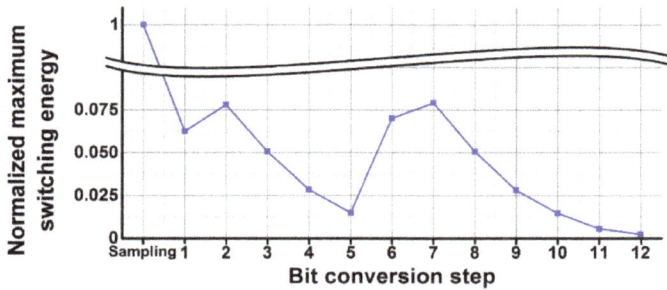

Figure 7. Normalized maximum switching energies of the capacitor DAC.

Figure 8 shows the theoretical normalized noise of the SAR ADC with the proposed FMS method and conventional DCMS method, which repeats the same operations for each A/D conversion, according to the total A/D conversion time. When the number of samplings is L, the noise for the proposed FMS method decreases by one over the square root of $(L - 1)$, according to Equation (6), and the noise for the conventional DCMS method decreases by one over the square root of L. However, when the number of samplings is 17, which is the maximum number for the developed CIS, the total A/D conversion time of the SAR ADC using the proposed FMS method is reduced to 8.4 µs from 19.2 µs, which is that of the SAR ADC using the conventional DCMS method.

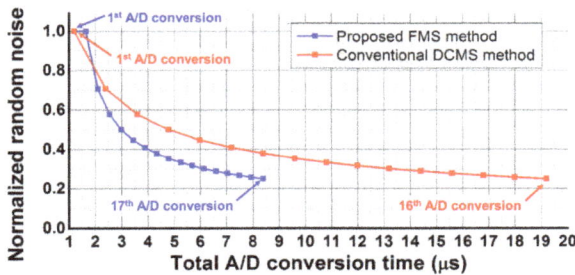

Figure 8. Theoretical normalized noise of the SAR ADC output.

3.2. Design of Digital Processing Logic

Figure 9 shows the block diagram of the proposed up/down counter-based digital processing logic for the 12-bit SAR ADC employing the proposed FMS method. The proposed digital processing logic consists of a digital-to-pulse converter (DPC) and 17 unit cells for the maximum number of samplings of 17, of which each unit cell consists of a T-F/F and 3-input MUX. The T-F/F output or its inverting output, Q or Qb, respectively, is transferred to the next T-F/F though the 3-input MUX according to the control signal, UP. The DPC, which consists of a 2-input MUX, selects a signal, GND or $PULSE$, according to the comparator output, CMP_OUT, and its output, DPC_OUT, is applied to one of the T-F/Fs through the 3-input MUX selected by the control signal, SEL. The control signal, EN_T, becomes low to maintain the output of the T-F/F when the control signal, UP or SEL, changes.

The digital processing logic subtracts the A/D conversion result of the PGA reset voltage from that of the amplified pixel signal voltage, and the developed CIS uses the upper 13-bit, from $D_{OUT}[16]$ to $D_{OUT}[5]$, among the 17-bit outputs of the digital processing logic for displaying the captured image. The MSB generated via subtraction is used as a sign bit.

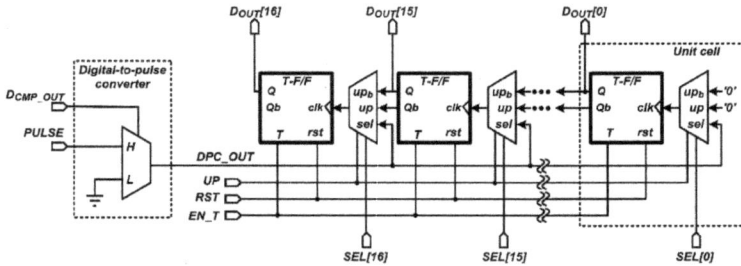

Figure 9. Block diagram of the digital processing logic.

Figure 10 shows the timing diagram of the digital processing logic. Before converting the PGA reset voltage, all T-F/Fs are reset by the control signal, RST, and the digital processing logic acts as a down counter by setting the control signal, UP, to low. At the first A/D conversion, the 12-bit SAR ADC converts the PGA reset voltage to 12-bit, and D_{CMP_OUT} from the MSB to the LSB is sequentially provided to the DPC. When D_{CMP_OUT} is low or high, DPC_OUT keeps or toggles the output of the T-F/F, respectively, starting from $D_{OUT}[15]$ to $D_{OUT}[4]$ sequentially. Therefore, D_{OUT} becomes two's complement of the first A/D conversion result. From the second to seventeenth A/D conversion, the 12-bit SAR ADC converts the PGA reset voltage to the lower 4-bit, and DPC_OUT is sequentially provided from the fourteenth to seventeenth T-F/F, which corresponds to the lower 4-bit. Therefore, the lower 4-bit conversion results are repeatedly subtracted from D_{OUT} 16 times, and then averaged by 16. Therefore, the upper 13-bit, from $D_{OUT}[16]$ to $D_{OUT}[4]$, becomes a final A/D conversion result of the PGA reset voltage. Afterwards, the 12-bit SAR ADC converts the amplified pixel signal voltage to 12-bit. The same operations as those in the PGA reset voltage conversion are repeated, with the exception of the digital processing logic, which acts as an up counter by setting UP to high. Therefore, the upper 13-bit becomes the difference between the A/D conversion results of the PGA reset voltage and the amplified pixel signal voltage.

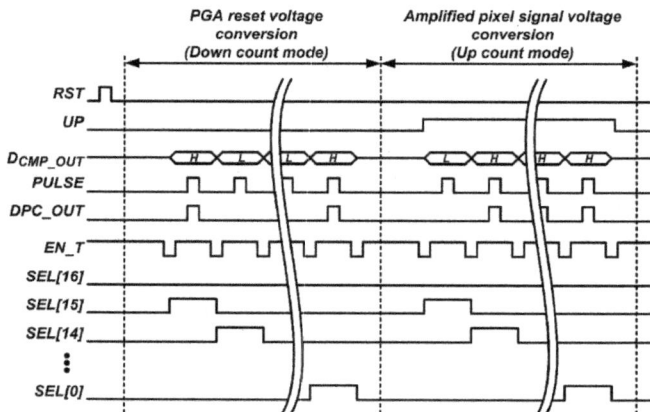

Figure 10. Timing diagram of the digital processing logic.

The proposed up/down counter-based digital processing logic uses a T-F/F and 3-input MUX per unit cell. As a result, the number of transistors per unit cell is reduced by 29% compared with that of the SAR ADC described in [3]. Moreover, the proposed digital processing logic is also applicable to the multi-bit cyclic ADC with the error correction algorithm in [18] by controlling the number of output pulses from the DPC without the use of additional circuits in the unit cell. The multi-bit cyclic ADC sequentially generates the multi-bit per clock cycle to compensate for the comparator offset.

4. Experimental Results

The proposed FMS method is verified using a 256×128 pixel array CIS with column-parallel 12-bit SAR ADCs. Figure 11 shows the chip photomicrograph and readout channel layout of the developed CIS which is fabricated using a 0.18 μm 1-poly 4-metal CMOS process. The developed CIS occupies an area of 2.35 mm $\times$ 2.35 mm.

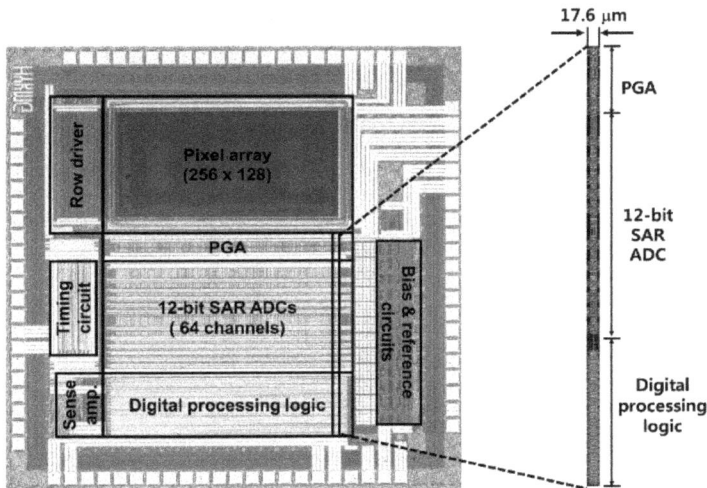

Figure 11. Photomicrograph and readout channel layout of the developed CIS.

The developed CIS uses supply voltages of 2.8 V for the pixel array, analog circuit, and 1.8 V for digital circuit. The total power consumption excluding PAD power is obtained from post layout simulation results under the operating conditions at the number of sampling of 17 and 90 frames/s. It has 4.4 mW which includes the power consumptions of the pixel array of 0.79 mW, reference and bias circuits of 1.14 mW, and readout channel array of 2.3 mW. In a readout channel, the PGA, SAR ADC, and digital processing logic consume 20.6 μW, 11.3 μW, and 4.3 μW, respectively.

Figure 12 shows the measured sensor output signal and random noise of the developed CIS according to light intensity, which is obtained from the average and standard deviation values of 100 images, respectively, at a PGA gain of $\times 1$ and the number of samplings of 17. The output signal and random noise increases linearly according to the light intensity. However, the nonlinearity of the conversion capacitor and source follower in the pixel causes the non-linear behavior [3,19]. The light sensitivity obtained from the slope of the output signal is 6.2 V/lx· s. Random noise has a minimum value of 1.2 LSB (270.4 μV) in near-dark condition, where flickers and thermal noises of the pixel and readout circuits are the dominant sources of noise. As the light intensity increases, random noise increases due to photon shot noise. At a light intensity of 40 lx, the measured output signal and random noise have 3110 LSB and a maximum value of 34 LSB, respectively. An output signal of 3110 LSB corresponds to the full well capacity of 11.4 ke$^-$ with a conversion gain of 60 μV/e$^-$ [19]. The dynamic range obtained from the ratio of the full well capacity to the minimum random noise measured in near-dark conditions is 68.1 dB. The signal-to-noise ratio (SNR) obtained from the ratio of the output signal to random noise is 39.2 dB at a light intensity of 40 lx.

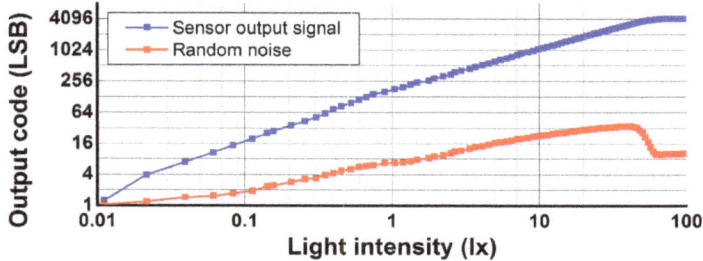

Figure 12. Measured output signal and random noise according to light intensity.

Figure 13 shows the measured differential non-linearity (DNL) and integral non-linearity (INL) of the 12-bit SAR ADC for the developed CIS. The measurement results of the DNL and INL are $-0.62/+1.37$ LSB and $-1.42/+3.55$ LSB, respectively. Since code saturations are caused by the parasitic capacitors connected to upper capacitors in the capacitor DAC, the peaks of the DNL and INL repeatedly occur. In addition, a mismatch of C_{ATT} repeatedly causes peaks of DNL and INL in each 128 LSB over the full code range. Linearity of the CIS is not affected by the INL of the ADC, but are determined by the photon shot noise and photo conversion nonlinearity [3]. In addition, the error of V_{ERR} is not occurred due to capacitance mismatch of the capacitor DAC since the D/A conversion error of the capacitor DAC is cancelled out while converting the input voltage to D_{1st_Nbit} and generating V_{ERR} from D_{1st_N-bit}. However, since the asymmetry of $+V_{REF}$ and $-V_{REF}$ causes an offset error in V_{ERR}, the A/D conversion range for the lower 4-bit conversion should be wider than the offset error.

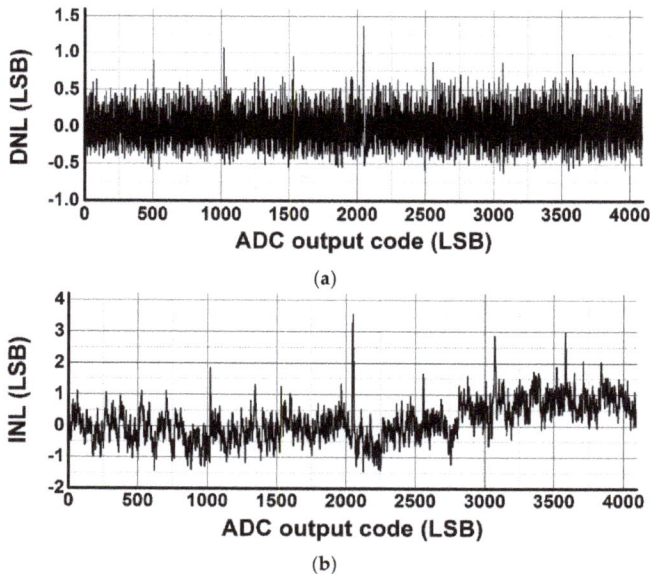

(a)

(b)

Figure 13. Measured (a) DNL and (b) INL of the SAR ADC.

Figure 14 shows the measured and theoretical input referred noises according to the number of samplings and PGA gain at the same exposure time. As the number of samplings increased from 1 to 17, the input referred noise decreases from 848.3 µV to 270.4 µV, 449.2 µV to 155.8 µV, and 255 µV to 96.5 µV, for a PGA gain of ×1, ×2, and ×4, respectively. V_{Kth_ERR}, which is the sum of input referred noises at the 1st and Kth conversions, is repeatedly converted to lower 4-bit after 1st A/D conversion.

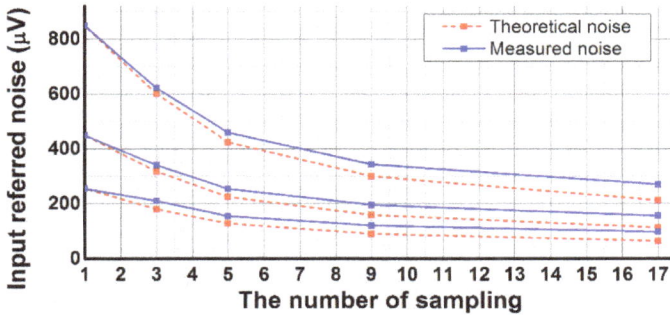

Figure 14. Measured and theoretical input referred noise.

Therefore, lower 4-bit for the fast conversion is derived by considering a measured input referred noise of 848.3 µV, which corresponds to 3.9 LSB, at the number of samplings of 1 and a PGA gain of ×1. Theoretically, for the number of samplings of L, the input referred noise should be decreased by one over the square root of $(L-1)$, but the measured input referred noise is greater than the theoretical noise due to the linearity error of the lower 4-bit conversion of the SAR ADC. The lower 4-bit conversion results after the first conversion exhibits a Gaussian distribution due to temporal noise, but linearity error distorts the distribution. In addition, the measured input referred noise proportionally decreases according to the PGA gain because, for the input referred noise, the noise caused by the readout circuit is dominant compared with what is caused by the pixel circuit [1]. The developed CIS achieves the lowest input referred noise of 96.5 µV at the number of samplings of 17 and a PGA gain of ×4.

Figure 15 shows the captured image of the developed CIS at the number of samplings of 17 and a PGA gain of ×1. The captured image exhibits a 12-bit resolution, but it is difficult to evaluate the noise performance of the CIS with the naked eye because the monitor system generally features an 8-bit resolution. To solve the above problem, images are captured in short exposure time and are displayed by using the lower 5-bits among the 12-bit CIS outputs, corresponding to a digital gain of ×128. Figure 16a,b shows the captured images for the following number of samplings, 1 and 17, respectively, at a PGA gain of ×1, in which Figure 16b has less noise than Figure 16a.

Figure 15. Captured 12-bit image at the number of samplings of 17.

(a)

(b)

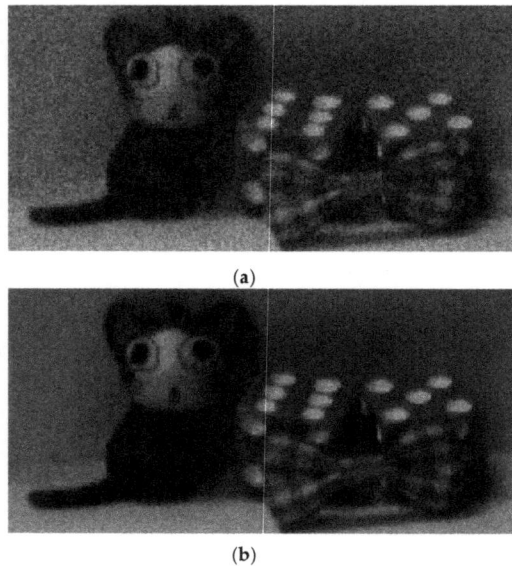

Figure 16. Captured lower 5-bit images in short exposure time for the following number of samplings: (a) 1 and (b) 17.

The performance of the developed CIS is summarized in Table 1 and compared with prior CISs using column-parallel SAR ADCs in Table 2. The figure of merit, representing noise and energy efficiency, was defined as:

$$FOM = \frac{Power \times Noise}{Pixel\_rate} \tag{7}$$

where *Pixel_rate* is a product of the total number of pixels and the frame rate. The CIS employing the proposed FMS method achieves the best FOM of 145 $\mu V \cdot nJ$ by simultaneously reducing the total A/D conversion time and random noise.

Table 1. Performance summary.

Parameter	Value
Process	0.18 μm 1-poly 4-metal CMOS process
Supply voltage	2.8 V/1.8 V
Chip size	2.35 mm $\times$ 2.35 mm
Pixel array size	256 (H) $\times$ 128 (V)
Maximum frame rate	90 frames/s
Pixel size	4.4 μm $\times$ 4.4 μm
Conversion gain	60 μV/e$^-$
Full well capacity	11.4 ke$^-$
Sensitivity	6.2 V/lx$\cdot$s
Column FPN at dark	0.17 LSB
SNR	39.2 dB
Dynamic range	68.1 dB
ADC input range	0.9 V
ADC resolution	12-bit
DNL	-0.62/+1.37 LSB
INL	-1.42/+3.55 LSB
Power consumption	4.4 mW

Sensors **2016**, *16*, 27

Table 2. Comparison with prior CISs with column-parallel SAR ADCs.

Parameter	This Work	[3]	[12]	[13]	[14]	[20]	[21]
Pixel array size	256 × 128	4112 × 2186	1280 × 800	920 × 256	644 × 488	54 × 50	64 × 45
Frame rate (frame/s)	90	60	35	9	120	7.4	21.2
ADC Resolution (bit)	12	14	11	9	14	10	8
Random noise (μV_{rms})	96.5 (0.44 LSB)	130.5	1500	5300	83	0.98 LSB	0.5 LSB
Power consumption (mW)	4.4	108.5	40	1.1	78	0.014	0.021
FOM ($\mu V \cdot nJ$)	145	265	1674	28147	171	-	-

5. Conclusions

In this paper, a fast multiple sampling method for CISs with column-parallel 12-bit SAR ADCs is proposed. The SAR ADC repeatedly converts a pixel output to 4-bit after the first 12-bit A/D conversion. As a result, each A/D conversion time after the first A/D conversion is reduced to 37.5% of the first A/D conversion time, and the total A/D conversion time at the number of samplings of 17 is reduced to 44% of that of the SAR ADC using conventional DCMS method. The 12-bit SAR ADC uses a 10-bit capacitor DAC with an attenuation capacitor and four scaled reference voltages to reduce the area. A simple up/down counter-based digital processing logic, consisting of a 3-input MUX and T-F/F is proposed to perform complex calculations for multiple sampling and DCDS. The measurement results shows that random noise decreases from 848.3 μV to 270.4 μV by using the proposed multiple sampling method, and the best FOM of 145 $\mu V \cdot nJ$ is achieved. Therefore, the proposed multiple sampling method is suitable for low-noise, high-frame rate CISs with column-parallel SAR ADCs.

Acknowledgments: This research was supported by the Industrial and Educational Cooperative R&D Program between Hanyang University and SK Hynix Semiconductor Inc. The authors would like to thank Jaseung Gou and Sang-Dong Yoo of SK Hynix Semi-conductor Inc. for their useful discussion and feedback.

Author Contributions: M.-K.K. and O.-K.K. proposed the idea and designed the circuits; S.-K.H. verified the circuits; M.-K.K. performed the experiments; S.-K.H. and O.-K.K. verified the experiments; M.-K.K. and S.-K.H. wrote the paper.

Conflicts of Interest: The authors declare no conflict of interest.

References

1. Kawai, N.; Kawahito, S. Noise analysis of high-gain, low-noise column readout circuits for CMOS image sensors. *IEEE Trans. Electron. Devices* **2004**, *51*, 185–194. [CrossRef]
2. Takahashi, H.; Noda, T.; Matsuda, T.; Watanabe, T.; Shinohara, M.; Endo, T.; Takimoto, S.; Mishima, R.; Nishimura, S.; Sakurai, K.; *et al.* A 1/2.7 inch low-noise CMOS image sensor for full HD camcorders. In Proceedings of the IEEE International Solid-State Circuits Conference. Digest of Technical Papers, San Francisco, CA, USA, 11–15 February 2007; pp. 510–511.
3. Matsuo, S.; Bales, T.J.; Shoda, M.; Osawa, S.; Kawamura, K.; Andersson, A.; Haque, M.; Honda, H.; Almond, B.; Mo, Y.; *et al.* 8.9-megapixel video image sensor with 14-b column parallel SA-ADC. *IEEE Trans. Electron. Devices* **2009**, *56*, 2380–2389. [CrossRef]
4. Suh, S.; Itoh, S.; Aoyama, S.; Kawahito, S. Column-parallel correlated multiple sampling circuits for CMOS image sensors and their noise reduction effects. *Sensors* **2010**, *10*, 9139–9154. [CrossRef] [PubMed]
5. Chen, Y.; Xu, Y.; Mierop, A.J.; Theuwissen, A.J.P. Column parallel digital correlated multiple sampling for low-noise CMOS image sensors. *IEEE Sens. J.* **2012**, *12*, 793–799. [CrossRef]
6. Lim, Y.; Koh, K.; Kim, K.; Yang, H.; Kim, J.; Jeong, Y.; Lee, S.; Lee, H.; Lim, S.-H.; Han, Y.; *et al.* A 1.1 e-temporal noise 1/3.2-inch 8 Mpixel CMOS image sensor using pseudo-multiple sampling. In Proceedings of the IEEE International Solid-State Circuits Conference. Digest of Technical Papers, San Fransisco, CA, USA, 7–11 February 2010; pp. 396–397.
7. Fowler, A.; Gatley, I. Noise reduction strategy for hybrid IR focal plane arrays. In Proceedings of the Infrared Sensors: Detectors, Electronics, and Signal Processing (SPIE), San Diego, CA, USA, 21 July 1991; pp. 127–133.

8. Eltoukhy, H.; Salama, K.; El-Gamal, A. A 0.18 μm CMOS Bioluminescence Detection Lab-on-Chip. *IEEE J. Solid State Circuits* **2006**, *41*, 651–662. [CrossRef]
9. Chae, Y.; Cheon, J.; Lim, S.; Kwon, M.; Yoo, K.; Jung, W.; Lee, D.-H.; Ham, S.; Han, G. A 2.1 M pixels, 120 frame/s CMOS image sensor with column-parallel ΔΣ ADC architecture. *IEEE J. Solid State Circuits* **2011**, *46*, 236–247. [CrossRef]
10. Seo, M.-W.; Suh, S.-H.; Iida, T.; Takasawa, T.; Isobe, K.; Watanabe, T.; Itoh, S.; Yasutomi, K.; Kawahito, S. A low-noise high intrascene dynamic range CMOS image sensor with a 13 to 19b variable-resolution column-parallel folding-integration/cyclic ADC. *IEEE J. Solid State Circuits* **2012**, *47*, 272–283. [CrossRef]
11. Kim, J.-H.; Jung, W.-K.; Lim, S.-H.; Park, Y.-J.; Choi, W.-H.; Kim, Y.-J.; Kang, C.-E.; Shin, J.-H.; Choo, K.-J.; Lee, W.-B.; *et al.* A 14b extended counting ADC implemented in a 24 Mpixel APS-C CMOS image sensor. In Proceedings of the IEEE International Solid-State Circuits Conference. Digest of Technical Papers, San Francisco, CA, USA, 19–23 February 2012; pp. 390–392.
12. Chen, D.G.; Tang, F.; Bermak, A. A Low-Power Pilot-DAC Based Column Parallel 8b SAR ADC With Forward Error Correction for CMOS Image Sensors. *IEEE Trans. Circuits. Syst. I Reg. Pap.* **2013**, *60*, 2572–2583. [CrossRef]
13. Chen, D.G.; Tang, F.; Law, M.K.; Zhong, X.; Bermak, A. A 64fJ/step 9-bit SAR ADC array with forward error correction and mixedsignal CDS for CMOS image sensors. *IEEE Trans. Circuits. Syst. I Reg. Pap.* **2014**, *61*, 3085–3092. [CrossRef]
14. Kim, J.-B.; Hong, S.-K.; Kwon, O.-K. A Low-Power CMOS Image Sensor With Area-Efficient 14-bit Two-Step SA ADCs Using Pseudomultiple Sampling Method. *IEEE Trans. Circuits Syst. II* **2015**, *62*, 451–455. [CrossRef]
15. Ginsburg, B.P.; Chandrakasan, A.P. An Energy-Efficient Charge Recycling Approach for a SAR Converter With Capacitive DAC. In Proceedings of the IEEE International Symposium on Circuits and Systems, Kobe, Japan, 23–26 May 2005; pp. 184–187.
16. White, M.H.; Lampe, D.R.; Blaha, F.C.; Mack, I.A. Charactarization of surface channel CCD image arrays at low light levels. *IEEE J. Solid State Circuits* **1974**, *9*, 1–13. [CrossRef]
17. Gowda, S.M.; Shin, H.J.; Wong, H.-S.P.; Xiao, P.H.; Yang, J. Image sensor with direct digital correlated sampling. US Patent #6,115,066, 5 September 2000.
18. Park, J.-H.; Aoyama, S.; Watanabe, T.; Isobe, K.; Kawahito, S. A High-Speed Low-Noise CMOS Image Sensor With 13-b Column-Parallel Single-Ended Cyclic ADCs. *IEEE Trans. Electron. Devices* **2009**, *56*, 2414–2422. [CrossRef]
19. Han, L.; Yao, S.; Xu, J.; Xu, C.; Gao, Z. Analysis of incomplete charge transfer effects in a CMOS image sensor. *J. Semicond.* **2013**, *34*, 054009. [CrossRef]
20. Cevik, I.; Ay, S. An Ultra-Low Power Energy Harvesting and Imaging (EHI) Type CMOS APS Imager with Self-Power Capability. *IEEE Trans. Circuits Syst. I* **2015**, *62*, 2177–2186. [CrossRef]
21. Ay, S. A 1.32pW/frame.pixel 1.2V CMOS Energy Harvesting and Imaging (EHI) APS Imager. In Proceedings of the 2011 International Solid- State Circuits Conference (ISSCC), San Francisco, CA, USA, 20–24 February 2011; pp. 116–117.

sensors

MDPI

Article

Long-Term Continuous Double Station Observation of Faint Meteor Showers

Stanislav Vítek [1],*, Petr Páta [1], Pavel Koten [2] and Karel Fliegel [1]

[1] Faculty of Electrical Engineering, Czech Technical University in Prague, Technická 2, 166 27 Prague, Czech Republic; pata@fel.cvut.cz (P.P.); fliegek@fel.cvut.cz (K.F.)

[2] Astronomical Institute of the Academy of Sciences of the Czech Republic, Fričova 298, 251 65 Ondřejov, Czech Republic; koten@asu.cas.cz

* Correspondence: viteks@fel.cvut.cz; Tel.: +420-224-352-232

Academic Editor: Gonzalo Pajares Martinsanz

Received: 13 July 2016; Accepted: 7 September 2016; Published: 14 September 2016

Abstract: Meteor detection and analysis is an essential topic in the field of astronomy. In this paper, a high-sensitivity and high-time-resolution imaging device for the detection of faint meteoric events is presented. The instrument is based on a fast CCD camera and an image intensifier. Two such instruments form a double-station observation network. The MAIA (Meteor Automatic Imager and Analyzer) system has been in continuous operation since 2013 and has successfully captured hundreds of meteors belonging to different meteor showers, as well as sporadic meteors. A data processing pipeline for the efficient processing and evaluation of the massive amount of video sequences is also introduced in this paper.

Keywords: faint meteor shower; meteoroid; CCD camera; image intensifier; image processing

---

1. Introduction

Modern image sensors used in astronomy provide high sensitivity and high frame-rates, allowing for the detection of weak and rapidly-changing events in the atmosphere. Among these phenomena are meteors, streaks of light that appear in the sky when an interplanetary dust particle ablates in the Earth's atmosphere. The study of meteors and meteoroids provides clues about their parent objects: comets [1] and asteroids [2]. The light curve of the meteor contains information about the mass of the original particle. Both the shape of this curve as well as the height interval where the meteor radiates correspond to the structure of the parent meteoroid. It is possible to investigate the above-mentioned and many other properties of meteors using video records. Video data has the advantage that if the meteor is recorded with high time resolution from at least two stations simultaneously, its atmospheric trajectory can be calculated. Moreover, the heliocentric orbit can be determined if we know the exact time of the event, which is common for video observation. It was shown that the properties of systems with image intensifiers enable the detection of meteors with masses down to fractions of a gram.

The multi-station observation of meteors using two or more video systems first appeared in the 1970s [3,4], and became a standard technique for the measurement of meteoroid trajectories. Current networks vary in the number of cameras and observation locations. Video systems within the framework of the SPanish Meteor Network (SPMN) use three cameras at three different locations [5]. The Cameras for Allsky Meteor Surveillance (CAMS) system operates 60 identical narrow-angle field of view (FOV, 30°) cameras at three locations [6]. The above-mentioned networks employ low-cost 1/3″ or 1/2″ security cameras [7] with a typical spatial resolution of 720 × 564 pixels and a frame rate between 20 and 25 frames per second (fps). Highly-sensitive E2V CCD still cameras [8] or cameras with custom-made electronic shutter systems [9] are also utilized. GigE (Gigabit Ethernet) cameras with 30 fps are used in the framework of the French initiative Fireball Recovery and InterPlanetary Observation Network (FRIPON) [10].

Double-station observation using S-VHS (Super Video Home System) camcorders coupled with image intensifiers started at the Ondřejov observatory about two decades ago [11]. The system had a horizontal resolution of 420 lines per picture, and video data were stored on S-VHS tapes, generally unsuitable for scientific purposes due to the mechanical movement of the tape causing video jitter. The MAIA (Meteor Automatic Imager and Analyzer) system introduced in this paper is a technological successor of the original analog one [12]. It consists of two identical stations placed in Ondřejov and Kunžak, Czech Republic (see Figure 1). The distance between stations is 92.5 km. The Ondřejov camera is aiming at the azimuth of 40°, elevation 45°, the Kunžak at azimuth of 120°, elevation 45°.

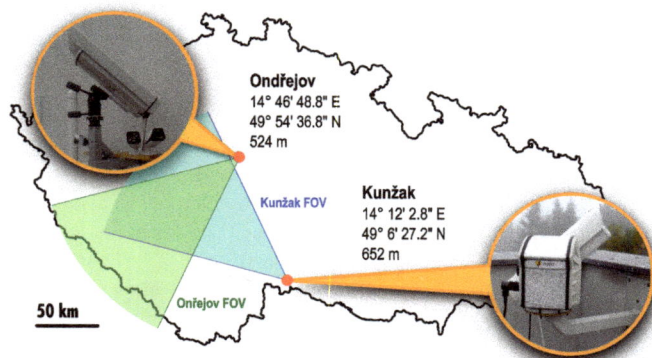

Figure 1. Location of the Meteor Automatic Imager and Analyzer (MAIA) stations. FOV: field of view.

With spatial resolution close to VGA and temporal resolution up to 61 fps, every single MAIA station produces almost 2 TB of raw video data per night. Since there is limited data storage at each MAIA station, image processing algorithms have to make a decision about the importance of the recorded data during daytime and free up the disk space for the following night. Due to restricted Internet access, the problem of the automatic processing of a massive amount of data has to be solved employing local computing power (e.g., Field-Programmable Gate Array, FPGA, or Graphical Processing Unit, GPU). These techniques make high-performance parallel processing tasks feasible. The computing power of a GPU can reach a thousandfold performance of a standard central processing unit (CPU) with an affordable price. The utilization of a GPU enables a substantial improvement in the performance of astronomical data processing algorithms. At this point, it is worth mentioning related solutions of N-body [13], radio-telescope signal correlation [14], adaptive mesh refinement [15], and gravitational microlensing [16]. These examples mostly focus on complicated numerical and cosmological problems, data mining [17], and the visualization of tera-scale astronomical datasets [18]. Classical image processing problems are addressed to a lesser extent.

The paper is organized as follows. Section 2 gives an overview of the technical properties of the MAIA system. Quality and formal aspects of acquired video data (significantly affecting the design of video processing pipeline) are addressed in Section 3. Section 4 concludes the paper.

2. Meteor Automatic Imager and Analyzer

The design of the MAIA system (see Figure 2) is based on our expertise gained with its previous analog version used in Ondřejov for many years. The electro-optical subsystem of MAIA consists of two main components: a second-generation image intensifier XX1332 and a GigE camera JAI CM-040GE. The image intensifier has a large diameter input (50 mm) and output (40 mm) apertures, high gain (typically 30,000 to 60,000 lm/lm), and a spatial resolution of 30 lp/mm [19]. Since the diameter of the photocathode in image intensifier is 50 mm, and the angle of view for meteor observation should be about 50°, then the most suitable focal length of the input lens comes at about 50 mm. The MAIA system uses a Pentax SMC FA 1.4/50 mm—this lens offers the angle of view of 47°. Due to the large aperture, a high input signal-to-noise ratio is achieved at the intensifier.

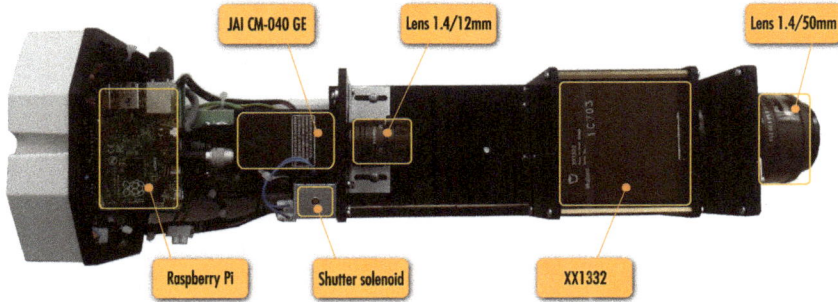

Figure 2. MAIA uncovered.

Camera JAI CM-040GE is equipped with a 1/2″ progressive scan CCD sensor. This sensor provides a resolution of 776 × 582 pixels (i.e., 0.45 megapixels), 8.3 μm square pixels, and 10- or 8-bit monochrome output. Maximum framerate of 61.5 fps can be increased if needed with vertical binning and partial scan. The exposure time can vary between 54.874 s to 16.353 ms (or preset electronic shutter 1/60 to 1/10,000 in 10 steps) in full frame scan. The camera has high sensitivity of 1.3 lux (on the sensor, maximum gain, shutter off, 50% of peak video level) and a signal-to-noise ratio greater than 50 dB at 0 dB gain setting. The focal length of the camera lens (Pentax H1214-M 1.4/12 mm) was selected to get a perfect match between the output screen of the image intensifier (diameter of 40 mm), the height of the CCD (4.83 mm), and a suitable distance between the camera and the image intensifier (about 10 cm).

The outer housing of the device was selected taking into account weatherproof requirements. The characteristics of the housing are very similar to those required for video surveillance. The body of the housing is made of extruded aluminum, and the end-cover plates of die-cast aluminum. The weatherproof feature is maintained by the rubber gaskets between the cover plates, and three cable glands. The housing is equipped with a heater kit and a sun shield. One-day exposure of sunlight through the fast lens reliably damages the sensitive layer of the image intensifier. Thus, the crucial part of the device is a mechanical shutter fulfilling the function of protection from the Sun. The solenoid opens the shutter for acquisition while a spring ensures that the shutter is closed in the case of power failure. The electro-mechanical design of the solenoid-operated mechanical shutter was a delicate matter. High reliability was required, but electromagnetic compatibility issues also had to be carefully treated, since the image intensifier is prone to electromagnetic interference.

MAIA has a local computer (Raspberry Pi) to handle data transfers between the main computer and the instrument. The data communication between the main computer and the instrument include video stream from the CCD camera, control signals (shutter, heating, local power supply), and environmental data (temperature, humidity). The distance between the device and the main computer is about 10–15 m. A fiber optic cable is used to ensure uncompromising protection of the instrument during thunderstorms.

2.1. Electro-Optical Characteristics

The XX1332 image intensifier has a highly nonlinear input–output conversion function as a result of the automatic gain control (AGC). This nonlinearity can be characterized by the dependence of the normalized gain (normalized ratio of the output pixel level in the captured image and the input power of the light, measured at a wavelength of 650 nm) on the normalized input power. The curve describing this dependency is depicted in Figure 3a. The image intensifier's AGC feature helps to accommodate extremely high dynamic range, and also brings high nonlinearity, which is especially critical for photometric measurements.

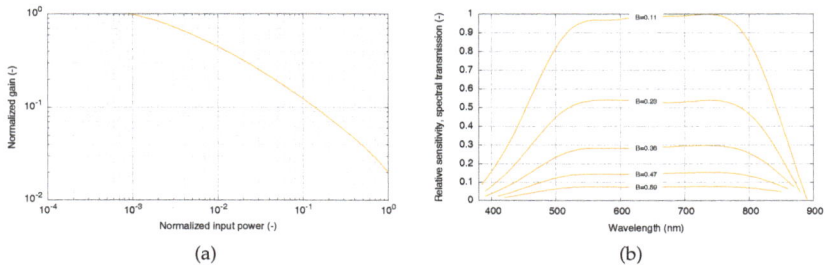

Figure 3. Properties of the MAIA system (**a**) Gain of the intensifier; (**b**) Spectral sensitivity.

The overall relative spectral sensitivity can be seen in Figure 3b. This characteristic takes into account the properties of the input lens, the image intensifier, the camera lens, and the camera itself. There are five curves of the relative spectral sensitivity for several digital values in the output image (B = 1 means white level; i.e., 255 in 8 bpp representation). This result applies to the particular setting of the camera; i.e., electronic shutter set to 1/100 s exposure time, gain of 0 dB, and zero black level. It is evident that the sensitivity is not constant for a chosen wavelength. This is the impact of AGC, as discussed above. The sensitivity is much higher for low-level light conditions in order to achieve sufficiently bright images on the image intensifier's output screen. However, the spectral dependence of the sensitivity does not change significantly with the variable gain set by the device's AGC. The FWHM (Full Width Half Maximum) spectral range of the system is approximately 455–845 nm; i.e., slightly shifted to the near-infrared (NIR) domain. This property is crucial, since meteors radiate significantly in this spectral region [20].

2.2. Spatiotemporal Characteristics

Intensified TV systems exhibit a prominent speckle noise component, caused by the intensifier's AGC. The level of individual bright spots in the video frame fluctuates significantly, while the overall signal level remains roughly constant (i.e., a couple of bright spots increase their level, while the level is decreased for other bright spots). This phenomenon affects conventional image processing algorithms concerning their scalability and performance. Figure 4a shows the dependence of the pixel intensity standard deviation on the pixel intensity value for a video sequence of 100 frames.

Aside from the high non-linearity discussed in the previous section, MAIA also exhibits features of a shift-variant imaging system. Figure 5 shows the shape of stellar objects in the FOV. It is clearly visible that the object in the middle of the image has a circular shape, whereas objects occurring on the border are heavily distorted. The shape of a stellar object is usually modeled by an asymmetric Gaussian function [21]. Then, the object's spatial distortion (i.e., its ellipticity) can be described as a ratio of sigma parameters in vertical and horizontal directions. Figure 4b shows the dependence of an object's ellipticity on the angular distance from the center of the image. The object's ellipticity plays a significant role in the efficiency of the object detection—for angular distances higher than 40° (i.e., close to the border of the FOV), efficiency decreases significantly.

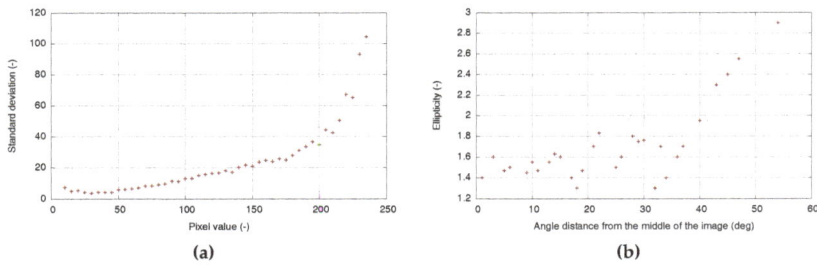

Figure 4. (**a**) Pixel value standard deviations; (**b**) Ellipticity of stellar objects.

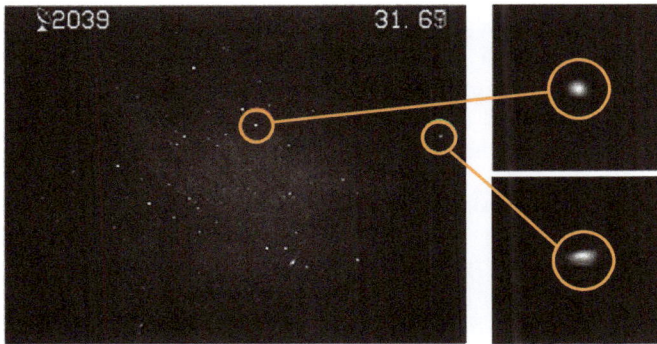

Figure 5. Shift-variant point spread function of the MAIA electro-optical system.

3. Video Data Processing Pipeline

The primary goal of the MAIA video processing pipeline is to find faint meteor showers in the recorded video sequence. A typical video sequence consists mostly of static stellar objects and noise. There is little or no change between the consecutive frames, due to the high frame rate of the camera. Thus, variations in the image data (e.g., meteors or optical transients) are detectable while using relatively simple algorithms based on comparison via image subtraction. Figure 6 summarizes the essential elements of the proposed video processing pipeline. The double-station system deals with simultaneous observation of the same astronomical events observed from two different locations. Therefore, it is important to ensure synchronous execution of the processes running on both stations. The internal clock of the computer is used as the time authority. An NTP (Network Time Protocol) service is employed to synchronize each computer. Achieved precision of approximately one second is sufficient for the successful identification of meteors recorded in double-station configuration. Precise alignment of the appropriate meteor frames is performed through calculation of the meteor's atmospheric trajectory.

As discussed in the previous sections, MAIA's electro-optical characteristics are far from those of an ideal imaging system. Currently, the conventional approach for obtaining the Point Spread Function (PSF) of the space-shift-variant system is based on modeling the wavefront aberrations using Zernike polynomials [22]. An efficient way of bypassing the impact of spatiotemporal fluctuations is described in [23]. The authors employed statistical analysis with nonlinear preprocessing of image intensity using Box-Cox and logarithmic transform. Our pipeline uses a more traditional method based on the detection and classification of the object while taking into account spatial relations between consecutive frames.

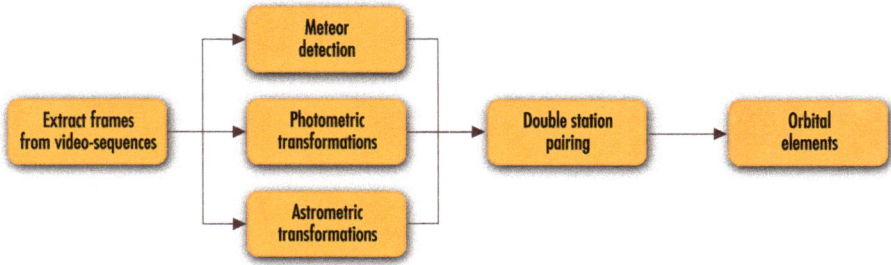

Figure 6. Double-station video processing pipeline.

3.1. Object Detection and Classification

A Canny algorithm [24] is used for object detection in the MAIA system. The algorithm detects edges, while static objects are identified as stars (see Figure 7a). The remaining moving objects can be classified as meteors. Linear motion between consecutive frames is then detected for such moving objects. If an object with linear motion is identified within a certain number of consecutive frames (usually at least five frames), then it is classified as a meteor. Finally, the positions of all detected objects are exported into an external text file (MAIA Object File, MOF). Moreover, the video sequence (approx. 80 frames, i.e. 35 frames before and 35 after the event) containing the meteor candidate is saved into a video file and further processed.

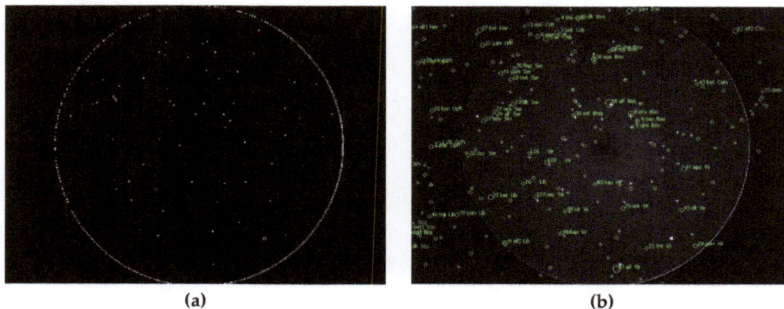

| (a) | (b) |

Figure 7. Calibration of the frame. (**a**) Edges detected by Canny filter; (**b**) Identified stars.

The first step in the video processing pipeline is the calculation of the calibration image. The calibration image is obtained as a time average of an odd number of video frames (five frames is usually sufficient). Then, the dark frame is subtracted, and the image is flat-fielded. Stars can be detected using values available in the MOF file. Another approach is based on repeated star detection using the Canny method. The star catalog information is loaded using the previously recorded meteor data (i.e., date, time, aiming point of the camera). Then, the catalog stars can be plotted over the acquired image (Figure 7b). Samples of the detected stars (x, y) and catalog stars (α, δ) are aligned. Then, parameters of the transformation between both coordinate systems are determined, and the corresponding pairs are identified. The signal of the stellar object is measured as a sum of the pixels in the box surrounding the star. Finally, the calibration curve is constructed (Figure 8a).

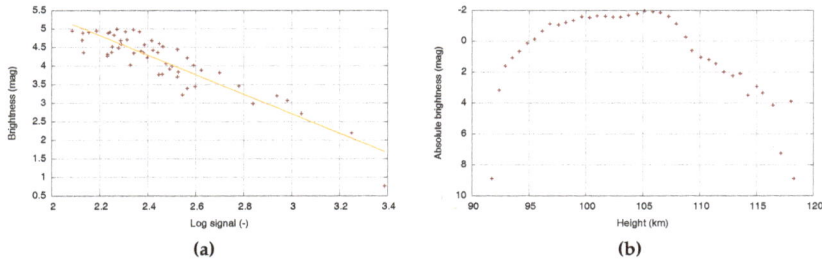

Figure 8. (**a**) Calibration curve; (**b**) Light curve of detected meteor.

The second step consists of the meteor's position and brightness measurements. The data stored in the MOF file are used. The particular point can be defined manually, or the MOF positions can be adjusted (Figure 9a). When finished, the boxes around the meteor are set for each frame. The meteor signal is measured as a sum of the pixels within the box (Figure 9b). The brightness of the meteor is determined using the calibration curve. Finally, the light curve of the meteor is calculated (Figure 8b).

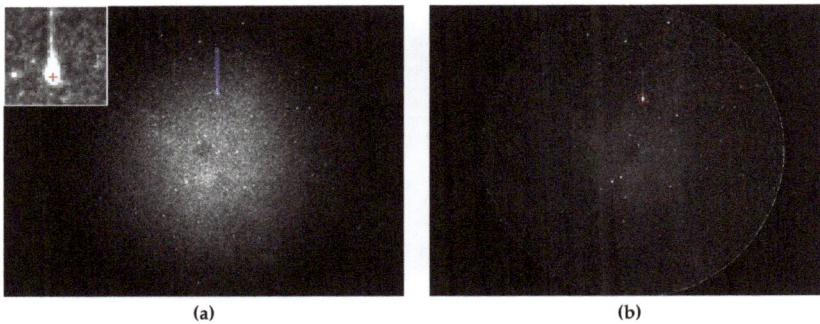

Figure 9. Measurement of meteor parameters. (**a**) Detected Meteor's path and its magnified detail; (**b**) Meteor signal measured within the box for each frame.

3.2. GPU Acceleration

The general-purpose CUDA (Compute Unified Device Architecture) GPU is a highly parallel multi-threaded architecture [25]. CUDA became the de facto standard software development kit (SDK) for astronomy computation [26]. One can find numerous studies on the acceleration of image processing for real-time applications, including techniques for real-time moving object detection, a topic related to the subject of this paper. The main bottleneck of GPU acceleration is inefficient data transfer between the host and the device—meaning that implementation of data transfers between the host and the GPU device can negatively affect the overall application performance [27]. The MAIA pipeline solves this issue by batching many small transfers into one larger transfer; i.e., the GPU simultaneously processes more frames.

The most time-consuming operation of the MAIA image processing pipeline is Canny detection. It is used for the detection of objects within particular video frames. This critical part of the pipeline is therefore implemented on the GPU. For the testing, we used video sequences acquired by the MAIA system with a duration of 10 min each (i.e., approximately 36,000 frames at the frame rate of 61 fps). It is worth noting that frame rate is not fully constant and depends highly on the bus workload. The resolution of a single frame is 776×582 pixels. Since a pixel takes 2 B of memory (10 b depth, monochrome), one frame requires approximately 1 MB of memory. Consequently, simultaneous processing of more frames brings substantial performance gain. Figure 10 shows the execution times required for processing, depending on the frame size. An image size of 3104 × 2328 pixels means that eight regular input frames are processed simultaneously. Assuming the speed of transfer of about 3000 MB/s between the host and the GPU, this transfer takes around 10 ms, including all necessary data (i.e., convolution kernels, both input and output). We compared execution times of the sequential code written in OpenCV, executed using an Intel Core i7-4790 machine with 16 GB of RAM to parallelized code executed using NVIDIA K4200 and NVIDIA Tesla K40 graphic cards.

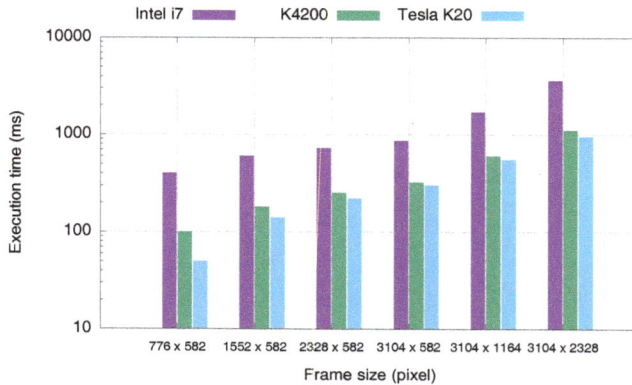

Figure 10. Comparison of the execution time of sequential and parallelized implementation of the Canny detection algorithm.

The time needed for the processing of a 60 s video sequence on the quad-core CPU is about 100 s. The current version of the GPU-optimized code can process the same sequence in 70 s. Our goal is to process a 60 s video sequence in 60 s or faster (i.e., in real time or faster).

4. Conclusions

Automatic double station observations using the MAIA system were carried out continuously during last three years, on each night with good weather conditions. For example, in the year 2015, the camera at Kunžak station was in operation for 202 nights and recorded more than 6500 meteors. The data for the Ondřejov station are slightly lower, since the observational conditions in Onřejov are worse, mostly due to light pollution. Currently, the observations and meteor detection are fully automatic, while the video data processing is semiautomatic, since the operator's input is still needed.

The comparison of the detection efficiency (which was done for the Perseid campaign) shows that the current MAIA pipeline can detect about 72% of meteors detected by the old system with a S-VHS camera. In the case of the old system, the detection software was usually run several times with parameters tuned manually by the operator. Such an approach is not applicable to the automatic MAIA system. On the other hand, the MAIA system successfully detected some meteors which were not detected by the old system.

The current implementation offers close to real-time video processing; i.e., the processing time is close to the duration of the captured video sequence. Once the development of the MAIA pipeline is finished, it will be released under GPL (General Public License) or a similar license (For information about the current status of the project, please visit http://maia-project.net).

Sensors **2016**, *16*, 1493

Acknowledgments: This work was supported by the grant No. 14-25251S Nonlinear imaging systems with spatially variant point spread function of the Czech Science Foundation. We also would like to thank namely Lukáš Fritsch for his help with the electronics, Petr Janout for his help with the optics, and Filip Kobrzek for his help with the mechanical part of the system.

Author Contributions: S. Vítek wrote the paper with contribution of all co-authors. All co-authors discussed results. S. Vítek wrote control software and GPU processing. P. Koten and P. Páta wrote image processing pipeline. K. Fliegel designed electro-optical part of MAIA system.

Conflicts of Interest: The authors declare no conflict of interest.

References

1. Trigo-Rodriguez, J.M.; Madiedo, J.M.; Williams, I.P.; Castro-Tirado, A.J. The outburst of the κ Cygnids in 2007: Clues about the catastrophic break up of a comet to produce an Earth-crossing meteoroid stream. *Mon. Not. R. Astron. Soc.* **2009**, *392*, 367–375.
2. Porubčan, V.; Williams, I.; Kornoš, L. Associations between asteroids and meteoroid streams. *Earth Moon Planets* **2004**, *95*, 697–712.
3. Clifton, K.S. Television studies of faint meteors. *J. Geophys. Res.* **1973**, *78*, 6511–6521.
4. Hawkes, R.L.; Jones, J. Two station television meteor studies. In *Symposium-International Astronomical Union*; Cambridge University Press: New York, NY, USA, 1980; Volume 90, pp. 117–120.
5. Madiedo, J.M.; Trigo-Rodriguez, J.M. Multi-station video orbits of minor meteor showers. *Earth Moon Planets* **2008**, *102*, 133–139.
6. Jenniskens, P.; Gural, P.; Dynneson, L.; Grigsby, B.; Newman, K.; Borden, M.; Koop, M.; Holman, D. CAMS: Cameras for Allsky Meteor Surveillance to establish minor meteor showers. *Icarus* **2011**, *216*, 40–61.
7. Samuels, D.; Wray, J.; Gural, P.S.; Jenniskens, P. Performance of new low-cost 1/3″ security cameras for meteor surveillance. In Proceedings of the 2014 International Meteor Conference, Giron, France, 18–21 September 2014; pp. 18–21.
8. Oberst, J.; Flohrer, J.; Elgner, S.; Maue, T.; Margonis, A.; Schrödter, R.; Tost, W.; Buhl, M.; Ehrich, J.; Christou, A.; et al. The smart panoramic optical sensor head (SPOSH)—A camera for observations of transient luminous events on planetary night sides. *Planet. Space Sci.* **2011**, *59*, 1–9.
9. Atreya, P.; Vaubaillon, J.; Colas, F.; Bouley, S.; Gaillard, B. CCD modification to obtain high-precision orbits of meteoroids. *Mon. Not. R. Astron. Soc.* **2012**, *423*, 2840–2844.
10. Colas, F.; Zanda, B.; Bouley, S.; Vaubaillon, J.; Vernazza, P.; Gattacceca, J.; Marmo, C.; Audureau, Y.; Kwon, M.K.; Maquet, L.; et al. The FRIPON and Vigie-Ciel networks. In Proceedings of the 2014 International Meteor Conference, Giron, France, 18–21 September 2014; pp. 34–38.
11. Koten, P. Software for processing of meteor video records. In Proceedings of the Asteroids, Comets, Meteors—ACM 2002, Berlin, Germany, 29 July–2 August 2002; pp. 197–200.
12. Koten, P.; Stork, R.; Páta, P.; Fliegel, K.; Vítek, S. Simultaneous analogue and digital observations and comparison of results. In Proceedings of the 2016 International Meteor Conference, Egmond, The Netherlands, 2–5 June 2016; pp. 133–136.
13. Hamada, T.; Nitadori, K.; Benkrid, K.; Ohno, Y.; Morimoto, G.; Masada, T.; Shibata, Y.; Oguri, K.; Taiji, M. A novel multiple-walk parallel algorithm for the Barnes–Hut treecode on GPUs–towards cost effective, high performance N-body simulation. *Comput. Sci.-Res. Dev.* **2009**, *24*, 21–31.
14. Wayth, R.B.; Greenhill, L.J.; Briggs, F.H. A GPU-based real-time software correlation system for the murchison widefield array prototype. *Publ. Astron. Soc. Pac.* **2009**, *121*, 857–865.
15. Schive, H.Y.; Tsai, Y.C.; Chiueh, T. Gamer: A GPU-accelerated adaptive mesh refinement code for astrophysics. *Astrophys. J. Suppl. Ser.* **2010**, *186*, 457–484.
16. Thompson, A.C.; Fluke, C.J.; Barnes, D.G.; Barsdell, B.R. Teraflop per second gravitational lensing ray-shooting using graphics processing units. *New Astron.* **2010**, *15*, 16–23.
17. Brescia, M.; Cavuoti, S.; Djorgovski, G.S.; Donalek, C.; Longo, G.; Paolillo, M. Extracting knowledge from massive astronomical data sets. In *Astrostatistics and Data Mining*; Springer: New York, NY, USA, 2012.
18. Hassan, A.; Fluke, C.J.; Barnes, D.G.; Kilborn, V. Tera-scale astronomical data analysis and visualization. *Mon. Not. R. Astron. Soc.* **2013**, doi:10.1093/mnras/sts513.
19. Fliegel, K.; Švihlík, J.; Páta, P.; Vítek, S.; Koten, P. Meteor automatic imager and analyzer: Current status and preprocessing of image data. *SPIE Opt. Eng. Appl.—Int. Soc. Opt. Photon.* **2011**, doi:10.1117/12.893700.

20. Borovička, J.; Koten, P.; Spurný, P.; Boček, J.; Štork, R. A survey of meteor spectra and orbits: Evidence for three populations of Na-free meteoroids. *Icarus* **2005**, *174*, 15–30.
21. Stetson, P.B. DAOPHOT: A computer program for crowded-field stellar photometry. *Publ. Astron. Soc. Pac.* **1987**, *99*, 191–222.
22. Anisimova, E.; Bednář, J.; Blažek, M.; Janout, P.; Fliegel, K.; Páta, P.; Vítek, S.; Švihlík, J. Estimation and measurement of space-variant features of imaging systems and influence of this knowledge on accuracy of astronomical measurement. *SPIE Opt. Eng. Appl.—Int. Soc. Opt. Photon.* **2014**, doi:10.1117/12.2061736.
23. Kukal, J.; Klimt, M.; Švihlík, J.; Fliegel, K. Meteor localization via statistical analysis of spatially temporal fluctuations in image sequences. *SPIE Opt. Eng. Appl.—Int. Soc. Opt. Photon.* **2015**, doi:10.1117/12.2188186.
24. Canny, J. A computational approach to edge detection. *IEEE Trans. Pattern Anal. Mach. Intell.* **1986**, *6*, 679–698.
25. Cook, S. *CUDA Programming: A Developer's Guide to Parallel Computing with GPUs*; Newnes: Sebastopol, CA, USA, 2012.
26. Fluke, C.J.; Barnes, D.G.; Barsdell, B.R.; Hassan, A.H. Astrophysical supercomputing with GPUs: Critical decisions for early adopters. *Publ. Astron. Soc. Aust.* **2011**, *28*, 15–27.
27. Harris, M. How to Optimize Data Transfers in CUDA C/C++. Available online: https://devblogs.nvidia.com/parallelforall/how-optimize-data-transfers-cuda-cc/ (accessed on 13 July 2016).

sensors   MDPI

Article

Evaluation of a Wobbling Method Applied to Correcting Defective Pixels of CZT Detectors in SPECT Imaging

Zhaoheng Xie [1,†], Suying Li [1,†], Kun Yang [2], Baixuan Xu [3] and Qiushi Ren [1,*]

[1] Department of Biomedical Engineering, Peking University, No. 5, Yiheyuan Road, Beijing 100871, China; xiezhaoheng@163.com (Z.X.); lisuying90@163.com (S.L.)
[2] Department of Control Technology and Instrument, Hebei University, No. 180, Wusi East Road, Baoding 071000, China; yangkun9999@hotmail.com
[3] General Hospital of Chinese People's Liberation Army, No. 28 Fuxing Road, Beijing 100039, China; xbx301@163.com
* Correspondence: renqsh@coe.pku.edu.cn; Tel.: +86-10-6276-7113
† These authors contributed equally to this work.

Academic Editor: Gonzalo Pajares Martinsanz
Received: 17 February 2016; Accepted: 16 May 2016; Published: 27 May 2016

Abstract: In this paper, we propose a wobbling method to correct bad pixels in cadmium zinc telluride (CZT) detectors, using information of related images. We build up an automated device that realizes the wobbling correction for small animal Single Photon Emission Computed Tomography (SPECT) imaging. The wobbling correction method is applied to various constellations of defective pixels. The corrected images are compared with the results of conventional interpolation method, and the correction effectiveness is evaluated quantitatively using the factor of peak signal-to-noise ratio (PSNR) and structural similarity (SSIM). In summary, the proposed wobbling method, equipped with the automatic mechanical system, provides a better image quality for correcting defective pixels, which could be used for all pixelated detectors for molecular imaging.

Keywords: semiconductor detector; CZT; defective points; calibration; wobbling method

1. Introduction

Radionuclide imaging has become one of the most advanced molecular imaging techniques to monitor physiological functions [1,2]. However, Single Photon Emission Computed Tomography (SPECT) images tend to be "noisy" because of the low amount of radiotracer per volume of interest (VOI) and the effect of Compton scattering in tissue and collimators. Over the past decades, researchers have been dedicated to developing gamma-cameras with improved spatial resolution and energy resolution. Semiconductor nuclear radiation detectors, especially cadmium zinc telluride ($Cd_{1-x}Zn_xTe$, CZT) materials, have been considered as alternatives to scintillator detectors [3–5] because of their good stopping power and low dark current [4,6,7]. Notably, the major advantage of semiconductor over traditional scintillator detectors is that they can directly convert the deposited photon energy into measurable signals, which could improve energy resolution and detection efficiency [4,5,7,8]. Hence, CZT detectors are regarded to be the most promising option for SPECT imaging.

The CZT detector consists of a semiconducting crystal that is bump-bonded to a large area ASIC and packaged with a high performance data acquisition system. Recently, great progress has been made in electronics [9–12] and crystal growing [13,14]. However, there are still some defective pixels, showing degraded features, such as split or broadening spectral peaks and extraordinarily low or high response, especially for pixels with smaller sizes or at the edge of module. Though the defective pixels only occupy a small proportion of each module, separate defects of each module may cluster to form continuous defective regions if several modules are arranged into a large detector, as shown in Figure 1. The continuous defective region in detector occurs in projections and introduces ring artefacts [15] after reconstruction, which degrades image quality and may lead to misinterpretation, e.g., misdiagnosis or overdiagnosis [16].

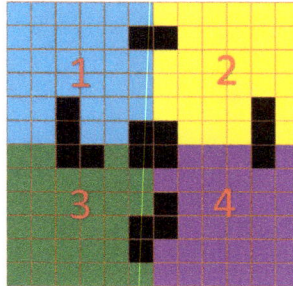

Figure 1. Example of different defective pixel patterns in multi-module pixelated detector; 1, 2, 3, 4 means four adjacent detector modules, black squares represent cluster of defective pixels.

Interpolation is a common correction method for defective pixels, which may work well for separate or individual bad pixels in homogeneous regions. For bad pixel cluster regions, however, simple interpolation often leads to inaccurate estimations. The determination of interpolation direction also poses more complexity since there are many variations for the constellation of defective pixels. Moreover, the advantage of interpolation method vanishes when the pixel size is small [17]. A proposed alternative plan is using sinogram-processing to eliminate the ring artifacts. Nonetheless, the 2D-wavelet-analysis [18] and polyphase decomposition in sinogram process [19] may introduce some noise or artifacts, and it has higher numerical complexity.

In this paper, an advanced wobbling method is proposed to correct the defective pixels. In order to accomplish the correction method and demonstrate its effects, we built a CZT SPECT system equipped with a simple mechanical device and conduct phantom experiment. The correction method is applied to various constellations of defective pixels. The correction effectiveness is evaluated quantitatively and is compared with conventional interpolation method. The results proved that the proposed wobbling method provides improvement in image quality of pixelated semiconductor detector, especially for the small object imaging with pinhole collimator.

2. Materials and Methods

2.1. Wobbling Correction Method

The wobbling method mainly includes following four steps:

2.1.1. 1st Step: Uniformity Correction and Wobbling Path Planning

The uniformity corrections are conducted by flood phantom first. From the flood image, some continuous bad pixels, which would affect image quality and cause ring artifacts after reconstruction, have been localized. Then all of defective pixels are recognized and the wobbling path is defined.

2.1.2. 2nd Step: Acquisition of the Wobbling Images

The basis of the wobbling correction method is the acquisition of wobbling images. Typically, it acquires different observations of the same object, *i.e.*, images with shifts of several pixel dimensions. Figure 2 shows an example of wobbling acquisition for the method. It is assumed that the projection on the detector is a "PKU"-shaped image and an "L"-shaped bad pixel pattern exists on CZT detector, as shown in Figure 2a. The detector performs a scan at Position 1 and subsequently acquires a wobbling image after a left shift of two pixel dimension in the horizontal direction (Figure 2b). In this way, we get the related images ("twin images", as shown in Figure 2c). Each of the wobbling images contributes useful information to the final corrected image and its correlation will be illustrated in the following 3rd Step.

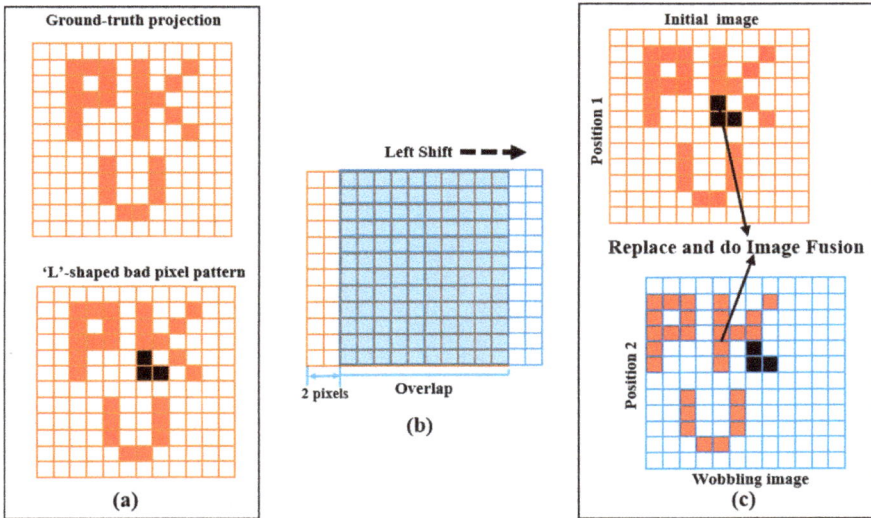

Figure 2. Description of the method for the wobbling technique. (**a**) Example of image without defectives and "L"-shaped bad-pixel pattern in detector; (**b**) The sketch of wobbling acquisition with Left shift; (**c**) Two images of different positions for the same object (Black points represent defective pixels in the fixed position of detector).

The wobbling acquisition can perform vertical and horizontal movement according to the defective-pixel patterns. In this paper, we mainly considered two-pixel shifts to correct defective pixels, so that there are 72.5% overlap between the two related images.

2.1.3. 3rd Step: Registration of Images Obtained from the Wobbling Method

After wobbling acquisition, each projection has two samples (images A and B), which respectively represent the initial and wobbling images, regarding the same imaging object. $f_A(x_A, y_A)$ and $f_B(x_B, y_B)$ are the number of emitted photons which are detected by each pixel. (x, y) means pixel index. The correlation between f_A and f_B can be expressed as a general rigid-body transformation that includes a combination of rotation and translation:

$$f_B(x_B, y_B) = R_{AB} \cdot f_A(x_A, y_A) + T_{AB} \tag{1}$$

where R_{AB} and T_{AB} is the translate transformation and rotate transformation between $f_A(x_A, y_A)$ and $f_B(x_B, y_B)$. A rigid-body transformation is applicable provided the pixel size of the CZT detector is negligible compared with the accuracy of the linear translation stage.

2.1.4. 4th Step: Replace the Defective Pixels and Apply an Image Fusion Algorithm

After establishing the point-to-point correspondence of f_A and f_B, we replace defective pixels using their counterparts and apply an image fusion algorithm to the two related images. The basic case of two roughly aligned images f_A and f_B with the overlapped area Ω is shown in Figure 3. In this paper, we mainly investigate the following two approaches to produce the mosaic image I_Ω:

(1)   Simple averaging function(Fusion method A):

$$I_\Omega = \frac{1}{2}f_A(x_A, y_A) + \frac{1}{2}f_B(x_B, y_B) \tag{2}$$

(2)   Distance weighted function(Fusion method B):

$$I_\Omega = \sigma f_A(x_A, y_A) + (1 - \sigma)f_B(x_B, y_B) \tag{3}$$

where $\sigma(0 \leqslant \sigma \leqslant 1)$ is the transition factor, as shown in Figure 3, based on the distance of the current pixel coordinate from its own boundary.

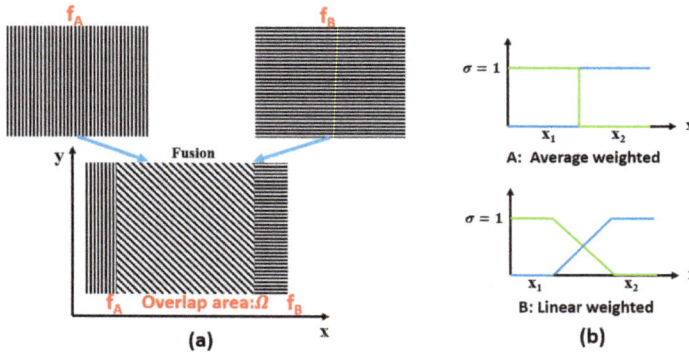

Figure 3. Image fusion algorithms and examples of different weighted function. (a) The overlapped regions are indicated by the slant dashed line. The two images on the left and right are to be stitched; (b) represents distinctive average features in overlapped region Ω. R and L are the region viewed exclusively in image f_A and f_B (L $\cap$ R = L $\cap$ Ω = R $\cap$ Ω = $\varnothing$).

The trade-off relationship between image quality and acquisition time is also considered here. Compared with the common acquisition mode, the acquisition time of each scan in the wobbling method is halved for two positions. In this way, the wobbling method can acquire sufficient counts without extending the data acquisition time. In this study, we are mainly interested in how to correct fixed defective pixels and edge-effects in CZT detectors, other defects such as geometric deformation, noise corruption and subsequent reconstruction algorithms, are not considered here.

2.2. System Description

The pinhole SPECT system is developed based on a CZT-detector and aiming for small animal imaging. The system, as shown in Figure 4, contains the detector, pinhole collimator and rotation stage where imaging objects are placed. The single pinhole collimator is made of tungsten alloy (ρ = 18.5 g/cm^3), with 0.8 mm aperture diameter, 1.38 mm channel height, and 60° opening angle. A linear translation stage is installed underneath the detector that provides support for the detector and the drive for wobble motion. It allows linear movement along the X, Y axes within a 50 mm range and its precision can reach 20 μm. The stage contains a grating ruler, which gives the feedback of motion position and guarantees the precision. With the translation stage, the wobbling range of detector can be adjusted from 0.25 to 50 mm in steps of 0.25 mm. The movement of linear stage is controlled by dedicated C++ codes, which are integrated into the acquisition software. After the first position the detector performs a scan and is subsequently wobbled for the following acquisition. Each scan is saved for post-processing according to the described wobbling method in Figure 2. A standard M1522 CZT detector module (Figure 4), has an active region of 40×40 mm^2 and 5 mm thick with Au contact (Redlen Technologies Inc., Saanichton, BC , Canada) was used in this work. The module is organized in a 16×16 array with a 2.46 mm pixel pitch and its acquisition software provides pixel position and energy value in a binary format. Energy resolution of this module is typical about 6.5% (Co-57 source).

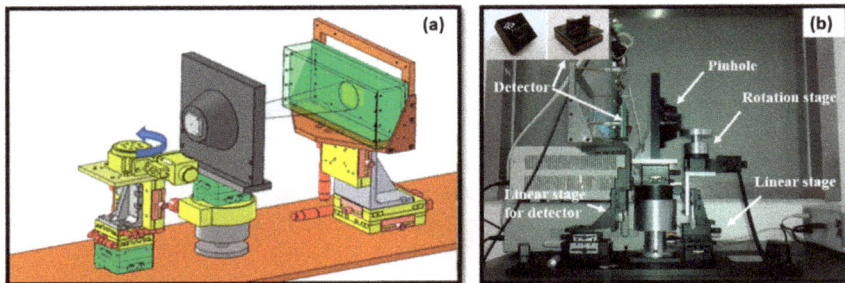

Figure 4. (a) Schematic diagram of the pinhole SPECT system design, which consists of rotary stage, collimator and CZT detector; **(b)** Close-up of the pinhole SPECT system.

2.3. Phantom Experiment

2.3.1. Flood Image Experiment

Flood images are obtained to check the performance of the CZT detector. We arranged four CZT modules into a 2×2 array, resulting in an 80×80 mm^2 detection area. It was exposed to a Tc-99m flood source with 2.22 mCi for 3600 s. The flood images are obtained with the 15% energy window (140 keV). The uniformity and energy spectrum of individual pixel as well as whole entire detector have been analyzed.

2.3.2. Line Source Phantom Experiment

To evaluate the performance of the proposed correction method, line source phantom experiment is designed as shown in Figure 5.

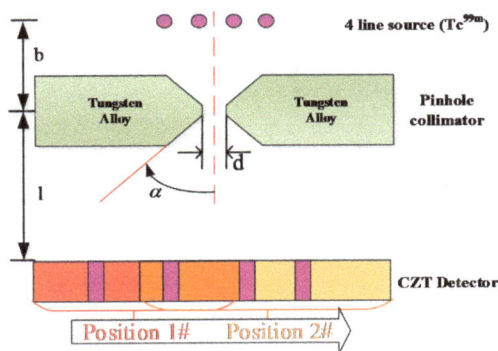

Figure 5. The geometry of keel-edge pinhole aperture and four linear sources phantom, where *b* is the perpendicular distance between source and the focal point (pinhole center), *l* is focal length, *d* is the physical diameter of pinhole, α is half of the pinhole opening angle.

Four line sources (2 mm inner diameter) are placed off-axis, symmetrically on the two sides of axis of rotation (AOR) which is 60 mm away from the aperture (120 mm from the detector surface). The center distance between each source is 4 mm. This geometric configuration approximately results in 2× magnification. Each line source is filled with 2.6 mCi/mL Tc-99m labelled medronate (MDP). After data acquisition, the correction and evaluation processes are performed by a Matlab-based program.

2.4. Image Quality Evaluation

Aiming at evaluating the effect of our wobbling correction method, we conducted phantom experiments and compared the results with a conventional correction method. Firstly, line sources are scanned statically as reference images. On the reference images, we deliberately defined some pixels as bad pixels and set the counts as zero, resulting in the image waiting to be corrected (identified as "bad image" in the following). Correction methods are applied to various bad-pixel patterns such as: (1) two bad pixels (vertical); (2) two bad pixels (horizontal); (3) three bad pixels; (4) four bad pixels. Here, we mainly investigate the effect of the wobbling method in the correction of bad pixels that appear in the region of interest (ROI) and appear in the background. We defined the effectiveness of a correction method as the similarity between the corrected image (C) and reference image (R). The similarity is assessed using peak signal-to-noise ratio (PSNR) and structural similarity (SSIM), which are the widely used full-reference quality metrics [20,21]. Given the reference image R and the corrected image C, both of size M × N, the PSNR between C and R is defined as:

$$PSNR(C,R) = 10\log_{10}(255^2/MSE(C,R)) \tag{4}$$

where:

$$MSE(C,R) = \frac{1}{MN}\sum_{i=1}^{M}\sum_{j=1}^{N}(C_{ij} - R_{ij}) \tag{5}$$

When MSE, which is short for mean squared error, approaches zero, the PSNR value approaches infinity, thus a higher PSNR value provides a higher image quality.

The SSIM is a well-known quality metric used to assess the similarity between two image, considering a combination of three factors that loss of correlation, luminance distortion and contrast distortion. The SSIM is defined as:

$$SSIM(C,R) = [l(C,R)]^{\alpha} \cdot [c(C,R)]^{\beta} \cdot [s(C,R)]^{\gamma} \tag{6}$$

where l(C, R) is the luminance comparison function, c(C, R) is contrast comparison function, and s(C, R) is structure comparison function and they are calculated as following equations:

$$l(C,R) = \frac{2\mu_C\mu_R + c1}{\mu_C^2 + \mu_R^2 + c1}, c(C,R) = \frac{2\sigma_C\sigma_R + c2}{\sigma_C^2 + \sigma_C^2 + c2}, s(C,R) = \frac{\sigma_{CR} + c3}{\sigma_C\sigma_R + c3} \tag{7}$$

Moreover, α, β, γ in Equation (7) are parameters used to adjust the relative importance of the three factors and the positive values constants c1, c2, c3 are used to avoid a null denominator. In order to simplify the expression, we set $\alpha = \beta = \gamma = 1$ and c1 = c2 = 2c3 in this paper. We use PSNR to assess the correction effect of pixels that appeared in ROI and use SSIM in terms of bad pixels appearing in the background. Here, we execute the wobbling correction method using averaging and distance-weighting fusion models (mentioned as "Method A" and "Method B") and compare the results with a conventional interpolation correction method.

3. Results

3.1. Flood Image Result

A typical image is shown in Figure 6. In the following pairs of spectra, we describe the different pixels' behavior. The black spots in Figure 6a represent dead pixels which have resulted in a large reduction of system sensitivity. Otherwise, pixels with split spectrum (247th pixel), broadening peaks (246th pixel) or non-standard response (65th pixel) are compared with "normal" pixels in Figure 6b,c. We found the unusual response pixels tend to bunch together, especially at the edge of the module. This may be attributed to the effects of local crystal structure variations. Overall, defective pixels represent less than 10% of all the pixels.

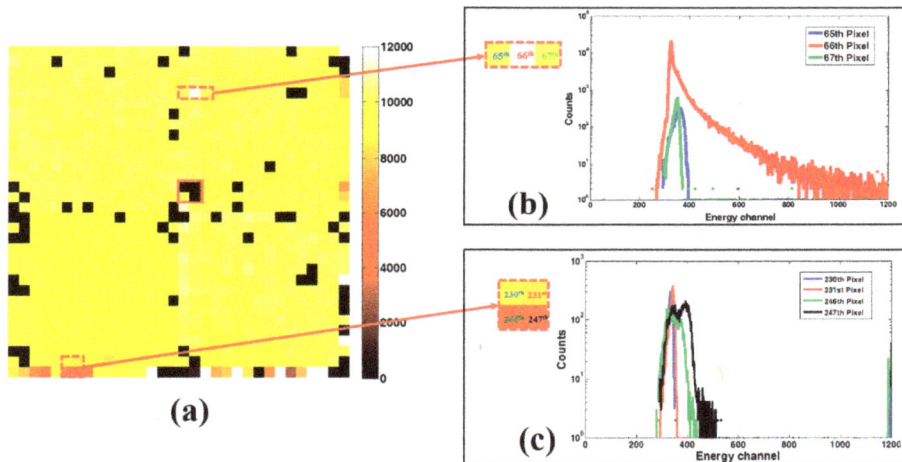

Figure 6. Spectra and counts are measured by flood phantom from a 2 × 2 modules of detector. (a) The Tc-99m flood image, the black spot (no counts) in the detector area (solid box) is caused by a cluster defect region. The white and brown spots (dashed box) in the image can be attributed to irregular response pixels; (b) Extreme high response pixels compared with adjacent normal pixels; (c) Pixels which show split or broadening spectral peak at the edge of detector.

3.2. Line Source Image Result

Figure 7 displays an example of the application of the wobbling correction method to four continuous bad pixels on the detector. A line source image is acquired statically as reference image (Figure 7a). Then we define four continuous bad pixels in ROI (8th and 9th row, 10th and 11th column) as Figure 7b. Figure 7c,d is the corrected image using a conventional interpolation method and the proposed wobbling correction method, respectively. Figure 8 compares line profiles of the pixel counts across the 8th row (the bad pixel location) after application of the different correction methods. The interpolation results in an obvious decrease of counts and information loss around the bad pixel region. Contrarily, the wobbling correction result shows similar counts as the reference image and recovers the missing information more effectively. The phantom results are valid examples to highlight the enhancement of the wobbling correction method.

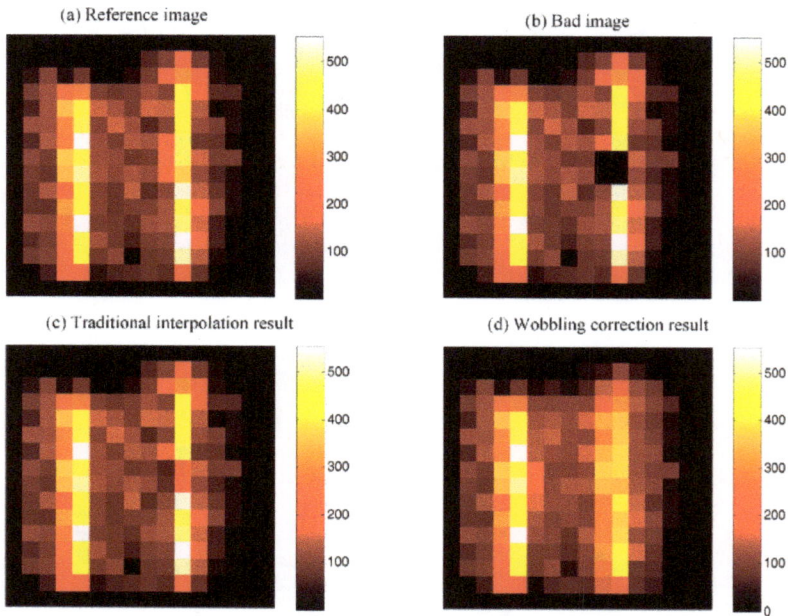

Figure 7. Results of the acquired line source phantom and correction results comparison. (**a**) Reference image: static imaging of line sources; (**b**) Bad image: image with 4 continuous bad pixels; (**c**) Conventional interpolation result: image using interpolation method; (**d**) Wobbling correction result: image using wobbling correction method.

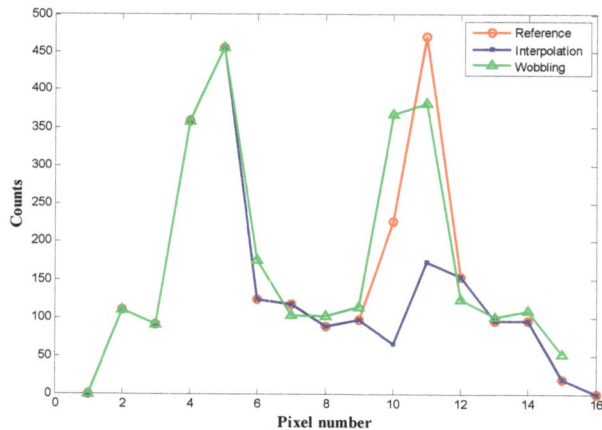

Figure 8. The pixel counts across row 8th of the corrected images. The blue line indicates the pixel counts after conventional interpolation and the green line indicates the counts after the proposed wobbling method is applied. The red line is the counts of the reference image.

4. Discussion

According to the evaluation factor defined in Section 2.4, the PSNR and SSIM are calculated for various bad pixel conditions with different pixel numbers and arrangements. The PSNR results of bad pixels appearing in the ROI are listed in Table 1. The SSIM results, assessing the correction effect for background bad pixels, are listed in Table 2. The improvements of the wobbling methods A and B compared with the conventional interpolation method are also marked in Tables 1 and 2. In order to demonstrate the improvement of the wobbling correction method explicitly, the PSNR and SSIM values of different correction methods for various bad-pixel numbers are plotted in Figures 9 and 10 respectively.

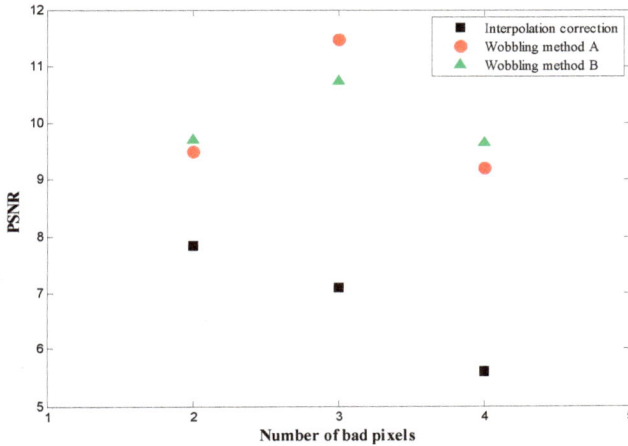

Figure 9. PSNR of different correction methods for various bad-pixel numbers.

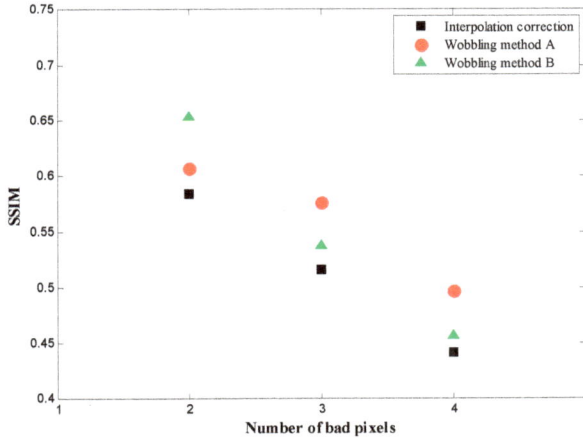

Figure 10. SSIM of different correction methods for various bad-pixel numbers.

The proposed wobbling correction method results in higher PSNR and SSIM, in particular when there are more than two continuous bad pixels. For the condition where bad pixels appear in the ROI, the improvement of the wobbling method results is around 10%~20% when the number of bad pixels is less than three. As the number of continuous bad pixels gets larger, he PSNR of the proposed wobbling method is significantly higher. For three continuous bad pixels, the PSNR of the wobbling method A and B is 1.6 and 1.5 times higher than the interpolation result. According to Table 2 of SSIM results, the overall improvement of the wobbling method is 10% for correction of background bad pixels, compared with conventional interpolation. As can be seen in Figures 8 and 9 defects up to 4 bad pixels can be corrected almost flawlessly when the wobbling correction is used. Conventional interpolation may already be inadequate for bad-pixel numbers up to four. This is because the conventional interpolation method is based on using the neighboring pixels to estimate the bad pixels' counts. When the defects cluster to form large regions of corrected pixels, the neighboring information is not sufficient for the correction, whereas the proposed method acquires two images of related wobbling positions, which can provide a comprehensive reference for correction, so the wobbling method is still effective when there are large defective regions. Moreover, the proper fusion model in the wobbling method can make effective use of the two wobbling images, producing desirable correction results.

Table 1. PSNR of correction for bad pixels appearing in the ROI.

Number of Bad Pixels	Shape	PSNR (Improvement %)		
		Conventional Interpolation	Method A	Method B
2	Vertical	6.3953	6.8457 (+7%)	7.3356 (+14.7)
2	Horizontal	7.8373	9.5113 (+21%)	9.7114 (+23.9%)
3	"L"-shaped	7.1087	11.4718 (+61.4%)	10.745 (+51.2%)
4	Square	5.6144	9.2081 (+64%)	9.6518 (+71.9%)

Table 2. SSIM of correction for background bad pixels.

Number of Bad Pixels	SSIM (Improvement %)		
	Conventional Interpolation	Method A	Method B
2	0.584	0.607 (+3.9%)	0.6532 (+11.8%)
3	0.5166	0.5767 (+11.6%)	0.5374 (+4%)
4	0.4419	0.497 (+12.5%)	0.4574 (+3.5%)

5. Conclusions

In this paper, we propose a novel method for correcting continuous bad pixels. Within the same acquisition time, images of two wobbling positions provide a reference for reasonable correction. We conduct corrections for various bad pixel conditions and use PSNR and SSIM estimators to evaluate the correction results. In the phantom experiment, the conventional interpolation method is used for comparison. The results show that wobbling method can correct continuous bad pixels effectively, no matter whether they appear in the ROI or background, whereas, a conventional interpolation method cannot effectively recover the original information, especially when there are more than three continuous bad pixels. In our future study, a proper image fusion algorithm of the wobbling images can be further investigated, resulting in a more effective correction. The wobbling method is conceptually simple, computationally efficient, and easy to use. Furthermore, this correction technique is potentially applicable to the standard pixelated detector, such as Si, GaAs and CdTe, with a proper motorized system.

Acknowledgments: This work is supported by the National Key Instrumentation Development Project Foundation of China (2011YQ030114), National Natural Science Foundation of China (11104058), Guangdong Innovative Research Team Program (2011S090), Foundation of Educational Commission of Hebei Province of China (ZD2015044), the Science Fund for Creative Research Groups of the National Natural Science Foundation of China (81421004) and the Natural Science Foundation of Hebei Province (A2011201155).

Author Contributions: The co-first author Zhaoheng Xie and Suying Li contributed to experiment conduction, data handling, paper writing; Kun Yang contributed to system motion control of linear stages and paper writing; Baixuan Xu contribute to preparation of 99mTc labelled MDP; Qiushi Ren contributed to experiment design and paper writing.

Conflicts of Interest: The authors declare no conflict of interest.

Abbreviations

The following abbreviations are used in this manuscript:

CZT   Cadmium zinc telluride
SPECT   Single Photon Emission Computed Tomography
VOI   Volume of interest
SNR   Signal-to-noise
PSNR   Peak signal-to-noise ratio
MSE   Mean squared error
SSIM   Structural similarity

References

1. Cherry, S.R.; Sorenson, J.A.; Phelps, M.E. *Physics in Nuclear Medicine*; Elsevier Health Sciences: London, UK, 2012.
2. Zaidi, H. *Quantitative Analysis in Nuclear Medicine Imaging*; Springer: New York, NY, USA, 2006.
3. Knoll, G.F. *Radiation Detection and Measurement*; John Wiley & Sons: New York, NY, USA, 2010.
4. Barber, H.; Apotovsky, B.; Augustine, F.; Barrett, H.; Dereniak, E.; Doty, F.; Eskin, J.; Hamilton, W.; Marks, D.; Matherson, K. Semiconductor pixel detectors for gamma-ray imaging in nuclear medicine. *Nucl. Instrum. Methods Phys. Res. Sect. A Accel. Spectrom. Detect. Assoc. Equip.* **1997**, *395*, 421–428. [CrossRef]
5. Awadalla, S. *Solid-State Radiation Detectors: Technology and Applications*; CRC Press: Boca Raton, FL, USA, 2015; Volume 41.
6. Barber, H.B. Application of II-VI materials to nuclear medicine. *J. Electron. Mater.* **1996**, *25*, 1232–1240. [CrossRef]
7. Scheiber, C.; Chambron, J. CdTe detectors in medicine: A review of current applications and future perspectives. *Nucl. Instrum. Methods Phys. Res. Sect. A Accel. Spectrom. Detect. Assoc. Equip.* **1992**, *322*, 604–614. [CrossRef]
8. Scheiber, C. CdTe and CdZnTe detectors in nuclear medicine. *Nucl. Instrum. Methods Phys. Res. Sect. A Accel. Spectrom. Detect. Assoc. Equip.* **2000**, *448*, 513–524. [CrossRef]
9. Matteson, J.L.; Skelton, R.T.; Pelling, M.R.; Suchy, S.; Cajipe, V.B.; Clajus, M.; Hayakawa, S.; Tümer, T.O. CZT detectors read out with the RENA-2 ASIC. In Proceedings of the Nuclear Science Symposium Conference Record, Fajardo, Puerto Rico, 23–29 October 2005; pp. 211–215.
10. Del Sordo, S.; Abbene, L.; Caroli, E.; Mancini, A.M.; Zappettini, A.; Ubertini, P. Progress in the development of CdTe and CdZnTe semiconductor radiation detectors for astrophysical and medical applications. *Sensors* **2009**, *9*, 3491–3526. [CrossRef] [PubMed]
11. Gao, W.; Gao, D.; Gan, B.; Wang, L.; Zheng, Q.; Xue, F.; Wei, T.-C.; Hu, Y. A novel data acquisition scheme based on a low-noise front-end asic and a high-speed adc for CZT-based pet imaging. In Proceedings of the 2012 18th IEEE-NPSS Real Time Conference (RT), Berkeley, CA, USA, 9–15 June 2012; pp. 1–4.
12. Zhang, Q.; Zhang, C.; Lu, Y.; Yang, K.; Ren, Q. Progress in the development of cdznte unipolar detectors for different anode geometries and data corrections. *Sensors* **2013**, *13*, 2447–2474. [CrossRef] [PubMed]
13. Amman, M.; Lee, J.S.; Luke, P.N.; Chen, H.; Awadalla, S.; Redden, R.; Bindley, G. Evaluation of thm-grown cdznte material for large-volume gamma-ray detector applications. *IEEE Trans. Nucl. Sci.* **2009**, *56*, 795–799. [CrossRef]
14. Shiraki, H.; Funaki, M.; Ando, Y.; Kominami, S.; Amemiya, K.; Ohno, R. Improvement of the productivity in the thm growth of CdTe single crystal as nuclear radiation detector. *IEEE Trans. Nucl. Sci.* **2010**, *57*, 395–399. [CrossRef]
15. Shepp, L.; Stein, J. Simulated reconstruction artifacts in computerized x-ray tomography. *Reconstr. Tomogr. Diagn. Radiol. Nucl. Med.* **1977**, *2*, 33–48.
16. Wischmann, H.-A.; Luijendijk, H.A.; Meulenbrugge, H.J.; Overdick, M.; Schmidt, R.; Kiani, K. Correction of amplifier nonlinearity, offset, gain, temporal artifacts, and defects for flat-panel digital imaging devices. In Proceedings of the International Society for Optics and Photonics Medical Imaging 2002, San Diego, CA, USA, 23 February 2002; pp. 427–437.

17. Orthen, A.; Wagner, H.; Martoiu, S.; Amenitsch, H.; Bernstorff, S.; Besch, H.J.; Menk, R.H.; Nurdan, K.; Rappolt, M.; Walenta, A.H. Development of a two-dimensional virtual-pixel x-ray imaging detector for time-resolved structure research. *J. Synchrotron Radiat.* **2004**, *11*, 177–186. [CrossRef] [PubMed]
18. Tang, X.; Ning, R.; Yu, R.; Conover, D. Cone beam volume ct image artifacts caused by defective cells in x-ray flat panel imagers and the artifact removal using a wavelet-analysis-based algorithm. *Med. Phys.* **2001**, *28*, 812–825. [CrossRef] [PubMed]
19. Anas, E.M.A.; Lee, S.Y.; Hasan, T. Removal of ring artifacts in x-ray micro tomography using polyphase decomposition and spline interpolation. In Proceedings of the 2010 International Conference on Electrical and Computer Engineering (ICECE), Dhaka, Bangladesh, 18–20 December 2010; pp. 638–641.
20. Pappas, T.N.; Safranek, R.J.; Chen, J. Perceptual criteria for image quality evaluation. In *Handbook of Image & Video Processing*; Academic Press: Cambridge, MA, USA, 2005; pp. 939–959.
21. Wang, Z.; Bovik, A.C. A universal image quality index. *IEEE Sign. Proc. Lett.* **2002**, *9*, 81–84. [CrossRef]

sensors

MDPI

Article

A Bevel Gear Quality Inspection System Based on Multi-Camera Vision Technology

Ruiling Liu *, Dexing Zhong, Hongqiang Lyu and Jiuqiang Han

School of Electronic and Information Engineering, Xi'an Jiaotong University, Xi'an 710049, China;
bell@xjtu.edu.cn (D.Z.); hongqianglv@xjtu.edu.cn (H.L.); jqhan@xjtu.edu.cn (J.H.)
* Correspondence: meggie@xjtu.edu.cn; Tel.: +86-29-8266-8665 (ext. 175)

Academic Editor: Gonzalo Pajares Martinsanz
Received: 6 July 2016; Accepted: 23 August 2016; Published: 25 August 2016

Abstract: Surface defect detection and dimension measurement of automotive bevel gears by manual inspection are costly, inefficient, low speed and low accuracy. In order to solve these problems, a synthetic bevel gear quality inspection system based on multi-camera vision technology is developed. The system can detect surface defects and measure gear dimensions simultaneously. Three efficient algorithms named Neighborhood Average Difference (NAD), Circle Approximation Method (CAM) and Fast Rotation-Position (FRP) are proposed. The system can detect knock damage, cracks, scratches, dents, gibbosity or repeated cutting of the spline, etc. The smallest detectable defect is 0.4 mm × 0.4 mm and the precision of dimension measurement is about 40–50 μm. One inspection process takes no more than 1.3 s. Both precision and speed meet the requirements of real-time online inspection in bevel gear production.

Keywords: multi-camera vision; bevel gear; defect detection; dimension measurement

1. Introduction

The bevel gear, also called the automobile shift gear, is an important part in an automobile transmission. Its quality directly affects the transmission system and the running state of the whole vehicle [1]. According to statistics, about one third of automobile chassis faults are transmission faults, in which gear problems account for the largest proportion [2]. Although the bevel gear's design, material and manufacturing technique are quite mature and standard, in today's high-speed production lines there are still inevitably defects such as dimension deviations, surface scratches, crushing, dents or insufficient filling, etc. Defective gears will not only reduce the performance of the car, but also cause security risks and potentially huge losses to manufacturers [3].

Figure 1a shows some kinds of bevel gears. The shape and structure of bevel gears are complicated and the types of defects various, so bevel gear quality inspection is very difficult. At present it is mainly completed manually, which is tedious and inefficient. It is hard to guarantee the quality stability and consistency. Recently Osaka Seimitsu Precision Machinery Corporation in Japan developed a high precision gear measuring instrument as shown in Figure 1b, but the inspection speed is very slow and cannot meet the requirements of high-speed production lines. There has been gear defect detection technology based on machine vision, but it's mainly used in the inspection of broken teeth and flat surfaces of general gears [4]. There is no precedent research for bevel gears with complicated shape and structure [5]. Because of the multiple measurement tasks and diverse defects, the traditional single vision system cannot obtain enough information. To measure the multiple dimensions quickly and accurately, to detect and classify the defects completely and precisely, a more efficient vision system is needed.

Machine vision is such an advanced quality inspection technology which uses intelligent cameras and computers to detect, identify, judge and measure objects [6–14]. To solve the above problems, this paper develops a multi-vision bevel gear quality inspection device which combines defect detection and dimension measurement in one system. Three efficient image processing algorithms named Neighborhood Average Difference (NAD), Circle Approximation Method (CAM) and Fast Rotation-Position (FRP) are proposed. The system successfully solves the problem of bevel gear defect detection and measurement at high speed and accuracy. One can find a short Chinese article describing this system in [15].

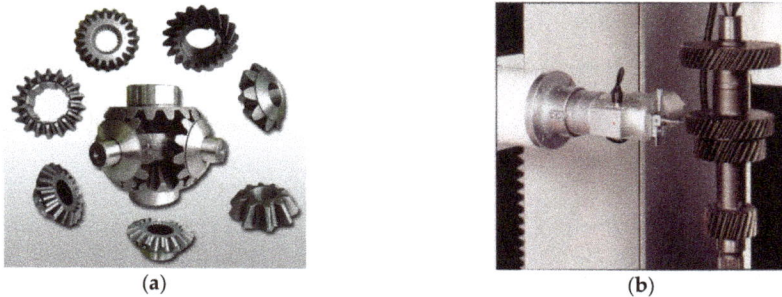

(a) (b)

Figure 1. (**a**) Samples of bevel gear; (**b**) Gear measuring instrument of Osaka Seimitsu Precision Machinery Corporation.

2. System Design

2.1. Inspection Project and Requirement Analysis

The defect detection and dimension measurement system developed in this paper is designed for two specific types of bevel gears, A_4 and F_6, as shown in Figure 2.

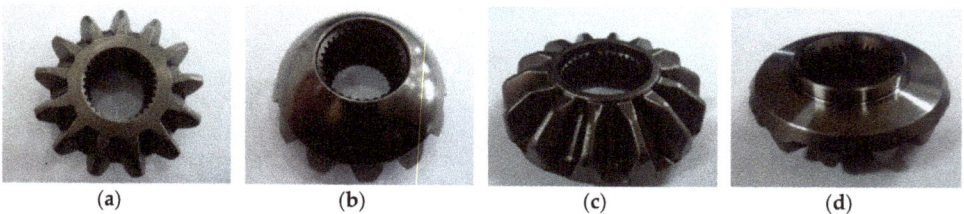

(a) (b) (c) (d)

Figure 2. Samples of gear A_4 and F_6: (**a**) Top surface of gear A_4; (**b**) Bottom surface of gear A_4; (**c**) Top surface of gear F_6; (**d**) Bottom surface of gear F_6.

The requirements of defect detection and measurement are as follows:

(1) Defect detection of tooth end surface, spherical surface and mounting surface, including knock damage, cracks, insufficient filling, folding, scratches, dents or gibbosity, repeated spline broaching, etc. As shown in Figure 3. Defects greater than 0.5 mm × 0.5 mm should be detected.
(2) Measurement of bevel gear's height, diameters of two circumcircles, spline's big and small diameters, end rabbet's diameter. Measurement accuracy is 80–100 μm.
(3) Defect detection speed is about 1 s for each surface.

2.2. Hardware Design

Bevel gears have many teeth and each tooth has different lateral surfaces. For example, gear A_4 has 13 teeth and 26 surfaces; gear F_6 has 14 teeth and 28 surfaces. The defects are diverse. How to realize the real-time online inspection of all the teeth and surfaces is a big challenge. One approach is using a single camera and rotational shooting mode, but it requires a high-precision rotary positioning system. The inspection speed is too low to meet the requirements. Because the bevel gear's teeth surfaces are uniformly distributed and rotationally symmetric, we propose a solution which can complete image acquisition and processing synchronously by using multiple high-speed cameras. The speed was an order of magnitude faster than single camera rotary positioning system.

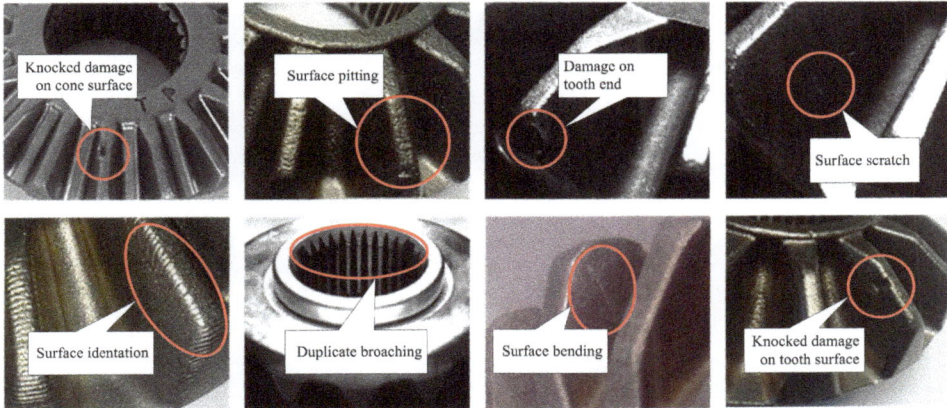

Figure 3. Examples of defects on bevel gear surface.

Figure 4 shows the total plan of our inspection system. Nine high speed USB cameras are used to capture different parts of the bevel gear simultaneously. The top camera is the core of the whole image acquisition system. It undertakes the positioning of the bevel gear, defect detection and measurement of the top and bottom surfaces.

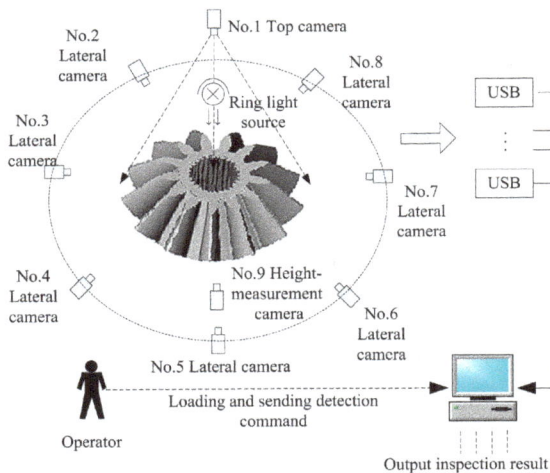

Figure 4. Diagram of bevel gear inspection system.

Seven side-view cameras are emplaced evenly around the bevel gear, with a 45° tilt angle with respect to the horizontal plane and perpendicular to gear tooth groove, detecting the lateral teeth surfaces omni-directionally. In addition, one more camera with a horizontal view is located on the platform to measure the bevel gear's height. After the gear is loaded, the operator gives a command that all of the cameras acquire images at the same time. An industrial computer receives and processes the images one by one and gives control signals according to the detected results. To meet the different accuracy requirements of bevel gear defect detection, three types of industrial cameras are adopted in this system. They are 3 megapixels, 2 megapixels and 1.3 megapixels, respectively. We can calculate that one pixel represents 40–50 μm according to the gear's actual size and image size. That is to say, the accuracy in the bevel gear's height and diameter measurements is 40–50 μm.

The system hardware includes four modules, which are camera module, control and process module, I/O module and gear rotary module, as shown in Figure 5. The camera module consists of nine high speed industrial cameras and a top ring light. The ring light illuminates the whole system. The cameras capture every part of the bevel gear. The control and processing module is responsible for image processing and output of instructions and inspection results. I/O module includes monitor, keyboard, mouse and pedal. The monitor displays the results. The software is operated by mouse and keyboard. The foot pedal starts the inspection of each surface. Software button starter and automatically trigger are also provided in this system. The gear rotation module is made up of a stepping motor and a reducer. It rotates the bevel gear precisely in the defect detection of special parts and template acquisition section.

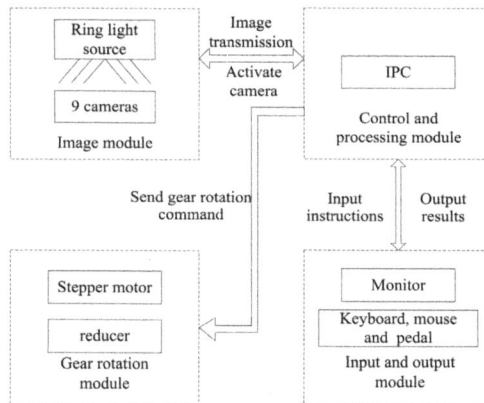

Figure 5. Hardware structure of the bevel gear inspection system.

Figure 6 shows the prototype of this multi-camera bevel gear quality inspection system. In order to adapt to various inspection tasks and objects, a series of position adjusting elements are designed. The height of top camera and illumination can be adjusted. Each camera can be adjusted slightly in the horizontal direction. For side-view cameras, the tilt angle and distance to the gear are adjustable in a large range, so the system can detect and measure the vast majority of bevel gears and other gears or parts of similar structure.

2.3. Software Design

Based on above hardware, a fast and multifunctional bevel gear inspection software is developed under the MFC and OpenCV platforms. The inspection includes two steps, top and bottom surface respectively. The software can automatically identify the current inspection is the top or the bottom of a bevel gear, and whether the bevel gear is in the right place. Top inspection includes two sub-modules, top surface and lateral surface, each one compose of size measurement and defect detection. The software flowchart is shown in Figure 7.

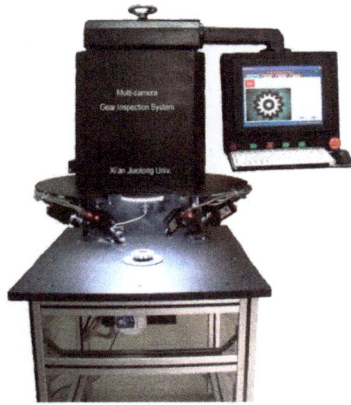

Figure 6. Prototype of multi-camera bevel gear inspection system.

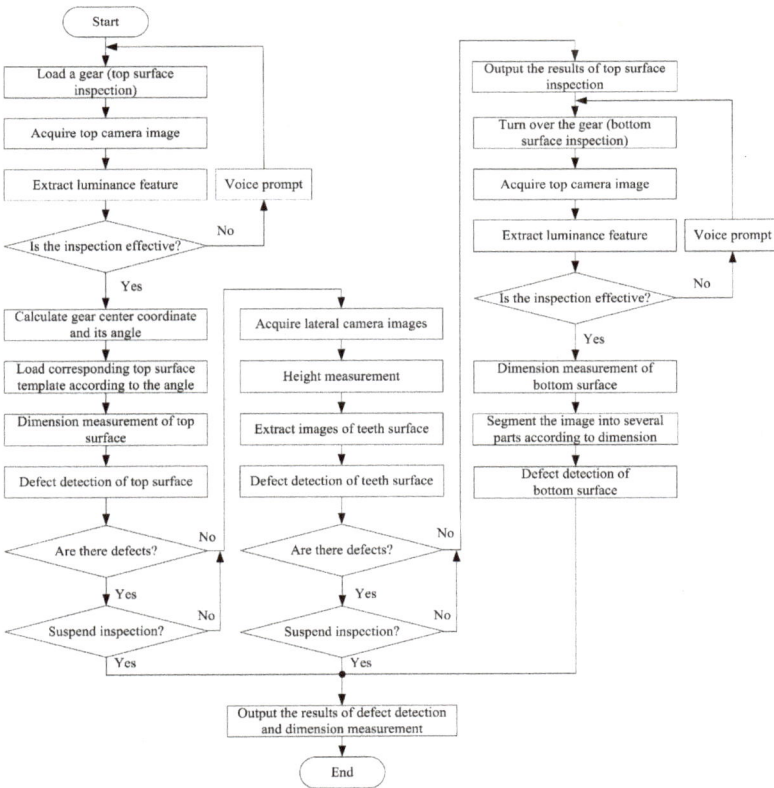

Figure 7. Software flowchart of bevel gear inspection system.

For a new inspection task, the software runs in order of top→lateral→bottom surface inspection. If any fault is found, the program marks it at once and outputs the defects and size measurement results, and then terminates. The software of the bevel gear inspection system employs a multi-thread technique. It includes four modules, which are defect detection, equipment calibration, template acquisition and parameter setting. The equipment calibration module calibrates the position, apertures and focuses of the nine cameras. The template acquisition module completes the sample collection and size calibration of a standard gear. The parameter setting module adjusts the related parameters for different kinds of inspection tasks. The defect detection module marks the faults in the gear image, and shows all the size and statistical data. Figure 8 shows the appearance of the software.

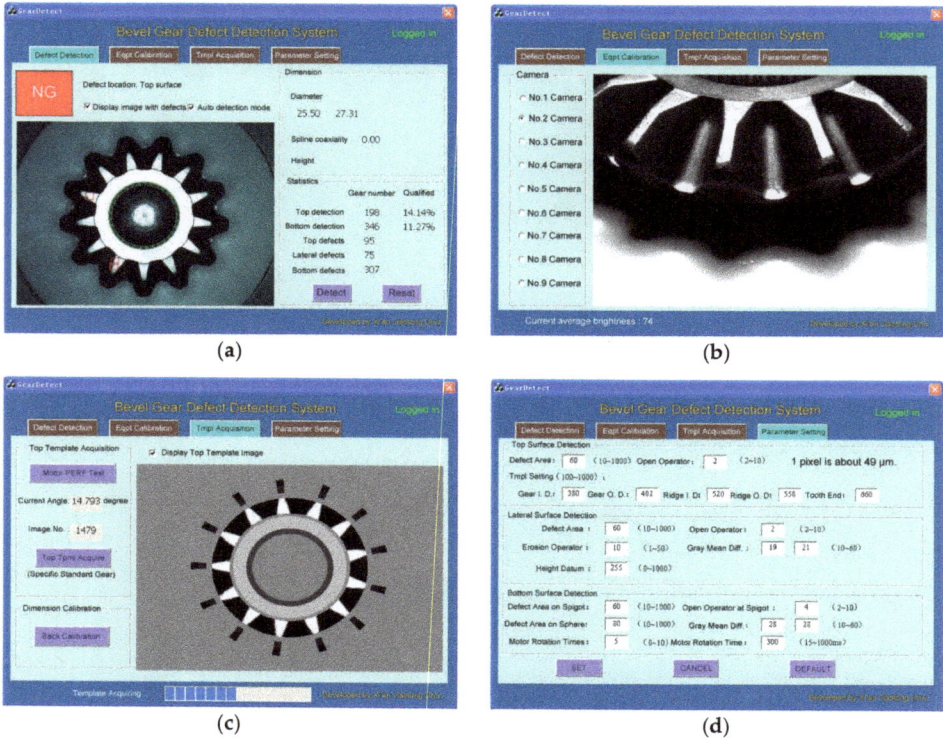

(a)

(b)

(c)

(d)

Figure 8. Soft interface of bevel gear inspection system: (**a**) Defect detection; (**b**) Equipment calibration; (**c**) Template acquisition; (**d**) Parameter setting.

3. Key Algorithms in Defect Detection

Because the defects of the bevel gears are complex and various, and the inspection time is short, the general algorithms cannot meet the requirements. In this system, we use multi-threading technology and propose several efficient image processing algorithms, which are NAD, CAM and FRP. The following introduces these key algorithms in detail.

3.1. Neighborhood Average Difference Method

From the point of texture, all kinds of defects are shown as waves, twists, undulations or roughness of the surface, thus defects can be detected by analyzing gear texture. Texture description operators based on gray-scale histograms include average, standard deviation, smoothness, third order moment, consistency and entropy, etc. A large number of experiments show that the combination of third order moment and smoothness do best in bevel gear defect detection, but these operators are sensitive to outside interference. The threshold setting is complicated and based on massive numbers of experiments and the results cannot show the characteristics of defects directly and comprehensively (such as area). In our system, we present an algorithm called Neighborhood Average Difference (NAD), which can extract the defects entirely and define larger ones as faults, ignoring tiny ones. The steps are as follows:

Step 1: Count the number of nonzero pixels in a neighborhood, denoted by N, given the neighborhood range is $L \times L$ pixels, here $L = 30$.

Step 2: Calculate the sum of gray values in the neighborhood: $S = \sum_{i=0}^{L-1} \sum_{j=0}^{L-1} p(i,j)$, in which $p(i,j)$ is the pixel value of point (i,j).

Step 3: Calculate the average pixel value in a neighborhood: $M = S/N$.

Step 4: For each point $p(i,j)$ in the neighborhood, if $|p(i,j) - M| > \delta$, δ is a threshold, then the point is marked by setting $p(i,j) = 255$.

Step 5: Scan the marked points in the image. If they are in connected region and the size of the region is greater than a threshold, then the region is a defect. Single marked points and tiny connected regions are ignored.

In addition, interference such as the impurities, dust and dirt on the surface can be removed partly by image dilation and erosion processing.

3.2. Circle Approximation Method

Circle detection is a classical problem in image processing [16–19]. To detect the actual and virtual circles on the bevel gear fast and accurately, an algorithm named CAM is presented. The main idea is: Draw a circle with initialized center and radius, then expands or contracts it slowly and moves the center continuously until it is tangent to the circle to be detected. Then keep adjusting its radius and center on condition that they remain tangent. The drawn circle will move closer and closer to the circle to be detected until they match perfectly. Figure 9 shows this process.

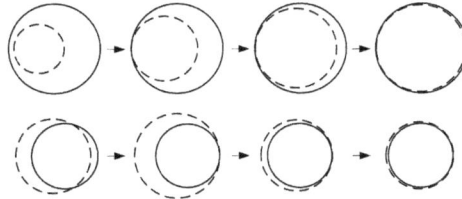

Figure 9. Circle Approximation Method: Each row shows an example of circle fitting process. The solid circles are to be detected and the dash circles are regulated to match the solid ones.

Circle center, radius and rotation angle step are three initial inputs of this method. The center and radius can be approximated only asking the drawn and the target circles have some common area. The more accurate the initial parameters are, the faster the method gets an idea result. It utilizes the positioning result in dimension measurement, and the steps are as follows:

Step 1: Extract the edge of the circle (actual or virtual) to be detected by threshold segmentation of the original image. Store the edge in target pixel set C.

Step 2: Set the initial center (x_0, y_0) and radius r_0 of the drawn circle. Set the rotation angle step θ.

Step 3: Increase or decrease radius r_0, $r_0 = r_0 + \delta_r$, or $r_0 = r_0 - \delta_r$, δ_r is the change value in each adjustment.

Step 4: Calculate the pixel set C' on the drawn circle with the current center and radius. Find the intersection of C and C', record as X. If X is empty, return to Step 3.

Step 5: If X is non-empty and the number of its elements is X_{num}, judge whether X_{num} satisfies the matching condition. If satisfies, output the center and radius information, otherwise move the center in its eight neighborhoods and find the position where X has the least elements, record and return to step 3.

The parameter θ is used in calculation of set X. Figure 10 shows how it is defined. R is the radius. A and B are the starting and ending point of a rotation. Line segment L connects A and B. For computational convenience, the center coordinate is set to (0, 0), B is set to $(R, 0)$, A is set to (x_1, y_1), then:

$$x_1 = R\cos\theta, \; y_1 = R\sin\theta \tag{1}$$

$$L^2 = (R - R\cos\theta)^2 + R^2\sin^2\theta \tag{2}$$

$$\theta = \arccos[(2R^2 - L^2)/2R^2] \tag{3}$$

Because the image is discrete, $L \geq 1$, then across $\left(\frac{2R^2-1}{2R^2}\right) \leq \theta \leq \pi$.

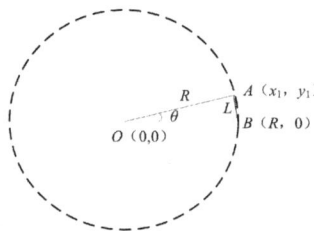

Figure 10. The calculation of rotation angle step θ in CAM.

The method counts target pixels on the drawn circle based on the parameter θ, it searches step by step like the second hand of a clock. The value of θ is closely related to the cost of computation. Theoretically, the greater the θ is, the faster the algorithm is, but the robustness of the algorithm will be decreased and the detection accuracy will be reduced at the same time, so the detection accuracy, the execution speed and the robustness of the algorithm should be all taken into account to set the angle step θ. In addition, θ can be set dynamically. In early of the algorithm a bigger θ is used to approach the detected circle quickly. θ becomes smaller as soon as the drawn circle touches the detected circle which is then searched and fitted in a small range. The algorithm is efficient and accurate with a dynamic theta.

Similar to Hough Transform, our method is robust to image noise and disturbance in virtue of cumulative sum. It has a simple searching idea and only requires a preprocessing of threshold segmentation. While in Hough Transform, preprocessing of edge detection or skeleton extraction is very time consuming.

In order to verify the detection results of our algorithm, we compare it with Hough transform by fitting the moon in Figure 11. The results of the two methods are almost identical and cannot be distinguished by naked eye. Details of the comparison are shown in Table 1. The data show that CAM is about 50 times faster than Hough Transform, with little difference in initial parameter setting. In addition, the classical Hough Transform cannot be used in high resolution images because the parameter space matrix will increase dramatically with the increase of the image size, resulting in a large amount of computation and storage space. However, our circle approximation algorithm is not restricted in this aspect. It is simple and fast with less computation, and is very suitable for the high precision circle detection situation.

CAM is suitable for fitting circles when a good initialization is available. Hough is suitable for fitting circles in the presence of noise and multiple models (multiple circle arcs). A solution from Hough transform will generally require refinement using an iterative method like CAM.

Our method is more suitable for detection of circles which have been positioned approximately. In actually, with the auxiliary system composed by sensors and cameras, the location of bevel gear won't deviate greatly. The gross location of the circle can be determined by software in advance. So our method can be successfully applied to most of the industrial circle detection occasions. The detection and production efficiency can be improved significantly with its speed advantage.

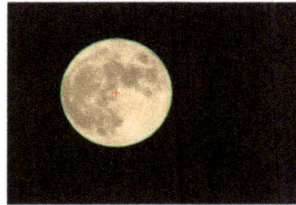

Figure 11. Moon picture (952 × 672 pixels) and circle fitting result. Red cross is the center.

Table 1. Compare Circle Approximation Method with Hough Transform in moon fitting.

Method	Min Radius/pixel	Max Radius/pixel	Center	Step Size of Angle/rad	Step Aize of Radius/pixel	Time/s
Hough Transform	150	250	Null	0.1	1	5.6623
CAM	150	250	Image center	0.006	0.5	0.1189

Like other circle detection methods, our method can detect solid circles and arcs. But it can also detect virtual circles such as gear spline circle, inscribed circle and circumscribed circle of a polygon. Figure 12 shows the fitting results of spline virtual circles on gear F_6 by our method.

(a) (b)

Figure 12. Fitting the spline circles by CAM: (**a**) Image of spline; (**b**) Detection results of inscribed and circumscribed circles of spline (denoted by green circles).

3.3. Fast Rotation-Position

In order to carry out accurate inspection of the bevel gear, the position of the workpiece should be determined quickly after loading. It includes horizontal, vertical and rotational position. This system uses a mechanical device to realize the horizontal and vertical position. Due to the complex gear structure and the short feeding time, it is difficult to achieve effective rotary position by hardware. We solve this problem by image processing method. Taking into account the center rotary symmetry of the bevel gear, the rotation position is divided into three steps which are center positioning, image acquisition of tooth surface and the rotary angle calculation:

(1)   Center positioning. To cut out the image of the inner circle, first segment the top camera image with a certain threshold, then remove interference of impurity and inherent defects by foreground and background area filtering. The circle center and diameter are extracted by area statistics and coordinates averaging. Figure 13a–c show this process.

(2)   Teeth-end image extraction. Segment Figure 13a again with a smaller threshold value and cut out the entire teeth-end image. Remove interference of impurity and inherent defects by foreground and background area filtering. Then draw two circles with the center calculated above and proper radiuses, extract only the teeth-end surface, as shown in Figure 13d–f.

(3)   Calculate the rotation angle. In order to implement template matching algorithm in defect detection, the original image of the bevel gear should be transformed to a standard position to extract the matched temple. The following shows how to calculate the rotary angle α.

For example, bevel gear A_4 has 13 teeth, which are distributed evenly. The angle between two teeth is $360°/13 = 27.69°$, as shown in Figure 14. O is the center of the gear and P_i is the geometric center of a tooth. The angle between x-axis and line OP_i is defined as α_i. For each tooth, α_i is converted into the first quadrant and within $0°$–$27.69°$. For gear A_4, $\alpha_i = \alpha_1, \alpha_1, \ldots, \alpha_{13}$. We eliminate any α_i that has great difference with others, and calculate the mean of the rest, which is defined as the rotation angle α of the gear to be detected. Experiments show that the rotation-position method provides excellent accuracy. The error is about $0.005°$, which can meet the requirements of the gear position determination.

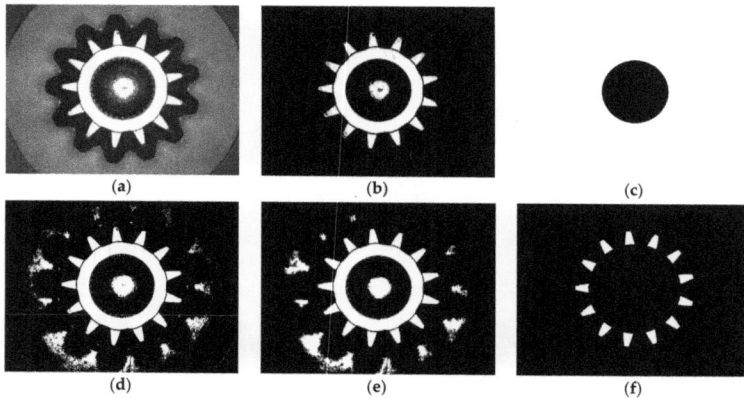

Figure 13. The image preprocess of FRP method: (**a**) Top surface image; (**b**) Segmentation for inner circle (with a larger threshold); (**c**) Inner circle detected; (**d**) Segmentation for teeth end (with a smaller threshold); (**e**) Result of area filting; (**f**) Teeth end image.

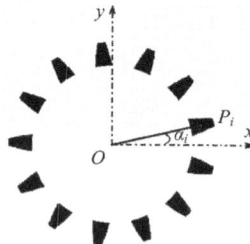

Figure 14. Calculation of gear rotation angle.

3.4. Template Matching and Collection

In machine vision, we often compare an input image with a standard image to find the difference or target, which is called template matching. It's fast and efficient. In this system we use template matching a lot to extract the area to be detected, eliminating the interference of non-inspection areas and improving the inspection speed.

Figure 15a is the top surface image of bevel gear A_4 and Figure 15b is the matching template. In Figure 15b, the value of background and edge of the gear are set to 150, which means they are a non-inspection area. Black, white and some other gray colors indicate different parts to be detected.

The templates are collected from a standard and perfect gear. Here is the acquisition process of top surface template. First extract the gear's center and rotation angle, and then divide target pixels into different parts according to their distance to the center and mark with different colors, remove background pixels completely. The templates are numbered and saved in order of rotation angle.

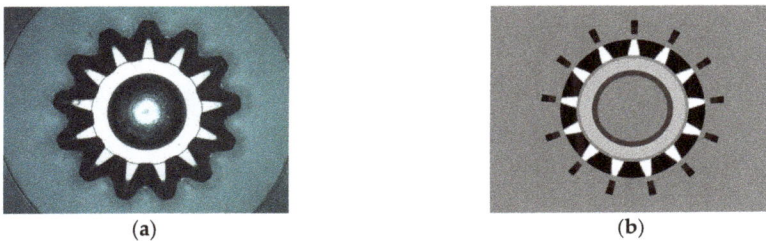

(a)   (b)

Figure 15. (a) Top surface of gear A4; (b) The matching template.

The lateral template is collected by the aid of a special gear painted with black and white matted coating, shown in Figure 16a. This ensures that a perfect template can be extracted through a simple image processing, shown in Figure 16b. Figure 16c is an original lateral image and Figure 16d is the extracted teeth surface by template matching.

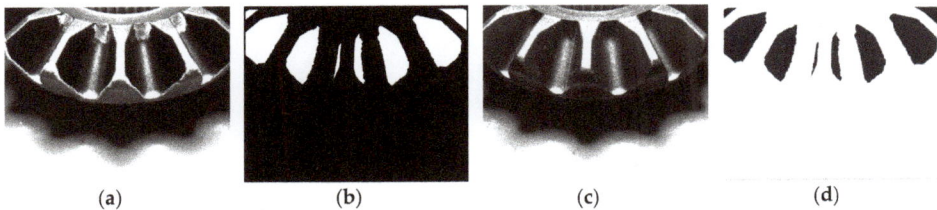

(a)   (b)   (c)   (d)

Figure 16. The collection and use of bevel gear lateral template. (a) Lateral image of template gear; (b) Extracted template; (c) Lateral image to be detected; (d) Extracted image by template matching.

In this system the templates collection and processing are completed automatically without artificial participation. Templates are automatically numbered and saved in order of rotation angle and camera. After a template is collected, the gear is rotated to another angle by the step-motor and the next collection begins. The angle interval is about $0.01°$–$0.02°$. The total template image size is as large as 25 GB, that is to say, the system achieves ideal inspection speed by sacrificing storage space.

4. Inspection Results

The inspection tasks and methods are different for bevel gears vary in structure. Figure 17a,b are bottom surfaces of gear F_6 and A_4 respectively. F_6 is flat and defects are obvious in the image. It is easy to detect. A_4 is spherical and the image is dark and weak in contrast. The defects are hard to be found in the image, especially for dints. The inspection algorithm of A_4 is much more complicated.

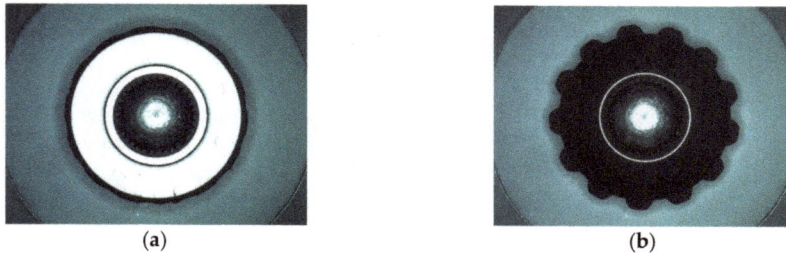

(a)

(b)

Figure 17. Bottom images of F_6 and A_4: (**a**) Bottom of F_6; (**b**) Bottom of A_4.

4.1. Example of Defect Detection and Dimension Measurement

In this section, we illustrate how the system works through a practical example, the inspection of bottom surface of bevel gear A_4.

4.1.1. Inspection of Top-Circle on Bottom Surface

The top of back surface is a standard sharp ring in the image of a perfect A_4 gear. But it is fragile and easily to be damaged in production process. Defects are shown as broken ring, burr or uneven color in the image. The system will detect and mark the defects as shown in Figure 18. The following illustrate the detail inspection process.

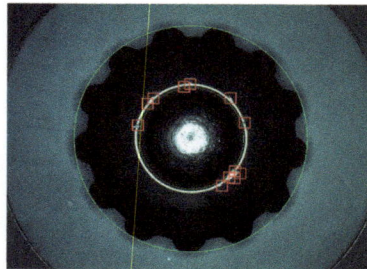

Figure 18. Inspection of bottom surface of gear A_4, red boxes mark defects and green circle marks circumcircle.

Step 1: Extract the ring and its neighborhood by threshold segmentation, and then remove small noise around the ring by area filter, the result is shown in Figure 19a.

Step 2: Fitting the ring boundaries. Generally the bevel gear is centrosymmetric, figure out the approximate center by average coordinate. Fit the inner and outer boundary circles of the ring by CAM. Red circles in Figure 19b are the fitting result. This is preparatory work for Step 3 and the following dimension measurement.

Step 3: Defect detection. The pixels of the top ring are highlighted for a qualified gear, so the dark pixels between two red circles are marked as defects. For the inside and outside neighborhood of the ring, defects are located by the combination of fixed and dynamic threshold segmentation. The merged result is shown as the white pixel groups in Figure 19c. Then eliminate additional arc by image open-operation and extract bigger defects by area filtering, as shown in Figure 19d. The end result is marked by red boxes in Figure 18.

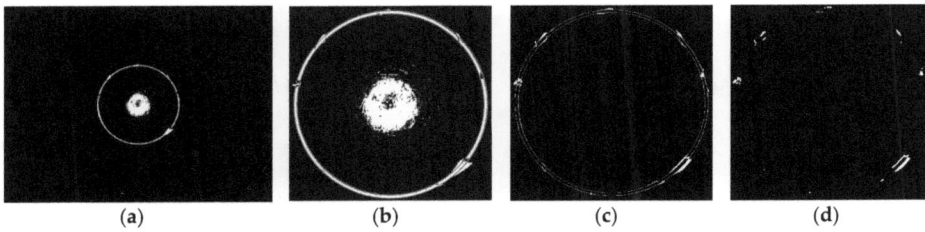

Figure 19. Defect detection of top ring on bottom surface: (**a**) Top ring extraction; (**b**) Top ring fitting (red circles); (**c**) Result of defect detection; (**d**) Defects after open-operation.

4.1.2. Inspection of Spherical Surface

The defects on spherical surface are not obvious in the top camera image and should be detected from lateral camera images. Figure 20a shows such an image. The region just facing the camera is highly lit and both sides are darker and darker. Experiments show that defects in the transition region are the most obvious, so the main inspection area is between the high light and the darkest regions.

To extract the target region fast, the system creates a template for each lateral camera, as shown in Figure 20b. Gray indicates the inspection region and black/white is the non-inspection region. The white stripe is used for device calibration and avoiding erroneous inspections. The whole sphere surface cannot be covered by seven cameras at a time. To ensure the defects can be detected at any angle, the gear is rotated 3 to 5 times by a stepper motor.

Figure 20. (**a**) The lateral image of sphere surface; (**b**) The template of sphere surface.

The motor rotation is time-consuming. The motor begins to rotate just at the end of a single image acquisition. The rotation completes when the image processing finishes, then the next image acquisition begins. Inspection time is saved greatly by this process. The defects are detected by the NAD method. Figure 21a shows an image with defects in it and Figure 21b shows the inspection result.

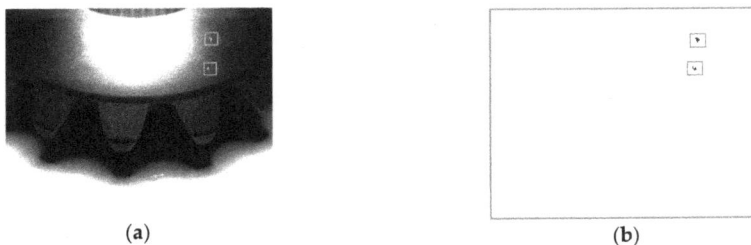

Figure 21. (**a**) A lateral image of the sphere surface; (**b**) Inspection result of Figure 21a.

4.1.3. Measurement of End Rabbet Diameter

Size measurement includes the diameters of large and small circumcircles and end rabbets. The two circumcircles are relatively easy to fit by threshold segmentation and CAM. The fitting results are shown as green circles in Figure 22a. The end rabbet is a shallow circular groove on the gear. Figure 22b shows its location and 22c is an amplified image. The division of two regions of different gray values in Figure 22c is the rabbet circle. It is dark in the image and hard to identify.

Because the rabbet circle is not distinct in the image, fixed or dynamic thresholding such as the Otsu algorithm or the interactive method are not applicable [20–22]. Figure 23a shows the gray histogram of the rabbet circle and its nearby pixels. We find that the pixels on the rabbet circle are the darkest. They are the left part of the histogram, about 3000–5000 pixels. We extract these pixels and then get the outline of the rabbet circle, as shown in Figure 23b.

We use CAM to fit the rabbet circle in Figure 23b. The result is shown as the red circle in Figure 22a. We prove again that our method is much faster than the Hough Transform and least-squares method while ensuring the accuracy.

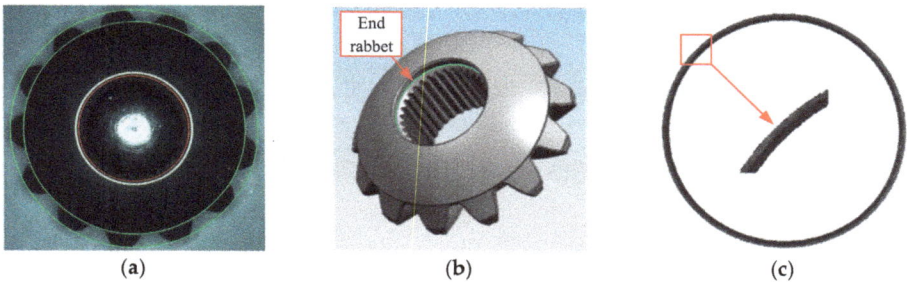

Figure 22. Size measurement of circumcircles and end rabbet: (**a**) Green circles are circumcircles and red is end rabbet circle. (**b**) Location of end rabbet; (**c**) Extracted rabbet and its amplified image.

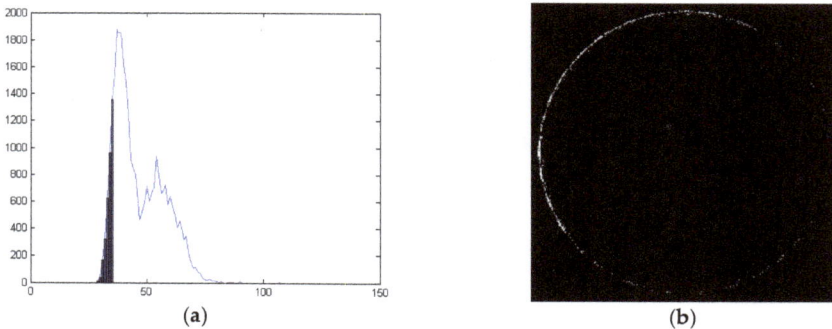

Figure 23. (**a**) Histogram of End rabbet and nearby pixels; (**b**) Extract the outline of rabbet circle.

4.2. Statistical Results of Bevel Gear Inspection

Using the inspection system developed in this paper, 84 bevel gears of type A_4 and 84 of type F_6 are tested. Tables 2–4 are the inspection results of A_4. Tables 5–7 are the inspection results of F_6. The data shows that size measurement of our system is 100% accurate under the current requirements. Measurement accuracy is about 45 μm and repeated error is no more than 90 μm. The minimum detectable defect is 0.4 mm × 0.4 mm. Inspection time of one surface is less than 1.3 s. Both the accuracy and the speed are suitable for the real-time on-line inspection in automatic production lines.

Table 2. Top surface inspection of gear A_4.

Results	Top Circle	Teeth Surface				Teeth Edge	
	Connected Fault	Knock	Dent	Knock	Insufficient Filling	Knock	Insufficient Filling
Number of defects	1	7	10	15	1	13	10
Detected defects	1	7	10	12	1	11	10
Accuracy/%	100	100	100	80	100	84.6	100

Table 3. Lateral surface inspection of gear A_4.

Results	Knock	Scratch	Dent	Gibbosity	Roughness	Peeling off
Number of defects	30	23	18	1	6	1
Detected defects	28	19	16	1	6	1
Accuracy/%	93.3	82.6	88.9	100	100	100

Table 4. Bottom surface inspection of gear A_4.

Results	Bottom Ring		Spherical Surface			Spline
	Knock	Scratch	Knock	Scratch	Dent	Duplicated Broaching
Number of defects	41	12	15	23	19	6
Detected defects	39	12	15	22	19	6
Accuracy/%	95.1	100	100	95.6	100	100

Table 5. Top surface inspection of gear F_6.

Results	Top Ring			Top Teeth Surface		Teeth Edge	
	Chamfer Lack	Knock	Dent	Knock	Dent	Knock	Insufficient Filling
Number of defects	2	7	8	17	9	34	18
Detected defects	2	7	8	16	9	30	18
Accuracy/%	100	100	100	94.1	100	88.2	100

Table 6. Lateral surface inspection of gear F_6.

Results	Lateral Teeth Surface				Lateral Teeth Small Surface		
	Knock	Dent	Scratch	Insufficient Filling	Knock	Dent	Scratch
Number of defects	42	8	11	3	13	5	4
Detected defects	37	8	8	3	11	5	3
Accuracy/%	88.1	100	72.7	100	84.6	100	75

Table 7. Bottom surface inspection of gear F_6.

Results	Bottom Ring		Mounting Surface			Spline
	Knock	Rough	Knock	Scratch	Size Error	Duplicated Broaching
Number of defects	3	3	2	5	4	3
Detected defects	3	3	2	4	4	3
Accuracy/%	100	100	100	80	100	100

5. Discussion

The presented multi-camera automobile bevel gear quality inspection system adopts advanced machine vision technology. It solves the problem of difficult inspection of complicated gears at high speed and with high precision. The research shows that machine vision can substitute most manual work in bevel gear inspection, improving the efficiency of production and the degree of automation.

The prototype of this system has been realized and applied in bevel gear production. There are still some technical problems to be solved. The edge of the gear is a non-detectable area in our system and some defects in low contrast images cannot be detected reliably. We are seeking better illumination schemes and new inspection ideas to solve these problems.

Acknowledgments: The work was supported by Natural Science Foundation of Shaanxi, China (No. 2012JQ8042) and the grants from National Natural Science Foundation of China (No. 61105021).

Author Contributions: All authors contributed extensively to the work presented in this paper and wrote the manuscript. Ruiling Liu reviewed the state-of-the-art and wrote the initial version of the paper. Dexing Zhong and Hongqiang Lyu participated the analysis of the vision techniques. Jiuqiang Han instructed the whole development of the system.

Conflicts of Interest: The authors declare no conflict of interest.

Abbreviations

NAD	Neighborhood Average Difference
CAM	Circle Approximation Method
FRP	Fast Rotation-Position

References

1. Shao, Y.M.; Liang, J.; Gu, F.S.; Chen, Z.G.; Ball, A. Fault prognosis and diagnosis of an automotive rear axle gear using a RBF-BP neural network. In *9th International Conference on Damage Assessment of Structures*; Ouyang, H., Ed.; IOP Publishing: Bristol, UK, 2011; Volume 305, pp. 613–623.
2. Parey, A.; Tandon, N. Impact velocity modelling and signal processing of spur gear vibration for the estimation of defect size. *Mech. Syst. Signal Process.* **2007**, *21*, 234–243. [CrossRef]
3. Saravanan, N.; Ramachandran, K.I. Fault diagnosis of spur bevel gear box using discrete wavelet features and decision tree classification. *Expert Syst. Appl.* **2009**, *36*, 9564–9573. [CrossRef]
4. Mark, W.D.; Lee, H.; Patrick, R.; Coker, J.D. A simple frequency-domain algorithm for early detection of damaged gear teeth. *Mech. Syst. Signal Process.* **2010**, *24*, 2807–2823. [CrossRef]
5. Hayes, M. Better Bevel Gear Production. Available online: http://www.gearsolutions.com/article/detail/6076/better-bevel-gear-production (accessed on 23 August 2016).
6. Golnabi, H.; Asadpour, A. Design and application of industrial machine vision systems. *Robot. Comput. Integr. Manuf.* **2007**, *23*, 630–637. [CrossRef]
7. Sun, T.H.; Tseng, C.C.; Chen, M.S. Electric contacts inspection using machine vision. *Image Vis. Comput.* **2010**, *28*, 890–901. [CrossRef]
8. Cubero, S.; Aleixos, N.; Molto, E.; Gomez-Sanchis, J.; Blasco, J. Advances in machine vision applications for automatic inspection and quality evaluation of fruits and vegetables. *Food Bioprocess Technol.* **2011**, *4*, 487–504. [CrossRef]
9. Fernandes, A.O.; Moreira, L.F.E.; Mata, J.M. Machine vision applications and development aspects. In Proceedings of the 2011 9th IEEE International Conference on Control and Automation (ICCA), Santiago, Chile, 19–21 December 2011.
10. Carrio, A.; Sampedro, C.; Sanchez-Lopez, J.L.; Pimienta, M.; Campoy, P. Automated low-cost smartphone-based lateral flow saliva test reader for drugs-of-abuse detection. *Sensors* **2015**, *15*, 29569–29593. [CrossRef] [PubMed]
11. Huang, K.Y.; Ye, Y.T. A novel machine vision system for the inspection of micro-spray nozzle. *Sensors* **2015**, *15*, 15326–15338. [CrossRef] [PubMed]
12. Liu, Z.; Li, X.J.; Li, F.J.; Wei, X.G.; Zhang, G.J. Fast and flexible movable vision measurement for the surface of a large-sized object. *Sensors* **2015**, *15*, 4643–4657. [CrossRef] [PubMed]
13. Fremont, V.; Bui, M.T.; Boukerroui, D.; Letort, P. Vision-based people detection system for heavy machine applications. *Sensors* **2016**, *16*, 128. [CrossRef] [PubMed]

14. Perez, L.; Rodriguez, I.; Rodriguez, N.; Usamentiaga, R.; Garcia, D.F. Robot guidance using machine vision techniques in industrial environments: A comparative review. *Sensors* **2016**, *16*, 335. [CrossRef] [PubMed]

15. Liu, R.; Zhong, D.; Han, J. Bevel gear detection system with multi-camera vision. *Hsi-An Chiao Tung Ta Hsueh/J. Xi'an Jiaotong Univ.* **2014**, *48*, 1–7.

16. Qiao, N.; Ye, Y.; Huang, Y.; Liu, L.; Wang, Y. Method of circle detection in pcb optics image based on improved point hough transform. In Proceedings of the 4th International Symposium on Advanced Optical Manufacturing and Testing Technologies: Optical Test and Measurement Technology and Equipment, Chengdu, China, 19 November 2008.

17. Wu, J.P.; Li, J.X.; Xiao, C.S.; Tan, F.Y.; Gu, C.D. Real-time robust algorithm for circle object detection. In Proceedings of the 9th International Conference for Young Computer Scientists, ICYCS 2008, Zhangjiajie, China, 18–21 November 2008.

18. Zhang, M.Z.; Cao, H.R. A new method of circle's center and radius detection in image processing. In Proceedings of the 2008 IEEE International Conference on Automation and Logistics, Qingdao, China, 1–3 September 2008.

19. Chiu, S.H.; Lin, K.H.; Wen, C.Y.; Lee, J.H.; Chen, H.M. A fast randomized method for efficient circle/arc detection. *Int. J. Innov. Comput. Inf. Control* **2012**, *8*, 151–166.

20. Blasco, J.; Aleixos, N.; Molto, E. Computer vision detection of peel defects in citrus by means of a region oriented segmentation algorithm. *J. Food Eng.* **2007**, *81*, 535–543. [CrossRef]

21. Tsneg, Y.H.; Tsai, D.M. Defect detection of uneven brightness in low-contrast images using basis image representation. *Pattern Recognit.* **2010**, *43*, 1129–1141. [CrossRef]

22. Jiang, H.H.; Yin, G.F. Surface defect inspection and classification of segment magnet by using machine vision technique. *Adv. Mater. Res.* **2011**, *339*, 32–35. [CrossRef]

![sensors logo] *sensors*

MDPI

Article

Substrate and Passivation Techniques for Flexible Amorphous Silicon-Based X-ray Detectors

Michael A. Marrs [1,*] **and Gregory B. Raupp** [2]

[1]   Flexible Electronics and Display Center, Arizona State University, Tempe, AZ 85284, USA
[2]   School for Engineering of Matter, Transport, and Energy, Arizona State University, Tempe, AZ 85287, USA;
       raupp@asu.edu
*   Correspondence: michael.marrs@asu.edu; Tel.: +1-480-727-6898

Academic Editor: Gonzalo Pajares Martinsanz
Received: 23 May 2016; Accepted: 19 July 2016; Published: 26 July 2016

Abstract: Flexible active matrix display technology has been adapted to create new flexible photo-sensing electronic devices, including flexible X-ray detectors. Monolithic integration of amorphous silicon (a-Si) PIN photodiodes on a flexible substrate poses significant challenges associated with the intrinsic film stress of amorphous silicon. This paper examines how altering device structuring and diode passivation layers can greatly improve the electrical performance and the mechanical reliability of the device, thereby eliminating one of the major weaknesses of a-Si PIN diodes in comparison to alternative photodetector technology, such as organic bulk heterojunction photodiodes and amorphous selenium. A dark current of 0.5 pA/mm^2 and photodiode quantum efficiency of 74% are possible with a pixelated diode structure with a silicon nitride/SU-8 bilayer passivation structure on a 20 µm-thick polyimide substrate.

Keywords: flexible electronics; flexible displays; flexible X-ray detectors; amorphous silicon PIN diodes; passivation

1. Introduction

1.1. Flexible Electronics

Flexible electronics are becoming more prevalent as previously developed flexible active matrix display technology is being implemented to produce a wide variety of photo-based biomedical sensing devices [1,2], including flexible X-ray detectors [3]. The primary immediate benefit of switching from conventional display glass to flexible substrates in digital radiography is the cost savings associated with the elimination of the significant ruggedization that must be incorporated to limit detector breakage.

Portable radiography has seen expansive growth into non-medical markets, such as security imaging and non-destructive testing for integrity analysis (i.e., looking for cracks in pipelines or inspection of aircraft structures). Though some portable X-ray panels do exist, they are comprised of glass thin film transistor (TFT) panels, and are rather bulky due to the ruggedization required to protect the costly panel from breakage. Given the cost of these panels, it would be advantageous to have a more rugged system that would be less prone to breakage and be much lighter for mobile users to carry.

Along with being both portable and lightweight, a digital X-ray panel with some degree of flexibility or bendability is also an appealing medical diagnostic or industrial imaging tool, especially if the overall digital X-ray detector thickness can be also minimized. A paramedic would be able to easily slide a thin and slightly flexible digital X-ray detector panel directly underneath the victim of a car accident at the scene, or an inspector could wrap the X-ray detector around a possibly cracked oil pipeline.

The relatively rapid development of flexible X-ray detectors and imagers has been enabled by the fact that existing active matrix thin film transistor flexible display technology can be easily ported to amorphous silicon (a-Si)-based indirect digital X-ray detectors, which have a similar underlying platform structure to liquid crystal displays (LCD). The principal new challenge to be addressed is flex-compatible fabrication of photodiodes in series with the thin film transistor at each pixel.

1.2. X-ray Source and Detector Structure and Operation

An X-ray source typically consists of a vacuum tube in which high voltage (i.e., 30–150 kV) electrons are accelerated into a metallic anode. The accelerated electrons pass close to the nuclei of the target material, or strike an inner shell electron in some cases, and produce an X-ray as the electron is slowed considerably by the opposing charge of the nucleus [4]. In the event of an inner electron shell collision, an electron from the outer shell moves into the vacancy in the inner shell and gives off an X-ray with energy that is characteristic of the anode material. The energy of the X-rays can be attenuated by a filter (which can be aluminum or carbon), producing a target energy range of 70–80 keV.

X-rays from the source pass through the patient or device under examination, with some of the X-rays attenuated by the nuclei of the intervening material. At the kilovolt energies typically employed by medical X-ray detectors, attenuation is usually the result of photoelectric absorption or Compton scattering. Photoelectric absorption produces meaningful data because an X-ray photon is absorbed and releases an electron from the absorbing atom. The X-ray absorption is proportional to the atomic number, the density, and the thickness of the absorbing media [5]. As an example, bone—which has a higher effective atomic number and is relatively dense—will absorb more X-rays than soft tissue, which results in a contrast difference in the completed radiograph. X-rays that do pass through the patient are absorbed by the detector.

A flat panel digital X-ray detector is usually classified as direct conversion or indirect conversion, depending on the process for converting the incident X-rays into charge. Direct conversion detectors utilize a photoconductive layer of amorphous selenium, which absorb the X-ray and produce an electron/hole pair. When a bias is applied across the selenium, the generated charge carriers are pulled towards the electrodes. When an X-ray is absorbed and generates an electron/hole pair, the charge carrier is drawn to the opposing electrode with limited scattering in comparison to indirect detection techniques [6]. However, the amorphous selenium has a smaller capture cross-section in comparison to the cesium iodide (CsI) screens used in indirect conversion, thus requiring the patient to be exposed to a higher dosage of X-rays to achieve the same resolution. In addition, the selenium film is usually 0.25–1.0 mm thick, requiring a large voltage on the order of 10 kV across the selenium to provide a sufficient electric field to extract the incident X-ray-generated charge carriers [6].

The basic components of an indirect digital X-ray detector include a scintillator phosphor conversion film that converts the incident X-rays into visible photons, a photodiode—usually amorphous silicon with organic photodiodes as an emerging competing technology—that detects the light emitted from the X-ray phosphor conversion film and converts the light into an electric charge, and a thin film transistor (TFT) that acts as a switch for the readout of the charge stored in the photodiode in between readout. In that sense, the operation of a digital X-ray detector can be compared to a large digital camera. The scintillator is typically composed of high capture cross section materials like cesium or the rare earth elements doped with a phosphor whose peak emission is in the visible spectrum usually between 500 and 550 nm for detectors manufactured with a-Si photodiodes. The a-Si photodiode is operated under reverse bias between -2 and -5 V so that the photocurrent dominates the total current flowing through the photodiode. The a-Si is grown to a thickness of 1.0 to 1.2 µm, which is more than adequate to capture most of the incident green light produced by the scintillator. The TFTs are laid out in a grid similar to an LCD display, where the gate lines are scanned sequentially and the charge stored on the source line is read out at the edge of the array.

1.3. Flexible Indirect X-ray Detectors

The simplest photodiode fabrication approach is to pattern the n-Si layer of the photodiode at each pixel, and to blanket deposit the i-Si, p-Si, and transparent conductor top metal contact continuously over the entire array. A major advantage of this full fill factor approach is that the entire pixel is covered by the i-Si absorption layer, thus eliminating any "dead" spots in the pixel where photon strikes will not be detected. This full fill factor approach also produces the smallest pixel size (and hence highest resolution) for a given set of design rules. For example, for X-ray detectors currently fabricated at the Flexible Electronics and Display Center, we produce 50 ppi full fill factor arrays, compared to much larger 83 ppi for pixelated diode arrays.

There are some significant processing drawbacks to the full fill factor design that are exacerbated by the use of polyethylene naphthalate (PEN) as a flexible substrate. Since there is a large coefficient of thermal expansion (CTE), mismatch between PEN and amorphous silicon (13 ppm/°C versus 3 ppm/°C), coupled with the high intrinsic compressive film stress of the i-Si layer detectors, arrays fabricated with the full fill factor approach have a tendency to curl significantly after post-fabrication debonding of the flexible substrate. In addition, the deposition process for the amorphous silicon diode layers takes place at 200 °C, which is near the melting temperature of the PEN (240 °C), and can take upwards of one hour to deposit the requisite 1.2 μm-thick film. The extended exposure at the elevated temperature of the PECVD (plasma enhanced chemical vapor deposition) process risks embrittling the PEN and ruining the detector. In spite of these drawbacks, full fill factor X-ray detectors on PEN substrates have been successfully demonstrated by multiple groups [3,7,8]. Previous results from the Flexible Electronics and Display Center are shared below in Figure 1. If the i-Si deposition process is not tightly controlled, stress-related failure, as shown in Figure 1, may occur.

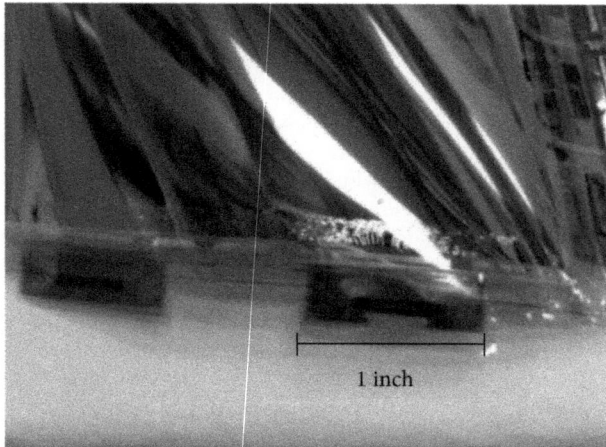

Figure 1. Full fill factor digital X-ray detector backplane showing stress related failure due to high intrinsic film stress in the i-Si layer.

In spite of these drawbacks, amorphous silicon photodiodes are more desirable for large area detectors than direct conversion amorphous selenium detectors because the amorphous silicon is significantly thinner (1.2 μm [3] vs. 100 μm [7]), enabling a tighter bending radius. Moreover, PECVD amorphous silicon can be deposited with better uniformity over large areas than the thermally evaporated amorphous selenium. Amorphous silicon photodiodes are preferable over organic photodiodes due to the air and moisture instability of organic photodiodes under irradiation, which requires a sufficiently robust moisture and oxidation barrier to prolong the life of the detector [9].

Another method for alleviating stress in the entire film stack is to use a flexible substrate whose coefficient of thermal expansion is a closer match to that of the deposited thin films. Polyimides are a class of flexible substrate materials receiving considerable attention because of their compatible thermal properties. The main advantages of polyimides over PEN as a substrate are that the coefficient of thermal expansion for polyimide is roughly four times smaller than PEN, while the safe operating temperature is roughly 200 °C higher. A comparison of the thermal properties of HD Microsystems PI 2611 [10], a polyimide, DuPont Teonex Q51 [11], and silicon [12,13] are shown below in Table 1.

Table 1. Comparison of structural, thermal, and electrical properties of polyimide, polyethylene naphthalate (PEN), and silicon.

	HD Microsystems PI 2611	DuPont Teonex Q51 PEN	Silicon	Units
Tensile Strength	350	274.6	7000	MPa
Glass Transition Temperature	360	121	-	°C
Continuous Operation Temperature	-	180	-	°C
Melting Temperature	620 (decomposes)	269	1415	°C
Coefficient of Thermal Expansion	3	13	2.6	ppm/°C
Dielectric Constant	2.9	3	11.9	

The increased operating temperature of the polyimide would allow for safe processing of a-Si at temperatures above 275 °C, which is a typical deposition temperature for high quality TFT and PIN diode a-Si films, without the risk of significant thermal damage to the substrate. In addition, the coefficient of thermal expansion of the PI 2611 is within 0.4 ppm/°C of silicon, meaning that it is less likely that intrinsic stress associated with a-Si deposition will cause polyimide substrates to physically curl once they are debonded from the carrier.

2. Materials and Methods

2.1. Substrate Preparation: Polyethylene Naphthalate (PEN) Bond/Debond

The bond/debond (or temporary bonding process) allows the processing of a flexible substrate bonded temporarily to a rigid carrier in conventional semiconductor or flat panel display (FPD) process tools that are built to handle rigid Si or glass substrates. The basic process flow for the bond process is shown in Figure 2a [14], where it is compared to the polyimide cast and debond process, which is described in the next sub-section. A temporary adhesive is spin-coated on to a rigid carrier. If the adhesive is thermally cured, a bake usually follows the spin process. The flexible substrate, in this case polyethylene naphthlate (PEN), is then mounted to the adhesive-coated carrier using a roll laminator. The adhesive is cured either by UV exposure, by baking, or with a combination of both.

The rigid carrier suppresses the bowing of the flexible substrate during processing to provide the requisite dimensional stability during device fabrication. Following device fabrication, the flexible substrate can be debonded from the rigid carrier to yield a flexible electronics device (display, sensor array, detector).

The main advantage of the bond/debond process is that the process requires little additional investment to an already existing display or semiconductor fabrication facility and is scalable to larger area substrates without significant alterations to the process. The main disadvantage of the bond/debond process is related to controlling substrate deformation (i.e., warp and bow), which can cause wafer handling and pattern alignment issues, substrate distortion, and in extreme cases delamination of the flexible substrate during processing. However, Haq, et al. have demonstrated a suitable distortion-free bond/debond process with controlled warp/bow that is capable of effective processing up to 200 °C [15].

(a) Bond/Debond

Plastic substrate bonded with
temporary adhesive to carrier

TFT Array Fabrication
130-180°C

Triggered Debond:
Mechanical or UV-induced

(b) Coat/Debond

Spin coated or slot dye coated
polyimide on glass carrier

TFT Array Fabrication
280-350°C

Manual Debond:
Substrate Edge is Cut
Polyimide is Manually Released

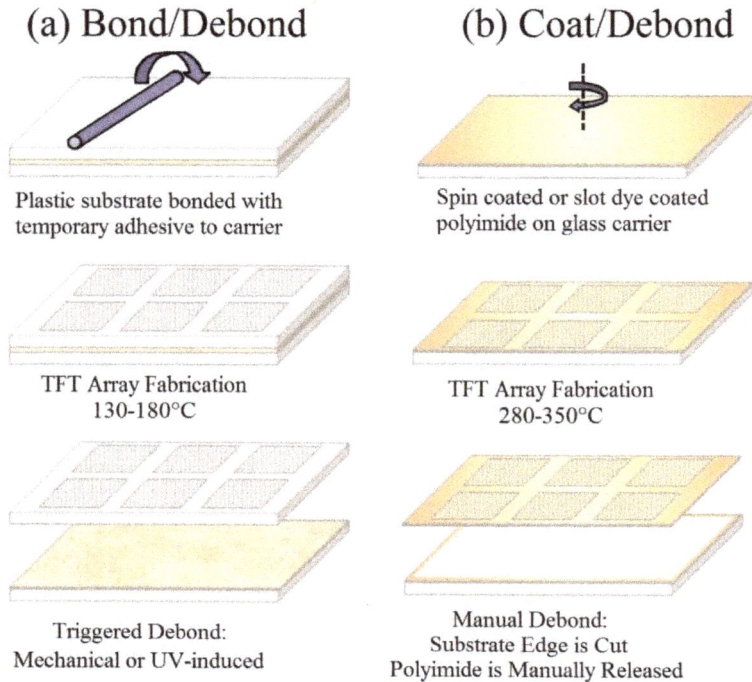

Figure 2. Bond/debond and polyimide coat/debond process flow. TFT: thin film transistor.

2.2. Substrate Preparation: Polyimide Cast and Debond Process

The polyimide cast and debond process is an attractive alternative to the EPLaR (Electronics on Plastic by Laser Release) [16] process, where slot die coating is used to cast polyimide as the base substrate onto a display glass substrate, as shown in Figure 2b. The adhesion strength of the polyimide is selectively modified at the edge of the carrier by applying an adhesion promoter such as VM652 from HD Microsystems. The increased adhesion strength at the edge significantly reduces edge-initiated debonding. However, the polyimide-to-glass adhesion strength has been tailored to be significantly smaller than adhesion strength of the polyimide to the adhesion promoter, thus eliminating the need for excimer laser melting of an interfacial layer to release polyimide, as in the EPLaR process. After the polyimide is cured at 375 °C in an inert atmosphere, a PECVD silicon nitride (SiN) barrier film is applied to protect the polyimide from future processing steps and to reduce the water vapor transmission rate (WVTR) of the polyimide. The TFT and PIN diode processing steps can proceed once the barrier is deposited.

After the device layers are fabricated, the backplane is cut inside the adhesion promotion layer picture frame and is removed manually. The adhesion promotion layer remains bonded to the glass carrier substrate along with a polyimide picture frame, which can now be reclaimed if desired. The curl of the polyimide substrate after debond depends on the formulation of the polyimide spin/spray casting solution and adhesion layer. The control of the post-debond curl is the biggest challenge of the polyimide process.

2.3. PIN Diode Processing and Device Structure

Figure 3 is a cross-sectional schematic that illustrates the basic full fill factor pixel design. The grey dashed line box highlights the PIN diode layers. The current full fill factor design places the PIN diode over the top of the TFT and uses an indium tin oxide (Siommon electrode for the V_{bias} connection on the p-doped side of the PIN diode. The i-Si, p-Si, and ITO blanket the entire X-ray detector array, and are etched outside the array in the field and over the driver connections. Processing of the full fill factor design was described previously [3,17]. The maximum processing temperature for the device is 200 °C.

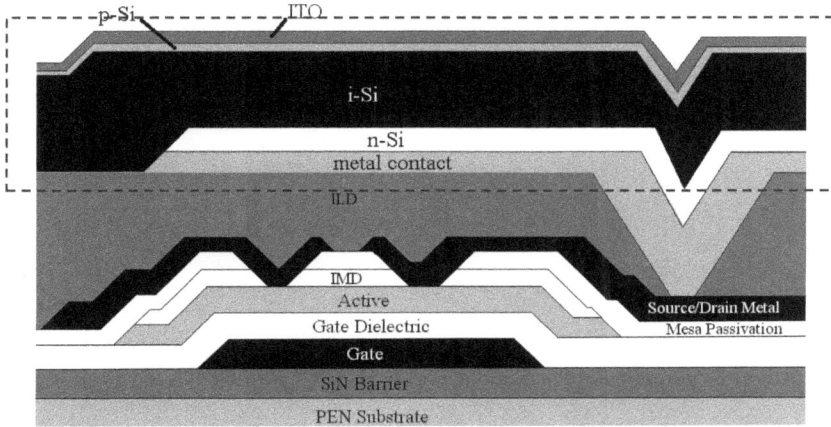

Figure 3. Full fill factor PIN photodiode connected in series to a TFT. The interlayer dielectric (ILD) is a combination of SU-8 (MicroChem) and SiN. The active layer is amorphous Si and the gate dielectric, mesa passivation, and intermetal dielectric (IMD) are SiN.

Although the full fill factor design provides a smaller-area, higher resolution pixel when subjected to the same design rules as alternative pixelated structures [18], there are some potential disadvantages as well. For example, since the i-Si is blanket deposited across the array, it imparts more stress on the substrate system than if the i-Si layer were patterned. In addition, there is also potential for crosstalk between the elements in the arrays, since photons absorbed physically over a certain pixel could instead be read out by a neighboring pixel because the i-Si film is continuous over the entire array.

Figure 4 is a cross-sectional schematic that illustrates the basic design for the alternative pixelated diode approach. The grey dashed line box again highlights the PIN diode layers. Pixelating the photodiode requires the V_{bias} connection to the p-doped side of the diode to be made using a subsequently deposited metal deposition. The readout lines can also be moved to this subsequent deposition step and then connected down to the TFT through a via at each pixel. This modification enables the maximum separation between the gate lines and the readout lines, which reduces the parasitic capacitance (noise) between the two. The separation of the diode and the TFT reduces the area available for detection by approximately 15% in a detector with a 207 μm by 207 μm pixel compared to the full fill factor design present in Figure 3 with the same pixel area.

The patterned PIN diode structure replaces the ITO V_{bias} connection with any opaque metal of choice. If a sputtered aluminum/1 wt% silicon alloy is used, the resistivity difference between the sputtered aluminum/silicon alloy and ITO is over a factor of 100 (4.1 μΩ-cm vs. 528 μΩ-cm), which should provide an immediate benefit by reducing the series resistance in the device. The ITO thickness and thus the sheet resistance is governed by optical constraints as well as electrical constraints. Specifically, the ITO layer needs to act as an anti-reflection coating at the target wavelength for the detection of incident light. In the case of a digital X-ray detector using a CsI scintillator, the target light wavelength for detection is the peak emission wavelength of the scintillator (roughly 525 nm). The ideal ITO thickness to produce minimum reflection at this wavelength is approximately 65 nm thick, as calculated from Snell's Law. However, the V_{bias} connection in the pixelated PIN diode structure depicted in Figure 4 does not have such optical constraints because the layer is so thick that it is sufficiently opaque to most visible light and only covers a small portion of the photodiode. This feature allows the separate control of the electrical and optical properties of the V_{bias} connection.

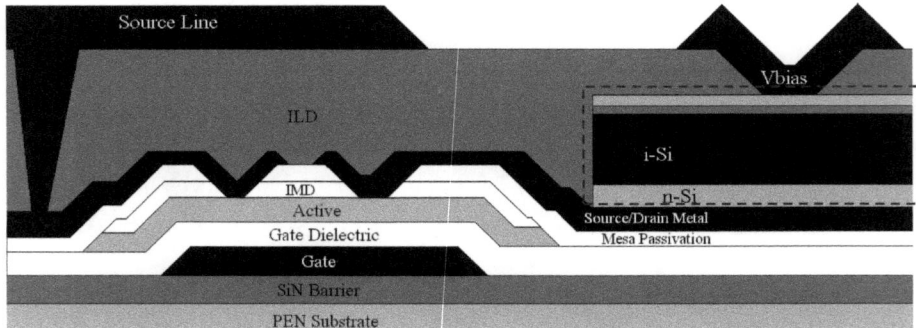

Figure 4. Patterned PIN photodiode connected in series to a TFT.

By using the aluminum/silicon alloy as the V_{bias} connection over ITO, the series resistance can be significantly reduced while still maintaining the optical properties necessary to maximize photon absorption in the photodiode. Series resistance is a parasitic, power consuming parameter whose effect on the diode performance is illustrated by equation (1), which is a modification of the ideal diode model:

$$I_D = I_S \left(e^{\frac{q(V_{bias} - I_D R_S)}{nkT}} - 1 \right) \tag{1}$$

where I_D is the diode current, I_S is the reverse saturation current, q is the charge of an electron, V_{bias} is the voltage applied across the diode, R_s is the series resistance, n is the diode ideality factor, k is Boltzmann's constant, and T is the temperature. The main contribution to series resistance comes from the diode contact resistance, bulk resistance of the diode layers, and the sheet resistance of the top metal layer [19]. The deleterious effects of series resistances are most noticeable at voltages near the open circuit voltage of the photodiode [20] as a deviation in the slope of the I–V curve from the vertical affecting the quantum efficiency in extreme cases [19].

2.4. PIN Diode Test Setup

2.4.1. L-I-V Sweeps of 1 mm^2 Diodes

PIN diode test structures with an area of 1 mm^2 were separately fabricated side-by-side with the main arrays to test and monitor the performance of the full fill factor and pixelated PIN diodes. A probe station fitted with a Keithley 4200 Semiconductor Characterization System was used to characterize the diode performance. The probe card on the probe station is fitted with a green LED illuminator (peak emission wavelength of 520 nm), which generates an irradiance 102 W/m^2 at the device surface. The diode is swept from −5 V to +1 V in the dark and again with the illuminator set to maximum current. The fill factor, open circuit voltage (V_{OC}), and short circuit current (I_{SC}) are extracted from the irradiated curve. The dark characteristics are fitted to the ideal diode equation, which yields the diode saturation current and the ideality factor. In addition, the minimum dark current, the dark current at −5 V, and the dark current at +1 V are extracted to indirectly track the shunt resistance and series resistance, as well as the magnitude of the dark current. After the sweeps are completed, the diode is held at −5 V in the dark as the prober continues to read the current an additional 120 times, with each reading commencing once the previous reading has collected sufficient charge to be safely above the noise floor threshold of the tester.

2.4.2. Diode Array Testing

A custom-built array test system with readout chips was built in-house to test the arrays. The tester was customizable so that arrays as small as 128 by 240 pixels or as large as 1024 by 720 pixels could be tested. The gate on time was set to 20 μs to allow for sufficient discharge of the photodiodes and the charge collected was recorded to an image with 12 bit range greyscale.

3. Results

3.1. L-I-V Sweep Comparison between Blanket and Pixelated Photodiodes

Figure 5A compares dark I-V characteristics of 1 mm^2 diodes fabricated with a blanket PIN diode structure with ITO top contact and no passivation (structure shown in Figure 3), with the characteristic behavior of diodes that are passivated and utilizing a metal strap as the top V_{bias} contact (structure shown in Figure 4). The passivated, pixelated PIN diode performs better than the blanket structure under both forward bias and reverse bias. The superior performance under forward bias is likely attributable to the reduced series resistance, which is expected to dominate the I-V characteristic as the voltage increases above 0.5 V. The primary advantage to decreasing the series resistance is to reduce the image lag and prevent inverse shadowing. A lower resistance allows charge to be drained faster and prevents the applied bias voltage from being pulled down when reading highly charged pixels.

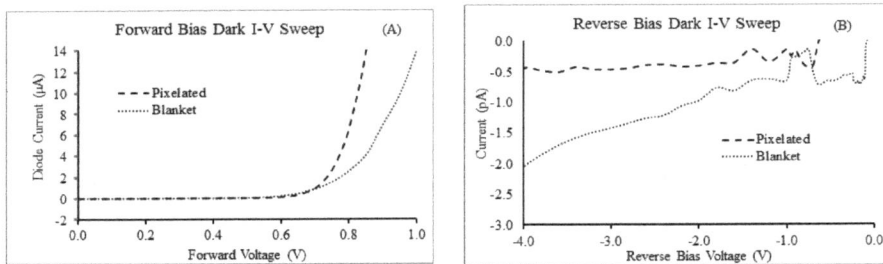

Figure 5. (**A**) Forward and (**B**) Reverse dark I-V sweep comparison between blanket and pixelated PIN diodes.

Under reverse bias operation—under which a photodiode will spend most of its operating lifetime—the leakage current for the pixelated diode is lower than the blanket coated diode and stays relatively flat with increasingly negative applied voltage (Figure 5B).

When irradiated with green light at an irradiance of 102 W/m^2, the quantum efficiency is calculated to be 78% for the blanket device and 74% for the pixelated device. The reduced quantum efficiency of the pixelated device is due to reflection at the passivation interfaces.

3.2. Effect of the Photodiode Passivation Layer

3.2.1. Dark Current vs. Time Stability

The nature of the passivation layer(s) affects the final dark current characteristics. When no passivation is used, the diodes are so leaky that it is difficult to consider them "diodes". In the example presented in Figure 5, the passivation is a two-layer stack comprising 2.0 μm of spin on SU-8 polymer and 300 nm of PECVD-deposited SiN. If only the SU-8 polymer is applied as the passivation layer, the initial dark current behavior is greatly improved. However, the dark current begins to degrade during continuous operation under reverse bias, as shown in Figure 6.

Figure 6 further shows that if only SiN is used as the passivation material, the initial dark current is significantly larger (~10 times) than the initial dark current for SU-8-coated diodes. However, the dark current exhibits a monotonically-decreasing current with time (eventually leveling off to a near constant value), consistent with thermally generated leakage associated with the depopulation of defects/traps in the i-Si [21]. This time-dependent behavior suggests that the PECVD SiN film deposition process may damage the exposed PIN diode layers, potentially increasing the number of surface trap sites. If only SU-8 is used as the passivation, the dark current increases with time, suggesting that SU-8 by itself may be an inadequate atmospheric contaminant barrier, and that atmospheric contaminants diffuse to the i-Si surface and react. The best results are observed when SU-8 is coated first to avoid photodiode damage, followed by SiN deposition.

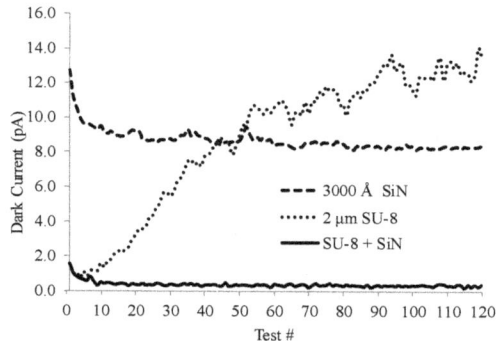

Figure 6. Dark current vs. test number for three different photodiode passivation schemes. The time interval between tests was not consistent due to the low current of the diodes and the charge integration time required to measure the current.

3.2.2. Effect of Diode Shape

The effect of surface generation/recombination can be explored by looking at the performance of photodiodes with the same detection area (i.e., the surface area of the side of the diode facing the incident light), but with a different perimeter. Square diodes have the minimum perimeter possible for a given area and thus have the smallest surface area of i-Si sidewalls exposed. Diodes with a rectangular shape have a larger area for potential surface recombination and generation to occur and thus should have a higher dark current than a square diode. To illustrate this point, two different diode geometries were investigated. Square Diode A has a 1 mm by 1 mm square cross-section exposed to the incident light, while rectangular Diode B has a 2 mm by 0.5 mm rectangular cross-section exposed to the incident light. Both diodes have the same capture area (1 mm^2).

A *t*-test reveals a small but statistically significant difference in the logarithm of the dark current, with the rectangular diodes having greater than two times the dark current of the square diodes.

3.2.3. Effect of a Backside Guard Ring on Humidity Sensitivity

The effect of the diode edge can also be illustrated by testing a structure where there is a diode completely surrounded by another diode as a protective backside guard ring, as shown schematically in Figure 7. The backside guard ring reduces the electric field at the edge of the device and prevents the depletion region from reaching the edge of the diode, thus mitigating surface leakage and enabling the capture of carriers generated outside of the active area [22]. The beneficial effects of the guard ring can be enhanced by grounding the n side of the diode, as opposed to having it float.

Figure 8 compares dark sweeps from diodes with and without guard rings, and clearly shows that the devices with the guard ring demonstrate a substantial reduction in the dark current. In addition, devices with the guard ring are less sensitive to environmental conditions, specifically humidity.

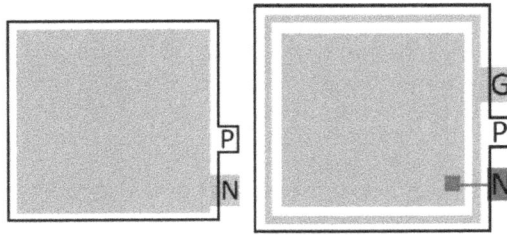

Figure 7. Top-down drawing comparing an unguarded PIN diode with only N and P terminals (**Left**); and guarded PIN diode structure with added guard ring (G) in addition to the N and P terminals (**Right**).

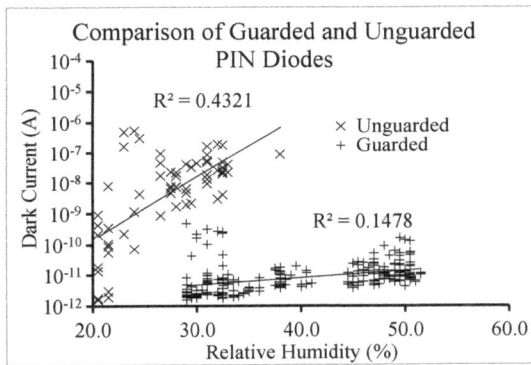

Figure 8. Comparison of the dark current vs. relative humidity for guarded and unguarded PIN diodes.

3.2.4. Array Performance

An advantage of the pixelated PIN diode approach is that the i-Si absorbing layer is not continuous over the entire array, making it extremely difficult for photons absorbed over one pixel to be read out by a neighboring pixel. This reduction in cross-talk between neighboring pixels results in increased contrast and improved resolution. Figure 9 shows a comparison between an array with a blanket i-Si layer and a pixel patterned i-Si when exposed to the same irradiance and with the same gate on time for the TFT array. The pixel patterned array demonstrates a notably higher contrast, thereby suggesting that it operates with a lower detection threshold.

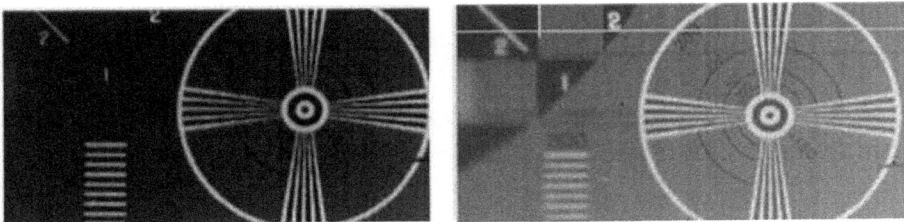

Figure 9. Full fill factor blanket i-Si photodiode array (**Left**) compared to pixel pattern photodiode array (**Right**).

3.2.5. PIN Diode Film Cracking

The pixelated PIN diode structure relieves the overall film stress of the entire device. With the blanket photodiode structure, the PIN diode layers have a tendency to crack and peel when the detector array is curled. Arrays with both the full fill factor blanket photodiode structure and the pixelated PIN diode structure were curled around a 3/8" stainless steel pipe, as shown in Figure 10.

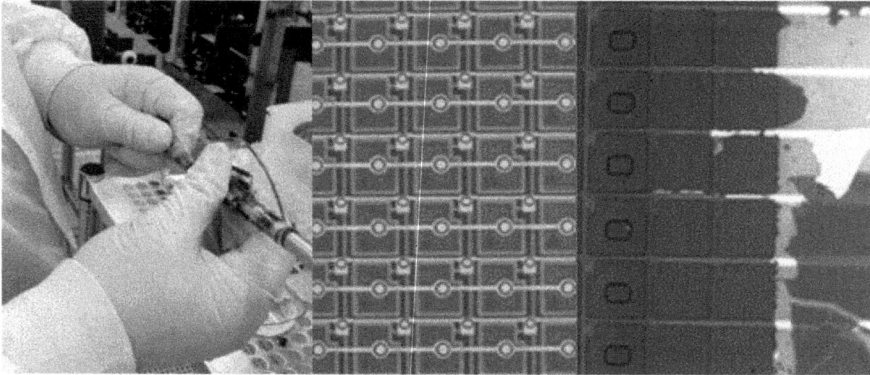

Figure 10. Flexible photodiode array wrapped around a 3/8" stainless steel pipe (**left**), microscope image of pixelated PIN diode structure (**center**), and full fill factor PIN diode structure (**right**) after wrapping around 3/8" stainless pipe.

Of the 24 arrays mechanically tested with the pixelated photodiode structure, none demonstrated any physical damage other than the embedding of particles from the pipe and manual handling of the arrays. Of the 24 arrays tested with the full fill factor PIN diode structure, 17 demonstrated significant cracking and flaking of the PIN diode layers, as shown in Figure 10.

3.2.6. Effect of the Substrate: PEN vs. Polyimide

Significant stress can be imparted on the polyimide if the peel strength is above 80 g/inch, resulting in the elongation or stretching of the polyimide. The polyimide monomer formulation is a factor in the coefficient of thermal expansion for the final film. If the peel strength is too high or the coefficient of thermal expansion is significantly different from the thin films deposited, significant curl will be present in the debonded polyimide, rendering the flexible X-ray detector unusable. Using HD microsystems 2611 series polyimide with an appropriate adhesion promotor at the edge of the rigid carrier substrate can yield a flexible substrate that is easily debonded post process and does not exhibit significant curl. The curl of the 2611 series polyimide is lower than PEN. A comparison of a debonded PEN X-ray detector, a debonded polyimide X-ray detector with large peel strength, and a debonded polyimide X-ray detector with appropriate peel strength and minimal coefficient of thermal expansion mismatch is shown in Figure 11.

Once the polyimide substrate is debonded, the user can be quite rough with the substrate and not be concerned with film cracking or peeling. One must be very cautious and slow while debonding the PEN, as perturbations during the debond process can cause the inorganic PIN diode layers to buckle under the sudden change in applied peeling force as the residual stress in the PEN is released.

The maximum processing temperature of the TFT and PIN diode processing is 200 °C, which is 20 °C above the recommended continuous operation temperature, but 69 °C below the melting temperature listed in Table 1. However, at 210 °C and above, the PEN crystallizes, leading to significant shrinkage, and is prone to fracture during the debond process [23]. In this dynamic situation, even minor perturbations in the temperature of the TFT and PIN diode thin film deposition processes could significantly affect the structure of the bonded PEN, potentially leading to film buckling during the debond process.

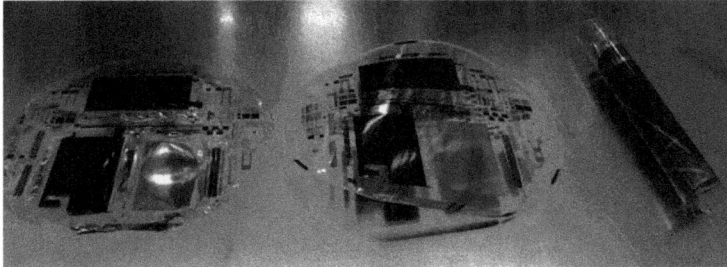

Figure 11. Curl comparison of low coefficient of thermal expansion (CTE) polyimide (**Left**), PEN (**Middle**), and high CTE polyimide (**Right**) debonded.

In addition to the reduced curl and the improved film adhesion post debond, another advantage of using polyimide over PEN is the significantly higher glass transition temperature of the polyimide (360 °C vs. 121 °C, respectively). The higher glass transition temperature allows for a higher deposition temperature during the PECVD steps. The added degree of freedom allows for a higher quality finished product, as the higher deposition temperature can lead to a lower dark current, higher quantum efficiency, and higher fill factor for the PIN diode films.

Simply increasing the deposition temperature of the PIN diode films from 200 °C to 250 °C, without making any other changes to the other deposition parameters, results in a significant increase in the quantum efficiency of the photodiodes at 520 nm from 57.8% to 61.8% when using a green LED light source with measured irradiance of 100 W/m^2. The increased deposition temperature of the PIN diode layers results in a denser, higher quality film that has fewer trap sites available to grab photogenerated electrons and holes before leaving the diode.

4. Conclusions

The benefits of a pixelated PIN diode structure were evaluated in comparison to full fill factor diodes with respect to flexible X-ray detectors. In addition to demonstrating lower dark current that is more stable with time and having less cross-talk between pixels, the pixelated structure is more robust with respect to flexing and bending of the substrate. The pixelated structure does not require additional masks in comparison to the blanket diode structure, but enables the passivation of the diode, allowing for the mitigation of surface edge-related leakage and reduction of the overall dark current of the device. The biggest downside to the structure is the extra layout spacing necessary to place the PIN diode next to the TFT, as opposed to on top of the TFT. However, this layout concern is only an issue with flexible detectors that must have a resolution less than 83 dpi.

Further benefits can be realized by utilizing polyimide as a substrate. HD Microsystems 2600 series polyimide has the right combination of low coefficient of thermal expansion and moderate adhesion strength to glass to enable a successful, low stress debond with minimal curl in the polyimide. In addition, the higher melting temperature of the polyimide allows for higher temperature PIN diode processing, leading to improved diode performance.

Acknowledgments: The research was sponsored by the Army Research Laboratory (ARL) and was accomplished under Cooperative Agreement W911NG-04-2-005 and DTRA MIPRs 11-2161M and 11-2308M. The views and conclusions contained in this document are those of the authors and should not be interpreted as representing the official policies, either expressed or implied, of the ARL, DTRA, or the U.S. Government. The U.S. Government is authorized to reproduce and distribute reprints for government purposes, notwithstanding any copyright notation herein.

Author Contributions: Marrs and Raupp collaborated on the writing of the paper, with both authors contributing significantly to the content. Raupp obtained the original funding for the work and helped interpret the experimental results. Marrs designed and processed the experiments associated with this work and interpreted the results.

Conflicts of Interest: The authors declare no conflict of interest. The founding sponsors had no role in the design of the study; in the collection, analyses, or interpretation of data; in the writing of the manuscript, and in the decision to publish the results.

References

1. Smith, J.; O'Brien, B.; Lee, Y.K.; Bawolek, E.; Christen, J. Application of Flexible OLED Display Technology for Electro-Optical Stimulation and/or Silencing of Neural Activity. *IEEE J. Disp. Technol.* **2014**, *10*, 514–520. [CrossRef]
2. Smith, J.T.; Katchman, B.A.; Kullman, D.; Obahiagbon, U.; Lee, Y.K.; O'Brien, B.; Madabhushi, K.P.; Surampudi, V.; Raupp, G.B.; Anderson, K.S.; et al. Application of Flexible OLED Display Technology to Point-of-Care Medical Diagnostic Testing. *IEEE J. Disp. Technol.* **2016**, *12*. [CrossRef]
3. Marrs, M.; Bawolek, E.; Smith, J.T.; Raupp, G.B.; Morton, D. Flexible Amorphous Silicon PIN Diode X-ray Detectors. *SPIE Def. Secur. Sens.* **2013**, *87300C*. [CrossRef]
4. Zink, F.E. The AAPM/RSNA Physics Tutorial for Residents: X-ray Tubes. *Radiographics* **1997**, *17*, 1259–1268. [CrossRef] [PubMed]
5. Carlton, R.R.; Adler, A.M. *Principles of Radiographic Imaging: An Art and A Science*, 5th ed.; Cengage Learning: Clifton Park, NY, USA, 2012; pp. 191–215.
6. Schaefer-Prokop, C.; Neitzel, U.; Venema, H.W.; Uffmann, M.; Prokop, M. Digital chest radiography: An update on modern technology, dose containment and control of image quality. *Eur. Radiol.* **2008**. [CrossRef] [PubMed]
7. Kuo, T.T.; Wu, C.M.; Lu, H.H.; Chan, I.; Wang, K.; Leou, K.C. Flexible X-ray Imaging Detector Based on Direct Conversion in Amorphous Selenium. *J. Vac. Sci. Technol. A* **2014**, *32*, 1–5. [CrossRef]
8. Lujan, R.A.; Street, R.A. Flexible X-ray Detector Array Fabricated with Oxide Thin-Film Transistors. *IEEE Electron Device Lett.* **2012**, *33*, 688–690. [CrossRef]
9. Ng, T.N.; Wong, W.S.; Chabinyc, M.L.; Sambandan, S.; Street, R.A. Flexible Image Sensor Array with Bulk Heterojunction Organic Photodiode. *Appl. Phys. Lett.* **2008**, *92*, 1–3. [CrossRef]
10. PI 2600 Series—Low Stress Applications. Product Bulletin. Available online: http://www.hdmicrosystems.com/HDMicroSystems/en_US/pdf/PI-2600_ProcessGuide.pdf (accessed on 4 April 2016).
11. Teonex®Q51 Datasheet. Available online: http://www.synflex.com/en/produkte_pdf/?id=19&areaid=flaechenisolierstoffe (accessed on 10 April 2016).
12. Petersen, K. Silicon as a Mechanical Material. *IEEE Proc.* **1982**, *70*, 420–456. [CrossRef]
13. General Properties of Si, Ge, SiGe, SiO_2 and Si_3N_4. Available online: http://www.virginiasemi.com/pdf/generalpropertiesSi62002.pdf (accessed on 7 April 2016).
14. Raupp, G.B.; O'Rourke, S.M.; O'Brien, B.P.; Ageno, S.K.; Loy, D.E.; Bawolek, E.J.; Allee, D.R.; Venugopal, S.M.; Kaminski, J.; Bottesch, D.; et al. Low-temperarature Amorphous-silicon Backplane Technology Development for Flexible Display in a Manufacturing Pilot-line Environment. *JSID* **2007**, *15*, 445–454. [CrossRef]
15. Haq, J.; Ageno, S.; Raupp, G.B.; Vogt, B.D.; Loy, D. Temporary Bond-debond Process for Manufacture of Flexible Electronics: Impact of Adhesive and Carrier Properties on Performance. *J. Appl. Phys.* **2010**, *108*, 114917. [CrossRef]
16. French, I.; George, D.; Kretz, T.; Templier, F.; Lifka, H. Flexible Displays and Electronics Made in AM-LCD Facilities by the EPLaR™ Process. *SID Symp. Dig. Tech. Pap.* **2007**, *38*, 1680–1683. [CrossRef]
17. Marrs, M.A.; Bawolek, E.J.; O'Brien, B.P.; Smith, J.T.; Strnad, M.; Morton, D.C. Flexible Amorphous Silicon PIN Diode Sensor Array Process Compatible with Indium Gallium Zinc Oxide Transistors. *SID Symp. Dig. Tech. Pap.* **2013**, *44*, 455–457. [CrossRef]
18. Theil, J.A.; Snyder, R.; Hula, D.; Lindahl, K.; Haddad, H.; Roland, J. a-Si:H Photodiode Technology for Advanced CMOS Active Pixel Sensor Imagers. *J. Non-Cryst. Solids* **2002**, *299*, 1234–1239. [CrossRef]
19. Dadu, M.; Kapoor, A.; Tripathi, K. Effect of Operating Current Dependent Series Resistance on the Fill Factor of a Solar Cell. *Sol. Energy Mater. Sol. Cells* **2002**, *71*, 213–218. [CrossRef]

Sensors **2016**, *16*, 1162

20. Series Resistance. Available online: http://www.pveducation.org/pvcdrom/solar-cell-operation/series-resistance (accessed on 18 May 2016).

21. Street, R.A. *Technology and Applications of Amorphous Silicon*; Springer-Verlag: Berlin, Germany, 2000; pp. 147–221.

22. Nam, H.G.; Shin, M.S.; Cha, K.H.; Cho, N.I. Fabrication of a Silicon PIN Diode for Radiation Detection. *J. Korean Phys. Soc.* **2006**, *48*, 1514–1519. [CrossRef]

23. Cygan, P.; Zheng, J.; Yen, S.; Jow, T. Thermal treatment of polyethylene-2,6-naphthalate (PEN) film and its influence on the morphology and dielectric strength. *Conf. Electr. Insul. Dielectr. Phenom.* **1993**, 630–635. [CrossRef]

sensors

MDPI

Article

Time-Resolved Synchronous Fluorescence for Biomedical Diagnosis

Xiaofeng Zhang [1,2,†,*], Andrew Fales [3] and Tuan Vo-Dinh [2,3,4,†,*]

[1] Department of Radiology, Duke University Medical Center, Durham, NC 27710, USA
[2] Fitzpatrick Institution for Photonics, Duke University, Durham, NC 27708, USA
[3] Department of Biomedical Engineering, Duke University, Durham, NC 27708, USA; Andrew.Fales@duke.edu
[4] Department of Chemistry, Duke University, Durham, NC 27708, USA
[*] Correspondence: Steve.Zhang@duke.edu (X.Z.); Tuan.Vodinh@duke.edu (T.V.-D.); Tel.: +1919-257-8564 (X.Z.); +1919-660-8520 (T.V.-D.); Fax: +1919-784-2711 (X.Z.); +1919-613-9145 (T.V.-D.)
[†] These authors contributed equally to this work.

Academic Editor: Gonzalo Pajares Martinsanz
Received: 28 July 2015; Accepted: 26 August 2015; Published: 31 August 2015

Abstract: This article presents our most recent advances in synchronous fluorescence (SF) methodology for biomedical diagnostics. The SF method is characterized by simultaneously scanning both the excitation and emission wavelengths while keeping a constant wavelength interval between them. Compared to conventional fluorescence spectroscopy, the SF method simplifies the emission spectrum while enabling greater selectivity, and has been successfully used to detect subtle differences in the fluorescence emission signatures of biochemical species in cells and tissues. The SF method can be used in imaging to analyze dysplastic cells *in vitro* and tissue *in vivo*. Based on the SF method, here we demonstrate the feasibility of a time-resolved synchronous fluorescence (TRSF) method, which incorporates the intrinsic fluorescent decay characteristics of the fluorophores. Our prototype TRSF system has clearly shown its advantage in spectro-temporal separation of the fluorophores that were otherwise difficult to spectrally separate in SF spectroscopy. We envision that our previously-tested SF imaging and the newly-developed TRSF methods will combine their proven diagnostic potentials in cancer diagnosis to further improve the efficacy of SF-based biomedical diagnostics.

Keywords: synchronous fluorescence; ultrafast; time-resolved; imaging; cancer diagnosis

1. Introduction

1.1. Synchronous Fluorescence

Optical technologies for rapid diagnosis of cancer and dysplasia are highly beneficial in early detection and timely treatment. In conventional diagnostics, a biopsy sample typically represents only a very limited area of the suspected tissue; and laboratory results using histopathology examinations are generally time-consuming. Fluorescence spectroscopy is a powerful technique that can be used to noninvasively analyze the fluorescent signatures of tissue. Autofluorescence of neoplastic and normal tissues using fixed-wavelength laser-induced fluorescence (LIF) have been investigated and used for cancer diagnosis in our laboratory as well as other laboratories [1–7]. Although studies have demonstrated reasonably good specificity and sensitivity in sample classification, fixed-wavelength excitation typically produces fluorescence spectra exhibiting featureless profiles or broad-band peaks, which do not fully exploit the diagnostic potentials of fluorescence spectroscopy. We have previously developed the theoretical foundations of the synchronous fluorescence (SF) method, which was characterized by simultaneously scanning both the excitation and emission wavelengths, while maintaining a constant wavelength interval [8,9]. The SF method has been coupled with multi-component analysis algorithms to obtain spectral signatures of environmental samples and to enhance selectivity in analyzing complex chemical and biological samples [9–14].

The SF method not only simplifies the emission spectrum, but is also less affected by Rayleigh and Raman scattering, compared to the conventional excitation-emission matrix fluorescence (EEMF) [15]. The SF methodology provides a simple way to rapidly measure fluorescence signals and spectral signatures of complex biological samples. Conventional fluorescence spectroscopy uses either a fixed-wavelength excitation (λ_{ex}) to produce an emission spectrum or a fixed-wavelength emission (λ_{em}) to record an excitation spectrum. With synchronous spectroscopy, the fluorescence signal is recorded while both λ_{em} and λ_{ex} are scanned simultaneously. A constant wavelength interval $\Delta\lambda$ is maintained between the excitation and emission wavelengths throughout the spectrum, as expressed by:

$$\lambda_{em} = \lambda_{ex} + \Delta\lambda \tag{1}$$

As a result, the intensity of synchronous signal I_s, can be written as the product of three functions [8,9]:

$$I_s(\lambda_{ex}, \lambda_{em}) = k\, c\, E_x(\lambda_{ex}) F_e(\lambda_{ex}, \lambda_{em}) E_m(\lambda_{em}) \tag{2}$$

where k is a constant dependent on the measurement geometry; c is the fluorophore concentration; E_x is the excitation absorption spectrum; E_m is the fluorescence emission spectrum; and F_e is the fluorescence efficiency describing the ratio of excitation light converted to fluorescence. If the wavelength interval $\Delta\lambda$ is chosen properly, the resulting SF spectrum will show one or a few features that are much more resolvable than those in the conventional fluorescence emission spectrum.

For a single molecular species, the observed SF signal intensity I_s is simplified (often to a single peak) and the bandwidth is narrower than in a conventional fluorescence emission spectrum. This feature significantly reduces spectral overlapping in multicomponent samples. This advantage can be well-demonstrated by the SF spectrum of a sample consisting of structurally-similar compounds: naphthalene, phenanthrene, anthracene, perylene, and tetracene, (Figure 1) [8], in which each compound gives practically only one "peak", similar to a chromatogram (*i.e.*, a spectrogram).

The SF method has found numerous applications in spectral analysis of complex samples, e.g., in environmental protection [16–18], food science [19–21], biological assays [22–28], and medical diagnosis [29–31]. In our laboratory, the SF method has been the basis for development of various instruments, including a portable field monitor [10] and an acousto-optic tunable filter (AOTF) system [11]. Most of the early applications of the SF method were focused on examining *in vitro* samples, e.g., analyzing air particulates [32], screening metabolites of a carcinogenic compound, benzo[a]pyrene, characterizing animal and human urine [12], determining carcinogen-macromolecular adducts [33,34], measuring cellular mitochondrial uptake of Rhodamine 123 [35], differentiating normal and neoplastic cells [36], investigating lysozyme-chitobioside interactions [37], studying non-calcium interactions of Fura-2 sensing dye [38], and detecting multiple fluorescent probes in DNA-sequencing [39]. Recently, the SF method has received increased interest in optical diagnosis of cancer. Several other research groups have carried out SF spectroscopy on *ex vivo* tissue samples, such as the cornea [40]. SF has been used to analyze normal and abnormal cervical tissues [41], and three-dimensional, total synchronous luminescence spectroscopy criteria for discrimination between normal and malignant breast tissue [42]. Moreover, the SF method has been used to characterize DMBA-TPA-induced squamous cell carcinoma in mice *in vivo* [43]. Data from total synchronous fluorescence spectroscopy measurements of normal and malignant breast tissue samples were introduced in supervised self-organizing maps, a type of artificial neural network, to obtain diagnosis [30]. Synchronous fluorescence spectroscopy was used for the detection and characterization of cervical cancers *in vitro* [31]. The SF technique has been applied on a variety types of skin tissue to show its narrow-band features and selectivity for *in vivo* analysis [44], and has been applied to both *in vitro* and *in vivo* imaging for cancer detection and diagnostics [30,31,41,45–47]. The recent surge of applications has highlighted the unique advantages of the SF method that offers a simple, yet effective, way to capture the fluorescent signatures of complex biochemical compounds in tissue for medical diagnostics.

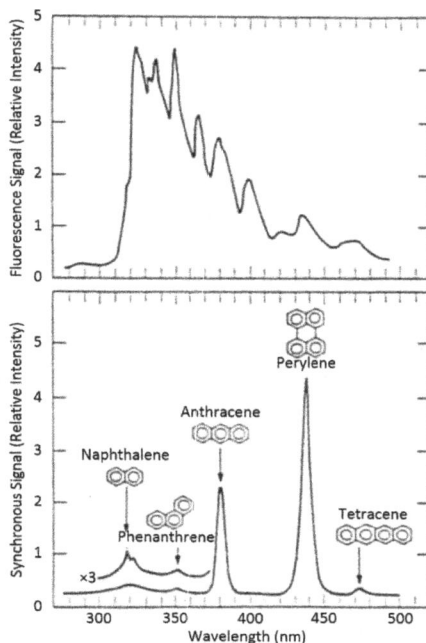

Figure 1. Conventional fixed-excitation fluorescence of a five-component mixture (upper curve, 258 nm excitation) and the SF spectrum of the same mixture showing improved spectral component separation (lower curve, 3 nm wavelength interval). Adapted from [8].

1.2. Synchronous Fluorescence Imaging

We have previously developed a synchronous fluorescence imaging (SFI) system to combine the great diagnostic potentials of the SF method and the large field-of-view of imaging in cancer diagnosis [45,46]. The SFI system can be incorporated into an endoscope for gastrointestinal cancer detection, as schematically shown in Figure 2.

A discriminant analysis and a multivariate statistical method were developed to differentiate pixels containing malignant and normal tissue. MATLAB (MathWorks, Natick, Massachusetts) programs were implemented to allow manual selection of the training pixels for a chosen classifier. The trained classifier would subsequently generate a diagnostic image indicating the condition of each pixel. Both the SF images and the whole set of fluorescence images were analyzed using these two methods. The specificity and sensitivity of the diagnostic image derived from the synchronous data were calculated, using the diagnostic image computed from the full spectral data as the gold standard. Classification results based on the SF data achieved good accuracy compared to the results based on the full spectra when the wavelength interval was chosen appropriately. We emphasize that an attractive advantage of SFI is that a much smaller data set is acquired without losing analytical specificity. Furthermore, the contrast between fluorescence peaks and background in the spectrum is significantly enhanced, which will benefit classification that relies on the subtle spectral differences between malignant and normal tissues. The SFI method, thus, dramatically reduces data acquisition time while maintaining a high classification accuracy, with diagnostic sensitivity from 82% to 97%, depending on the experimental conditions [45,46].

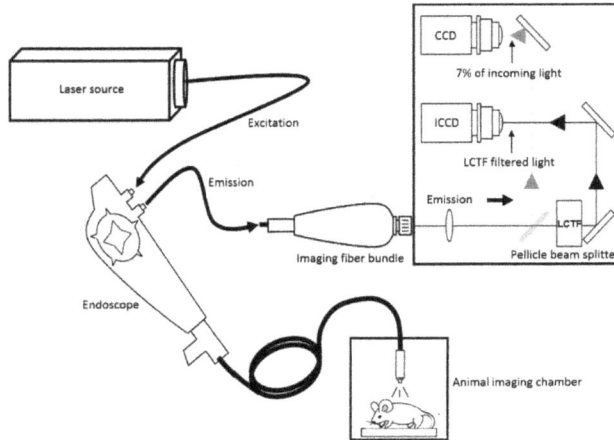

Figure 2. Schematic of the synchronous fluorescence imaging system. Adapted from [45].

2. Methods

2.1. Time-Resolved Synchronous Fluorescence

Despite its proven success in comparison to conventional fluorescent methods, the SF method still encounters difficulties in analyzing highly-complex fluorescent samples, where spectral overlapping of multiple fluorophores is severe. This limitation is especially detrimental when both absorption and emission spectra of the fluorophores are closely located. We propose to solve this problem by adding an additional dimension of measurement to the SF method: *i.e.*, using the time-domain information to further separate spectrally-similar fluorophores. The time-domain information of fluorophores is obtained by measuring time-resolved fluorescence decay, which is characteristic to the fluorophores, and, more importantly, is practically uncoupled to the spectral domain information. In our previous work, using an analog time-gating device, we showed improved spectral separation of complex samples [48]. The proposed method significantly improves spectral separation by obtaining the complete time-domain data in just one acquisition cycle.

Time-resolved photo-detection can be achieved using either analog or photon-counting methods. The analog method uses a photo-sensitive device to convert the photon flux to an electric current or voltage, which is subsequently quantized by an analog-digital convertor. Representative analog photodetectors are alkaline-metal photomultiplier tubes and semiconductor photodiodes. The analog method is easy to implement and has been widely used in many applications. However, it has notable limitations in detection sensitivity, response time, output linearity, and device aging. The photon-counting method is established on a completely different principle: it detects the arrival time of individual photons, also known as photon-tagging, and builds the temporal response curve by repeating the photon-tagging operation for a large number of times to create a histogram of the accumulated events of single-photon detection. In this method, the gain of the photosensor is set to a very high level disregarding the linearity range of the device. By doing so, it achieves single-photon sensitivity. The photon-tagging module determines the arrival time of photons with respect to the reference signal fed directly from the pulsed laser source. Compared to the analog method, the photon-counting method dramatically mitigates the limitations on detection sensitivity, response time, output linearity, and device aging.

In the present work, we used the photon-counting method with the single-photon avalanche diodes (SPADs) as the photosensors. With single-photon sensitivity, this configuration was well suited for low-light photodetection. Another important advantage of the photon-counting method is that its temporal resolution is no longer limited by the response time of the photosensor because the temporal response curve is the statistic of the arrival time of the photons, *i.e.*, the electric signal pulses generated by the photosensor in response to the incoming photons. Therefore, the delay due to response time of the photosensor is constant and can be subtracted from the final result collectively with other sources of delay, e.g., optical and electric transmission. Since the significant event is the arrival of photons, it is the timing and number of pulses, rather than the amplitude, that determine the output in photon-counting. Therefore, the linearity range of the photosensor is no longer a concern. In addition, the output linearity is highly stable compared to the analog method and do not require any calibration by the users. Finally, device aging would have virtually no impact on the performance of the photon-counting method because the photosensor produces binary signals whose amplitude is insignificant for the downstream photon-counter. It should be noted that a common limitation of the photon-counting method is the so-called "photon pile-up" effect, which is negligible under the low-light condition.

2.2. System Configuration

Here we report the first prototype system for a TRSF spectroscopy study. The system is schematically shown in Figure 3. A pulsed supercontinuum laser (SC400-2, Fianium, Southampton, UK) was used as a broadband light source: Laser emission between 400 and 2400 nm with an average power density of ~1 mW/nm with a beam diameter of ~1.5 mm in the visible range of 400–700 nm, at a repetition rate of 20 MHz and a typical pulse width of 10 ps. For fluorescence excitation, a short-pass filter (cutoff at 850 nm) was inserted to prevent infrared sample heating. A monochromator (DMC1-03, Optometrics, Littleton, MA, USA) was used to adjust the excitation wavelength. On the emission side, a liquid-crystal tunable filter (LCTF, VariSpec VIS-20, PerkinElmer, Waltham, MA, USA) was used as a band-pass filter tunable between 400 and 720 nm. Time-resolved photodetection was achieved using a time-correlated single-photon counting (TCSPC) module (SPC-150, Becker and Hickl, Berlin, Germany) coupled with an SPAD (Micro Photon Devices, Bolzano, Italy).

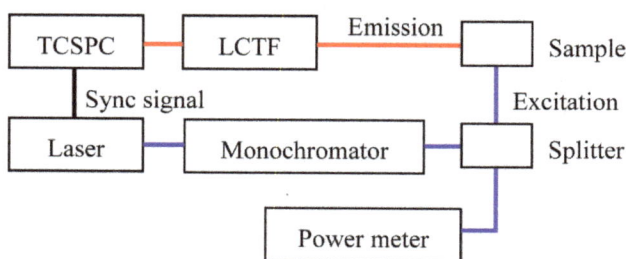

Figure 3. Schematic of the time-resolved synchronous fluorescence (TRSF) system.

It is noteworthy that the proposed TRSF methodology can be easily translated to imaging applications, for example, by coupling the photodetector to a wide-field imaging lens and performing raster scans in the imaging plane of the lens.

3. Results and Discussion

To demonstrate the proposed TRSF system, we used a sample consisting of methylene blue and oxazine 170 (both 10 μM) in deionized water. Methylene blue is frequently used as a dye to stain certain types of tissue during surgery, and can be used as an injectable to treat methemoglobinemia [49]. Oxazine 170 is an important laser dye and has recently been demonstrated in a feasibility study as a nerve-specific marker for neurosurgery [50]. These two fluorophores have substantially overlapping absorption and emission spectra that render SF spectral separation ineffective, shown in Figure 4, which were acquired using a commercially available spectrometer (Jobin Yvon FluoroMax 3, Horiba Scientific, Edison, NJ) with both excitation and emission slit widths set to 1 nm. The spectral sweeping step size was 5 nm for fluorescence (Figure 4a,c) and 1 nm for SF (Figure 4b,d). The wavelength interval of the SF scans was 20 nm.

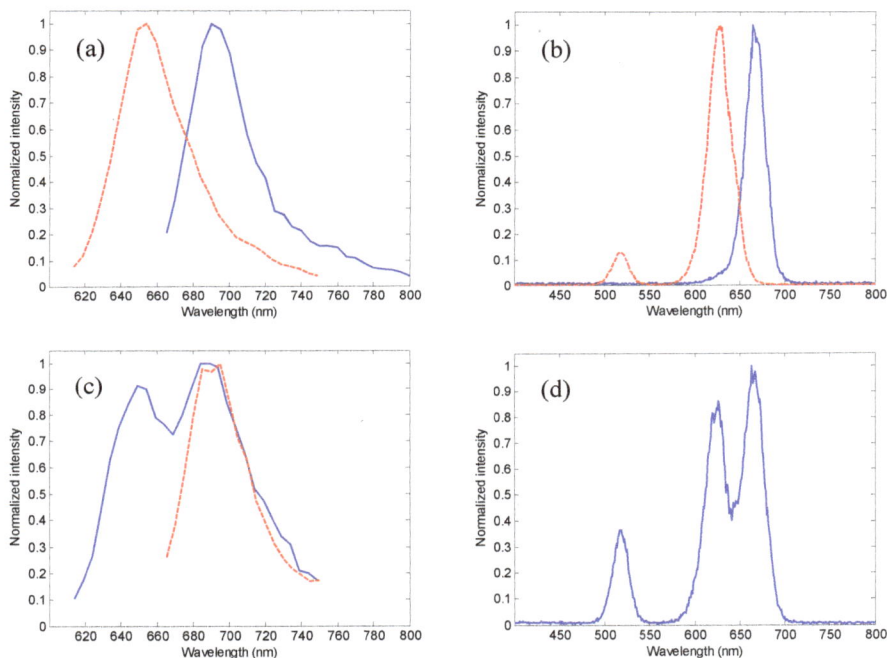

Figure 4. Fluorescence and standard SF spectra of methylene blue, oxazine 170, and their mixture. (**a**) conventional fluorescence emission spectra of methylene blue (blue solid line) and oxazine 170 (red broken line); (**b**) standard SF spectra of methylene blue (blue solid line) and oxazine 170 (red broken line); (**c**) conventional fluorescence spectra of the mixture of methylene blue and oxazine 170, excited at 614 nm (blue solid line) and 655 nm (red broken line); and (**d**) standard SF spectrum of the mixture of methylene blue and oxazine 170 (wavelength interval of 20 nm).

It is apparent in Figure 4c,d that spectral separation of methylene blue and oxazine 170 in the SF spectrum is significantly better than in conventional fluorescence spectrum. Nonetheless, we emphasize that the SF spectrum was acquired with a slit width of 1 nm and a sweeping step of 1 nm, which may be impractical in many applications, e.g., multispectral imaging or fast acquisition, when the spectral pass-band widths and sweeping step are significantly broader (typically 10 nm or higher). Under these circumstances, spectral separation may become unsatisfactory, which is shown below.

Using our newly-developed prototype system, we obtained both SF and TRSF spectra of the same set of samples consisting of methylene blue and oxazine 170 (freshly prepared). The excitation and emission wavelengths were adjusted by tuning the monochromator and the LCTF simultaneously between consecutive data acquisitions with a wavelength interval of 40 nm. We chose this wavelength interval because the optical density of this particular LCTF is the smallest at 40 nm within the range of typical interval values (20–60 nm). Note that the bandwidths of the LCTF and monochromator were 20 and 10 nm, respectively, limiting the smallest allowable wavelength interval for this configuration to 15 nm. Considering the transition bandwidth of the devices, the minimal practical value of the wavelength interval would be ~20 nm.

The TRSF and SF spectra acquired by the new system are shown in Figure 5. The TRSF spectra (Figure 5a–c) were acquired by plotting the photon count with respect to the photodetection/emission wavelength (the x-axis) and the photon time-tag (the y-axis). The SF spectra (Figure 5d–f) were obtained by disregarding the photon time-tag (*i.e.*, collapsing the y-axis). The spectral range (excitation) was 400–680 nm, which was jointly defined by the laser spectrum (400–850 nm) and the tuning range of LCTF (400–720 nm). The corresponding spectral range of the TRSF and SF scans was 440–720 nm.

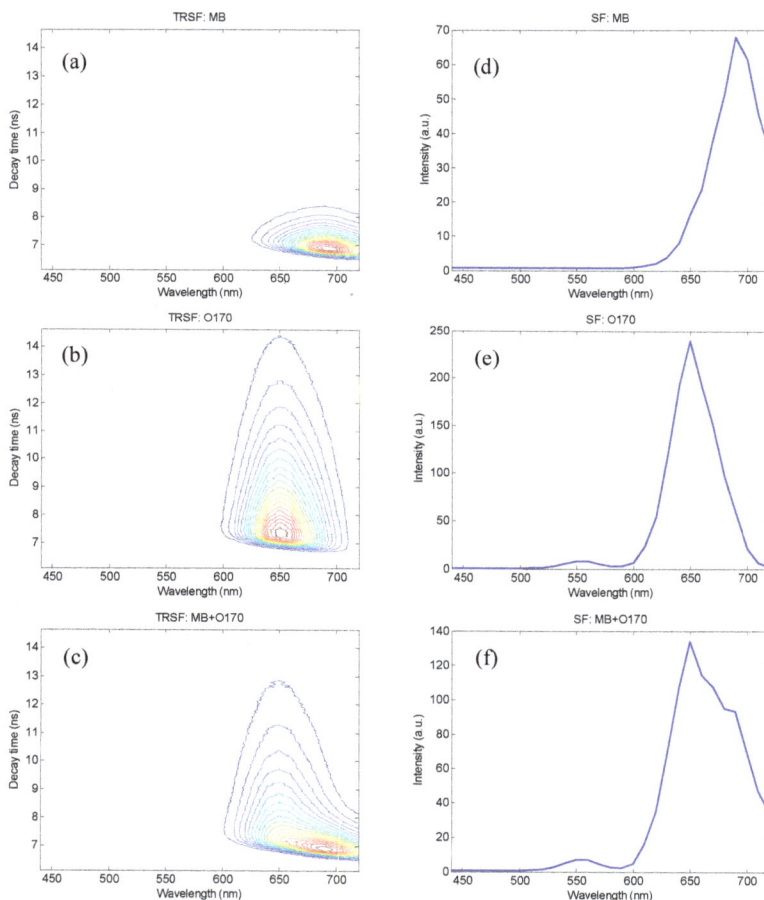

Figure 5. SF and TRSF spectra acquired using the TRSF system: linear contour plots of the TRSF spectra of (**a**) methylene blue (MB); (**b**) oxazine 170 (O170); (**c**) their mixture (MB + O170); and (**d–f**) SF spectra derived from the same data sets, respectively.

It was apparent that, using the SF method, methylene blue and oxazine 170 could not be spectrally resolved with the given excitation/emission configuration, Figure 5f. This was mainly because of peak-broadening due to differences in system configuration and device specifications, comparing Figures 4d and 5f. The difference in fluorescence quantum efficiency is another complicating factor, referring to the spectra obtained by the same instrument shown in Figure 5d–f. Both of these factors have made effective spectral separation of the two fluorophores impractical. With the time-resolved data shown in Figure 5c, however, clear distinction between the two fluorescent species is evident, comparing Figure 5c to Figure 5a,b. Specifically, single-component samples exhibit highly symmetric patterns (Figure 5a,b) because of the symmetric fluorescent emission peaks of the individual fluorophores. In contrast, the two-component sample produced a highly asymmetric pattern (Figure 5c) because fluorescence lifetime constants of the constituent fluorophores are different. In other words, one observes a shifting fluorescent emission peak at different time points of fluorescence decay, which can be better illustrated by plotting the SF spectra of the set of samples at different time points, Figure 6.

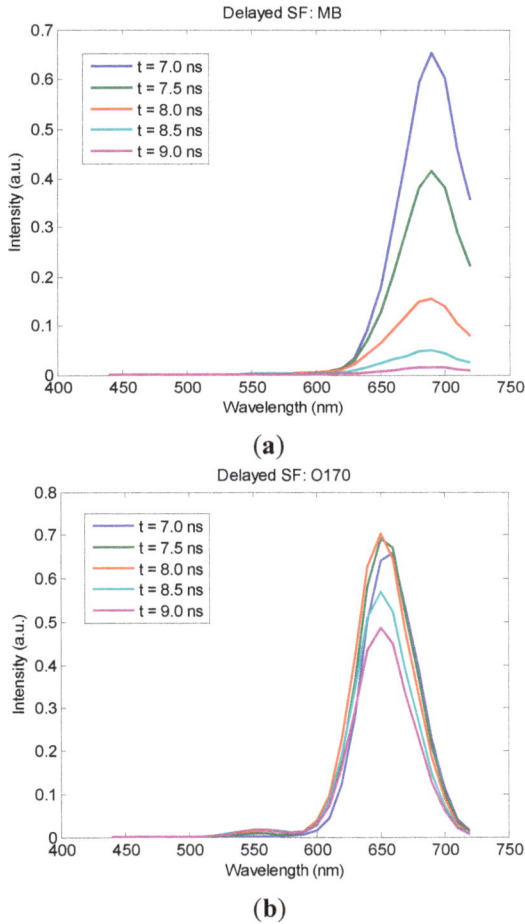

(a)

(b)

Figure 6. *Cont.*

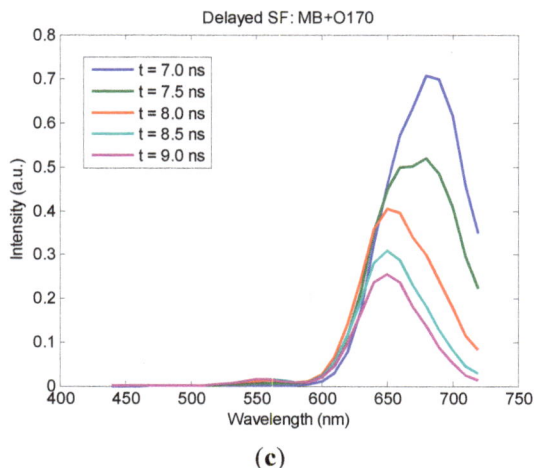

Figure 6. Time-delayed SF ("Delayed SF") plots derived from the TRSF data set of (**a**) methylene blue (MB); (**b**) oxazine 170 (O170); and (**c**) the mixture of the two fluorophores (MB + O170).

The existence of shifting peak wavelength in the "delayed SF" spectra can be used as a quantitative indicator in automatic multi-component fluorophore identification. With the additional time-resolved information, the lifetime constants of the fluorophores can be used in spectro-temporal analysis. Similar to the extension of the SF method to SFI, the concept of TRSF can also potentially be extended to biomedical imaging. Note that the TCSPC acquisition method is based on photon-counting and that it does not perform additional scans (e.g., time-binning) in the time-domain. For this reason, TRSF imaging will not increase acquisition time except for the intrinsic signal-to-noise limit. Furthermore, in analyzing complex biomedical specimens, high-dimensional analysis that involves spectral, temporal, and spatial information will likely give more robust classification results than low-dimensional analysis does. We are currently working toward TRSF imaging instrumentation and multi-dimensional data analysis algorithms for biomedical applications.

4. Conclusions

Synchronous fluorescence (SF) is a fast and information-rich spectroscopy method. It has unique advantages compared to the conventional fixed-excitation fluorescence spectroscopy. Based on the SF method, we demonstrated the feasibility of a new time-resolved SF (TRSF) method and instrument. This method is a significant advancement from our previous work that revealed differences in spectral data with different acquisition time-delay. In the present work, a complete dimension in the time-domain is added to the spectral data, which is conveniently obtained in one acquisition cycle. Compared to the standard SF method and the previous method, the proposed TRSF method better captures the full characteristics of fluorophores in complex samples. Using a two-fluorophore liquid sample as an example, our prototype TRSF system clearly showed its advantage in spectro-temporal separation of the fluorophores that were otherwise difficult to resolve with conventional SF spectroscopy. Although application of the proposed method on actual biological samples is beyond the scope of the present work, we envision that our previously-tested SFI and the newly-developed TRSF methods can combine the impressive diagnostic potentials of the SF spectroscopy, wide-field imaging, and the new ultrafast time-resolved acquisition, and that the new technologies will further improve the selectivity of fluorescence-based analysis and diagnostics. Lastly, we would like to emphasize that SF is a methodology for rapid data acquisition, which can be potentially applied to a number of analytical and imaging technologies to achieve superior performance that were previously unattainable.

Acknowledgments: This work was sponsored by the Duke Faculty Exploratory Fund.

Author Contributions: Xiaofeng Zhang and Tuan Vo-Dinh jointly conceived the idea of time-resolved synchronous fluorescence and authored the manuscript. Xiaofeng Zhang implemented the instrumentation and performed the experiments. Tuan Vo-Dinh provided critical intellectual and logistic supports for the work. Andrew Fales was a PhD student who assisted with sample preparation and initial testing of samples.

Conflicts of Interest: The authors declare no conflict of interest.

References

1. Georgakoudi, I.; Feld, M.S. The combined use of fluorescence, reflectance, and light-scattering spectroscopy for evaluating dysplasia in Barrett's esophagus. *Gastrointest. Endosc. Clin. North Am.* **2004**, *14*, 519–537. [CrossRef] [PubMed]
2. Ramanujam, N. Fluorescence spectroscopy of neoplastic and non-neoplastic tissues. *Neoplasia* **2000**, *2*, 89–117. [CrossRef] [PubMed]
3. Richards-Kortum, R.; Mitchell, M.F.; Ramanujam, N.; Mahadevan, A.; Thomsen, S. *In vivo* fluorescence spectroscopy: Potential for non-invasive, automated diagnosis of cervical intraepithelial neoplasia and use as a surrogate endpoint biomarker. *J. Cell. Biochem. Suppl.* **1994**, *19*, 111–119. [PubMed]
4. Harris, D.M.; Werkhaven, J. Endogenous porphyrin fluorescence in tumors. *Lasers Surg. Med.* **1987**, *7*, 467–472. [CrossRef] [PubMed]
5. Vo-Dinh, T.; Panjehpour, M.; Overholt, B.F.; Farris, C.; Buckley, F.P., III; Sneed, R. *In vivo* cancer diagnosis of the esophagus using differential normalized fluorescence (DNF) indices. *Lasers Surg. Med.* **1995**, *16*, 41–47. [PubMed]
6. Alfano, R.R. *Advances in Optical Biopsy and Optical Mammography*; Wiley: Hoboken, NJ, USA, 1997; pp. 1–203.
7. Lam, S.; Palcic, B.; McLean, D.; Hung, J.; Korbelik, M.; Profio, A.E. Detection of early lung cancer using low dose Photofrin II. *Chest* **1990**, *97*, 333–337. [CrossRef] [PubMed]
8. Vo-Dinh, T. Multicomponent analysis by synchronous luminescence spectrometry. *Anal. Chem.* **1978**, *50*, 396–401. [CrossRef]
9. Vo-Dinh, T. Synchronous luminescence spectroscopy: Methodology and applicability. *Appl. Spectrosc.* **1982**, *36*, 576–581. [CrossRef]
10. Alarie, J.P.; Vo-Dinh, T.; Miller, G.; Ericson, M.N.; Eastwood, D.; Lidberg, R.; Dominguez, M. Development of a battery-operated portable synchronous luminescence spectrofluorometer. *Rev. Sci. Instrum.* **1993**, *64*, 2541–2546. [CrossRef]
11. Hueber, D.M.; Stevenson, C.L.; Vo-Dinh, T. Fast scanning synchronous luminescence spectrometer based on acousto-optic tunable filters. *Appl. Spectrosc.* **1995**, *49*, 1624–1631. [CrossRef]
12. Vo-Dinh, T.; Uziel, M. Laser-induced room-temperature phosphorescence detection of benzo[a]pyrene-DNA adducts. *Anal. Chem.* **1987**, *59*, 1093–1095. [CrossRef] [PubMed]
13. Uziel, M.; Ward, R.J.; Vodinh, T. Synchronous fluorescence measurement of bap metabolites in human and animal urine. *Anal. Lett.* **1987**, *20*, 761–776. [CrossRef]
14. Vo-Dinh, T.; Gammage, R.B. Singlet-Triplet energy difference as a parameter of selectivity in synchronous phosphorimetry. *Anal. Chem.* **1978**, *50*, 2054–2058. [CrossRef]
15. Kumar, K.; Mishra, A.K. Analysis of dilute aqueous multifluorophoric mixtures using excitation-emission matrix fluorescence (EEMF) and total synchronous fluorescence (TSF) spectroscopy: A comparative evaluation. *Talanta* **2013**, *117*, 209–220. [CrossRef] [PubMed]
16. Chen, W.; Habibul, N.; Liu, X.Y.; Sheng, G.P.; Yu, H.Q. FTIR and synchronous fluorescence heterospectral two-dimensional correlation analyses on the binding characteristics of copper onto sissolved organic matter. *Environ. Sci. Technol.* **2015**, *49*, 2052–2058. [CrossRef] [PubMed]
17. Bauer, A.E.; Frank, R.A.; Headley, J.V.; Peru, K.M.; Hewitt, L.M.; Dixon, D.G. Enhanced characterization of oil sands acid-extractable organics fractions using electrospray ionization-high resolution mass spectrometry and synchronous fluorescence spectroscopy. *Environ. Toxicol. Chem.* **2015**, *34*, 1001–1008. [CrossRef] [PubMed]

18. Hur, J.; Lee, B.M.; Lee, T.H.; Park, D.H. Estimation of biological oxygen demand and chemical oxygen demand for combined sewer systems using synchronous fluorescence spectra. *Sensors* **2010**, *10*, 2460–2471. [CrossRef] [PubMed]

19. Ziak, L.; Majek, P.; Hrobonova, K.; Cacho, F.; Sadecka, J. Simultaneous determination of caffeine, caramel and riboflavin in cola-type and energy drinks by synchronous fluorescence technique coupled with partial least squares. *Food Chem.* **2014**, *159*, 282–286. [CrossRef] [PubMed]

20. Sergiel, I.; Pohl, P.; Biesaga, M.; Mironczyk, A. Suitability of three-dimensional synchronous fluorescence spectroscopy for fingerprint analysis of honey samples with reference to their phenolic profiles. *Food Chem.* **2014**, *145*, 319–326. [CrossRef] [PubMed]

21. Lenhardt, L.; Zekovic, I.; Dramicanin, T.; Dramicanin, M.D.; Bro, R. Determination of the botanical origin of honey by front-face synchronous fluorescence spectroscopy. *Appl. Spectrosc.* **2014**, *68*, 557–563. [CrossRef] [PubMed]

22. Ye, T.; Liu, Y.; Luo, M.; Xiang, X.; Ji, X.; Zhou, G.; He, Z. Metal-organic framework-based molecular beacons for multiplexed DNA detection by synchronous fluorescence analysis. *Analyst* **2014**, *139*, 1721–1725. [CrossRef] [PubMed]

23. Pagani, A.P.; Ibanez, G.A. Second-order multivariate models for the processing of standard-addition synchronous fluorescence-pH data. Application to the analysis of salicylic acid and its major metabolite in human urine. *Talanta* **2014**, *122*, 1–7. [CrossRef] [PubMed]

24. Madrakian, T.; Bagheri, H.; Afkhami, A. Determination of human albumin in serum and urine samples by constant-energy synchronous fluorescence method. *Luminescence* **2015**, *30*, 576–582. [CrossRef] [PubMed]

25. Schenone, A.V.; Culzoni, M.J.; Campiglia, A.D.; Goicoechea, H.C. Total synchronous fluorescence spectroscopic data modeled with first- and second-order algorithms for the determination of doxorubicin in human plasma. *Anal. Bioanal. Chem.* **2013**, *405*, 8515–8523. [CrossRef] [PubMed]

26. Kaur, K.; Saini, S.; Singh, B.; Malik, A.K. Highly sensitive synchronous fluorescence measurement of danofloxacin in pharmaceutical and milk samples using aluminium (III) enhanced fluorescence. *J. Fluoresc.* **2012**, *22*, 1407–1413. [CrossRef] [PubMed]

27. Teixeira, A.P.; Duarte, T.M.; Carrondo, M.J.; Alves, P.M. Synchronous fluorescence spectroscopy as a novel tool to enable PAT applications in bioprocesses. *Biotechnol. Bioeng.* **2011**, *108*, 1852–1861. [CrossRef] [PubMed]

28. Pulgarin, J.A.M.; Molina, A.A.; Robles, I.S.-F. Rapid simultaneous determination of four non-steroidal anti-inflammatory drugs by means of derivative nonlinear variable-angle synchronous fluorescence spectrometry. *Appl. Spectrosc.* **2010**, *64*, 949–955. [CrossRef] [PubMed]

29. Huang, L.; Guo, L.; Wan, Y.; Pan, P.; Feng, L. Simultaneous determination of three potential cancer biomarkers in rat urine by synchronous fluorescence spectroscopy. *Spectrochim. Acta Part A Mol. Biomol. Spectrosc.* **2014**, *120*, 595–601. [CrossRef] [PubMed]

30. Dramicanin, T.; Dimitrijevic, B.; Dramicanin, M.D. Application of supervised self-organizing maps in breast cancer diagnosis by total synchronous fluorescence spectroscopy. *Appl. Spectrosc.* **2011**, *65*, 293–297. [CrossRef] [PubMed]

31. Ebenezar, J.; Aruna, P.; Ganesan, S. Synchronous fluorescence spectroscopy for the detection and characterization of cervical cancers *in vitro*. *Photochem. Photobiol.* **2010**, *86*, 77–86. [CrossRef] [PubMed]

32. Vo-Dinh, T.; Gammage, R.B.; Martinez, P.R. Analysis of a workplace air particulate sample by synchronous luminescence and room-temperature phosphorescence. *Anal. Chem.* **1981**, *53*, 253–258. [CrossRef] [PubMed]

33. Vahakangas, K.; Trivers, G.; Rowe, M.; Harris, C.C. Benzo(a)pyrene diolepoxide-DNA adducts detected by synchronous fluorescence spectrophotometry. *Environ. Health Perspect.* **1985**, *62*, 101–104. [CrossRef] [PubMed]

34. Shields, P.G.; Kato, S.; Bowman, E.D.; Petruzzelli, S.; Cooper, D.P.; Povey, A.C.; Weston, A. Combined micropreparative techniques with synchronous fluorescence spectroscopy or 32P-postlabelling assay for carcinogen-DNA adduct determination. *IARC Sci. Publ.* **1993**, *124*, 243–254. [PubMed]

35. Askari, M.D.; Vo-Dinh, T. Implication of mitochondrial involvement in apoptotic activity of fragile histidine triad gene: Application of synchronous luminescence spectroscopy. *Biopolymers* **2004**, *73*, 510–523. [CrossRef] [PubMed]

36. Watts, W.E.; Isola, N.R.; Frazier, D.; Vo-Dinh, T. Differentiation of normal and neoplastic cells by synchronous fluorescence: Rat liver epithelial and rat hepatoma cell models. *Anal. Lett.* **1999**, *32*, 2583–2594. [CrossRef]

37. Viallet, P.M.; Vo-Dinh, T.; Vigo, J.; Salmon, J.M. Investigation of lysozyme-chitobioside interactions using synchronous luminescence and lifetime measurements. *J. Fluoresc.* **2002**, *12*, 57–63. [CrossRef]

38. Baucel, F.; Salmon, J.M.; Vigo, J.; Vo-Dinh, T.; Viallet, F. Investigation of noncalcium interactions of fura-2 by classical and synchronous fluorescence spectroscopy. *Anal. Biochem.* **1992**, *204*, 231–238.

39. Stevenson, C.L.; Johnson, R.W.; Vo-Dinh, T. Synchronous luminescence: A new detection technique for multiple fluorescent-probes used for DNA-sequencing. *Biotechniques.* **1994**, *16*, 1104–1111. [PubMed]

40. Uma, L.; Sharma, Y.; Balasubramanian, D. Fluorescence properties of isolated intact normal human corneas. *Photochem. Photobiol.* **1996**, *63*, 213–216. [CrossRef] [PubMed]

41. Vengadesan, N.; Anbupalam, T.; Hemamalini, S.; Ebenezar, J.; Muthuvelu, K.; Koteeswaran, D.; Aruna, P.R.; Ganesan, S. Characterization of cervical normal and abnormal tissues by synchronous luminescence spectroscopy. *Opt. Biopsy IV* **2002**, *4613*, 13–17.

42. Dramicanin, T.; Dramicanin, M.D.; Jokanovic, V.; Nikolic-Vukosavljevic, D.; Dimitrijevic, B. Three-dimensional total synchronous luminescence spectroscopy criteria for discrimination between normal and malignant breast tissues. *Photochem. Photobiol.* **2005**, *81*, 1554–1558. [CrossRef] [PubMed]

43. Diagaradjane, P.; Yaseen, M.A.; Yu, J.; Wong, M.S.; Anvari, B. Synchronous fluorescence spectroscopic characterization of DMBA-TPA-induced squamous cell carcinoma in mice. *J. Biomed. Opt.* **2006**, *11*. [CrossRef] [PubMed]

44. Vo-Dinh, T. Principle of synchronous luminescence (SL) technique for biomedical diagnostics. In Proceedings of the Biomedical Diagnostic, Guidance, and Surgical-Assist Systems II, San Jose, CA, USA, 3 May 2000; pp. 42–49.

45. Liu, Q.; Chen, K.; Martin, M.; Wintenberg, A.; Lenarduzzi, R.; Panjehpour, M.; Overholt, B.F.; Vo-Dinh, T. Development of a synchronous fluorescence imaging system and data analysis methods. *Opt. Express* **2007**, *15*, 12583–12594. [CrossRef] [PubMed]

46. Liu, Q.; Grant, G.; Vo-Dinh, T. Investigation of synchronous fluorescence method in multi-component analysis in tissue. *IEEE J. Sel. Top. Quantum Electron.* **2010**, *16*, 927–940.

47. Patra, D.; Mishra, A.K. Recent developments in multi-component synchronous fluorescence scan analysis. *TrAC Trends Anal. Chem.* **2002**, *21*, 787–798. [CrossRef]

48. Stevenson, C.L.; Vo-Dinh, T. Analysis of polynuclear aromatic compounds using laser-excited synchronous fluorescence. *Anal. Chim. Acta* **1995**, *303*, 247–253. [CrossRef]

49. Wendel, W.B. The control of methemoglobinemia with methylene blue. *J. Clin. Investig.* **1939**, *18*, 179–185. [CrossRef] [PubMed]

50. Park, M.H.; Hyun, H.; Ashitate, Y.; Wada, H.; Park, G.; Lee, J.H.; Njiojob, C.; Henary, M.; Frangioni, J.V.; Choi, H.S. Prototype nerve-specific near-infrared fluorophores. *Theranostics* **2014**, *4*, 823–833. [CrossRef] [PubMed]

sensors

MDPI

Article

A High Performance Banknote Recognition System Based on a One-Dimensional Visible Light Line Sensor

Young Ho Park [1], Seung Yong Kwon [1], Tuyen Danh Pham [1], Kang Ryoung Park [1,*], Dae Sik Jeong [1] and Sungsoo Yoon [2]

[1] Division of Electronics and Electrical Engineering, Dongguk University, 26 Pil-dong 3-ga, Jung-gu, Seoul 100-715, Korea; fdsarew@hanafos.com (Y.H.P.); sbaru07@dgu.edu (S.Y.K.); phamdanhtuyen@dongguk.edu (T.D.P.); jungsoft97@dongguk.edu (D.S.J.)

[2] Kisan Electronics, Sungsoo 2-ga 3-dong, Sungdong-gu, Seoul 133-831, Korea; ssyoon@kisane.com

* Correspondence: parkgr@dongguk.edu; Tel.: +82-10-3111-7022; Fax: +82-2-2277-8735

Academic Editor: Gonzalo Pajares Martinsanz
Received: 16 April 2015; Accepted: 8 June 2015; Published: 15 June 2015

Abstract: An algorithm for recognizing banknotes is required in many fields, such as banknote-counting machines and automatic teller machines (ATM). Due to the size and cost limitations of banknote-counting machines and ATMs, the banknote image is usually captured by a one-dimensional (line) sensor instead of a conventional two-dimensional (area) sensor. Because the banknote image is captured by the line sensor while it is moved at fast speed through the rollers inside the banknote-counting machine or ATM, misalignment, geometric distortion, and non-uniform illumination of the captured images frequently occur, which degrades the banknote recognition accuracy. To overcome these problems, we propose a new method for recognizing banknotes. The experimental results using two-fold cross-validation for 61,240 United States dollar (USD) images show that the pre-classification error rate is 0%, and the average error rate for the final recognition of the USD banknotes is 0.114%.

Keywords: banknote recognition; one-dimensional (line) sensor; pre-classification; USD banknote

1. Introduction

Functionality for detecting counterfeit banknotes and recognizing banknotes are required in various machines, such as banknote-counting machines and automatic teller machines (ATMs). Banknote recognition is defined as the recognition of the type of banknote (e.g., $1, $2, and $5), the direction of the banknote, and the date of issue of the banknote (e.g., least recent, recent, and most recent). Furthermore, banknote recognition facilitates the detection of counterfeit banknotes and allows the condition of banknotes to be monitored. Thus, research on banknote recognition has developed rapidly [1–8].

In general, a banknote's obverse and reverse image differ. Hence, a banknote is constituted by four different patterns (obverse-forward, obverse-backward, reverse-forward, and reverse-backward) according to the direction of the banknote received by the machine. This increases the number of banknote classes four-fold. Therefore, the complexity of banknote recognition increases accordingly, which, in turn, increases the processing time and the banknote recognition error rate.

Previous studies can be divided into those that address the recognition of a banknote's orientation [9] and those concerned with banknote recognition [10–17]. Wu *et al.* proposed a banknote-orientation recognition method using a back-propagation (BP) network [9]. They classified the input direction of banknotes by using a three-layer BP network. The performance of their method is high, however, they used only one type of banknote—viz., renminbi (RMB) 100 Yuan—in their experiment.

Kagehiro *et al.* proposed a hierarchical classification method for United States Dollar (USD) banknotes. Their method consists of three stages, using a generalized learning-vector quantization (GLVQ) algorithm to achieve high-speed processing with a high degree of accuracy [10]. The performance of their method was such that 99% of the banknotes were correctly recognized. Hasanuzzaman *et al.* proposed a banknote-recognition method based on speeded-up robust features (SURF) features [11]. Because it uses SURF features, their method is robust to illumination and scaling changes, as well as image rotation. At two seconds, however, the processing time required for a 3 GHz CPU computer is excessively long. Gai *et al.* proposed a feature-extraction method based on the quaternion wavelet transform (QWT) and generalized Gaussian distribution (GGD) for banknote classification [12]. With their method, USD banknotes are classified using a BP neural network. Ahmadi *et al.* and Omatu *et al.* proposed a banknote-classification method using principal component analysis (PCA), self-organizing map (SOM) clustering, and a learning-vector quantization (LVQ) classifier [13–17]. They presented a method to improve the reliability of banknote classification using local PCA [13–17] for data-feature extraction.

In other research [18], Yeh *et al.* proposed a method for detecting counterfeit banknotes based on multiple-kernel support vector machines (SVMs). They segmented a banknote into partitions, and took the luminance histograms of the partitions (with its own kernels) as the inputs to the system. In order to fuse the multiple kernels, they used semi-definite programming (SDP) learning. Unlike our research, their method is not concerned with recognizing the type of USD banknote. Rather, it is designed to detect counterfeit Taiwanese banknotes. This is the central difference between their research and ours. Two classes are needed with their method [18]: genuine and counterfeit banknotes. Our research, however, requires 64 classes: four directions and 16 banknote types ($1, $2, $5, recent $5, most recent $5, $10, recent $10, most recent $10, $20, recent $20, most recent $20, $50, recent $50, most recent $50, $100, and recent $100).

In previous research [19,20], Bruna *et al.* proposed a method to detect various types of counterfeit Euro banknotes. They used a near-infrared (NIR) light illuminator and a camera to capture the banknote image. This differs from our method, which uses a visible-light USD image. In order to discriminate counterfeit from genuine banknotes, they used a percentage of the pixels that satisfy predetermined conditions inside predefined regions. In addition, they used a correlation measure between the patches learned at the training stage and the corresponding pixels in the searching regions to recognize of the banknote type.

Hasanuzzaman *et al.* proposed a method for recognizing banknotes for the visually impaired based on SURF and the spatial relationship of matched SURF features [21]. Although they obtained a correct-recognition rate of 100% with USD banknotes, the number of classes and images for testing, at seven and 140, respectively, was too small. As noted above, by contrast, our experiments involved 64 classes and 61,240 images, as shown in Table 2.

In previous studies on banknote recognition, each banknote type was considered as constituted by four classes (obverse-forward, obverse-backward, reverse-forward, and reverse-backward), rather than a single class. Consequently, the number of banknote classes is considerably large. To overcome these problems, we propose a new method for recognizing banknotes. When compared with previous studies, the proposed method is novel in the following four ways:

(1)   The region-of-interest (ROI) area on the captured banknote image is located by corner detection algorithm, but there still exist the effect of the misalignment, geometric distortion, and non-uniform illumination on the ROI image. Therefore, a sub-sampled image of 32 × 6 pixels was used in our research in order to extract features more efficiently by reducing this effect.

(2)   In order to further reduce this effect on the recognition accuracy, we propose the feature extraction method by PCA with the sub-sampled image.

(3) Pre-classification is performed hierarchically: the first classification demarcates the obverse and reverse side, and the second demarcates the forward and backward direction. This pre-classification process reduces classification errors.

(4) Pre-classification is performed using a SVM based on the optimal feature vector extracted using PCA on the sub-sampled image. Then, the final type of banknote is recognized using the classifier based on a K-means algorithm.

Table 1 compares the methods proposed in previous studies with ours. The remainder of our paper is organized as follows: we describe the proposed method in Section 2. In addition, we present experimental results, discussions, and concluding remarks in Sections 3–5, respectively.

2. Proposed Method

2.1. Overall Procedure

An overview of our pre-classification method for the recognition of USD banknotes is provided in Figure 1. We defined four directions in our study: the obverse side in a forward direction is "Direction A", the obverse side in a backward direction is "Direction B", the reverse side in a forward direction is "Direction C", and the reverse side in a backward direction is "Direction D".

Table 1. Comparison of the proposed method with previous methods.

Category	Methods	Strengths	Weakness
Non-pre-classification-based method	-Using GLVQ [10]	Additional processing time for pre-classification is not required	Accuracy enhancement is limited because of the large number of classes of banknotes, including two sides (obverse and reverse) and two directions (forward and backward)
	-Using SURF features [11]		
	-Using QWT, generalized Gaussian distribution, and BP neural network [12]		
	-Using local PCA, SOM, and LVQ [13–17]		
	-Using correlation measure [19,20]		
	-Using SURF and the spatial relationship of matched SURF features [21]		
Pre-classification-based method	-Using BP neural network [9]	The number of classes of banknotes can be reduced four-fold, because of the pre-classification of the two sides and two directions	The classification accuracy of banknote type is not presented
	-Using SVM classifier with PCA features (**proposed method**)		Additional processing time is required for pre-classification

As shown in the figure, our method consists of four steps. First, to reduce the processing time, the input image is sub-sampled to yield a single image that is 32×6 pixels in size. Second, the optimal feature vector is extracted using PCA to classify the side of the banknote. Third, the side of the input image is determined as either obverse or reverse using an SVM. Fourth, the optimal feature vector for classifying the banknote's direction is extracted using PCA. Finally, the forward or backward direction is determined using the SVM. Then, the final kind of banknote is determined by K-means algorithm of 16 classes using the PCA features of the sub-sampled image.

2.2. Image Acquisition and Pre-Processing

In this study, the banknote image was captured using a commercial banknote-counting machine. Due to the size and cost limitation of the banknote-counting machine, the banknote image of visible light is captured by one-dimensional (line) sensor instead of conventional two-dimensional (area) sensor. That is, one line (row) image is acquired instead of a space (row by column) image at each time. The line image is acquired with visible light emitting diode (LED) while the input banknote is moving through the roller inside the banknote-counting machine at fast speed. The resolution of one line image is 1584 pixels, and 464 line images are acquired by this system. Based on this, the two-dimensional 1584 × 464 pixels image of the banknote is finally acquired by sequentially combining the 464 line images.

Because the banknote image is captured by the line sensor while it is moved through the roller at fast speed misalignment, geometric distortion, and non-uniform illumination effects frequently occur in the captured images, as shown in Figure 2, and the images with these effects are included in our database for experiments.

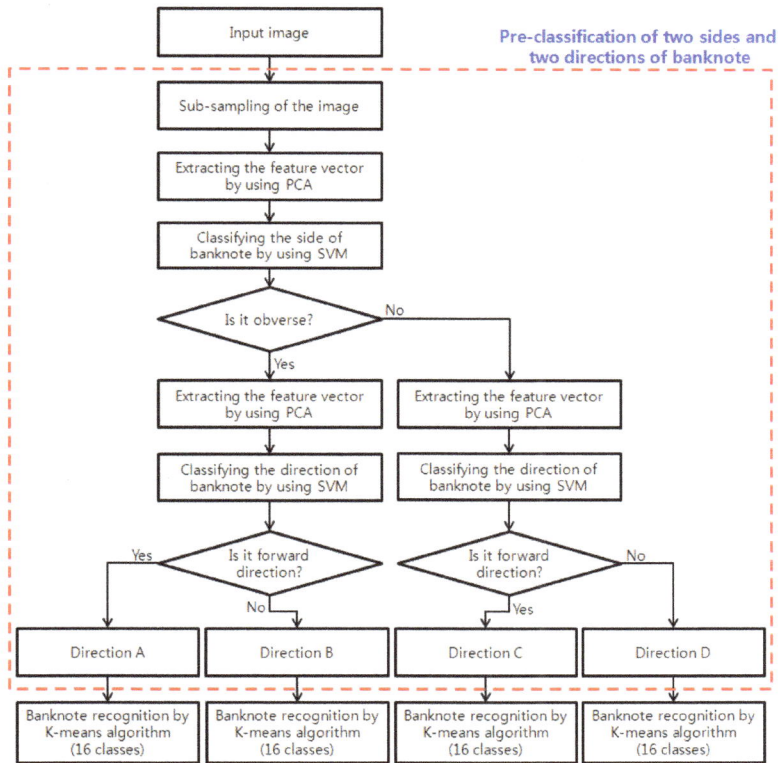

Figure 1. Overview of the proposed method.

To overcome these problems, the ROI area on the banknote is located by corner detection algorithm, but there still exist the positional variations on the ROI image as shown in Figure 3. In addition, the image resolution for the ROI area on banknote is as high as 1212 × 246 pixels, which leads to considerable processing time and the inclusion of noise and redundant data. Therefore, a sub-sampled image of 32 × 6 pixels was used in our research in order to extract features more efficiently by reducing the processing time and any unnecessary data without the positional variations, as shown in Figure 3.

(a)

(b)

(c)

(d)

Figure 2. Examples of misalignment, geometric distortion, and non-uniform illumination on the captured images: the cases of (**a**); (**b**) misalignment; (**c**) geometric distortion; and (**d**) non-uniform illumination.

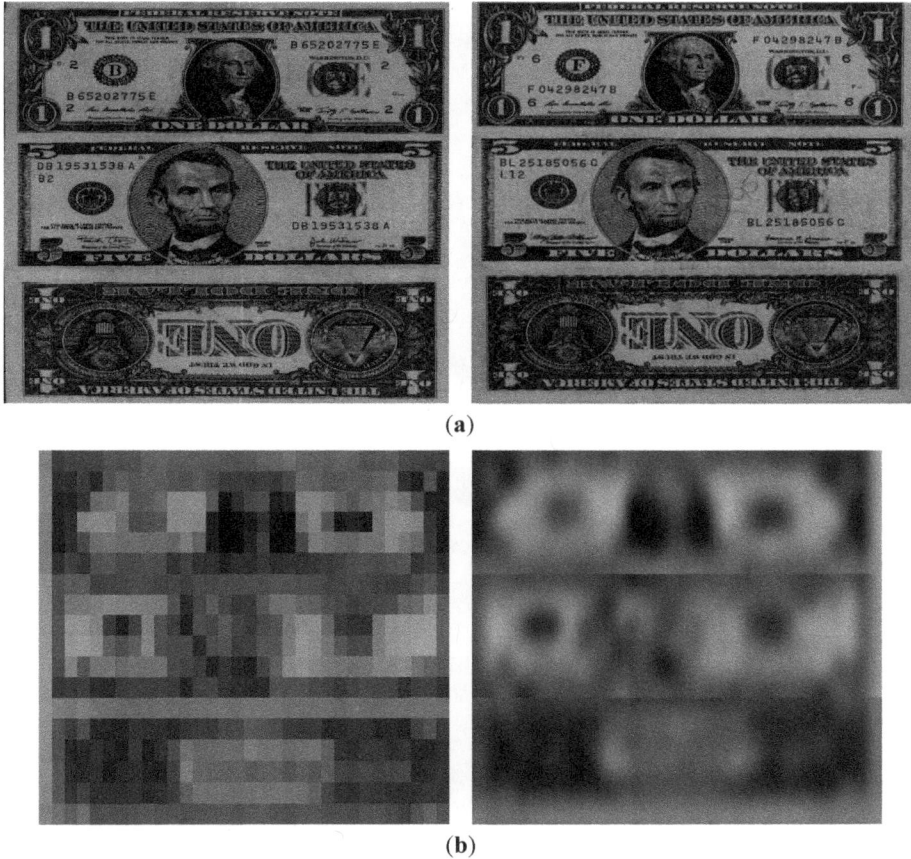

Figure 3. Examples of sub-sampled images: (a) original ROI areas of the banknote; (b) sub-sampled images.

The images of the 1st row of Figure 3b are the sub-sampled ones of the 1st row of Figure 3a. Like this, the images of the 2nd, and 3rd rows of Figure 3b are the sub-sampled ones of the 2nd, and 3rd rows of Figure 3a, respectively. If we show the sub-sampled image as its actual size (32×6 pixels), it is difficult for readers to discriminate the image because the size of sub-sampled image is much smaller than that (1212×246 pixels) of original ROI area of Figure 3a. Therefore, we show the magnified sub-sampled images whose size is larger than actual one (32×6 pixels) for higher visibility to readers in Figure 3b. Even with the sub-sampled image, there still exist the positional variations as shown in Figure 3b. Therefore, we propose the feature extraction method by PCA in order to reduce this effect on the recognition performance, and detail explanations are shown in the next section.

2.3. Feature Extraction by PCA, and Classification with an SVM

In general, as the dimensionality of data increases, feature extraction and pattern classification require much processing. In addition, an increase in the dimensionality can degrade the classification accuracy. PCA is a popular stochastic method that facilitates the analysis of high-dimensional data by dimensionality reduction [13,22]. This characteristic in PCA allows it to analyze the patterns of data more easily by reducing the dimensions of the data with minimal data loss. The procedure for conducting PCA is as follows. First, the covariance matrix Σ of the data is calculated:

$$\Sigma = \frac{1}{N} \sum_{n=1}^{N} (x_n - \mu)(x_n - \mu)^{T} \tag{1}$$

where N is the amount of data, μ is the mean of x_n, and x_n is the input data. Then, the eigenvalues and eigenvectors for the covariance matrix are calculated. In this study, the data represented by PCA is used as a feature vector for the classification of the banknote's direction ("Direction A", "Direction B", "Direction C" and "Direction D" in Figure 1).

To classify the direction of the banknote, we used an SVM. In general, the decision function of an SVM is defined as [22,23]:

$$f(x) = \text{sgn}\left(\sum_{1}^{l} y_i \alpha_i K(x, x_i) + b \right) \tag{2}$$

where l is the amount of data and $y_i \in [1, -1]$ is the indicator vector. In our study, the value "1" is assigned to the correct class and the value "−1" to the incorrect class. α_i and b are the weight value to $K(x, x_i)$ and off-set used in the decision function of SVM, respectively [22,23]. $K(x, x_i)$ is a kernel function. In our experiments, LibSVM software [24] was used to determine the optimal parameters. This software provides various kernel functions. Using training data to experiment, the linear kernel was selected as the optimal kernel for classifying the banknote's direction. The classification step using the SVM classifier consists of two sub-steps, as shown in Figure 1. The first sub-step involves classifying the group of obverse sides (Direction A or B) and the group of reverse sides (Direction C or D). Finally, the direction of the banknote is determined in the second sub-step. For example, if the result of the first sub-step is that the banknote is determined to belong in the obverse-sided group, the direction of the banknote is determined as either Direction A or B in the second sub-step, as shown in Figure 1.

3. Experimental Results

We collected a database consisting of 61,240 USD banknote images for our experiments. The images in the database consisted of the four directions for the 16 types of banknotes ($1, $2, $5, recent $5, most recent $5, $10, recent $10, most recent $10, $20, recent $20, most recent $20, $50, recent $50, most recent $50, $100, recent $100). As shown in Table 2, the number of images used in our study is similar to that of [10], but much larger than in most previous studies [11–17]. In addition, our study uses more classes than any previous study [10–17]. Because there is no open database of USD banknote images, it is difficult to compare the performance of our method with those of previous researches on same condition. Therefore, we show that the number and classes of banknote images in our experiment are comparatively larger than those of previous researches in Table 2.

Table 2. Comparison of the number of images and classes used in previous studies with that of our study.

Method	The Number of Images	The Number of Classes (Including Two Sides and Two Directions)
[10]	65,700	48
[11]	140	28
[12]	15,000	24
[13,16]	2400	24
[14,15]	3600	24
[17]	3570	24
Our study	61,240	64

Figure 4 shows example images of a $1 banknote. In this study, the database of 61,240 images was randomly divided into two subsets, Group 1 and Group 2, for training and testing, respectively, as shown in Table 3.

(a)

(b)

(c)

(d)

Figure 4. Examples of images: (**a**) Direction A; (**b**) Direction B; (**c**) Direction C; and (**d**) Direction D.

Table 3. Number of images in each group and direction.

Category	Group 1	Group 2	Total Number
Direction A	7692	7689	15,381
Direction B	7618	7621	15,239
Direction C	7692	7689	15,381
Direction D	7618	7621	15,239
Total	30,620	30,620	61,240

On the basis of a two-fold cross-validation scheme, the experiments were performed with Groups 1 and 2 from Table 3. First, we used Group 1 for the training process, and Group 2 for testing. Then, the training and testing processes were performed repeatedly, alternating between Group 1 and Group 2.

In this study, we used PCA to extract features, as explained in Section 2.3. The classification performance is affected by the number of PCA dimensions. Therefore, the optimal PCA dimensionality that produces the best classification accuracy is obtained by experimentation with the training database. This involves measuring the classification accuracy with various numbers of PCA dimensions. Figure 5 shows the experimental classification results for Group 1 and 2 using a Bayesian classifier with various PCA dimensions. The classification error is calculated as the summed value of Type-1 and -2 errors. A Type-1 error means that the obverse side (or forward direction) was incorrectly determined as the reverse side (or backward direction). A Type-2 error means that the reverse side (or backward direction) was incorrectly determined as the obverse side (or forward direction).

In our research, we defined four directions in our study: the obverse side in a forward direction is "Direction A," the obverse side in a backward direction is "Direction B," the reverse side in a forward direction is "Direction C," and the reverse side in a backward direction is "Direction D." Therefore, "Direction AB" means the obverse side in a forward or backward direction in Figure 5. "Direction CD" means the reverse side in a forward or backward direction.

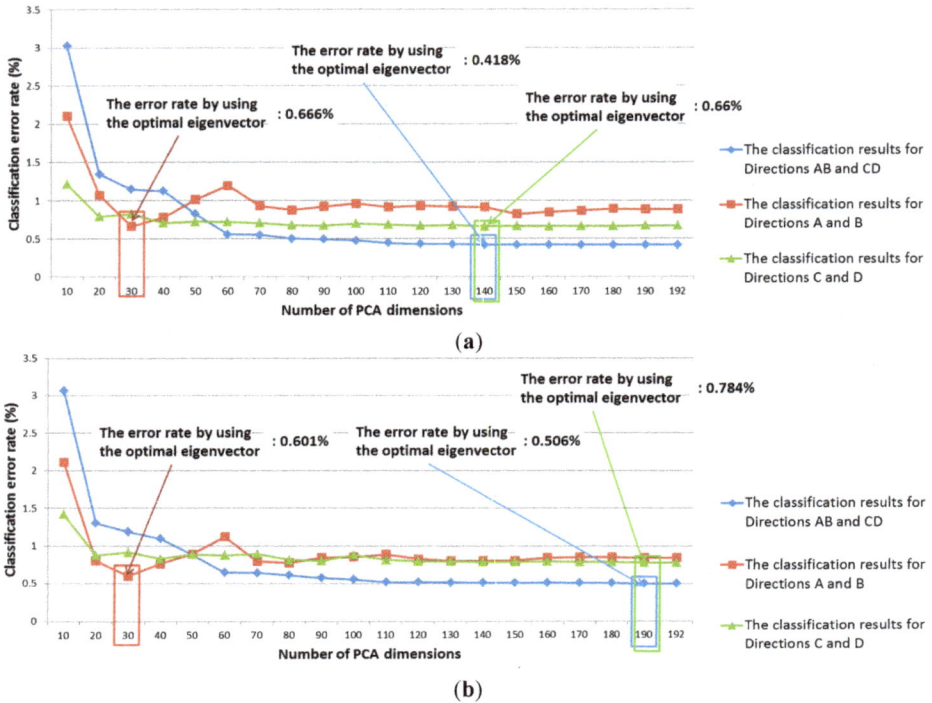

Figure 5. Classification accuracy of PCA according to number of PCA dimensions: (a) Group 1; (b) Group 2.

For comparison, we also measured the classification accuracy using linear-discriminant analysis (LDA) instead of PCA. The classification results from using the training data are shown in Figure 6.

Table 4 shows the respective classification results from PCA and LDA using the testing data. As shown in Table 4, although LDA outperforms PCA for classifying Direction A and Direction C, the average classification accuracy for the four directions with PCA is higher than it is with LDA.

Table 4. Average error rate from two testing sub-databases using LDA and PCA (unit: %).

Method	Direction A	Direction B	Direction C	Direction D	Average Error
LDA	0.241	3.012	1.138	3.708	2.025
PCA	0.689	0.564	1.931	1.352	1.134

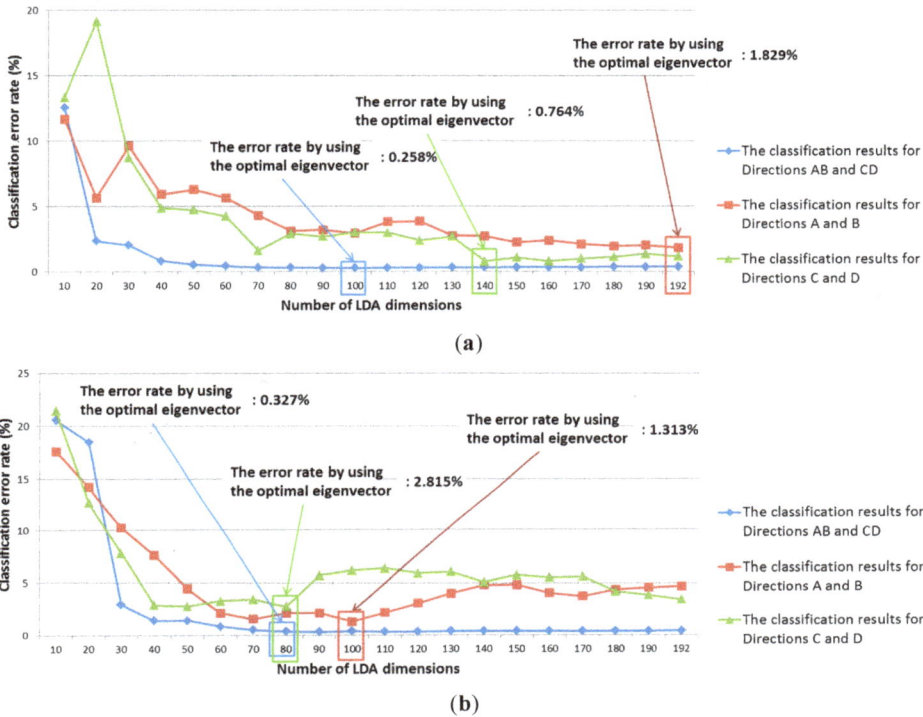

Figure 6. Classification accuracy of LDA according to number of LDA dimensions: (**a**) Group 1; (**b**) Group 2.

To improve the pre-classification performance, we used an SVM classifier instead of a Bayesian classifier for testing by implementing the PCA features with the optimal PCA dimensionality obtained with the training data. In this paper, four SVM kernels were considered: Linear, Polynomial, RBF, and Sigmoid. Using the SVM classifier, the classification accuracy for all four kernels was 100% when using the training database. Therefore, we used the linear kernel in our proposed method because it required processing time is shorter than that of the other kernels. Table 5 and Figure 7 show the results from using the testing database. In Figure 7, equal error rate (EER) means the error rate when the difference between Type-1 and Type-2 errors is minimized.

When using PCA with the SVM classifier, the pre-classification error is 0%. This indicates that no misclassification occurred when our proposed method was applied using the testing database. In addition, the classification accuracy of the SVM is higher than that of the Bayesian classifier using the PCA features. As shown in Table 5 and Figure 7, we compared the accuracy from using Bayesian classifiers with both LDA and PCA. The latter (PCA with Bayesian classifiers) is demonstrably superior to the former (LDA with Bayesian classifiers). As a result, we can confirm that the usability of PCA is higher than LDA. In addition, PCA with an SVM is superior to PCA with Bayesian classifiers, as shown in Table 5 and Figure 7. Hence, we can confirm that the usability of an SVM is higher than that of the Bayesian method. Table 5 and Figure 7 show that PCA with an SVM is superior to merely using an SVM, and we can confirm that the usability of PCA with an SVM is higher than that of an SVM without PCA.

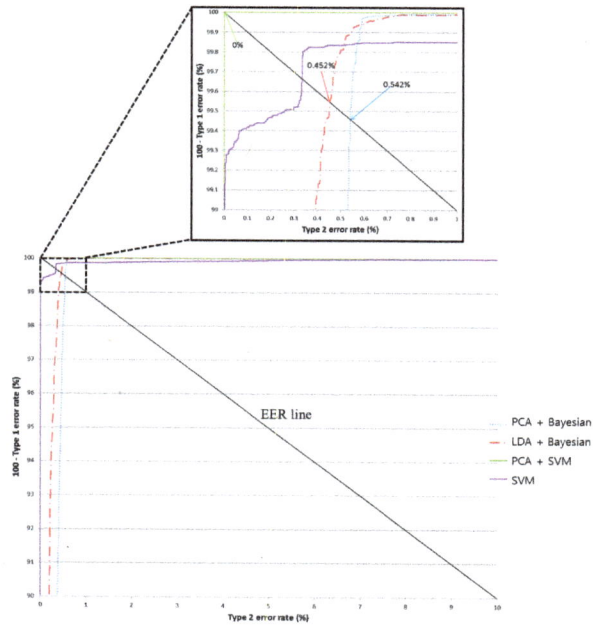

(a)

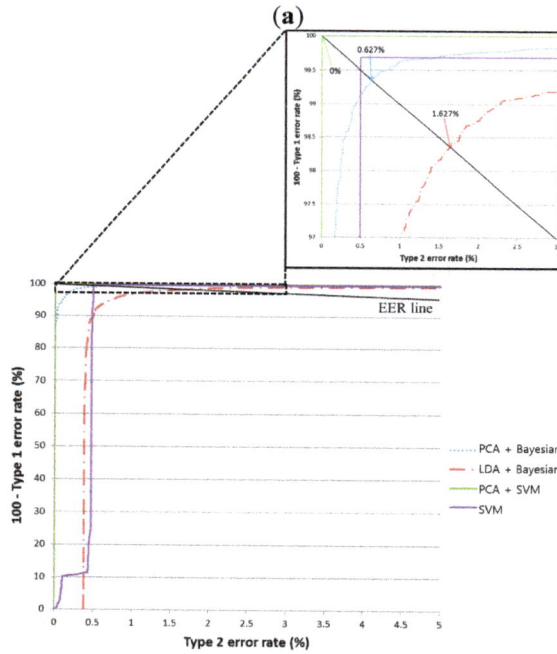

(b)

Figure 7. *Cont.*

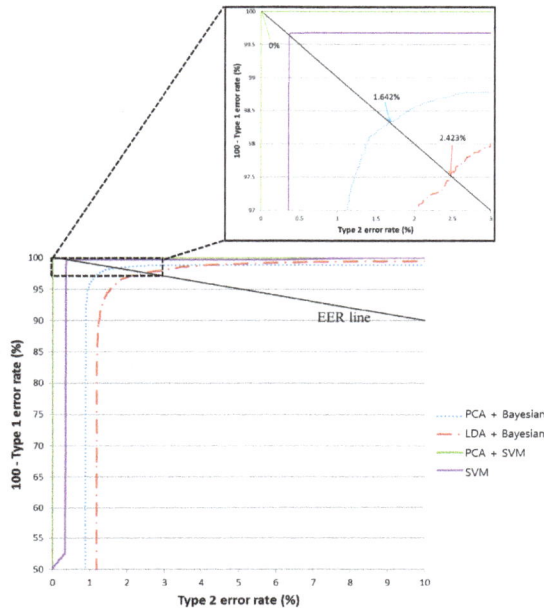

Figure 7. ROC curve from the pre-classification stage using two testing sub-databases (In each figure, PCA+SVM is the proposed method): (**a**) the average classification results for the obverse side (Direction A or B) and the reverse side (Direction C or D); (**b**) the average classification results for Direction A and Direction B; and (**c**) the average classification results for Direction C and Direction D.

Table 5. Average error rates during pre-classification using two testing sub-databases (unit: %).

Method	Direction A	Direction B	Direction C	Direction D	Average Error
LDA + Bayesian	0.241	3.012	1.138	3.708	2.025
PCA + Bayesian	0.689	0.564	1.931	1.352	1.134
SVM	0.419	0.579	0.325	0.441	0.441
PCA + SVM (proposed method)	0	0	0	0	0

In addition, we performed banknote recognition using a K-means algorithm after the pre-classification of the banknote's direction. The features for K-means were extracted using PCA. For banknote recognition, the optimal dimensionality of the PCA was experimentally obtained in the same way as it was during the pre-classification step. Thus, the optimal dimensionality was determined to be that which produces the smallest recognition error among various possible numbers of PCA dimensions. Here, the recognition error was calculated at a 100% correct-classification rate. Correct classification means that the input banknote was correctly recognized in its corresponding class—That is, a $1 banknote was correctly recognized as belonging to the $1 class. The recognition results for the training data according to PCA dimensionality are shown in Figure 8. Using the optimal PCA dimensionality for each direction class, the image features were extracted and classified using the K-means method. As shown in Figures 5 and 8, we apply PCA at the three stages such as the pre-classification of obverse (AB) and reverse (CD) sides, that of forward and backward directions (A and B, or C and D), and the banknote recognition by K-means algorithm. The optimal numbers of dimension of the feature vectors after applying PCA are different according to the PCA application for the three stages as shown in Figures 5 and 8.

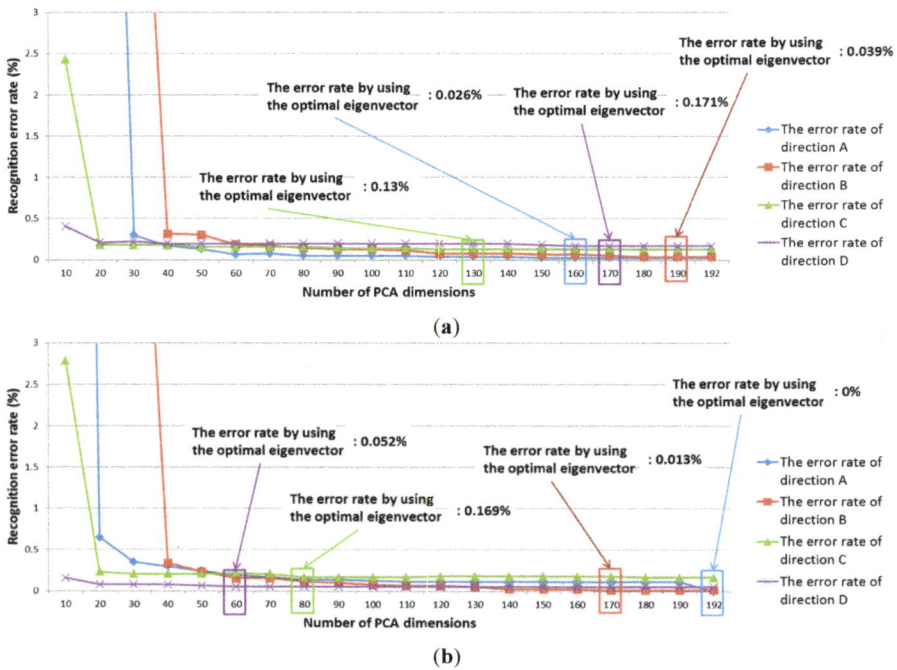

Figure 8. Recognition results for training data using the K-means method: (**a**) Group 1; (**b**) Group 2.

Table 6 presents the average banknote-recognition results obtained using K-means with the two testing sub-databases. As shown in Table 6, the error rate for Directions A and B is lower than that of Directions C and D. This is because the reverse side of a USD banknote is less easily discriminated than the obverse, as shown in Figures 9 and 10. On average, the recognition error for the four directions is approximately 0.114%.

Because there is no open source for banknote recognition, it is difficult to compare our method with previous ones. However, because one previous method [12] measured its accuracy with the USD database like our research, we compared the accuracy of [12] to that of our method, as shown in Table 7. In [12], feature extraction is done with QWT and GGD. In our paper, feature extraction is done with PCA. In [12], the classifier used is a BP neural network without pre-classification, while in our paper SVM is used instead.

Table 6. Results from performing banknote recognition after pre-classifying the banknote's direction (unit: %).

Method	Direction A	Direction B	Direction C	Direction D	Average Error
K-means	0.091	0.052	0.176	0.138	0.114

In Table 7, the false-recognition rate refers to the frequency of one banknote image being recognized as an incorrect banknote—e.g., a $10 bill in Direction A recognized as a $10 bill in Direction B, or a $20 bill in Direction A recognized as a $50 bill in Direction A. The reject rate refers to the frequency with which either the type or the class of a banknote is indeterminate. As shown in Table 7, we can confirm that the accuracy of our method is higher than that of previous methods [12] for a greater number of classes and images. Our dataset consists of 61,240 USD banknote images. There was no mistreated banknote included in the dataset. As shown in Table 7, the false recognition rate was 0.114% with the rejection rate of 0% with all the images of 61,240 by our method. Therefore, the correct recognition rate was 99.886%.

Table 7. Comparison of the accuracy of the proposed method with previous methods.

Category	Previous Methods [12]	Proposed Method
Number of classes	24	64
Number of USD images	15,000	61,240
False recognition rate (%)	0.12	0.114
Reject rate (%)	0.58	0

Figure 9 illustrates cases where a banknote was correctly recognized with our method. In each pair of images, the lower image is the image sub-sampled at 32 × 6 pixels.

(a)

(b)

(c)

Figure 9. *Cont.*

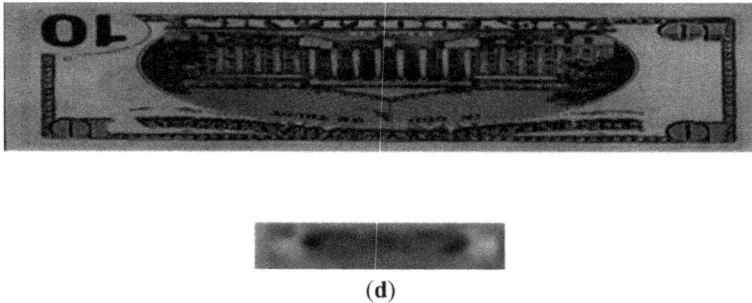

(**d**)

Figure 9. Accurately recognized banknotes and their sub-sampled images: (**a**) Direction A with $100;
(**b**) Direction B with recent $50; (**c**) Direction C with recent $100; and (**d**) Direction D with recent $10.

As explained in Section 2.2, the four corner positions of the ROI area on the banknote are detected by corner detection algorithm. Then, the ROI area defined by these four corner positions is segmented from the original input image, and this area is rotated based on the left-upper corner position. From that, the rotation compensated area is obtained as shown in the middle images of Figure 10a–c, respectively. Then, the sub-sampled image of 32 × 6 pixels is acquired based on this area as shown in the bottom images of Figure 10a-c, respectively. Therefore, the image rotation (the top images of Figure 10a-c) does not affect the correct acquisition of the ROI area and sub-sampled image, and the consequent accuracy of banknote recognition is not affected by the image rotation.

Figure 11 illustrates cases where a banknote was incorrectly recognized with our method. In the image pairs, the lower image is the image sub-sampled at 32 × 6 pixels. Each image in the left column was falsely recognized as belonging to the class in the right column. For example, the image in the first row ($100, Direction A) was falsely recognized as belonging the class in the right column ($5, Direction A). As shown in Figure 11, incorrect recognitions occurred when some part of the banknote was damaged (e.g., in the $50-Direction-B image from the second row in the left column) or when a contaminant was present in the upper or lower part of the banknote (e.g., the images in the other rows of the left columns). In particular, as seen in the images from the first, third, and fourth rows of Figure 11, the banknote was falsely recognized because the upper and lower parts of banknote were stained with dark ink. Thus, these cases illustrate a contaminated banknote. In addition, banknotes are falsely recognized when thick vertical lines of discontinuity appear in the mid part of the banknote as a result of being folded, as shown in the images from the second row in Figure 11. This case illustrates a damaged banknote.

In the next experiment, we measured the processing time for our method using a desktop computer (a 3.5 GHz CPU with 8 GB RAM). Most previous research on banknote recognition does not measure the processing time, focusing exclusively on the recognition accuracy instead [9,10,12–17]. Therefore, it is difficult to compare the processing time of our method with others. In addition, one extant method measured the processing time during testing but not training [11]. Therefore, we measured only the processing time for our method during testing. As shown in Table 8, our method recognizes a banknote image at approximately 172 images/sec (1000/5.83). The method that measured the processing time [11] required as much as two seconds with a 3 GHz CPU desktop computer and is therefore much slower than our method.

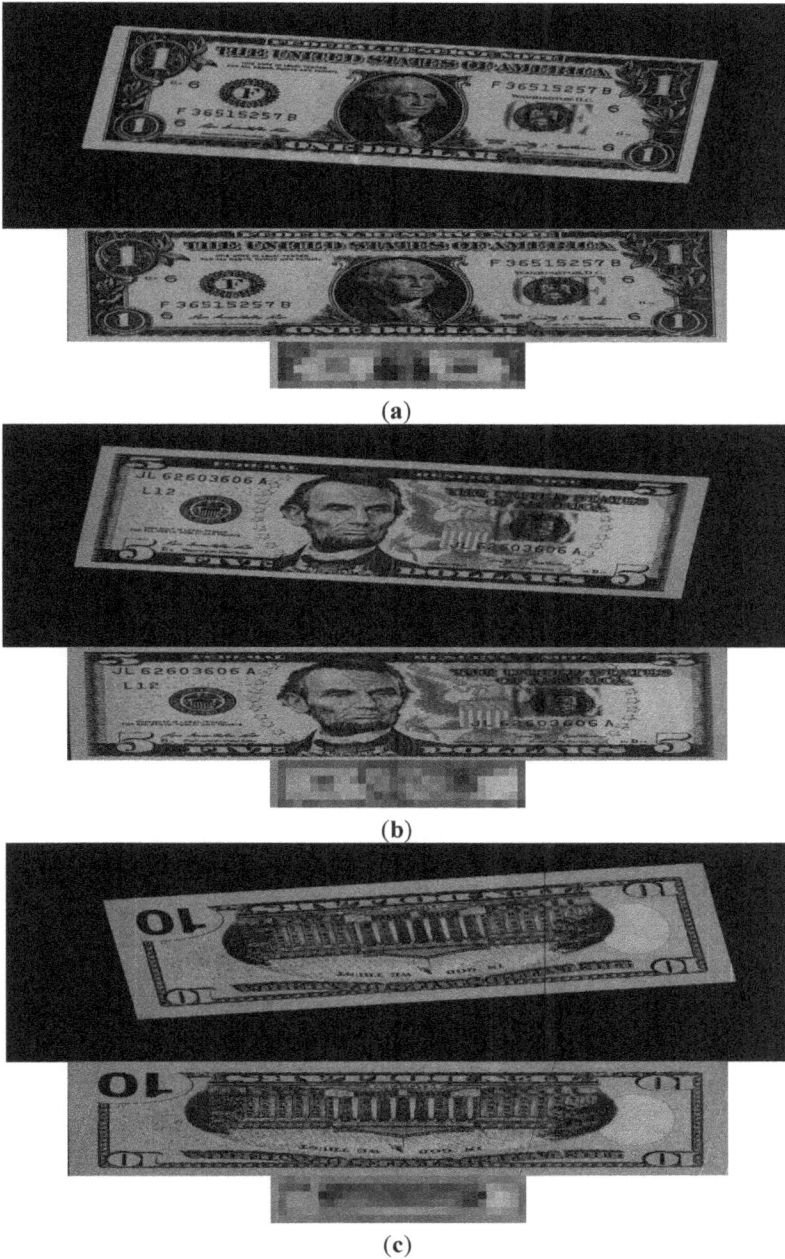

Figure 10. Examples of rotated banknote, the corresponding ROI area, and sub-sampled images. In (**a**)–(**c**), top, middle, and bottom figures are the original input, ROI area, and sub-sampled images, respectively: (**a**) the 1st example; (**b**) the 2nd example; and (**c**) the 3rd example.

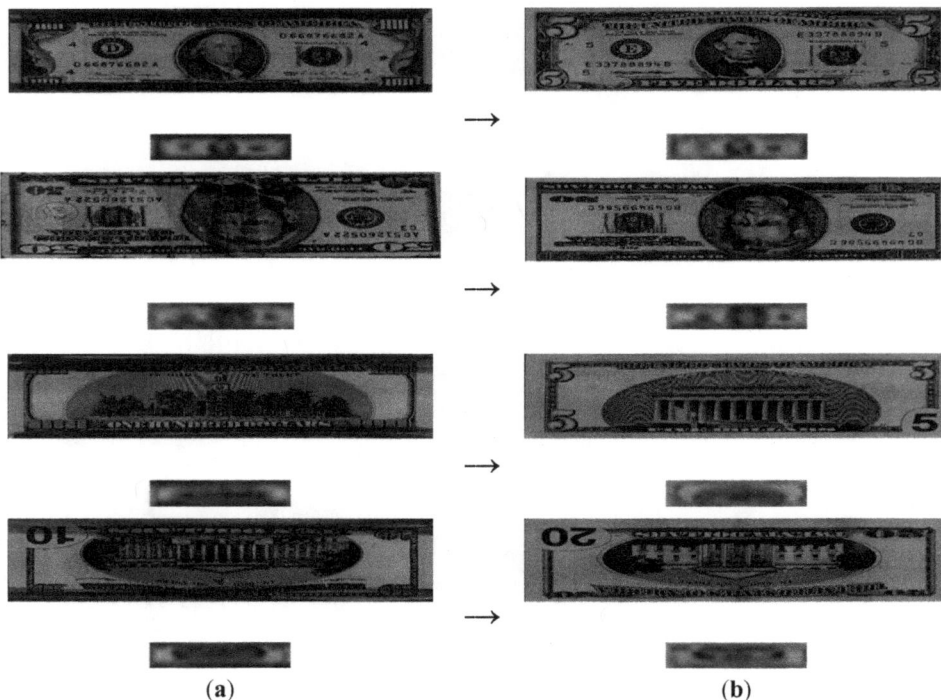

Figure 11. Cases of incorrect banknote recognition: (**a**) input images; (**b**) falsely recognized classes.

Table 8. Processing time of the proposed method on PC (unit: ms).

Image Sub-Sampling	Obverse/Reverse Side Classification		Forward/Backward Direction Classification		K-Means Matching	Total
	Feature Vector Extraction	Side Classification	Feature Vector Extraction	Direction Classification		
2.06	1.27	0.06	1.95	0.04	0.45	5.83

In addition, we measured the processing time on TI DSP. As shown in Table 9, the total processing time is about 15.6 ms and our method recognizes a banknote image at approximately 64 images/sec (1000/15.6) on banknote counting machine.

Table 9. Processing time of the proposed method on banknote counting machine (unit: ms).

Image Sub-Sampling	Obverse/Reverse Side Classification		Forward/Backward Direction Classification		K-Means Matching	Total
	Feature Vector Extraction	Side Classification	Feature Vector Extraction	Direction Classification		
4.21	3.24	1.89	3.31	1.91	1.04	15.6

In addition, we measured the total memory usage on banknote counting machine. The total memory usage is about 1.6 MB. It is a total of 734,976 Bytes (1584 × 464) for original input image, 298,152 Bytes (1212 × 246) for the ROI image, 192 Bytes (32 × 6) for the sub-sampled image, 442,368 Bytes

(192 × 192 × 3 × 4 (float number type)) for the PCA transform matrix, 2304 Bytes (192 × 3 × 4 (float number type)) for the PCA eigenvalues, 124,716 Bytes for the SVM models, and 49,152 Bytes (192 × 16 (class centers) × 4 (directions of A, B, C, D) × 4 (float number type)) for the K-means algorithm.

4. Discussions of Experimental Results

For our research, we obtained the global features from the sub-sampled 32 × 6-pixel image. When comparing the sub-sampled images (of different directions and banknote types) in Figures 3 and 9, we found that each sub-sampled image can be discriminated. As a result, it is clear that the global features (useful for discriminating Directions A–D and classifying the 16 classes, as shown in Figure 1) still remain in the sub-sampled image despite losing the detailed (local) texture patterns. This is confirmed by the accuracy in pre-classification and the recognition rates in Tables 5 and 7. In addition, using the sub-sampled image enables real-time processing, as shown in Table 8 (172 images/sec (1000/5.83)), with lower memory usage. In general, the misalignment of a banknote often occurs because banknotes are cut inconsistently during production, and this issue is aggravated by variations in the capturing position of the banknote image in banknote-counting machines and ATMs. Misalignment degrades the recognition accuracy, but by using the sub-sampled image, the recognition accuracy is affected less by a misaligned banknote image.

As shown in Figures 3b and 9, the texture components are so blurred in the sub-sampled image that its features are difficult to be extracted by gradient-direction method or Gabor filtering. In addition, because of the limitations to the processing time in a banknote-counting machine or an ATM with low processing power, Gabor filtering—which requires high processing time—is difficult to implement with our method. Therefore, we used the method of extracting features with PCA, and its performance is validated by the results in Table 5, Table 7, and Table 8.

As shown in Table 5 and Figure 7, we compared the accuracy of both LDA and PCA with Bayesian classifiers, and PCA with an SVM for pre-classification. The accuracy from using LDA and PCA with Bayesian classifiers did not reach 100% (the error rates for the former and latter were 2.025% and 1.134%, respectively). In addition, we measured the accuracy of an SVM without PCA, and it did not reach 100%, either, as shown in Table 5 and Figure 7. Based on these results, we are assured that this problem is not simple, and that the usability of PCA with an SVM is higher than in other methods. Because the experiments were done on such a large classes and numbers of USD banknotes compared with previous research (as shown in Table 2), we expect that these results are sufficiently validated.

In our research, we aim at developing an algorithm that can be applied to actual banknote-counting machines and ATMs with low processing power. Therefore, more sophisticated algorithms with higher processing times are unfeasible for these machines due to the limited processing time.

In our research, the final classification of the banknote into 16 classes is also based on PCA features. By using a K-means clustering algorithm, 16 class centers (vectors) are obtained in each Direction A–D, as illustrated in Figure 1. The class mean (center) for each class can be estimated by calculating the average positions (geometric centers) of the data (in each class) in each axis of dimension. However, the accuracy of this method can be affected by the outliers (the error data whose position are far from the center of the class). That is, if the number of the outliers is large, the center of the class is calculated as that close to the outliers, and the incorrect center position can be obtained consequently. However, K-means algorithm is the unsupervised (iterative) learning method which can group the K classes data automatically based on the minimum distance between each sample data and class center. Therefore, its consequent accuracy of determining the correct class center is less affected by the outlier data than that by calculating the geometric center.

With the extracted PCA features from the inputted banknote image, the Euclidean distance between these PCA features and each center (vector) of the 16 classes (which are determined by K-means method) is calculated after pre-classification, and one class center (with the smallest Euclidean distance) is selected as the final class for the banknote (nearest class mean (NCM) method).

In our method, an SVM is used to pre-classify the four classes (Directions A–D) in a hierarchical manner, and a K-means method is used to recognize the banknote from the 16 classes, as illustrated in Figure 1. Because a conventional SVM has been used to discriminate between two classes, and because accurate banknote recognition ultimately requires 16 classes, the SVM was not used for banknote recognition. In order to adopt the SVM for banknote recognition in 16 classes, either a multi-class SVM would be necessary or the conventional SVM would need to be applied repeatedly in a hierarchical way. Both options require too much processing time for use in a banknote-counting machine or an ATM with low processing power. Therefore, we used the K-means method for the banknote recognition in 16 classes. In other words, we performed banknote recognition based on the minimum Euclidean distance with the class center by a K-means clustering method. In addition, because an SVM is usually superior to the K-means method or others for discriminating between two classes (as shown in Table 5), we used the SVM hierarchically to pre-classify the four classes (Direction A–D).

If the ROI area on banknote of 1212×246 pixels is used for PCA training, it is difficult to obtain the covariance matrix for PCA training because the size of covariance matrix is so high as $298,152(1212 \times 246) \times 298,152(1212 \times 246)$ (which requires the memory usage larger than 88 GBs). PCA usually has the functionalities of both dimension reduction and acquisition of optimal features. Therefore, as shown in Figure 7 and Table 5, the average error by PCA + SVM is smaller than that by SVM without PCA. In addition, as shown in Figure 5a, the dimensionality reduction is large (from 192 to 140) in case of the second trial of training when exchanging the training and testing data based on two-fold cross validation. Therefore, although the dimensionality reduction in some cases (Figure 5b) is minimal, the PCA transform with the sub-sampled images is necessary in our research.

Our method can recognize the kind of banknote through the pre-classification of obverse (AB) and reverse (CD) sides, and forward and backward directions (A and B, or C and D) only with the visible light image. Our method cannot identify counterfeits because the detection of counterfeit requires additional information such as magnetic, infrared sensors, *et al.*, in addition to the visible light sensor.

5. Conclusions

We propose in this paper a novel method for pre-classifying banknotes' direction for implementation in banknote-recognition systems. The results of our experiments showed that the error rate for the proposed pre-classification method was lower than that of other methods. In addition, the banknote-recognition error rate after pre-classifying the banknote's direction was as low as 0.114%. However, incorrect recognition occurred when part of the banknote was damaged or when contaminants were present in the upper or lower region of the banknote.

Although banknote images with a limited degree of mistreatment are included in our database, those including tears or hand written notes are not included in our database. As the future work, we would propose to test these images and research a method of enhancing the recognition accuracy with these kinds of poor quality images. In addition, we intend to study an algorithm for rejecting poor-quality banknote images based on a quality measure or the confidence level of the matching score and we plan to apply our method for pre-classifying the banknote's direction to other currencies, and to compare the accuracy of the method according to the banknote type.

Acknowledgments: This research was supported by a grant from the Advanced Technology Center R&D Program funded by the Ministry of Trade, Industry& Energy of Korea (10039011).

Author Contributions: Young Ho Park and Kang Ryoung Park designed the overall system and wrote the pre-classification algorithm. In addition, they wrote and revised the paper. Seung Yong Kwon and Tuyen Danh Pham implemented the banknote recognition algorithm. Dae Sik Jeong and Sungsoo Yoon helped with the dataset collection and experiments.

Conflicts of Interest: The authors declare no conflict of interest.

References

1. Aoba, M.; Kikuchi, T.; Takefuji, Y. Euro banknote recognition system using a three-layered perceptron and RBF networks. *IPSJ Trans. Math. Model. Appl.* **2003**, *44*, 99–109.
2. Takeda, F.; Omatu, S. A neuro-money recognition using optimized masks by GA. *Lect. Notes Comput. Sci.* **1995**, *1011*, 190–201.
3. Kosaka, T.; Omatu, S. Classification of the Italian liras using the LVQ method. In Proceedings of IEEE International Conference on Systems, Man, and Cybernetics, Tokyo, Japan, 12–15 October 1999; pp. 845–850.
4. Liu, J.F.; Liu, S.B.; Tang, X.L. An algorithm of real-time paper currency recognition. *J. Comput. Res. Dev.* **2003**, *40*, 1057–1061.
5. Hassanpour, H.; Farahabadi, P.M. Using hidden markov models for paper currency recognition. *Experts Syst. Appl.* **2009**, *36*, 10105–10111. [CrossRef]
6. García-Lamont, F.; Cervantes, J.; López, A. Recognition of Mexican banknotes via their color and texture features. *Experts Syst. Appl.* **2012**, *39*, 9651–9660. [CrossRef]
7. Choi, E.; Lee, J.; Yoon, J. Feature extraction for banknote classification using wavelet transform. In Proceedings of International Conference on Pattern Recognition, Hong Kong, China, 20–24 August 2006; pp. 934–937.
8. Ahangaryan, F.P.; Mohammadpour, T.; Kianisarkaleh, A. Persian banknote recognition using wavelet and neural network. In Proceedings of the International Conference on Computer Science and Electronics Engineering, Hangzhou, China, 23–25 March 2012; pp. 679–684.
9. Wu, Q.; Zhang, Y.; Ma, Z.; Wang, Z.; Jin, B. A banknote orientation recognition method with BP network. In Proceedings of WRI Global Congress on Intelligent Systems, Xiamen, China, 19–21 May 2009; pp. 3–7.
10. Kagehiro, T.; Nagayoshi, H.; Sako, H. A hierarchical classification method for US bank notes. In Proceedings of IAPR Conference on Machine Vision Applications, Tsukuba Science City, Japan, 16–18 May 2005; pp. 206–209.
11. Hasanuzzaman, F.M.; Yang, X.; Tian, Y. Robust and effective component-based banknote recognition by SURF features. In Proceedings of the 20th Annual Wireless and Optical Communications Conference, Newark, NJ, USA, 15–16 April 2011; pp. 1–6.
12. Gai, S.; Yang, G.; Wan, M. Employing quaternion wavelet transform for banknote classification. *Neurocomputing* **2013**, *118*, 171–178. [CrossRef]
13. Ahmadi, A.; Omatu, S.; Kosaka, T. A PCA based method for improving the reliability of bank note classifier machines. In Proceedings of the 3rd International Symposium on Image and Signal Processing and Analysis, Rome, Italy, 18–20 September 2003; pp. 494–499.
14. Ahmadi, A.; Omatu, S.; Kosaka, T. A Study on evaluating and improving the reliability of bank note neuro-classifiers. In Proceedings of the SICE Annual Conference, Fukui, Japan, 4–6 August 2003; pp. 2550–2554.
15. Ahmadi, A.; Omatu, S.; Fujinaka, T.; Kosaka, T. Improvement of reliability in banknote classification using reject option and local PCA. *Inf. Sci.* **2004**, *168*, 277–293. [CrossRef]
16. Omatu, S.; Yoshioka, M.; Kosaka, Y. Bank note classification using neural networks. In Proceedings of IEEE Conference on Emerging Technologies and Factory Automation, Patras, Greece, 25–28 September 2007; pp. 413–417.
17. Omatu, S.; Yoshioka, M.; Kosaka, Y. Reliable banknote classification using neural networks. In Proceedings of the 3rd International Conference on Advanced Engineering Computing and Applications in Sciences, Sliema, Malta, 11–16 October 2009; pp. 35–40.
18. Yeh, C.Y.; Su, W.P.; Lee, S.J. Employing Multiple-kernel support vector machines for counterfeit banknote recognition. *Appl. Soft Comput.* **2011**, *11*, 1439–1447. [CrossRef]
19. Bruna, A.; Farinella, G.M.; Guarnera, G.C.; Battiato, S. Forgery detection and value identification of Euro banknotes. *Sensors* **2013**, *13*, 2515–2529. [CrossRef] [PubMed]
20. Battiato, S.; Farinella, G.M.; Bruna, A.; Guarnera, G.C. Counterfeit detection and value recognition of Euro banknotes. In Proceedings of International Conference on Computer Vision Theory and Applications, Barcelona, Spain, 21–24 February 2013; pp. 63–66.
21. Hasanuzzaman, F.M.; Yang, X.; Tian, Y. Robust and effective component-based banknote recognition for the blind. *IEEE Trans. Syst. Man Cybern. Part. C Appl. Rev.* **2012**, *42*, 1021–1030. [CrossRef] [PubMed]

22. Murphy, K.P. *Machine Learning: A Probabilistic Perspective*; The MIT Press: Cambrige, MA, USA, 2012.
23. Vapnik, V.N. *Statistical Learning Theory*; John Wiley & Sons, Inc.: New York, NY, USA, 1998.
24. LIBSVM—A Library for Support Vector Machines. Available online: http://www.csie.ntu.edu.tw/~cjlin/libsvm (accessed on 3 December 2014).

sensors

MDPI

Article

Uncertainty Comparison of Visual Sensing in Adverse Weather Conditions [†]

Shi-Wei Lo [1],*, Jyh-Horng Wu [1], Lun-Chi Chen [1], Chien-Hao Tseng [1], Fang-Pang Lin [1] and Ching-Han Hsu [2],*

[1] National Center for High-Performance Computing, No. 7, R&D 6th Rd., Hsinchu Science Park, Hsinchu 30076, Taiwan; jhwu@nchc.narl.org.tw (J.-H.W.); casper@nchc.narl.org.tw (L.-C.C.); 0903049@nchc.narl.org.tw (C.-H.T.); fplin@nchc.narl.org.tw (F.-P.L.)

[2] Department of Biomedical Engineering and Environmental Sciences, National Tsing Hua University, No. 101, Section 2, Kuang-Fu Road, Hsinchu 30013, Taiwan

* Correspondence: LSW@nchc.narl.org.tw (S.-W.L.); cghsu@mx.nthu.edu.tw (C.-H.H.); Tel.: +886-3-57-76-085 (ext. 354) (S.-W.L.); Tel.: +886-3-57-15-131 (ext. 80-815) (C.-H.H.)

[†] This paper is an extended version of our paper published in Lo, S.-W.; Wu, J.-H.; Chen, L.-C.; Tseng, C.-H.; Lin, F.-P. Flood Tracking in Severe Weather. In Proceedings of the International Symposium on Computer, Consumer and Control, Taichung, Taiwan, 10–12 June 2014; pp. 27–30.

Academic Editor: Gonzalo Pajares Martinsanz
Received: 28 April 2016; Accepted: 15 July 2016; Published: 20 July 2016

Abstract: This paper focuses on flood-region detection using monitoring images. However, adverse weather affects the outcome of image segmentation methods. In this paper, we present an experimental comparison of an outdoor visual sensing system using region-growing methods with two different growing rules—namely, GrowCut and RegGro. For each growing rule, several tests on adverse weather and lens-stained scenes were performed, taking into account and analyzing different weather conditions with the outdoor visual sensing system. The influence of several weather conditions was analyzed, highlighting their effect on the outdoor visual sensing system with different growing rules. Furthermore, experimental errors and uncertainties obtained with the growing rules were compared. The segmentation accuracy of flood regions yielded by the GrowCut, RegGro, and hybrid methods was 75%, 85%, and 87.7%, respectively.

Keywords: vision application; outdoor imaging; visual sensing; flood detection

1. Introduction

In the summer and during typhoon season, the western coast of Taiwan is particularly vulnerable to flooding, especially during the period between May and October. Every year, abundant rainfall causes numerous deaths and serious damage to the economy [1–7]. One of the most challenging problems with regard to flood response is the precise localization of flood risk. This task is performed by early warning systems (EWSs) for flood prevention and disaster management. EWSs are extensively applied to mitigate flood risk, and they work by detecting abnormalities and predicting the onset of flooding with remote sensors. They can also provide real-time information during floods [8–11]. Traditionally, EWSs monitor flooding with remote sensing technology such as satellite imaging and electronic sensors installed nearby rivers and seaports. Satellite images cover hundreds of kilometers, generally providing only the broadest outlines of potential risk. On the other hand, using electronic sensors to measure water levels remains unfeasible, owing to the sheer number of sensors needed. These devices have a limited geographic range and extensive power requirements. Moreover, they incur massive costs in installation and maintenance. Therefore, the development of a long-term sustainable EWS with the ability to precisely localize areas of risk is crucial to the field of flood monitoring and early warning.

Visual sensing techniques are widely employed in various fields for vision applications such as inspection, surveillance, and monitoring. Unlike active sensors, vision sensing techniques indirectly measure physical information from captured images and video. Such systems record particular behavior, activity, and other changes in the field scene. The use of a visual sensing system to perform an indirect estimate of the region, position, velocity, and attitude of a monitored object is well known. However, the influence of weather phenomena on visual sensing systems remains an open research issue with many unanswered questions [12,13]. Conventional imaging systems are designed to capture scenes in ideal atmospheric conditions, such as indoors. However, outdoor vision applications must be capable of capturing images even in adverse weather conditions [14,15]. Such conditions limit the accuracy of the estimated attitude of a monitored object [16–19].

Fog and stained lenses are the most pernicious phenomena for outdoor vision systems. The image intensity, color, and shape are altered by interactions between light and the atmosphere. First, fog results from suspended particles, mist, raindrops, rain streaks, and heavy spray rain. Another major source with fog are raindrop streaks. Consider a camera system capturing the volume of raindrops; this volume comprises randomly-distributed and high-velocity raindrops. Raindrop streaks are projected in non-uniform stripes onto the scene. They produce sharp changes to the intensity during image acquisition. Subsequent imaging processes are also affected by concentrated rain streaks. Relevant research regarding raindrop detection and removal can be found in [20–24].

Second, raindrop stains tend to adhere to the lens of imaging devices. Each stain refracts and reflects light, generating shape and intensity changes in images. Figure 1a,b shows an example of a stain on a camera lens, and Figure 1c,d shows the detection result of a time-varying flood region. In imaging systems, the projections of a raindrop stain on an image are a non-uniform refraction mask on pixels. Due to the composite raindrop stains, the image intensity is randomly nebulized. However, the effects of raindrop stains on camera lenses have not been thoroughly investigated. This study also focused on rain stains, which are a common atmospheric condition in vision systems.

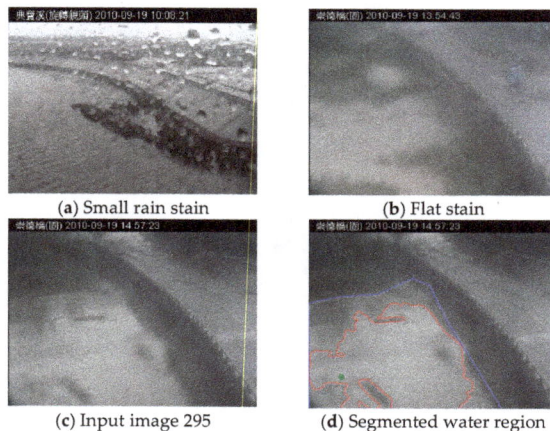

(a) Small rain stain   (b) Flat stain

(c) Input image 295   (d) Segmented water region

Figure 1. Two stain types: (**a**) a small stain and (**b**) a flat stain overlapping the camera lens. The effect of stains when applied to flood detection for (**c**) a stained outdoor image and (**d**) the concave region affected by stains on an image. Image 295's flood region segmented with RegGro is represented by a red contour. The blue contour represents the ground truth, and the green dot is the location of the seed used with RegGro. (Note: The Traditional Chinese in header of (**a**) and (**b**–**d**) are represented the location in the Dianbao River and the Changed Bridge, respectively).

To recover clear images in adverse weather, associative restoration techniques should be introduced. Fog removal techniques can be applied during preprocessing, before visual sensing applications. Fog removal (or image dehazing) techniques restore image clarity by eliminating the medium effects of fog. The basic principle behind recovering fog-free images involves estimating the transmission of light in the medium of fog scenes, and then eliminating the scattered light caused by the medium, in order to provide a clear image. This topic has been discussed in previous literature [25–29]. However, in long-term video sequencing, image dehazing remains challenging, because it is independent from the atmospheric changes in each frame. Variations in luminosity, the fog level, fog distribution, and light scattering randomly affect video sequencing. A single image clarification filter with constant parameters is insufficient for estimating entire video sequences.

To understand the influence of adverse weather conditions, an image segmentation application has been employed to manipulate videos of rainy conditions captured using an outdoor imaging system. The vision application scenario involved flood-tide detection. Flood regions must be segmented into precise shapes in order to determine hazard levels and provide automatic flood warnings to support EWSs [14].

The remainder of this paper is organized as follows: Section 2 reviews the interactive segmentation problem and the advantages of region-based segmentation. Section 3 describes the two region-based rules and the image set in detail. The experimental results are given in Section 4. Finally, a discussion and conclusions are provided respectively in Sections 5 and 6.

2. Image Segmentation in Environmental Application

This paper focuses on flood detection using small-scale monitoring images to identify the part of flow in a water region, surrounding buildings, and geographic background. However, interference introduced from elements—such as variance to the water region, raindrops on the camera screen, blurred scenes from water atomization, and fierce wind—negatively affects traditional image segmentation methods, such as background subtraction, thresholding, and watershed processing.

Image segmentation has been widely applied in industry and medicine. More recently, the process has been used in environmental object analysis [30–34]. For outdoor images, simple segmentation process parameters, such as threshold values, cannot be established for precise flood region segmentation [35–40]. This is because region colors, region shapes, scene illumination, fog distribution, rain, and other atmospheric conditions vary over time. Visual information is somewhat independent between frames in video sequences comprised in a single shot.

Interactive image segmentation schemes with a few simple user inputs provide a better solution for natural images than fully-automatic schemes [41]. First, users indicate the location of the object and background using strokes as markers or seeds. Then, images are initially over-segmented into several small contiguous and perceptually similar regions (or superpixels), using mean shift [42], Bayesian flooding [43], graph-based [44,45], or contour-based [46] methods, among others. Finally, the region-merging stage automatically merges the initial regions with constraints to the boundary, shape, region, and topology [43]. The object is obtained from the background following a merging task. However, most interactive schemes require pre-segmentation to divide the image into small regions. Furthermore, most such schemes require the user to draw the specific shape of the initial markers, in order to fit the location, boundary, and features of the object and background. For flood detection, however, the location, boundary, and features of objects are time-varying. By comparison, both region-based segmentation methods analyzed in this paper do not require pre-segmentation; rather, the user roughly places a few seeds on the flow surface without deliberate selection. Region-based segmentation involves selecting seed points in the region of interest and using an algorithm to grow the region from the seed point according to the seed-pixel intensity and previously set criteria. The seed intensity depends on the pixels in each frame, rather than constrained values. This facilitates the successful deployment of the growing process in various frames with differing intensities. Therefore, region-based segmentation is suitable for time-varying intensity and shape conditions.

Based on the above reasons, region-based image segmentation was selected as the most suitable method for identifying flood regions and estimating the degree of hazard. In addition, region-based image segmentation exhibits properties that increase coupling to the seed location, rather than limit the set intensity. Therefore, temporal shape transformations and size variations to the flood regions can be traced.

3. Material and Methods

Region-based segmentation involves the assumption that the pixels within a region possess similar properties, such as color, intensity (gray level), and texture. Based on this, criteria for a similarity test were designed to determine whether neighboring pixels in a region are similar. If a similarity criterion is satisfied, neighboring pixels can be inferred to belong to the same growing region as their neighbors. Similarity criteria are crucial factors that shape growing patterns and result in differing final regions. In this study, we used two region-based algorithms to trace flood regions: RegGro (a modified region-growing algorithm), and GrowCut. The criteria used in the growing process differ between these algorithms. Details for these algorithms are provided in the following sections.

3.1. RegGro Method

The purpose of RegGro is to group pixels into meaningful regions, starting from a specific seed pixel and spreading to neighbors that satisfy the growing rule [40]. The growing rule is a set of criteria used to determine whether neighboring pixels should be added to the region. The fundamental disadvantage to intensity-based region segmentation is that the intensity provides no spatial information. The established threshold is a single value or a set gray level. Hence, to implement the growing rule, RegGro uses the dynamic mean intensity with a threshold window (where the window size is $\pm$ the intensity distance). The dynamic mean intensity is the sum mean intensity of all pixels that belong to a specific region. This mean intensity is updated each time a new pixel is added to the region. Specifically, the mean intensity is a dynamic statistic that depends on the current region, rather than the established intensity of the initial seed pixel. Thus, the mean intensity is more suitable for spreading over the blurred boundary when the region and background have not been determined. The RegGro rule pseudocode can be described as follows (Algorithm 1):

Algorithm 1: RegGro rule pseudocode.

```
//For each pixel p
for all p in image A
    //Copy previous state of p and the mean intensity of all q
    labels_new=labels; Mean_Intensity(q);
    //All current q try to spread to neighbors p
    for all neighbors of p
        if (Intensity (p) ∈ Mean_Intensity (q))
            //Successful growth spreads out to the current p.
            labels_new(p)=labels(q)
        end if
    end for
end for
```

Here, p represents the set of background pixels. Before segmentation, then, p denotes all of the pixels in an image. The image segmentation process can be understood as a process that partitions p into two subregions: the foreground and the background. Initially, all p in the image are labeled as the background region, where q—the pixels belonging to the foreground—are seeded pixels. Initially, the mean intensity of all q, Mean_Intensity(q), is the intensity of only one seeded pixel. This seeded pixel is the first q. Then, q attempts to spread to the neighbors (8-connected pixels). The region-growing process involves labeling a neighboring pixel p as the foreground in a larger region when intensity(p) falls within the Mean_Intensity(q), as shown in Figure 2a. The region-growing or spreading rule of the foreground region's pixel q and neighbors p, hereafter referred to as the δ function, is defined as follows:

$$p(x,y) = \begin{cases} Foreground\ if\ Intensity\,(p) \in Window\langle Mean\_Intensity\,(q)\rangle \\ Background\ otherwise \end{cases} \tag{1}$$

where the intensity of pixel p is the V-channel value of an image in the HSV color space. The Mean_Intensity is the pixel mean intensity of q with a window of the intensity distance. This is dynamically updated with each new pixel added to q. The constant intensity distance is a window of the Mean_Intensity set to 0.065, because all images are converted to a floating format ranging between 0 and 1, as shown in Figure 2b.

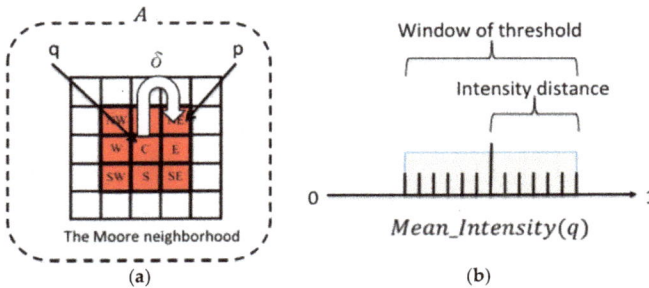

Figure 2. (**a**) Region-growing process from q to 8-connected neighbors p that satisfy the δ function; and (**b**) threshold for the window of intensity. The center of the window is the value of Mean_Intensity, and the window size is $\pm$ the intensity distance (± 0.065).

3.2. GrowCut Method

GrowCut provides an alternative to region-based methods. GrowCut applies cellular automation as the region-growing rule [47]. In automata evolution models, each pixel is treated as a cell that grows and struggles with other cells. Region growing begins from the seed pixels, spreads outward, and attempts to occupy the entire image. Here, the region growth criteria are called the local transition function, known as the δ function. This function defines the rule for calculating the state of a current cell coupled with the state of neighboring cells. Moreover, unlike traditional region growing in only one direction, the state of the region pixels can reverse-grow with neighboring pixels. Thus, the automation evolution can grow the region bi-directionally until all criteria have been satisfied (Figure 3). The pseudocode of the automata evolution rule is described as follows (Algorithm 2):

Algorithm 2: GrowCut rule pseudocode.

```
//For each cell p
for all p in image A
    //Copy the previous state of p
    l_p^{t+1} = l_p^t; θ_p^{t+1} = θ_p^t;
    //All current cells q try to attack p
    for all neighbors p
        if g ( || C_p - C_q || ) · θ_q^t > θ_p^t
            //Successful attacks spread to neighbors p
            l_p^{t+1} = l_q^t; θ_p^{t+1} = g ( || C_p - C_q || ) · θ_q^t;
        end if
    end for
end for
```

---

where the label l_q denotes a foreground pixel, label l_p denotes a background pixel, θ is the strength of the pixels, and $\theta \in [0,1]$. Here, $\vec{C}$ is the intensity of the pixel, and $g\left(\|\vec{C_p} - \vec{C_q}\|\right)$ is the absolute difference between p and q. In the initial states, all q are set to $l_q = 0$ ($0 = background$, $1 = foreground$), $\theta_q = 0$, $\vec{C_q} = Seed(x,y)$. The growing rule for GrowCut—i.e., the δ function—is defined as follows:

$$p(x,y) = \begin{cases} Foreground\ if\ g\left(\|\vec{C_p} - \vec{C_q}\|\right) \cdot \theta_q^t > \theta_p^t \\ Background\ Otherwise \end{cases} \tag{2}$$

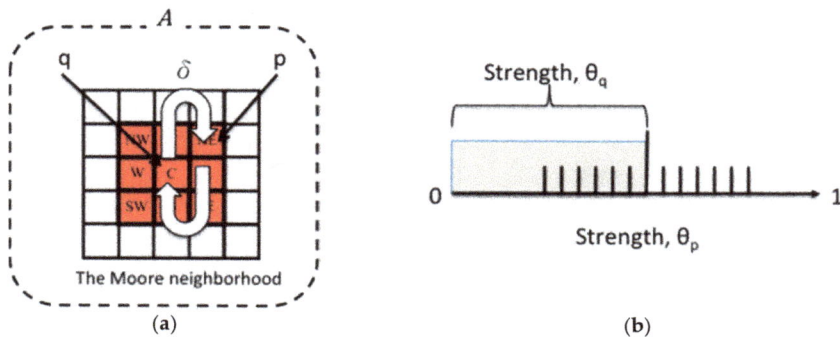

Figure 3. (a) Region-growing process from cell q to its neighbors or reverse-growing from its neighbors, with the δ function; and (b) the strength threshold for the growing rule. The region grows when $\theta_q > \theta_p$; otherwise the region reverse-grows.

3.3. Hybrid RegGro and GrowCut

We proposed a hybrid RgGc that employs a neural network model in order to combine these two growing methods. The hybrid RgGc applies a neural network to classify the input image as fog, stained, or normal scenes. Then, RegGro and GrowCut were applied to process fog and stain images, respectively. The GrowCut method has also been used to segment images of normal scenes.

Detecting fog and stain scenes is a difficult task for image recognition. It is also unclear how the properties of fog and stain should be described. Currently, neural networks have been central to the largest advances in image recognition performance in recent years. The network model learns what distinguishes images, rather than relying on manually-specified differences. To automatically recognize the fog and stain images, a neural network model is presented as a classifier. This model is trained using TensorFlow [48]. Following the training instruction [49], the model is trained with the Typhoon Image Set, as described in Section 3.4, to distinguish between three labels (viz., fog, stained, and normal). This model uses 4000 training steps. Each step chooses ten random images from the Typhoon Image Set, and feeds them into the final layer in order to derive predictions. Those predictions were then compared to the actual labels in order to update the final layer's weights through the back-propagation process. This test evaluation is the best estimate of how the trained model will perform with regard to the classification task. Model evaluations were performed using a running average of the parameters computed over time. After the model was fully trained, its accuracy was approximately 99%.

Figure 4 shows the workflow of the training model. The trained model classifies an input image. A decision is made regarding whether an input image is foggy, stained, or normal. The hybrid RgGc then automatically switches to the RegGro and GrowCut methods to process fog and stained images separately.

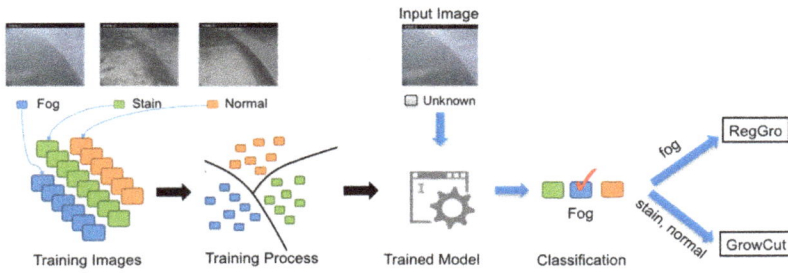

Figure 4. Flowchart of the training model and classification input images. During the training process, the training images are labeled as fog, stained, or normal. The hybrid RgGc method classifies input images and then pipes to different growing methods. (Note: The Traditional Chinese in header of all images is represented the location in the Changed Bridge).

3.4. Image Set and Ground Truth

3.4.1. Typhoon Image Set

In this case study, two region-based segmentation algorithms were employed to identify flood regions. Historical outdoor images were recorded during a typhoon-induced rainstorm that occurred in September 2010 in Taiwan. The capture period was between 12:00 p.m. and 5:50 p.m., 19 September 2010. The outdoor imaging system replayed real-time videos streamed to Internet applications. For our evaluation, we extracted one image each minute, for a total of 350 images in the test image set. The video stream was decomposed to a spatial resolution of 352×288 in JPEG format. Part of the test image set is shown as thumbnails in Figure 5. The images were captured between noon (when the raining began) and nightfall (at the flood tide).

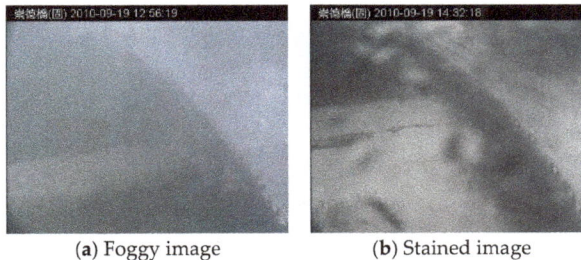

(**a**) Foggy image   (**b**) Stained image

Figure 5. Image set and weather conditions. The image set was captured in adverse weather conditions. The selected sample images show fog (**a**) and stained (**b**) patterns. (Note: The Traditional Chinese in header of all images is represented the location in the Changed Bridge).

3.4.2. Ground Truth of Flood Segments

To evaluate the segmentation results of previous algorithms, a ground truth that yields accurate segmentation results is needed. The ground truth also provides a statistical basis for evaluating region segmentation and boundary detection, as shown in Figure 6. Therefore, flood regions in 450 outdoor images were labeled manually. Examples of these manually-labeled flood regions are shown in Figure 7, where the red boundaries represent the flood region coverage in the original images.

The "true detection" and "false detection" of detected flood segments in each image are described as follows:

$$if \ \{(rT > 70\% \ of \ g.t.) \ \&\& \ (rO < 30\% \ of \ g.t.) \ \&\& \ (rU < 30\% \ of \ g.t.)\} \ ;$$
$$True \ Detection \ else \ Fault \ Detection \tag{3}$$

where, *g.t.* denotes the pixels of the ground truth, *rT* is the resulting region pixel of the algorithm that matches the *g.t.*, *rO* denotes over-segmenting that grows to non-*g.t.*, and *rU* denotes under-segmenting that misses the *g.t.*

The algorithms' respective accuracy for the whole image set is derived as follows:

$$Algorithms' \ Accuracy = \left(\frac{\sum TrueOfDetection}{\sum ImageSet} \right) \times 100\% \tag{4}$$

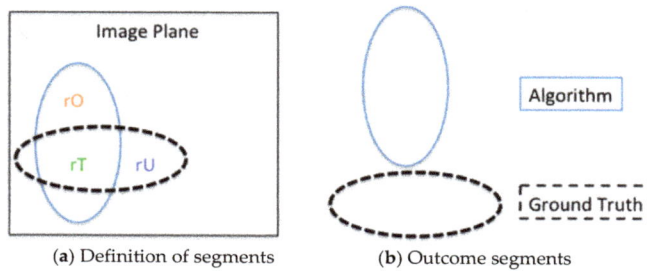

(a) Definition of segments   (b) Outcome segments

Figure 6. Accuracy determined according to the ground truth. (**a**) the resulting region pixels of the algorithm in the image plane are classified as: rT, matching the ground truth; rO, over-segmented; and rU, under-segmented; (**b**) the outcome segments produced with algorithm and ground truth.

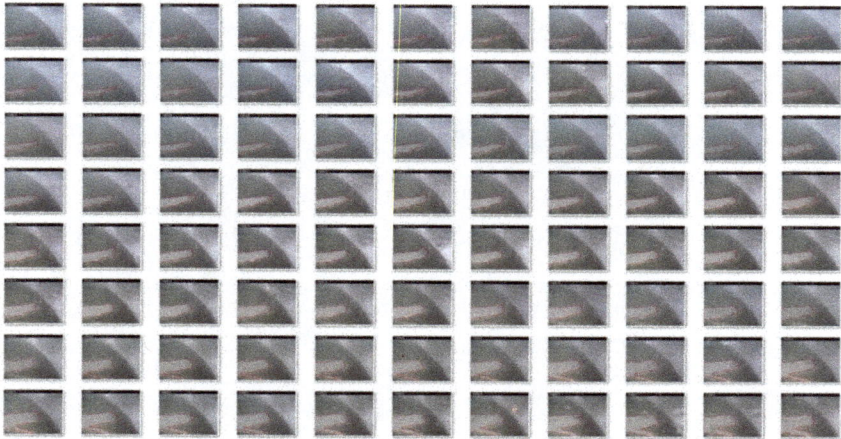

Figure 7. Part of the ground truth of the image set. The red boundaries are manually-labeled segments of flood regions. (Note: The Traditional Chinese in header of all images is represented the location in the Changed Bridge).

4. Results

RegGro and GrowCut were employed to determine flood regions in outdoor images. Various seed-location and image-filtering settings were tested to determine the optimal set that resulted in superior flood regions. The results of flood segments were evaluated according to the ground truth.

4.1. Performance of RegGro

The accuracy of the flood regions identified with RegGro is shown in Figure 8. The intensity distance ranged from 0.025 to 0.15. The highest accuracy achieved using RegGro was 85.7%, with an intensity distance of 0.065. Images without flooding were excluded to avoid the problem of selecting seed points in non-object regions. The remaining 335 valid images were used to evaluate the segmentation algorithms. In a prior experiment, we found that image filtering cannot substantially improve the segmentation accuracy of the RegGro algorithm. We examined several image filters, including the mean, median, bottom-hat, and histogram equalization. However, the maximum accuracy of RegGro was achieved using non-filtered images. To thoroughly understand the segmentation performance, the comparison results of image sequences are presented in Figure 9. The data in Figure 9 show the flood region accuracy evaluated within a time series. This process is crucial for an EWS in order to trace flood variations precisely during the tide process. Inconsistent segments were set as False (1), and consistent segments were set as True (0). This clearly indicates that the segmentation accuracy for the initial period of rain was insufficient. Specifically, before Image 40, the majority of flood segments were not consistent with the ground truth. The remaining flood segments exhibited accurate regions, excluding a few failures in subsequent images. The results of the flood-region segmentation are partially shown in Figure 10 with a step of 10 frames.

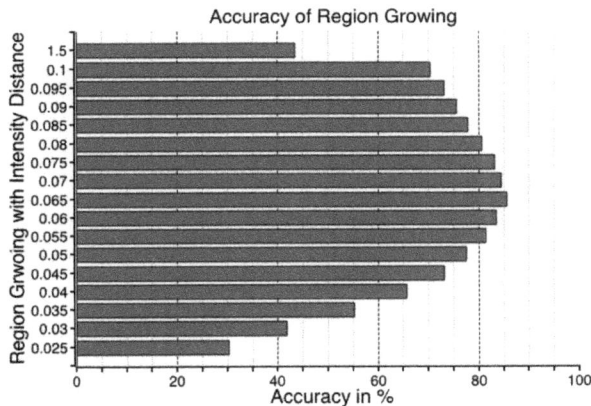

Figure 8. RegGro accuracy. The accuracy was determined according to the ground truth. Each horizontal bar shows the accuracy with a different intensity distance, ranging from 0.025 (RegGro_025) to 0.15 (RegGto_150). The highest accuracy was 85.7% with an intensity distance of 0.065.

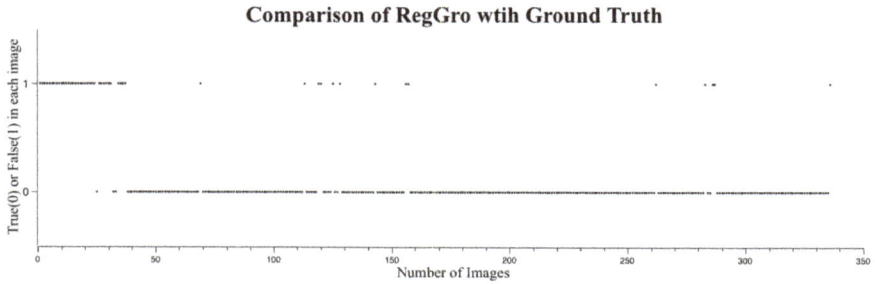

Figure 9. Segmentation success or failure with RegGro. True (0) indicates success, and False (1) indicates failure. Most false detections occurred in the first 40 images with heavy rain and fog.

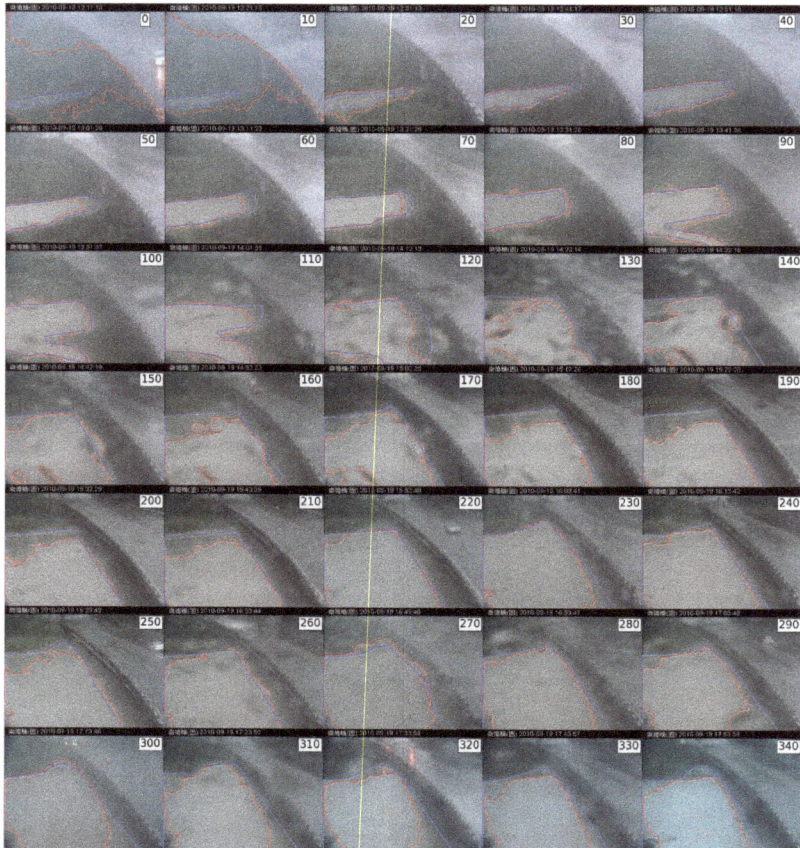

Figure 10. Part of RegGro's results with red segments from growing methods. The blue line is the ground truth, and the green marker is the initial seed for the growing methods. There were few flood segmentation failures in heavy rain and fog. Some failures occurred with raindrop stains on the CCTV screen. (Note: The Traditional Chinese in header of all images is represented the location in the Changed Bridge).

4.2. Performance of GrowCut

The accuracy of flood regions identified with GrowCut is shown in Figure 11. Unlike RegGro, some image filters in GrowCut can improve flood detection in various segments. The maximum accuracy was 75.2% when using the mean filter with 16×16 or 18×18 masks. After testing several image filters, the experimental results showed that the mean filter is superior for enhancing the outcome provided by GrowCut. Specifically, the mean filter increased the accuracy of the GrowCut algorithm from 68.1% to 75.2%. To thoroughly understand the segmenting performance, a comparison of image sequences is shown in Figure 12. The data in Figure 12 indicate the flood region accuracy evaluated within a time series. Inconsistent segments were set to False (1), and consistent segments were set to True (0). This clearly indicates that the segmenting accuracy for the initial rain period failed during two periods of rain. The first period was the same as that using RegGro. That is, before Image 40, most flood segments were not consistent with the ground truth. The second period was between Images 70 and 100, and exhibited more failed segments than RegGro. In the remaining flood segments, however, GrowCut yielded only a few failures in subsequent images. In other words, GrowCut provided nearly perfect segmentation from Image 100 onward. The results of flood region segmentation are partially shown in Figure 13 with a step of 10 frames.

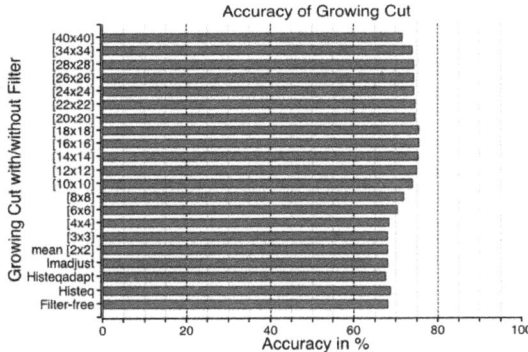

Figure 11. GrowCut accuracy. The accuracy was determined according to the ground truth. Each horizontal bar shows the accuracy with a different filter: viz., the mean, imadjust, histeqadapt, histeq, and filter-free GrowCut. The highest accuracy is 75.2% with the mean filter and both 18×18 and 16×16 masks.

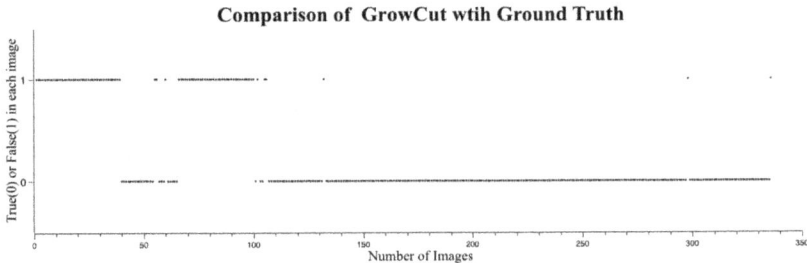

Figure 12. Segmentation success or failure with GrowCut. True (0) indicates success, and False (1) indicates failure. Most false detections occurred between Images 0–40 and Images 70–100 with heavy rain and fog. However, GrowCut was more robust to raindrop stains on the CCTV screen (after Image 100).

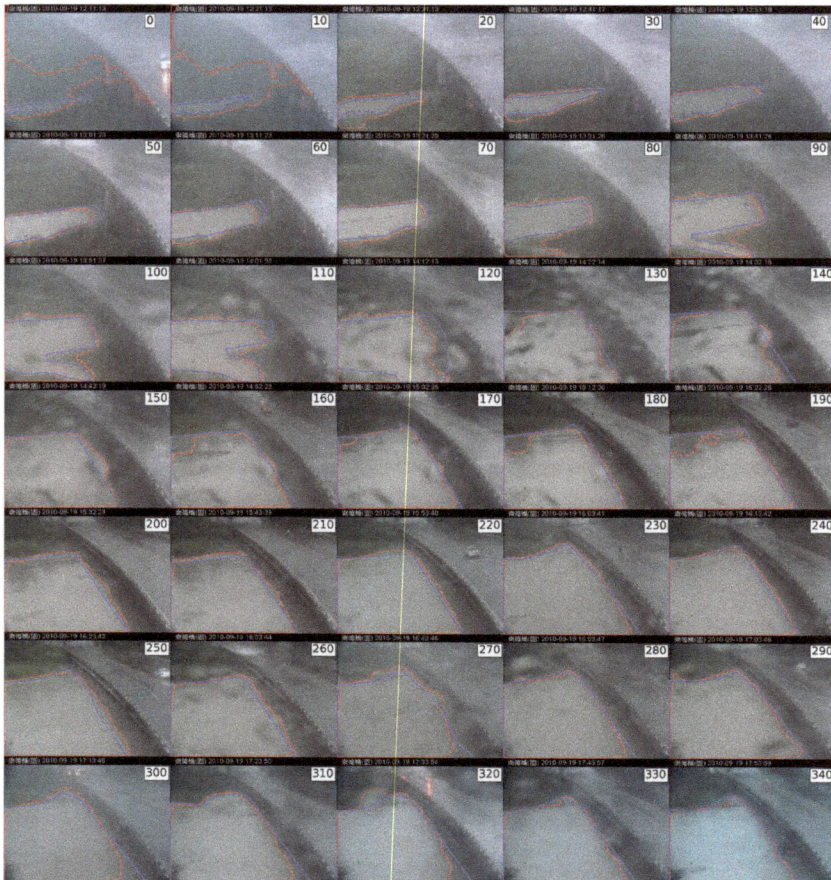

Figure 13. Part GrowCut's results with red segments from growing methods. The blue line is the ground truth, and the green marker is the initial seed for the growing methods. Most flood segmentation failures occurred during heavy rain and fog. GrowCut is robust to raindrop stains on the CCTV screen. (Note: The Traditional Chinese in header of all images is represented the location in the Changed Bridge).

4.3. Performance of Hybrid RegGro and GrowCut

When combining RegGro and GrowCut, the hybrid RgGc was 87.7% accurate. The accuracy of the flood regions identified with the hybrid RgGc is shown in Figure 14. To thoroughly understand the segmentation performance, the comparison results of image sequences are presented in Figure 14. The data in Figure 15 show the flood region accuracy evaluated within a time series. Inconsistent segments were set as False (1), and consistent segments were set as True (0). This result also clearly indicates that both RegGro and GrowCut failed to segment the flood regions as well as the hybrid RgGc during the initial period of heavy rain and fog. The results are consistent with the previous observations in Sections 4.2 and 4.3. The hybrid RgGc exploited the strength of both growing methods, with more accurate detections than GrowCut for Images 65~100, and RegGro for Images 110–150. The results of the flood-region segmentation, ground truth, and seed marker are partially shown in Figure 16.

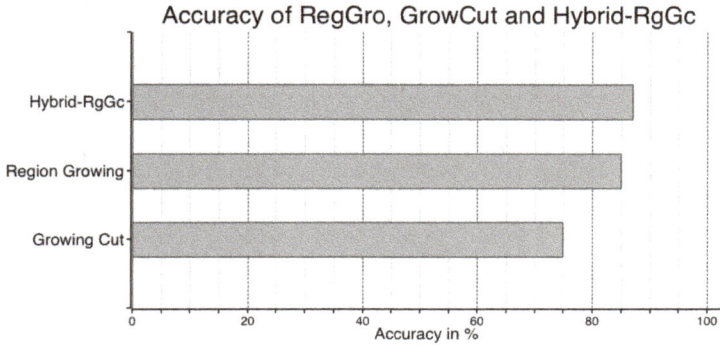

Figure 14. Hybrid RgGc accuracy. The accuracy was determined according to the ground truth. Each horizontal bar shows the accuracy with a different growing method. Outperforming both RegGro and GrowCut, the hybrid RgGc was 87.7% accurate.

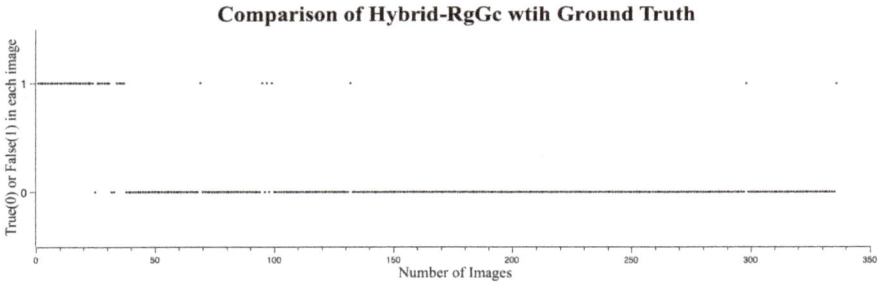

Figure 15. Segmentation success or failure with the hybrid RgGc. True (0) indicates success and False (1) indicates failure. Most false detections occurred in the first 40 images with heavy rain and fog. Both methods failed to segment the flood regions as well as the hybrid RgGc.

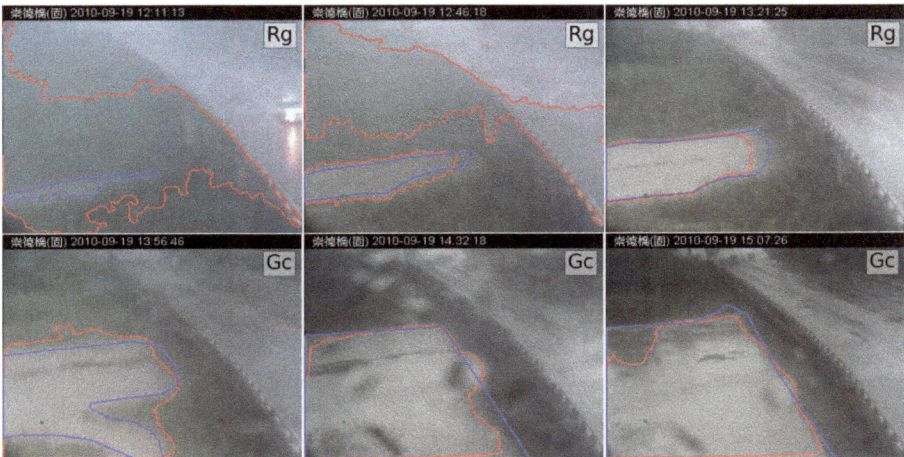

Figure 16. *Cont.*

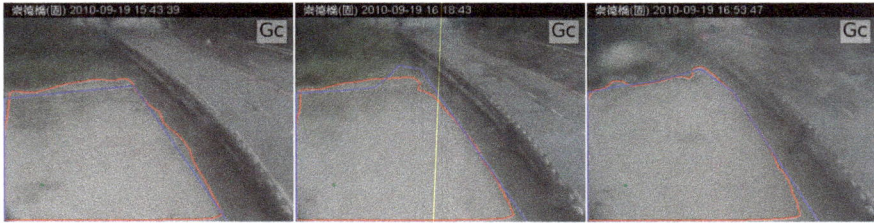

Figure 16. Hybrid RgGc results with red segments from growing methods. The blue line is the ground truth, and the green marker is the initial seed for the growing methods. This figure only shows Images 0, 35, 70, 105, 140, 175, 210, 245, and 280 from the image set, in order to clearly show the contours and seeded marker. The text in the upper-right corner distinguishes between images processed with RegGro (Rg) and GrowCut (Gc). (Note: The Traditional Chinese in header of all images is represented the location in the Changed Bridge).

5. Discussion

In this section, we discuss how poor atmospheric conditions affect segmentation outcomes and ground-truth proceedings. As stated in the introduction, fog and stains are the primary factors that affect the outcome. Moreover, the ground truth also serves as a crucial evaluation factor.

5.1. Influence of Fog and Heavy Rainfall

Fog and haze are the main factors that disturb light reflection in scenes, causing unexpected variations in image intensity. In our image set, fog and haze occurred for only a short period when the rainstorms began—specifically, the period between Images 0 and 40. Since rainstorms involve suspended particles, mist, raindrops, raindrop streaks, and heavy rain spray, they render scenes extremely unclear. An image of the fog that formed before the initial rainstorm at noon is shown in Figure 17. The resulting blurry image was difficult to segment using the proposed region-based method. We enhanced the image using filters and equalization to improve the histogram distribution. However, this enhancement procedure also affected the remaining images, causing the region-based segmentation method to tend toward overestimations and underestimations. Both algorithms failed to segment the initial period of torrential rainfall. Therefore, we infer that the presence of fog and haze influence the final outcome of flood segmentation. Furthermore, during heavy rainfall, CCTV cameras that rely exclusively on visible light are more easily blocked by raindrops and fog. Multispectral image sensors should be used to address this issue [50]. For example, infrared cameras use infrared light to capture scenes. Infrared light has a longer wavelength than visible light, and it can penetrate heavy fog and rainfall to form clearer images.

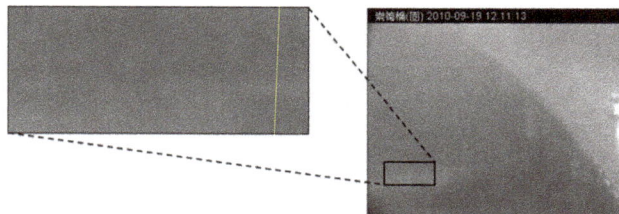

Figure 17. Fog example. Fog and haze regress pixel intensity. (Note: The Traditional Chinese in header of all images is represented the location in the Changed Bridge).

5.2. Influence of Stains on Lenses

Various stains on the camera lens also exert critical effects on image processing. Figure 18 shows an example of the differing outcomes of these algorithms with these effects. Briefly, image segmentation is a process that involves segmenting objects of interest from the background. Although extracted segments have discriminatory boundaries, they are sensitive to minor foreground and background boundaries. However, the presence of stains on the camera lens directly disturbs the overall image intensity. Occasionally, the segmentation process stalls on stained areas (see Image 270 in Figure 18b). This is the reason why RegGro yielded numerous segmentation failures from Images 100–350 (see Figure 9). GrowCut exhibited a superior ability to resist rain stains on the camera lens during the specific period between Images 100 and 350 (Figure 12). We compared crucial periods, at the start (Image 152) and middle (Image 270). Although RegGro provided superior segmentation accuracy during the initial period, the effects of stained areas stalled region growth. By contrast, GrowCut exhibited superior resistance to stains on the camera lens (see Image 270 in Figure 18a), yet it tended to overestimate flood regions at the start (see Image 152 in Figure 18a). Thus, GrowCut primarily yielded segmentation failures before Image 100 (see Figure 12).

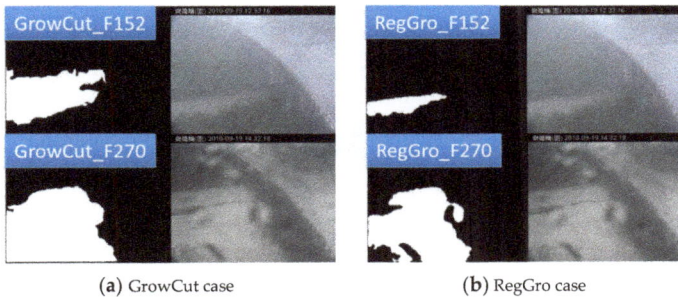

(**a**) GrowCut case   (**b**) RegGro case

Figure 18. Comparison of the two algorithms with example images from the start (Image 152) and middle (Image 270). (**a**) GrowCut covering the stains (Image 152) and overestimating them (Image 270); (**b**) RegGro identifying the flood region (Image 152) and affected by stains (Image 270). (Note: The Traditional Chinese in header of all images is represented the location in the Changed Bridge).

5.3. Ground Truth

We also inquired as to whether a ground truth is the only method for estimating the performance of an algorithm. However, it must be asked whether the ground truth exactly represents the flood region. In Figure 19, compared with the ground truth segment, the RegGro segment determines segmentation failure. In this case, the RegGro segment is not considered a flood region, ostensibly leading to a false detection. In fact, this RegGro segment can be treated as an assembly of various flood regions. Moreover, manually-labeled flood regions are sometimes underestimated, because the subjects sedulously avoid solid boundaries in order to ensure that segments remain isolated from the background. Thus, the manually-labeled region might be smaller than the actual region. To evaluate the performance success or failure, an intelligent and flexible evaluation of the ground truth should be conducted in future studies.

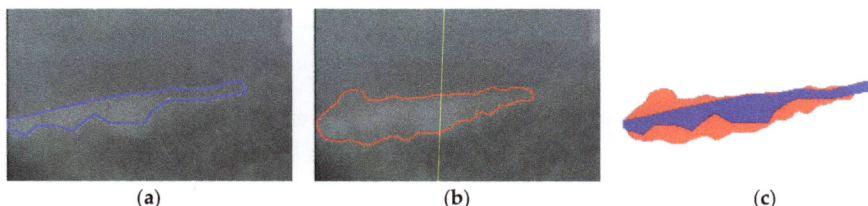

Figure 19. Example of a failed identification. The RegGro region (in red) overlaps with the ground truth (in blue). (**a**) ground truth; (**b**) RegGro region; and (**c**) comparison of regions.

6. Conclusions and Future Work

Vision systems are used for various image applications, such as feature detection, stereo vision, segmentation, recognition, and tracking. These image processes are affected by weather conditions, particularly fog and stains that occur on lenses. However, such visual effects barely affect human vision, because the original scene behind the stain can, nevertheless, be approximated. In most image processing applications—particularly processes that use the pixel intensity and graph information—stained regions belong to neither the foreground nor the background. Therefore, outdoor imaging and monitoring applications are limited by the visual effects generated by adverse weather conditions. This was the primary motivation for this study. We aimed to understand not only the influence of adverse weather on outdoor imaging, but also how the performance of image applications can be improved.

In this study, two region-based segmentation algorithms and a hybrid method that combines both algorithms were applied to flood detection in adverse weather. A case study was presented to demonstrate the performance of these algorithms and their bottleneck for low-cost vision-based flood monitoring. The experimental results indicate the advantages and disadvantages of both algorithms, and the effects that poor atmospheric conditions have on segmentation outcomes. Both methods have unique advantages and disadvantages for fog and stained conditions, respectively. The segmentation accuracy of flood regions yielded by GrowCut and RegGro was 75% and 85%, respectively. Although RegGro was more accurate, it was inadequate for stained images. If the ability to resist stains were incorporated into RegGro, it could achieve more accurate results (see Figure 15 for the hybrid's results). Thus, we have combined the advantages of both RegGro and GrowCut into a hybrid RgGc with a network classifier. In doing so, we improved the results by approximately 2.7%. Moreover, we shall investigate the feasibility of multispectral cameras in terms of improving the accuracy of image segmentation and preserving visual information in outdoor images with scenes of fog.

Acknowledgments: The authors would like to thank all three reviewers for their comments that help improve the manuscript.

Author Contributions: Shi-Wei Lo and Ching-Han Hsu supervised the research and contributed to manuscript organization. Jyh-Horng Wu, Lun-Chi Chen, Chien-Hao Tseng and Fang-Pang Lin contributed in in situ equipment, monitoring data acquisition and visual sensor network. Shi-Wei Lo developed the research framework and wrote the manuscript.

Conflicts of Interest: The authors declare that there is no conflict.

References

1. Lin, M.L.; Jeng, F.S. Characteristics of hazards induced by extremely heavy rainfall in Central Taiwan—Typhoon Herb. *Eng. Geol.* **2000**, *58*, 191–207. [CrossRef]
2. Tsou, C.Y.; Feng, Z.Y.; Chigira, M. Catastrophic landslide induced by Typhoon Morakot, Shiaolin, Taiwan. *Geomorphology* **2011**, *127*, 166–178. [CrossRef]

3. Guo, X.Y.; Zhang, H.Y.; Wang, Y.Q.; Clark, J. Mapping and assessing typhoon-induced forest disturbance in Changbai Mountain National Nature Reserve using time series Landsat imagery. *J. Mt. Sci.* **2015**, *12*, 404–416. [CrossRef]

4. Chen, S.C.; Lin, T.W.; Chen, C.Y. Modeling of natural dam failure modes and downstream riverbed morphological changes with different dam materials in a flume test. *Eng. Geol.* **2015**, *188*, 148–158. [CrossRef]

5. Zhuang, J.Q.; Peng, J.B. A coupled slope cutting—A prolonged rainfall-induced loess landslide: A 17 October 2011 case study. *Bull. Eng. Geol. Environ.* **2014**, *73*, 997–1011. [CrossRef]

6. Tsou, C.Y.; Chigira, M.; Matsushi, Y.; Chen, S.C. Fluvial incision history that controlled the distribution of landslides in the Central Range of Taiwan. *Geomorphology* **2014**, *226*, 175–192. [CrossRef]

7. Chigira, M. Geological and geomorphological features of deep-seated catastrophic landslides in tectonically active regions of Asia and implications for hazard mapping. *Episodes* **2014**, *37*, 284–294.

8. Lo, S.-W.; Wu, J.-H.; Lin, F.-P.; Hsu, C.-H. Cyber Surveillance for Flood Disasters. *Sensors* **2015**, *15*, 2369–2387. [CrossRef] [PubMed]

9. Massari, C.; Tarpanelli, A.; Moramarco, T. A fast simplified model for predicting river flood inundation probabilities in poorly gauged areas. *Hydrol. Process.* **2015**, *29*, 2275–2289. [CrossRef]

10. Holcer, N.J.; Jelicic, P.; Bujevic, M.G.; Vazanic, D. Health protection and risks for rescuers in cases of floods. *Arh. Za Hig. Rada I Toksikol. Arch. Ind. Hyg. Toxicol.* **2015**, *66*, 9–13.

11. Fang, S.F.; Xu, L.D.; Zhu, Y.Q.; Liu, Y.Q.; Liu, Z.H.; Pei, H.; Yan, J.W.; Zhang, H.F. An integrated information system for snowmelt flood early-warning based on internet of things. *Inf. Syst. Front.* **2015**, *17*, 321–335. [CrossRef]

12. Lo, S.W.; Wu, J.H.; Chen, L.C.; Tseng, C.H.; Lin, F.P. Fluvial Monitoring and Flood Response. In Proceedings of the 2014 IEEE Sensors Applications Symposium (SAS), Queenstown, New Zealand, 18–20 February; pp. 378–381.

13. Lo, S.W.; Wu, J.H.; Chen, L.C.; Tseng, C.H.; Lin, F.P. Flood Tracking in Severe Weather. In Proceedings of the 2014 International Symposium on Computer, Consumer and Control (Is3c 2014), Taichung, Taiwan, 10–12 June 2014; pp. 27–30.

14. Krzhizhanovskaya, V.V.; Shirshov, G.S.; Melnikova, N.B.; Belleman, R.G.; Rusadi, F.I.; Broekhuijsen, B.J.; Gouldby, B.P.; Lhomme, J.; Balis, B.; Bubak, M.; et al. Flood early warning system: Design, implementation and computational modules. *Procedia Comput. Sci.* **2011**, *4*, 106–115. [CrossRef]

15. Castillo-Effer, M.; Quintela, D.H.; Moreno, W.; Jordan, R.; Westhoff, W. Wireless sensor networks for flash-flood alerting. In Proceedings of The Fifth IEEE International Caracas Conference On Devices, Circuits and Systems, Punta Cana, Dominican Republic, 3–5 November 2004; Volume 1, pp. 142–146.

16. Chen, Z.; Di, L.; Yu, G.; Chen, N. Real-Time On-Demand Motion Video Change Detection in the Sensor Web Environment. *Comput. J.* **2011**, *54*, 2000–2016. [CrossRef]

17. Kim, J.; Han, Y.; Hahn, H. Embedded implementation of image-based water-level measurement system. *IET Comput. Vis.* **2011**, *5*, 125–133. [CrossRef]

18. Nguyen, L.S.; Schaeli, B.; Sage, D.; Kayal, S.; Jeanbourquin, D.; Barry, D.A.; Rossi, L. Vision-based system for the control and measurement of wastewater flow rate in sewer systems. *Water Sci. Technol.* **2009**, *60*, 2281–2289. [CrossRef] [PubMed]

19. Lo, S.-W.; Wu, J.-H.; Lin, F.-P.; Hsu, C.-H. Visual Sensing for Urban Flood Monitoring. *Sensors* **2015**, *15*, 20006–20029. [CrossRef] [PubMed]

20. Garg, K.; Nayar, S.K. Detection and removal of rain from videos. In Proceedings of the 2004 IEEE Computer Society Conference on Computer Vision and Pattern Recognition, Washington, DC, USA, 27 June–2 July 2004; Volume 1, pp. 528–535.

21. Tripathi, A.K.; Mukhopadhyay, S. A Probabilistic Approach for Detection and Removal of Rain from Videos. *IETE J. Res.* **2011**, *57*, 82–91. [CrossRef]

22. Tripathi, A.K.; Mukhopadhyay, S. Meteorological approach for detection and removal of rain from videos. *IET Comput. Vis.* **2013**, *7*, 36–47. [CrossRef]

23. Adler, W.F. Rain impact retrospective and vision for the future. *Wear* **1999**, *233*, 25–38. [CrossRef]

24. Garg, K.; Nayar, S.K. Vision and rain. *Int. J. Comput. Vis.* **2007**, *75*, 3–27. [CrossRef]

25. Pang, J.; Au, O.C.; Guo, Z. Improved Single Image Dehazing Using Guided Filter. In Proceedings of the APSIPAASX, Xi'an, China, 17–21 October 2011; pp. 1–4.

26. Shwartz, S.; Namer, E.; Schechner, Y.Y. Blind Haze Separation. In Proceedings of the 2006 IEEE Computer Society Conference on Computer Vision and Pattern Recognition, New York, NY, USA, 17–22 June 2006; pp. 1984–1991.
27. Kopf, J.; Neubert, B.; Chen, B.; Cohen, M.; Cohen-Or, D.; Deussen, O.; Uyttendaele, M.; Lischinski, D. Deep Photo: Model-Based Photograph Enhancement and Viewing. In Proceedings of the ACM SIGGRAPH Asia 2008, Singapore, 11–13 December 2008.
28. Xiao, C.X.; Gan, J.J. Fast image dehazing using guided joint bilateral filter. *Vis. Comput.* **2012**, *28*, 713–721. [CrossRef]
29. Kim, J.H.; Jang, W.D.; Sim, J.Y.; Kim, C.S. Optimized contrast enhancement for real-time image and video dehazing. *J. Vis. Commun. Image Represent.* **2013**, *24*, 410–425. [CrossRef]
30. Blaschke, T. Object based image analysis for remote sensing. *Isprs J. Photogramm. Remote Sens.* **2010**, *65*, 2–16. [CrossRef]
31. Dos Santos, P.P.; Tavares, A.O. Basin Flood Risk Management: A Territorial Data-Driven Approach to Support Decision-Making. *Water* **2015**, *7*, 480–502. [CrossRef]
32. Mason, D.C.; Giustarini, L.; Garcia-Pintado, J.; Cloke, H.L. Detection of flooded urban areas in high resolution Synthetic Aperture Radar images using double scattering. *Int. J. Appl. Earth Obs. Geoinf.* **2014**, *28*, 150–159. [CrossRef]
33. Long, S.; Fatoyinbo, T.E.; Policelli, F. Flood extent mapping for Namibia using change detection and thresholding with SAR. *Environ. Res. Lett.* **2014**, *9*. [CrossRef]
34. Chen, S.; Liu, H.J.; You, Y.L.; Mullens, E.; Hu, J.J.; Yuan, Y.; Huang, M.Y.; He, L.; Luo, Y.M.; Zeng, X.J.; et al. Evaluation of High-Resolution Precipitation Estimates from Satellites during July 2012 Beijing Flood Event Using Dense Rain Gauge Observations. *PLoS ONE* **2014**, *9*. [CrossRef] [PubMed]
35. Oliva, D.; Osuna-Enciso, V.; Cuevas, E.; Pajares, G.; Pérez-Cisneros, M.; Zaldívar, D. Improving segmentation velocity using an evolutionary method. *Expert Syst. Appl.* **2015**, *42*, 5874–5886. [CrossRef]
36. Foggia, P.; Percannella, G.; Vento, M. Graph Matching and Learning in Pattern Recognition in the Last 10 Years. *Int. J. Pattern Recognit. Artif. Intell.* **2014**, *28*, 1450001. [CrossRef]
37. Ducournau, A.; Bretto, A. Random walks in directed hypergraphs and application to semi-supervised image segmentation. *Comput. Vis. Image Underst.* **2014**, *120*, 91–102. [CrossRef]
38. Oliva, D.; Cuevas, E.; Pajares, G.; Zaldivar, D.; Perez-Cisneros, M. Multilevel Thresholding Segmentation Based on Harmony Search Optimization. *J. Appl. Math.* **2013**, *2013*. [CrossRef]
39. Vantaram, S.R.; Saber, E. Survey of contemporary trends in color image segmentation. *J. Electron. Imaging* **2012**, *21*, 040901. [CrossRef]
40. Gonzalez, R.C.; Woods, R.E. *Digital Image Processing*, 3rd ed.; Prentice Hall: Upper Saddle River, NJ, USA, 2008.
41. Peng, B.; Zhang, L.; Zhang, D. A survey of graph theoretical approaches to image segmentation. *Pattern Recognit.* **2013**, *46*, 1020–1038. [CrossRef]
42. Ning, J.; Zhang, L.; Zhang, D.; Wu, C. Interactive image segmentation by maximal similarity based region merging. *Pattern Recognit.* **2010**, *43*, 445–456. [CrossRef]
43. Panagiotakis, C.; Grinias, I.; Tziritas, G. Natural Image Segmentation Based on Tree Equipartition, Bayesian Flooding and Region Merging. *IEEE Trans. Image Process.* **2011**, *20*, 2276–2287. [CrossRef] [PubMed]
44. Couprie, C.; Grady, L.; Najman, L.; Talbot, H. Power watersheds: A new image segmentation framework extending graph cuts, random walker and optimal spanning forest. In Proceedings of the 2009 IEEE 12th International Conference on Computer Vision, Kyoto, Japan, 29 September–2 October 2009; pp. 731–738.
45. Panagiotakis, C.; Papadakis, H.; Grinias, E.; Komodakis, N.; Fragopoulou, P.; Tziritas, G. Interactive image segmentation based on synthetic graph coordinates. *Pattern Recognit.* **2013**, *46*, 2940–2952. [CrossRef]
46. Arbelaez, P.; Maire, M.; Fowlkes, C.; Malik, J. Contour Detection and Hierarchical Image Segmentation. *IEEE Trans. Pattern Anal. Mach. Intell.* **2011**, *33*, 898–916. [CrossRef] [PubMed]
47. Vezhnevets, V.; Konouchine, V. GrowCut: Interactive multi-label ND image segmentation by cellular automata. *Proc. Graphicon* **2005**, *1*, 150–156.
48. Abadi, M.; Agarwal, A.; Barham, P.; Brevdo, E.; Chen, Z.; Citro, C.; Corrado, G.; Davis, A.; Dean, J.; Devin, M.; et al. TensorFlow: Large-Scale Machine Learning on Heterogeneous Distributed Systems. 2016, arXiv:1603.04467.

Sensors **2016**, *16*, 1125

49. TensorFlow. Available online: https://www.tensorflow.org/ (accessed on 28 April 2016).
50. NASA Spinoff. Available online: https://spinoff.nasa.gov/ (accessed on 28 April 2016).

sensors

MDPI

Article

Object Occlusion Detection Using Automatic Camera Calibration for a Wide-Area Video Surveillance System

Jaehoon Jung [1], Inhye Yoon [1,2] and Joonki Paik [1,*]

[1] Department of Image, Chung-Ang University, 84 Heukseok-ro, Dongjak-gu, Seoul 06974, Korea; gjslkjs@gmail.com (J.J.); inhyey@gmail.com (I.Y.)

[2] ADAS Camera Team, LG Electronics, 322 Gyeongmyeong-daero, Seo-gu, Incheon 22744, Korea

*   Correspondence: paikj@cau.ac.kr; Tel.: +82-10-7123-6846

Academic Editor: Gonzalo Pajares Martinsanz
Received: 3 May 2016; Accepted: 23 June 2016; Published: 25 June 2016

Abstract: This paper presents an object occlusion detection algorithm using object depth information that is estimated by automatic camera calibration. The object occlusion problem is a major factor to degrade the performance of object tracking and recognition. To detect an object occlusion, the proposed algorithm consists of three steps: (i) automatic camera calibration using both moving objects and a background structure; (ii) object depth estimation; and (iii) detection of occluded regions. The proposed algorithm estimates the depth of the object without extra sensors but with a generic red, green and blue (RGB) camera. As a result, the proposed algorithm can be applied to improve the performance of object tracking and object recognition algorithms for video surveillance systems.

Keywords: occlusion detection; automatic camera calibration; depth estimation; moving object detection; video surveillance system

1. Introduction

Recently, the demand for object tracking and recognition algorithms is increasing due to video surveillance. An object occlusion is a major factor for the performance degradation of a video surveillance system. For this reason, various object occlusion detection and handling methods were studied.

Mei et al. proposed an object tracking with consideration of occlusion that is detected using the occlusion map [1]. Since this method uses a target template to obtain the occlusion map, it is difficult to detect the object occlusion when the target template is unavailable. Zitnick et al. generated a depth map using a stereo camera and detected object occlusion regions [2]. However, this method needs two cameras for stereo matching to generate the depth map. Sun et al. proposed an optimization approach using the visibility constraint for the stereo matching and then generated the depth map by minimizing the energy function [3]. Since a stereo camera-based occlusion detection method needs an additional camera, it is not easy to implement in an already installed wide-area surveillance system.

To solve this problem, single camera based depth map estimation methods were proposed. Matyunin et al. estimated the depth using an infrared sensor [4]. However, this method cannot work in the outdoor scene since an infrared sensor is interrupted by sunlight. Im et al. proposed a single red, green, and blue (RGB) camera-based object depth estimation method using multiple color-filter apertures (MCA) [5]. However, this method needs a special aperture for the object depth estimation, and produces color distortion at boundary of the out-focused objects. Zonglei et al. used a patterned box for semi-automatic camera calibration [6]. Lin et al. estimated vanishing points using traffic lanes, and estimated the distance of a frontal vehicle using a single RGB camera for a collision warning system [7]. Since this method uses the traffic lane for vanishing point estimation, distance estimation is impossible when an input image does not contain a traffic lane. Song et al. detected features from a moving object, and automatically calibrated the camera [8]. However, Song's method cannot avoid the camera calibration error when feature points change while the object is moving.

To solve these problems, the proposed method first performs automatic camera calibration using both moving objects and background structures to estimate camera parameters. Given the camera parameters, the proposed algorithm estimates the object depth with regard to a reference plane, and then detects the object occlusion. To estimate vanishing points and lines, the proposed algorithm detects parallel lines in the input image. Bo et al. detected straight lines from background structures using the one-dimensional (1D) Hough transform for automatic camera calibration [9]. Since this method uses only a single image, it is impossible to automatically calibrate the camera when the background does not have line components. Moreover, accuracy of the camera calibration is degraded by with non-parallel lines. Lv et al. detected human foot and head points for automatic camera calibration [10]. However, if the estimated foot and head points are not sufficiently accurate or if the object motion is linear, the camera calibration is impossible. Moreover, accuracy of the camera calibration result depends on the object detection results. To solve these problems, the proposed algorithm combines the background structure lines with human foot and head information to estimate vanishing points and lines.

This paper is organized as follows: Section 2 describes background theory of the camera geometry, and Section 3 presents the proposed object occlusion detection algorithm. Experimental results of the proposed algorithm are shown in Section 4, and Section 5 concludes the paper.

2. Theoretical Background of Camera Geometry

Estimation of the depth needs a 3D space information. To obtain the projective relationship between the 2D image and 3D space information, a camera geometry is used with camera parameter that describe camera sensor, lens, optical axis, and a position of the camera in the world coordinate. In the pin-hole camera model [11], a point in the 3D space is projected onto a point in the 2D image as

$$
s \begin{bmatrix} x \\ y \\ 1 \end{bmatrix} = \begin{bmatrix} f_x & skew & p_x \\ 0 & f_y & p_y \\ 0 & 0 & a \end{bmatrix} \begin{bmatrix} r_{11} & r_{12} & r_{13} & t_1 \\ r_{21} & r_{22} & r_{23} & t_2 \\ r_{31} & r_{32} & r_{33} & t_3 \end{bmatrix} \begin{bmatrix} X \\ Y \\ Z \\ 1 \end{bmatrix} = A[R|t] \begin{bmatrix} X \\ Y \\ Z \\ 1 \end{bmatrix} \tag{1}
$$

where s represents the scale, $[\, x \quad y \quad 1 \,]^T$ a point in the 2D image, matrix A consists of intrinsic camera parameters, f_x and f_y focal length in the x- and y-axis focal length, respectively, *skew* the skewness, a the aspect ratio, camera rotation matrix R consists of camera rotation parameters r_{ij}, $[\, t_1 \quad t_2 \quad t_3 \,]^T$ the camera translation vector, and $[\, X \quad Y \quad Z \quad 1 \,]^T$ a point in the 3D space.

To simplify the description without loss of generality, we assume that the focal lengths f_x and f_y are equivalent, the principal point is at the image center, the skewness is equal to zero, and the aspect ratio is equal to 1. In the same manner, we also assume that the camera rotation angle with regard to the Z-axis is equal to zero, and the camera translation with regard to both X- and Y-axis is equal to zero to calculate the extrinsic matrix $[R|t]$ as

$$
[R|t] = [R_Z(\rho)R_X(\theta)T(0,0,h_c)] \tag{2}
$$

where R_Z represents the rotation matrix with regard to the Z-axis, R_X the rotation matrix with regard to the X-axis, T the transformation for a translation, h_c the camera height.

The 2D image is generated by the light that is reflected by an object and than arrives at the camera sensor. In this process, a single object is projected onto the 2D image plane with different sizes according to the distance from the camera as shown in Figure 1. For this reason, parallel lines in the 3D space are projected on the 2D image plane as non-parallel lines depending on the depth. Using the apparent non-parallel lines in the image, the vanishing points can be estimated as a intersected points of those lines. Since the projective camera transformation model projects a point in the 3D space onto the camera sensor, the camera parameters can be estimated using vanishing points.

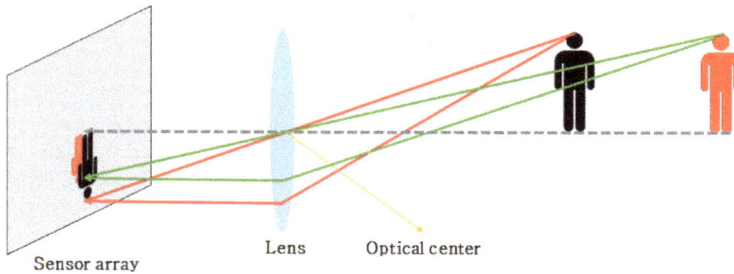

Figure 1. Projective model of the camera.

The point in the 3D space is projected onto a camera sensor corresponding to a point in the 2D image using a projective transform. However, a point in the 2D image cannot be inversely projected onto a unique point in the 3D space since the camera projection transform is not a one-to-one function. On the other hand, if there is a reference plane, a point in the 2D image can be inversely projected onto a point on the reference plane that is defined in the 3D space. As a result, the proposed algorithm estimates the object depth using the 2D image based on a pre-specified reference plane.

3. Automatic Calibration-Based Occlusion Detection

The proposed object occlusion detection algorithm consists of three steps: (i) automatic camera calibration; (ii) object depth estimation; and (iii) occlusion detection. Figure 2 shows the block diagram of the proposed algorithm, where I_k represents the k-th input frame, L represents the extracted lines, P represents the projective matrix, D represents the object depth information, and O represents the detected region of an occluded object.

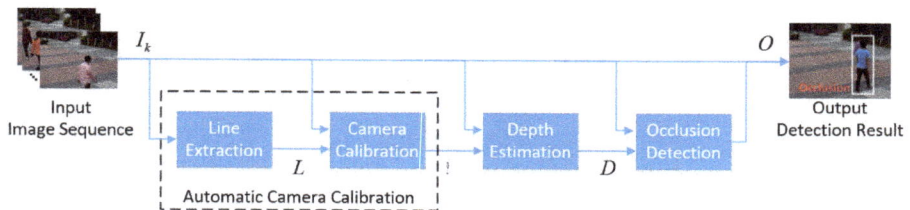

Figure 2. Block diagram of the proposed occlusion detection algorithm.

3.1. Automatic Camera Calibration

The proposed algorithm estimates the camera parameters for the object depth estimation followed by object occlusion detection. Since semi-automatic camera calibration is the simplest way to estimate parameters using a synthetic calibration pattern [6], its performance is highly dependent on the experience of a user. To solve this problem, the proposed algorithm uses an automatic camera calibration method that extracts lines from the image, and then estimates vanishing points and lines [12].

To detect human foot and head points, the proposed algorithm detects the foreground by modeling the background using the Gaussian mixture model (GMM) [13]. The object region is then detected by labeling a sufficiently large object. Given an object region, the vertically highest point is determined as the head point. On the other hand, the average of the bottom 20 percent points is determined as the foot point.

A pair of parallel lines that connect head points and foot points are used to detect vanishing points and lines. Since the lines connecting head points and foot points are non-parallel when the height of an object changes while walking, the proposed algorithm detects the uniform height of the object only when pedestrian's legs are crossing as

$$\frac{1}{n} \sum_{i=1}^{n} (p_i - p_f)^2 < T_C \tag{3}$$

where p_i represents the candidate foot points, n the number of candidate foot points, p_f the detected foot point, and T_C the threshold value.

To combine object foot and head information with the background structure lines, the proposed algorithm detects edges that are used to detect vanishing points and lines [14]. The detected foot and head points and background structure lines are shown in Figure 3.

(a)	(b)

Figure 3. Results of the line detection: (**a**) foot-to-head line and (**b**) background structure lines.

The vanishing points and lines are estimated using the detected foot-to-head line and background structure lines. For the robust vanishing points and lines estimation, a sufficient number of foot and head points are required. For that reason, the proposed algorithm estimates the vanishing points and lines depending on the number of the human foot and head points according to the following three cases:

Theoretically, homography estimation for camera calibration requires four 2D coordinates that can solve eight linear equations. However, a practical random sample consensus (RANSAC) based robust camera calibration needs at least eight points such that the calibration error is minimized as shown in Figure 4.

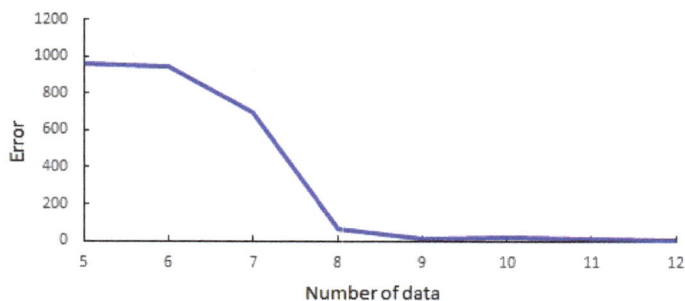

Figure 4. Focal length estimation error depending on the number of foot-head data sets with 30% outliers.

Case 1. If the number of detected foot and head point sets is less than N, the vanishing points and lines are estimated using background structure lines. More specifically, three vanishing points are selected from background lines intersecting points using the RANSAC algorithm. Among three vanishing points, the lowest one is determined as the vertical vanishing point. The line connecting the remaining two vanishing point is determined as the horizontal vanishing line.

Case 2. If the number of detected foot and head sets is more than N but the object motion is linear, the vertical vanishing point can be estimated only using the object foot and head points. The vertical vanishing point is determined at the intersected point of foot-to-head lines as shown in Figure 5. However, if the object moves linearly, estimation of a horizontal vanishing line is impossible since only one horizontal vanishing point is estimated. In this case, the vanishing line is estimated using background structure lines.

Case 3. If the number of detected foot and head sets is more than N and object motion is not linear, vanishing points and lines can be estimated using foot and head points. A foot-to-foot line that connects two foot points and the corresponding head-to-head line that connects two head points are used to estimate the horizontal vanishing points. As a result, the horizontal vanishing line can be estimated using the two horizontal vanishing points as shown in Figure 5.

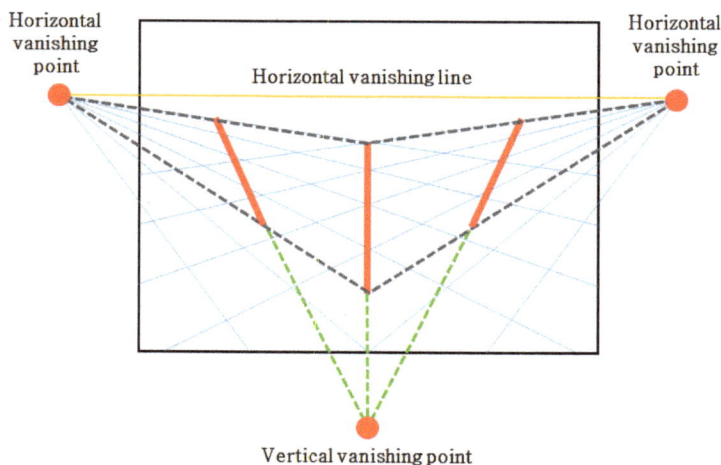

Figure 5. Definition of vanishing points and the horizontal vanishing line.

Camera parameters are calculated using the estimated vertical vanishing point and the horizontal vanishing line as [15]

$$
\begin{aligned}
f &= \sqrt{(a_3/a_2 - p_y)(v_y - p_y)} \\
\rho &= \operatorname{atan}(-v_x/v_y) \\
\theta &= \operatorname{atan}(-\sqrt{v_x^2 + v_y^2}/f) \\
h_c &= h_o / (1 - \tfrac{d(o_h, v_l)\|o_f - v\|}{d(o_f, v_l)\|o_h - v\|})
\end{aligned}
\tag{4}
$$

where f represents the focal length, ρ the roll angle, θ the tilt angle, h_c the camera height, v_l the horizontal vanishing line $a_1 x + a_2 y + a_3 = 0$, $v = \begin{bmatrix} v_x & v_y \end{bmatrix}^T$ the vertical vanishing point, h_o the object height, o_f the object foot point, o_h the object head point, and $d(A, B)$ the distance between a point A and a line B.

3.2. Object Depth Estimation and Occluded Region Detection

The proposed algorithm uses object depth information to detect an occluded region. To estimate the depth of an object, the 2D image coordinate is projected onto the reference plane in the 3D space using a projective matrix. Since the object foot points should be on the ground plane, the proposed algorithm uses the ground plane as the reference plane, which means that the ground plane is considered as the XY plane because the camera height is calculated as the distance between the ground plane and the camera. Using an object foot point in the 2D image, the object foot point on the ground plane in the 3D space can be calculated. To detect the foot point in the 3D space, a foot point in the 2D image is inversely projected onto the 3D space using a projective matrix as

$$
X = \left(P^T P\right)^{-1} P^T x_f
\tag{5}
$$

where x_f represents the foot point in the 2D image, matrix P the projective matrix, and X the inversely projected coordinate of x_f. Inversely projected coordinate X is normalized by the Z-axis value to detect the foot point in the 3D space as

$$
X_f = \frac{X}{Z}
\tag{6}
$$

where Z represents the Z-axis value of X, and X_f the foot point on the ground plane in the 3D space.

An object depth is estimated by computing the distance between the object and camera. However, the foot point appears in the finite position in the input image. For this reason, the proposed algorithm uses the nearest foot point as a pivot point for the object depth estimation. The estimated depth is then normalized using the farthest distance. If an object is far enough from the camera, depth of the object foot point is assumed to be the Y-axis coordinate since the camera pan angle is equal to zero and the pivot point is on the ground plane. If the object depth is equal to the object foot point depth, the object depth is calculated as

$$
d = \frac{\left| Y_f - Y_p \right|}{d_F}
\tag{7}
$$

where d represents the object depth, Y_p the Y-axis value of the pivot point, Y_f the object foot point, and d_F the farthest distance. Figure 6 shows the proposed object depth estimation model, where d_N represents the nearest distance, (X_p, Y_p, Z_p) the pivot point, and (X_f, Y_f, Z_f) the object foot point.

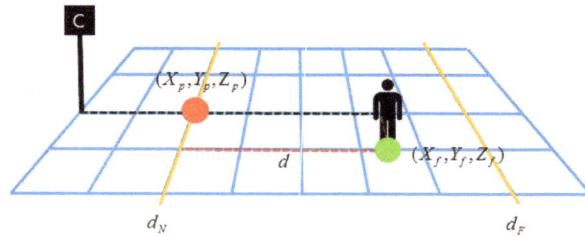

Figure 6. The proposed object depth estimation model.

The proposed algorithm detects the object occlusion using the estimated object depth. The depth of the same object in adjacent frames slowly changes. On the other hand, if the object is occluded, the estimated depth of the object rapidly changes. Based on the observation, object occlusion is detected as

$$O = \begin{cases} \text{true,} & \text{if } |d_{t-1} - d_t| \geq T_O \\ \text{false,} & \text{otherwise} \end{cases} \tag{8}$$

where O represents the object occlusion detection result, d_t the depth of the object at time t, and T_O the threshold value for the object occlusion detection. We can only estimate depth from standing human objects whose feet lie on the reference plane assuming that each object has a uniform depth.

4. Experimental Results

This section shows the results of the proposed automatic camera calibration and object occlusion detection algorithms. For the experiment, test video sequences of resolution 1280 × 720 were acquired using a camera installed at 2.2 to 7.2 m high. In addition to the in-house test sets, Vision and Autonomous System Center's (VASC) stereo dataset is also used to compare the performance of the proposed method with existing stereo matching-based methods [16].

Figure 7 shows the result of three different methods for automatic camera calibration. A ground plane is drawn on the image using grid lines with a 0.5 m interval to show the accuracy of the camera calibration. The background structure-based method makes a poor calibration result because of insufficient, non-parallel line segments and random textures of natural objects as shown in Figure 7a. The moving object-based method degrades the calibration performance because of the incompletely detected moving object and only linear motion of the object as shown in Figure 7b. On the other hand, the proposed method significantly improves the accuracy of camera calibration because it uses neither incomplete background structures nor multiple object positions in the same line as shown in Figure 7c.

Figure 8 shows the result of object depth estimation using the proposed method. Figure 8b shows the calibration result with the superimposed ground plane. The estimated depths of moving objects are shown in Figure 8c.

Figure 9 compares depth estimation results using the stereo matching-based and the proposed methods. Figure 9a,b respectively show the left and right images of the "Toy" in the VASC stereo dataset. Figure 9c shows the stereo matching-based depth estimation result. Figure 9d shows guidelines to detect an region, where red lines represent the objects bottom boundary and blue lines the object's top boundary. Figure 9e shows the superimposed grid of the reference plane that is the camera calibration result. The calibration result using the proposed object depth estimation method is shown in Figure 9f. Although the stereo matching-based method generated many holes in textureless regions without features, the proposed method successfully estimated the continuous depth map.

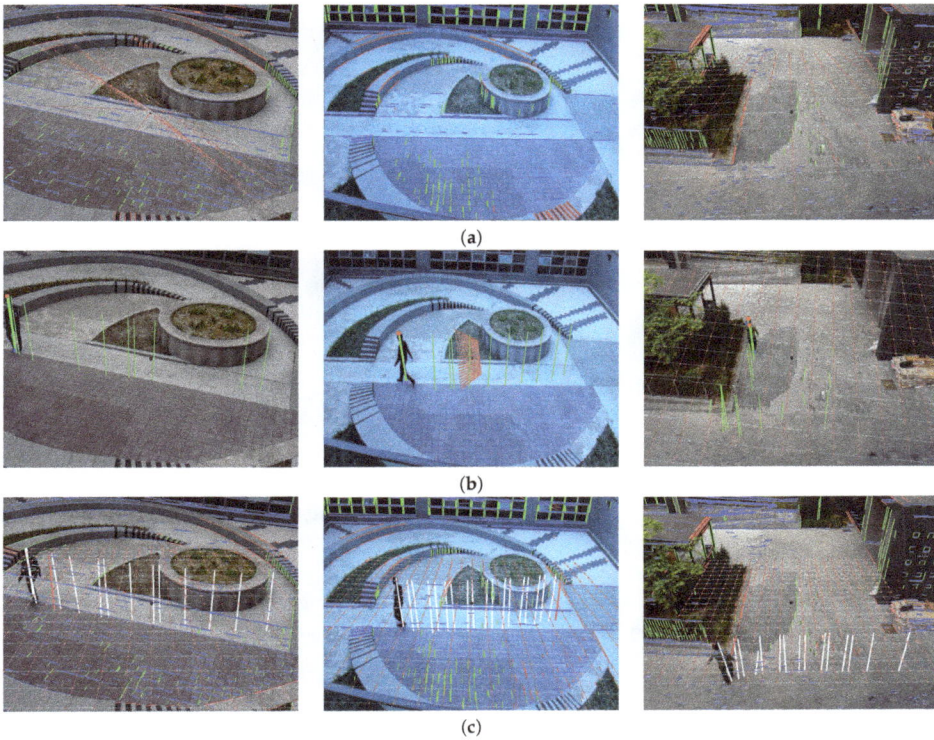

Figure 7. Results of three different method for camera calibration: (**a**) background structure-based method; (**b**) moving object-based method; and (**c**) the proposed camera calibration method.

Figure 8. The results of the proposed object depth estimation: (**a**) input image; (**b**) calibration result in the form of the superimposed grid representing the ground plane; and (**c**) the depth estimation result.

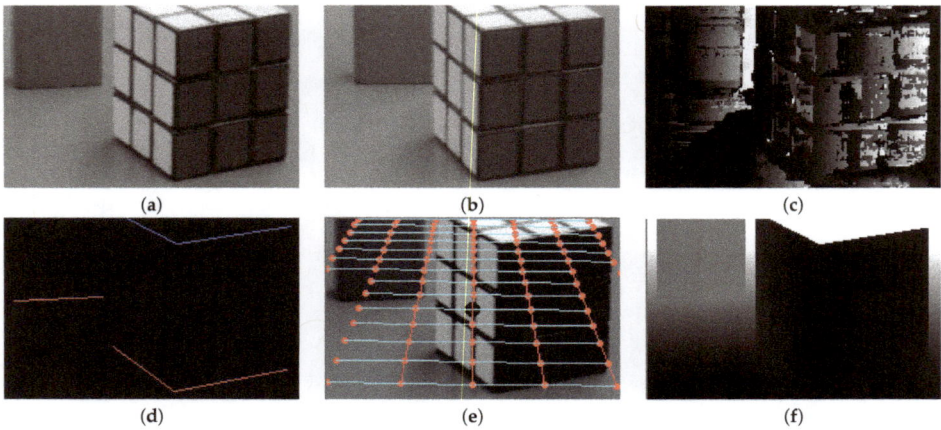

Figure 9. Comparison of depth estimation results using the stereo-based and the proposed methods: (**a**) the left stereo image; (**b**) the right stereo image; (**c**) estimated depth map using the stereo matching-based method; (**d**) guide lines of the left image; (**e**) estimated ground plane of the left image; and (**f**) estimated depth map of the left image using the proposed method.

Figure 10 shows the detection result of an occluded object using the proposed occlusion detection algorithm with the threshold distance of 1.0 m. Figure 10a shows the detection result of the occluded object by a background structure. Figure 10b shows the detection result of the occluded object by another object, and Figure 10c shows the detection result in a different test video. The proposed method can successfully detect occlusion in various test videos. As shown in Figure 10d detection of the y-axis value of an object foot position may results in erroneous detection of occlusion. However, the proposed method can correctly detect occlusion in the scene-invariant manner since it uses the depth in formation in the 3D space.

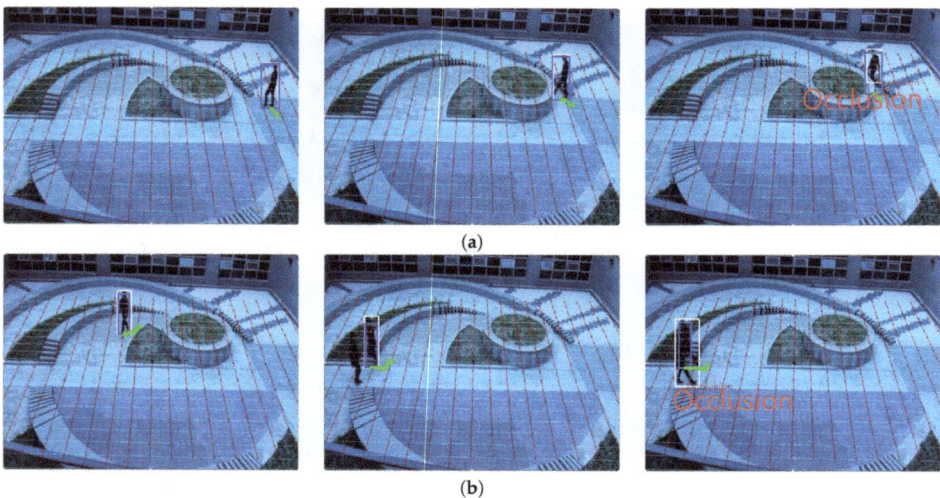

Figure 10. *Cont.*

Figure 10. Results of occlusion detection three selected frames in each video: (**a**) occlusion by background; (**b**) occlusion by another object; (**c**) result of occlusion detection detection in another video and (**d**) result occlusion detection without depth information.

5. Conclusions

In this paper, we presented a fully automatic object occlusion detection method by estimating the object depth from a single uncalibrated camera. The proposed algorithm can robustly calibrate a camera by combining the background structure line components and moving object information. In addition, object depth is estimated using a single RGB camera. As a result, the object occlusion is successfully detected by analyzing the object depth information. The proposed method can be applied to object detection and tracking in a multiple-view surveillance system.

The fundamental assumption of the proposed occlusion detection algorithm is that there is a single, flat ground on which all objects move around. If the ground is not flat or slanted, the estimated depth becomes inaccurate, and as a result, object detection may fail. In that case, the nonflat ground can be approximated by piece-wise flat one, and the slanting ground can be taken care of in the calibration process. In spite of the restrictions, the proposed method is suitable for a wide range of surveillance applications, such as multiple camera video tracking with object handover and normalized metadata generation-based video indexing and retrieval because of its economical implementation.

Acknowledgments: This research was supported by the MSIP (Ministry of Science, ICT and Future Planning), Korea, under the ITRC (Information Technology Research Center) support program (IITP-2016-H8501-16-1018) supervised by the IITP (Institute for Information & Communications Technology Promotion), by Institute for Information & Communications Technology Promotion (IITP) grant funded by the Korea government (MSIP) (B0101-15-0525, Development of global multi-target tracking and event prediction techniques based on real-time large-scale video analysis), and by the Technology Innovation Program (Development of Smart Video/Audio Surveillance SoC & Core Component for Onsite Decision Security System) under Grant 10047788.

Author Contributions: Jaehoon Jung and Inhye Yoon initiated the research and designed the experiments, and Joonki Paik wrote the paper.

Conflicts of Interest: The authors declare no conflict of interest.

References

1. Mei, X.; Ling, H.; Wu, Y.; Blasch, E.; Bai, L. Minimum error bounded efficient l1 tracker with occlusion detection. In Proceedings of the 2011 IEEE Conference on Computer Vision and Pattern Recognition, Providence, RI, USA, 20–25 June 2011; pp. 1257–1264.
2. Zitnick, C.L.; Kanade, T. A cooperative algorithm for stereo matching and occlusion detection. *Pattern Anal. Mach. Intell. IEEE Trans.* **2000**, *22*, 675–684.
3. Sun, J.; Li, Y.; Kang, S.B.; Shum, H.Y. Symmetric stereo matching for occlusion handling. In Proceedings of the 2015 IEEE Computer Society Conference on Computer Vision and Pattern Recognition, San Diego, CA, USA, 20–25 June 2005; Volume 2, pp. 399–406.
4. Matyunin, S.; Vatolin, D.; Berdnikov, Y.; Smirnov, M. Temporal filtering for depth maps generated by Kinect depth camera. In Proceedings of the 3DTV Conference: The True Vision-Capture, Transmission and Display of 3D Video, Antalya, Turkey, 16–18 May 2011; pp. 16–18.
5. Lm, J.; Jung, J.; Paik, J. Single camera-based depth estimation and improved continuously adaptive mean shift algorithm for tracking occluded objects. In Proceedings of the 16th Pacific-Rim Conference on Advances in Multimedia Information Processing, Gwangju, Korea, 16–18 September 2015; pp. 246–252.
6. Huang, Z.; Boufama, B. A semi-automatic camera calibration method for augmented reality. In Proceedings of the IEEE International Conference on System, Man and Cybernetics, Yasmine Hammamet, Tunisia, 6–9 October 2002.
7. Lin, H.Y.; Chen, L.Q.; Lin, Y.H.; Yu, M.S. Lane departure and front collision warning using a single camera. In Proceedings of the 2012 IEEE International Symposium on Intelligent Signal Processing and Communications Systems, New Taipei, Taiwan, 4–7 November 2012; pp. 64–69.
8. Song, Y.; Wang, F.; Yang, H.; Gao, S. Easy to calib: Auto-calibration of camera from sequential images based on VP and EKF. In Proceedings of the 2014 International Conference on Innovative Computing Technology, Luton, UK, 13–15 August 2014; pp. 41–45.
9. Li, B.; Peng, K.; Ying, X.; Zha, H. Vanishing point detection using cascaded 1D Hough transform single images. *Pattern Recognit. Lett.* **2012**, *33*, 1–8.
10. Lv, F.; Zhao, T.; Nevatia, R. Camera calibration from video of a walking human. *Pattern Anal. Mach. Intell. IEEE Trans.* **2006**, *28*, 1513–1518.
11. Cipolla, R.; Drummond, T.; Robertson, D.P. Camera calibration from vanishing points in image of architectural scenes. In Proceedings of the British Machine Vision Conference, Nottingham, UK, 13–16 September 1999; Volume 2, pp. 382–391.
12. Beardsley, P.; Murray, D. Camera calibration using vanishing points. In Proceedings of the British Machine Vision Conference, Leeds, UK, 22–24 September 1992; pp. 416–425.
13. Stauffer, C.; Grimson, W.E.L. Adaptive background mixture models for real–time tracking. In Proceedings of the 1999 IEEE Computer Society Conference on Computer Vision and Pattern Recognition, Fort Collins, CO, USA, 23–25 June 1999; Volume 2.
14. Lutton, E.; Maitre, H.; Lopez-Krahe, J. Contribution to the determination of vanishing points using Hough transform. *Pattern Anal. Mach. Intell. IEEE Trans.* **2002**, *16*, 430–438.
15. Liu, J.; Collins, R.T.; Liu, Y. Surveillance camera autocalibration based on pedestrian height distributions. In Proceedings of the British Machine Vision Conference, Dundee, UK, 29 August–2 September 2011; pp. 1–11.
16. Vision and Autonomous System Center's Image Dataset. Available online: http://vasc.ri.cmu.edu/idb/ (accessed on 28 July 1997).

sensors

MDPI

Article

A Crowd-Sourcing Indoor Localization Algorithm via Optical Camera on a Smartphone Assisted by Wi-Fi Fingerprint RSSI

Wei Chen, Weiping Wang, Qun Li *, Qiang Chang and Hongtao Hou

College of Information Systems and Management, National University of Defense Technology, Changsha 410073, China; weichen@nudt.edu.cn (W.C.); wangwp@nudt.edu.cn (W.W.); qchang@telin.ugent.be (Q.C.); houhongtao@nudt.edu.cn (H.H.)
* Correspondence: liqunnudt@163.com; Tel.: +86-731-8457-3558

Academic Editor: Gonzalo Pajares Martinsanz
Received: 24 January 2016; Accepted: 16 March 2016; Published: 19 March 2016

Abstract: Indoor positioning based on existing Wi-Fi fingerprints is becoming more and more common. Unfortunately, the Wi-Fi fingerprint is susceptible to multiple path interferences, signal attenuation, and environmental changes, which leads to low accuracy. Meanwhile, with the recent advances in charge-coupled device (CCD) technologies and the processing speed of smartphones, indoor positioning using the optical camera on a smartphone has become an attractive research topic; however, the major challenge is its high computational complexity; as a result, real-time positioning cannot be achieved. In this paper we introduce a crowd-sourcing indoor localization algorithm via an optical camera and orientation sensor on a smartphone to address these issues. First, we use Wi-Fi fingerprint based on the K Weighted Nearest Neighbor (KWNN) algorithm to make a coarse estimation. Second, we adopt a mean-weighted exponent algorithm to fuse optical image features and orientation sensor data as well as KWNN in the smartphone to refine the result. Furthermore, a crowd-sourcing approach is utilized to update and supplement the positioning database. We perform several experiments comparing our approach with other positioning algorithms on a common smartphone to evaluate the performance of the proposed sensor-calibrated algorithm, and the results demonstrate that the proposed algorithm could significantly improve accuracy, stability, and applicability of positioning.

Keywords: fingerprint localization; optical camera; image processing; orientation sensor; crowd-sourcing; smartphone

---

1. Introduction

As the demands on location-based services (LBSs) increase, developing positioning systems (PSs) has generated great concern over the last decade; in particular, location information is an important basic feature in LBSs, and it is of great significance to get access to the users' location information anytime and anywhere; in addition, users are more and more concerned about the performance of PSs, such as accuracy and stability. Outdoors, the Global Navigation Satellite System (GNSS), including GPS, GLONASS, Galileo, and BDS, which works well in most outdoor applications, is a major way to get users' location information, and the accuracy is adequate for civilian use. Unfortunately, when users are in indoor environments, there exist some limitations, for example, lack of line of sight and multiple paths, both of which will cause satellite signal attenuation, so GNSS is not capable of providing services with sufficient positioning accuracy. At the same time, the widespread use and popularity of mobile devices have spurred extensive demands on indoor LBSs in recent years, which have greatly promoted the rapid development of entertainment, health, business, and other sectors, for example, users can easily get local information from the LBS when they are arriving at a place that is totally unfamiliar to them; likewise, when consumers want to go shopping in a big mall, when they arrive on-site equipment could send electronic coupons to them for advertising, two applications that without doubt, need to be based on precise indoor location information. To this end, indoor positioning systems have been widely developed, including Assisted GNSS (AGNSS) [1,2], Difference GNSS (DGNSS) [3,4], radio frequency identification (RFID) [5–7], WLANs [8–10], WSNs [11–13], optical sensors [14–16], and cooperative localization [17,18]. On the whole, no system exists that can adapt to the changing environment to meet different demands for localization due to their particular advantages and disadvantages.

With their great international popularity and embedded hardware sensors, smartphones have become the ideal personal and mobile navigator used to implement indoor positioning due to their low cost and self-contained properties. Meanwhile, Wi-Fi infrastructures are more and more generalized, indoor positioning systems based on Wi-Fi fingerprint and smartphones are widely applied, which includes the offline acquisition and online positioning stages. In the offline acquisition stage, the radio signal strength of reference points (RPs) is sampled to build a database, which is always time consuming and laborious; in the online positioning stage, the real-time received radio signal strength is used to locate with methods such as K-nearest neighbors (KNN) [19], K-weighted nearest neighbors (KWNN) [20], Bayesian [21,22], or artificial neural network (ANN) [23]. However, a Wi-Fi signal fluctuates because it is influenced by the complex indoor environment, which results in a low accuracy; besides, if the database of the Wi-Fi fingerprint cannot be updated in time, that is, the stored data cannot remain consistent with the actual signal, this will reduce the positioning accuracy.

Researchers have compared Wi-Fi fingerprint positioning with data from the built-in microelectronic mechanical systems (MEMSs) in smartphones, e.g., accelerometer, gyroscope, magnetometer, to improve Wi-Fi fingerprint positioning accuracy; these algorithms are well known as pedestrian dead reckoning (PDR) methods [20,24]. However, this technique has two drawbacks: on the one hand, it is only able to provide accurate data for a limited time because of the sensor errors arising from bias and noise, especially for the low-cost inertial sensors found in mobile phones, so the accumulating errors grow rapidly with the walking distance of pedestrians, on the other hand, we have tested a common smartphone and found that those sensors' output is unstable and low accuracy, which will influence the positioning performance. Other researchers have also proposed positioning approaches combining Wi-Fi fingerprints and image matching [25–28]; these approaches can solve the problems mentioned above, however, a large-scale images database is required for those approaches, and the image processing and matching task is so time consuming that the required instantaneity is hard to guarantee in general.

Taking those advantages and disadvantages of stand-alone Wi-Fi fingerprints and stand-alone image matching positioning into account, this paper proposes a crowd-sourcing indoor localization algorithm via an optical camera on a smartphone to improve accuracy and instantaneity. First, we use the Wi-Fi signal fingerprint based on the KWNN algorithm to get the K nearest neighbors to make a coarse estimation, Second, we adopt a mean-weighted exponent algorithm with image features extracted by the Scale-Invariant Feature Transform (SIFT) operator and orientation sensor, whose output will constraint random images choice for a mobile node on a smartphone to refine the results based on a multithread mechanism. Compared with the stand-alone SIFT-based image matching positioning method, the proposed algorithm can reduce the search space of images and reduce the position estimation time. Compared with stand-alone Wi-Fi fingerprint positioning, the proposed algorithm can improve the accuracy, moreover, to prove effect of proposed algorithm further, we compare it with the visual positioning system based on the Speed-up Robust Feature (SURF) and KWNN algorithms used in literature [25]. Equally important, we utilize a crowd-sourcing method to update and supplement the database, which will retain consistency with the actual signal and obviously improve the stability of the positioning system. Experimental results show that the proposed algorithm could significantly overcome the disadvantages of stand-alone Wi-Fi fingerprint and stand-alone SIFT-based image matching positioning, and is more effective than SURF combined with KWNN positioning with respect to accuracy and stability. The proposed algorithm has been realized on an Android operating system smartphone, which has good research and application prospects.

The rest of the paper is organized as follows: Section 2 will review recent related works and summarize our contributions. In Section 3, we will present an overview of the proposed sensors-calibrated positioning system, including the offline acquisition and online positioning stages, then the proposed sensors-calibrated algorithm will be introduced in detail in Section 4, and the experimental results and evaluations will be shown in Section 5. Section 6 summarizes the whole paper, then provides our conclusions and an outlook of future work.

2. Related Works

Indoor positioning plays an important role in providing the location of people and objects. There are three principles used in indoor positioning: triangulation, scene analysis, and proximity. Triangulation locates the target point's coordinates by using distances and angles between the target point and three other points with known coordinates, this principle includes Time of Arrival (TOA), Time Difference of Arrival (TDOA), Angle of Arrival (AOA), and Time of Flight (TOF). Scene analysis is a principle of positioning by analyzing the characteristics of the measured values, for example, received signal strength indicator (RSSI) is a popular scene analysis method to realize localization; available signals include Wi-Fi, radio frequency, geomagnetic, visible light, and so on. The proximity principle is mainly used in radio frequency-based positioning systems; in this technique, for a grid of antennas with fixed locations within a building, when a mobile node is detected, the closest antenna or the one receiving the strongest signal is selected to calculate the mobile node's location [29], for example, cellular localization based on Cell-ID is one of the popular applications.

To meet the demanded high accuracy, some researchers have used hardware-based solutions to get a precise distance to locate objects accurately, which is a range-based method. For instance, Seybold [30] calculated the true distances between router and smartphone with a logarithmic-distance path-loss model. Zou [31] applied a weighted path-loss model to RFID signals to implement indoor positioning, on the basis of Zou, Chen [32] adopted a Kalman filter algorithm fusing received strength of Wi-Fi fingerprints, PDR, and landmarks to locate the position indoors. Fidan [33] used a geometric cooperative technique and path-loss model to estimate in real time the time-consuming coefficient of path-loss in ranging sensors, the simulation results indicated that the proposed approach can effectively adapt to uncertainty of the loss factor in signal propagation. Yang [34] and Kotaru [35] measured the phase–distance relationship of the radio signal to locate position, the experimental results show that those methods could reach an accuracy of decimeter level.

However, all of the mentioned solutions based on hardware above have some disadvantages and limitations from the aspect of practicability on a smartphone, for example, it is hard to synchronize the clocks between many common measuring devices, and the outputs of sensors on smartphones have low accuracy and stability in general, which will directly influence positioning performance. Finally, high-quality equipment will lead to increased cost, it is obviously not suitable for a smartphone.

In contrast, other researchers, including Google, Microsoft, Apple, *etc.* used software-based solutions to improve positioning performance, whose advantages are more flexibility, high accuracy, low cost, and short development cycle, and have developed some software-based positioning systems. Among those, Wi-Fi fingerprint is a common range-free way with great popularity, and various methods and theories about it have been researched to solve those issues and limitations mentioned in the Introduction. Schussel [36] and Li [37] used the actual signal and Gaussian distribution process to generate some fingerprints to fill the whole database. Through this approach, the positioning problem in different densities and environments was solved. To reduce computational complexity and positioning accuracy on the online stage, clustering theory has been used; Liu [38] proposed a dynamic constraint KNN algorithm by using indoor layout geometry information to cluster, however, the proposed algorithm required the extraction of features of the specific indoor area, so it is hard to promote its use. Saha [39] used the relative size of RSSI as a metric for clustering the fingerprint and tried to solve the equipment diversity issue in crowd-sourcing. The proposed algorithm can achieve a higher accuracy, but its processing involving clustering is also complicated. Lee [40] proposed a support vector machine-based clustering approach and used the margin between two canonical hyperplanes for classification; it reserved the fingerprint distribution property and obtained a superior decision boundary, but did not take instantaneity into consideration. In addition, the drawback of a time-consuming and laborious database prebuilding in the offline stage also needs to be overcome, the accuracy will decrease if the database is not being updated in time, thus, some researchers proposed a crowd-sourcing mechanism [41,42] to update and supplement the database.

To reduce development cost for the positioning system and increase its practicability, researchers have used smartphones' built-in sensors to locate cooperatively. Keller [43] focused on the gyroscope and barometer in current smartphone to realize the indoor navigation with storey detection, because of their work, a position in three dimensions could be interpolated by the determination of a change of detected height with the knowledge of the relative motion distance or a prior known gradient of the stairs/ramp. Chen [32] utilized a Kalman filter to fuse Wi-Fi and sensors integrated in a phone as well as landmarks to finish locating; Galván-Tejada [44] adopted microphone, magnetometer, and light sensor to locate cooperatively; Shin [45] used Wi-Fi fingerprint and PDR to develop SmartSLAM, which locates pedestrians and constructs an indoor map at the same time. By using off-the-shelf smartphones and the methods mentioned above, they have realized indoor positioning, however, we have tested the output of sensors in common smartphones and found that the output values have low accuracy; even if we put the phone in a stationary state, the output fluctuated, therefore, we think that its low accuracy and stability will influence the performance of positioning systems.

Recent advances in charge-coupled device (CCD) sensor technologies have been applied to positioning and several visual positioning systems have been developed. Considering Wi-Fi fingerprint may have shadow areas where the Wi-Fi signal is weak, Song [46] proposed a new positioning technique by which captured images' features and their coordinate information are stored to form a fingerprint, then they used the SURF algorithm along with Random Sample Consensus (RANSAC) to realize image similarity comparisons. Their method worked both indoors and outdoors, including the shadow areas where Wi-Fi fingerprint signals are weak. However, the time consumption was not discussed. Kawaji [47] utilized omnidirectional panoramic images and a PCA–SIFT algorithm to locate, with the proposed confidence factor locality sensitive hashing; this could improve the image matching performance and speed up image retrieval, however, the method required omnidirectional image acquisition and a higher performance processor, and could not be implemented on a mobile phone. Considering positioning accuracy, time consumption, automation and scalability, Levchev [28] proposed a simultaneous fingerprinting and mapping system, which used a particle filtering algorithm to fuse data sampled from sensors in smart mobile devices. Its Android application could achieve an average localization error of under 2 m, but its map and database generation process was complex and time-consuming; Further, a so-called Mobile Visual Indoor Positioning System (MoVIPS) using smartphone cameras was proposed by Werner [26] in 2011 and extended in 2014 [27]. MoVIPS used a dead reckoning approach and estimated orientation factors from smartphone compasses to reduce the amount of images, then quantized and clustered the SURF feature points, and used a transformation matrix to improve the distance estimation, a making it therefore faster, more precise and more efficient.

Visual positioning systems with image matching theories can improve the accuracy, unfortunately, it is generally necessary to construct an image database to be recognized, and the matching process is always time-consuming; in particular, these algorithms should be improved for use on a smartphone with ideal accuracy and instantaneity. Inspired by the researched literature, we propose our algorithm with the following contributions:

(1) In the preliminary positioning stage, we use a KWNN-based Wi-Fi fingerprint positioning algorithm to make a coarse estimation, then select K images corresponding to K-nearest neighbors as retrievable images to reduce the time spent in retrieving images.

(2)   After Wi-Fi fingerprint positioning, we introduce an orientation matching factor calculated using a three-dimensional orientation sensor on a smartphone to constrain random images choice for a coarse estimated location, then, we use the SIFT operator to extract image features to calculate the image matching factor on a multithread mechanism.

(3)   We propose a mean-weighted exponent algorithm to improve accuracy by fusing the KWNN algorithm, orientation matching factor, and image matching factor.

(4)   We use crowd-sourcing to update the database; users can upload their positioning results to a server, which can aids others' positioning the next time if the database has not recorded some RPs information in the offline stage.

(5)   Our algorithm has been realized in an Android operating system smartphone, and is easy to implement and apply. There is no need to use additional hardware assistance and the costs are greatly reduced.

3. Overview of the Proposed Sensors-Calibrated Positioning System

We propose a crowd-sourcing indoor localization algorithm via an optical camera and orientation sensor on a smartphone, with the purpose of simplifying indoor localization with high accuracy through the use of a Wi-Fi module and camera as well as the orientation sensor in the smartphone, and it does not need to be assisted by other hardware. The proposed system includes two stages: offline acquisition and online positioning stages. In the offline stage, the smartphone will scan available access points (APs) at different RPs, then record the RP's coordinates, AP's MAC address, and its RSSI values, while recording the front image of RPs and output values of the three-dimensional orientation sensor. In this paper, we use a relational database whose format is shown in Table 1 to store those data, and those images will be preprocessed in a server using the SIFT operator and extracting the features vector.

Table 1. Storage format of the database.

Coordinate	AP Number	AP_1	AP_2	...	AP_m	Image	Orientation Sensor Data
(x_1, y_1)	m_1	$MAC_{m1} =$ $RSSI_{m1}$	$MAC_{m1} =$ $RSSI_{m1}$	...	$MAC_{m1} =$ $RSSI_{m1}$	1	$\{\theta_{x1}, \theta_{y1}, \theta_{z1}\}$
(x_2, y_2)	m_2	$MAC_{m2} =$ $RSSI_{m2}$	$MAC_{m2} =$ $RSSI_{m2}$	...	$MAC_{m2} =$ $RSSI_{m2}$	2	$\{\theta_{x2}, \theta_{y2}, \theta_{z2}\}$
(x_3, y_3)	m_3	$MAC_{m3} =$ $RSSI_{m3}$	$MAC_{m3} =$ $RSSI_{m3}$	...	$MAC_{m3} =$ $RSSI_{m3}$	3	$\{\theta_{x3}, \theta_{y3}, \theta_{z3}\}$
$\vdots$	$\vdots$	$\vdots$	$\vdots$	$\vdots$	$\vdots$	$\vdots$	$\vdots$
(x_n, y_n)	m_n	$MAC_{mn} =$ $RSSI_{mn}$	$MAC_{mn} =$ $RSSI_{mn}$	...	$MAC_{mn} =$ $RSSI_{mn}$	n	$\{\theta_{xn}, \theta_{yn}, \theta_{zn}\}$

In the online positioning stage, users open the Wi-Fi and scan accessible APs, record the MAC address and RSSI values in the cache of the phone, then users can choose to use Wi-Fi fingerprint positioning based on the KWNN algorithm to perform rough localization. In order to achieve higher accuracy, we use the built-in sensor to calibrate the positioning result of Wi-Fi fingerprint. The framework of this approach is shown in Figure 1. Users can open the camera to take a picture, and the three-dimensional orientation values will be recorded in the cache simultaneously. The features vector are extracted from the captured images and recorded temporarily in the phone cache. The orientation values will be used to calculate the orientation factor, which can restrain random image choices at a position. The temporary features vector and stored features vector corresponding to the K nearest neighbors are used to calculate the image matching factor, then the mean-weighted exponent algorithm will fuse KWNN, orientation matching, and image matching results to improve the accuracy further. At the same time, the users can upload the results and those temporarily stored data including MAC address, RSSI, image feature, and orientation values into a server to update the database and aid better location estimate of others the next time.

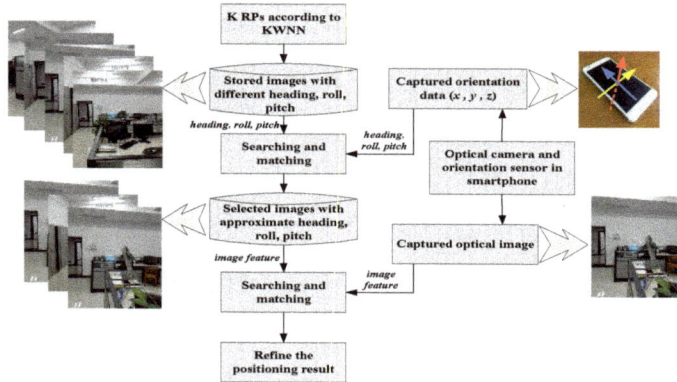

Figure 1. Framework of positioning calibration using built-in sensors in a smartphone.

Table 1 shows the recording format of RPs' coordinates, Wi-Fi fingerprint, captured image sequence number, and output values of the orientation sensor. The coordinates are specified by the user when building the database in the offline stage, but when using crowd-sourcing to update, they are calculated by our proposed algorithm. The AP number means the total number of available APs when users scan Wi-Fi, it is worth noting that the AP number of different RPs are not all equal because of the changing environment at different places, thus, m_1, m_2, ..., m_n in Table 1 are not all equal. AP_1–AP_m record the received APs' MAC and RSSI; when making a coarse estimation, the KWNN algorithm only calculates the distance of RSSI with the same MAC address. Image sequence number is recorded for 1 to n, and the image features are stored in another data table named as the corresponding sequence number of images. Because of the random direction when users take a picture, namely, it is likely that users take pictures at the same place but the direction of the image is different, so in this paper we use orientation factor to restrain this case of randomness.

4. Crowd-Sourcing Localization Assisted by Wi-Fi Fingerprints

4.1. Wi-Fi Fingerprint Positioning Based on KWNN

K-weighted nearest neighbors is a version of the K-nearest neighbors (KNN) algorithm improved by introducing a weighted distance factor. If the RPs' distribution is sparse or non-uniform, after calculating the distances and selecting K nearest neighbors, the classification result of KNN is worse than that of KWNN. Torres-Sospedra's [8] extensive experimental results showed that Euclidean distance is not the optimal choice to describe the similarity between fingerprints. On the contrary, the Sorensen distance function can achieve a better accuracy than Euclidean distance. Therefore, in this paper, we choose the Sorensen function for the KWNN algorithm. The definition of the Sorensen distance function is:

$$L_{(d,i)} = \frac{\sum_{i=1}^{n} \left| RSSI_{MN}^{(i,n)} - RSSI_{RP}^{(i,n)} \right|}{\sum_{i=1}^{n} \left| RSSI_{MN}^{(i,n)} + RSSI_{RP}^{(i,n)} \right|} \tag{1}$$

where i is a sequence of RPs, n denotes the number of APs with the same MAC between mobile node and RPs, MN denotes mobile node, RP denotes a reference point, then, the K-nearest distance will be selected as the nearest neighbors. Finally, the distances corresponding to the K neighbors are used to calculate the normalized weights using [20]:

$$\omega_i = \frac{1/L_{(d,i)}}{\sum\limits_{i=1}^{K}(1/L_{(d,i)})} \tag{2}$$

According to the normalized weight and RPs' coordinates, a coarse estimation using the KWNN-based Wi-Fi fingerprint can use the following formula [20]:

$$P_i(x,y)_{KWNN} = \left\{ \sum_{i=1}^{K} x_i \cdot \omega_i, \sum_{i=1}^{K} y_i \cdot \omega_i \right\} \tag{3}$$

In KWNN, it is well known that K is a critical parameter that will directly influence the positioning accuracy, so we have explored that and the results are shown in Figure 2. Each K was simulated 1000 times to calculate the mean error of positioning; we can see from Figure 2 that K = 3 is the optimal choice whose mean error is about 1.59 m, so we set K as 3 in this paper, which will determine the number of retrievable images.

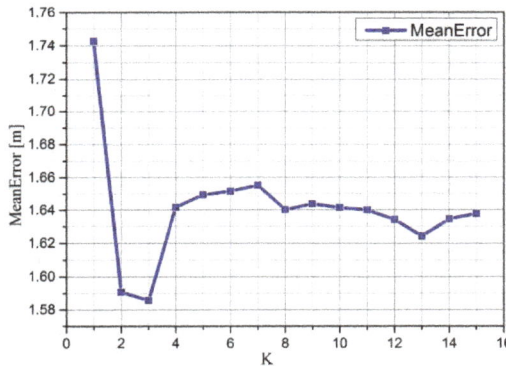

Figure 2. The K value's impact on positioning.

4.2. Precise Localization Calibrated with Built-in Sensors on a Smartphone

At the image matching positioning stage, the images and three-dimensional orientation sensor data captured via the user's smartphone will be matched with the stored data according to the K nearest neighbors. This process can be described like this: first, we calculate the angle difference between real-time obtained orientation values and stored values, and the results are expressed as orientation matching factors, afterwards, the captured image will be processed to calculate the image matching factor. It is worth noting that the orientation matching factors are three-dimensional vectors, but the image matching factors are 1 × K vectors, where K is the parameter in the KWNN algorithm. Finally, we will utilize the mean-weighted exponent algorithm to correct further the estimated results of the Wi-Fi fingerprint. More pertinent details will be introduced in the following sections.

4.2.1. Three-Dimensional Orientation Matching

The orientation sensor is used to detect motion pose such as azimuth, pitch and roll. We show the 3D model of orientation sensor in Figure 3, it is software-based and derives its three-dimensional floating data from the accelerometer sensor and the geomagnetic field sensor using rotation matrix. Since this whole process has been packaged into the Android API, we can use it directly.

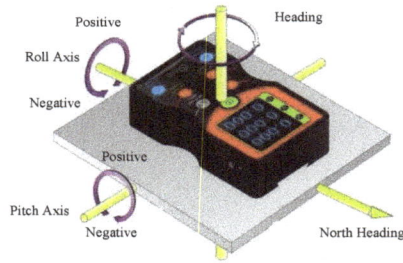

Figure 3. 3D model of an orientation sensor.

We use the orientation sensor in a smartphone to improve the image matching performance, so we should know its coordinate system shown in Figure 4a, where the z-axis represents azimuth or heading. When the phone rotates around the z-axis, the first element in the three-dimensional floating data, whose range is 0–360°, indicates the angle between the geomagnetic field and the phone's y-axis, so if the value is equal to 0°, 90°, 180°, or 270°, it means the heading is north, east, south, or west, respectively, and we show this case in Figure 4b. The x-axis represents pitch. When the phone rotates around the x-axis, the second element in the three-dimensional floating data, whose range is −180° to +180° indicates the angle between the geomagnetic field and the phone's z-axis, specifically, when the top of the phone is up, the angle changes from 0° to −180°, when the bottom of the phone is up, the value changes from 0° to 180°. This case is shown in Figure 4c. The x-axis represents roll. When the phone rotates around the y-axis, the second element in the three-dimensional float data, whose range is −90° to +90° indicates the angle between the geomagnetic field and the phone's x-axis, specifically, when the left of the phone is up, the angle changes from 0° to −90°, when the right of phone is up, the value changes from 0° to +90°. This case is shown in Figure 4d.

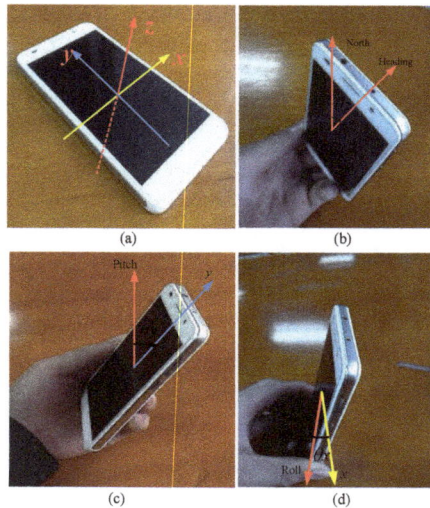

Figure 4. (**a**) The coordinate system in a smartphone; (**b**) heading indication of a smartphone; (**c**) pitch indication of a smartphone; (**d**) roll indication of a smartphone.

In our algorithm, when the user takes a picture, the output values of the orientation sensor and the image will be recorded into the database simultaneously. The angle difference between real-time obtained orientation values and stored values can improve the stability of image matching performance, namely, the values are used to constrain random image choices for a coarse estimated location. A smaller angle difference indicates that the direction of the user and the pose of the smartphone are closer to the corresponding RPs. In this paper, we define the orientation matching factor to measure the closeness as follows: assume that the real-time obtained three-dimensional orientation values are represented as $\{\theta x, \theta y, \theta z\}$, the stored values acquired from the K nearest nodes are denoted as $\{\}$, then, the angle difference can be calculated using the following formula:

$$\Delta\theta_i = \sqrt{(\theta x - \theta x_i)^2 + (\theta y - \theta y_i)^2 + (\theta z - \theta z_i)^2} \tag{4}$$

Finally, the orientation matching factor is the normalized weights of the reciprocal value of the angle differences:

$$\zeta_i = \frac{1/\Delta\theta_i}{\sum\limits_{i=1}^{K} (1/\Delta\theta_i)} \tag{5}$$

where $i = 1, 2, \ldots , K$ indicates the sequence of RPs in Equations (4) and (5).

4.2.2. Image Feature Matching by an Optical Camera

Image feature matching has been researched for indoor localization for years, and image features can be extracted with many algorithms, for example, SIFT, PCA-SIFT, SURF. Niu [25] and Luo [48] compared SIFT and SURF from the perspective of time, scale, rotation, blur, illumination and affine. We have tested this in our experiments and it proves that SURF is more than three times faster than SIFT, however, it is not optimal to handle scale changes, rotation transformation and blur; a performance comparison of those three changes is shown in Figure 5.

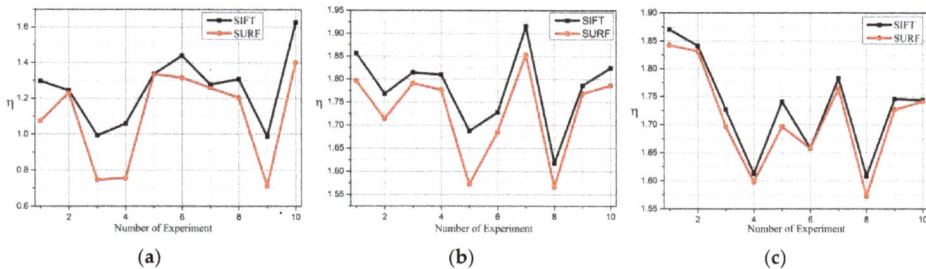

Figure 5. (**a**) the impact of scale changes on image matching factor; (**b**) the impact of rotation on image matching factor; (**c**) the impact of blur on image matching factor.

Considering that there exists scale, rotation and blur changes when users take photos in many cases, thus, we choose the SIFT algorithm proposed by Lowe [49] to extract image features. Extracted features can be expressed as $N \times 128$-dimensional feature vectors and stored in a database, where N denotes the total number of obtained feature points. The image captured on a smartphone by users is processed by the same algorithm, then we calculate one by one the Sorensen distances between the stored feature vectors and the real-time captured ones, we calculate the ratio by comparing the distance of the closest neighbor to that of the second-closest neighbor as the metric for similarity of images. Lowe demonstrated that a metric threshold of "feature distance" between 0.4 and 0.6 is optimal when the two matching images change with scaling change and rotation transformation [49], further, we have found in our many experiments that when the threshold is equal to 0.55, the positioning mean error is smallest, the result is shown in Section 5.2.1. The process of image matching is as follows: it is based on the condition that we have obtained K images after coarse positioning with the KWNN-based Wi-Fi fingerprint, where K is the number of nearest neighbors. As mentioned above, the amount of database images in the candidate set has been reduced and the randomness of images choice have been restricted with the orientation sensor. To decrease the computational time, K images will be processed in a multithread mechanism.

Those K images corresponding to K nearest neighbors in the database can be expressed as $I_{(DB,i)}$ ($i = 1, 2, ..., K$), and feature vectors are indicated as $F_{(DB,i)}$ ($i = 1, 2, ..., K$), the image captured by users is represented as Is and its corresponding feature vectors can be donated as Fs, then we use the following formula to calculate the common feature vectors Fc between the captured image and stored images:

$$Fc = (F_{(DB,i)} \cap Fs) \in [0, min(F_{(DB,i)}, Fs)] \tag{6}$$

where $i = 1, 2, ..., K$. After that, two ratios between those images can be calculated, there are $((F_{(DB,i)} \cap F_s)/F_{(DB,i)})$ and $((F_{(DB,i)} \cap F_s)/F_S)$. Accordingly, we use the two ratios to calculate the image matching factor. This factor can be defined as:

$$\eta_i \Rightarrow arg\ max(\frac{F_{(DB,i)} \cap Fs}{F_{(DB,i)}} + \frac{F_{(DB,i)} \cap Fs}{Fs}) \tag{7}$$

where $i = 1, 2, 3$ denotes the index of the selected images. The image in the database that should be picked up if it maximizes the image matching factor, as we can see, if the two images don't contain common feature points, the η is 0, if the two images match totally, the η is 2, so the range of η values is between $0.0 \leqslant \eta \leqslant 2.0$, and a larger η indicates a greater similarity between two images.

4.2.3. Precise Localization Using Mean-Weighted Exponent Algorithm

In this section we will introduce our fusion method, which we name the mean-weighted exponent algorithm, to calibrate the estimated location through the KWNN algorithm. The reason why we chose an exponent function is due to its nonlinearity, namely, an exponent function has a more significant change rate compared with a linear function.

According to the selected K RPs, we have calculated the distance factor ω by KWNN, and the orientation matching factor ξ as well as image matching factor η through the proposed algorithm. With those critical parameters, we fuse them by calculating the sum, which will be used as the power of the exponent function. Similar to the previous principle, the normalized correction factor after fusion is represented as:

$$\xi'_i = \frac{e^{(\omega_i + \lambda\eta_i + \gamma\zeta_i)}}{\sum\limits_{i=1}^{K} e^{(\omega_i + \lambda\eta_i + \gamma\zeta_i)}} \tag{8}$$

where K is the number of nearest neighbors, $i = 1, 2, \ldots, K$, λ and γ denote adjustment parameters. In this paper, we set both λ and γ to 1. Finally, the location of the user improved by the camera and orientation sensor on the smartphone is estimated by utilizing the following formula, similar to Equation (3):

$$P_i(x, y)_{Fianl} = \left\{ \sum_{i=1}^{K} x_i \cdot \xi'_i, \sum_{i=1}^{K} y_i \cdot \xi'_i \right\} \tag{9}$$

Equally important, after calculating the orientation matching and image matching factors, it should be noted that the orientation matching factors are three-dimensional vectors, but the image matching factors are $1 \times K$ vectors, where K is the parameter in the KWNN algorithm. Then, we first set thresholds to judge the maximum value of those factors, if the maximum values are both greater than the two thresholds, this indicates that the location of the reference point corresponding to that maximum value can be used as the final location of the user. In this paper, they are equal to 1.8 and 0.8.

4.3. Updating the Database by Crowd-Sourcing

One of the disadvantages of existing Wi-Fi fingerprint positioning is the very high costs in time and manpower during the offline acquisition and calibration phases. To overcome these limitations, our proposed algorithm supports users in uploadiang their scanned Wi-Fi data and captured images into a server to update the database, that is, crowd-sourcing, Estelles-Arolas [50] overviewed this and defined the idea as "a new online distributed production model in which people collaborate and may be rewarded to complete tasks" [51], in our design, we always assume the users are willing to upload the right results into the database. Apart from the fact that it can greatly reduce the complexity of building the database, we can find other advantages from our experiment in that the accuracy of both the Wi-Fi fingerprint and proposed sensors-calibrated system positioning results are higher and more stable as the amount of data in the database increases, as the results in Section 5.2.3 show, so users contributing to crowd-sourcing can get a more precise and stable results. Moreover, it can aid others to locate the next time if the database has not recorded some RPs information in the offline stage, in this way, it also decreases the manpower cost.

5. Experimental Results and Evaluation

5.1. Experimental Setup

To validate our proposed algorithm, we have developed an application on an Android smartphone. Our experiment was performed on the fourth floor at the College of Information System and Management of NUDT, with an area of around 1250 m^2 containing six fixed APs, the layout of floor and RPs installation are shown in Figure 6. During the experiment, we scan the Wi-Fi signal RSSI and record the captured front images and three-dimensional orientation values to build the database shown in Figure 7. We did our best to try to capture fixed icons, such as signs, house numbers, *etc.* We collected 150 RPs' information in the offline stage in the first round. In our experiment, our experimental hardware included: a TP-link router, ThinkPad E420 with Win 7 64-bit operating system, Intel(R) i3-2310M CPU 2.10 GHz 4 cores and 4 G memory. The smartphone is a Huawei Honor 6 (H60-L01) mobile 4G, 16 G ROM whose sensors parameters are listed in Tables 2 and 3 moreover, we used a Kingston 8 GB micro SD memory card to store data in the phone. We developed our software using Android Developer Tools v 21.1.0.

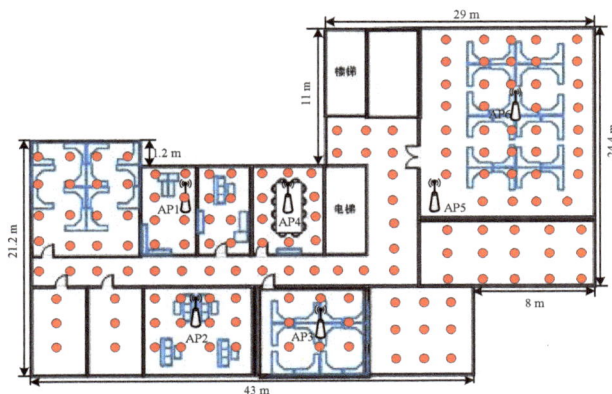

Figure 6. Layout and RPs.

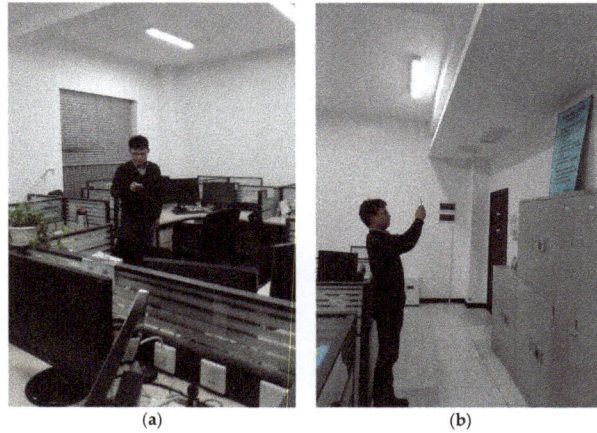

(a)   (b)

Figure 7. Scanning WiFi fingerprint (**a**) and capturing images (**b**).

Table 2. Parameters for the 3D orientation sensor in the Huawei Honor 6.

Name	Supplier	Resolution	Heading		Pitch		Roll	
			Range	Accuracy	Range	Accuracy	Range	Accuracy
Inemo	STmicroelectronics	0.01°	0~360°	0.1°	−90°~+90°	0.1°	−180°~+180°	0.1°

Table 3. Parameters for the optical camera in the Huawei Honor 6.

Type	Orientation	Focal Length	View Angle	Size
BSI CMOS	90°	3.79 mm	63°	1920 × 1080

The simple interface of the app and positioning result are displayed in Figure 8. The red point represents the current location, and a prompt message is at the top.

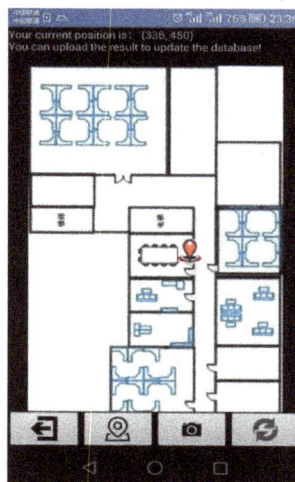

Figure 8. Display interface and positioning result.

Four buttons are shown in the bottom, from left to right, their function is: exit, Wi-Fi positioning, image matching positioning, and upload data to the server. Clicking the Wi-Fi positioning button uses the KWNN algorithm to estimate the position for users, After the KWNN algorithm implements the positioning, clicking the image matching positioning opens the camera automatically and further corrects the roughly estimated result using our proposed algorithm. Clicking the upload data button uploads the data and the refined localization in the phone cache into the database. Clicking exit exits our positioning application software. Initially, the user can point to a location on the map and upload the current data into the server to build a database.

5.2. Experimental Results and Performance Evaluation

5.2.1. Image Matching Evaluation

An experiment is conducted in the meeting room shown in Figure 9, in which we display the captured image and matched images (a), (b), and (c) through KWNN, as well as corresponding orientation vectors. The red pentagram indicates the current position and its captured image, scanned APs' RSSI value are $\{-63,-44,-85,-94,-89,-86\}$, the other three RSSI values stored in the database are $\{-67,-46,-90,-80,-81,-86\}$, $\{-73,-38,-85,-80,-83,-80\}$, and $\{-75,-45,-88,-82,-88,-81\}$. The captured orientation sensor data is $\{25.62,-70.69,-1.69\}$, the other three stored in the database are $\{29.18,-65.32,-3.98\}$, $\{68.61,-84.61,-0.51\}$ and $\{347.52,-81.46,0.2\}$. In our experiment, we compressed all images with the size of 1034×731, the image captured by the user at the online stage is processed with the SIFT operator and the number of extracted features $Fs = 714$. We list the other three images' features and common features in Table 4, meanwhile, we calculate those factors using the methodology described in Section 4.

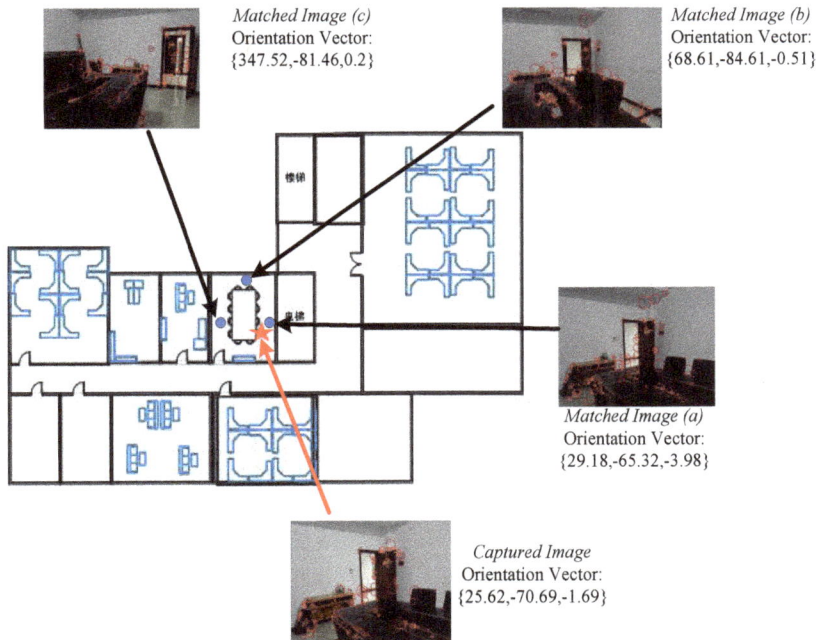

Matched Image (c) Orientation Vector: $\{347.52,-81.46,0.2\}$

Matched Image (b) Orientation Vector: $\{68.61,-84.61,-0.51\}$

Matched Image (a) Orientation Vector: $\{29.18,-65.32,-3.98\}$

Captured Image Orientation Vector: $\{25.62,-70.69,-1.69\}$

Figure 9. Comparison of image and orientation matching.

Table 4. Feature number and matching factors.

index	$F_{(DB,i)}$	F_c	η_i	ζ_i
(a)	659	647	1.888	0.853
(b)	643	563	1.664	0.128
(c)	628	489	1.464	0.018

It is well known that Lowe demonstrated that a metric threshold for Euclidean distance between 0.4 and 0.6 is optimal [49,52]. In this case the probability density functions for correct matches are obviously higher than for incorrect matches. In our experiment, we have explored the metric threshold's impact on the positioning accuracy of SURF and SIFT algorithm, it worth noting that "Wi-Fi+SURF" means the method using Wi-Fi fingerprint to make a coarse estimation and SURF-based image matching algorithm to refine the positioning results, which was used in [25], "Wi-Fi+SURF" is the proposed algorithm in our paper. The results are shown in Figure 10, where we selected 150 RPs in the database, and located each group 200 times with different thresholds. As we can see clearly, the mean error is lowest when the threshold is set to 0.55 both of the two algorithm, however, the mean error increases significantly when the threshold is greater than 0.75, the accuracy becomes worse, consequently, we set the threshold to 0.55 when using feature vector to match images and realize positioning in the online stage.

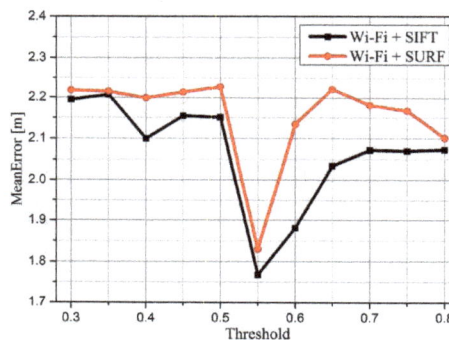

Figure 10. Different thresholds' impact on positioning accuracy.

5.2.2. Positioning Evaluation

To validate the accuracy and correctness of the proposed algorithm with calibrated sensors, we have performed an experiment with 150 RPs in the database. The Cumulative Distribution Function (CDF) and root mean square error (RMSE) of the stand-alone KWNN-based Wi-Fi fingerprint algorithm, stand-alone SIFT-based positioning algorithm, SURF combined with Wi-Fi positioning algorithm and proposed sensors-calibrated algorithm are shown in Figure 11. In Figure 11a, we count the errors of 200 positioning events and we can see that positioning accuracy could be greatly improved by the sensors-calibrated algorithm compared with other approaches, for example, the error probability distribution of the KWNN-based algorithm under 1 m is about 12%, under 2 m is about 53.5%, these results demonstrate that the accuracy of the KWNN-based algorithm is not high enough. However, with correction of optical image and orientation sensor matching, this probability distribution increases to 43%, the probability under 2 m is about 80%. The stand-alone SIFT-based algorithm also has a high error probability distribution, and these results demonstrate that the accuracy of the KWNN-based algorithm is improved by visual positioning algorithm obviously (*i.e.*, the sensors-calibrated algorithm). Significantly, the correctness of the KWNN directly impacts the accuracy of the sensors-calibrated algorithm. The RMSEs of different algorithms are shown in Figure 11b, where the RMSE of the sensors-calibrated algorithm is between 1.2 m and 1.5 m, less than the RMSE of the other three positioning algorithms. The stand-alone SIFT-based and "SURF+Wi-Fi" method have close RMSE results, while the stand-alone KWNN-based algorithm is worse, ranging from 2.1 to 2.4. We can conclude that our proposed sensors-calibrated algorithm has better performance than the others and the estimated results are closer to the true location after calibration.

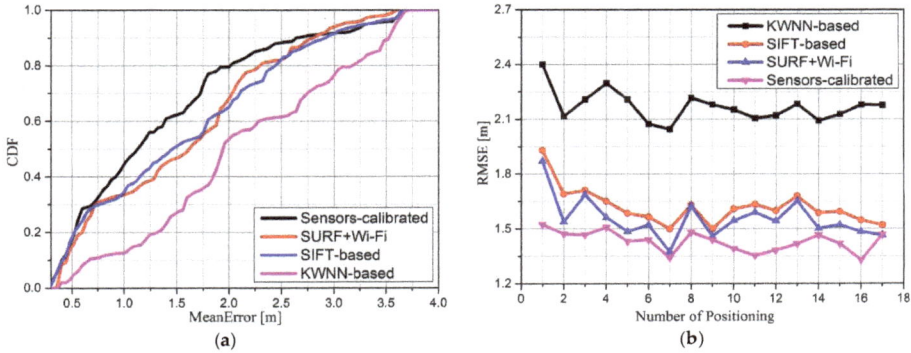

Figure 11. Performance comparison of our sensors-calibrated algorithm with other approaches.

5.2.3. Performance of Crowd-Sourcing for Positioning

We use the error CDF and accuracy as the standard to evaluate the performance of crowd-sourcing. In our small-scale experiment, five people participated in uploading data on the fourth floor at the College of Information System and Management of NUDT. We assume users participate actively, uploading positioning results freely and honestly, that is, there is no unreliable data in the experiment. In this section, we select compare the performance of the stand-alone KWNN algorithm and our proposed algorithm. The number of items in the database is set to 50, 100, 150, 200, 250, 300, and 350 to simulate the increase in the database size through crowd-sourcing. The results are shown in Figure 12.

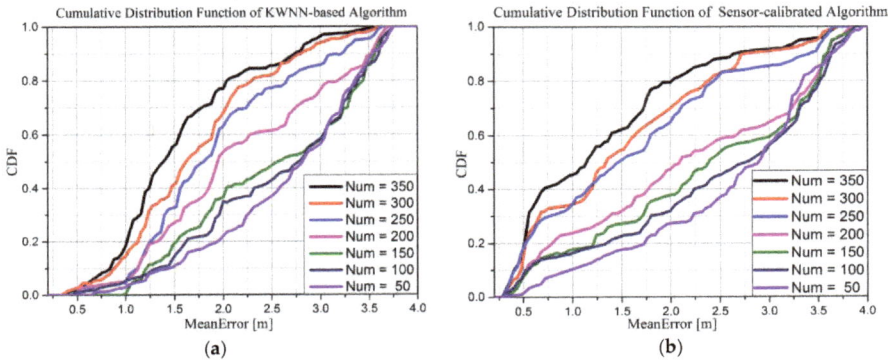

Figure 12. Performance comparison of crowd-sourcing.

Figure 12a,b show the error CDF of the KWNN-based and the proposed sensors-calibrated algorithm, respectively. From Figure 12 we can conclude that:

(1) As the number of RPs in the database increases, the accuracy of the KWNN-based and our sensors-calibrated algorithm are both improved.

(2) The sensors-calibrated algorithm has a better performance. For example, when the number of RPs is equal to 350, the mean error under 1 m of KWNN-based algorithm is about 20%, but, the mean error under 1 m of the proposed algorithm is about 46%, an obvious improvement.

(3) For the same database, if the estimated results of the KWNN-based algorithm are more precise and more stable, which will choose better nearest neighbors, so finally, the accuracy of our sensors-calibrated algorithm will be better.

The RMSE and mean error of the stand-alone KWNN-based Wi-Fi fingerprint algorithm, stand-alone SIFT-based positioning algorithm, SURF combined with Wi-Fi positioning algorithm and proposed sensors-calibrated algorithm are shown in Figure 13, where the mean error indicates the closeness between the average value of a number of measurements and an accepted reference value, namely, trueness.

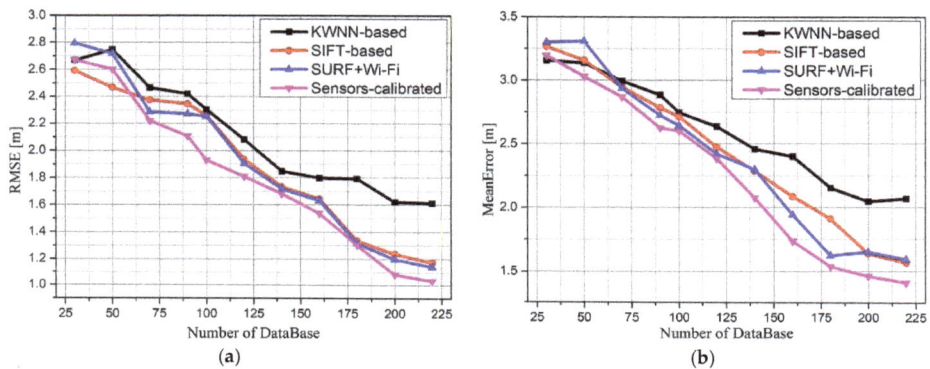

Figure 13. RMSE and mean error comparison of crowd-sourcing with different algorithms.

The RMSE indicates a measure of the system's combined trueness and precision, namely, accuracy. Each group was performed 200 times. Figure 13a compares the RMSE of these four algorithms, Figure 13b compares the mean errors of these four algorithms. We can conclude the following from the shown results: as the number of RPs increases, the RMSE of the KWNN-based algorithm decreases from 2.8 m to 1.6 m, reducing by 42.9%. The sensors-calibrated algorithm decreases from 2.7 m to 1.02 m, reducing by 62%, stand-alone SIFT image matching algorithm and SURF combinated with Wi-Fi algorithm decrease 54.2% and 57.9%, respectively.

The RMSE of the sensors-calibrated algorithm is less than that of the other three algorithms, which means the position from the sensors-calibrated algorithm is closer to the true location, while the stand-alone SIFT-based algorithm and SURF combined with Wi-Fi algorithm results are approximate.

From Figure 13b, we can see that the mean error of visual positioning algorithm is less than that of KWNN-based Wi-Fi fingerprint positioning, and with the coarse estimation by stand-alone KWNN-based algorithm, the proposed sensors-calibrated algorithm is less than those of the other three algorithms as the number of RPs increases, and the rate of increase of the proposed algorithm is larger than that of the others, which indicates the sensors-calibrated algorithm has a better performance.

We can use data variance to evaluate the stability of positioning. Positioning variance comparisons of the different algorithms are shown in Figure 14. As we can see, when the number of RPs increase, the variance of the four algorithms decreases, however, the positioning variance of the proposed sensors-calibrated algorithm is less than the variance of other algorithms, stand-alone SIFT-based algorithm and SURF combined with Wi-Fi algorithm are approximate, therefore, the results demonstrate that the proposed algorithm is more stable than KWNN.

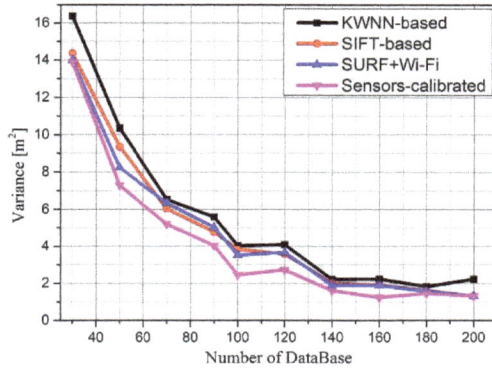

Figure 14. Positioning variance comparison.

5.2.4. Performance Comparison of Time Consume for Positioning

Time consumption is an important aspect for positioning systems, and it's well known that the SURF algorithm is faster than the SIFT algorithm. In our proposed sensors-calibrated algorithm, we use a multithread mechanism to extract features of the K stored images corresponding to the K nearest neighbors in parallel by using the SIFT operator, so the proposed algorithm's time consumption for estimating positions of is not worse than SURF, but the time consumption of a stand-alone SIFT-based algorithm is almost three times that of our proposed sensors-calibrated algorithm, and the position estimation time increases greatly as the image database increases in size, thus, we show the results of stand-alone KWNN-based, SURF combined with Wi-Fi algorithm and the proposed algorithm (*i.e.*, SURF+Wi-Fi) in Table 5.

Table 5. Time-consuming comparison of different algorithms.

Method	1	2	3	4	5	6	7	8
KWNN-based/ms	18.33	16.14	22.57	25.44	28.16	27.29	28.19	24.18
SURF + Wi-Fi/ms	1834.23	1898.58	1833.12	1777.30	1747.72	1716.91	1747.72	1908.65
Proposed/ms	1739.45	1734.18	1731.09	1699.88	1702.02	1691.47	1688.45	1743.09

As we can see, the time consumption of the KWNN-based algorithm is the least; the average is 23.79 ms. As we put more emphasis on the time consumption of image matching positioning after coarse estimation by the KWNN-based algorithm, with the help of the multithread mechanism, the average time consumed by our proposed algorithm is 1716.20 ms. At the same time, the time consumption of the image matching approach based on SURF is 1808.03 ms, so the results demonstrate that high computational cost of image matching positioning have been reduced greatly by our sensors-calibrated algorithm.

6. Conclusions

Aiming at solving the imprecise and time-consuming outcomes in Wi-Fi fingerprint and image matching positioning, respectively, in this paper we propose a crowd-sourcing indoor localization algorithm via the optical camera and orientation sensor in a smartphone. Real-case experimental results of different methods demonstrate that the sensors-calibrated algorithm has a better accuracy and stability performance, compared with three other algorithms (*i.e.*, stand-alone KWNN-based positioning, stand-alone SIFT image matching positioning and SURF combined with Wi-Fi positioning).

In summary, the proposed sensors-calibrated crowd-sourcing indoor localization algorithm via the optical camera and orientation sensor on a smartphone has three innovative aspects:

(1)   We explore the impact of SIFT and SURF thresholds on positioning when we choose the SIFT operator to extract image features. We conclude from our experiments that the positioning error is the least when the threshold is set as 0.55 in both the SIFT and SURF algorithms.

(2)   We propose a mean-weighted exponent algorithm to fuse the output of the built-in sensors in a smartphone. Small scale real-case scenarios results demonstrate that the proposed algorithm has a better performance because of its nonlinearity to calibrate the positioning accuracy.

(3)   We use crowd-sourcing to update and supplement the database. We also explored its influence on positioning. The experimental results show that the accuracy and stability of both algorithms have been improved, and the sensors-calibrated algorithm is obviously better than other three algorithms in this paper.

Compared with the other positioning systems, the proposed system has the following advantages:

(1)   *Higher performance.* We apply the mean-weighted exponent algorithm to calculate the correction factors based on the output of the camera and orientation sensor. These factors will improve the performance of positioning system greatly in accuracy, stability, and applicability.

(2)   *Easier usage.* All required hardware is widely available in common smartphones. It is unnecessary to use deploy additional installations and users can use their smartphones for positioning in two modes: if high accuracy is unnecessary, they can choose Wi-Fi fingerprint positioning, otherwise, they can open the camera to improve the positioning results through the proposed sensors-calibrated algorithm.

(3)   *Lower cost.* The proposed algorithm can be implemented on a smartphone. Firstly, we use the smartphone to locate, and on the other hand, we adopt crowd-sourcing via the smartphone to update the database. Equally important, we do not need to preset tags for image identification in the indoor scenario. However, if we do preset tags for image identification, it will be better for our algorithm.

For indoor positioning systems, it is ideal to achieve the optimal performance with the simplest equipment or method; however, there are many issues to be solved. Our proposed algorithm still has some shortcomings; fortunately, those shortcomings will be addressed in our future work:

Firstly, we should consider combining other built-in sensors in the smartphone. We have performed experiments under the assumption that the light is stable in indoor environments. In practice, the light might change as the weather changes. Smartphone cameras are sensitive to lighting and susceptible to this lowering the quality of the images. We will handle this aspect in the future by combining other sensors or other methods such as MoVIPS [26,27].

Secondly, the influence of equipment diversity should be considered. It is well known that the same built-in sensors in different smartphones have different performances, such as accuracy and stability, which will influence the positioning performance. We have tested the output value of the orientation sensor on the Huawei Honor 6 mobile phone and Motorola Moto X Pro mobile phone. The stability of the sensors in these two devices is obviously different. For simplification, we have done all experiments with the same brand of smartphone. Exploring the influence of equipment diversity for positioning will be a part of our future work.

Finally, the crowd-sourcing selection mechanism should also be considered. Crowd-sourcing is a suitable method to build a database for positioning. On the one hand, it can reduce the time needed to build the database. On the other hand, it can improve positioning performance as demonstrated by our experiments. However, the thing we should take into account is how can we filter out the unreliable data from all the uploaded data. At the same time, we will explore the quality control and management of crowd-sourcing to select better and more useful data to realize localization.

Acknowledgments: The authors wish to thank the reviewers and editors for the careful review and the effort in processing this paper. This study was financially supported by the National Nature Science Foundation of China (71031007) and (61273198).

Author Contributions: The work presented in this paper correspond to a collaborative development by all authors. Weiping Wang read and revised the manuscript for many times, and gave the comprehensive guidance. Qun Li and Hongtao Hou helped us design the system and revised our manuscript. Qiang Chang programmed and wrote the manuscript. Wei Chen programmed, performed the experiment, analyzed data and wrote the manuscript.

Conflicts of Interest: The authors declare no conflict of interest.

References

1. Monnerat, M.; Couty, R.; Vincent, N.; Huez, O.; Chatre, E. The assisted GNSS, technology and applications. In Proceedings of the 17th International Technical Meeting of the Satellite Division of the Institute of Navigation, Long Beach, CA, USA, 21–24 September 2004; pp. 2479–2488.
2. Deng, Z.; Zou, D.; Huang, J.; Chen, X.; Yu, Y.P. The assisted GNSS boomed up location based services. In Proceedings of the 5th International Conference on Wireless Communications, Networking and Mobile Computing, Beijing, China, 24–26 September 2009; pp. 1–4.
3. Wang, J.; Xu, G. Move Difference Relative Positioning Method Based on GNSS. *Modern Navig.* **2015**, *3*, 250–255.
4. Chen, R.; Chu, T.; Li, J.; Li, X.; Chen, Y. DGNSS-C: A differential solution for enhancing smartphone GNSS performance. In Proceedings of the 27th International Technical Meeting of the Satellite Division of the Institute of Navigation, Tampa, FL, USA, 8–12 September 2014; pp. 490–497.
5. Xu, C.; Firner, B.; Zhang, Y.; Howard, R.; Li, J.; Lin, X. Improving RF-Based Device-Free Passive Localization in Cluttered Indoor Environments Through Probabilistic Classification Methods. In Proceedings of the 11th International Conference on Information Processing in Sensor Networks, Beijing, China, 16–20 April 2012; pp. 209–220.
6. Schmitt, S.; Adler, S.; Kyas, M. The Effects of Human Body Shadowing in RF-based Indoor Localization. In Proceedings of the International Conference on Indoor Positioning and Indoor Navigation, Busan, Korea, 27–30 October 2014; pp. 307–313.
7. Pelka, M.; Bollmeyer, C.; Hellbruck, H. Accurate Radio Distance Estimation by Phase Measurements with Multiple Frequencies. In Proceedings of the International Conference on Indoor Positioning and Indoor Navigation, Busan, Korea, 27–30 October 2014; pp. 142–151.
8. Torres-Sospedra, J.; Montoliu, R.; Martinez-Uso, A.; Avariento, J.P. UJIIndoorLoc: A New Multi-building and Multi-floor Database for WLAN Fingerprint-based Indoor Localization Problems. In Proceedings of the International Conference on Indoor Positioning and Indoor Navigation, Busan, Korea, 27–30 October 2014; pp. 261–270.
9. Zhou, M.; Zhan, Q.; Xu, K.; Tian, Z.; Wang, Y.; He, W. PRIMAL: Page Rank-Based Indoor Mapping and Localization Using Gene-Sequenced Unlabeled WLAN Received Signal Strength. *Sensors* **2015**, *15*, 24791–24817. [CrossRef] [PubMed]
10. Ma, L.; Xu, Y. Received Signal Strength Recovery in Green WLAN Indoor Positioning System Using Singular Value Thresholding. *Sensors* **2015**, *15*, 1292–1311. [CrossRef] [PubMed]
11. Zhang, S.; Xing, T. Open WSN indoor localization platform design. In Proceedings of the 2nd International Symposium on Instrumentation and Measurement, Sensor Network and Automation, Toronto, ON, Canada, 23–24 December 2013; pp. 845–848.
12. Laoudias, C.; Michaelides, M.P.; Panayiotou, C. Fault Tolerant Target Localization and Tracking in Binary WSNs using Sensor Health State Estimation. In Proceedings of International Conference on Communications, Budapest, Hungary, 9–13 June 2013; pp. 1469–1473.
13. Pak, J.M.; Ahn, C.K.; Shmaliy, Y.S.; Lim, M.T. Improving Reliability of Particle Filter-Based Localization in Wireless Sensor Networks via Hybrid Particle/FIR Filtering. *IEEE Trans. Ind. Inform.* **2015**, *11*, 1089–1098. [CrossRef]
14. Lim, H.; Sudipta, S.N.; Cohen, M.F.; Uyttendaele, M.; Kim, H.J. Real-time monocular image-based 6-DoF localization. *Int. J. Robot. Res.* **2015**, *34*, 476–492. [CrossRef]

15. Ifthekhar, M.S.; Saha, N.; Jang, Y.M. Neural Network Based Indoor Positioning Technique in Optical Camera Communication System. In Proceedings of the International Conference on Indoor Positioning and Indoor Navigation, Busan, Korea, 27–30 October 2014; pp. 431–435.

16. Yang, S.H.; Kim, H.S.; Son, Y.H.; Han, S.K. Three-Dimensional Visible Light Indoor Localization Using AOA and RSS with Multiple Optical Receivers. *J. Lightwave Technol.* **2014**, *32*, 2480–2485. [CrossRef]

17. Taniuchi, D.; Liu, X.; Nakai, D.; Maekawa, T. Spring Model Based Collaborative Indoor Position Estimation with Neighbor Mobile Devices. *IEEE J. Sel. Top. Signal Process.* **2015**, *9*, 268–277. [CrossRef]

18. Chang, Q.; Hou, H.T.; Zeng, X.H.; Li, Q.; Wang, W.P. A Survey of GNSS based Cooperative Positioning. *J. Astronaut.* **2014**, *35*, 13–20.

19. Zhao, Y.; Liu, K.H.; Ma, Y.T.; Li, Z. An improved k-NN algorithm for localization in multipath environments. *J. Wirel. Commun. Netw.* **2014**, *1*, 1–10.

20. Chang, Q.; Velde, S.V.D.; Wang, W.; Li, Q.; Hou, H.T. Wi-Fi fingerprint positioning updated by pedestrian dead reckoning for mobile phone indoor localization. In *China Satellite Navigation Conference (CSNC) 2015 Proceedings: Volume III*; Springer: Berlin, Germany, 2015.

21. Wang, D.; Zhou, Y.; Wei, Y. A Bayesian Compressed Sensing Approach to Robust Object Localization in Wireless Sensor Networks. In Proceedings of the International Conference on Mobile Services, Anchorage, AK, USA, 27 June–2 July 2014; pp. 24–30.

22. Hejc, G.; Seitz, J.; Vaupel, T. Bayesian Sensor Fusion of Wi-Fi Signal Strengths and GNSS Code and Carrier Phases for Positioning in Urban Environments. In Proceedings of the International Conference on Position Location and Navigation Symposium, Monterey, CA, USA, 5–8 May 2014; pp. 1026–1032.

23. Soltani, M.M.; Motamedi, A.; Hammad, A. Enhancing Cluster-based RFID Tag Localization Using Artificial Neural Networks and Virtual Reference Tags. In Proceedings of the International Conference on Indoor Positioning and Indoor Navigation, Montbeliard-Belfort, France, 28–31 October 2013; pp. 93–105.

24. Chiang, K.W.; Liao, J.K.; Tsai, G.J.; Chang, H.W. The Performance Analysis of the Map-Aided Fuzzy Decision Tree Based on the Pedestrian Dead Reckoning Algorithm in an Indoor Environment. *Sensors* **2016**, *16*, 34.

25. Niu, J.; Ramana, K.V.; Wang, B.; Rodrigued, J.J.P.C. A Robust Method for Indoor Localization using WiFi and Surf based on image fingerprint registration. *Ad-HOC Mob. Wirel. Netw.* **2014**, *8487*, 346–359.

26. Werner, M.; Kessel, M.; Marouane, C. Indoor Positioning Using Smartphone Camera. In Proceedings of the International Conference on Indoor Positioning and Indoor Navigation, Guimarães, Portugal, 21–23 September 2011; pp. 1–6.

27. Marouane, C.; Maier, M.; Feld, S.; Werner, M. Visual Positioning Systems—An Extension to MoVIPS. In Proceedings of the International Conference on Indoor Positioning and Indoor Navigation, Guimaraes, Portugal, 21–23 September 2011; pp. 95–104.

28. Levchev, P.; Krishnan, M.N.; Yu, C.; Menke, J. Simultaneous Fingerprinting and Mapping for Multimodal Image and WiFi Indoor Positioning. In Proceedings of the International Conference on Indoor Positioning and Indoor Navigation, Busan, Korea, 27–30 October 2014; pp. 442–450.

29. Gu, Y.; Anthony, L.; Niemegeers, I. A Survey of Indoor Positioning Systems for Wireless Personal Networks. *IEEE Commun. Surv. Tutor.* **2009**, *11*, 13–32. [CrossRef]

30. Seybold, J.S. *Introduction to RF Propagation*; John Wiley & Sons: Hoboken, NJ, USA, 2005.

31. Zou, H.; Wang, H.; Xie, L.; Jia, Q.S. An RFID indoor positioning system by using weighted path loss and extreme learning machine. In Proceedings of the International Conference on Cyber-Physical Systems, Networks, and Applications, Taipei, Taiwan, 19–20 August 2013; pp. 66–71.

32. Chen, Z.; Zou, H.; Jiang, H.; Zhu, Q.; Soh, Y.C.; Xie, L. Fusion of WiFi, Smartphone Sensors and Landmarks Using the Kalman Filter for Indoor Localization. *Sensors* **2015**, *15*, 715–732. [PubMed]

33. Fidan, B.; Uamy, I. Adaptive Environmental Source Localization and Tracking with Unknown Permittivity and Path Loss Coefficients. *Sensors* **2015**, *15*, 31125–31141. [CrossRef] [PubMed]

34. Yang, L.; Chen, Y.; Li, X.Y.; Xiao, C.; Li, M.; Liu, Y.H. Tagoram: Real-Time Tracking of Mobile RFID Tags to High Precision Using COTS Devices. In Proceedings of the Annual International Conference on Mobile Computing and Networking, Maui, HI, USA, 7–11 September 2014; pp. 237–248.

35. Kotaru, M.; Joshi, K.; Bharadia, D.; Katti, S. SpotFi: Decimeter Level Localization Using WiFi. In Proceedings of the Annual Conference of the Special Interest Group on Data Communication, London, UK, 7–21 August 2015; pp. 269–282.

36. Schussel, M.; Pregizer, F. Coverage Gaps in Fingerprinting Based Indoor Positioning: The Use of Hybrid Gaussian Processes. In Proceedings of the International Conference on Indoor Positioning & Indoor Navigation, Banff, AB, Canada, 13–16 October 2015; pp. 1–9.

37. Li, L.; Shen, G.; Zhao, C.; Moscibroda, T.; Lin, J.H.; Zhao, F. Experiencing and Handling the Diversity in Data Density and Environmental Locality in an Indoor Positioning Service. In Proceedings of International Conference on Mobile Computing & Networking, Maui, HI, USA, 7–11 September 2014; pp. 459–470.

38. Liu, C.; Wang, J. A Constrained KNN Indoor Positioning Model Based on a Geometric Clustering Fingerprinting Technique. *Geomat. Inf. Sci. Wuhan Univ.* **2014**, *39*, 1287–1292.

39. Saha, A.; Sadhukhan, P. A Novel Clustering Strategy for Fingerprinting based Localization System to Reduce the Searching Time. In Proceedings of the 2nd International Conference on Recent Trends in Information Systems, Kolkata, India, 9–11 July 2015; pp. 538–543.

40. Lee, C.W.; Lin, T.N.; Fang, S.H.; Chou, Y.C. A Novel Clustering-Based Approach of Indoor Location Fingerprinting. In Proceedings of the 24th International Symposium on Personal, Indoor and Mobile Radio Communications: Mobile and Wireless Networks, London, UK, 8–11 September 2013; pp. 3191–3196.

41. Philipp, D.; Baier, P.; Dibak, C.; Durr, F.; Rothermel, K.; Becker, S.; Peter, M.; Fritsch, D. MapGENIE: Grammar-enhanced Indoor Map Construction from crowd-sourcing Data. In Proceedings of the 12th International Conference on Pervasive Computing and Communication, Budapest, Spain, 24–28 March 2014; pp. 139–147.

42. Zhou, M.; Wong, K.S.; Tian, Z.; Luo, X. Personal Mobility Map Construction for crowd-sourcing Wi-Fi Based Indoor Mapping. *IEEE Commun. Lett.* **2014**, *18*, 1427–1430.

43. Keller, F.; Willemsen, T.; Sternberg, H. Calibration of smartphones for the use in indoor navigation. In Proceedings of the International Conference on Indoor Positioning and Indoor Navigation, Sydney, NSW, Australia, 13–15 November 2012; pp. 1–8.

44. Galván-Tejada, C.E.; García-Vázquez, J.P.; Galvan-Tejada, J.I.; Delgado-Contreras, J.R.; Brena, R.F. Infrastructure-Less Indoor Localization Using the Microphone, Magnetometer and Light Sensor of a Smartphone. *Sensors* **2015**, *15*, 20355–20372. [PubMed]

45. Shin, H.; Chon, Y.; Cha, H. Unsupervised Construction of Indoor Floor Plan Using Smartphone. *IEEE Trans. Syst. Man Cybern. Part C.* **2012**, *42*, 889–898. [CrossRef]

46. Song, J.; Hur, S.; Park, Y. Fingerprint-Based User Positioning Method Using Image Data of Single Camera. In Proceedings of the International Conference on Indoor Positioning and Indoor Navigation, Banff, AB, Canada, 13–16 October 2015.

47. Kawaji, H.; Hatada, K.; Yamasaki, T.; Aizawa, K. Image-based Indoor Positioning System: Fast Image Matching using Omnidirectional Panoramic Images. In Proceedings of the ACM Workshop on Multimodal Pervasive Video Analysis, Florence, Italy, 29 October 2010; pp. 1–4.

48. Luo, J.; Oubong, G. A Comparison of SIFT, PCA-SIFT and SURF. *Int. J. Image Process.* **2009**, *3*, 143–151.

49. Lowe, D.G. Distinctive Image Features from Scale-invariant Keypoints. *Int. J. Comput. Vis.* **2004**, *60*, 91–110. [CrossRef]

50. Estellés-Arolas, E. Towards an Integrated Crowd-sourcing Definition. *J. Inf. Sci.* **2012**, *38*, 189–200. [CrossRef]

51. Vukovic, M.; Lopez, M.; Laredo, J. Peoplecloud for the Globally Integrated Enterprise. In Proceedings of the International Conference on Service Oriented Computing, San Francisco, CA, USA, 7–10 December 2010; pp. 109–114.

52. Qian, S.; Zhu, J. Improved SIFT-based Bidirectional Image Matching Algorithm. *Mech. Sci. Technol. Aerosp. Eng.* **2007**, *26*, 1179–1182.

sensors

MDPI

Article

Parallax-Robust Surveillance Video Stitching

Botao He [1,*] and Shaohua Yu [2]

[1] School of Optical and Electronic Information, Huazhong University of Science and Technology,
Wuhan 430074, China
[2] Wuhan Research Institute of Posts and Telecommunications, Wuhan 430074, China; shyu@fhzz.com.cn
* Correspondence: hebotao@hust.edu.cn; Tel.:+86-27-87693908

Academic Editor: Gonzalo Pajares Martinsanz
Received: 7 October 2015; Accepted: 17 December 2015; Published: 25 December 2015

Abstract: This paper presents a parallax-robust video stitching technique for timely synchronized surveillance video. An efficient two-stage video stitching procedure is proposed in this paper to build wide Field-of-View (FOV) videos for surveillance applications. In the stitching model calculation stage, we develop a layered warping algorithm to align the background scenes, which is location-dependent and turned out to be more robust to parallax than the traditional global projective warping methods. On the selective seam updating stage, we propose a change-detection based optimal seam selection approach to avert ghosting and artifacts caused by moving foregrounds. Experimental results demonstrate that our procedure can efficiently stitch multi-view videos into a wide FOV video output without ghosting and noticeable seams.

Keywords: video stitching; video surveillance; layered warping; parallax

1. Introduction

Image stitching, also called image mosaicing or panorama stitching, has received a great deal of attention in computer vision [1–8]. After decades of development, the fundamentals of image stitching are well studied and relatively mature now. There are many research works on image stitching [1–8], and it is typically solved by estimating a global 2D projective warp to align the input images. A 2D projective warp uses a homography parameterized by 3×3 matrices [1–3,9], which can preserve global image structures, but cannot handle parallax. It is correct only if the scene is planar or if the views differ purely by rotation. However, in practice, such conditions are usually hard to satisfy, thus ghosting and seams yield (see Figure 1). If there is parallax in input images, no global homograhpy exists that can be used to align these images. When a global warp is used to stitch these images, ghosting like Figure 1a would appear. Some advanced image composition techniques such as seam cutting [10–12] can be used to relieve these artifacts. However, if there are moving objects across the seams, another kind of ghosting like Figure 1b would yield.

Previous research indicates that one of the most challenging problems to create seamless and drift-free panoramas is performing a correct image alignment rather than using a simple global projective model and then fix the alignment error [6,8,9]. Thus, some recent image stitching methods focus on using spatially-varying warping algorithms to align the images [6–8], these methods can handle parallax and allow for local deviation to some extent but require more computation.

With wide applications in robotics, industrial inspection, surveillance and navigation, video stitching faces all the problems as image stitching does and can be more challenging due to moving objects in videos. Some researchers tried to build panoramic images by aligning video sequences [13,14], which is panoramic image generation rather than expansion of the FOV of dynamic videos. Other works focus on freely moving devices [15–18], especially mobile devices, which include techniques such as efficient computation of temporal varying homography [15], optimal seam selection for blending [16], and so on. However, due to complex computation and low resolution, they may not be suitable for surveillance application.

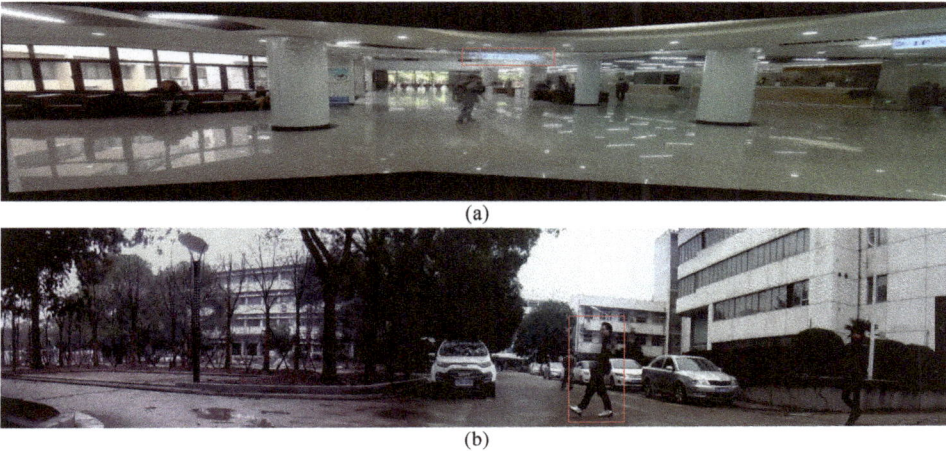

(a)

(b)

Figure 1. The main reason for ghosting comes from misalignment error. Here we show two types of ghosting in red boxes: (**a**) ghosting caused by using a global projective warping; (**b**) ghosting caused by persons moving across seams.

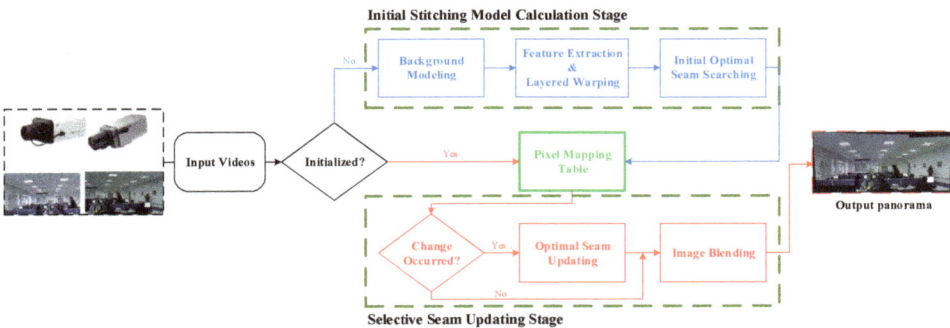

Figure 2. Outline of the proposed video stitching procedure. Our method consists of two stages: initial stitching model calculation stage and selective seam updating stage. In the initial stitching model calculation stage, we first use background modeling algorithm to generate still background of each input source video, then utilize layered warping algorithm to align background images, finally, we perform optimal seam searching and image blending to generate panorama backgrounds. The resulting stitching model is a mapping table in which each entry indicates the correspondence between the pixel index of source images and that of panorama image. To relieve the ghosting effect caused by moving objects, at selective seam updating stage, we perform seam updating according to whether there are objects moving across previous seams or not. Our method reaches a balance between suppressing ghosting artifacts and real-time requirement.

In this paper, we present an efficient parallax-robust surveillance video stitching procedure that combines layered warping and the change-detection based seam updating approach. As the alignment validity of video stitching is still crucial as it is in seamless image stitching, we use a layered warping method for video registration instead of a simple global projective warp.

Xiao *et al.* [19] provides a similar layer-based video registration algorithm, but it aims at aligning a single mission video sequence to the reference image via layer mosaics and region expansion, rather than building a dynamic panoramic video for motion monitoring. Through dividing matched feature pairs into multiple layers (or planes) and local alignment based on these layers, the layered warping method seems to be more robust to parallax. Moreover, the warping data for fixed surveillance videos are stored in an index table for "recycling-use" in subsequent frames to avoid repeated registration and interpolations. This index table is referred to as the initial stitching model in this paper. Aside from layered warping, a local change-detection based seam updating method for overlapping regions is performed to disambiguate the ghosting caused by moving foregrounds. Figure 2 shows the video stitching process presented in this paper.

2. Related Works

Recently, video stitching has drawn a lot of attentions [11,20,21] due to its wide usage in public security. Generally speaking, surveillance video stitching can be regarded as image stitching for every individual frame since the camera positions are always fixed. Different from image stitching technologies, video stitching requires more strict real-time processing ability, and large parallax and dynamic foregrounds must be carefully considered to obtain consistent wide field of view videos. Image stitching is relatively a well studied problem in computer vision [1,9,22]. Several freewares and commercial softwares are also available for performing image stitching, like AutoStitch [23], Microsoft's Image Compositing Editor [24], and Adobe's Photoshop CS5 [25] mosaicing feature. However, these approaches all work under the assumption that the input images contains little or no parallax, which implies that the scene is either sufficiently far away from the camera to be considered planar, or that the images have been taken from a camera carefully rotated about its center of projection. This assumption is too strict to be satisfied in real surveillance scenarios. Thus, misalignment artifacts like ghosting or broken image structures will make the final panorama visually unacceptable (see Figure 1).

Some advanced image composition techniques, such as seam cutting [10–12] and blending [26,27], have been employed to reduce the artifacts. However, these methods alone still cannot handle significant parallax. In this paper, we also use seam cutting and blending as the final steps to suppress artifacts. Recent studies on spatially-varying warping are another way out [6,7,28]. As-projective-as-possible (APAP) warps [7] employed local projective warps within the overlapping regions and performed moving direct linear transformation [29] to smoothly extrapolate local projective warps into the non-overlapping regions. Shape-Preserving Half-Projective (SPHP) warp [28] spatially combines a projective transformation and a similarity transformation and has the strengths of both. However, instead of improving alignment accuracy, its main concern is to decrease distortion of non-overlapping area caused by the projective transformation. So even if it introduces APAP [7] into their warp, they cannot solve the problem of structure distortion in the overlapping area in APAP [7]. Gao *et al.* [6] proposed to uses a dual homography warp (DHW) algorithm for scenes containing two dominant planes (ground plane and distant plane). While it performs well if the required setting is true, it may fail when there are more than two planes in the source images. Inspired by DHW [6], we propose to use a layered warping algorithm to align the background scenes, which is location-dependent and turned out to be more robust to parallax than the traditional global projective warping methods and more flexible than DHW [6] which can only process images with two planes.

Apart from parallax, moving foregrounds are another reason for ghosting in video stitching. Although we propose to use layered warping to align images as accurate as possible and to utilize seam cutting to composite source images, some artifacts may still exist when objects move across the seam. Liu *et al.* [20] only used the stitching model calculated with first few frames to stitch following frames and didn't consider the moving foregrounds. In contrast, Tennoe *et al.* [30] and Hu *et al.* [11] update the seam in every frame, which is very time-consuming. To balance between suppressing the artifacts and the real-time requirement, we propose to first detect changes around the previous seams, and only perform seam update when there are moving objects across seams. Since the price of change detection around seams is much lower than that of updating seams, artifacts caused by moving foregrounds can be suppressed with acceptable time consumption by our method.

3. Initial Stitching Model Calculation

Since our focus is on improving image alignment accuracy and reducing artifacts caused by moving objects, we do not change the conventional pipeline [9] of image stitching with different number of input sources. For ease of illustration, in the following text, we only describe the layered warping algorithm and selective seam updating algorithm with two input videos. In the experiment section, we provide stitching results with both two and more than two input videos.

3.1. Background Image Generation and Feature Extraction

For fixed surveillance cameras, we only perform alignment at the stitching model calculation stage because the computation of temporal varying homography may result in palpable jitter of background scenes in the panoramic video. Background modeling is essential to avoid volatile foreground [31,32]. We take the first N_{gmm} frames of input videos to establish the background frame utilizing the Gaussian Mixture Model (GMM) [33]. SIFT [3,34] features of the background frame is extracted and matched into pairs through Best-Bin-First (BBF) algorithm [35].

3.2. Layered Warping

Inspired by DHW [6], we assume that different objects in a scene usually lie in different depth layers, the objects in the same layer (plane) shall be consistent with each other in spatial transformation. Compared to warping using a global homography or warping using dual homography, layered warping may be more adequate and robust for abundant scenes.

Layer registration. We denote the input images as I_1 and I_2 respectively, and the matching feature pairs as $\mathcal{F} = \{(\mathbf{p}_i^1, \mathbf{p}_i^2)\}_{i=1}^{N}$, where $\mathbf{p}_i^k$ is the coordinate of the i-th matching point from I_k ($k = 1, 2$). Given the matching feature pairs, we first utilize Random Sample Consensus (RANSAC) algorithm [36] to robustly group the feature pairs into different layers, then estimate the homography for each layer. Denote the consistent matching feature pairs of layer k as $\mathcal{L}_k$, the number of matches in $\mathcal{L}_k$ as $|\mathcal{L}_k|$ and its corresponding homography as $\mathbf{H}_k$. To guarantee the grouped layer to be representative, we introduce a threshold N_{min} which denotes the minimum number of matching pairs in $|\mathcal{L}_k|$. Layers whose number of matching pairs is smaller than N_{min} are simply dropped. The detailed layer registration process is presented in Algorithm 1.

Algorithm 1 Layer registration utilizing multiple-layer RANSAC

Input: Initial pair set $\mathcal{F}_1 = \mathcal{F}$, threshold N_{min} and iteration index $k = 0$;
Output: Each layer's matching pair set $\mathcal{L}_k$ and its corresponding homography $\mathbf{H}_k$;
   repeat
      $k \leftarrow k + 1$
      RANSAC in pair set $\mathcal{F}_k$ for model $\mathbf{p}_{k_n}^1 \times \mathbf{H}\mathbf{p}_{k_n}^2 = 0$, where $(\mathbf{p}_{k_n}^1, \mathbf{p}_{k_n}^2) \in \mathcal{F}_k$;
      Divide outliers $\mathbf{V}_{out}$ and inliers $\mathbf{V}_{in}$ according to $\mathbf{H}$;
      if $|\mathbf{V}_{in}| \geq N_{min}$ **then**

         Set matching pair set of the k-th layer as $\mathcal{L}_k = \mathbf{V}_{in}$;
         Set homography of the k-th layer as $\mathbf{H}_k = \mathbf{H}$;
      end if
      Set the pair set of next iteration as $\mathcal{F}_{k+1} = \mathbf{V}_{out}$;
   until $|\mathcal{F}_{k+1}| < N_{min}$

Through the layer registration process, we divide the matched feature pairs to multiple layers, each of which contains a set of feature pairs that are consistent with a common homography. Figure 3 shows two examples of layer registration, in which feature points are illustrated as circles and points with the same color are from the same layer. From Figure 3 we can see that the grouped matching pairs of the same layer are almost from the same plane or with the same depth, which is in accordance with our expectation.

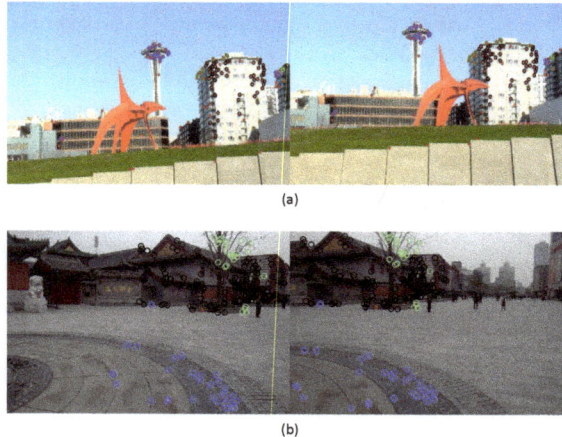

(a)

(b)

Figure 3. Examples of matched feature points in multiple layers, different color indicate different layers: (a) multiple-layer matched feature pairs of images with large parallax; (b) multiple-layer matched feature pairs of images with a distant plane and ground plane

Local alignment. To simplify calculations of local alignment, the source image is divided into $M \times N$ grids. Since a feature point is not usually coincident with any grid vertex, we use the distances between the grid center and its nearest neighbors in different layers to vote on the warp of the due grid. A grid g_j is represented by its center point $\mathbf{c}_j$. The homography of the grid g_j, denoted by $\mathbf{H}_j^*$, is computed by fusing $\mathbf{H}_k$ of multiple layers using a weight w_k^j by Equation (1)

$$\mathbf{H}_j^* = \sum_k w_k^j \mathbf{H}_k \tag{1}$$

where $w_k^j = a_k^j / \sum_i a_i^j$ and a_k^j is a position dependent Gaussian weight:

$$a_k^j = \exp\left(-\frac{||\mathbf{c}_j - \mathbf{p}_k^*||^2}{\sigma^2}\right) \tag{2}$$

Here $\mathbf{p}_k^*$ denotes the nearest neighboring feature point of $\mathbf{c}_j$ in layer k and σ is a scale constant.

After deriving the local homography for each grid, the target pixel position $\mathbf{p}'$ in the reference image for the source pixel at position $\mathbf{p}$ in grid g_j can be easily obtained by Equation (3):

$$\mathbf{p}' = \mathbf{H}_j^* \mathbf{p} \tag{3}$$

This process is referred to as forward mapping [9]. To accelerate the computation, we only perform forward mapping once, and store the correspondence between source pixel positions and target pixel positions in the pixel mapping table. This pixel mapping table is exactly the stitching model. The index tables are stored so that the warped image can be obtained by looking up each corresponding pixel in the source image instead of repeated transformation when new frames arrive.

Figure 4a,b shows the panorama images with global projective warp and layered warp respectively. It is clear that the building in Figure 4b is better aligned than that in Figure 4a.

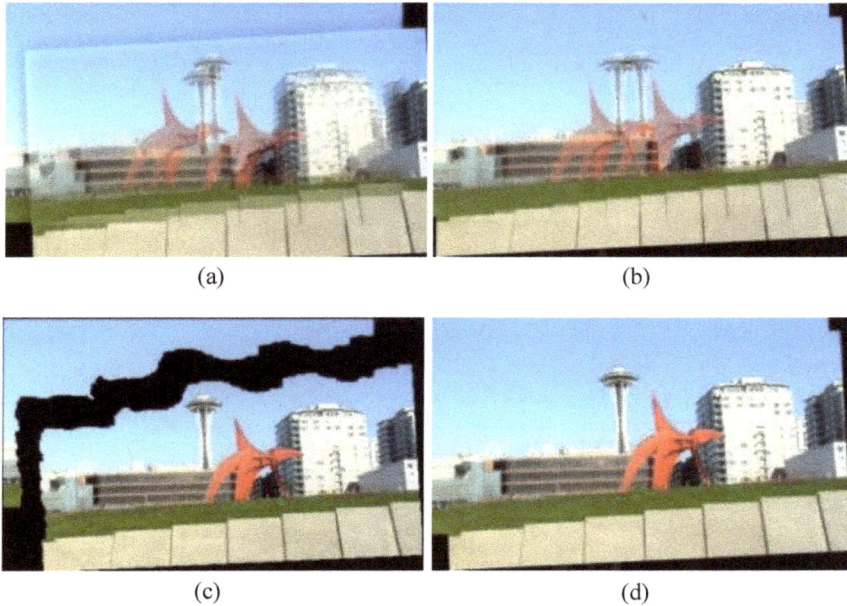

Figure 4. Examples of panoramas with (**a**) global projective warp and (**b**) layered warp, and seam selection results in (**c**). (**d**) is the final fused image according to selected seam in (**c**). The seam selection result (**c**) is based on stitched image (**b**) with layered warp.

3.3. Optimal Seam Cutting

Though layered warping is parallax-robust, it is applied only in the stitching model calculation stage on the extracted background scene. We perform the seam selection method to disambiguate the ghosting caused by the moving foreground.

Optimal seam searching method, also called optimal seam selection [5] or seam cutting, is to search for an optimal seam path which is a pixel-formed continuous curve in the overlapping area to connect pairwise warped images.

The seam should neither introduce inconsistent scene elements nor intensity differences. Therefore, two criteria are applied in this paper to form the difference map of overlaps: the intensity energy E_{ij}^C and gradient energy E_{ij}^g which are defined as:

$$\begin{cases} E_{ij}^C &= \frac{||I_A(i,j) - I_B(i,j)||}{\max(I_A(i,j), I_B(i,j))} \\ E_{ij}^g &= ||\nabla I_A(i,j) - \nabla I_B(i,j)||^2 \end{cases} \tag{4}$$

Here $I_A(i,j)$, $I_B(i,j)$, $\nabla I_A(i,j)$ and $\nabla I_B(i,j)$ are the intensity and gradient of pixel (i,j) in image A and B respectively. Finding the optimal stitching seam is an energy minimization problem (see Equation (5)) and can be converted to a binary Markov Random Field (MRF) labeling problem [37]:

$$\arg\min_{ij} \sum_{ij} (E_{ij}^C + \lambda E_{ij}^g) \tag{5}$$

To accelerate the computing process, the warped images and background images are down sampled before seam selection and restored to the original size after seam selection in our procedure. Figure 4c and d show an example of seam selection. From Figure 4c we can see, the selected seam mainly crosses flat areas with little gradients or intensity differences, thus the resulted panorama is visually consistent with no noticeable ghosting.

4. Selective Seam Updating

After initial stitching model calculation, the videos are overall pre-aligned. However, for moving foregrounds, the previous seam may lose its optimality or even miss information. Since the seam cutting algorithm requires complex computation even on the down-sampled frames, it is difficult to be used to update video frames in real-time. So we perform a change detection based seam updating method instead of real-time seam selection.

4.1. Change Detection around Previous Seams

First of all, we resize the new warped frames to a smaller scale as we did in the optimal seam selection section. Even if the images have been scaled down, direct calculation of the gradient of the two images and evaluation of the change may be expensive. However, we observed that compared with changes in non-overlapping area, those in the overlapping area are more likely to violate the optimal seam. Furthermore, only changes across the optimal seam may result in the failure of it (see Figure 5), which cam be measured by gradient difference.

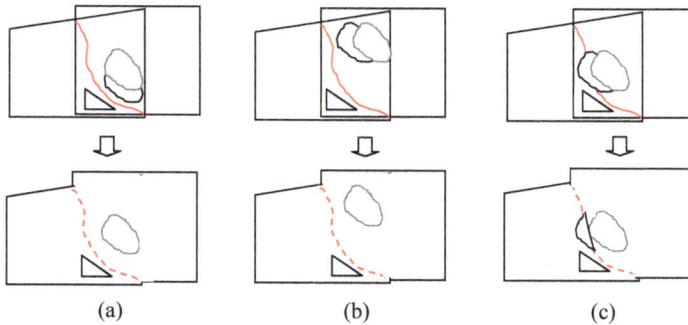

Figure 5. (**a**) The original aligned frame with a stitching line; (**b**) newly aligned frame with changes not cross the line; (**c**) frame with changes cross the line.

As an optimal seam is searched, the original gradient value set $\tilde{G}_0$ of the seam is stored. We set $\tilde{g}_{i0}$ as the original gradient of pixel p_i in the present seam, and $\tilde{g}_{it}$ as the gradient in time t. To calculate the number of pixels that have large gradient variations, we use the following rule to judge if changes occured at pixel p_i or not:

$$C_t = \{p_i | \frac{\tilde{g}_{it} - \tilde{g}_{i0}}{\tilde{g}_{i0}} > \delta\} \tag{6}$$

If the total pixel number in C_t is bigger than N_{cd}, we consider that there are new moving objects in the overlapping area and the optimal seam shall be updated. In Equation (6), δ is a constant chosen empirically, whereas N_{cd} depends on $\tilde{N}$, which is the total pixel number of the seam. We set it as $0.3\tilde{N}$ in our experiments.

4.2. Seam Updating

For each new warped frame, change detection, as described in the previous section, is performed to see if an alteration of the seam is needed, if so, we select a new seam by the seam selection algorithm presented in Section 3.3, otherwise, we continue to use the previous seam and move on to the image blending process.

4.3. Blending

Seam cutting can eliminate ghosting, but it provides images without overlaps which may result in noticeable seams. We expand the seamed image with a spherical dilating kernel, and use a simple weighted linear blending method [9] to blend the images.

5. Experiments

To evaluate the performance of our video stitching procedure, we conduct some experiments on both still images with parallax and actual surveillance videos.

5.1. Experimental Settings

There are several empirical parameters which should be manually tuned for different cases. At the background modeling stage, we use the initial 20 frames to construct a GMM model with 5 components for each pixel. The frame number may be set larger if the scenes are more cluttered. In our experiment, N_{min} is set as 12, which is the minimum number of matching pairs for each layer, and the source 720P image is divided into 80×45 grids for layered warping. We provide stitching results with both two, three and four channels as input.

5.2. Stitching Still Images

To evaluate the proposed layered warp algorithm, we first conduct experiments on still images with parallax and compare our results with other image stitching methods in Figures 6 and 7.

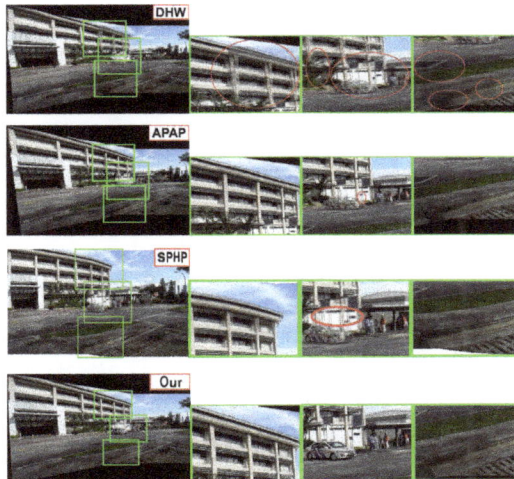

Figure 6. Comparisons among dual homography warp (DHW) [6], as-projective-as-possible (APAP) [18], Shape-Preserving Half-Projective (SPHP) [28] and Our algorithm, Red circles highlight errors.

DHW [6] only considers two layers (a distant plane and a ground plane), so it fails when the scene has multiple depth layers or contains complex objects. Our algorithm is more adequate and robust than DHW for abundant scenes (see Figure 6). Figure 7 shows the comparison among our algorithm and some state-of-art algorithms on a tough scene from [8]. As APAP tries to align two images over the whole overlapping region, it distorts the salient image structure, such as the pillar indicated by the red rectangle. SPHP also fails when the overlapping area covers most of the image and there is large parallax in it (see Figure 7c). From Figures 6 and 7 we can see, the stitching results of our algorithm contain no noticeable distortions and are more visually acceptable than other methods.

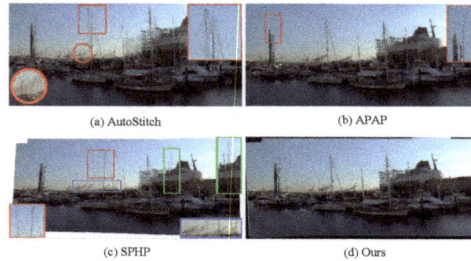

Figure 7. Comparisons among various stitching methods. colored rectangles and their zoom-in rectangles illustrate the false misalignment.

5.3. Stitching Fixed Surveillance Videos

To evaluate the effectiveness of selective seam updating strategy, we perform extensive experiments on fixed surveillance videos, and compare our results with that of without seam updating. The results are shown in Figure 8, from which we can see, when we do not perform seam updating in the video stitching stage, some obvious artifacts caused by moving objects would make the stitched video visually unacceptable. The comparison results indicate that the change detection based seam updating approach is rather helpful to avert ghosting and mis-alignments caused by moving targets, especially when there are moving objects across previous seams.

Figure 8. Frames taken from the output Field-of-View (FOV) videos: (**a1–d1**) are taken from the output videos using initialized seam without updating. (**a2–d2**) are taken from the output video using change-detection based seam updating approach. (**a1**), (**a2**), (**b1**) and (**b2**) are the stitching results of two channels of 720P videos. (**c1**) and (**c2**) are the stitching results of three channels of 720P videos. While for (**d1**) and (**d2**), they are the stitching results of four 720P videos.

5.4. Time Analysis

Apart from surpressing ghosting caused by parallax and moving targets, the real-time requirement of video stitching task should also be seriously considered when developing algorithms. To evaluate the speed of the proposed method, we take two videos as input and test the time consumption for different video resolutions. To make a fair comparison, the speed of stitching with temporal varying homography, layered warping without selective seam updating and layered warping with selective seam updating are all listed in Table 1. These experiments are all conducted on a PC with Intel Core i3 2.33 GHz CPU with two cores and 2GB RAM. All tested algorithms are implemented with C++.

From Table 1 we can see, as the index table has already been established according to the registration result in the stitching model calculation stage, frames can be projected to the panoramic frame by directly indexing instead of alignment and mapping for every new frame, thus the stitching efficiency of the proposed method is greatly improved. Table 1 also shows that the selective seam updating process slows down the frame stitching process to some extent.

Table 1. Comparisons of the speed among different video stitching methods on input-videos with different resolutions.

Resolution	Stitching with Temporal Varying Homography	Proposed Algorithm without Seam Updating	Proposed Algorithm with Seam Updating
720p: 1280 × 720	4.764 s	0.051 s	0.083 s
480P: 720 × 480	2.326 s	0.035 s	0.045 s
CIF: 352 × 288	1.025 s	0.021 s	0.032 s

6. Conclusions

This paper presents an efficient parallax-robust stitching algorithm for fixed surveillance videos. The proposed method consists of two parts: alignment work done at the stitching model calculation stage and change detection based frame updating at the video stitching stage. The algorithm uses layered warping method to pre-align the background scene which is robust to scene parallax. As for each new frame, an efficient change detection based seam updating method is adopted to avert ghosting and artifacts caused by moving foregrounds. Thus, the algorithm can provide good stitching performance with no ghosting and artifacts for dynamic scenes efficiently.

Author Contributions: The work was realized with the collaboration of all of the authors. Botao He and Shaohua Yu contributed to the algorithm design and code implementation. Botao He contributed to the data collection and baseline algorithm implementation and drafted the early version of the manuscript. Shaohua Yu organized the work and provided the funding, supervised the research and critically reviewed the draft of the paper. All authors discussed the results and implications, commented on the manuscript at all stages and approved the final version.

Conflicts of Interest: The authors declare no conflict of interest.

References

1. Brown, M.; Lowe, D. Recognising Panoramas. In Proceedings of the Ninth IEEE International Conference on Computer Vision, Nice, France, 13–16 October 2003; pp. 1218–1225.
2. Szeliski, R. Video mosaics for virtual environments. *IEEE Comput. Graph. Appl.* **1996**, *16*, 22–30.
3. Eden, A.; Uyttendaele, M.; Szeliski, R. Seamless image stitching of scenes with large motions and exposure differences. *Comput. Vis. Pattern Recognit.* **2006**, doi:10.1109/CVPR.2006.268.
4. Levin, A.; Zomet, A.; Peleg, S.; Weiss, Y. *Seamless Image Stitching in the Gradient Domain*; Springer Berlin Heidelberg: Berlin, Germany, 2004; pp. 377–389.
5. Mills, A.; Dudek, G. Image stitching with dynamic elements. *Image Vis. Comput.* **2009**, *27*, 1593–1602.
6. Gao, J.; Kim, S.J.; Brown, M.S. Constructing Image Panoramas Using Dual-Homography Warping. In Proceedings of the EEE Conference on Computer Vision and Pattern Recognition (CVPR), Providence, RI, USA, 20–25 June 2011; pp. 49–56.

7. Zaragoza, J.; Chin, T.J.; Tran, Q.H.; Brown, M.S.; Suter, D. As-projective-as-possible image stitching with moving DLT. *IEEE Trans. Pattern Anal. Mach. Intell.* **2014**, *36*, 1285–1298.

8. Zhang, F.; Liu, F. Parallax-Tolerant Image Stitching. In Proceedings of the IEEE Conference on Computer Vision and Pattern Recognition (CVPR), Columbus, OH, USA, 23–28 June 2014; pp. 3262–3269.

9. Szeliski, R. Image alignment and stitching: A tutorial. *Found. Trends Comput. Graph. Vis.* **2006**, *2*, 1–104.

10. Agarwala, A.; Dontcheva, M.; Agrawala, M.; Drucker, S.; Colburn, A.; Curless, B.; Salesin, D.; Cohen, M. Interactive digital photomontage. *ACM Trans. Graph.* **2004**, *23*, 294–302.

11. Hu, J.; Zhang, D.Q.; Yu, H.; Chen, C.W. Discontinuous Seam Cutting for Enhanced Video Stitching. In Proceedings of the IEEE International Conference on Multimedia and Expo (ICME), 29 June–3 July 2015; pp. 1–6.

12. Kwatra, V.; Schödl, A.; Essa, I.; Turk, G.; Bobick, A. Graphcut textures: Image and video synthesis using graph cuts. *ACM Trans. Graph.* **2003**, *22*, 277–286.

13. Steedly, D.; Pal, C.; Szeliski, R. Efficiently Registering Video into Panoramic Mosaics. In Proceedings of the Tenth IEEE International Conference on Computer Vision, Beijing, China, 17–21 October 2005; pp. 1300–1307.

14. Hsu, C.T.; Tsan, Y.C. Mosaics of video sequences with moving objects. *Signal Process. Image Comm.* **2004**, *19*, 81–98.

15. El-Saban, M.; Izz, M.; Kaheel, A. Fast Stitching of Videos Captured from Freely Moving Devices by Exploiting Temporal Redundancy. In Proceedings of the 17th IEEE International Conference on Image Processing (ICIP), Hong Kong, China, 26–29 September 2010; pp. 1193–1196.

16. El-Saban, M.; Izz, M.; Kaheel, A.; Refaat, M. Improved Optimal Seam Selection Blending for Fast Video Stitching of Videos Captured from Freely Moving Devices. In Proceedings of 18th IEEE International Conference on the Image Processing (ICIP), Brussels, Belgium, 11–14 September 2011; pp. 1481–1484.

17. Okumura, K.I.; Raut, S.; Gu, Q.; Aoyama, T.; Takaki, T.; Ishii, I. Real-Time Feature-Based Video Mosaicing at 500 fps. In Proceedings of the IEEE/RSJ International Conference on Intelligent Robots and Systems (IROS), Tokyo, Japan, 3–7 November 2013; pp. 2665–2670.

18. Au, A.; Liang, J. Ztitch: A Mobile Phone Application for Immersive Panorama Creation, Navigation, and Social Sharing. In Proceedings of the IEEE 14th International Workshop on Multimedia Signal (MMSP), Banff, AB, USA, 17–19 September 2012; pp. 13–18.

19. Xiao, J.; Shah, M. Layer-based video registration. *Mach. Vis. Appl.* **2005**, *16*, 75–84.

20. Liu, H.; Tang, C.; Wu, S.; Wang, H. Real-Time Video Surveillance for Large Scenes. In Proceedings of the International Conference on Wireless Communications and Signal Processing (WCSP), Nanjing, China, 9–11 November 2011; pp. 1–4.

21. Zeng, W.; Zhang, H. Depth Adaptive Video Stitching. In Proceedings of the Eighth IEEE/ACIS International Conference on Computer and Information Science, Shanghai, China, 1–3 June 2009; pp. 1100–1105.

22. Brown, M.; Lowe, D.G. Automatic panoramic image stitching using invariant features. *Int. J. Comput. Vis.* **2007**, *74*, 59–73.

23. AutoStitch. Available online: http://cvlab.epfl.ch/ brown/autostitch/autostitch.html (accessed on 18 December 2015).

24. Microsoft Image Composite Editor. Available online: http://research.microsoft.com/en-us/um/ redmond/groups/ivm/ICE/ (accessed on 18 December 2015).

25. Adobe Photoshop CS5. Available online: http://www.adobe.com/products/photoshop (accessed on 18 December 2015).

26. Burt, P.J.; Adelson, E.H. A multiresolution spline with application to image mosaics. *ACM Trans. Graph.* **1983**, *2*, 217–236.

27. Pérez, P.; Gangnet, M.; Blake, A. Poisson image editing. *ACM Trans. Graph.* **2003**, *22*, 313–318.

28. Chang, C.H.; Sato, Y.; Chuang, Y.Y. Shape-Preserving Half-Projective Warps for Image Stitching. In Proceedings of the IEEE Conference on Computer Vision and Pattern Recognition (CVPR), Columbus, OH, USA, 23–28 June 2014; pp. 3254–3261.

29. Alexa, M.; Behr, J.; Cohen-Or, D.; Fleishman, S.; Levin, D.; Silva, C.T. Computing and rendering point set surfaces. *IEEE Trans. Vis. Comput. Graph.* **2003**, *9*, 3–15.

30. Tennoe, M.; Helgedagsrud, E.; Næss, M.; Alstad, H.K.; Stensland, H.K.; Gaddam, V.R.; Johansen, D.; Griwodz, C.; Halvorsen, P. Efficient Implementation and Processing of a Real-Time Panorama Video Pipeline. In Proceedings of the IEEE International Symposium on Multimedia (ISM), Anaheim, CA, USA, 9–11 December 2013; pp. 76–83.

31. Kumar, P.; Dick, A.; Brooks, M.J. Integrated Bayesian Multi-Cue Tracker for Objects Observed from Moving Cameras. In Proceedings of the 23rd International Conference on Image and Vision Computing New Zealand, Christchurch, New Zealand, 26–28 November 2008; pp. 1–6.

32. Kumar, P.; Dick, A.; Brooks, M.J. Multiple Target Tracking with an Efficient Compact Colour Correlogram. In Proceedings of the 10th International Conference on Control, Automation, Robotics and Vision, Hanoi, Vietnam, 17–20 December 2008; pp. 699–704.

33. Zivkovic, Z. Improved adaptive Gaussian mixture model for background subtraction. *Proc. Int. Conf. ICPR Pattern Recognit.* **2004**, *2*, 28–31.

34. Kumar, P.; Ranganath, S.; Huang, W. Queue Based Fast Background Modelling and Fast Hysteresis Thresholding for Better Foreground Segmentation. In Proceedings of the Fourth Pacific Rim Conference on Multimedia Information, Communications and Signal, Singapore, 15–18 December 2003; pp. 743–747.

35. Beis, J.S.; Lowe, D.G. Shape Indexing Using Approximate Nearest-Neighbour Search in High-Dimensional Spaces. In Proceedings of the IEEE Computer Society Conference on Computer Vision and Pattern Recognition, San Juan, Territory, 17–19 June 1997; pp. 1000–1006.

36. Fischler, M.A.; Bolles, R.C. Random sample consensus: A paradigm for model fitting with applications to image analysis and automated cartography. *Comm. ACM* **1981**, *24*, 381–395.

37. Szeliski, R.; Zabih, R.; Scharstein, D.; Veksler, O.; Kolmogorov, V.; Agarwala, A.; Tappen, M.; Rother, C. A comparative study of energy minimization methods for markov random fields with smoothness-based priors. *IEEE Trans. Pattern Anal. Mach. Intell.* **2008**, *30*, 1068–1080.

![sensors logo] *sensors*

MDPI

Article

Monocular-Vision-Based Autonomous Hovering for a Miniature Flying Ball

Junqin Lin, Baoling Han * and Qingsheng Luo

School of Mechanical Engineering, Beijing Institute of Technology, 5 Zhongguancun South Street, Haidian District, Beijing 100081, China; E-Mails linjunqin2010@163.com (J.L.); luoqsh@bit.edu.cn (Q.L.)
* Correspondence: hanbl@bit.edu.cn; Tel./Fax: +86-10-6891-8856.

Academic Editor: Gonzalo Pajares Martinsanz
Received: 14 April 2015; Accepted: 1 June 2015; Published: 5 June 2015

Abstract: This paper presents a method for detecting and controlling the autonomous hovering of a miniature flying ball (MFB) based on monocular vision. A camera is employed to estimate the three-dimensional position of the vehicle relative to the ground without auxiliary sensors, such as inertial measurement units (IMUs). An image of the ground captured by the camera mounted directly under the miniature flying ball is set as a reference. The position variations between the subsequent frames and the reference image are calculated by comparing their correspondence points. The Kalman filter is used to predict the position of the miniature flying ball to handle situations, such as a lost or wrong frame. Finally, a PID controller is designed, and the performance of the entire system is tested experimentally. The results show that the proposed method can keep the aircraft in a stable hover.

Keywords: monocular-vision sensor; vision measurement; flying height detecting; MAVs; hovering

1. Introduction

The problem of hovering for a micro air vehicle (MAV) is considered here. It would sense the environment based on the information of the vision sensor, breaking the dependence on the Global Navigation Satellite Systems (GNSS), such as GPS and GLONASS, which will be selectively unavailable and will be disabled in cluttered or indoor environments. The commonly-used range measuring technologies to control the height in hovering, such as radar or laser range finder, are simply impractical for MAVs, because of the size and the weight barriers [1]. In addition to the small volume and light weight, the vision system also has advantages of passive illumination, low power consumption and low cost.

Many research works have been done on controlling the hovering of a vehicle based on vision. In the early years, helicopters were used in experiments because the MAV had not been widely studied. For example, a 67-kg Yamaha R-50 helicopter used a combination of stereo vision and feature tracking to control the hovering [2], and a fly-by-wire helicopter ATTHeS (Advanced Technology Testing Helicopter System) used the camera to track a two-dimensional template of *a priori* unknown features [3]. Both of them have the advantages of the sufficient space and carrying capacity of the helicopter. MAVs are characterized as small volume and limited payloads, so it is difficult to miniaturize the equipment for use on MAVs, such as the stereo vision needing a minimum baseline between two cameras.

Vision combined with structured environments has been applied successfully in hovering control [4–6]. For instance, an autonomous quadrotor aircraft can perform stable hovering above a pattern glyph with a data-fusion algorithm using both visual system measurements and inertial sensors [6]. However, the vision system relied on the pattern glyph, which is specially designed to align the aircraft with the glyph. Hence, these methods are still restrained by artificially structured environments in practical applications.

How to sense the height of MAVs by vision for hovering control using natural terrain is also a challenge. In Cherian's work, height is measured by the machine learning to analyze the texture of the image captured by a downwards looking camera [7]. The reliance on the sufficient texture of images limits the application of this method. Optical flow can be used in several movement-control situations, such as terrain following [8], flight in the vicinity of obstacles or over flat terrain [9] and obstacle avoidance [10,11]. As for the control of hovering, optic flow is difficult with respect to measuring the height because it cannot produce a range unless there is a significant motion that goes against the objective of staying at a near perfect hover. Vision has achieved impressive results for the hovering of MAVs, which is fused with other sensors, especially the inertial measurement units (IMUs) [12–14]. These data fusion techniques are usually computationally complex, and the detecting reliability depends on several sensors.

While most of reviewed works have been done using vision assisted with other sensors, the vision-based hovering control proposed in this paper uses only one vision sensor. The MAV does not have any other sensors to measure its position or posture. Our method is inspired by honey bees, which utilize a stored snapshot as a visual marker and keep stable by the disparity between the current retinal image and the snapshot [15,16]. A camera is mounted directly under the MAV, and it takes an image of the ground underneath the MAV as a reference position. An error between the current positions and given reference position in the image plane is computed. When the height of the MFB changes, the distance of two points on the ground is invariable, but the distance of imaging points in the camera changes correspondingly. There is a negative correlation between these two distances. Therefore, we can achieve hovering control of the MFB by maintaining the distance variation near zero.

2. Vision-Based Position Estimation

For the purpose of achieving stable hovering, the positon of the MAV, which is relative to an external reference, such as the ground, should be determined at first. We estimate the position of the MAV based on visual information. In addition, a high-speed image processing algorithm also plays an important role to control the performance. Therefore, the image processing algorithm should be as fast as possible.

In this paper, we present an embedded MAV that uses information extracted from real-time images for hovering control. The ball-shaped outer shell of the MFB is lifted and propelled by the coaxial contra rotating twin rotors, which are mounted at the top of the ball (Figure 1). Four control surfaces at the button of the ball are used to control the motion direction of the ball freely. Therefore, the MFB almost has no restraints on its movement and can fly omnidirectionally. For the control of the hovering, the aircraft just needs the thrust of the coaxial contra rotating twin rotors, owing to its special power distribution, and it will always keep straight down to the ground.

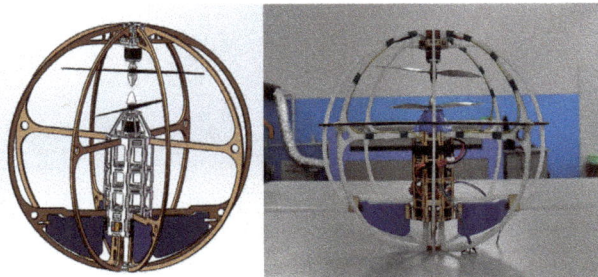

Figure 1. Miniature flying ball.

2.1. Monocular Vision Analysis

The camera is mounted directly under the MFB. In the camera coordinate system, the optical center of the camera is set as the origin, and the Z axis is parallel to the optical axis. The X and Y axes are parallel to the image plane, as shown in Figure 2a. A perspective projection model of a pinhole camera allows the position in a 2D image plane to be inferred from the 3D position [17]. The model projects an arbitrary point $A = (x_w, y_w, z_w)^T$ on the ground to point $a = (u, v)^T$ on the image plane (the camera image) expressed by Equation (1):

$$z_c \bullet \begin{bmatrix} u \\ v \\ 1 \end{bmatrix} = \begin{bmatrix} \frac{f}{d_x} & -\frac{f}{d_x}\cot\theta & u_0 & 0 \\ 0 & \frac{f}{d_y\sin\theta} & v_0 & 0 \\ 0 & 0 & 1 & 0 \end{bmatrix} \begin{bmatrix} R & t \\ 0^T & 1 \end{bmatrix} \begin{bmatrix} x_w \\ y_w \\ z_w \\ 1 \end{bmatrix} \tag{1}$$

where Z_c is an arbitrary positive scalar. The six intrinsic parameters $(f, d_x, d_y, \theta, u_0, v_0)$ can be derived from the calibration of the camera. The extrinsic parameters, the R and T metrics, denote the rotation and translation of the camera. For the control of hovering, the aircraft just needs the thrust upward. It will always keep nearly straight down to the ground, and the image plane of the camera will keep approximately parallel to the ground. Therefore, the extrinsic parameters can be simplified as:

$$R \doteq \begin{bmatrix} 1 & 0 & 0 \\ 0 & 1 & 0 \\ 0 & 0 & 1 \end{bmatrix}, T \doteq \begin{bmatrix} 0 \\ 0 \\ h \end{bmatrix} \tag{2}$$

where h denotes the height of the camera. Then, we can get Equation (3):

$$u - u_0 = \frac{fx_w}{h}, v - v_0 = \frac{fy_w}{h} \tag{3}$$

Consider the situation of a line AB on the ground and the corresponding line ab on the image plane. When the height of the MFB changes into h', as shown in Figure 2b, ab on the image plane will change into $a'b'$:

$$h = \frac{f \bullet l_{AB}}{l_{ab}}, h' = \frac{f \bullet l_{AB}}{l_{a'b'}} \tag{4}$$

From Equation (4), the measurement formula is obtained:

$$\frac{h}{h'} = \frac{l_{a'b'}}{l_{ab}} \tag{5}$$

The essential problem of the system is to achieve stable hovering of the MFB by maintaining $h' = h$. According to Equation (5), this condition can be transferred into keeping $l_{a'b'}$ nearly equal to l_{ab}. Therefore, the problem of controlling the height of the MFB is turned into finding the corresponding points ab and $a'b'$.

2.2. Measurement from Images

In order to reduce computing requirements, it is desirable to keep the detection region of images as small as possible. This also indicates that not all of the visual angles in the field of view have the same relationship for flight control. Visual angles pointing at the margin of images correspond to regions that may disappear during flight and, thus, do not require computation. For visual angles close to zero (*i.e.*, close to the center of the visual field), the measurable magnitude tends to be zero. These two limits suggest that the region of interest lies around the middle (Figure 3), where measurements are both reliable and relevant for controlling the aircraft.

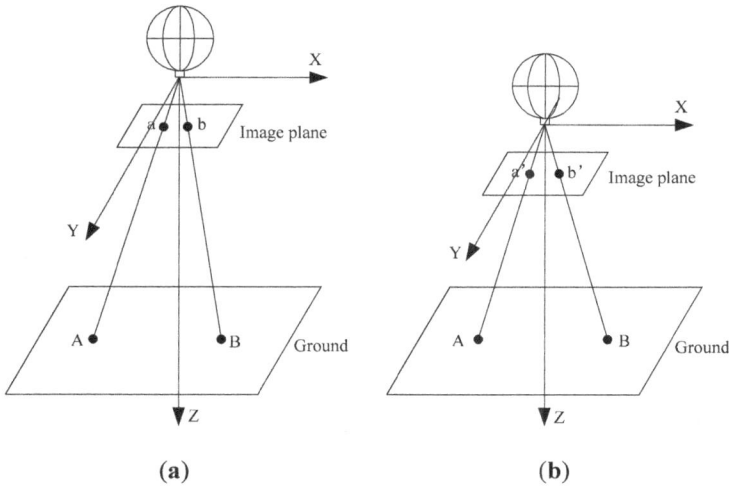

Figure 2. The monocular vision model and coordinate system used.

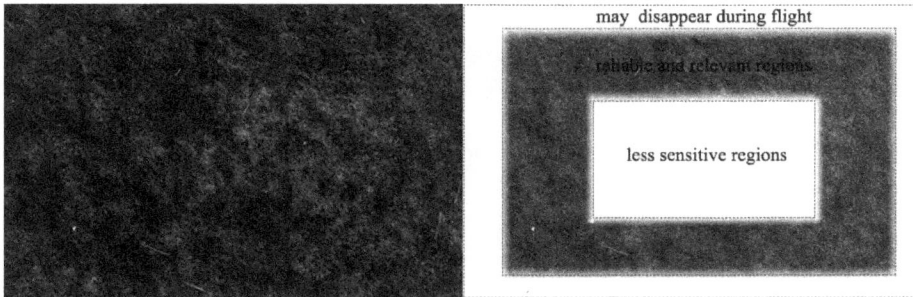

Figure 3. (**Left**) An image captured from a downwards looking camera of the miniature flying ball (MFB); (**Right**) Representation of the regions where measurements are both reliable and relevant.

2.2.1. Feature Extraction

The camera was calibrated to get the intrinsic parameters before use. Then, we can use the parameters to remove the geometrical distortion of the camera and obtain the accurate distances on the image plane.

In the follow-up process, we should find out the corresponding points between subsequence frames and the reference frame. The first step is to extract the features in the frames. There are some common features, such as point, region and contour. Generally speaking, feature point extractions, such as Harris, SUSAN (Smallest Univalue Segment Assimilating Nucleus) and DOG (Difference of Gaussian), are relatively easy in their calculation and are invariable against the rotation, translation and illumination variation.

In the application of our system, the sequence images may change not only in the rotation, translation and illumination variation, but also in the scale and the viewpoint. Therefore, we adopt the SIFT algorithm proposed by Lowe [18], which can deal with these situations. However, the algorithm of SIFT is too time-consuming and is impractical in our system. Therefore, we use the SURF algorithm [19], the improved version of SIFT, to achieve quicker speed and more feature points. The extraction result of SURF is shown in Figure 4.

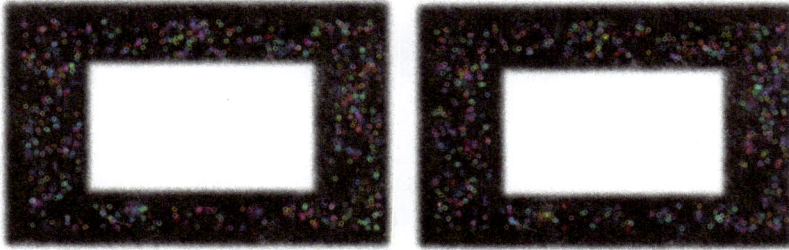

Figure 4. Feature extraction results of the reference frame (**Left**) and one of sequence frames (**Right**).

2.2.2. Feature Matching

After feature extraction, the corresponding points between frames should be matched. Conventional techniques include the correlation coefficient method, the Hausdorff distance method, measurements of similarity, *etc.* We can obtain a multitude of information, such as position, scale, principal orientation and feature vector, about the points when the feature extraction is executed. The feature vectors contain the neighborhood information of the feature points, so we can use the nearest neighbor method to find out the potential matching points to achieve a faster processing speed. The calculation formula can be expressed as follows:

$$D = \left\{ \min_{j=1,2,\dots,N_2} \sqrt{\sum_{i=1}^{k} \left(f_{ik} - f_{jk} \right)^2} \middle| i = 1, 2, \dots, N_1 \right\} \tag{6}$$

where N_1 and N_2 are the number of feature points extracted from two frames, respectively. In order to get a robust matching result, we pick out two minimal candidates denoted as d_{ij} and d_{ij}'. If $d_{ij} \leq \alpha \cdot d_{ij}'$ (α is set as 0.7 in the experiment), then d_{ij} is considered as the optimal matching point, and the wrong matching can be rejected effectively. The matching results are shown in Figure 5.

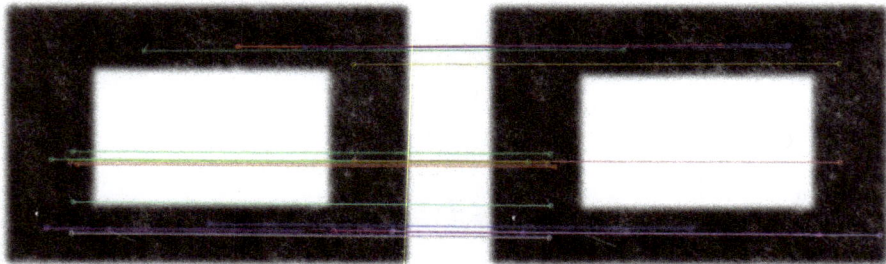

Figure 5. The matching result of two frames. The lines connect the match points between two frames.

2.2.3. Distance of Frames

In order to reduce measurement errors and get accurate data, we use the statistical method to calculate the distance between frames. Assume that we have found N pairs of matching points, and then, we can get the height and x, y variations between two frames from:

$$\begin{cases} \Delta x = \frac{\sum_{i=1}^{n} x_i}{n} - \frac{\sum_{i=1}^{n} x_i'}{n} \\ \Delta y = \frac{\sum_{i=1}^{n} y_i}{n} - \frac{\sum_{i=1}^{n} y_i'}{n} \\ \frac{h}{h'} = \frac{\sum_{i=1}^{n} a'b_i'}{\sum_{i=1}^{n} ab_i} \end{cases} \tag{7}$$

where (x_i, y_i) is the coordinate of the feature on the image plane at the target height and (x_i', y_i') is the coordinate at other moments. $a'b_i'$ and ab_i denote the Euclidean distances of each of the two features on the image plane of two moments. Then, we can get the variable $(\Delta x, \Delta y)$ in the x and y directions and variable $\frac{h}{h'}$ in the z direction.

From Equation (7), we can figure out that the position estimation of the MFB depends on the feature position on the image plane, but is unrelated to the intrinsic parameters of the camera. This is because the position parameters we consider here are relatively variable. The intrinsic parameters of the camera have been eliminated during the calculation. Thus, our method can avoid the measurement error introduced by the calibration error of the camera.

2.3. Kalman Filter

A Kalman filter can utilize the basic physics model of the MFB to estimate a dot from its previous position when it is not sensed correctly by the vision system. The Kalman filter will smooth the sensing results and prevent drastic changes in the control commands, which will cause the MFB to become unstable.

In the implementation, when a datum is considered as an outlier, the Kalman filter will respond by modifying the measurement update covariance matrix in order to reduce the confidence in that data. The state vector (T_j) of the Kalman filter is composed of h and the x and y position variations of the frames, which are being filtered (Equation (8)). The measurement input (Z_j) of the filter is also h and the x and y position variations of each frame, which make the H matrix of the update equations in the basic Kalman filter equations the identity matrix.

$$T_j = \begin{bmatrix} h \\ x \\ y \end{bmatrix} \tag{8}$$

Three variations are independent of each other, so the equations can be even further transformed, as shown in Equation (10). The equations are then simplified by breaking all matrix operations into single element operations.

$$\begin{aligned} \hat{u}(k)_j &= z(k)_j - \hat{t}(k)_{j|j+1} \\ \hat{s}(k)_j &= p(k,k)_{j|j-1} + r(k)_j \\ k(k)_j &= p(k,k)_{j|j-1} s(k)_j^{-1} \\ \hat{t}_{j|j} &= \hat{t}(k)_{j|j-1} + k(k)_j \hat{u}(k)_j \\ p(k,k)_{j|j} &= p(k,k)_{j|j-1} - k(k)_j p(k,k)_{j|j-1} \\ k &= 1, 2, 3. \end{aligned} \tag{9}$$

2.4. Controller Design

In order to achieve the desired position, three variables h, x and y of the MFB should be controlled with a specific controller. Couplings between the three channels are ignored for simplification. The h (height) and the x and y variations are controlled by three off-board PID controllers independently.

For the position x/y of the MFB, the two controllers are similar. P gain is limited to ensure the stability due to the latency of the system. The I term is used to reduce the steady-state errors. The vision data are used to calculate the D terms, since there are no other sensors to measure the speed. Because the field of view (FOV) of the camera is limited, it is necessary to set a threshold value as the output of the PID controller. Otherwise, the corresponding feature points will be out of the FOV, and then, the MFB will be uncontrollable.

For the h-controller, the processing method is basically the same as with x/y. The main difference is that the I term is adjusted to provide enough thrust for the hovering MFB according to different payloads and changing battery status.

3. Materials

To validate the proposed hovering control strategy, we ran a series of experiments in a real flying platform developed in our laboratory (Figure 1). In this section, we describe the hardware and software architecture that we used.

3.1. Hardware Description

The whole system (Figure 6) consists of two parts: the MFB aircraft, consisting of coaxial contra rotating twin rotors, control surfaces, an onboard controller and sensors, such as the camera and IMUs; and the ground station, which receives and processes sensor data, computes the pose of the aircraft, performs the controller strategy and sends the control commands back to the aircraft.

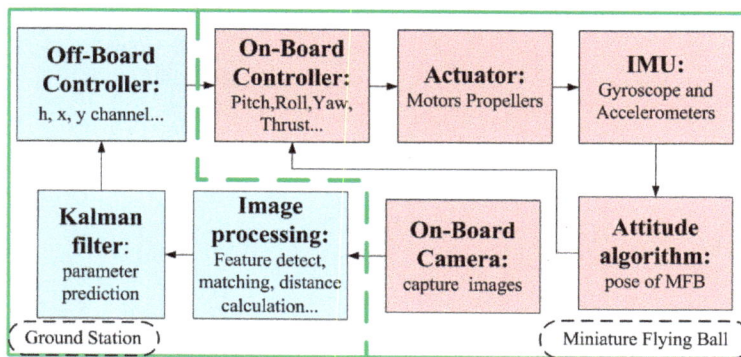

Figure 6. System overview.

3.1.1. MFB Aircraft

The MFB can perform at 200-Hz control frequency to drive the high-torque DC brushless motors, which enables rapid reaction to changes in the environment. A pair of coaxial rotors spin clockwise and anticlockwise, respectively, which counteracts the rotors' torque and provide thrust. This is different from similar works, such as the flying robot in [20] and the gimball in [21]. For them, the control surfaces are needed to balance the rotors torque, and this will cause the aircrafts to tilt toward the ground. Our aircraft just needs the thrust of the coaxial contra rotating twin rotors, and it will always keep straight down to the ground during hovering.

The IMU on the MFB aircraft consists of three gyroscopes sensing the angular velocity of each rotation axis and three accelerometers sensing the acceleration of each translation axis. It is used to provide the reference pose of the MFB. Limited by the space and payloads of the MFB, a miniature camera is used in the vision system. The specification of the camera will be discussed in more detail later in the article.

3.1.2. Ground Station

To avoid putting a heavy and powerful computer on the aircraft, images are wirelessly transmitted to a personal computer (PC), through our custom software. An off-board PC (Intel i5, 2 GB RAM) is used for data processing, pose estimation and performing the controller strategy in our system. The aircraft is equipped with the wireless image transmission subsystem and the command transmission subsystem modules, which can transmit image data and control commands at a rate of 20 Hz wirelessly.

3.2. Software Architecture

The main software on the ground station consists of two major parts: vision-based position estimation and a controller. Besides, there are two minor blocks in the entire software structure: the GUI block and the communication block. As shown in Figure 7, the GUI block can provide the interface between the entire system and the user for sensory data displaying, parameter controlling and parameter tuning. The communication block is used to integrate the hardware and software between the aircraft and the ground station.

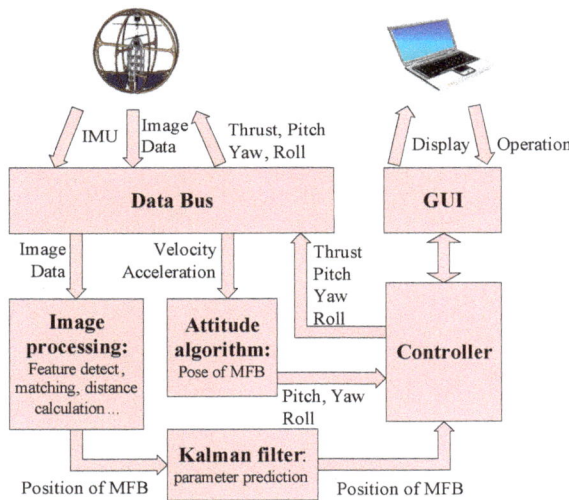

Figure 7. Software architecture.

4. Experiments

The performance of the vision-based hovering method is demonstrated using the MFB described in the previous sections. The diameter of the platform is 42 cm, and the total mass is 625 g. The parameters of the selected camera are as follows: the focus length is $8 \times 10^{-3}\,m$; the resolution of CCD is 640×480; the size of the CCD is $1/3''$; all algorithm development is in C++, which utilizes the OpenCV library for feature extraction. The experiment environment is a football field shown in Figure 8. The entire system was evaluated in four parts: (1) evaluating the solution of the system; (2) evaluating the static measurement precision of the system; (3) testing the dynamic hovering precision of the MFB; and (4) testing the robustness of the system.

Figure 8. The flying test on the football field.

4.1. Evaluating the Solution of the System

Through the derivation of Equation (4) with respect to the length on the image plane, we can get:

$$\Delta h = -\frac{f \cdot l}{x^2} \cdot \Delta x = -\frac{h^2}{f \cdot l} \cdot \Delta x \tag{10}$$

where Δh represents the height resolution according to the resolution of the CCD Δx. This means that we can sense the minimum change Δh with a specific CCD. Here, assume the height of MFB is set as 3 m. The parameters are substituted into Equation (10), and we can get the relation of the resolution (Δh) with the distance (l) from the center. As can be seen from Figure 9, the resolution near the center (less than 0.8 m) is too coarse, and it will become stable after a certain distance (large than 1.4 m). We can choose the appropriate distance and resolution as we need.

In Equation (10), the $-\frac{h \cdot \Delta x}{f \cdot l}$ is constant for the system. Therefore, the height resolution Δh is linearly related to the height h. This means that the higher the aircraft hover, the lower the detection precision will be. Therefore, our method is better applicable within a proper altitude according to the sensitivity of the detector.

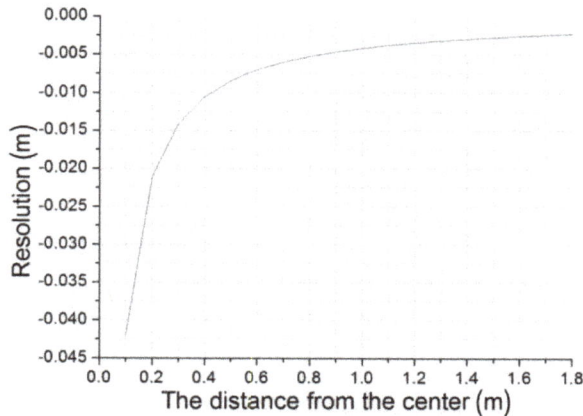

Figure 9. The relation of the resolution with the distance from the center.

4.2. Static Measurement Precision of the Vision System of MFB

In this part, we designed experiments to show the validity of our measurement method. The MFB was hung up by a rope, and its height can be accurately measured by measuring tape. The initial height of the MFB was set at 2 m, and the height could be moved up and down freely. The camera on the MFB looked straight down to the ground, and it took several pictures every time when the height of the MFB changed as planned. According to the schematic diagram presented in Figure 4 and the basic Equation (7), we can get the relationship of $\frac{h}{h'}$ and the actual distance variation (Figure 10).

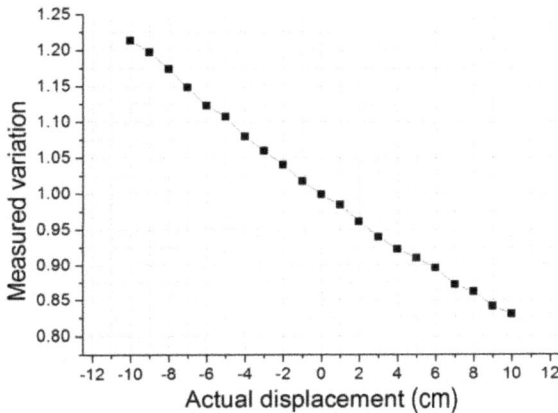

Figure 10. The relation of actual displacement and the measured variation.

The graph's horizontal axis shows the actual distance variation from the hovering positon, and the vertical axis shows the $\frac{h}{h'}$ calculated by Equation (7). Therefore, in the actual flight, we can figure out the actual distance variation from the hovering positon by $\frac{h}{h'}$, which we get from monocular vision. For example, when the $\frac{h}{h'}$ equal to 1, this means that the actual positon is at the hovering positon. When the $\frac{h}{h'}$ is less than 1, this shows that the actual positon is above the hovering positon. Otherwise, the actual positon is beneath the hovering positon. Therefore, we can maintain the value of $\frac{h}{h'}$ equal to 1 by controlling the motion of the MFB to achieve stable hovering.

4.3. Dynamic Hovering Precision of MFB

When hovering, the quadrotor is very stable. The 6 DOF of the MFB can be traced by an optical tracking system from NDI Polaris, which uses precise marks to track. The 6 DOF pose of the system obtained by NDI Polaris is used as the standard to inspect the hovering accuracy of the MFB. It can work at a speed of 60 frames per second and recover the position with a precision of around 0.3 mm RMS.

The main experimental processes are as follows: Firstly, the MFB is controlled by the remote control handle to an arbitrary height. Then enable the proposed hovering method is enabled to keep the MFB stable. During the hovering control, NDI Polaris is used to record the trajectory of the MFB. In Figure 11, the height was set at 3 m, and the flight path for a 60-s hover can be recorded and displayed. Overall, the position error during 60 s of hovering has an RMS value of 1.54 cm in the X position, 1.04 cm in the Y position and 0.92 cm in Z position, which yields an absolute RMS error value of 1.36 cm, as shown in Figure 12.

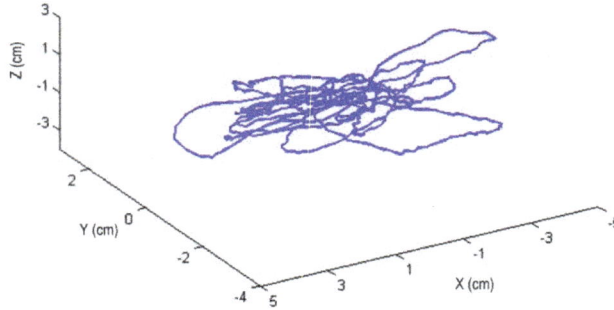

Figure 11. Position error while hovering during 60 s. The RMS value of the position error is 1.54 cm in *X*, 1.04 cm in *Y* and 0.92 cm in *Z*.

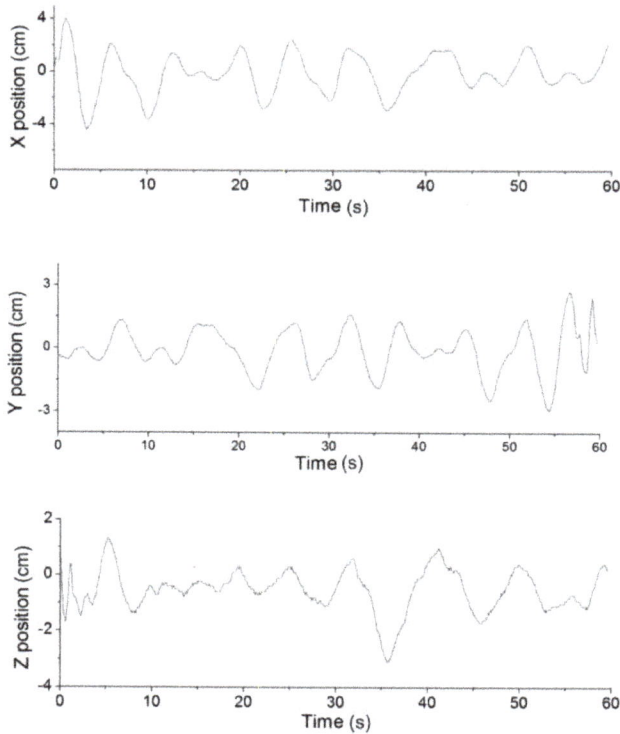

Figure 12. Position error in the *X*, *Y* and *Z* positions. The value remains ±4 cm. The *Z* position is more accurate than the *X* and *Y* positions.

4.4. Testing the Robustness of the System

We conduct this experiment to demonstrate the method's feasibility when the system is affected by a strong disturbance; that is to demonstrate if the MFB can go back to the original hovering altitude when it suffers a sudden change in height, so as to verify the robustness of the system. The experimental setup is the same as that of the previous experiments in Section 4.3. The only difference here is that a sudden height change is employed. To be specific, after the MFB is hovering for a while, we pull it down to certain height (24 cm in this trial) and observe the MFB's position variation.

The recorded data are shown in Figure 13. The blue line shows the height of the MFB; the red line is the desired height; and the black bars present the corresponding output of the vision system with the proposed method. Note that the outputs are too small to display, so they have been magnified 100-times here. In the experiment, during the first 37 s, the MFB was kept hovering at the expected position with a small deviation. Right at 38 s, it was pulled 24 cm down promptly and released soon. The output of the vision system reached the maximum at this moment. Note that a threshold is set to prevent drastic changes of the output, which will cause the MFB to become unstable. Then, during 38 to 48 s, the MFB returned back to the original height, as expected. In this process, the output decreases with the rising height of the MFB. After 48 s, the ball was hovering around the original height again. The graph above shows that our proposed vision system can always detect the MFB's orientation and distance and generate negative feedback to control the MFB hovering at the required height.

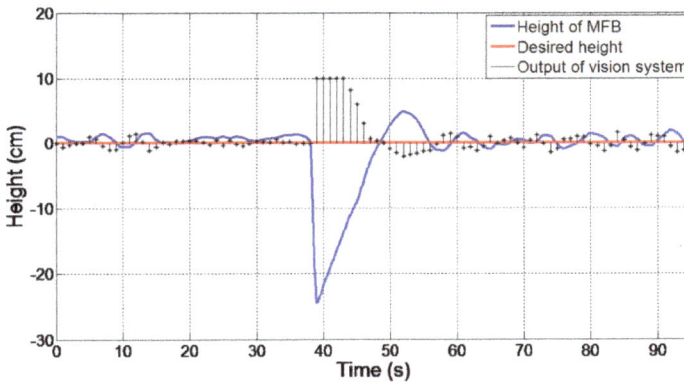

Figure 13. The response to the sudden height change.

4.5. Discussion

The experiments are based on the assumption that the MFB can always keep straight down to the ground, so that the image planes are completely parallel to each other. Of course, it is impossible to keep the image plane completely parallel without error during the flight. Here, we discuss the error caused by the vibration of the MFB.

Figure 14 shows the altitude variation recorded by the IMU of the MFB during one flight with coaxial rotors operating only. At the stage of preparing for taking off, the MFB was kept straight downward, and three Euler angles were initialized to 0. During the flight, the pitch and roll in the horizontal direction varied slightly and kept within a range of less than 1.5°. The yaw in the vertical direction deflected with small angles. When landed, the MFB rolled and bounced to slow down until stopping. This caused the three Euler angles to change drastically. After stopping, the three Euler angles kept steady.

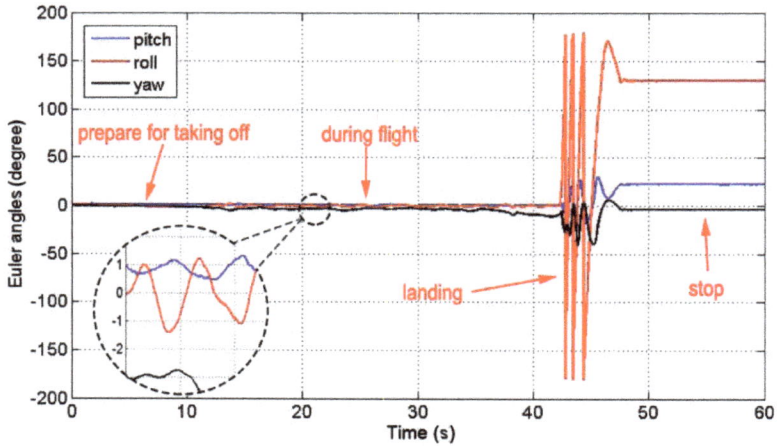

Figure 14. IMU data during one flight.

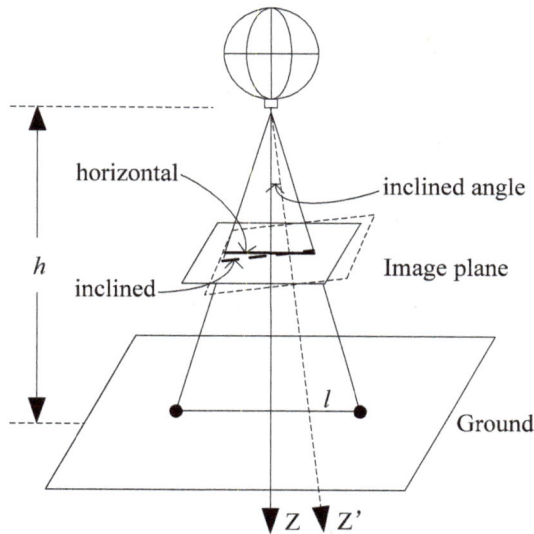

Figure 15. Inclined model of the MFB.

The schematic of the inclined image plane is shown in Figure 15. The relative error caused by the inclined angle to the measurement result can be expressed as:

$$\delta = \frac{((\tan(\arctan(l/h) + \Delta\alpha)) + (\tan(\arctan(l/h) - \Delta\alpha))) - 2l/h}{2l/h} \times 100\% \qquad (11)$$

where l is the distance of the feature to the center on the ground, h is the height of the MFB, $\Delta\alpha$ is the inclined angle and δ is the relative error. Plugging in the conventional value ($l = 1\,\text{m}, h = 3\,\text{m}$), then we can get the relationship of the relative error δ and the inclined angle $\Delta\alpha$, as shown in Figure 16.

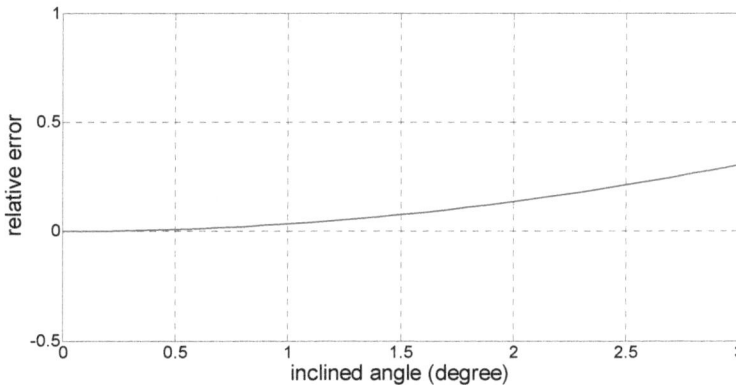

Figure 16. The relationship of the inclined angle with the relative error.

In Figure 16, we find that the pitch and roll in the horizontal direction keep within a range of less than 1.5° during normal flight. Then, the relative error caused by the inclined angle will not exceed 0.076%. Therefore, the effect of the inclined angle caused by the tilt of the MFB during flight has little impact on the measurement results. In some other cases, the MFB may be unstable with a great tilt angle caused by the wind or other unpredictable events.

5. Conclusions and Future Work

An onboard camera is utilized to measure the position of an MFB to maintain a stable hover. The position variation between the subsequent frames and the reference image can be acquired from analyzing their correspondence points. Then, the measurement is processed by the Kalman filter and sent to the PID controller to control the MFB. The experimental results show that the visual method is able to estimate the aircraft's positon in unknown environments and to guide the MFB to achieve stable hovering.

Several improvements will be made in the future work. First, we will develop a robust control algorithm, which allows the aircraft to adapt itself to the uncertainty state estimate. Secondly, our work in this paper lacks consideration of bad weather, especially wind, so we plan to expand the functionality of the system by visual techniques in order to adapt to more applications. Thirdly, from Figure 14, we can figure out that the maximum rotation of the MFB is about 10.6° during the 30-s flight. Although this has no effect on our hovering control method, we can derive the rotation of the system from Equation (7) and apply it in the precise control of the MFB in future work.

Author Contributions: Junqin Lin and Baoling Han conceived of and designed the experiments. Junqin Lin and Qingsheng Luo performed the experiments and analyzed the data. Junqin Lin wrote the paper.

Conflicts of Interest: The authors declare no conflict of interest.

References

1. Garratt, M.A.; Lambert, A.J.; Teimoori, H. Design of a 3D snapshot based visual flight control system using a single camera in hover. *Auton. Robot.* **2013**, *34*, 19–34. [CrossRef]
2. Amidi, O.; Kanade, T.; Fujiita, K. A visual odometer for autonomous helicopter flight. *Rob. Auton. Syst.* **1999**, *28*, 185–193. [CrossRef]
3. Oertel, C.-H. Machine vision-based sensing for helicopter flight control. *Robotica* **2000**, *18*, 299–303. [CrossRef]
4. Mejias, L.; Saripalli, S.; Campoy, P.; Sukhatme, G.S. Visual Servoing of an Autonomous Helicopter in Urban Areas Using Feature Tracking. *IFAC Proc. Vol.* **2007**, *7*, 81–86. [CrossRef]

5.   Bonak, M.; Matko, D.; Blai, S. Quadrocopter control using an on-board video system with off-board processing. *Rob. Auton. Syst.* **2012**, *60*, 657–667. [CrossRef]
6.   Gomez-Balderas, J.E.; Salazar, S.; Guerrero, J.A.; Lozano, R. Vision-based autonomous hovering for a miniature quad-rotor. *Robotica* **2014**, *32*, 43–61. [CrossRef]
7.   Cherian, A.; Andersh, J.; Morellas, V.; Papanikolopoulos, N.; Mettler, B. Autonomous altitude estimation of a UAV using a single onboard camera. In Proceedings of the IEEE/RSJ International Conference on Intelligent Robots and Systems, St. Louis, MO, USA, 10–15 October 2009; pp. 3900–3905.
8.   Garratt, M.A.; Chahl, J.S. Vision-Based Terrain Following for an Unmanned Rotorcraft. *IFAC Proc. Vol.* **2007**, *7*, 81–86. [CrossRef]
9.   Beyeler, A.; Dario, J.Z. Vision-based control of near-obstacle flight. **2009**, *27*, 201–219. [CrossRef]
10.   Zingg, S.; Scaramuzza, D.; Weiss, S.; Siegwart, R. MAV navigation through indoor corridors using optical flow. In Proceedings of the IEEE International Conference on Robotics and Automation (ICRA), Anchorage, AK, USA, 3–7 May 2010; pp. 3361–3368.
11.   Beyeler, A.; Zufferey, J.C.; Floreano, D. 3D vision-based navigation for indoor microflyers. In Proceedings of the IEEE International Conference on Robotics and Automation, Roma, Italy, 10–14 April 2007; pp. 1336–1341.
12.   Chowdhary, G.; Johnson, E.N.; Magree, D.; Wu, A.; Shein, A. GPS-denied Indoor and Outdoor Monocular Vision Aided Navigation and Control of Unmanned Aircraft. *J. F. Robot.* **2013**, *30*, 415–438. [CrossRef]
13.   Shen, S.; Mulgaonkar, Y.; Michael, N.; Kumar, V. Vision-Based State Estimation and Trajectory Control Towards High-Speed Flight with A Quadrotor. Available online: http://www.kumarrobotics.org/wp-content/uploads/2014/01/2013-vision-based.pdf (accessed on 5 June 2015).
14.   Zhang, T.; Li, W.; Kühnlenz, K.; Buss, M. Multi-Sensory Motion Estimation and Control of an Autonomous Quadrotor. *Adv. Robot.* **2011**, *25*, 1493–1514. [CrossRef]
15.   Cartwright, B.A.; Collett, T.S. How honey bees use landmarks to guide their return to a food source. *Nature* **1982**, *295*, 560–564. [CrossRef]
16.   Cartwright, B.A.; Collett, T.S. Landmark learning in bees. *J. Comp. Physiol. Psychol.* **1983**, *151*, 521–543. [CrossRef]
17.   Wu, A.D.; Johnson, E.N. Methods for localization and mapping using vision and inertial sensors. In Proceedings of the AIAA Guidance, Navigation and Control Conference and Exhibit, Honolulu, HI, USA, 18–21 August 2008.
18.   Lowe, D.G. Distinctive Image Features from Scale-Invariant Keypoints. *Int. J. Comput. Vis.* **2004**, *60*, 91–110. [CrossRef]
19.   Bay, H.; Ess, A.; Tuytelaars, T.; Van Gool, L. Speeded-Up Robust Features (SURF). *Comput. Vis. Image Underst.* **2008**, *110*, 346–359. [CrossRef]
20.   Klaptocz, A. Design of Flying Robots for Collision Absorption and Self-Recovery. Ph.D Thesis, École Polytechneque Fédérale de Lausanne, Lausanne, Switzerland, 2012.
21.   Briod, A.; Kornatowski, P.; Jean-Christophe, Z.; Dario, F. A Collision-resilient Flying Robot. *IFAC Proc. Vol.* **2007**, *7*, 81–86. [CrossRef]

![sensors logo] *sensors*

MDPI

Review

Driver Distraction Using Visual-Based Sensors and Algorithms

Alberto Fernández [1,*], Rubén Usamentiaga [2], Juan Luis Carús [1] and Rubén Casado [2]

[1]   Grupo TSK, Technological Scientific Park of Gijón, 33203 Gijón, Asturias, Spain; juanluis.carus@grupotsk.com

[2]   Department of Computer Science and Engineering, University of Oviedo, Campus de Viesques, 33204 Gijón, Asturias, Spain; rusamentiaga@uniovi.es (R.U.); rcasado@lsi.uniovi.es (R.C.)

*   Corrospondence: alberto.fernandez@grupotsk.com; Tel.: +34-984-29-12-12; Fax: +34-984-39-06-12

Academic Editor: Gonzalo Pajares Martinsanz

Received: 14 July 2016; Accepted: 24 October 2016; Published: 28 October 2016

Abstract: Driver distraction, defined as the diversion of attention away from activities critical for safe driving toward a competing activity, is increasingly recognized as a significant source of injuries and fatalities on the roadway. Additionally, the trend towards increasing the use of in-vehicle information systems is critical because they induce visual, biomechanical and cognitive distraction and may affect driving performance in qualitatively different ways. Non-intrusive methods are strongly preferred for monitoring distraction, and vision-based systems have appeared to be attractive for both drivers and researchers. Biomechanical, visual and cognitive distractions are the most commonly detected types in video-based algorithms. Many distraction detection systems only use a single visual cue and therefore, they may be easily disturbed when occlusion or illumination changes appear. Moreover, the combination of these visual cues is a key and challenging aspect in the development of robust distraction detection systems. These visual cues can be extracted mainly by using face monitoring systems but they should be completed with more visual cues (e.g., hands or body information) or even, distraction detection from specific actions (e.g., phone usage). Additionally, these algorithms should be included in an embedded device or system inside a car. This is not a trivial task and several requirements must be taken into account: reliability, real-time performance, low cost, small size, low power consumption, flexibility and short time-to-market. The key points for the development and implementation of sensors to carry out the detection of distraction will also be reviewed. This paper shows a review of the role of computer vision technology applied to the development of monitoring systems to detect distraction. Some key points considered as both future work and challenges ahead yet to be solved will also be addressed.

Keywords: driver distraction detection; visual-based sensors; image processing

1. Introduction

According to the most recent published World Health Organization (WHO) report, it was estimated that, in 2013, 1.25 million people were killed on the roads worldwide, making road traffic injuries a leading cause of death globally [1]. Most of these deaths happened in low- and middle-income countries, where rapid economic growth has been accompanied by an increased motorization and therefore, road traffic injuries. In addition to deaths on the roads, up to 50 million people incur non-fatal injuries each year as a result of road traffic crashes, while there are additional indirect health consequences associated with this growing epidemic. Road traffic injuries are currently estimated to be the ninth leading cause of death across all age groups globally, and are predicted to become the seventh leading cause of death by 2030 [1].

Distracted driving is a serious and growing threat to road safety [1]. Collisions caused by distracted driving have captured the attention of the US Government and professional medical organizations during the last years [2]. The prevalence and identification as a contributing factor in crashes is seen as an *epidemic* of American roadways, in words of Ray LaHood, when he was US Secretary of Transportation in 2012 [3]. There is not an exact figure regarding statistics about accidents caused by inattention (and its subtypes) since studies are made in different places, different time frames and therefore, different conditions. The studies referenced below show both the different statistics about inattention in general and those recorded when produced by distraction and fatigue in particular. These authors have estimated that distraction and inattention account for somewhere between 25% and 75% of all crashes and near crashes [4–8].

The trend towards increasing the use of in-vehicle information systems (IVISs) is critical [9] because they induce visual, manual and cognitive distraction [10] and may affect driving performance in qualitatively different ways [11]. Additionally, the advancement and prevalence of personal communication devices has exacerbated the problem during these last years [12]. All these factors can lead to the increment of the number of tasks subordinate to driving activity. These tasks, namely secondary tasks, which may lead to distraction [13], include eating, drinking, the act of taking something or tuning the radio or the use of cell phones and other technologies. The secondary tasks that take drivers' eyes off the forward roadway [14,15] reduce visual scan [16] and increase cognitive load may be particularly dangerous [13]. For example, the use of cell phones while driving, according to naturalistic studies [17], causes thousands of fatalities in the United States every year [18,19].

The purpose of this paper is the analysis of the state-of-the-art regarding the detection of drivers' distraction. The scope of the paper can be seen in Figure 1 and is commented as follows. The main methods for face detection, face tracking and detection of facial landmarks are summarized in Section 2 because they are a key component in many of the video-based inattention monitoring systems. In Sections 3–5, the main algorithms for biomechanical, visual and cognitive distraction detection are reviewed, respectively. Additionally, in Section 6, there are some algorithms detecting mixed types of distraction and, hence, are also reviewed. The relationship between facial expressions and distraction is also explored in Section 7. The key points for the development and implementation of sensors to carry out the detection of distraction will be considered in Section 8. In Section 9, the key ones to test and train driving monitoring systems are summarized. Privacy issues related to camera sensors are commented in Section 10. Lastly, conclusions, future aspects and challenges ahead will be considered in Section 11.

With the objective of introducing the scope and limitations of this review, some key aspects have been briefly introduced as follows. Driver distraction is just one form of inattention, which occurs when drivers divert their attention away from the driving task to focus on another activity. Therefore, a "complete" solution should consider all aspects of inattention. At least, the system should detect both distraction and drowsiness as the main contributing factors in crashes and near-crashes. As stated before, in this work, only distraction algorithms are summarized but one must not forget that other forms of inattention should be taken into account. Moreover, the use of on-board sensors already available in the vehicle to analyze driver behaviour is a low-cost and powerful alternative to the vision-based monitoring systems [20,21]. However, these systems should not be treated like different alternatives, because they can be used together (fusioned) in order to obtain indicators for monitoring [22]. Hence, for the sake of completeness, in this paper review only "purely" vision-based monitoring systems have been reviewed.

One of the challenges in decreasing the prevalence of distracted drivers is that many of them report that they believe they can drive safely while distracted [23]. However, for example, in connection with the use of mobile phones while driving, there is a great deal of evidence interacting with mobile devices, such as sending messages or engaging in conversations, which can impair driving performance because this interaction can create distraction. Moreover, a recent research showed that phone notifications alone significantly disrupted performance, even when drivers did not directly interact with a mobile device during the task [24]. Another study suggests that people in general can reduce both inattention and hyperactivity symptoms simply by silencing the smartphones and avoiding notifications [25]. Therefore, it is clear that drivers should not use and notice the presence of the smartphones inside the car while driving. It should be pointed out that distraction generation is a very complex process and is scarcely addressed here. We recommend some research papers that focused on driver distraction generation: Angell et al. [26] focused on the process of cognitive load in naturalistic driving; Liang et al. [27] addressed the adaptive behaviour of the driver under task engagement and their results on visual, cognitive and combined distraction; Caird analyzed the effects of texting on driving [28]. In the context of intelligent vehicles, Ohn et al. [29] highlights the role of humans by means of computer vision techniques.

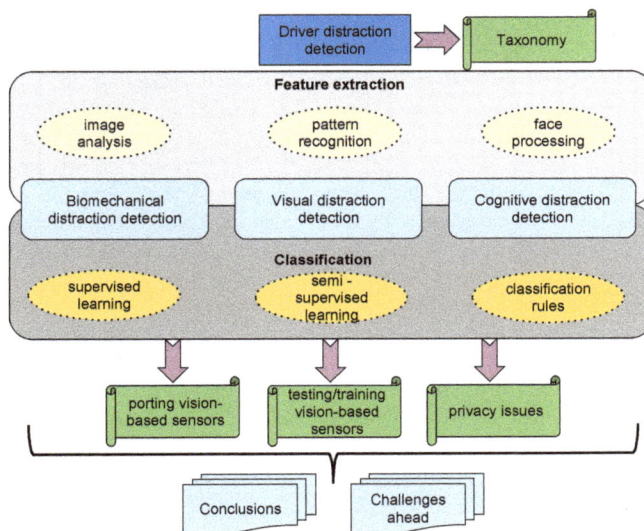

Figure 1. Scope of the present work.

1.1. Taxonomy

Both distraction and inattention have been inconsistently defined and the relationship between them remains unclear [30]. The use of different, and sometimes inconsistent, definitions of driver distraction can create a number of problems for researchers and road safety professionals [31]. Inconsistent definitions across studies can make the comparison of research findings difficult or impossible, can also lead to different interpretations of crash data and, therefore, to conclude different estimates of the role of distraction in crashes. This problem can be further seen in these recent works [32–35]. Many definitions have been proposed in order to define distraction [5,7,8,31]. Regan et al. [35] proposed a taxonomy of both driver distraction and inattention in which distraction is conceptualized as just one of several factors that may give rise to inattention. They concluded that driver inattention means *"insufficient or no attention to activities critical for safe driving"*. They defined driver distraction as *"the diversion of attention away from activities critical for safe driving toward a competing activity, which may result in insufficient or no attention to activities critical for safe driving"*. The definition proposed here is almost identical to that coined for driver distraction by Lee et al. [31].

It is acknowledged that the taxonomy proposed by Reagan et al. [35] suffers from "hindsight bias", that is, the forms of driver inattention proposed are derived from studies of crashes and critical incidents in which judgements have been made after the fact about whether or not a driver was attentive to an activity critical for safe driving [35]. Driving consists of a variety of sub-tasks and it may not be possible to attend to all at the same time. Determining which sub-task is more important (and the driver, thus, should attend to) can often only be determined after the fact (i.e., after a crash or incident occurred) and, hence, this attribution of inattention is somewhat arbitrary [36]. Additionally, the dynamics of distraction [37], which identifies breakdowns on interruption as an important contributor to distraction should also be considered as part of this taxonomy, and hence, timing and context have implications on the algorithm design that should be taken into account.

1.2. Methodology

Papers addressed in this review are within the topic of distraction detection using vision-based systems. The search and review strategy is described below. A comprehensive review of the English language scientific literature was performed. It encompassed the period from 1 January 1980 to 31 August 2016. The following databases were used: EBSCO, ResearchGate, ScienceDirect, Scopus, Pubmed, Google Scholar and Web of Knowledge. Search terms related to driver distraction were employed combining all of them: driver, visual, cognitive, manual, biomechanicall, vision, vision-based, impairment, distraction, distractions, review, task, tasks, inattention, performance, phone, sms, vehicle, problem, looking, face, head, pose, glasses, illumination, self-driving, tracking, sensors, image, traffic, safety, facts, privacy, issues, porting, taxonomy. Many items were returned from the search criteria shown before. These were, then, reviewed using the following criteria. Exclusion criteria were obviously non-relevant papers or from medical, electronic, networking, marketing and patent topics. Only publications from peer-reviewed English language journals were considered for inclusion. Additionally, reviewed papers were ordered by the number of references in order to include all relevant papers. Finally, in order to get the latest published papers, search filters were applied for this purpose. Search filters were applied to get publications only from years 2015 and 2016. References and bibliographies from the selected papers identified were examined to determine potentially additional papers. A total of approximately 1500 publications were revised in the review process.

2. Face and Facial Landmarks Detection

A common face processing scheme in many inattention monitoring systems, which can be seen in Figure 2, includes the following steps:

- Face detection and head tracking. In many cases a face detection algorithm is used as a face tracking one. In other cases, a face detection algorithm is used as an input for a more robust face tracking algorithm. When the tracking is lost, a face detection call is usually involved (that is why in Figure 2 these steps are placed inside the same external orange box).
- Localization of facial features (e.g., eyes). Facial landmarks localization is usually performed, but it should be noted that, in some cases, no specific landmarks are localized. So, in such cases, estimation of specific cues are extracted based on anthropometric measures from both face and head.

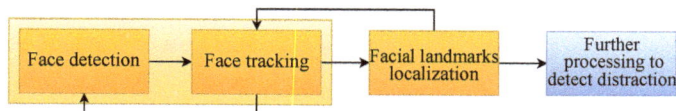

Figure 2. Common steps in most distraction monitoring systems.

2.1. Face Detection

Viola-Jones [38] have made object detection practically feasible in real world applications, which contains three main ideas that make possible to build and run in real time: the integral image, classifier learning with AdaBoost, and the attentional cascade structure [39]. This framework is used to create state-of-the-art detectors (e.g., face detector [40]), available, for example, in Opencv library.

However, this framework turned out to be really time-consuming [41]. Moreover, cascade detectors work well on frontal faces but sometimes, they fail to detect profile or partially occluded faces.

One possible solution is to use the standard approach for human detection [42], which can also be used for face detection [39]. This approach is based on the Histogram of Oriented Gradients (HOG), which is a feature descriptor used in computer vision and image processing for the purpose of object detection. This approach can be trained with less images and faster [43]. Deep Learning approaches can also be used for face detection. For example, in [44], a deep learning approach, called DP2MFD, is used. DP2MFD detects faces at multiples scales, poses and occlusion by integrating deep pyramid features with Deformable Parts Models (DPMs). Experiments were carried out on four publicly available unconstrained face detection datasets, which demonstrated the effectiveness of the approach. However, this face detector was tested on a machine with 4 cores, 12 GB RAM, 1.6 GHz processing speed and it took about 26 s. Consequently, complex features may provide better discrimination power than Haar-like features for the face detection task. However, they generally increase the computational cost [44].

Some modifications to the Viola-Jones algorithm have been proposed [45,46] to speed up the algorithm. For example, in [45], different optimization techniques to speed up the Viola-Jones detector for embedded smart camera applications have been discussed. In their paper, skin colour information is integrated with the Viola-Jones detector in order to reduce the computation time. PICO (Pixel Intensity Comparison-based Object detection) is another modification of the standard Viola-Jones object detection framework, which scans the image with a cascade of binary classifiers at all reasonable positions and scales [46]. This algorithm can achieve competitive results at high processing speed. This is especially evident on devices with limited hardware support for floating point operations. PICO outperforms the other two OpenCV detectors in terms of accuracy and processing speed.

Since driver face monitoring system should work in all light conditions, lighting and camera selection is one of the most important stage in the design of the system. Lighting devices not only should provide enough light in environment, but they also should not hurt his/her eyes. For example, learning-based methods (e.g., Viola-Jones algorithm or PICO) can also be used for face detection in Infrared (IR) images [47].

Position of the camera inside the car is another key factor in the detection rate. For example, in [48], if the camera is installed under the front mirror of the car, face detection has 85% accuracy. But if it is installed on the dashboard, face detection reaches up to 93%. This is because they used the Viola-Jones face detector, which is trained to distinguish faces that are tilted up to about 45° out of plane (towards a profile view) and up to about 15° in plane. Therefore, if the camera is installed on the dashboard, the captured images will contain frontal or near-frontal faces. In [49], the camera was placed over the steering wheel column for two reasons: a) it facilitates the estimation of gaze angles, such as pitch, which is relevant for detecting distraction, and b) from a production point of view, it is convenient to integrate a camera into the dashboard. On the downside, when the wheel is turning, there will be some frames in which the drivers face may be occluded by the steering wheel. However, the driver is seldom very sleepy or inattentive to traffic while turning the steering wheel.

2.2. Face Tracking

Head pose estimation can be defined as the ability to infer the orientation of a person's head relative to the view of a camera and different studies have reported statistics showing consistent range of head motion [50], which (see Figure 3) can be decomposed in:

- Saggital flexion/extension, i.e., forward to backward movement of the neck usually from −60° to 70°, which can be characterized by pitch angle.
- Axial rotation, i.e., right to left rotation of the head usually from −80° to 75°, which can be characterized by yaw angle.
- Lateral bending, i.e., right to left bending of the neck usually from −41° to 36°, which can be characterized by roll angle.

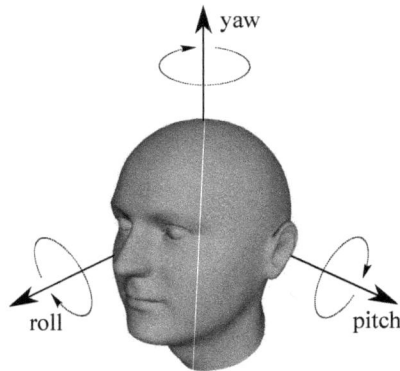

Figure 3. Head pose can be decomposed in pitch, yaw and roll angles.

Many vision-based algorithms for pose estimation have shown good performance when the head is near frontal, which is 95% of the time. But it is during those 5% of the time when interesting events, which are critical for safety, will occur [51]. Furthermore, as face orientation while driving is normally frontal, if the driver faces in other directions for a long period of time, this is probably due to fatigue or inattention [52]. Hence, a key component for a monitoring system based on face tracking is the ability to robustly and continuously operate even during large head movements. However, face tracking remains a challenging vision problem and, hence, a system for a continuous estimation of head movement is needed. On the other hand, as many head tracking algorithms have shown good performance when the head is near frontal, it can be concluded that the driver is looking away when tracking is unavailable. This information could be an alternative approach instead of adding more cameras to increase the range of the tracker.

Accordingly, numerous research works and publications have been trying to perform face tracking using a single camera and they are discussed as follows. Head pose estimation methods based on geometric approach using facial landmark and its 3D correspondences [49,53–56] can provide a good estimation and operate in real-time. For example, in [53], 3D pose estimation is achieved based on the position of the eyes and the mouth. A similar approach is proposed in [54], where only three points (eye centers and the middle point between the nostrils) are used to estimate continuous head orientation and gaze direction. Very closed to this approach, in [55], at least four prominent facial features are extracted from the face. After that, their correspondence on a 3D generic-face model is used to estimate head orientation. Oyini et al. [56] proposed the visual analysis of head position using a single camera aligning and scaling the 3D head model of the face according to the position and distance between the two eyes of the face in the 2D image. Another interesting approach recently published is [49], where a 3D head pose estimation system is proposed. This system is based on the 49 tracked 2D facial landmarks from Supervised Descent Method (SDM) tracker [57].

Other options include the combination of information [58–60], using for example, several classifiers [58,59] or combining 2D and 3D algorithms [60]. Asthana et al. [58] developed a system able to handle 3D pose variations up to ±45° in yaw and ±30° in pitch angles combining four different face detectors based on Viola-Jones framework. The drawback of this approach is that it requires four classifiers in order to track the face so it increases the execution time and memory requirements. In [59], the system consists of three interconnected modules, which detects drivers' head, provides initial estimates of head pose, and continuously tracks its position and orientation in six degrees of freedom. Pelaez et al. [60], combined 2D and 3D algorithms to provide head pose estimation and regions of interest identification based on 3D information from a range imaging camera.

Alternatively, more than a camera can be used to implement the tracking [51,61–63], that is, a distributed camera system is commonly used, where two or more cameras can be located inside the car cockpit. Following this line of research, in [61], they proposed a distributed camera framework for gaze estimation using head pose dynamics based on the algorithm proposed in [51]. They predict three gaze zones: right, front and left. In [51], a continuous head movement estimator (CoHMEt) is proposed, which independently tracks the head in each camera, and their outputs are further analyzed to choose the best perspective and corresponding head pose. When tracking is lost, due to either the loss of facial point detection or the rejection of the estimated points, reinitialization is performed using a scoring criterion. In [62], they also used a two-camera system to overcome challenges in head pose estimation, which allows for continuous tracking even under large head movements, as proposed in [51]. Therefore, following the setup of [51], a two-camera system can provide a simple solution in order to improve tracking during large head movements. Two cameras are also used in [63] for head pose estimation. Head pose is tracked over a wide operational range in the yaw rotation angle using both camera perspectives.

For a quantitative evaluation over the algorithms the Mean Absolute Error (MAE) is commonly used. Best results for the different algorithms can be seen in Table 1, where different databases are used. For example, in [49,56] the public database of Boston University (BU) is used to evaluate the performance of the proposed head pose estimation scheme. Some other algorithms used naturalistic on-road data set [59]. Moreover, some algorithms achieving good performance did not indicate any database [60]. LISA-P Head Pose database [55] introduces head pose data from on-road daytime and nighttime drivers of different age, race and gender, with continuous ground truth measurements and manual annotation of facial features. Therefore, this database can be used to compare head pose algorithms and head behaviour studies. The LISA-P Head Pose Database consists of 14 video sequences of drivers in on-road driving environment in natural and spontaneous conditions. The video sequences were collected at a frame rate of 30 frames per second, with a 640 × 480 pixel resolution.

Table 1. Mean Absolute Error (MAE) (in degrees) of face tracking algorithms comparison working in an automobile environment.

Algorithm	Roll(°)	Yaw(°)	Pitch(°)
La Cascia et al. [64]	9.8	4.7	2.4
Oyini et al. [56] average results 1 camera	5.3	3.9	5.2
Oyini et al. [56] uniform illumination 1 camera	4.8	3.8	3.9
Oyini et al. [56] varying illumination 1 camera	5.3	5.1	6.3
Vicente et al. [49] 1 camera	3.2	4.3	6.2
Pelaez et al. [60] 1 Kinect device	2.7	3.8	2.5
Murphy et al. [59] 1 camera	2.4	4.7	3.4
Tawari et al. [51] (MPS + POS) 1 camera	3.0	8.2	7.6
Tawari et al. [51] (MPS + POS) 2 cameras	3.8	7.0	8.6
Tawari et al. [51] (MPS + POS) 3 cameras	3.5	5.9	9.0
Tawari et al. [51] (CLM + POS) 1 camera	3.4	6.9	9.3
Tawari et al. [51] (CLM + POS) 2 cameras	3.6	5.7	8.8
Tawari et al. [51] (CLM + POS) 3 cameras	2.7	5.5	8.5

Based on the results from Table 1, in [56], the MAE decreased by an average of 1.3° due to illumination variations. In [51], the best performance of 3.9% failure rate, which is the percentage of the time that the system output is unreliable, is achieved with the three-camera view compared with that of over 15% for the single view, which is a significant improvement.

2.3. Location of Facial Features

The detection of facial features (also called landmarks) is an essential part of many face monitoring systems. The problem of the precise and robust detection of facial landmarks has drawn a lot of attention during this decade. State-of-the-art methods include tree models [65,66], DPM [67], SDM [57], explicit shape regression [68] or learning local binary features [69]. A comprehensive survey of facial feature point detection can be seen here [70]. All the above listed research suffers more or less from a lack of verification and performance analysis with a realistic variation in lighting conditions. Therefore, further research should be performed in order to adapt these algorithms to the traffic research in general and to the drivers' monitoring systems in particular. Difficulties for proper detection of drivers' facial features are mainly due to the non-uniformity of light sources, asymmetric shades on their face and eye regions, or rapid changes in light intensity during real-world driving due to shadows caused by buildings, bridges, trees, or, for example, when entering or leaving a tunnel [71].

Eyes, as one of the most salient facial features reflecting individuals' affective states and focus of attention [72], have become one of the most remarkable information sources in face analysis. Eye tracking serves as the first step in order to get glance behaviour, which is of most interest because it is a good indicator of the direction of the driver's attention [73]. Glance behaviour can be used to detect both visual and cognitive distraction [74]. It has also been used by many studies as an indicator of distraction while driving [75] and has been evaluated in numerous ways [73]. Therefore, both eye detection and tracking form the basis for further analysis to get glance behaviour, which can be used for both cognitive and visual distraction.

Eye tracking data is typically captured through the use of a vehicle instrumented with an in-vehicle eye tracker system. On one hand, complex systems consist of single or multiple cameras directed at the driver's face. As the number of face cameras increases, so does the ability of the system to capture larger and more dramatic head movements of the driver. On the other hand, simpler systems consisting of one or two cameras are usually less expensive and easier to install than more complex systems. For example, in [76], a comparison of eye tracking systems with one and three cameras using Smart Eye technology [77] is performed. The system uses a single standard camera of VGA resolution together with IR flash illuminators. The three-camera system used is the Smart Eye Pro [77], which has similar properties as the one-camera system, but it also facilitates gaze direction in full 3D.

Eye detection is required before eye region processing. Eye detection methods can be divided into two general categories: (1) methods based on imaging in IR spectrum; and (2) feature-based methods. A literature survey on robust and efficient eye localization in real-life scenarios can be seen in [72], and a review on eye localization in car environment can be seen in [78].

Methods based on imaging in IR spectrum, which are commonly called "hardware-based" approaches, rely on IR illuminators to generate the bright pupil effect to driver head pose and gaze estimation. These methods use two ring-type IR light-emitting diodes: one located near the camera optical axis and the other located far from it. This approach is often used to detect visual distraction. In contrast to these methods, in [79], the authors use a progressive camera and only one on-axis lighting source [80]. In this situation, the camera always produces images with bright pupils and image processing techniques are applied to detect pupils. Based on thresholding techniques, the possible pupils can be selected. An appearance model, trained using Principal Component Analysis (PCA) and Support Vector Machine (SVM), is exploited to verify the final pupils. To increase the robustness against eyeglasses, the Generalized Symmetry Transform (GST) is incorporated achieving a recognition rate of 99.4% and 88.3% for users not wearing and wearing eyeglasses, respectively.

Regarding feature-based methods, different techniques are commonly applied. Image binarization [81], projection [82,83], face anthropometric properties of the face [84], individual classifiers [85] or particle filtering [86] can be used to detect driver's eyes. For example, in [86], an algorithm for eyes tracking based on particle filtering is proposed. Their method works with a low-cost IR camera device at a low frame rate. They used a single particle filter to track both eyes at the same time. Evaluation was carried out in a driving simulator with five users achieving an average accuracy of 93.25%. In [85], two individual classifiers based on Haar-like features, one for the head and another for both eyes, were used. They tested face and eye detection in their research vehicle in daylight conditions achieving a hit rate of 97.2% for eye detection and a false alarmn of 4.6%.

All in all, the task of accurate eye localization is challenging due to the high degree of eyes appearance variability: facial expression variations, occlusion, pose, lighting and other imaging conditions and quality [72], are frequently encountered in car environments. Another problem that is scarcely addressed in the literature is that, in strong sunlight, the driver tends to squint, which makes, even more difficult to track the eyes. To mitigate these deficiencies, different approaches can be adopted. Sigari et al. [82] proposed to extract symptoms of hypo-vigilance based on eye-region processing but without explicit eye detection stage. Flores et al. [84] proposed a combination of algorithms in order to deal with illumination conditions for both day and night. Rezaei et al. [71] used a methodology to enhance the accuracy, performance and effectiveness of Haar-like classifiers, especially for complicated lighting conditions. These authors also proposed ASSAM [87], which is based on the asymmetric properties of the driver's face due to illumination variations. A good solution is also to use a "divide and conquer" strategy to handle different variations at different stages [72].

3. Biomechanical Distraction

In connection with biomechanical detection and recognition using computer vision techniques, we can find two approaches. The first one involves hands secondary tasks recognition involving hands action, while the second one is based on hands tracking and information.

3.1. Secondary Tasks Involving Biomechanical Distraction

Zhao et al. [88–91] proposed different maching learning approaches to detect predefined driving postures, where four predefined postures were considered: (1) grasping the steering wheel; (2) operating the shift lever; (3) eating; and (4) talking on a cellular phone, which are recorded from the passenger seat, that is, from the right profile view of the driver. Yan et al. [92] proposed a combination of the Motion History Image (MHI) and POHG, and the application of Random Forest (RF) classifier for driving actions recognition. Trying to improve the accuracy of the aforementioned approach, the same authors included a Convolutional Neural Network (CNN) [93], which was tested over three datasets covering four driving postures: (1) normal driving; (2) responding to a cell phone call; (3) eating; and (4) smoking. For fair comparison, Yan et al. [93] re-implemented aforementioned state-of-the-art approaches [88–91] and carried out experiments on other two popular vision descriptor approaches (PHOG [94] and SIFT [95]). Classification accuracy of all of these methods can be seen in Table 2 evaluated on the Southeast University (SEU) driving posture dataset [88].

In connection with secondary tasks recognition, different computer vision algorithms have been proposed in order to detect cell phone usage of the driver while driving [96–100]. High recognition rates are usually obtained (from 86.19% to 95%) using very different approaches. Computer vision techniques seem to be the best approach for this task, whose results can be seen in Table 3, compared to other non-computer vision algorithms relying on inertial sensors of the mobile phone [101]. Best results are obtained by the algorithm proposed by Xu et al. [99], which consists of two stages: first, the frontal windshield region localization using DPM; next, they utilized Fisher vectors (FV) representation to classify the driver's side of the windshield into cell phone usage violation and non-violation classes. The proposed method achieved about 95% accuracy with a dataset of more than 100 images with drivers in a variety of challenging poses with or without cell phones.

It can be concluded that many different computer vision and machine learning techniques can be used to recognize predefined postures involving hand gestures. The CNN model offered a better performance than other approaches but with some limitations. The algorithm needs high computational resources making difficult to be applied in some conditions with common hardware architecture (e.g., embedded systems). Moreover, training a CNN needs a large amount of data, which is also difficult to obtain in some scenarios.

Table 2. Classification accuracy evaluated on the Southeast University (SEU) driving posture dataset [88].

Algorithm	Features	Classifier	Average Accuracy (%)
Zhao et al. [88]	Homomorphic filtering, skin-like regions segmentation and Contourlet Transform (CT)	RF	90.63
Zhao et al. [89]	Geronimo-Hardin-Massopust (GHM) multiwavelet transform	Multiwavelet Transform	89.23
Zhao et al. [90]	Histogram-based feature description by Pyramid Histogram of Oriented Gradients (PHOG) and spatial scale-based feature description	Perceptron classifiers	94.20
Zhao et al. [91]	Homomorphic filter, skin-like regions segmentation, canny edge detection, connected regions detection, small connected regions deletion and spatial scale ratio calculation	Bayes classifier	95.11
Bosch et al. approach [94]	PHOG	SVM	91.56
Lowe et al. approach [95]	SIFT	SVM	96.12
Yan et al. [93]		CNN	99.78

Table 3. Computer vision algorithms to detect cell phone usage. High recognition rates are usually obtained using very different approaches.

Algorithm	Features	Classifier	Recognition Rate (%)
Zhang et al. [96]	Features from the driver's face, mouth and hand	Hidden Conditional Random Fields (HCRF)	91.20
Artan et al. [97]	Image descriptors extracted from a region of interest around the face	SVM	86.19
Berri et al. [98]	Percentage of the Hand and Moment of Inertia	FV	91.57
Xu et al. [99]	DPM	FV	95
Seshadri et al. [100]	Raw pixels and HOG features	Real AdaBoost, SVM, RF	93.86

3.2. Hands Information

Hand detection is a challenging problem as human hands are highly deformable and are also exposed to different illumination conditions [102]. One approach for object detection relies on a sliding-window, where a model is learned based on positive samples (i.e., hands in different poses) of fixed size and negative samples with no hands. A classifier is then used to learn a classification rule. In order to detect hands at different scales, this scheme can be applied on hand images at different sizes. But a sliding window-based approach trained on hand instances was shown to be prone to false positive detection rates [103]. A recent common approach to improve the results is the assumption that hands can only be found in a small and predefined set of regions [103,104].

As opposed to training a model for hand shape or appearance and running a sliding window detector, two different approaches are analyzed in [103] taking into account three activity classes:

(1) two hands on the wheel; (2) hands on the instrument panel and (3) hand on the gear shift. The motion-cue-based hand approach uses temporal accumulated edges in order to maintain the most reliable and relevant information motion and then, it is fitted with ellipses in order to produce the location of the hands. The static-cue-based approach uses features in each frame in order to learn a hand presence model for each of the three regions and a second-stage classifier (SVM) produces the final activity classification. Martin et al. [104] also constraint the problem of hands detection to a number of regions of interest. They used HOG at different scales. Afterwards, a SVM is used to learn a hand presence in each of the three regions and 'two hands on the wheel' model for the wheel region. A similar approach is proposed in [102], training a linear SVM model for each region using a different set of descriptors.Ohn et al. [62] incorporated hand gestures in order to study preparatory motions before a maneuver had been performed, training a hand detector using fast feature pyramids. Gradient and colour channels are extracted for each patch image. They used CIE-LUV colour channels because they worked better compared to RGB and HSV. Afterwards, an AdaBoost classifier was applied in order to learn the features from the hands and finally, they trained a SVM-based detector using HOG features to differentiate the left hand from the right one. Later on, Ohn et al. [105] also explored the use of a pyramidal representation for each region of interest using HOG finding that edge features are particularly successful in the task of hands detection.

In order to compare these algorithms, a dataset of synchronized RGB and depth videos collected in an operating vehicle was proposed [106]. The CVRR-HANDS 3D dataset was designed in order to study natural human activity under difficult settings (background, illumination, occlusion) containing three subsets: (1) hand localization; (2) hand and objects localization; and (3) 19 hand gestures for occupant-vehicle interaction. Five regions of interest were considered: (1) wheel; (2) lap; (3) hand rest; (4) gear; and (5) instrument panel. Recognition rates from some of these previous algorithms using this database can be seen in Table 4.

Table 4. Hands recognition in different regions inside the car using CVRR-HANDS 3D dataset [106].

Algorithm	Features	Classifier	Regions	Recognition Rate (%)
Ohn et al. [106]	RGB data	SVM	5	52.1
Ohn et al. [106]	RGB combined with depth data	SVM	5	69.4
Martin et al. [104]	Hands cues	SVM	3	83
Martin et al. [104]	Hands and head cues	SVM	3	91
Ohn et al. [105]	Hands cues	SVM	3	90
Ohn et al. [105]	Hands and head cues	SVM	3	94

Summarizing, a common approach is to recognize if the hands are positioned in one of the established areas (wheel, gearbox and so on) and to track them over time. It could be considered that the steering wheel is the critical area because it is where hands should remain most of the time while driving. If hands remained in a non-critical zone for a certain period of time, which could be different for each of the non-critical areas, an alarm would be created to warn drivers to lay their hands in the correct position.

Hand Disambiguation

There is another interesting problem to solve related to hands detection that needs further research: hand disambiguation [107]. Once hands are detected, it is crucial to ensure that the hands belong to the driver. Both hand disambiguation and hand activity detection should be studied and considered together in order to infer final, clear and unambiguous results.

4. Visual Distraction

Visual distraction is often related to the on-board presence of electronic devices such as mobile phones, navigation or multimedia systems, requiring active control from the driver. It can also be related to the presence of salient visual information away from the road causing spontaneous off-road eye glances and momentary rotation of the head. A 2006 report on the results of a 100-car field experiment [4] showed that almost 80% of all crashes and 65% of all near-crashes involved drivers looking away from the forward roadway just prior to the incident.

Engagements in visually distracting activities divert drivers' attention from the road and cause occasional lapses, such as imprecise control of the vehicle [108], missed events [28], and increasing reaction times [108]. Visual time sharing between the driving task and a secondary task reveals that the glance frequency to in-car devices is correlated to the task duration, but the average glance duration does not change with task time or glance frequency [109]. Drivers do not usually increase the glance duration for more difficult or longer tasks but rather increase the accumulated visual time sharing duration by increasing the number of glances away from the road [110]. As both single long glances and accumulated glance duration have been found to be detrimental for safety [110–112], a driver distraction detection algorithm based on visual behaviour should take both glance duration and repeated glances into account [113].

One one hand, high-resolution cameras placed throughout the cabin are needed to view the driver's eyes from all head positions and at all times. Several economic and technical challenges of integrating and calibrating multiple cameras should be tackled to achieve this. Technically, eye orientation cannot always be measured in vehicular environments because eye region can be occluded by (1) sunlight reflections on eyeglasses; (2) the eye blink of the driver; (3) a large head rotation; (4) sunglasses; (5) wearing some kind of mascaras; (6) direct sunlight; (7) hats, caps, scarves; or (8) varying real world illumination conditions.

On the other hand, many security systems do not require such detailed gaze direction but they need coarse gaze direction to reduce false warnings [114,115]. For example, forward collision warning (FCW) systems need not only exterior observations but interior observations of the driver's attention as well to reduce false warnings (distracting and bothering the driver), that is, coarse gaze direction can be used in order to control the timing of warning emission when the system detects that the driver is not facing forwards.

Taking into account that errors in facial feature detection greatly affect gaze estimation [116], many researchers have measured coarse gaze direction by using only head orientation with the assumption that coarse gaze direction can be approximated by head orientation [117]. Head pose is a strong indicator of a driver's field of view and his/her focus of attention [59]. It is intrinsically linked to visual gaze estimation, which is the ability to characterize the direction in which a person is looking [118]. However, it also should be noted that drivers use a time-sharing strategy when engaged in a visual-manual task where the gaze is constantly shifted between the secondary task and the driving scene for short intervals of time [119] and often position the head in between the two involved gaze targets and only uses the eyes to quickly move between the two targets. In this situation, a face tracking algorithm would recognize this as a distracted situation based on head position, but the driver is constantly looking the road ahead. Therefore, in an ideal situation, both driver gaze tracking and eyes-off-road should be detected together [49].

In short, visual distraction can be categorized into two main approaches as it can be seen in Figure 4. In the first approach, which can be called "coarse", researchers measured the coarse gaze direction and the focus of attention by using only head orientation with the assumption that the coarse gaze direction can be approximated by the head orientation. In the second approach, which can be called "fine", researchers considered both head and eye orientation in order to estimate detailed and local gaze direction.

Moreover, considering its operating principles, visual distraction systems can be grouped in two main categories: hardware- and software-based methods. Additionally, some systems can combine these two approaches and therefore, a third category can also be considered, as seen in Figure 4.

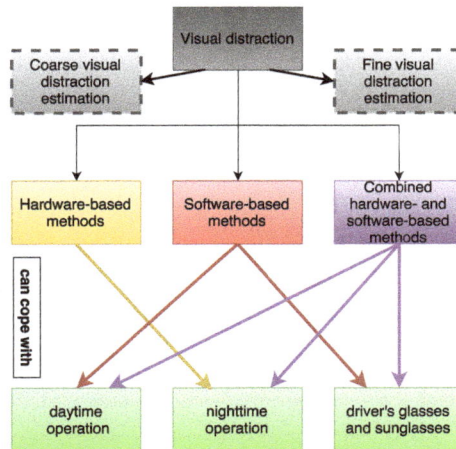

Figure 4. Visual distraction algorithms categorization.

4.1. Hardware-Based Methods to Extract Gaze Direction

Hardware-based approaches to head pose and gaze estimation rely on Near Infrared (NIR) illuminators to generate the bright pupil effect. These methods use two ring-type IR light-emitting diodes: one located near the camera's optical axis and the other located far from it [120–126]. The light source near the camera optical axis makes a bright pupil image caused by the red-eye effect, and the other light source makes a normal dark pupil image. The pupil was, then, easily localized by using the difference between bright and dark pupil images. Ji et al. used the size, shape, and intensity of pupils, as well as the distance between the left and right pupil, to estimate the head orientation. Specifically, the authors used the pupil-glint displacement to estimate nine discrete gaze zones [121,122], a geometric disposition of the IR LEDs similar to that of Morimoto et al. [120] and two Charge Coupled Device (CCD) cameras embedded on the dashboard of the vehicle. In connection with the CCD cameras, the first one is a narrow angle camera, focusing on the driver's eyes to monitor eyelid movement while the second one is a wide angle camera focusing on his/her head to track and monitor head movement. Based on this work, Gu et al. [124] proposed a combination of the Kalman filtering with the head motion to predict the features localization and used Gabor wavelet in order to detect the eyes constrained to the vicinity of predicted location. Another existent approach proposed by Batista et al. used dual Purkinje images to estimate a driver's discrete gaze direction [125]. A rough estimation of the head-eye gaze was described based on the position of the pupils. The shape of the face is modeled with an ellipse and the 3D face pose is recovered from a single image assuming a ratio of the major and minor axes obtained through anthropometric face statistics. In this method, further research is necessary in order to improve the accuracy of the face orientation estimation, which is highly dependent on the image face ellipse detection.

The aforementioned NIR illumination systems work particularly well at night. The major advantage of these methods is the exact and rapid localization of the pupil. However, performance can drop dramatically due to the contamination introduced by external light sources [126,127]. In addition, during daytime, sunlight is usually far stronger than NIR light sources and hence, the red-eye effect may not occur. Moreover, these methods could not work with drivers wearing glasses because the lenses create large specular reflections and scatter NIR illumination [127–129]. While the contamination due to artificial lights can easily be filtered with a narrow band pass filter, sunlight contamination will still exist [126]. Furthermore, such systems are vulnerable to eye occlusion caused by head rotation and blinking [114].

4.2. Software-Based Methods to Extract Gaze Direction

Combining facial feature locations with statistical elliptical face modelling, Batista et al. [83] presented a framework to determine the gaze of a driver. To determine the gaze of the face, an elliptical face modelling was used taking the eye's pupil locations to constraint the shape, size and location of the ellipse. The proposed solution can measure yaw head rotation over $[-30°, +30°]$ interval and pitch head rotation over $[-20°, +20°]$ interval.

Furthermore, despite the technical challenges of integrating multiple cameras, Bergasa et al. [130] proposed a a subspace-based tracker based on head pose estimation using two cameras. More specifically, the initialization phase was performed using the Viola and Jones algorithm [40] and a 3D model of the face was constructed and tracked. In this work, head pose algorithm, which was the base for visual distraction estimation, could track the face correctly up to $[-40°, +40°]$.

A limitation of the software-based methods is the fact that they cannot often be applied at night [126,131]. This has motivated some researchers to use active illumination based on IR LEDs, exploiting the bright pupil effect, which constitutes the basis of these systems [126,131] (explained in previous section), or combine both methods, which can be seen in the next section.

4.3. Hardware- and Software-Based Methods to Extract Gaze Direction

Lee et al. [114] proposed a system for both day and night conditions. A vision-based real-time gaze zone estimator based on a driver's head orientation composed of yaw and pitch is proposed. The authors focused on estimating a driver's gaze zone on the basis of his/her head orientation, which is essential in determining a driver's inattention level. For night conditions, additional illumination to capture the driver's facial image was provided. The face detection rate was higher than 99% for both daytime and nighttime.

The use of face salient points to track the head was introduced by Jimenez et al. [132], instead of attempting to directly find the eyes using object recognition methods or the analysis of image intensities around the eyes. The camera was modified to include an 850 nm band-pass filter lens covering both the image sensor and the IR LEDs in order: (a) to improve the rejection of external sources of IR radiation and reduce changes in illumination and (b) to facilitate the detection of the pupils, because the retina is highly reflective of the NIR illumination of the LEDs. An advantage of salient points tracking is that the approach is more robust to the eyes occlusion whenever they occur, due to the driver's head or body motion.

Later on, the same authors extended their prior work in order to improve non-invasive systems for sensing a driver's state of alert [133]. They used a kinematic model of the driver's motion and a grid of salient points tracked using the Lukas-Kanade optical flow method [132]. The advantage of this approach is that it does not require one to directly detect the eyes, and therefore, if the eyes are occluded or not visible from the camera when the head turns, the system does not loose the tracking of the eyes or the face, because it relies on the grid of salient points and the knowledge of the driver's motion model. Experiments involving fifteen people showed the effectiveness of the approach with a correct eyes detection rate of 99.41% on average. It should be noted that this work is focused on sensing the drivers' state of alert, which is calculated measuring the percentage of eyelid closure over time (PERCLOS), and it is not focused on distraction detection.

Eyes Off the Road (EOR) detection system is proposed in [49]. The system collects videos from a CCD camera installed on the steering wheel column and tracks facial features. Using a 3D head model, the system estimates the head pose and gaze direction. For night time operation, the system requires an IR illumination. The proposed system does not suffer from the common drawbacks of NIR based systems [121,122,125], because it does not rely on the bright pupil effect. The system works reliably with drivers of different ethnicities wearing different types of glasses. However, if the driver is wearing sunglasses, it is not possible to robustly detect the pupil. Thus, to produce a reliable EOR estimation in this situation, only head pose angles are taken into account.

Cyganek et al. [134] proposed a setup of two cameras operating in the visible and near infra-red spectra for monitoring inattention. In each case (visible and IR) two cascade of classifiers are used. The first one is used for the detection of the eye regions and the other for the verification stage.

Murphy-Chutorian et al. used Local Gradient Orientation (LGO) and Support Vector Regression (SVR) to estimate the driver's continuous yaw and pitch [135]. They used head pose information extracted from a LGO and SVR to recognize drivers' awareness. The algorithm was further developed in [59] by introducing a head tracking module built upon 3D motion estimation and a mesh model of the driver's head. There is a general weakness here as the tracking module may easily diverge from face shapes that are highly different to the given mesh model.

4.4. Driver Distraction Algorithms Based on Gaze Direction

In these previous Sections 4.1–4.3, gaze direction is extracted using different methods. The next step is to detect distraction using gaze direction regardless of the type of method used to extract this information, and hence, is commented as follows.

Many software-based methods have been proposed in order to detect visual distraction, many of which, rely on "course" information extracted from visual cues [114,136–139]. Hattori et al. [136] introduced a FCW system using drivers' behavioural information. Their system determines distraction when it detects that the driver is not looking straight ahead. Following this approximation, an Android app [137] has been developed to detect and alert drivers of dangerous driving conditions and behaviour. Images from the front camera of the mobile phone are scanned to find the relative position of the driver's face. By means of a trained model [38] four face related categories were detected: (1) no face is present; (2) facing forwards, towards the road; (3) facing to the left and (4) facing to the right. Another related system is proposed by Flores et al. [138] where, in order to detect distraction, if the system detects that the face position is not frontal, an alarm cue is issued to alert the driver of a danger situation. Lee et al. [114] proposed a vision-based real-time gaze zone estimator based on a driver's head orientation composed of yaw and pitch. This algorithm is based on normalized histograms of horizontal and vertical edge projections combined with an ellipsoidal face model and a SVM classifier for gaze estimation. In the same research line but in a more elaborated fashion, Yuging et al. [139] used machine vision techniques to monitor the driver's state. The face detection algorithm is based on detection of facial parts. Afterwards, the facial rotation angle is calculated based on the analysis of the driver's head rotation angles. When the angle of facial orientation is not in a reasonable range and lasts for a relatively long time, it can be thought that the driver is distracted and warning information will be provided.

Additionally, other software-based approaches rely on "fine" information considering both head and eye orientation in order to estimate distraction [83,130,140,141]. Pohl et al. [140] focused on estimating the driver's visual distraction level using head pose and eye gaze information with the assumption that the visual distraction level is non-linear: visual distraction increased with time (the driver looked away from the road scene) but nearly instantaneously decreased (the driver re-focused on the road scene). Based on the pose and eye signals, they established their algorithm for visual distraction detection. Firstly, they used a Distraction Calculation (DC) to compute the instantaneous distraction level. Secondly, a Distraction Decision-Maker (DDM) determined whether the current distraction level represented a potentially distracted driver. However, to increase the robustness of the method, also the robustness of the eye and head tracking device to adverse lighting conditions has to be improved.

Bergasa et al. [126] presented a hardware- and software-based approach for monitoring driver vigilance. It is based on a hardware system, for real time acquisition of driver's images using an active IR illuminator and a software implementation for real time pupil tracking, ocular measures and face pose estimation is proposed. Finally, driver's vigilance level is determined from the fusion of the measured parameters into a fuzzy system. The authors yielded an accuracy percentage close to 100% both at night and for users not wearing glasses. However, the performance of the system decreases during daytime, especially in bright days, and at the moment, the system does not work with drivers wearing glasses [126].

Recently, Lee et al. [141] evaluated four different vision-based algorithms for distraction under different driving conditions. These algorithms were chosen for their ability to distinguish between distracted and non-distracted states using eye-tracking data [141]. The resulting four algorithms, summarized in Table 5, are commented next:

1. Eyes off forward roadway (EOFR) estimates distraction based on the cumulative glances away from the road within a 6-s window [7].
2. Risky Visual Scanning Pattern (RVSP) estimates distraction by combining the current glance and the cumulative glance durations [142].
3. "AttenD" estimates distraction associated with three categories of glances (glances to the forward roadway, glances necessary for safe driving (i.e., at the speedometer or mirrors), and glances not related to driving), and it uses a buffer to represent the amount of road information the driver possesses [143–145].
4. Multi distraction detection (MDD) estimates both visual distraction using the percent of glances to the middle of the road and long glances away from the road, and cognitive distraction by means of the concentration of the gaze on the middle of the road. The implemented algorithm was modified from Victor et al. [146] to include additional sensor inputs (head and seat sensors) and adjust the thresholds for the algorithm variables to improve robustness with potential loss of tracking.

Table 5. AUC comparisons by algorithm across tasks.

Task	Algorithm			
	RVSP	**EOFR**	**AttenD**	**MDD**
Arrows	0.67	0.75	0.71	0.87
Bug	0.78	0.87	0.80	0.86

Considering the results of the ROC curves, AUC values, accuracy and precision, it is apparent that a trade-off exists between ensuring distraction detection and avoiding false alarms, which complicates determining the most promising algorithm. More specifically, the MDD algorithm showed the best performance across all evaluation metrics (accuracy, precision, AUC). Although the EOFR algorithm had promising AUC values, the AttenD algorithm often yielded better accuracy and precision. Additionally, the RVSP algorithm consistently yielded the lowest values for both accuracy and precision, but yielded a slightly higher AUC value than AttenD. All of the algorithms succeeded in detecting distraction well above chance detection (AUC = 0.5). The performance of the algorithms varied by task, with little difference in performance for the looking and reaching task (bug) but more stark differences for the looking and touching (arrows). The AUC for each task for each algorithm is provided in Table 5.

5. Cognitive Distraction

Cognitive distraction is a critical area of concern with regard to driver distraction, particularly as related to tasks of listening and conversing, but also, as related to spontaneously occurring processes like daydreaming or becoming lost in thought, which may occur really often on long drives. The term "cognitive load" can be defined as any workload imposed on a driver's cognitive processes [26]. There are several types (and subtypes) of scenarios where cognitive load may occur during (see Figure 5), and therefore, affect driving. For further information, the reader may refer to [26]. These include:

1. Cognitive load imposed by secondary tasks undertaken while driving.
2. Cognitive load associated with the driver's internal activity.
3. Cognitive load arising from the driving task itself.

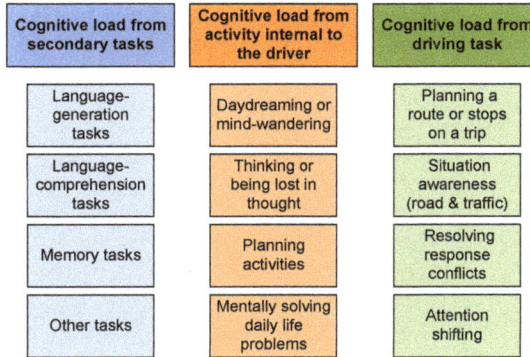

Cognitive load from secondary tasks	Cognitive load from activity internal to the driver	Cognitive load from driving task
Language-generation tasks	Daydreaming or mind-wandering	Planning a route or stops on a trip
Language-comprehension tasks	Thinking or being lost in thought	Situation awareness (road & traffic)
Memory tasks	Planning activities	Resolving response conflicts
Other tasks	Mentally solving daily life problems	Attention shifting

Figure 5. Classification of main types and subtypes of cognitive load while driving.

5.1. Behavioral and Physiological Indicators of Cognitive Load

The research literature documents several types of measures associated with periods of cognitive load. Secondary tasks imposing cognitive load lead to: (1) a high percentage of glances on the forward road and; (2) unusually long glances on the forward road. These two metrics together have been found to be uniquely indicative of cognitive loads [110,147]. Moreover, a narrowing of the spatial extent of scanning [147] is also produced, which is reflected in slightly fewer glances to locations where the mirrors, the speedometer and the areas peripheral to the road centre are located [26,148]. As a result, cognitive load may cause an increasing gaze concentration towards the middle of the road [11].

An eye-gaze pattern could be used to differentiate the action of only driving from driving under the influence of any cognitive task [147]. Drivers under cognitive distraction had fewer saccades per unit time, which was consistent with less exploration of the driving environment [149]. Saccades may be a valuable index of mental workload [150]. In fact, the standard deviations of both eye and head movement could be suitable for detecting cognitive distraction causing gaze concentration and slow saccades when drivers are looking forward [151]. A higher blink rate and a shrink in visual searching range were observed when the driver was cognitively distracted [152]. Kircher et al. [144] indicated the percentage of time the driver spent observing the road ahead, which is called the percentage road center (PRC) of gaze direction, was more than 92% under cognitive distraction in a field study.

Therefore, both glance and blink measures can be used to detect cognitive distraction. For example, He et al. [153] have observed that mind-wandering has effects on glance patterns and blink rates similar to those observed in periods of cognitive secondary task load. During mind-wandering, there is also an increasing concentration of gaze on the forward road with concomitant narrowing of scanning, longer glances on the forward road, and changes in blink rate [26]. Results from [154] suggested that performance data and oculomotor scanning behavior may allow the detection of drivers' mind wandering episodes before they are recognized by the driver himself/herself, potentially providing interventions to detect inattentiveness and alert drivers. Blink rate seems to be a promising indicator of cognitive processing [27]. However, there are measurement issues that may affect how successfully it can be applied in discriminating different types of task loading during driving. For example, there are some questions left about whether it offers sufficient sensitivity when extracted from real world data acquired from a complex task like driving wherein there are inherent temporal variations in driving task load [26,27].

Physiological measures can also be used to detect cognitive load. The average value of pupil diameter is suitably used as a physiological feature for detection of cognitive distraction [155]. When cognitive loads such as arithmetic or conversation were imposed to the subjects, dilation of pupils occurred by acceleration of the sympathetic nerve, resulting in an increase of diameter of pupils [156]. The average value of pupil diameter caused by cognitive loads, such as arithmetic, increased by 13.1% compared with ordinary driving [156]. The tests were performed in a driver simulator in controlled settings. Further experiments are required in a naturalistic setting. Moreover, additional works highlighted the difficulty in estimating cognitive load using pupil diameter during a dialogue task [157] or in different lighting conditions [158].

The same limitation applies to other physiological measures, as Heart Rate (HR), which tends to increase as cognitive task load raises [159]. The traditional method to quantify these physiological measures is by wearing physiological sensors. However, HR measurements can be acquired using computer vision techniques, and consequently, special care has been taken reviewing HR information. Additionally, it is considered a good indicator of fatigue, stress and cognitive load.

By means of the use of HR information the cognitive state of a driver can be monitored [160] in controlled settings. Changes in HR have been noted during certain driving tasks [161]. Similarly, Apparies et al. [162] showed that HR and Heart Rate Variability (HRV) may serve as early indicators of fatigue. In general, HRV specifically measures mental workload, while HR measures physical one [163]. HRV analysis is a strong indicator of mental stress or workload caused by driving tasks [162,164,165]. Experiments carried out in a driving simulator by Zhao et al. [166] found that human heart rates violently fluctuate during a mental stress situation. Ostlund et al. [167] and Miller et al. [165] identified both HR and HRV as promising indicators of the driver's stress level, by increasing HR and decreasing HRV [165,167]. Physiological measures, such as HR and skin conductance level, tend to increase as cognitive task load increases [159].

There are some research works able to extract HR and HRV from face video images in real time from human faces [168–170]. Eulerian Video Magnification framework [171] can be also used to obtain human pulse from a video sequence [172]. In [173], the authors described an approach offering a non-invasive, non-contact means of cardiac monitoring. Once the HRV time series are extracted, feature generation, feature selection and classification should be performed. The conventional method that uses Fast Fourier Transform (FFT) analysis on HRV is 2-min long. In [174], a new method developed by using wavelet-based feature and SVM for classification uses only 1-min HRV signals. Moreover, this method increases accuracy, sensitivity and specificity compared to FFT-based results.

Therefore, concerning cardiovascular measures, they have been reported to be sensitive to mental workload changes and both HR and HRV are widely adopted mental workload measures because they are easy to use and provide fundamental information about the autonomic nervous system [175]. Most methods [168–170] enable low-cost, non-contact cardiovascular activity monitoring using regular RGB cameras by analyzing minute skin color changes caused by periodic blood flow. Nevertheless, for automotive applications, these methods can encounter difficulties under different illumination conditions [176]. In [176], the authors proposed an artifact reduction method, which is caused by lighting variation. Another option is to use an IR-based camera system suitable for automotive applications [177].

To conclude this section, the use of physiological parameters can be used to monitor the cognitive state of the driver. Many of these parameters have been described in controlled settings, but further experiments are required to validate their capability in naturalistic conditions. The main algorithms in this matter are included in the next section.

5.2. Algorithms

Zhang et al. [178] used a decision tree approach to estimate drivers' cognitive workload from eye gaze-related features and driving performance. Liang, Reyes, et al. [179] showed that the SVM models can also detect cognitive distraction. The model's accuracy and sensitivity increased with window size, suggesting that using longer periods to summarize the data made the distraction signal easier for the models to detect. The conclusion was that the best models were obtained using 40-s window size. Additionally, Liang, Lee, et al. [180] also used Bayesian Network (BN) models and found that they could identify cognitive load reliably for simulator data, and also found that Dynamic Bayesian Networks (DBNs), which considered time dependencies of driver's behaviour, gave a better performance than static BN models. This fact suggests that time-dependent relationship is critical in estimating the cognitive state of the driver. However, to train DBN models, longer training sequences are necessary to obtain more accurate and sensitive models. The results obtained in [180] using BNs, which stated that window size did not affect model performance, clearly conflict with those of Liang et al. [179], which found that larger window sizes improved the detection of cognitive distraction, although another data mining method, SVM, was applied in that study. An additional work from Liang et al. [181] compared SVMs, SBNs, and DBNs in detecting cognitive distraction using the best parameter settings from the same dataset used in the previous two studies [179,180]. DBNs produced the most accurate and sensitive models compared to SBN and SVM. Based on the comparisons of SVMs and BNs, Liang et al. [27,182,183] used a hierarchical layered algorithm, which incorporated both a DBN and a supervised clustering algorithm, to identify feature behaviors when drivers were in different cognitive states. This layered algorithm includes a DBN algorithm at the higher level to model the time-dependent relationship of driver behavior and a supervised clustering algorithm at the lower level to identify feature behaviors. The layered algorithm overcomes the disadvantages of DBNs and significantly improves computational efficiency in training and prediction. Miyaji et al. [184] proposed an approach to detect eye and head movement tracked via standard deviation and categorized features for pattern recognition by using AdaBoost method to detect distraction. The authors compared the performance achieved by both SVM and AdaBoost in estimating cognitive workload, finding that AdaBoost could achieve higher accuracy. Additionally, Miyaji et al. [156] introduced a mixed method by applying a SVM and an AdaBoost classifier for three parameters: (1) heart rate; (2) visual information (standard deviation of both gaze and head rotation angles) and (3) pupil diameter to assess the level of the driver's vigilance. Recently, a new machine learning tool, Extreme Learning Machine (ELM) [185,186]), has gained much attention due to its simple structure, high generalization capability, and fast computational speed. For example, in [187], ELM and SVM were applied to detect drivers' workload using eye movement, as well as eye movement combined with driving performance data. The results suggested that both methods can detect drivers' workload at high accuracy, but ELM outperformed SVM in most cases.

The results of all the works mentioned so far can be summarized in Table 6. Common features include the use of eye gaze-related features, driving performance, pupil diameter features and HR. It should also be noted that very good results can be obtained using only eye gaze-related features. Additionally, many supervised machine learning techniques have been proposed so far: decision trees, SVM, BN, DBN, AdaBoost or ELM.

All these distraction detection systems are based on supervised learning, meaning that the training of such systems need to be "supervised" by human experts by providing a target set for training data containing distraction status. The supervised learning paradigm is only suitable for early stage research and may not be suitable for implementation in real driving cases because of the huge cost and difficulty of creating a target distraction status set, which would require additional subjective ratings by the driver [115], post-processing by the experimentalists [56], or additional computation based on data from other sources [179]. For example, in a recent study [188], labeling drivers' distraction state involves the development of Graphical User Interface (GUI), the training of external evaluators, and the actual labeling time, which is approximately 21.5 h of manpower (43 min per one of the 30 evaluator) to label the entire video segments. For naturalistic driving, where the driver voluntarily decides which tasks to perform at any time, the labeling process can become infeasible. On the other hand, data without known distraction states (unlabeled data) can be collected readily, e.g., from drivers' naturalistic driving records.

Table 6. Supervised algorithms for cognitive distraction detection.

Algorithm	Features	Classifier	Accuracy (%)
Zhang et al. [178]	Eye gaze-related features and driving performance	Decision Tree	81
Zhang et al. [178]	Eye gaze-related features	Decision Tree	80
Zhang et al. [178]	Pupil-diameter features	Decision Tree	61
Zhang et al. [178]	Driving performance	Decision Tree	60
Liang, Reyes, et al. [179]	Eye gaze-related features and driving performance	SVM	83.15
Liang, Reyes, et al. [179]	Eye gaze-related features	SVM	81.38
Liang, Reyes, et al. [179]	driving performance	SVM	54.37
Liang, Lee, et al. [180]	Eye gaze-related features and driving performance data	DBNs	80.1
Miyaji et al. [156]	Heart rate, Eye gaze-related features and pupil diameter	AdaBoost	91.5
Miyaji et al. [156]	Eye gaze-related features	SVM	77.1 (arithmetic task)
Miyaji et al. [156]	Eye gaze-related features	SVM	84.2 (conversation task)
Miyaji et al. [156]	Eye gaze-related features	AdaBoost	81.6 (arithmetic task)
Miyaji et al. [156]	Eye gaze-related features	AdaBoost	86.1 (conversation task)
Yang et al. [187]	Eye gaze-related features and driving performance data	ELM	87.0
Yang et al. [187]	Eye gaze-related features and driving performance data	SVM	82.9

With the purpose of tackle these deficiencies, Unsupervised and Semi-Supervised algorithms can be used. For example, in [12], Semi-Supervised Extreme Learning Machine (SS-ELM) is proposed for drivers' distraction detection. SS-ELM outperformed supervised ELM in both accuracy and model sensitivity, suggesting that the proposed semi-supervised detection system can extract information from unlabeled data effectively to improve the performance. SS-ELM based detection system has the potential of improving accuracy and alleviating the cost of adapting distraction detection systems to new drivers, and thus, more promising for real world applications. However, several points are unclear from these preliminary results [12] further explored in [189], where the Semi-Supervised Learning (SSL) paradigm is introduced to real time detection of distraction based on eye and head movements.

In [189], two graph-based SSL methods were compared with supervised learning methods. These algorithms are detailed as follows. Laplacian Support Vector Machine (LapSVM), which is an extension of SVMs to SSL under manifold regularization framework [190], and SS-ELM were compared with three supervised learning methods (static BN with Supervised Clustering (SBN-SC) [180,183], ELM and SVM) and one low-density-separation-based method (Transductive SVM (TSVM) [191]). To capture realistic eye and head movements patterns, data from an on-road experiment were used. By utilizing unlabeled data, the graph-based semi-supervised methods reduced the labeling cost and improved the detection accuracy. The highest accuracy of 97.2% and G-mean of 0.959 were achieved by SS-ELM. The benefits of using SSL methods increased with the size of unlabeled data set showing that by exploring the data structure without actually labeling them, extra information to improve models performance can be obtained.

It is worth noting that cognitive distraction detection is only performed in "laboratory settings" and not in real conditions. In real life situations, when the driver is under cognitive load (e.g., mind wandering): (1) he is alone and does not interact with anybody; (2) he is also the only one who can decide whether or not to activate the attentional processing of distractive thoughts [192];

and (3) drivers are likely to be performing multiple tasks at the same time (e.g., talking on the mobile phone and listening to music). Moreover, there are two main limitations intrinsic to laboratory-based studies. First of all, most of these studies require that the execution of predefined tasks last for no more than some minutes. In our opinion, such experiments make it very difficult, if not impossible, to infer, for instance, the long-term effectiveness of for example, warning signals, monotonous driving (in general, real driving), on the basis of the results of experiments that are typically so short; And secondly, the drivers are abnormally vigilant to the driving task because they are being observed [193]. In connection with this point, the use of physiological parameters, which form the basis for cognitive distraction detection, have also been extracted in controlled settings and not in real conditions.

6. Mixing Types of Distraction

There are some algorithms trying to detect mixing types of distraction, whose results can be seen in Table 7. In [194], facial features are extracted to detect both visual and cognitive distractions. Binary classifiers (normal vs distracted) are built for visual and cognitive distraction detection. Gaze and Action Units (AU) features are useful in order to detect visual distractions, while AU features are particularly important for cognitive distractions. It should be pointed out that the cognitive tasks considered in this study are closely related to talking activities.

Table 7. Mixing types of distraction detection algorithms.

Algorithm	Features	Classifier	Average Accuracy (%)
Li et al. [194]	AU and head pose	LDC (visual distraction) and SVM (cognitive distraction)	80.8 (LDC), 73.8 (SVM)
Craye et al. [195]	eye behaviour, arm position, head orientation and facial expressions using both color and depth images	Adaboot and HMM	89.84 (Adaboot), 89.64 (HMM)
Liu et al. [196]	Head and eye movements	SVM, ELM and CR-ELM	85.65 (SVM), 85.98 (ELM), 86.95 (CR-ELM)
Ragab et al. [197]	arm position, eye closure, eye gaze, facial expressions and head orientation using depth images	Adaboost, HMM, RF, SVM, CRF, NN	82.9 (RF—type of distraction detection), 90 (RF—distraction detection)

Liu et al. [196] applied Cluster Regularized Extreme Learning Machine (CR-ELM) for detecting mixing types of distraction. Compared with the traditional ELM, CR-ELM introduces an additional regularization term penalizing large covariance of training data within the same clusters in the output space. CR-ELM, ELM and SVM were compared to predict mixing types of distraction. They simulated the mixing types of distraction by combining two types of distracting driving activities (a visual task and a cognitive one). CR-ELM showed lower error rate on most of the 11 subjects (see Table 7).

There are other approaches trying to merge both RGB and depth images to get the features to be used by the algorithms [195,197]. Craye et al. [195] extracted features from face and body using both color and depth images in order to build a distraction system, which is composed of four sub-modules: eye behaviour (gaze and blinking), arm position, head orientation and facial expressions. The information from these modules are merged together using two different classification techniques: Adaboost classifier and Hidden Markov Model (HMM). A set of video sequences was collected to test the system. Five distractive tasks were recorded and manually labelled for training and evaluation. HMM outperforms Adaboost for most drivers. Finally, a vision-based driver distraction is investigated using several machine learning techniques in [197]. IR and Kinect cameras were used in this system, where five visual cues were calculated: arm position, eye closure, eye gaze, facial expressions and head orientation. These cues were fed into a classifier, such as Adaboost, HMM, RF, SVM, Conditional Random Field (CRF) or NN, in order to detect and recognize the type of distraction.

7. The Relationship between Facial Expressions and Distraction

Facial expressions can be described at different levels [198]. A widely used description is Facial Action Coding System (FACS) [199], which is a human-observer-based system developed to capture subtle changes in facial expressions. With FACS, these expressions are decomposed into one or more AUs [200]. AU recognition and detection have attracted much attention recently [201]. Meanwhile, psychophysical studies indicate that basic emotions have corresponding universal facial expressions across all cultures [202]. This is reflected by most current facial expression recognition systems attempting to recognize a set of prototypic emotional expressions including disgust, fear, joy, surprise, sadness and anger [201], which can be helpful in predicting driving behaviour [203].

Therefore, in this work, main facial expression works in the driving environment are described in accordance with the two aforementioned levels (FACS and prototypic emotional expressions) and how they are related with distraction.

On one hand, in connection with FACS and distraction while driving, the reference work is the one proposed by Li et al. [194]. The authors performed the analysis of driver's facial features under cognitive and visual distractions. In addition to the obvious facial movement associated with secondary tasks such as talking, they hypothesized that facial expression can play an important role in cognitive distraction detection. They studied the top five features (from a total of 186 features) to predict both cognitive and visual distraction. For cognitive distraction, the most important features to consider are: (1) head yaw; (2) Lip Corner Depressor (AU15); (3) Lip Puckerer (AU18); (4) Lip Tightener (AU23) and (5) head roll. For visual distraction, the most important features to consider are: (1) Lip Tightener (AU23); (2) jaw drop (AU26); (3) head yaw; (4) Lip Suck (AU28) and (5) Blink (AU45). The results indicated that gaze and AU features are useful for detecting visual distractions, while AU features are particularly important for cognitive distractions. It should be pointed out that since the cognitive tasks considered in this study are closely related to talking activities, their future work will include the analysis of other cognitive tasks (e.g., thinking or solving math problems).

On the other hand, in connection with prototypic emotional expressions, there are some works trying to study how these emotions affect behaviour.

The relationship between emotion and cognition is complex, but it is widely accepted that human performance is altered when a person is in any emotional state. It is really important to fully understand the impact of emotion on driving performance because, for example, roadways are lined with billboard advertisements and messages containing a lot of different emotional information. Moreover, the distracting effects of emotions may come in other forms, such as cell phone, passenger conversations, radio information or texting information [204]. For example, Chan et al. [204] conducted a study to examine the potential for distraction from emotional information presented on roadside billboards. The findings in this study showed that emotional distraction: (a) can seriously modulate attention and decision-making abilities and have adverse impacts on driving behavior for several reasons and (b) can impact driving performance by reorienting attention away from the primary driving task towards the emotional content and negatively influence the decision-making process. In another study with a similar line of work, Chan et al. [205] showed that emotion-related auditory distraction can modulate attention to differentially influence driving performance. Specifically, negative distractions reduced lateral control and slowed driving speeds compared to positive and neutral distractions.

Some studies have shown that drivers who are more likely to become angry (e.g., those with high trait anger rates) tend to engage in more aggressive behavior on the road, which can result in negative outcomes such as crashes [206]. Moreover, anger negatively influences several driving performance and risky behaviors such as infractions, lane deviations, speed, and collisions [207].

In conclusion, aggressiveness and anger are emotional states that extremely influence driving behaviour and increase the risk of accident. However, a too low level of activation (e.g., resulting from emotional states like sadness) also leads to reduced attention and distraction as well as prolonged reaction time and, therefore, lowers driving performance [208]. On this basis, research and experience have demonstrated that being in a good mood is the best precondition for safe driving and that happy drivers produce fewer accidents [209]. In other words, *happy drivers are better drivers* [208,210]. Facial expression and emotion recognition can be used in advanced car safety systems, which, on one hand, can identify hazardous emotional drivers' states that can lead to distraction and, on the other, can provide tailored (according to each state and associated hazards) suggestions and warnings to the driver [211].

8. Sensors

Once the algorithms for distraction detection have been designed and implemented, the next step is to port them to an embedded device or system to be executed inside the car. However, porting a vision-based algorithm is not a straightforward step and some key factors should be taken into account. Furthermore, there is not a standard implementation platform, so different alternatives have been proposed by both the scientific community and the industry.

8.1. Porting a Vision Algorithm to an Embedded Automotive System

The implementation of computer vision applications in automotive environments is not straightforward because several requirements must be taken into account: reliability [212,213], real-time performance [213–215], low-cost [216–219], spatial constraints [217,219], low power consumption [220], flexibility [219], rapid prototyping [215,221], design requirements [217] and short time to market [217]. Therefore, there must be a trade-off among these design requisites [217]. Moreover, there is not a commonly accepted hardware and software platform, so different solutions have been proposed by the industry and the scientific community. Last but not least, some driver distraction guidelines and test procedures for all applications to be used while driving should be considered [222], and so should ADAs.

One approach can rely on the use of microprocessors, which incorporates the functions of a computer's central processing unit (CPU) on a single integrated circuit (IC). For example, in [223], a vision-based system for monitoring the loss of attention, tested under day and night driving conditions, is proposed. The algorithm was cross-validated using brain signals and finally, implemented on a Single Board Computer (SBC). Another example is presented in [224], where a vehicle was equipped with a USB camera connected to the system in order to track the driver's eyes for fatigue detection.

A similar approach is the use of digital signal processors (DSPs) [225], which can perform multiplications and additions in a single cycle and have parallel processing capabilities. DSPs have been used in image and audio signal processing when the use of microcontrollers was not enough. These processors were used in [215], where an optimized vision library approach for embedded systems was presented. VLIB is a software library that accelerates computer vision applications for high-performance embedded systems. By significantly speeding up pixel-intensive operations, the library provides more headroom for innovative algorithms, and enables processing of more channels at higher resolutions. Authors optimized the library for the Texas Instruments C64x/C64x+ DSP cores. Karuppusamy et al. [226] proposed an embedded implementation of facial landmarks detection based on both Viola-Jones face detector and facial landmarks detection using extended Active Shape Model (ASM) [227]. However, DSPs imply a much higher cost compared with other options such as field-programmable gate arrays (FPGAs) [228].

Another option is to use hardware implementation, since it can achieve a much better computational performance, where two types are commonly used namely, FPGA and ASIC. A FPGA is an integrated circuit designed to be configured by a customer or a designer after manufacture.

FPGAs take advantage of high speed operations, especially for parallelizable operations achieving good performance in face monitoring applications [229–232]. For example, several well-known algorithms have been used and optimized for this field of application, such as: (a) spatial and temporal filtering, motion detection and optical flow analysis [229] or (b) gray scale projection, edge detection with Prewitt operator and complexity functions [230]. Additionally, the use of Application-Specific Integrated Circuits (ASIC), which is an IC customized for a particular use rather than intended for general-purpose use, has also been considered [233]. FPGAs have an important advantage over ASICs: they are reconfigurable, which gives them some of the flexibility of software. ASICs are only used for high volume manufacturing and long series due to higher initial engineering cost.

Developing the whole application in hardware is a cumbersome task, so hybrid solutions have appeared combining both software and hardware implementations. The work in [234] describes a System on a Chip (SOC) designed to support a family of vision algorithms. However, this system uses an ASIC, so it cannot be completely reconfigured. This important drawback makes impossible to update the device. A generic embedded hardware and software architecture was proposed to design and evaluate ADAS vision applications [221]. Although the system is useful to test some vision applications, the performance obtained in the case study showed that the system is not powerful enough to run more complex applications including I/Os management, vehicle communications or more demanding vision applications. In [219], a reconfigurable embedded vision system reaching the requirements of ADAS applications is presented. A SOC, which is formed by an FPGA with a dual core ARM, is prepared to be easily reconfigured. A lane departure warning system was implemented in the case study obtaining a good computational performance. The obtained computational time allows the system to include another more complex vision algorithm running in parallel. In [235], they proposed an approach to predict performances of image processing algorithms on different computing units of a given heterogeneous SOC.

Despite the fact that in recent years some authors have been trying to propose some architectures in order to achieve some key factors in embedded ADAS systems inside a vehicle [219,221,235], these efforts do not seem to be sufficient to reach the requirements stated before. The use of microprocessors in embedded computer vision-based systems has experienced a significant growth in recent years. Moreover, current premium cars implement more than 90 Electronic Control Units (ECU) with close to 1 Gigabyte embedded software code [236]. In 2018, 30% of the overall vehicle cost is predicted to stem from vehicle electronics [237]. The independence of different applications (with different criticality levels) running on the same platform must be made evident. Therefore, the development of embedded automotive systems has become quite complex. To that end, the use of standards and frameworks is indispensable.

8.2. Commercial Sensors

8.2.1. Smart Eye

The Smart Eye [77] system is a well-suited head and gaze tracking method for the demanding environment of a vehicle cabin and flexible to cope with most research projects. It consists of a multi-camera system running on a single PC and on a single algorithm. The system is scalable from 2 up to 8 cameras allowing 360° head and eye tracking. A typical configuration inside a vehicle cabin is composed of four cameras with two IR lightings, located on the dashboard on either side of the steering wheel. Smart Eye offers a sampling rate of 60 Hz (up to 8 cameras) or 120 Hz (up to 4 cameras). The field of view, depending on the number of cameras, is in the range of 90°–360°. The data output includes over 145 values covering, among others, gaze, eyelid, pupilometry and head tracking, raw and filtered gaze, blinks, fixations and saccades. Smart Eye has been used in several driver assistance and inattention systems, such as [76,143–145,238].

8.2.2. EyeAlert

EyeAlert [239], cited in several publications [128,240], has been conceived to detect driver inattention using computer vision and to generate a warning signal in case of dangerous situation. The EyeAlert system focuses entirely on the driver's alertness levels or inattention to the road ahead, regardless of the time of the day or the weather conditions. Three models are available:

- EyeAlert EA410 detects both distracted and fatigue driving. The EA410 has a highly integrated IR camera, a computer, an image processing unit and an alarm. The EA410 technology is protected by over ten patents. The system will also respond in case the driver does not focus on driving.
- EyeAlert EA430 with GPS detects both distracted and fatigue driving. Moreover, a minimum speed threshold is programmed into the internal GPS to prevent false alarms in urban environments.
- EyeAlert EA450 with Data detects both distracted and fatigue driving. Additionally, minimum speed threshold, sensitivity, volume and data can be remotely programmed. The minimum speed and sensitivity controls allow the reduction of false alarms in highway and urban environments.

8.2.3. Seeing Machines

Seeing Machines [241] builds image-processing technology that tracks the movement of a person's eyes, face, head, and facial expressions. It monitors fatigue and distraction events in real-time and uses IR technology to provide fatigue and distraction monitoring at any time of the day. The system can also combine multiple camera sensors to detect a wider range of movements. The Seeing Machines' system continuously measures operator eye and eyelid behaviour to determine the onset of fatigue and micro sleeps and delivers real-time detection and alerts. The system has been used in many different driver assistance and inattention systems [11,142,151,156,242–244].

8.2.4. Visage Technologies AB

Visage Technologies AB [245] provides a state-of-the-art commercial head tracker based on feature-point detection and tracking of the nose boundary and eye regions. Visage SDK finds and tracks the face and facial features, including gaze direction, in video sequences in real time. It provides pupil coordinates, 3D gaze direction as well as (with a calibration step) screen-space gaze point. Visage Technologies also features support for embedded systems like FPGA and IR light tracking for poor lighting conditions.

8.2.5. Delphi Electronics Driver Status Monitor

Delphi Electronics, a major automotive electronics manufacturer, developed a single camera Driver Status Monitor (DSM) [246]. By detecting and tracking the driver's facial features, the system analyzes eye-closures and head pose to infer his/her fatigue or distraction. This information is used to warn the driver and to modulate the actions of other safety systems. The system includes the use of NIR illumination, an embedded processing unit, as well as the camera (resolution of 640×480 pixels).

8.2.6. Tobii Technologies

Tobii Technologies develops Tobii's eye-tracking technology for integration into volume products such as computers, computer games, virtual reality and cars. The Tobii platform consists of two-camera sensors, placed at different angles, and operating at IR frequencies to eliminate interference from external light. The system can distinguish whether the driver's eyes are open or closed or if the driver has turned his/her head. The sensors work even when the driver is wearing glasses or sunglasses. By observing the specifics of eyelid closure, in combination with eye gaze patterns, an active safety system powered by Tobii's eye tracking sensor can reliably detect if a driver is falling asleep and warns him/her properly. Moreover, Tobbi Technologies provides the Tobii EyeChip, which is a dedicated eye tracking SOC ASIC.

8.2.7. SensoMotoric Instruments

SensoMotoric Instruments GmbH (SMI) [247] is a German company, whose eye tracking solutions can measure head position and orientation, gaze direction, eyelid opening, and pupil position and diameter. Eye trackers use a sampling rate of 120 Hz for head pose and gaze measurement, 120 Hz for eyelid closure and blink measurement, and 60 Hz for combined gaze, head pose, and eyelid measurement. It also provides PERCLOS information for drowsiness detection. It is a computer-based system and needs user calibration. In [248], SensoMotoric was used to recognize the pupil in each image in order to measure horizontal and vertical eye movements.

8.2.8. Automobile Manufacturers

Nissan introduces its new Driver Attention Alert system with the 2016 Nissan Maxima, which was unveiled at the New York International Auto Show [249]. The automaker has highlighted that the new system will be able to detect drowsy and inattentive driving and it will alert the driver about the situation by giving visual and audible warning. Ford's Driver Alert [250] seems only to detect drowsiness but not distraction. The Driver Alert system comprises a small forward-facing camera connected to an on-board computer. The camera is mounted on the back of the rear view mirror and is trained to identify lane markings on both sides of the vehicle. When the vehicle is on the move, the computer looks at the road ahead and predicts where the car should be positioned relative to the lane markings. Then, it measures where the vehicle actually is and, if the difference is significant, the system issues a warning. The Saab Driver Attention Warning System [251] detects visual inattention and drowsy driving. The system uses two miniature IR cameras integrated with Smart Eye technology [77] to accurately estimate head pose, gaze, and eyelid status. When a driver's gaze is not located inside the primary attention zone (which covers the central part of the frontal windshield) for a predefined period, an alarm is triggered. Toyota has equipped their luxury Lexus models with their Driver Monitoring System [252]. The system permanently monitors the movement of the driver's head when looking from side to side using a NIR camera installed at the top of the steering wheel column. The system is integrated into Toyota's pre-crash system, which warns the driver when a collision is likely to happen. In 2007, Volvo Cars introduced Driver Alert Control to alert tired and non-concentrating drivers [253,254]. Based on the idea that the technology for monitoring eyes is not yet sufficiently mature and human behavior varies from one person to another, Volvo Cars developed the system based on car progress on the road. It is reported that Driver Alert Control monitors the car movements and assesses whether the vehicle is driven in a controlled or uncontrolled way. More recently, a Hyundai concept car (the Hyundai HCD-14) incorporates Tobbi Technologies to track the eyes [255].

9. Simulated vs. Real Environment to Test and Train Driving Monitoring Systems

The development of the computer vision algorithm only represents one part of all the cycle of the product design. One of the hardest tasks is to validate the whole system with the wide variety of driving scenarios [256]. In order to complete the whole "process development" of the vision-based ADAS, some key points are presented.

In order to monitor both the driver and his/her driving behaviour, several hardware and software algorithms are being developed, but they are tested mostly in simulated environments instead of in real driving ones. This is due to the danger of testing inattention in real driving environments [21]. Experimental control, efficiency, safety, and ease of data collection are the main advantages of using simulators [257,258]. Some researches have validated that driving simulators can create driving environment relatively similar to road experiments [259–261]. However, some considerations should be taken into account since simulators can produce inconsistent, contradictory and conflicting results. For example, low-fidelity simulators may evoke unrealistic driving behavior and, therefore, produce invalid research outcomes. One common issue is that real danger and the real consequences of actions do not occur in a driving simulator, giving rise to a false sense of safety, responsibility, or competence [262]. Moreover, simulator sickness symptoms may undermine training effectiveness and negatively affect the usability of simulators [262].

A study on distraction in both simulated and real environment was conducted in [11] and it was found out that the driver's physiological activity showed significant difference. Engstorm et al. [11] stated that physiological workload and steering activity were both higher under real driving conditions compared to simulated environments. In [257], the authors compared the impact of a narrower lane using both a simulator and real data, showing that the speed was higher in the simulated roads, consistent with other studies. In [263], controlled driving yielded more frequent and longer eye glances than the simulated driving setting, while driving errors were more common in simulated driving. In [167], the driver's heart rate changed significantly while performing the visual task in real-word driving relative to a baseline condition, suggesting that visual task performance in real driving was more stressful.

After the system is properly validated in a driver simulator, it should be validated in real conditions as well, because various factors including light variations and noise can also affect the driver's attention. The application on a real moving vehicle presents new challenges like changing backgrounds and sudden variations of lighting [264]. Moreover, a useful system should guarantee real time performance and quick adaptability to a variable set of users and to natural movements performed during driving [264]. Thus, it is necessary to make simulated environments appear more realistic [203].

To conclude, in most previous studies, independent evaluations using different equipment and conditions (mainly simulated environments) resulted in time-consuming and redundant efforts. Moreover, inconsistency in the algorithm performance metrics makes it difficult to compare algorithms. Hence, the only way to compare most algorithms and systems is the metrics provided by each author when comparing their values, but with scarce information about the used images and conditions. Public data sets covering simulated and real driving environments should be released in the near future, as stated by some authors previously [203].

10. Privacy Issues Related to Camera Sensors

Although there is a widespread agreement for intelligent vehicles to improve safety, the study of driver behaviour to design and evaluate intelligent vehicles requires large amounts of naturalistic driving data [265]. However, in current literature, there is a lack of publicly available naturalistic driving data largely due to concerns over individual privacy. It also should be noted that a real-time visual-based distraction detection system does not have to save the video stream. Therefore, privacy issues are mostly relevant in research works were video feed is collected and stored to be studied at a later stage, for example in the large naturalistic studies conducted in the US.

Typical protection of the individuals' privacy in a video sequence is commonly referred as "de-identification" [266]. Although this fact will help protect the identities of individual drivers, it impedes the purpose of sensorizing vehicles to control both drivers and their behaviour. In an ideal situation, a de-identification algorithm would protect the identity of drivers while preserving sufficient details to infer their behaviour (e.g., eye gaze, head pose or hand activity) [265].

Martin et al. [265,267] proposed the use of de-identification filters to protect the privacy of drivers while preserving sufficient details to infer their behaviour. Following this idea, a de-identification filter preserving only the mouth region can be used for monitoring yawning or talking and a de-identification filter preserving eye regions can be used for detecting fatigue or gaze direction, which is precisely proposed by Martin et al. [265,267]. More specifically, the authors implemented and compared de-identification filters made up of a combination of preserving eye regions for fine gaze estimation, superimposing head pose encoded face masks for providing spatial context and replacing background with black pixels for ensuring privacy protection. A two-part study revealed that human facial recognition experiment had a success rate well below the chance while gaze zone estimation accuracy disclosed 65%, 71% and 85% for One-Eye, Two-Eyes and Mask with Two-Eyes, respectively.

Fernando et al. [268] proposed to use video de-identification in the automobile environment using personalized Facial Action Transfer (FAT), which has recently attracted a lot of attention in computer vision due to its diverse applications in the movie industry, computer games, and privacy protection. The goal of FAT is to "clone" the facial actions from the videos of a person (source) to another one (target) following a two-step approach. In the first step, their method transfers the shape of the source person to the target subject using the triangle-based deformation transfer method. In the second step, it generates the appearance of the target person using a personalized mapping from shape changes to appearance changes. In this approach video de-identification is used to pursue two objectives: (1) to remove person-specific facial features and (2) to preserve head pose, gaze and facial expression.

11. General Discussion and Challenges Ahead

The main visual-based approaches reviewed in this paper are summarized in Table 8 according to some key factors.

A major finding emerging from two recent research works reveals that just-driving baselines may, in fact, not be "just driving" [26,269], containing a considerable amount of cognitive activity in the form of daydreaming and lost-in-thought activity. Moreover, eye-gaze patterns are somewhat idiosyncratic when visual scanning is disrupted by cognitive workload [27]. Additionally, "look-but-failed-to-see" impairment under cognitive workload is an obvious detriment to traffic safety. For example, Strayer et al. [270] found that recognition memory for objects in the driving environment was reduced by 50% when the driver was talking on a handsfree cell phone, inducing failures of visual attention during driving. Indeed, visual, manual and cognitive distraction often occur simultaneously while driving (e.g., texting while driving and other cell-phone reading and writing activities). Therefore, the estimates of crash risk based on comparisons of activities to just-driving baselines may need to be reconsidered in light of the possible finding that just-driving baselines may contain the aforementioned frequent cognitive activity. As a result, for example, secondary tasks effects while driving should be revised [269]. Accordingly, as detecting driver distraction depends on how distraction changes his/her behavior compared to normal driving without distraction, periods with minimal or no cognitive activity should be identified in order to train the distraction detection algorithms.

Additionally, computer vision techniques can be used, not only for extracting information inside the car, but also for extracting information outside the car, such as traffic, road hazards, external conditions of the road ahead, intersections, or even position regarding other cars. The final step should be the correlation between the driver's understanding and the traffic context. One of the first works trying to fuse "out" information (visual lane analysis) and "in" information (driver monitoring) is the one proposed by Apostoloff et al. [271], pointing out the benefits of this approach. Indeed, visual lane analysis can be used for "higher-order tasks", which are defined by interacting with other modules in a complex driver assistance system (e.g., understanding the driver's attentiveness—distraction—to the lane-keeping task [272]). Hirayama et al [273] focused on temporal relationships between the driver's eye gaze and the peripheral vehicles behaviour. In particular, they concluded that the timing when a driver gazes towards the overtaking event under cognitive distraction is later than that under the neutral state. Therefore, they showed that the temporal factor, that is, timing, of a reaction is important for understanding the state by focusing on cognitive distraction in a car-driving situation. Additionally, Rezaei et al. [87] proposed a system correlating the driver's head pose to road hazards (vehicle detection and distance estimation) by analyzing both simultaneously. Ohn et al. [274] proposed a framework for early detection of driving maneuvers using cues from the driver, the environment and the vehicle. Tawari et al. [275] provided early detection of driver distraction by continuously monitoring driver and surround traffic situation. Martin et al. [276] focused on intersections and studied the interaction of head, eyes and hands as the driver approaches a stop-controlled intersection. In this line work of research, Jain et al. [277] deal with the problem of anticipating driving maneuvers a few seconds before the driver performs them.

Table 8. Summary of visual-based approaches to detect different types of driver distraction.

Approach	Distraction Detection Approaches			Real Conditions	Operation	
	Manual	Visual	Cognitive		Daytime	Nighttime
Zhao et al. [88]	✔	✗	✗	✗	✔	✗
Zhao et al. [89]	✔	✗	✗	✗	✔	✗
Zhao et al. [90]	✔	✗	✗	✗	✔	✗
Zhao et al. [91]	✔	✗	✗	✗	✔	✗
Bosch et al. [94]	✔	✗	✗	✗	✔	✗
Lowe et al. [95]	✔	✗	✗	✗	✔	✗
Yan et al. [92]	✔	✗	✗	✗	✔	✗
Yan et al. [93]	✔	✗	✗	✗	✔	✗
Zhang et al. [96]	✔	✗	✗	✗	✔	✗
Artan et al. [97]	✔	✗	✗	✔	✔	✔
Berri et al. [98]	✔	✗	✗	✔	✔	✗
Xu et al. [99]	✔	✗	✗	✔	✔	✔
Seshadri et al. [100]	✔	✗	✗	✔	✔	✗
Ohn et al. [106]	✔	✗	✗	✔	✔	✗
Martin et al. [104]	✔	✗	✗	✔	✔	✗
Ohn et al. [105]	✔	✗	✗	✔	✔	✗
Morimoto et al. [120]	✗	✔	✗	✔	✗	✔
Ji et al. [121]	✗	✔	✗	✔	✗	✔
Ji et al. [122]	✗	✔	✗	✔	✗	✔
Ji et al. [123]	✗	✔	✗	✔	✗	✔
Gu et al. [124]	✗	✔	✗	✔	✗	✔
Batista el al. [125]	✗	✔	✗	✔	✗	✔
Bergasa et al. [126]	✗	✔	✗	✔	✔	✔
Lee et al. [114]	✗	✔	✗	✔	✔	✔
Vicente et al. [49]	✗	✔	✗	✔	✔	✔
Cyganek et al. [134]	✗	✔	✗	✔	✔	✔
Donmez et al. [142]	✗	✔	✗	✗	✔	✗
Klauer et al. [7]	✗	✔	✗	✔	✔	✔
Kircher et al. [143]	✗	✔	✗	✔	✔	✔
Kircher et al. [144]	✗	✔	✗	✔	✔	✔
Kircher et al. [145]	✗	✔	✗	✔	✔	✔
Victor et al. [146]	✗	✔	✗	✔	✔	✗
Zhang et al. [178]	✗	✗	✔	✗	✔	✗
Liang et al. [179]	✗	✗	✔	✗	✔	✗
Liang et al. [180]	✗	✗	✔	✗	✔	✗
Liang et al. [181]	✗	✗	✔	✗	✔	✗
Liang et al. [27]	✗	✗	✔	✗	✔	✗
Liang et al. [182]	✗	✗	✔	✗	✔	✗
Liang et al. [183]	✗	✗	✔	✗	✔	✗
Miyaji et al. [184]	✗	✗	✔	✗	✔	✗
Miyaji et al. [156]	✗	✗	✔	✗	✔	✗
Yang et al. [187]	✗	✗	✔	✗	✔	✗
Li et al. [194]	✗	✔	✔	✗	✔	✗
Craye et al. [195]	✔	✔	✗	✗	✔	✗
Liu et al. [196]	✗	✔	✔	✗	✔	✗
Ragab et al. [197]	✔	✔	✗	✗	✔	✗

There are many factors that can modulate distraction. For example, as discussed in Section 7, emotional information can modulate attention and decision-making abilities. Additionally, numerous studies link highly aroused stress states with impaired decision-making capabilities [278], decreased situational awareness [279], and degraded performance, which could impair driving ability [280]. Another driver state, often responsible for traffic violations and even road accidents that can lead to distraction, is confusion or irritation, as it is related to loss of self-control and, therefore, loss of vehicle control, which can be provoked by non-intuitive user interfaces or defective navigation systems as well as by complex traffic conditions, mistakable signs and complicated routing. Moreover, the amount of information that needs to be processed simultaneously during driving is a source of confusion especially for older people [281], who have slower perception and reaction times. Just like stress, confusion or irritation leads to impairment of driving capabilities including driver's perception, attention, decision making, and strategic planning. Nervousness corresponds to a level of arousal above the "normal" one, which best suits to the driving task [211]. It is an affective state with negative impact both on decision-making process and strategic planning. Nervousness can be induced by a variety of reasons either directly related to the driving task like novice drivers or by other factors like personal/physical conditions [211].

The system should be validated, firstly, in a driver simulator and afterwards, in real conditions, where various factors including variations in lighting and noise can also affect both the driver's attention and the performance of the developed algorithms. Therefore, public data sets covering simulated and real driving environments should be released. The driver's physiological responses could be different in a driver simulator from those in real conditions [11,167,257,263]. Hence, while developing an inattention detection system, the simulated environment must be a perfect replica of the real environment. However, they are normally used in research and simulated scenarios, but not in real ones, due to the problems of vision systems working in outdoor environments (lighting changes, sudden movements, etc.). Moreover, they do not work properly with users wearing glasses and may need high computational requirements.

Data-driven applications will require large amount of labeled images for both training and testing the system. Both manual data reduction and labeling of data are time-consuming and they are also subject to interpretation of the reductionist. Therefore, to deal with this problem, two approaches are emerging from the literature: (1) unsupervised or semi-supervised learning and (2) automatic data reduction. For example, in connection with the first approach, Liu et al. [189] commented the benefits of SSL methods. Specifically, the explained the benefits of using SSL increased with the size of unlabeled data set showing that by exploring the data structure without actually labeling them, extra information to improve models performance can be obtained. On the other hand, there has been a hype in data reduction using vehicle dynamics and looking outside on large scale naturalistic driving data [282–284], and looking in at the driver [285].

In many distraction detection systems, the use of commercial sensors is usually performed [77,239,241,245–247]. We understand that the reason from this is twofold: these systems are well-established solutions offering both head and gaze tracking in the car environment and the efforts of the investigation can be focused to detect and predict distraction from the outputs from these commercial sensors instead of developing a new sensor from the very beginning. These commercial sensors can operate using one camera [239,245–247], two cameras [241] or even up to 8 cameras [77] placed all over the vehicle cabin. What we find missing is some research works trying to compare these commercial sensors in order to highlight the pros and cons of each one. Also, missing from the literature is the comparison between a new sensor and a commercial one trying to offer a competitive solution from the sake of the research community.

Author Contributions: Alberto Fernández, Rubén Usamentiaga, Juan Luis Carús have made the review of the already existing methods and have written the paper as well. Alberto Fernández, Rubén Usamentiaga and Rubén Casado have designed paper concept and structure and have revised its content.

Conflicts of Interest: The authors declare no conflict of interest.

References

1. World Health Organization. Global Status Report on Road Safety 2015. Available online: http://apps.who.int/iris/bitstream/10665/189242/1/9789241565066_eng.pdf?ua=1 (accessed on 2 July 2016).
2. Llerena, L.E.; Aronow, K.V.; Macleod, J.; Bard, M.; Salzman, S.; Greene, W.; Haider, A.; Schupper, A. An evidence-based review: Distracted driver. *J. Trauma Acute Care Surg.* **2015**, *78*, 147–152.
3. Carsten, O.; Merat, N. Protective or not? (visual distraction). In Proceedings of the 2015 4th International Conference on Driver Distraction and Inattention, Sydney, Australia, 9–11 November 2015.
4. Dingus, T.A.; Klauer, S.; Neale, V.; Petersen, A.; Lee, S.; Sudweeks, J.; Perez, M.; Hankey, J.; Ramsey, D.; Gupta, S.; et al. *The 100-Car Naturalistic Driving Study, Phase II—Results of the 100-Car Field Experiment*; National Highway Traffic Safety Administration: Washington, DC, USA, 2006.
5. Ranney, T.A.; Mazzae, E.; Garrott, R.; Goodman, M.J. NHTSA driver distraction research: Past, present, and future. In *Driver Distraction Internet Forum*; National Highway Traffic Safety Administration: Washington, DC, USA, 2000; Volume 2000.
6. Klauer, S.G.; Neale, V.L.; Dingus, T.A.; Ramsey, D.; Sudweeks, J. Driver inattention: A contributing factor to crashes and near-crashes. In Proceedings of the Human Factors and Ergonomics Society Annual Meeting, Orlando, FL, USA, 26–30 September 2005; SAGE Publications: Thousand Oaks, CA, USA; Volume 49, pp. 1922–1926.
7. Klauer, S.G.; Dingus, T.A.; Neale, V.L.; Sudweeks, J.D.; Ramsey, D.J. *The Impact of Driver Inattention on Near-Crash/crash Risk: An Analysis Using the 100-Car Naturalistic Driving Study Data*; National Highway Traffic Safety Administration: Washington, DC, USA, 2006.
8. Talbot, R.; Fagerlind, H. Exploring inattention and distraction in the SafetyNet accident causation database. *Accid. Anal. Prev.* **2009**, *60*, 445–455.
9. Bennakhi, A.; Safar, M. Ambient Technology in Vehicles: The Benefits and Risks. *Procedia Comput. Sci.* **2016**, *83*, 1056–1063.
10. Lee, J.D. Driving safety. In *Review of Human Factors*; Nickerson, R.S., Ed.; Human Factors and Ergonomics Society: San Francisco, CA, USA, 2006; pp. 172–218.
11. Engström, J.; Johansson, E.; Östlund, J. Effects of visual and cognitive load in real and simulated motorway driving. *Transp. Res. Part F Traffic Psychol. Behav.* **2005**, *8*, 97–120.
12. Liu, T.; Yang, Y.; Huang, G.B.; Lin, Z. Detection of Drivers' Distraction Using Semi-Supervised Extreme Learning Machine. In *Proceedings of ELM-2014*; Springer: Berlin/Heidelberg, Germany, 2015; Volume 2, pp. 379–387.
13. Simons-Morton, B.G.; Guo, F.; Klauer, S.G.; Ehsani, J.P.; Pradhan, A.K. Keep your eyes on the road: Young driver crash risk increases according to duration of distraction. *J. Adolesc. Health* **2014**, *54*, S61–S67.
14. Dingus, T.A.; Neale, V.L.; Klauer, S.G.; Petersen, A.D.; Carroll, R.J. The development of a naturalistic data collection system to perform critical incident analysis: An investigation of safety and fatigue issues in long-haul trucking. *Accid. Anal. Prev.* **2006**, *38*, 1127–1136.
15. Harbluk, J.L.; Noy, Y.I.; Trbovich, P.L.; Eizenman, M. An on-road assessment of cognitive distraction: Impacts on drivers' visual behavior and braking performance. *Accid. Anal. Prev.* **2007**, *39*, 372–379.
16. Recarte, M.A.; Nunes, L.M. Mental workload while driving: Effects on visual search, discrimination, and decision making. *J. Exp. Psychol. Appl.* **2003**, *9*, 119–137.
17. Klauer, S.G.; Guo, F.; Simons-Morton, B.G.; Ouimet, M.C.; Lee, S.E.; Dingus, T.A. Distracted driving and risk of road crashes among novice and experienced drivers. *N. Engl. J. Med.* **2014**, *370*, 54–59.
18. Bergmark, R.W.; Gliklich, E.; Guo, R.; Gliklich, R.E. Texting while driving: The development and validation of the distracted driving survey and risk score among young adults. *Inj. Epidemiol.* **2016**, *3*, doi:10.1186/s40621-016-0073-8.
19. Administration, N.H.T.S. Traffic Safety Facts: Distracted Driving 2014. Available online: https://crashstats.nhtsa.dot.gov/Api/Public/ViewPublication/812260 (accessed on 2 July 2016).
20. Carmona, J.; García, F.; Martín, D.; Escalera, A.D.L.; Armingol, J.M. Data fusion for driver behaviour analysis. *Sensors* **2015**, *15*, 25968–25991.
21. Sahayadhas, A.; Sundaraj, K.; Murugappan, M. Detecting driver drowsiness based on sensors: A review. *Sensors* **2012**, *12*, 16937–16953.

22. Daza, I.G.; Bergasa, L.M.; Bronte, S.; Yebes, J.J.; Almazán, J.; Arroyo, R. Fusion of optimized indicators from Advanced Driver Assistance Systems (ADAS) for driver drowsiness detection. *Sensors* **2014**, *14*, 1106–1131.

23. Hoff, J.; Grell, J.; Lohrman, N.; Stehly, C.; Stoltzfus, J.; Wainwright, G.; Hoff, W.S. Distracted driving and implications for injury prevention in adults. *J. Trauma Nurs.* **2013**, *20*, 31–34.

24. Stothart, C.; Mitchum, A.; Yehnert, C. The attentional cost of receiving a cell phone notification. *J. Exp. Psychol. Hum. Percept. Perform.* **2015**, *41*, 893–897.

25. Kushlev, K.; Proulx, J.; Dunn, E.W. Silence Your Phones: Smartphone Notifications Increase Inattention and Hyperactivity Symptoms. In Proceedings of the 2016 CHI Conference on Human Factors in Computing Systems, San Jose, CA, USA, 7–12 May 2016; pp. 1011–1020.

26. Angell, L.; Perez, M.; Soccolich, S. *Identification of Cognitive Load in Naturalistic Driving*; Virginia Tech Transportation Institute: Blacksburg, VA, USA, 2015.

27. Liang, Y. Detecting Driver Distraction. Ph.D. Thesis, University of Iowa, Iowa City, IA, USA, May 2009.

28. Caird, J.K.; Johnston, K.A.; Willness, C.R.; Asbridge, M.; Steel, P. A meta-analysis of the effects of texting on driving. *Accid. Anal. Prev.* **2014**, *71*, 311–318.

29. Ohn-Bar, E.; Trivedi, M.M. Looking at Humans in the Age of Self-Driving and Highly Automated Vehicles. *IEEE Trans. Intell. Veh.* **2016**, *1*, 90–104.

30. Toole, L.M. Crash Risk and Mobile Device Use Based on Fatigue and Drowsiness Factors in Truck Drivers. Ph.D. Thesis, Virginia Tech, Blacksburg, VA, USA, October 2013.

31. Lee, J.D.; Young, K.L.; Regan, M.A. *Defining Driver Distraction*; CRC Press: Boca Raton, FL, USA, 2008.

32. Foley, J.P.; Young, R.; Angell, L.; Domeyer, J.E. Towards operationalizing driver distraction. In Proceedings of the 7th International Symposium on Human Factors in Driver Assessment, Training, and Vehicle Design, Bolton Landing, NY, USA, 17–20 June 2013; pp. 57–63.

33. Young, R.A. A Tabulation of Driver Distraction Definitions. Available online: https://www.yumpu.com/en/document/view/22871401/a-tabulation-of-driver-distraction-definitions-toyota/5 (accessed on 27 October 2016).

34. Young, R. Cognitive distraction while driving: A critical review of definitions and prevalence in crashes. *SAE Int. J. Passeng. Cars Electron. Electr. Syst.* **2012**, *5*, 326–342.

35. Regan, M.A.; Hallett, C.; Gordon, C.P. Driver distraction and driver inattention: Definition, relationship and taxonomy. *Accid. Anal. Prev.* **2011**, *43*, 1771–1781.

36. Engström, J.; Monk, C.; Hanowski, R.; Horrey, W.; Lee, J.; McGehee, D.; Regan, M.; Stevens, A.; Traube, E.; Tuukkanen, M.; et al. *A Conceptual Framework and Taxonomy for Understanding and Categorizing Driver Inattention*; European Commission: Brussels, Belgium, 2013.

37. Lee, J.D. Dynamics of driver distraction: The process of engaging and disengaging. *Ann. Adv. Autom. Med.* **2014**, *58*, 24–32.

38. Viola, P.; Jones, M. Rapid object detection using a boosted cascade of simple features. In Proceedings of the 2001 IEEE Computer Society Conference on Computer Vision and Pattern Recognition (CVPR 2001), Kauai, HI, USA, 8–14 December 2001; Volume 1.

39. Zafeiriou, S.; Zhang, C.; Zhang, Z. A Survey on Face Detection in the wild: Past, present and future. *Comput. Vis. Image Underst.* **2015**, *138*, 1–24.

40. Viola, P.; Jones, M.J. Robust real-time face detection. *Int. J. Comput. Vis.* **2004**, *57*, 137–154.

41. Jensen, O.H. Implementing the Viola-Jones Face Detection Algorithm. Ph.D. Thesis, Technical University of Denmark, Lyngby, Denmark, September 2008.

42. Dalal, N.; Triggs, B. Histograms of oriented gradients for human detection. In Proceedings of the 2005 IEEE Computer Society Conference on Computer Vision and Pattern Recognition (CVPR 2005), Miami, FL, USA, 20–25 June 2005; Volume 1, pp. 886–893.

43. Dlib. Make Your Own Object Detector. Available online: http://blog.dlib.net/2014/02/dlib-186-released-make-your-own-object.html (accessed on 25 November 2015).

44. Ranjan, R.; Patel, V.M.; Chellappa, R. A Deep Pyramid Deformable Part Model for Face Detection. 2015, arXiv preprint arXiv:1508.04389.

45. Wang, Q.; Wu, J.; Long, C.; Li, B. P-FAD: Real-time face detection scheme on embedded smart cameras. *IEEE J. Emerg. Sel. Top. Circuits Syst.* **2013**, *3*, 210–222.

46. Markuš, N.; Frljak, M.; Pandžić, I.S.; Ahlberg, J.; Forchheimer, R. A Method for Object Detection Based on Pixel Intensity Comparisons Organized in Decision Trees. 2013, arXiv preprint arXiv:1305.4537

47. Reese, K.; Zheng, Y.; Elmaghraby, A. A comparison of face detection algorithms in visible and thermal spectrums. In Proceedings of the International Conference on Advances in Computer Science and Application, New Delhi, India, 25–27 May 2012.

48. Abtahi, S.; Omidyeganeh, M.; Shirmohammadi, S.; Hariri, B. YawDD: A yawning detection dataset. In Proceedings of the 5th ACM Multimedia Systems Conference, Singapore, 19–27 March 2014; pp. 24–28.

49. Vicente, F.; Huang, Z.; Xiong, X.; de la Torre, F.; Zhang, W.; Levi, D. Driver Gaze Tracking and Eyes Off the Road Detection System. *IEEE Trans. Intell. Transp. Syst.* **2015**, *16*, 2014–2027.

50. Murphy-Chutorian, E.; Trivedi, M.M. Head pose estimation in computer vision: A survey. *IEEE Trans. Pattern Anal. Mach. Intell.* **2009**, *31*, 607–626.

51. Tawari, A.; Martin, S.; Trivedi, M.M. Continuous Head Movement Estimator for Driver Assistance: Issues, Algorithms, and On-Road Evaluations. *IEEE Trans. Intell. Transp. Syst.* **2014**, *15*, 818–830.

52. Zhu, Z.; Ji, Q. Real time and non-intrusive driver fatigue monitoring. In Proceedings of the 2004 7th International IEEE Conference on Intelligent Transportation Systems, Washington, DC, USA, 3–6 October 2004; pp. 657–662.

53. Garcia-Mateos, G.; Ruiz, A.; Lopez-de Teruel, P.E.; Rodriguez, A.L.; Fernandez, L. Estimating 3D facial pose in video with just three points. In Proceedings of the 2008 IEEE Computer Society Conference on Computer Vision and Pattern Recognition Workshops (CVPRW'08), Anchorage, AK, USA, 23–28 June 2008; pp. 1–8.

54. Kaminski, J.Y.; Knaan, D.; Shavit, A. Single image face orientation and gaze detection. *Mach. Vis. Appl.* **2009**, *21*, 85–98.

55. Martin, S.; Tawari, A.; Murphy-Chutorian, E.; Cheng, S.Y.; Trivedi, M. On the design and evaluation of robust head pose for visual user interfaces: Algorithms, databases, and comparisons. In Proceedings of the 4th International Conference on Automotive User Interfaces and Interactive Vehicular Applications, Portsmouth, NH, USA, 17–19 October 2012; pp. 149–154.

56. Oyini Mbouna, R.; Kong, S.G.; Chun, M.G. Visual analysis of eye state and head pose for driver alertness monitoring. *IEEE Trans. Intell. Transp. Syst.* **2013**, *14*, 1462–1469.

57. Xiong, X.; de la Torre, F. Supervised descent method and its applications to face alignment. In Proceedings of the 2013 IEEE Conference on Computer Vision and Pattern Recognition (CVPR), Portland, OR, USA, 25–27 June 2013; pp. 532–539.

58. Asthana, A.; Marks, T.K.; Jones, M.J.; Tieu, K.H.; Rohith, M. Fully automatic pose-invariant face recognition via 3D pose normalization. In Proceedings of the 2011 IEEE International Conference on Computer Vision (ICCV), Barcelona, Spain, 6–13 November 2011; pp. 937–944.

59. Murphy-Chutorian, E.; Trivedi, M.M. Head pose estimation and augmented reality tracking: An integrated system and evaluation for monitoring driver awareness. *IEEE Trans. Intell. Transp. Syst.* **2010**, *11*, 300–311.

60. Peláez, G.; de la Escalera, A.; Armingol, J. Head Pose Estimation Based on 2D and 3D Information. *Phys. Procedia* **2014**, *22*, 420–427.

61. Tawari, A.; Trivedi, M.M. Robust and continuous estimation of driver gaze zone by dynamic analysis of multiple face videos. In Proceedings of the 2014 IEEE Intelligent Vehicles Symposium Proceedings, Ypsilanti, MI, USA, 8–11 June 2014; pp. 344–349.

62. Ohn-Bar, E.; Tawari, A.; Martin, S.; Trivedi, M.M. Predicting driver maneuvers by learning holistic features. In Proceedings of the 2014 IEEE Intelligent Vehicles Symposium Proceedings, Ypsilanti, MI, USA, 8–11 June 2014; pp. 719–724.

63. Ohn-Bar, E.; Tawari, A.; Martin, S.; Trivedi, M.M. On surveillance for safety critical events: In-vehicle video networks for predictive driver assistance systems. *Comput. Vis. Image Underst.* **2015**, *134*, 130–140.

64. La Cascia, M.; Sclaroff, S.; Athitsos, V. Fast, reliable head tracking under varying illumination: An approach based on registration of texture-mapped 3D models. *IEEE Trans. Pattern Anal. Mach. Intell.* **2000**, *22*, 322–336.

65. Zhu, X.; Ramanan, D. Face detection, pose estimation, and landmark localization in the wild. In Proceedings of the 2012 IEEE Conference on Computer Vision and Pattern Recognition (CVPR), Providence, RI, USA, 16–21 June 2012; pp. 2879–2886.

66. Uricár, M.; Franc, V.; Hlavác, V. Facial Landmark Tracking by Tree-based Deformable Part Model Based Detector. In Proceedings of the IEEE International Conference on Computer Vision Workshops, Santiago, Chile, 11–18 December 2015; pp. 10–17.

67. Uricár, M.; Franc, V.; Thomas, D.; Sugimoto, A.; Hlavác, V. Real-time multi-view facial landmark detector learned by the structured output SVM. In Proceedings of the 2015 11th IEEE International Conference and Workshops on Automatic Face and Gesture Recognition (FG), Ljubljana, Slovenia, 4–8 May 2015; Volume 2, pp. 1–8.

68. Cao, X.; Wei, Y.; Wen, F.; Sun, J. Face alignment by explicit shape regression. *Int. J. Comput. Vis.* **2014**, *107*, 177–190.

69. Ren, S.; Cao, X.; Wei, Y.; Sun, J. Face alignment at 3000 fps via regressing local binary features. In Proceedings of the IEEE Conference on Computer Vision and Pattern Recognition, Columbus, OH, USA, 23–28 June 2014; pp. 1685–1692.

70. Wang, N.; Gao, X.; Tao, D.; Li, X. Facial feature point detection: A comprehensive survey. 2014, arXiv preprint arXiv:1410.1037

71. Rezaei, M.; Klette, R. Adaptive Haar-like classifier for eye status detection under non-ideal lighting conditions. In Proceedings of the 27th Conference on Image and Vision Computing New Zealand, Dunedin, New Zealand, 26–28 November 2012; pp. 521–526.

72. Song, F.; Tan, X.; Chen, S.; Zhou, Z.H. A literature survey on robust and efficient eye localization in real-life scenarios. *Pattern Recognit.* **2013**, *46*, 3157–3173.

73. Kircher, K.; Ahlström, C. *Evaluation of Methods for the Assessment of Minimum Required Attention*; Swedish National Road and Transport Research Institute (VTI): Linkoping, Sweden, 2015.

74. Victor, T. Keeping Eye and Mind on the Road. Ph.D. Thesis, Uppsala University, Uppsala, Sweden, 2005.

75. Flannagan, C.A.; Bao, S.; Klinich, K.D. *Driver Distraction From Cell Phone Use and Potential for Self-Limiting Behavior*; University of Michigan: Ann Arbor, MI, USA, 2012.

76. Ahlstrom, C.; Dukic, T. Comparison of eye tracking systems with one and three cameras. In Proceedings of the 7th International Conference on Methods and Techniques in Behavioral Research, Eindhoven, The Netherlands, 24–27 August 2010.

77. SmartEye. Smart Eye Pro. Available online: http://smarteye.se/products/smart-eye-pro/ (accessed on 3 January 2016).

78. Sigari, M.H.; Pourshahabi, M.R.; Soryani, M.; Fathy, M. A Review on Driver Face Monitoring Systems for Fatigue and Distraction Detection. *Int. J. Adv. Sci. Technol.* **2014**, *64*, 73–100.

79. Zhao, S.; Grigat, R.R. Robust eye detection under active infrared illumination. In Proceedings of the 2006 IEEE 18th International Conference on Pattern Recognition (ICPR 2006), Washington, DC, USA, 20–24 August 2006; Volume 4, pp. 481–484.

80. Zhao, S.; Grigat, R.R. An automatic face recognition system in the near infrared spectrum. In *Machine Learning and Data Mining in Pattern Recognition*; Springer: Berlin/Heidelberg, Germany, 2005; pp. 437–444.

81. Smith, P.; Shah, M.; Lobo, N.D.V. Determining driver visual attention with one camera. *IEEE Trans. Intell. Transp. Syst.* **2003**, *4*, 205–218.

82. Sigari, M.H. Driver hypo-vigilance detection based on eyelid behavior. In Proceedings of the IEEE 7th International Conference on Advances in Pattern Recognition (ICAPR'09), Kolkata, India, 4–6 Feburary 2009; pp. 426–429.

83. Batista, J. A drowsiness and point of attention monitoring system for driver vigilance. In Proceedings of the 2007 IEEE Intelligent Transportation Systems Conference (ITSC 2007), 30 September–3 October 2007; pp. 702–708.

84. Flores, M.J.; Armingol, J.M.; de la Escalera, A. Driver drowsiness warning system using visual information for both diurnal and nocturnal illumination conditions. *EURASIP J. Adv. Signal Process.* **2010**, *2010*, 438205.

85. Rezaei, M.; Klette, R. Simultaneous analysis of driver behaviour and road condition for driver distraction detection. *Int. J. Image Data Fusion* **2011**, *2*, 217–236.

86. Craye, C.; Karray, F. Multi-distributions particle filter for eye tracking inside a vehicle. In *Image Analysis and Recognition*; Springer: Berlin/Heidelberg, Germany, 2013; pp. 407–416.

87. Rezaei, M.; Klette, R. Look at the driver, look at the road: No distraction! No accident! In Proceedings of the 2014 IEEE Conference on Computer Vision and Pattern Recognition (CVPR), Columbus, OH, USA, 23–28 June 2014; pp. 129–136.

88. Zhao, C.; Zhang, B.; He, J.; Lian, J. Recognition of driving postures by contourlet transform and random forests. *Intell. Transp. Syst. IET* **2012**, *6*, 161–168.
89. Zhao, C.; Gao, Y.; He, J.; Lian, J. Recognition of driving postures by multiwavelet transform and multilayer perceptron classifier. *Eng. Appl. Artif. Intell.* **2012**, *25*, 1677–1686.
90. Zhao, C.H.; Zhang, B.L.; Zhang, X.Z.; Zhao, S.Q.; Li, H.X. Recognition of driving postures by combined features and random subspace ensemble of multilayer perceptron classifiers. *Neural Comput. Appl.* **2013**, *22*, 175–184.
91. Zhao, C.; Zhang, B.; He, J. Vision-based classification of driving postures by efficient feature extraction and bayesian approach. *J. Intell. Robot. Syst.* **2013**, *72*, 483–495.
92. Yan, C.; Coenen, F.; Zhang, B. Driving posture recognition by joint application of motion history image and pyramid histogram of oriented gradients. *Int. J. Veh. Technol.* **2014**, *2014*, 719413.
93. Yan, C.; Coenen, F.; Zhang, B. Driving posture recognition by convolutional neural networks. *IET Comput. Vis.* **2016**, *10*, 103–114.
94. Bosch, A.; Zisserman, A.; Munoz, X. Representing shape with a spatial pyramid kernel. In Proceedings of the 6th ACM International Conference on Image and Video Retrieval, Hyderabad, India, 6–12 January 2007; pp. 401–408.
95. Lowe, D.G. Distinctive image features from scale-invariant keypoints. *Int. J. Comput. Vis.* **2004**, *60*, 91–110.
96. Zhang, X.; Zheng, N.; Wang, F.; He, Y. Visual recognition of driver hand-held cell phone use based on hidden CRF. In Proceedings of the 2011 IEEE International Conference on Vehicular Electronics and Safety (ICVES), Beijing, China, 10–12 July 2011; pp. 248–251.
97. Artan, Y.; Bulan, O.; Loce, R.P.; Paul, P. Driver cell phone usage detection from HOV/HOT NIR images. In Proceedings of the 2014 IEEE Conference on Computer Vision and Pattern Recognition Workshops (CVPRW), Columbus, OH, USA, 23–28 June 2014; pp. 225–230.
98. Berri, R.A.; Silva, A.G.; Parpinelli, R.S.; Girardi, E.; Arthur, R. A Pattern Recognition System for Detecting Use of Mobile Phones While Driving. 2014, arXiv preprint arXiv:1408.0680.
99. Xu, B.; Loce, R.P. A machine learning approach for detecting cell phone usage. In Proceedings of the IS&T/SPIE Electronic Imaging, International Society for Optics and Photonics, San Francisco, CA, USA, 4 March 2015.
100. Seshadri, K.; Juefei-Xu, F.; Pal, D.K.; Savvides, M.; Thor, C.P. Driver Cell Phone Usage Detection on Strategic Highway Research Program (SHRP2) Face View Videos. In Proceedings of the IEEE Conference on Computer Vision and Pattern Recognition (CVPR) Workshops, Boston, MA, USA, 7–12 June 2015.
101. Li, Y.; Zhou, G.; Li, Y.; Shen, D. Determining driver phone use leveraging smartphone sensors. *Multimed. Tools Appl.* **2015**, *1*, doi:10.1007/s11042-015-2969-7.
102. Ohn-Bar, E.; Trivedi, M. In-vehicle hand activity recognition using integration of regions. In Proceedings of the 2013 IEEE Intelligent Vehicles Symposium (IV), Gold Coast City, Australia, 23–26 June 2013; pp. 1034–1039.
103. Ohn-Bar, E.; Martin, S.; Trivedi, M.M. Driver hand activity analysis in naturalistic driving studies: Challenges, algorithms, and experimental studies. *J. Electron. Imaging* **2013**, *22*, 041119.
104. Martin, S.; Ohn-Bar, E.; Tawari, A.; Trivedi, M.M. Understanding head and hand activities and coordination in naturalistic driving videos. In Proceedings of the 2014 IEEE Intelligent Vehicles Symposium Proceedings, Ypsilanti, MI, USA, 8–11 June 2014; pp. 884–889.
105. Ohn-Bar, E.; Martin, S.; Tawari, A.; Trivedi, M. Head, eye, and hand patterns for driver activity recognition. In Proceedings of the IEEE 2014 22nd International Conference on Pattern Recognition (ICPR), Stockholm, Sweden, 24–28 August 2014; pp. 660–665.
106. Ohn-Bar, E.; Trivedi, M.M. The power is in your hands: 3D analysis of hand gestures in naturalistic video. In Proceedings of the 2013 IEEE Conference on Computer Vision and Pattern Recognition Workshops (CVPRW), Portland, OR, USA, 23–28 June 2013; pp. 912–917.
107. Lee, S.R.; Bambach, S.; Crandall, D.J.; Franchak, J.M.; Yu, C. This hand is my hand: A probabilistic approach to hand disambiguation in egocentric video. In Proceedings of the 2014 IEEE Conference on Computer Vision and Pattern Recognition Workshops (CVPRW), Columbus, OH, USA, 23–28 June 2014; pp. 557–564.
108. Drews, F.A.; Yazdani, H.; Godfrey, C.N.; Cooper, J.M.; Strayer, D.L. Text messaging during simulated driving. *Hum. Factors J. Hum. Factors Ergon. Soc.* **2009**, *51*, 762–770.

109. Tsimhoni, O.; Arbor, A. *Time-Sharing of a Visual in-Vehicle Task While Driving: Findings from the Task Occlusion Method*; University of Michigan, Transportation Research Institute: Ann Arbor, MI, USA, 2003.

110. Victor, T.W.; Harbluk, J.L.; Engström, J.A. Sensitivity of eye-movement measures to in-vehicle task difficulty. *Transp. Res. Part F Traffic Psychol. Behav.* **2005**, *8*, 167–190.

111. Donmez, B.; Boyle, L.N.; Lee, J.D. Safety implications of providing real-time feedback to distracted drivers. *Accid. Anal. Prev.* **2007**, *39*, 581–590.

112. Klauer, S.G.; Guo, F.; Sudweeks, J.; Dingus, T.A. *An Analysis of Driver Inattention Using a Case-Crossover Approach on 100-Car Data: Final Report*; National Highway Traffic Safety Administration: Washington, DC, USA, 2010.

113. Ahlstrom, C.; Kircher, K.; Kircher, A. A gaze-based driver distraction warning system and its effect on visual behavior. *IEEE Trans. Intell. Transp. Syst.* **2013**, *14*, 965–973.

114. Lee, S.J.; Jo, J.; Jung, H.G.; Park, K.R.; Kim, J. Real-time gaze estimator based on driver's head orientation for forward collision warning system. *IEEE Trans. Intell. Transp. Syst.* **2011**, *12*, 254–267.

115. Wollmer, M.; Blaschke, C.; Schindl, T.; Schuller, B.; Farber, B.; Mayer, S.; Trefflich, B. Online driver distraction detection using long short-term memory. *IEEE Trans. Intell. Transp. Syst.* **2011**, *12*, 574–582.

116. Neale, V.L.; Dingus, T.A.; Klauer, S.G.; Sudweeks, J.; Goodman, M. *An Overview of the 100-Car Naturalistic Study and Findings*; National Highway Traffic Safety Administration: Washington, DC, USA, 2005.

117. Boyraz, P.; Yang, X.; Hansen, J.H. Computer vision systems for context-aware active vehicle safety and driver assistance. In *Digital Signal Processing for in-Vehicle Systems and Safety*; Springer: Berlin/Heidelberg, Germany, 2012; pp. 217–227.

118. Hammoud, R.I.; Wilhelm, A.; Malawey, P.; Witt, G.J. Efficient real-time algorithms for eye state and head pose tracking in advanced driver support systems. In Proceedings of the 2005 IEEE Computer Society Conference on Computer Vision and Pattern Recognition (CVPR 2005), San Diego, CA, USA, 20–25 June 2005; Volume 2.

119. Tivesten, E.; Dozza, M. Driving context and visual-manual phone tasks influence glance behavior in naturalistic driving. *Transp. Res. Part F Traffic Psychol. Behav.* **2014**, *26*, 258–272.

120. Morimoto, C.H.; Koons, D.; Amir, A.; Flickner, M. Pupil detection and tracking using multiple light sources. *Image Vis. Comput.* **2000**, *18*, 331–335.

121. Ji, Q.; Yang, X. Real time visual cues extraction for monitoring driver vigilance. In *Computer Vision Systems*; Springer: Berlin/Heidelberg, Germany, 2001; pp. 107–124.

122. Ji, Q.; Yang, X. Real-time eye, gaze, and face pose tracking for monitoring driver vigilance. *Real Time Imaging* **2002**, *8*, 357–377.

123. Ji, Q. 3D Face pose estimation and tracking from a monocular camera. *Image Vis. Comput.* **2002**, *20*, 499–511.

124. Gu, H.; Ji, Q.; Zhu, Z. Active facial tracking for fatigue detection. In Proceedings of the 6th IEEE Workshop on Applications of Computer Vision (WACV 2002), Orlando, FL, USA, 3–4 December 2002; pp. 137–142.

125. Batista, J.P. A real-time driver visual attention monitoring system. In *Pattern Recognition and Image Analysis*; Springer: Berlin/Heidelberg, Germany, 2005; pp. 200–208.

126. Bergasa, L.M.; Nuevo, J.; Sotelo, M.; Barea, R.; Lopez, M.E. Real-time system for monitoring driver vigilance. *IEEE Trans. Intell. Transp. Syst.* **2006**, *7*, 63–77.

127. Hansen, D.W.; Ji, Q. In the eye of the beholder: A survey of models for eyes and gaze. *IEEE Trans. Pattern Anal. Mach. Intell.* **2010**, *32*, 478–500.

128. D'Orazio, T.; Leo, M.; Guaragnella, C.; Distante, A. A visual approach for driver inattention detection. *Pattern Recognit.* **2007**, *40*, 2341–2355.

129. Nuevo, J.; Bergasa, L.M.; Sotelo, M.; Ocaña, M. Real-time robust face tracking for driver monitoring. In Proceedings of the IEEE Intelligent Transportation Systems Conference (ITSC'06), Toronto, ON, Canada, 17–20 September 2006; pp. 1346–1351.

130. Bergasa, L.M.; Buenaposada, J.M.; Nuevo, J.; Jimenez, P.; Baumela, L. Analysing driver's attention level using computer vision. In Proceedings of the 11th International IEEE Conference on Intelligent Transportation Systems (ITSC 2008), Beijing, China, 12–15 October 2008; pp. 1149–1154.

131. Flores, M.J.; Armingol, J.M.; De la Escalera, A. Driver drowsiness detection system under infrared illumination for an intelligent vehicle. *IET Intell. Transp. Syst.* **2011**, *5*, 241–251.

132. Jimenez-Pinto, J.; Torres-Torriti, M. Face salient points and eyes tracking for robust drowsiness detection. *Robotica* **2012**, *30*, 731–741.

133. Jiménez-Pinto, J.; Torres-Torriti, M. Optical flow and driver's kinematics analysis for state of alert sensing. *Sensors* **2013**, *13*, 4225–4257.

134. Cyganek, B.; Gruszczyński, S. Hybrid computer vision system for drivers' eye recognition and fatigue monitoring. *Neurocomputing* **2014**, *126*, 78–94.

135. Murphy-Chutorian, E.; Doshi, A.; Trivedi, M.M. Head pose estimation for driver assistance systems: A robust algorithm and experimental evaluation. In Proceedings of the 2007 IEEE Intelligent Transportation Systems Conference (ITSC 2007), Bellevue, WA, USA, 30 September–3 October 2007; pp. 709–714.

136. Hattori, A.; Tokoro, S.; Miyashita, M.; Tanaka, I.; Ohue, K.; Uozumi, S. Development of Forward Collision Warning System Using the Driver Behavioral Information. Available online: http://papers.sae.org/2006-01-1462/ (accessed on 26 October 2016).

137. You, C.W.; Lane, N.D.; Chen, F.; Wang, R.; Chen, Z.; Bao, T.J.; Montes-de Oca, M.; Cheng, Y.; Lin, M.; Torresani, L.; et al. Carsafe app: Alerting drowsy and distracted drivers using dual cameras on smartphones. In Proceedings of the 11th Annual International Conference on Mobile Systems, Applications, and Services, Taipei, Taiwan, 25–28 June 2013; pp. 13–26.

138. Flores, M.J.; Armingol, J.M.; de la Escalera, A. Real-time warning system for driver drowsiness detection using visual information. *J. Intell. Robot. Syst.* **2010**, *59*, 103–125.

139. Yuying, J.; Yazhen, W.; Haitao, X. A surveillance method for driver's fatigue and distraction based on machine vision. In Proceedings of the 2011 International Conference on Transportation, Mechanical, and Electrical Engineering (TMEE), Changchun, China, 16–18 December 2011; pp. 727–730.

140. Pohl, J.; Birk, W.; Westervall, L. A driver-distraction-based lane-keeping assistance system. *Proc. Inst. Mech. Eng. Part I J. Syst. Control Eng.* **2007**, *221*, 541–552.

141. Lee, J.; Moeckli, J.; Brown, T.; Roberts, S.; Victor, T.; Marshall, D.; Schwarz, C.; Nadler, E. Detection of driver distraction using vision-based algorithms. In Proceedings of the 23rd International Conference on Enhanced Safety of Vehicles, Seoul, Korea, 27–30 May 2013.

142. Donmez, B.; Boyle, L.N.; Lee, J.D. Mitigating driver distraction with retrospective and concurrent feedback. *Accid. Anal. Prev.* **2008**, *40*, 776–786.

143. Kircher, K.; Kircher, A.; Ahlström, C. *Results of a Field Study on a Driver Distraction Warning System*; Swedish National Road and Transport Research Institute (VTI): Linkoping, Sweden, 2009.

144. Kircher, K.; Ahlstrom, C.; Kircher, A. Comparison of two eye-gaze based real-time driver distraction detection algorithms in a small-scale field operational test. In Proceedings of the 5th International Symposium on Human Factors in Driver Assessment, Training and Vehicle Design, Big Sky, MT, USA, 22–25 June 2009; pp. 16–23.

145. Kircher, K.; Kircher, A.; Claezon, F. *Distraction and Drowsiness—A Field Study*; Swedish National Road and Transport Research Institute (VTI): Linköping, Sweden, 2009.

146. Victor, T. The Victor and Larsson (2010) distraction detection algorithm and warning strategy. *Volvo Technol.* **2010**, *1*, 0–6.

147. Angell, L.S.; Auflick, J.; Austria, P.; Kochhar, D.S.; Tijerina, L.; Biever, W.; Diptiman, T.; Hogsett, J.; Kiger, S. *Driver Workload Metrics Task 2 Final Report*; National Highway Traffic Safety Administration: Washington, DC, USA, 2006.

148. Recarte, M.A.; Nunes, L.M. Effects of verbal and spatial-imagery tasks on eye fixations while driving. *J. Exp. Psychol. Appl.* **2000**, *6*, 31.

149. Harbluk, J.L.; Noy, Y.I.; Eizenman, M. *The Impact of Cognitive Distraction on Driver Visual Behaviour and Vehicle Control*; Transport Canada: Ottawa, ON, Canada, 2002.

150. May, J.G.; Kennedy, R.S.; Williams, M.C.; Dunlap, W.P.; Brannan, J.R. Eye movement indices of mental workload. *Acta Psychol.* **1990**, *75*, 75–89.

151. Miyaji, M.; Kawanaka, H.; Oguri, K. Driver's cognitive distraction detection using physiological features by the adaboost. In Proceedings of the IEEE 12th International IEEE Conference on Intelligent Transportation Systems (ITSC'09), St. Louis, MO, USA, 4–7 October 2009; pp. 1–6.

152. Yang, Y. The Effects of Increased Workload on Driving Performance and Visual Behaviour. Ph.D. Thesis, University of Southampton, Southampton, UK, December 2011.

153. He, J.; Becic, E.; Lee, Y.C.; McCarley, J.S. Mind wandering behind the wheel performance and oculomotor correlates. *Hum. Factors J. Hum. Factors Ergon. Soc.* **2011**, *53*, 13–21.

154. He, J. Identify Mind-Wandering Behind the Wheel. Ph.D. Thesis, University of Illinois at Urbana-Champaign, Champaign, IL, USA, December 2010.

155. Kahneman, D.; Tursky, B.; Shapiro, D.; Crider, A. Pupillary, heart rate, and skin resistance changes during a mental task. *J. Exp. Psychol.* **1969**, *79*, 164–167.

156. Miyaji, M.; Kawanaka, H.; Oguri, K. Effect of pattern recognition features on detection for driver's cognitive distraction. In Proceedings of the 2010 13th International IEEE Conference on Intelligent Transportation Systems (ITSC), Funchal, Portugal, 19–22 September 2010; pp. 605–610.

157. Heeman, P.A.; Meshorer, T.; Kun, A.L.; Palinko, O.; Medenica, Z. Estimating cognitive load using pupil diameter during a spoken dialogue task. In Proceedings of the 5th International Conference on Automotive User Interfaces and Interactive Vehicular Applications, Eindhoven, The Netherlands, 27–30 October 2013; pp. 242–245.

158. Pfleging, B.; Fekety, D.K.; Schmidt, A.; Kun, A.L. A Model Relating Pupil Diameter to Mental Workload and Lighting Conditions. In Proceedings of the 2016 CHI Conference on Human Factors in Computing Systems, San Jose, CA, USA, 7–12 May 2016; pp. 5776–5788.

159. Mehler, B.; Reimer, B.; Coughlin, J.F. Physiological reactivity to graded levels of cognitive workload across three age groups: An on-road evaluation. In Proceedings of the Human Factors and Ergonomics Society Annual Meeting, San Francisco, CA, USA, 27 September–1 October 2010; SAGE Publications: Thousand Oaks, CA, USA; Volume 54, pp. 2062–2066.

160. Partin, D.L.; Sultan, M.F.; Thrush, C.M.; Prieto, R.; Wagner, S.J. Monitoring Driver Physiological Parameters for Improved Safety. Available online: http://papers.sae.org/2006-01-1322/ (accessed on 26 October 2016).

161. Liu, S.H.; Lin, C.T.; Chao, W.H. The short-time fractal scaling of heart rate variability to estimate the mental stress of driver. In Proceedings of the 2004 IEEE International Conference on Networking, Sensing and Control, Taipei, Taiwan, 21–23 March 2004; Volume 2, pp. 829–833.

162. Apparies, R.J.; Riniolo, T.C.; Porges, S.W. A psychophysiological investigation of the effects of driving longer-combination vehicles. *Ergonomics* **1998**, *41*, 581–592.

163. Wickens, C.D.; Lee, J.D.; Liu, Y.; Gordon-Becker, S. *Introduction to Human Factors Engineering*; Pearson Education: Upper Saddle River, NJ, USA, 1998.

164. Mulder, L. Measurement and analysis methods of heart rate and respiration for use in applied environments. *Biol. Psychol.* **1992**, *34*, 205–236.

165. Miller, E.E. Effects of Roadway on Driver Stress: An On-Road Study using Physiological Measures. Ph.D. Thesis, University of Washington, Seattle, WA, USA, July 2013.

166. Zhao, C.; Zhao, M.; Liu, J.; Zheng, C. Electroencephalogram and electrocardiograph assessment of mental fatigue in a driving simulator. *Accid. Anal. Prev.* **2012**, *45*, 83–90.

167. Östlund, J.; Nilsson, L.; Törnros, J.; Forsman, Å. Effects of cognitive and visual load in real and simulated driving. *Transp. Res. Part F Traffic Psychol. Behav.* **2006**, *8*, 97–120.

168. Poh, M.Z.; McDuff, D.J.; Picard, R.W. Non-contact, automated cardiac pulse measurements using video imaging and blind source separation. *Opt. Express* **2010**, *18*, 10762–10774.

169. Poh, M.Z.; McDuff, D.J.; Picard, R.W. Advancements in noncontact, multiparameter physiological measurements using a webcam. *IEEE Trans. Biomed. Eng.* **2011**, *58*, 7–11.

170. Fernandez, A.; Carus, J.L.; Usamentiaga, R.; Alvarez, E.; Casado, R. Unobtrusive health monitoring system using video-based physiological information and activity measurements. In Proceedings of the IEEE 2015 International Conference on Computer, Information and Telecommunication Systems (CITS), Paris, France, 11–13 June 2015; pp. 1–5.

171. Wu, H.Y.; Rubinstein, M.; Shih, E.; Guttag, J.V.; Durand, F.; Freeman, W.T. Eulerian video magnification for revealing subtle changes in the world. *ACM Trans. Graph.* **2012**, *31*, 65.

172. Chambino, P. Android-Based Implementation of Eulerian Video Magnification for Vital Signs Monitoring. Ph.D. Thesis, Faculdade de Engenharia da Universidade do Porto, Porto, Portugal, July 2013.

173. Balakrishnan, G.; Durand, F.; Guttag, J. Detecting pulse from head motions in video. In Proceedings of the 2013 IEEE Conference on Computer Vision and Pattern Recognition (CVPR), Washington, DC, USA, 23–28 June 2013; pp. 3430–3437.

174. Li, G.; Chung, W.Y. Detection of driver drowsiness using wavelet analysis of heart rate variability and a support vector machine classifier. *Sensors* **2013**, *13*, 16494–16511.

175. Solovey, E.T.; Zec, M.; Garcia Perez, E.A.; Reimer, B.; Mehler, B. Classifying driver workload using physiological and driving performance data: Two field studies. In Proceedings of the SIGCHI Conference on Human Factors in Computing Systems, Toronto, ON, Canada, 26 April –1 May 2014; pp. 4057–4066.

176. Lee, D.; Kim, J.; Kwon, S.; Park, K. Heart rate estimation from facial photoplethysmography during dynamic illuminance changes. In Proceedings of the 2015 37th Annual International Conference of the IEEE Engineering in Medicine and Biology Society (EMBC), Milano, Italy, 25–29 August 2015; pp. 2758–2761.

177. Jeanne, V.; Asselman, M.; den Brinker, B.; Bulut, M. Camera-based heart rate monitoring in highly dynamic light conditions. In Proceedings of the 2013 IEEE International Conference on Connected Vehicles and Expo (ICCVE), Las Vegas, NV, USA, 2–6 December 2013; pp. 798–799.

178. Zhang, Y.; Owechko, Y.; Zhang, J. Driver cognitive workload estimation: A data-driven perspective. In Proceedings of the 7th International IEEE Conference on Intelligent Transportation Systems, Washington, DC, USA, 3–6 October 2004; pp. 642–647.

179. Liang, Y.; Reyes, M.L.; Lee, J.D. Real-time detection of driver cognitive distraction using support vector machines. *IEEE Trans. Intell. Transp. Syst.* **2007**, *8*, 340–350.

180. Liang, Y.; Lee, J.D.; Reyes, M.L. Nonintrusive detection of driver cognitive distraction in real time using Bayesian networks. *Transp. Res. Rec. J. Transp. Res. Board* **2007**, *2018*, 1–8.

181. Liang, Y.; Lee, J.D. Comparing Support Vector Machines (SVMs) and Bayesian Networks (BNs) in detecting driver cognitive distraction using eye movements. In *Passive Eye Monitoring: Algorithms, Applications and Experiments*; Springer: Berlin/Heidelberg, Germany, 2008; pp. 285–300.

182. Liang, Y.; Lee, J.D. Using a Layered Algorithm to Detect Driver Cognitive Distraction. In Proceedings of the seventh International Driving Symposium on Human Factors in Driver assessment, Training, and Vehicle Design, New York, NY, USA, 17–20 June 2013; pp. 327–333.

183. Liang, Y.; Lee, J.D. A hybrid Bayesian Network approach to detect driver cognitive distraction. *Transp. Res. Part C Emerg. Technol.* **2014**, *38*, 146–155.

184. Miyaji, M.; Danno, M.; Kawanaka, H.; Oguri, K. Driver's cognitive distraction detection using AdaBoost on pattern recognition basis. In Proceedings of the IEEE International Conference on Vehicular Electronics and Safety (ICVES 2008), Columbus Ohio, OH, USA, 22–24 September 2008; pp. 51–56.

185. Huang, G.B.; Zhu, Q.Y.; Siew, C.K. Extreme learning machine: Theory and applications. *Neurocomputing* **2006**, *70*, 489–501.

186. Huang, G.B.; Zhou, H.; Ding, X.; Zhang, R. Extreme learning machine for regression and multiclass classification. *IEEE Trans. Syst. Man Cybern. Part B Cybern.* **2012**, *42*, 513–529.

187. Yang, Y.; Sun, H.; Liu, T.; Huang, G.B.; Sourina, O. Driver Workload Detection in On-Road Driving Environment Using Machine Learning. In *Proceedings of ELM-2014*; Springer: Berlin/Heidelberg, Germany, 2015; Volume 2, pp. 389–398.

188. Li, N.; Busso, C. Predicting Perceived Visual and Cognitive Distractions of Drivers With Multimodal Features. *IEEE Trans. Intell. Transp. Syst.* **2015**, *16*, 51–65.

189. Liu, T.; Yang, Y.; Huang, G.B.; Yeo, Y.; Lin, Z. Driver Distraction Detection Using Semi-Supervised Machine Learning. *IEEE Trans. Intell. Transp. Syst.* **2015**, *17*, 1–13.

190. Belkin, M.; Niyogi, P.; Sindhwani, V. Manifold regularization: A geometric framework for learning from labeled and unlabeled examples. *J. Mach. Learn. Res.* **2006**, *7*, 2399–2434.

191. Joachims, T. Transductive inference for text classification using support vector machines. In Proceedings of the Sixteenth International Conference on Machine Learning (ICML), Bled, Slovenia, 27–30 June 1999; Volume 99, pp. 200–209.

192. Lemercier, C.; Pêcher, C.; Berthié, G.; Valéry, B.; Vidal, V.; Paubel, P.V.; Cour, M.; Fort, A.; Galéra, C.; Gabaude, C.; et al. Inattention behind the wheel: How factual internal thoughts impact attentional control while driving. *Saf. Sci.* **2014**, *62*, 279–285.

193. Ho, C.; Gray, R.; Spence, C. To what extent do the findings of laboratory-based spatial attention research apply to the real-world setting of driving? *IEEE Trans. Hum. Mach. Syst.* **2014**, *44*, 524–530.

194. Li, N.; Busso, C. Analysis of facial features of drivers under cognitive and visual distractions. In Proceedings of the 2013 IEEE International Conference on Multimedia and Expo (ICME), San Jose, CA, USA; 15–19 July 2013; pp. 1–6.

195. Craye, C.; Karray, F. Driver distraction detection and recognition using RGB-D sensor. 2015, arXiv preprint arXiv:1502.00250.

196. Liu, T.; Yang, Y.; Huang, G.B.; Lin, Z.; Klanner, F.; Denk, C.; Rasshofer, R.H. Cluster Regularized Extreme Learning Machine for Detecting Mixed-Type Distraction in Driving. In Proceedings of the 2015 IEEE 18th International Conference on Intelligent Transportation Systems (ITSC), Las Palmas de Gran Canaria, Spain, 15–18 September 2015; pp. 1323–1326.
197. Ragab, A.; Craye, C.; Kamel, M.S.; Karray, F. A Visual-Based Driver Distraction Recognition and Detection Using Random Forest. In *Image Analysis and Recognition*; Springer: Berlin/Heidelberg, Germany, 2014; pp. 256–265.
198. Tian, Y.L.; Kanade, T.; Cohn, J.F. Handbook of face recognition. In *Ch Facial Expression Analysis*; Springer: Berlin/Heidelberg, Germany, 2005; pp. 487–519.
199. Ekman, P.; Friesen, W. *Facial Action Coding System: A Technique for the Measurement of Facial Movement*; Consulting Psychologists: San Francisco, CA, USA, 1978.
200. Tian, Y.; Kanade, T.; Cohn, J.F. Facial expression recognition. In *Handbook of Face Recognition*; Springer: Berlin/Heidelberg, Germany, 2011; pp. 487–519.
201. Shan, C.; Gong, S.; McOwan, P.W. Facial expression recognition based on local binary patterns: A comprehensive study. *Image Vis. Comput.* **2009**, *27*, 803–816.
202. Ekman, P.; Friesen, W.V.; Press, C.P. *Pictures of Facial Affect*; Consulting Psychologists Press: Palo Alto, CA, USA, 1975.
203. Kang, H.B. Various approaches for driver and driving behavior monitoring: A review. In Proceedings of the 2013 IEEE International Conference on Computer Vision Workshops (ICCVW), Sydney, Australia, 1–8 December 2013; pp. 616–623.
204. Chan, M.; Singhal, A. The emotional side of cognitive distraction: Implications for road safety. *Accid. Anal. Prev.* **2013**, *50*, 147–154.
205. Chan, M.; Singhal, A. Emotion matters: Implications for distracted driving. *Saf. Sci.* **2015**, *72*, 302–309.
206. Deffenbacher, J.L.; Lynch, R.S.; Filetti, L.B.; Dahlen, E.R.; Oetting, E.R. Anger, aggression, risky behavior, and crash-related outcomes in three groups of drivers. *Behav. Res. Ther.* **2003**, *41*, 333–349.
207. Jeon, M.; Walker, B.N.; Gable, T.M. Anger effects on driver situation awareness and driving performance. *Presence Teleoper. Virtual Environ.* **2014**, *23*, 71–89.
208. Eyben, F.; Wöllmer, M.; Poitschke, T.; Schuller, B.; Blaschke, C.; Färber, B.; Nguyen-Thien, N. Emotion on the road—Necessity, acceptance, and feasibility of affective computing in the car. *Adv. Hum. Comput. Interact.* **2010**, *2010*, 263593.
209. James, L. *Road Rage and Aggressive Driving: Steering Clear of Highway Warfare*; Prometheus Books: Amherst, NY, USA, 2000.
210. Grimm, M.; Kroschel, K.; Harris, H.; Nass, C.; Schuller, B.; Rigoll, G.; Moosmayr, T. On the necessity and feasibility of detecting a driver's emotional state while driving. In *Affective Computing and Intelligent Interaction*; Springer: Berlin/Heidelberg, Germany, 2007; pp. 126–138.
211. Katsis, C.D.; Rigas, G.; Goletsis, Y.; Fotiadis, D.I. Emotion Recognition in Car Industry. In *Emotion Recognition: A Pattern Analysis Approach*; Wiley: New York, NY, USA, 2015; pp. 515–544.
212. Stein, F. The challenge of putting vision algorithms into a car. In Proceedings of the 2012 IEEE Computer Society Conference on Computer Vision and Pattern Recognition Workshops (CVPRW), Providence, RI, USA, 16–21 June 2012; pp. 89–94.
213. Nieto Doncel, M.; Arróspide Laborda, J.; Salgado Álvarez de Sotomayor, L.; García Santos, N. Video-Based Driver Assistance Systems. Ph.D. Thesis, Robert Bosch SRL, Cluj Napoca, Romania, October 2008.
214. Kim, K.; Choi, K. SoC Architecture for Automobile Vision System. In *Algorithm & SoC Design for Automotive Vision Systems*; Springer: Berlin/Heidelberg, Germany, 2014; pp. 163–195.
215. Dedeoğlu, G.; Kisačanin, B.; Moore, D.; Sharma, V.; Miller, A. An optimized vision library approach for embedded systems. In Proceedings of the 2011 IEEE Computer Society Conference on Computer Vision and Pattern Recognition Workshops (CVPRW), Colorado Springs, CO, USA, 20–25 June 2011; pp. 8–13.
216. Satzoda, R.K.; Lee, S.; Lu, F.; Trivedi, M.M. Snap-DAS: A Vision-based Driver Assistance System on a Snapdragon TM Embedded Platform. In Proceedings of the 2015 IEEE Intelligent Vehicles Symposium (IV), Seoul, Korea, 28 June–1 July 2015.
217. Nieto, M.; Otaegui, O.; Vélez, G.; Ortega, J.D.; Cortés, A. On creating vision-based advanced driver assistance systems. *IET Intell. Transp. Syst.* **2014**, *9*, 59–66.

218. Pelaez, C.; Garcia, F.; de la Escalera, A.; Armingol, J. Driver Monitoring Based on Low-Cost 3-D Sensors. *IEEE Trans. Intell. Transp. Syst.* **2014**, *15*, 1855–1860.

219. Velez, G.; Cortés, A.; Nieto, M.; Vélez, I.; Otaegui, O. A reconfigurable embedded vision system for advanced driver assistance. *J. Real Time Image Process.* **2014**, *10*, 1–15.

220. Forster, F. Heterogeneous Processors for Advanced Driver Assistance Systems. *ATZelektronik Worldw.* **2014**, *9*, 14–18.

221. Anders, J.; Mefenza, M.; Bobda, C.; Yonga, F.; Aklah, Z.; Gunn, K. A hardware/software prototyping system for driving assistance investigations. *J. Real Time Image Process.* **2013**, *11*, 1–11.

222. Young, R.; Zhang, J. Safe Interaction for Drivers: A Review of Driver Distraction Guidelines and Design Implications. Available online: http://papers.sae.org/2015-01-1384/ (accessed on 26 October 2016).

223. Dasgupta, A.; George, A.; Happy, S.; Routray, A. A vision-based system for monitoring the loss of attention in automotive drivers. *IEEE Trans. Intell. Transp. Syst.* **2013**, *14*, 1825–1838.

224. Krishnasree, V.; Balaji, N.; Rao, P.S. A Real Time Improved Driver Fatigue Monitoring System. *WSEAS Trans. Signal Process.* **2014**, *10*, 146.

225. Veeraraghavan, H.; Papanikolopoulos, N.P. *Detecting Driver Fatigue Through the Use of Advanced Face Monitoring Techniques*; University of Minnesota: Minneapolis, MN, USA, 2001.

226. Karuppusamy, S.; Jerome, J.; Shankar, N. Embedded implementation of facial landmarks detection using extended active shape model approach. In Proceedings of the IEEE 2014 International Conference on Embedded Systems (ICES), Coimbatore, India, 3–5 July 2014; pp. 265–270.

227. De Marsico, M.; Nappi, M.; Riccio, D.; Wechsler, H. Robust face recognition for uncontrolled pose and illumination changes. *IEEE Trans. Syst. Man Cybern. Syst.* **2013**, *43*, 149–163.

228. Malinowski, A.; Yu, H. Comparison of embedded system design for industrial applications. *IEEE Trans. Ind. Inform.* **2011**, *7*, 244–254.

229. Moreno, F.; Aparicio, F.; Hernández, W.; Paez, J. A low-cost real-time FPGA solution for driver drowsiness detection. In Proceedings of the 29th Annual Conference of the IEEE Industrial Electronics Society, Roanoke, VA, USA, 2–6 June 2003; Volume 2, pp. 1396–1401.

230. Wang, F.; Qin, H. A FPGA based driver drowsiness detecting system. In Proceedings of the IEEE International Conference on Vehicular Electronics and Safety, Xian, China, 14–16 October 2005; pp. 358–363.

231. Sanz, R.; Salvador, R.; Alarcon, J.; Moreno, F.; López, I. Embedded Intelligence on Chip: Some FPGA-Based Design Experiences. Available online: http://www.intechopen.com/books/pattern-recognition-recent-adv ances/embedded-intelligence-on-chip-some-fpgabased-design-experiences (accessed on 26 October 2016).

232. Samarawickrama, M.; Pasqual, A.; Rodrigo, R. FPGA-based compact and flexible architecture for real-time embedded vision systems. In Proceedings of the 2009 International Conference on Industrial and Information Systems (ICIIS), Peradeniya, Sri Lanka, 28–31 December 2009; pp. 337–342.

233. Mielke, M.; Schafer, A.; Bruck, R. Asic implementation of a gaussian pyramid for use in autonomous mobile robotics. In Proceedings of the 2011 IEEE 54th International Midwest Symposium on Circuits and Systems (MWSCAS), Seoul, Korea, 7–10 August 2011; pp. 1–4.

234. Stein, G.P.; Rushinek, E.; Hayun, G.; Shashua, A. A computer vision system on a chip: A case study from the automotive domain. In Proceedings of the IEEE Computer Society Conference on Computer Vision and Pattern Recognition-Workshops, Miami, FL, USA, 20–25 June 2005; p. 130.

235. Saussard, R.; Bouzid, B.; Vasiliu, M.; Reynaud, R. Towards an Automatic Prediction of Image Processing Algorithms Performances on Embedded Heterogeneous Architectures. In Proceedings of the 2015 44th International Conference on Parallel Processing Workshops (ICPPW), Beijing, China, 1–4 September 2015; pp. 27–36.

236. Ebert, C.; Jones, C. Embedded software: Facts, figures, and future. *Computer* **2009**, *42*, 42–52.

237. Macher, G.; Stolz, M.; Armengaud, E.; Kreiner, C. Filling the gap between automotive systems, safety, and software engineering. *E I Elektrotech. Informationstech.* **2015**, *132*, 1–7.

238. Jin, L.; Niu, Q.; Jiang, Y.; Xian, H.; Qin, Y.; Xu, M. Driver sleepiness detection system based on eye movements variables. *Adv. Mech. Eng.* **2013**, *5*, 648431.

239. Lumeway. EyeAlert Distracted Driving and Fatigue Warning Systems. Available online: http://www.lumeway.com/EA.htm (accessed on 3 January 2016).

240. Craye, C.; Rashwan, A.; Kamel, M.S.; Karray, F. A Multi-Modal Driver Fatigue and Distraction Assessment System. *Int. J. Intell. Transp. Syst. Res.* **2015**, *14*, 1–22.
241. SeeingMachines. Advanced Driver Fatigue and Distraction Detection. Available online: http://www.seeingmachines.com/ (accessed on 6 January 2016).
242. Fletcher, L.; Apostoloff, N.; Petersson, L.; Zelinsky, A. Vision in and out of vehicles. *IEEE Intell. Syst.* **2003**, *18*, 12–17.
243. Fletcher, L.; Loy, G.; Barnes, N.; Zelinsky, A. Correlating driver gaze with the road scene for driver assistance systems. *Robot. Auton. Syst.* **2005**, *52*, 71–84.
244. Friedrichs, F.; Yang, B. Camera-based drowsiness reference for driver state classification under real driving conditions. In Proceedings of the 2010 IEEE Intelligent Vehicles Symposium (IV), San Diego, CA, USA, 21–24 June 2010; pp. 101–106.
245. VisageTechnologies. Face Tracking and Analysis. Available online: http://www.visagetechnologies.com/ (accessed on 6 January 2016).
246. Edenborough, N.; Hammoud, R.; Harbach, A.; Ingold, A.; Kisačanin, B.; Malawey, P.; Newman, T.; Scharenbroch, G.; Skiver, S.; Smith, M.; et al. Driver state monitor from delphi. In Proceedings of the IEEE Computer Society Conference on Computer Vision and Pattern Recognition, Miami, FL, USA, 20–25 June 2005; Volume 2, pp. 1206–1207.
247. SMI. Eye and gaze tracking systems. Available online: www.smivision.com (accessed on 6 January 2016).
248. Juhola, M.; Aalto, H.; Joutsijoki, H.; Hirvonen, T.P. The classification of valid and invalid beats of three-dimensional nystagmus eye movement signals using machine learning methods. *Adv. Artif. Neural Syst.* **2013**, *2013*, doi:10.1155/2013/972412.
249. Nissan. Nissan Introduces Driver Attention Alert for Passenger Safety. Available online: http://automobiletechnology.automotive-business-review.com/news/nissan-introduces-driver-attention-alert-for-passenger-safety-020415-4546074 (accessed on 6 March 2016).
250. Ford. Ford Technology: Driver Alert, Lane Departure Warning. Available online: http://technology.fordmedia.eu/documents/newsletter/FordTechnologyNewsletter082010.pdf (accessed on 6 March 2016).
251. Nabo, A. Driver attention-dealing with drowsiness and distraction. Available online: http://smarteye.se/wp-content/uploads/2015/01/Nabo-Arne-IVSS-Report.pdf (accessed on 26 October 2016).
252. Ishiguro, H.; Hayashi, T.; Naito, T.; Kasugai, J.; Ogawa, K.; Ohue, K.; Uozumi, S. Development of facial-direction detection sensor. In Proceedings of the 13th Its World Congress, London, UK, 8–12 October 2006.
253. Volvo. Volvo Cars Introduces New Systems for Alerting Tired And Unconcentrated Drivers. Available online: http://www.mobileye.com/wp-content/uploads/2011/09/Volvo.DriverAlert.pdf (accessed on 6 January 2016).
254. MobileEye. MobileEye Advanced Vehicle Technologies Power Volvo Car's Driver Alert Control (DAC) System. Available online: http://www.mobileye.com/wp-content/uploads/2011/09/ MobileyeAdvanceVehicleTechnologiesPowerVolvo.pdf (accessed on 6 Janyary 2016).
255. SAE. Hyundai HCD-14 Genesis Concept Previews Eye-Tracking, Gesture-Recognition Technologies. Available online: http://articles.sae.org/11727/ (accessed on 17 January 2016).
256. Velez, G.; Otaegui, O. Embedded Platforms for Computer Vision-based Advanced Driver Assistance Systems: A Survey. 2015, arXiv preprint arXiv:1504.07442.
257. Rosey, F.; Auberlet, J.M.; Moisan, O.; Dupré, G. Impact of Narrower Lane Width. *Transp. Res. Rec. J. Transp. Res. Board* **2009**, *2138*, 112–119.
258. Konstantopoulos, P.; Chapman, P.; Crundall, D. Driver's visual attention as a function of driving experience and visibility. Using a driving simulator to explore drivers' eye movements in day, night and rain driving. *Accid. Anal. Prev.* **2010**, *42*, 827–834.
259. Auberlet, J.M.; Rosey, F.; Anceaux, F.; Aubin, S.; Briand, P.; Pacaux, M.P.; Plainchault, P. The impact of perceptual treatments on driver's behavior: From driving simulator studies to field tests-First results. *Accid. Anal. Prev.* **2012**, *45*, 91–98.
260. Johnson, M.J.; Chahal, T.; Stinchcombe, A.; Mullen, N.; Weaver, B.; Bedard, M. Physiological responses to simulated and on-road driving. *Int. J. Psychophysiol.* **2011**, *81*, 203–208.

261. Mayhew, D.R.; Simpson, H.M.; Wood, K.M.; Lonero, L.; Clinton, K.M.; Johnson, A.G. On-road and simulated driving: Concurrent and discriminant validation. *J. Saf. Res.* **2011**, *42*, 267–275.

262. De Winter, J.; Happee, P. Advantages and Disadvantages of Driving Simulators: A Discussion. In Proceedings of the Measuring Behavior, Utrecht, The Netherlands, 28–31 August 2012; pp. 47–50.

263. Bach, K.M.; Jæger, M.G.; Skov, M.B.; Thomassen, N.G. Evaluating driver attention and driving behaviour: Comparing controlled driving and simulated driving. In Proceedings of the 22nd British HCI Group Annual Conference on People and Computers: Culture, Creativity, Interaction, Swindon, UK, 1–5 September 2008; pp. 193–201.

264. Masala, G.; Grosso, E. Real time detection of driver attention: Emerging solutions based on robust iconic classifiers and dictionary of poses. *Transp. Res. Part C Emerg. Technol.* **2014**, *49*, 32–42.

265. Martin, S.; Tawari, A.; Trivedi, M.M. Toward privacy-protecting safety systems for naturalistic driving videos. *IEEE Trans. Intell. Transp. Syst.* **2014**, *15*, 1811–1822.

266. Newton, E.M.; Sweeney, L.; Malin, B. Preserving privacy by de-identifying face images. *IEEE Trans. Knowl. Data Eng.* **2005**, *17*, 232–243.

267. Martin, S.; Tawari, A.; Trivedi, M.M. Balancing Privacy and Safety: Protecting Driver Identity in Naturalistic Driving Video Data. In Proceedings of the 6th International Conference on Automotive User Interfaces and Interactive Vehicular Applications, New York, NY, USA, 17–19 September 2014; pp. 1–7.

268. la Torre, F.D. Vision-Based Systems for Driver Monitoring and De-Identification. Available online: http://fot-net.eu/wp-content/uploads/sites/7/2015/09/Fernando-de-la-Torre.pdf (accessed on 30 April 2016).

269. Young, R. Revised Odds Ratio Estimates of Secondary Tasks: A Re-Analysis of the 100-Car Naturalistic Driving Study Data. Available online: http://papers.sae.org/2015-01-1387/ (accessed on 26 October 2016).

270. Strayer, D.L.; Drews, F.A.; Johnston, W.A. Cell phone-induced failures of visual attention during simulated driving. *J. Exp. Psychol. Appl.* **2003**, *9*, 23–32.

271. Apostoloff, N.; Zelinsky, A. Vision in and out of vehicles: Integrated driver and road scene monitoring. *Int. J. Robot. Res.* **2004**, *23*, 513–538.

272. Shin, B.S.; Xu, Z.; Klette, R. Visual lane analysis and higher-order tasks: A concise review. *Mach. Vis. Appl.* **2014**, *25*, 1519–1547.

273. Hirayama, T.; Mase, K.; Takeda, K. Analysis of temporal relationships between eye gaze and peripheral vehicle behavior for detecting driver distraction. *Int. J. Veh. Technol.* **2013**, *2013*, doi:10.1155/2013/285927.

274. Ohn-Bar, E.; Tawari, A.; Martin, S.; Trivedi, M.M. Vision on wheels: Looking at driver, vehicle, and surround for on-road maneuver analysis. In Proceedings of the IEEE Conference on Computer Vision and Pattern Recognition Workshops, Columbus, OH, USA, 24–27 June 2014; pp. 185–190.

275. Tawari, A.; Sivaraman, S.; Trivedi, M.M.; Shannon, T.; Tippelhofer, M. Looking-in and looking-out vision for urban intelligent assistance: Estimation of driver attentive state and dynamic surround for safe merging and braking. In Proceedings of the 2014 IEEE Intelligent Vehicles Symposium Proceedings, Dearborn, MI, USA, 8–11 June 2014; pp. 115–120.

276. Martin, S.; Rangesh, A.; Ohn-Bar, E.; Trivedi, M.M. The rhythms of head, eyes and hands at intersections. In Proceedings of the 2016 IEEE Intelligent Vehicles Symposium (IV), Gothenburg, Sweden, 19–22 June 2016; pp. 1410–1415.

277. Jain, A.; Koppula, H.S.; Raghavan, B.; Soh, S.; Saxena, A. Car that knows before you do: Anticipating maneuvers via learning temporal driving models. In Proceedings of the IEEE International Conference on Computer Vision, Santiago, Chile, 13–16 December 2015; pp. 3182–3190.

278. Baddeley, A.D. Selective attention and performance in dangerous environments. *Br. J. Psychol.* **1972**, *63*, 537–546.

279. Vidulich, M.A.; Stratton, M.; Crabtree, M.; Wilson, G. Performance-based and physiological measures of situational awareness. *Aviat. Space Environ. Med.* **1994**, *65*, 7–12.

280. Helmreich, R.L.; Chidester, T.R.; Foushee, H.C.; Gregorich, S.; Wilhelm, J.A. How effective is cockpit resource management training. *Flight Saf. Dig.* **1990**, *9*, 1–17.

281. Ball, K.; Rebok, G. Evaluating the driving ability of older adults. *J. Appl. Gerontol.* **1994**, *13*, 20–38.

282. Satzoda, R.K.; Trivedi, M.M. Drive analysis using vehicle dynamics and vision-based lane semantics. *IEEE Trans. Intell. Transp. Syst.* **2015**, *16*, 9–18.

Sensors **2016**, *16*, 1805

283. Satzoda, R.K.; Gunaratne, P.; Trivedi, M.M. Drive quality analysis of lane change maneuvers for naturalistic driving studies. In Proceedings of the 2015 IEEE Intelligent Vehicles Symposium (IV), Seoul, Korea, 18 June–1 July 2015; pp. 654–659.

284. Kusano, K.D.; Montgomery, J.; Gabler, H.C. Methodology for identifying car following events from naturalistic data. In Proceedings of the 2014 IEEE Intelligent Vehicles Symposium Proceedings, Dearborn, MI, USA, 8–11 June 2014; pp. 281–285.

285. Martin, S.; Ohn-Bar, E.; Trivedi, M.M. Automatic Critical Event Extraction and Semantic Interpretation by Looking-Inside. In Proceedings of the 2015 IEEE 18th International Conference on Intelligent Transportation Systems, Las Palmas, Spain, 15–18 September 2015; pp. 2274–2279.

MDPI AG

St. Alban-Anlage 66

4052 Basel, Switzerland

Tel. +41 61 683 77 34

Fax +41 61 302 89 18

http://www.mdpi.com

Sensors Editorial Office

E-mail: sensors@mdpi.com

http://www.mdpi.com/journal/sensors

www.ingramcontent.com/pod-product-compliance
Lightning Source LLC
Chambersburg PA
CBHW041213220326
41597CB00032BA/5313